# LASERS

Anthony E. Siegman

Professor of Electrical Engineering
Stanford University

University Science Books
Sausalito, California

University Science Books
55D Gate Five Road
Sausalito, CA 94965

Manuscript Editor: Aidan Kelly
Designer: Robert Ishi
Production: Miller/Scheier Associates, Palo Alto, CA
TEXpert: Laura Poplin
Printer and Binder: The Maple-Vail Book Manufacturing Group

Copyright © 1986 by University Science Books
Reproduction or translation of any part of this work
beyond that permitted by Sections 107 or 108 of the
1976 United States Copyright Act without the permission of
the copyright owner is unlawful. Requests for permission
or further information should be addressed to
the Permissions Department, University Science Books.

Library of Congress Catalog Card Number: 86-050346

ISBN 0-935702-11-3

Printed in the United States of America

This manuscript was prepared at Stanford University using the text editing facilities of the Context and Sierra DEC-20 computers and Professor Donald Knuth's TEX typesetting system. Camera-ready copy was printed on an Autologic APS-$\mu$5 phototypesetter.

# CONTENTS

Preface — xiii
Units and Notation — xv
List of Symbols — xvii

## BASIC LASER PHYSICS

1. An Introduction to Lasers — 1
2. Stimulated Transitions: The Classical Oscillator Model — 80
3. Electric Dipole Transitions in Real Atoms — 118
4. Atomic Rate Equations — 176
5. The Rabi Frequency — 221
6. Laser Pumping and Population Inversion — 243
7. Laser Amplification — 264
8. More On Laser Amplification — 307
9. Linear Pulse Propagation — 331
10. Nonlinear Optical Pulse Propagation — 362
11. Laser Mirrors and Regenerative Feedback — 398
12. Fundamentals of Laser Oscillation — 457
13. Oscillation Dynamics and Oscillation Threshold — 491

## OPTICAL BEAMS AND RESONATORS

14. Optical Beams and Resonators: An Introduction — 558
15. Ray Optics and Ray Matrices — 581
16. Wave Optics and Gaussian Beams — 626
17. Physical Properties of Gaussian Beams — 663
18. Beam Perturbation and Diffraction — 698
19. Stable Two-Mirror Resonators — 744
20. Complex Paraxial Wave Optics — 777
21. Generalized Paraxial Resonator Theory — 815
22. Unstable Optical Resonators — 858
23. More on Unstable Resonators — 891

## LASER DYNAMICS AND ADVANCED TOPICS

24. Laser Dynamics: The Laser Cavity Equations — 923
25. Laser Spiking and Mode Competition — 954
26. Laser $Q$-Switching — 1004
27. Active Laser Mode Coupling — 1041
28. Passive Mode Locking — 1104
29. Laser Injection Locking — 1129
30. Hole Burning and Saturation Spectroscopy — 1171
31. Magnetic-Dipole Transitions — 1213

# LIST OF TOPICS

|  |  |
|---|---|
| Preface | xiii |
| Units and Notation | xv |
| List of Symbols | xvii |

## BASIC LASER PHYSICS

### Chapter 1  An Introduction to Lasers

| | | |
|---|---|---|
| 1.1 | What Is a Laser? | 2 |
| 1.2 | Atomic Energy Levels and Spontaneous Emission | 6 |
| 1.3 | Stimulated Atomic Transitions | 18 |
| 1.4 | Laser Amplification | 30 |
| 1.5 | Laser Pumping and Population Inversion | 35 |
| 1.6 | Laser Oscillation and Laser Cavity Modes | 39 |
| 1.7 | Laser Output-Beam Properties | 49 |
| 1.8 | A Few Practical Examples | 60 |
| 1.9 | Other Properties of Real Lasers | 66 |
| 1.10 | Historical Background of the Laser | 74 |
| 1.11 | Additional Problems for Chapter 1 | 76 |

### Chapter 2  Stimulated Transitions: The Classical Oscillator Model

| | | |
|---|---|---|
| 2.1 | The Classical Electron Oscillator | 80 |
| 2.2 | Collisions and Dephasing Processes | 89 |
| 2.3 | More on Atomic Dynamics and Dephasing | 97 |
| 2.4 | Steady-State Response: The Atomic Susceptibility | 102 |
| 2.5 | Conversion to Real Atomic Transitions | 110 |

### Chapter 3  Electric Dipole Transitions in Real Atoms

| | | |
|---|---|---|
| 3.1 | Decay Rates and Transition Strengths in Real Atoms | 118 |
| 3.2 | Line Broadening Mechanisms in Real Atoms | 126 |
| 3.3 | Polarization Properties of Atomic Transitions | 135 |
| 3.4 | Tensor Susceptibilities | 143 |
| 3.5 | The "Factor of Three" | 150 |
| 3.6 | Degenerate Energy Levels and Degeneracy Factors | 153 |
| 3.7 | Inhomogeneous Line Broadening | 157 |

### Chapter 4  Atomic Rate Equations

| | | |
|---|---|---|
| 4.1 | Power Transfer From Signals to Atoms | 176 |

| | | |
|---|---|---|
| 4.2 | Stimulated Transition Probability | 181 |
| 4.3 | Blackbody Radiation and Radiative Relaxation | 187 |
| 4.4 | Nonradiative Relaxation | 195 |
| 4.5 | Two-Level Rate Equations and Saturation | 204 |
| 4.6 | Multilevel Rate Equations | 211 |

**Chapter 5  The Rabi Frequency**

| | | |
|---|---|---|
| 5.1 | Validity of the Rate Equation Model | 221 |
| 5.2 | Strong Signal Behavior: The Rabi Frequency | 229 |

**Chapter 6  Laser Pumping and Population Inversion**

| | | |
|---|---|---|
| 6.1 | Steady-State Laser Pumping and Population Inversion | 243 |
| 6.2 | Laser Gain Saturation | 252 |
| 6.3 | Transient Laser Pumping | 257 |

**Chapter 7  Laser Amplification**

| | | |
|---|---|---|
| 7.1 | Practical Aspects of Laser Amplifiers | 264 |
| 7.2 | Wave Propagation in an Atomic Medium | 266 |
| 7.3 | The Paraxial Wave Equation | 276 |
| 7.4 | Single-Pass Laser Amplification | 279 |
| 7.5 | Stimulated Transition Cross Sections | 286 |
| 7.6 | Saturation Intensities in Laser Materials | 292 |
| 7.7 | Homogeneous Saturation in Laser Amplifiers | 297 |

**Chapter 8  More On Laser Amplification**

| | | |
|---|---|---|
| 8.1 | Transient Response of Laser Amplifiers | 307 |
| 8.2 | Spatial Hole Burning, and Standing-Wave Grating Effects | 316 |
| 8.3 | More on Laser Amplifier Saturation | 323 |

**Chapter 9  Linear Pulse Propagation**

| | | |
|---|---|---|
| 9.1 | Phase and Group Velocities | 331 |
| 9.2 | The Parabolic Equation | 339 |
| 9.3 | Group Velocity Dispersion and Pulse Compression | 343 |
| 9.4 | Phase and Group Velocities in Resonant Atomic Media | 351 |
| 9.5 | Pulse Broadening and Gain Dispersion | 356 |

**Chapter 10  Nonlinear Optical Pulse Propagation**

| | | |
|---|---|---|
| 10.1 | Pulse Amplification With Homogeneous Gain Saturation | 362 |
| 10.2 | Pulse Propagation in Nonlinear Dispersive Systems | 375 |
| 10.3 | The Nonlinear Schrödinger Equation | 387 |
| 10.4 | Nonlinear Pulse Broadening in Optical Fibers | 388 |
| 10.5 | Solitons in Optical Fibers | 392 |

**Chapter 11  Laser Mirrors and Regenerative Feedback**

| | | |
|---|---|---|
| 11.1 | Laser Mirrors and Beam Splitters | 398 |
| 11.2 | Interferometers and Resonant Optical Cavities | 408 |
| 11.3 | Resonance Properties of Passive Optical Cavities | 413 |

| | | | |
|---|---|---|---|
| | 11.4 | "Delta Notation" for Cavity Gains and Losses | 428 |
| | 11.5 | Optical-Cavity Mode Frequencies | 432 |
| | 11.6 | Regenerative Laser Amplification | 440 |
| | 11.7 | Approaching Threshold: The Highly Regenerative Limit | 447 |

### Chapter 12  Fundamentals of Laser Oscillation

| | | |
|---|---|---|
| 12.1 | Oscillation Threshold Conditions | 457 |
| 12.2 | Oscillation Frequency and Frequency Pulling | 462 |
| 12.3 | Laser Output Power | 473 |
| 12.4 | The Large Output Coupling Case | 485 |

### Chapter 13  Oscillation Dynamics and Oscillation Threshold

| | | |
|---|---|---|
| 13.1 | Laser Oscillation Buildup | 491 |
| 13.2 | Derivation of the Cavity Rate Equation | 497 |
| 13.3 | Coupled Cavity and Atomic Rate Equations | 505 |
| 13.4 | The Laser Threshold Region | 510 |
| 13.5 | Multiple-Mirror Cavities and Etalon Effects | 524 |
| 13.6 | Unidirectional Ring-Laser Oscillators | 532 |
| 13.7 | Bistable Optical Systems | 538 |
| 13.8 | Amplified Spontaneous Emission and Mirrorless Lasers | 547 |

## OPTICAL BEAMS AND RESONATORS

### Chapter 14  Optical Beams and Resonators: An Introduction

| | | |
|---|---|---|
| 14.1 | Transverse Modes in Optical Resonators | 559 |
| 14.2 | The Mathematics of Optical Resonator Modes | 565 |
| 14.3 | Build-Up and Oscillation of Optical Resonator Modes | 569 |

### Chapter 15  Ray Optics and Ray Matrices

| | | |
|---|---|---|
| 15.1 | Paraxial Optical Rays and Ray Matrices | 581 |
| 15.2 | Ray Propagation Through Cascaded Elements | 593 |
| 15.3 | Rays in Periodic Focusing Systems | 599 |
| 15.4 | Ray Optics With Misaligned Elements | 607 |
| 15.5 | Ray Matrices in Curved Ducts | 614 |
| 15.6 | Nonorthogonal Ray Matrices | 616 |

### Chapter 16  Wave Optics and Gaussian Beams

| | | |
|---|---|---|
| 16.1 | The Paraxial Wave Equation | 626 |
| 16.2 | Huygens' Integral | 630 |
| 16.3 | Gaussian Spherical Waves | 637 |
| 16.4 | Higher-Order Gaussian Modes | 642 |
| 16.5 | Complex-Argument Gaussian Modes | 649 |
| 16.6 | Gaussian Beam Propagation in Ducts | 652 |
| 16.7 | Numerical Beam Propagation Methods | 656 |

# PREFACE

This book presents a detailed and comprehensive treatment of laser physics and laser theory which can serve a number of purposes for a number of different groups. It can provide, first of all, a textbook for graduate students, or even well-prepared seniors in science or engineering, describing in detail how lasers work, and a bit about the applications for which lasers can be used. Problems, references and illustrations are included throughout the book.

Second, it can also provide a solid and detailed description of laser physics and the operational properties of lasers for the practicing engineer or scientist who needs to learn about lasers in order to work on or with them.

Finally, the advanced sections of this text are sufficiently detailed that this book will provide a useful one-volume reference for the experienced laser engineer or laser researcher's bookshelf. The discussions of advanced laser topics, such as optical resonators, Q-switching, mode locking, and injection locking, extend far enough into the current state of the art to provide a working reference on these and similar topics. References for further reading in the recent literature are included in nearly every section.

One unique feature of this book is that it removes much of the quantum mystique from "quantum electronics" (the generic label often applied to lasers and laser applications). Many people think of lasers as quantum devices. In fact, however, most of the basic concepts of laser physics, and virtually all the practical details, are classical in nature. Lasers (and masers) of all types and in all frequency ranges are simply electronic devices, of great interest and importance to the electronics engineer.

In the analogous case of semiconductor electronics, for example, the transistor is not usually thought of as a quantum device. Mental images of holes and electrons as classical charged particles which accelerate, drift, diffuse and recombine are used both by semiconductor device engineers to do practical device engineering, and by solid-state physics researchers to understand sophisticated physics experiments. These classical concepts serve to explain and make understandable what is otherwise a complex quantum picture of energy bands, Bloch wavefunctions, Fermi-Dirac distributions, and occupied or unoccupied quantum states. The same simplification can be accomplished for lasers, and laser devices can then be very well understood from a primarily classical viewpoint, with only limited appeals to quantum terms or concepts.

The approach in this book is to build primarily upon the classical electron oscillator model, appropriately extended with a descriptive picture of atomic energy levels and level populations, in order to provide a *fully accurate, detailed and physically meaningful* understanding of lasers. This can be accomplished without requiring a previous formal background in quantum theory, and also without attempting to teach an abbreviated and inadequate course in this subject on the spot. A thorough understanding of laser devices is readily available through this book, in terms of classical and descriptively quantum-mechanical concepts, without a prior course in quantum theory.

I have also attempted to review, at least briefly, relevant and necessary background material for each successive topic in each section of this book. Students will find the material most understandable, however, if they come to the book with some background in electromagnetic theory, including Maxwell's equations; some understanding of the concept of electromagnetic polarization in an atomic medium; and some familiarity with the fundamentals of electromagnetic wave propagation. An undergraduate-level background in optics and in Fourier transform concepts will certainly help; and although familiarity with quantum theory is *not* required, the student must have at least enough introduction to atomic physics to be prepared to accept that atoms do have quantum properties, especially quantum energy levels and transitions between these levels.

The discussions in this book begin with simple physical descriptions and then go into considerable analytical detail on the stimulated transition process in atoms and molecules; the basic amplification and oscillation processes in laser devices; the analysis and design of laser beams and resonators; and the complexities of laser dynamics (including spiking, Q-switching, mode locking, and injection locking) common to all types of lasers. We illustrate the general principles with specific examples from a number of important common laser systems, although this book does not attempt to provide a detailed handbook of different laser systems. Extensive references to the current literature will, however, guide the reader to this kind of information.

There is obviously a large amount of material in this book. The author has taught an introductory one-quarter "breadth" course on basic laser concepts for engineering and applied physics students using most of the material from the first part of the book on "Basic Laser Physics" (see the Table of Contents), especially Chapters 1–4, 6–8 and 11–13. A second-quarter "depth" course then adds more advanced material from Chapters 5, 9, 10, 30, 31 and selected sections from Chapters 24–29. A complete course on optical beams and resonators can be taught from Chapters 14 through 23.

I am very much indebted to many colleagues for help during the many years while this book was being written. I wish it were possible to thank by name all the students in my classes and my research group who lived through too many years of drafts and class notes. Special thanks must go to Judy Clark, who became a TEX and computer expert and did so much of the editing and manuscript preparation; to the Air Force Office of Scientific Research for supporting my laser research activities over many years; to Stanford University, and especially to Donald Knuth, for providing the environment, and the computerized text preparation tools, in which this book could be written; and to the Alexander von Humboldt Foundation and the Max Planck Institute for Quantum Optics in Munich, who supplied the opportunity for the manuscript at last to be completed. Finally, there are my wife Jeannie, and my family, who made it all worthwhile.

*Anthony E. Siegman*

# UNITS AND NOTATION

The units and dimensions in this book are almost entirely mks, or SI, except for a few concessions to long-established habits such as expressing atomic densities $N$ in atoms/cm$^3$ and cross sections $\sigma$ in cm$^2$. Such non-mks values should of course always be converted to mks units before plugging them into formulas.

In general, lower-case symbols in bold-face type such as $\boldsymbol{\mathcal{E}}(\boldsymbol{r},t)$, $\mathbf{b}(\boldsymbol{r},\mathbf{t})$, $\mathbf{h}(\boldsymbol{r},\mathbf{t})$, and so on refer to electromagnetic field quantities as real vector functions of space and time, while $\mathcal{E}(r,t)$, $b(r,t)$, $h(r,t)$, etc., refer to the scalar counterparts of the same quantities. Bold-face capital letters $\mathbf{E}$, $\mathbf{B}$, $\mathbf{H}$, etc., refer to the complex phasor amplitudes of the same vector quantities with $e^{j\omega t}$ variations, while $\tilde{E}$, $\tilde{B}$, $\tilde{H}$, etc., are the complex phasor amplitudes of the corresponding scalars. As illustrated here, complex quantities are sometimes, but not always, identified by a superposed tilde.

In writing sinusoidal signals and waves, waves propagating toward positive $z$ are written in the "electrical engineer's form" of $\exp j(\omega t - \beta z)$ rather than the "physicist's form" of $\exp i(kz - \nu t)$. (This of course does *not* imply that $i \equiv -j$!) Linewidths $\Delta f$, $\Delta\omega$, $\Delta\lambda$ and pulsewidths $\Delta t$, $\tau$ or $T$, unless specifically noted, always mean the full width at half maximum (FWHM).

In contrast to much of the published literature, an attenuation or gain coefficient $\alpha$ in this book always refers to an *amplitude* or *voltage* growth rate, such as for example $\mathcal{E}(z) = \mathcal{E}(0)\exp \pm \alpha z$. Signal powers or intensities in this book, therefore, always grow or attenuate with exponential growth coefficients $2\alpha$ rather than $\alpha$.

The notation in the book has a few other minor idiosyncrasies. First, we are often concerned with signals and waves inside laser crystals, in which the host crystal itself has a dielectric constant $\epsilon$ and an index of refraction $n$ even without any atomic transition present. To take the dielectric properties of a possible host medium into account, the symbols $\epsilon$, $c$ and $\lambda$ in formulas in this text always refer to the dielectric permeability, velocity of light and wavelength of the radiation *in the dielectric medium* if there is one. We then use $c_0$ and $\lambda_0$ in the few cases where it is necessary to refer to these same quantities specifically in vacuum. The advantage of this choice is that all our formulas involving $\epsilon$, $c$ and $\lambda$ remain correct with or without a dielectric host medium, without needing to clutter these formulas with different powers of the refractive index $n$.

The other special convention peculiar to this book is the nonstandard manner in which we define the complex susceptibility $\tilde{\chi}_{at}$ associated with a resonant atomic transition. In brief, we define the linear relationship between the induced polarization $\tilde{P}_{at}$ on an atomic transition in a laser medium and the electric field $\tilde{E}$ that produces this polarization by the convention that $\tilde{P}_{at} = \tilde{\chi}_{at}\epsilon\tilde{E}$ where $\epsilon$ is the dielectric permeability *of the host laser crystal* rather than the vacuum value $\epsilon_0$ usually used in this definition. The merits of this nonstandard approach are argued in Chapter 2.

# LIST OF SYMBOLS

Throughout this text we attempt to follow a consistent notation for subscripts, using the conventions that:

$a$ = either *atomic*, as in atomic transition frequency $\omega_a$ or homogeneous atomic linewidth $\Delta\omega_a$; or sometimes *absorption*, as in absorption coefficient $\alpha_a$.

$c$ = *cavity*, as in cavity decay time $\tau_c$ or cavity energy decay rate $\gamma_c$; also, *carrier*, as in carrier frequency $\omega_c$.

$d$ = *doppler*, as in doppler broadening with linewidth $\Delta\omega_d$, and by extension any other kind of inhomogeneous broadening.

$e$ = *external*, as in cavity external coupling factor $\delta_e$ or external decay rate $\gamma_e$; also, sometimes, *effective*, as in effective lifetime or pumping rate.

$m$ = *molecular* or *maser*, generally used to refer to atomic or maser or laser quantities, e.g., laser gain coefficient $\alpha_m$ or laser growth rate $\gamma_m$.

$o$ = *ohmic*, referring generally to internal ohmic and/or scattering losses, as in the ohmic loss coefficient $\alpha_0$ or ohmic cavity decay rate $\gamma_0$. Also used in several other ways, generally to indicate an initial value; a thermal equilibrium value; a small-signal or unsaturated value; a midband value; or a free-space (vacuum) values, as in $c_0$, $\epsilon_0$, and $\lambda_0$.

$p$ = *pump*, as in pumping rate $R_p$ or pump transition probability $W_p$.

We also frequently use $ax \equiv$ *axial*; $avail \equiv$ *available*; $circ \equiv$ *circulating*; $eff \equiv$ *effective*; $eq \equiv$ *equivalent*; $inc \equiv$ *incident*; $opt \equiv$ *optimum*; $out \equiv$ *output*; $refl \equiv$ *reflected*; $rt \equiv$ *round-trip*; $sat \equiv$ *saturation*; $sp \equiv$ *spontaneous* or *spiking*; $ss \equiv$ *small-signal* or *steady-state*; and $th \equiv$ *threshold* as compound subscripts.

A partial list of symbols used in the text then includes:

$\alpha$ = exponential gain or loss coefficient for amplitude (or voltage); also, amplitude parameter for gaussian optical pulse

$\alpha''$ = second derivative of $\alpha(\omega)$ with respect to $\omega$

$\tilde{\alpha}_n$ = complex amplitude of $n$-th order Hermite-gaussian mode

$\alpha_m$ = maser/laser/molecular gain (or loss) coefficient

$\alpha_0$ = ohmic and/or scattering loss coefficient

$\beta$ = propagation constant, including host dielectric effects, but usually not loss or atomic transition effects; also, chirp parameter for gaussian pulse; relaxation-time ratio in multilevel laser pumping systems; Bohr magneton

$\beta_I$ = Nuclear magneton

$\beta', \beta''$ = first and second derivatives of $\beta(\omega)$ with respect to $\omega$

$\Delta\beta_m$ = added propagation constant term due to reactive part of an atomic transition

## LIST OF SYMBOLS

$\gamma =$ in general, an energy or population decay rate

$\gamma_c =$ decay rate for cavity stored energy ($\equiv 1/\tau_c$)

$\gamma_i =$ total downward population decay rate from energy level $E_i$

$\gamma_{ij} =$ population decay rate from upper level $E_i$ to lower level $E_j$

$\gamma_{nr} =$ nonradiative part of total decay rate for a classical oscillator or an atomic transition

$\gamma_{rad} =$ radiative decay rate for classical electron oscillator or real atomic transition

$\tilde{\gamma} =$ complex eigenvalue for optical resonator or lensguide

$\tilde{\gamma}_{mn} =$ complex eigenvalue for $mn$-th order transverse eigenmode

$\Gamma = \alpha + j\beta =$ complex propagation constant for an optical wave

$\Gamma = \alpha - j\beta =$ complex gaussian pulse parameter

$\delta =$ coefficient of (logarithmic) fractional power gain or loss, per bounce or per round trip

$\delta_c =$ total (round-trip) power loss coefficient due to cavity losses plus external coupling

$\delta_e =$ cavity loss coefficient due to external coupling only

$\delta_m =$ power gain coefficient due to laser atoms

$\delta_0 =$ cavity loss coefficient due to internal (ohmic) losses only

$\Delta_m =$ AM or FM modulation index

$\epsilon =$ dielectric permeability of a medium

$\epsilon_0 =$ dielectric permeability of free space (vacuum)

$\eta =$ efficiencies of various sorts; also, characteristic impedance $\sqrt{\mu/\epsilon}$ of a dielectric medium

$\eta_0 =$ characteristic impedance of free space (vacuum)

$\lambda =$ optical wavelength (in a medium); also, eigenvalue for optical ray matrix

$\lambda_0 =$ optical wavelength in vacuum

$\lambda_a, \lambda_b =$ eigenvalues of periodic lensguide or $ABCD$ matrix

$\Lambda =$ spatial period of optical grating

$\mu =$ electric or magnetic dipole moment; also, magnetic permeability of a magnetic medium

$\mu_e =$ electric dipole moment

$\mu_m =$ magnetic dipole moment

$\mu_0 =$ magnetic permeability of free space

$\rho =$ amplitude reflection or transmission of optical mirror or beamsplitter; also, distance between two points; $\rho(\omega) =$ cavity mode density

$\tilde{\rho} =$ complex amplitude reflection or transmission of optical mirror or beamsplitter

$\sigma =$ ohmic conductivity; also, transition cross section, standard deviation

$\sigma_{ij} =$ cross section for stimulated transition from level $E_i$ to $E_j$

$\tau =$ lifetime or decay time

$\tau_c =$ cavity decay time due to all internal losses plus external coupling

$\tau_i =$ total lifetime (energy decay time) for energy level $E_i$

$\theta, \phi, \psi =$ phase shifts and phase angles of various sorts

$\psi(\mathbf{r}, t) =$ Schrödinger wave function

## LIST OF SYMBOLS

$\psi_{mn}$ = Guoy phase shift for an $mn$-th order gaussian beam
$\tilde{\chi}$ = susceptibility of a dielectric or magnetic medium = $\chi' + j\chi''$
$\chi', \chi''$ = real and imaginary parts of $\tilde{\chi}$
$\tilde{\chi}_{at}$ = susceptibility of a resonant atomic transition
$\tilde{\chi}_e, \tilde{\chi}_m$ = electric (magnetic) dipole susceptibilities
$\omega$ = frequency (in radians/second)
$\omega'$ = in general, a frequency that has been shifted, pulled, or modified in some small manner
$\omega_a$ = atomic transition frequency
$\omega_b$ = a beat frequency (between two signals)
$\omega_c$ = cavity or circuit resonant frequency; also, carrier frequency
$\omega_i(t)$ = instantaneous frequency of a phase-modulated signal
$\omega_m$ = generally, a modulation frequency of some sort
$\omega_q$ = resonant frequency of $q$-th axial mode
$\omega_R$ = Rabi frequency on an atomic transition
$\omega_{sp}$ = Spiking or relaxation-oscillation frequency
$\delta\omega_q$ = frequency pulling of axial mode frequency $\omega_q$
$\Delta\omega$ = linewidth, or frequency tuning, in radians/sec
$\Delta\omega_a$ = atomic linewidth (FWHM)nin radians/sec
$\Delta\omega_{ax}$ = axial mode spacing between adjacent axial modes
$\Omega$ = solid angle; also, radian frequency or rotation rate

$\tilde{a}_i, \tilde{b}_i$ = normalized wave amplitudes
$A$ = area
$A_{ji}$ = Einstein $A$ coefficient on $E_j \to E_i$ transition
$ABCD$ = matrix elements for optical ray matrix or paraxial optical system
$b$ = magnetic field as real function of space and time; also, confocal parameter for gaussian beam
$\boldsymbol{b}$ = magnetic field as real vector function of space and time; also, confocal parameter for gaussian beam
$B$ = magnetic field; also, pressure-broadening coefficient or "$B$ integral" for nonlinear interaction
$\tilde{B}$ = phasor amplitude of sinusoidal $B$ field
$c$ = velocity of light in a material medium
$c_0$ = velocity of light in vacuum
$C$ = in general, an unspecified constant; also, electrical capacitance; coupling coefficient in mode competition analysis
$CC$ = complex conjugate (of preceding term)
CEO = classical electron oscillator model
$d$ = electric displacement as real function of space and time; also, distance or displacement
$\boldsymbol{d}$ = electric displacement as real vector function of space and time
$D$ = dimensionless dispersion parameter
$\tilde{D}$ = phasor amplitude of sinusoidal electric displacement
$e$ = magnitude of electronic charge
$\mathcal{E}$ = electric field; usually, real field $\mathcal{E}(x,t)$ as function of space and time

# LIST OF SYMBOLS

$\tilde{E}$ = phasor amplitude of sinusoidal $E$ field

$E_n(t)$ = amplitude of $n$-th mode in a normal mode expansion

$f$ = frequency in Hz ($\equiv$ cycles/sec); also, lens focal length

$f^\#$ = lens $f$-number

$\Delta f$ = linewidth, or frequency detuning, in Hz

$\Delta f_a$ = atomic transition linewidth (FWHM) in Hz

$\Delta f_d$ = doppler or inhomogeneous linewidth (FWHM) in Hz

$F$ = oscillator strength for an atomic transition; also, lens $f$-number

$\mathcal{F}$ = finesse, of interferometer or laser cavity

$\tilde{F}(x)$ = Fresnel integral function

$F_{ji}$ = oscillator strength of $E_j \rightarrow E_i$ atomic transition $\equiv \gamma_{\text{rad},ji}/3\gamma_{\text{rad,ceo}}$

$g$ = amplitude (or voltage) gain, as a number; also, gaussian stable resonator parameter; magnetic resonance $g$ value

$g(v), g(\omega)$ = normalized lineshapes

$\tilde{g}$ = complex amplitude (or voltage) gain, as a (complex) number

$g_i, g_j$ = degeneracy factors for quantum energy levels $E_i$ and $E_j$

$g_I$ = nuclear magnetic resonance $g$ value

$\tilde{g}_{rt}$ = round-trip voltage gain inside an optical cavity

$G$ = power gain (as a number); also, electrical conductance

$G_{dB}$ = power gain in decibels

$h$ = magnetic intensity as real function of space and time; also, Planck's constant

$\hbar$ = $h/2\pi$

$\boldsymbol{h}$ = magnetic $H$ field as real vector function of space and time

$h_n$ = $n$-th order polynomial function

$\tilde{H}$ = phasor amplitude of sinusoidal $H$ field

$H_n$ = $n$-th order hermite polynomial

$I$ = intensity (power/unit area) of an optical wave; also sometimes, loosely, total power in the wave

$I_m$ = modified Bessel function of order $m$

$I_{sat}$ = amplifier (or absorber) saturation intensity

$j$ = current density as real function of space and time; also, $\sqrt{-1}$

$\boldsymbol{j}$ = current density as real vector function of space and time

$\tilde{J}$ = phasor amplitude of sinusoidal current density

$J_m$ = Bessel function of order $m$

$k$ = propagation vector of optical wave = $\omega/c$

$K$ = scalar constant in various equations (especially coupled rate equations); also, spring constant in classical oscillator model

$L$ = length; electrical inductance

$m$ = electron mass; also, magnetization (magnetic dipole moment per unit volume) as real function of time

$\boldsymbol{m}$ = magnetization (magnetic dipole moment per unit volume) as real vector function of space and time

$m, \tilde{m}$ = half-trace parameter for ray or $ABCD$ matrix

$M$ = proton mass; molecular mass

## LIST OF SYMBOLS

$\tilde{M}$ = phasor amplitude of sinusoidal magnetic dipole moment

$M$ = optical ray matrix or $ABCD$ matrix

$n$ = refractive index; also, photon number $n(t)$ (number of photons per cavity mode)

$n_2$ = optical Kerr coefficient $n_{2E}$ or $n_{2I}$

$N$ = atomic number or level population; usually interpreted as atoms per unit volume, sometimes as total number of atoms

$\Delta N$ = population difference, or population difference density, on an atomic transition ($\Delta N_{ij} \equiv N_i - N_j$)

$N$ = Fresnel number $a^2/L\lambda$ for an optical beam or resonator

$N_c$ = collimated Fresnel number for an unstable optical resonator

$N_{eq}$ = equivalent Fresnel number for an unstable optical resonator

$N_i$ = population, or population density, in atomic energy level $E_i$

$p$ = perimeter, period or round-trip path length, for cavities or periodic lensguides; also, electric polarization (electric dipole moment per unit volume) as real function of time, and laser mode density or mode number

$\boldsymbol{p}$ = electric polarization (electric dipole moment per unit volume) as real vector function of space and time

$p_m$ = path length (round-trip) through an atomic or laser gain medium

$P$ = power, in watts; also, pressure, in torr

$P_n(t)$ = polarization driving term for $n$-th order cavity mode in coupled-mode expansion

$\tilde{P}$ = phasor amplitude of sinusoidal electric polarization

$q$ = axial mode index

$\tilde{q}$ = complex gaussian beam parameter or complex radius of curvature

$\hat{q}$ = reduced gaussian beam parameter, $\tilde{q}/n$

$r$ = amplitude reflectivity of mirror or beamsplitter; also, dimensionless or normalized pumping rate; displacement off axis of optical ray

$r'$ = reduced slope $n\, dr/dz$ for optical ray

$\boldsymbol{r}$ = shorthand for spatial coordinates $x, y, z$

$\tilde{r}_{ij}$ = complex scattering matrix element, or mirror or beamsplitter reflection coefficient

$r_p$ = dimensionless pumping rate or inversion ratio, relative to threshold pumping rate or threshold inversion density

$d\boldsymbol{r}$ = volume element, $dV$ or $dx\, dy\, dz$

$R$ = power reflectivity of mirror or beamsplitter ($\equiv |r|^2$); also, electrical resistance; radius of curvature for mirror, dielectric interface, or optical wave

$\hat{R}$ = reduced radius of curvature $R/n$

$R_p$ = pumping rate in atoms per second and, usually, per unit volume

$s$ = spatial frequency (cycles/unit length)

$\boldsymbol{s}$ = shorthand for transverse spatial coordinates $x, y$

$d\boldsymbol{s}$ = transverse area element $dA$ or $dx\, dy$

$\boldsymbol{S}$ = multiport scattering matrix (matrix elements $S_{ij}$)

## LIST OF SYMBOLS

$t$ = time; also, amplitude transmission through mirror, beamsplitter, or light modulator

$\tilde{t}$ = complex amplitude transmission coefficient through mirror, beamsplitter or light modulator

$\tilde{t}_{ij}$ = complex scattering matrix element, or mirror/beamsplitter transmission coefficient

$T$ = power transmission of mirror or beamsplitter ($\equiv |t|^2$); also, cavity round-trip transit time, or temperature (K)

$\boldsymbol{T}$ = dimensionless susceptibility tensor

$T_b$ = laser oscillation build-up time

$T_{nr}$ = temperature of "nonradiative" surroundings

$T_{rad}$ = temperature of radiative surroundings

$T_1$ = energy decay time, population recovery time, longitudinal relaxation time

$T_2$ = dephasing time, collision time, transverse relaxation time

$T_2^*$ = effective $T_2$ or dephasing time for inhomogeneous (gaussian) transition

$\tilde{u}$ = complex (and usually normalized) optical wave amplitude

$U$ = energy or, more commonly, energy density (energy per unit volume)

$U_a$ = energy density in a collection of atoms or atomic energy level populations

$U_{bbr}$ = energy density of blackbody radiation

$v$ = velocity of an atom, an electron, or a wave

$\tilde{v}$ = complex spot size for Hermite-gaussian modes

$v_g$ = group velocity

$v_\phi$ = phase velocity

$V, V_c$ = volume (of a cavity mode or field pattern)

$w$ = gaussian spot size parameter ($1/e$ amplitude point)

$w_{ij}$ = total relaxation transition probability (per atom, per second) from level $E_i$ to level $E_j$

$W_{ij}$ = stimulated transition probability (per atom, per second) from level $E_i$ to level $E_j$

$W_p$ = pumping transition probability (per atom, per second)

$x(t)$ = displacement of electronic charge in classical electron oscillator model

$z_D$ = dispersion length for dispersive pulse broadening

$z_R$ = Rayleigh range for a gaussian or collimated optical beam

$Z$ = atomic number

$2^*$ = dimensionless population saturation factor, with values between $2^* = 1$ (lower level empties out rapidly) and $2^* = 2$ (lower level bottlenecked)

$3^*$ = dimensionless polarization overlap factor for atomic interactions, with numerical value between 0 and 3

# LASERS

# 1
# AN INTRODUCTION TO LASERS

Lasers are devices that generate or amplify coherent radiation at frequencies in the infrared, visible, or ultraviolet regions of the electromagnetic spectrum. Lasers operate by using a general principle that was originally invented at microwave frequencies, where it was called *m*icrowave *a*mplification by *s*timulated *e*mission of *r*adiation, or *maser* action. When extended to optical frequencies this naturally becomes *l*ight *a*mplification by *s*timulated *e*mission of *r*adiation, or *laser* action.

This basic laser or maser principle is now used in an enormous variety of devices operating in many parts of the electromagnetic spectrum, from audio to ultraviolet. Practical laser devices in particular employ an extraordinary variety of materials, pumping methods, and design approaches, and find a great variety of applications. The study of laser and maser devices and their scientific applications is often referred to as the field of *quantum electronics*.

From an electronics-engineering viewpoint, the developments that followed the operation of the first ruby laser in 1960 suddenly pushed the upper limit of coherent electronics from the millimeter-wave range, using microwave tubes and transistors, out to include the submillimeter, infrared, visible, and ultraviolet spectral regions (and soft X-ray lasers are now on the horizon). All the familiar functions of coherent signal generation, amplification, modulation, information transmission, and detection are now possible at frequencies up to a million times higher, or wavelengths down to a million times shorter, than previously. But it has also become possible for engineers and scientists, in fields of technology ranging from microbiology to auto manufacture, to perform an almost unlimited variety of new and unexpected functions made possible by the short wavelengths, high powers, ultrashort pulsewidths, and other unique characteristics of these laser devices.

In the twenty-odd years since the first appearance of coherent light, lasers have become widespread and almost commonplace devices. The importance and the excitement of the laser and its applications, however, still can hardly be overestimated. The objective of this book is to explain in detail how lasers work, what the performance characteristics of typical lasers are, and how lasers are employed in a wide variety of applications. Our goal in this opening chapter is

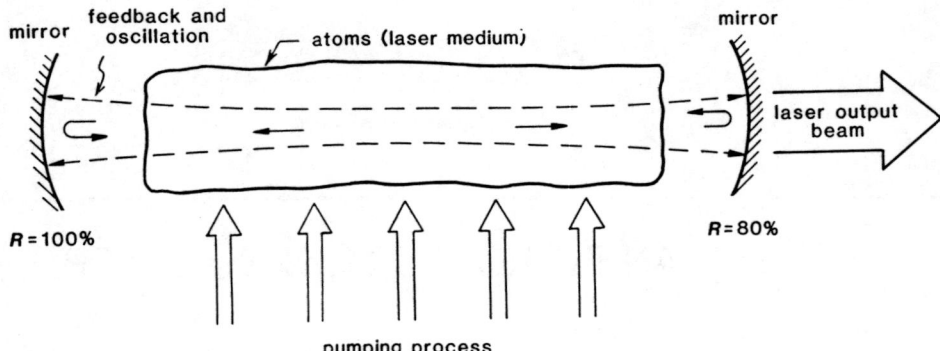

FIGURE 1.1
Elements of a typical laser oscillator.

to give an abbreviated overview of these same points, as a synopsis of what will be presented in much more detail in the remainder of the book.

## 1.1 WHAT IS A LASER?

Lasers, broadly speaking, are devices that generate or amplify light, just as transistors and vacuum tubes generate and amplify electronic signals at audio, radio, or microwave frequencies. Here "light" must be understood broadly, since different kinds of lasers can amplify radiation at wavelengths ranging from the very long infrared region, merging with millimeter waves or microwaves, up through the visible region and extending now to the vacuum ultraviolet and even X-ray regions. Lasers come in a great variety of forms, using many different laser materials, many different atomic systems, and many different kinds of pumping or excitation techniques. The beams of radiation that lasers emit or amplify have remarkable properties of directionality, spectral purity, and intensity. These properties have already led to an enormous variety of applications, and others undoubtedly have yet to be discovered and developed.

### Essential Elements of a Laser

The essential elements of a laser device, as shown in Figure 1.1, are thus: (i) a *laser medium* consisting of an appropriate collection of atoms, molecules, ions, or in some instances a semiconducting crystal; (ii) a *pumping process* to excite these atoms (molecules, etc.) into higher quantum-mechanical energy levels; and (iii) suitable *optical feedback elements* that allow a beam of radiation to either pass once through the laser medium (as in a laser amplifier) or bounce back and forth repeatedly through the laser medium (as in a laser oscillator).

These elements come in a great variety of forms and fashions, as we will see when we begin to examine each of them in more detail.

### Laser Atoms and Laser Pumping

For simplicity we will from now on use "atoms" as a general term for whatever kind of atoms or molecules or ions or semiconductor electrons may be used as the laser medium. A pumping process is then required to excite

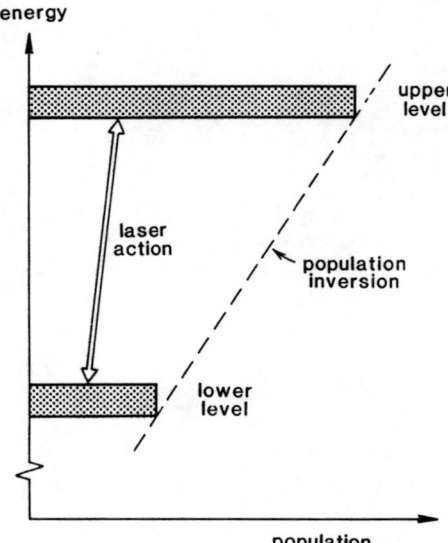

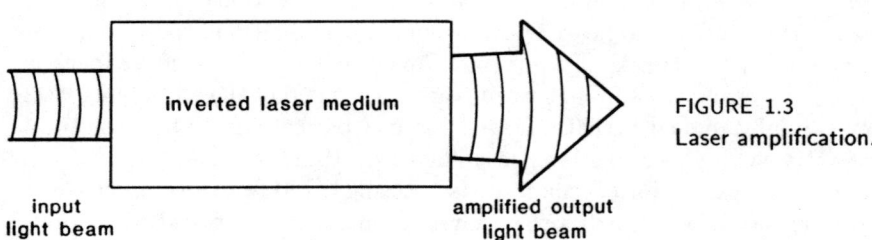

FIGURE 1.2
Population inversion between two quantum-mechanical energy levels.

FIGURE 1.3
Laser amplification.

these atoms into their higher quantum-mechanical energy levels. Practical laser materials can be pumped in many ways, as we will describe later in this text.

For laser action to occur, the pumping process must produce not merely excited atoms, but a condition of *population inversion* (Figure 1.2), in which more atoms are excited into some higher quantum energy level than are in some lower energy level in the laser medium. It turns out that we can obtain this essential condition of population inversion in many ways and with a wide variety of laser materials—though sometimes only with substantial care and effort.

### Laser Amplification

Once population inversion is obtained, electromagnetic radiation within a certain narrow band of frequencies can be coherently amplified if it passes through the laser medium (Figure 1.3). This amplification bandwidth will extend over the range of frequencies within about one atomic linewidth or so on either side of the quantum transition frequency from the more heavily populated upper energy level to the less heavily populated lower energy level.

Coherent amplification means in this context that the output signal after being amplified will more or less exactly reproduce the input signal, except for a substantial increase in amplitude. The amplification process may also add some small phase shift, a certain amount of distortion, and a small amount of

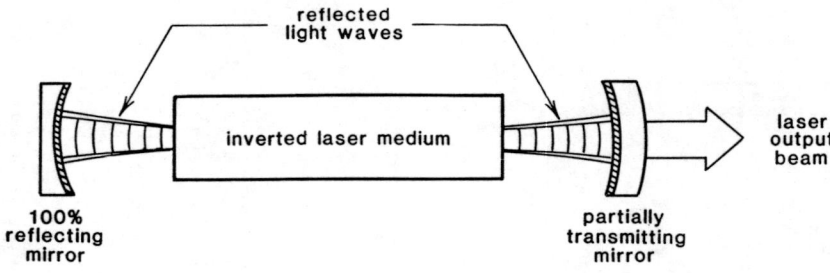

FIGURE 1.4
Laser oscillation.

amplifier noise. Basically, however, the amplified output signal will be a coherent reproduction of the input optical signal, just as in any other coherent electronic amplification process.

Laser Oscillation

Coherent amplification combined with feedback is, of course, a formula for producing oscillation, as is well known to anyone who has turned up the gain on a public-address system and heard the loud squeal of oscillation produced by the feedback from the loudspeaker output to the microphone input. The feedback in a laser oscillator is usually supplied by mirrors at each end of the amplifying laser medium, carefully aligned so that waves can bounce back and forth between these mirrors with very small loss per bounce (Figure 1.4). If the net laser amplification between mirrors, taking into account any scattering or other losses, exceeds the net reflection loss at the mirrors themselves, then coherent optical oscillations will build up in this system, just as in any other electronic feedback oscillator.

When such coherent oscillation does occur, an output beam that is both highly directional and highly monochromatic can be coupled out of the laser oscillator, either through a partially transmitting mirror on either end, or by some other technique. This output in essentially all lasers will be both extremely bright and highly coherent. The output beam may also in some cases be extremely powerful. Just what we mean by "bright" and by "coherent" we will explain later.

## REFERENCES

The first stimulated emission devices, before lasers, were various kinds of masers, which operated on essentially the same basic physical principles, but at much lower frequencies and with much different experimental techniques. For an overview and unified approach to all these devices, see my earlier texts *Microwave Solid-State Masers* (McGraw-Hill, 1964) and *An Introduction to Lasers and Masers* (McGraw-Hill, 1971).

Some other good books on lasers can be found. A more elementary introduction, with good illustrations, is D. C. O'Shea, W. R. Callen, and W. T. Rhodes, *Introduction to Lasers and Their Applications* (Addison-Wesley, 1977). A good general coverage is also given in O. Svelto, *Principles of Lasers* (Plenum Press, 1982). Two well-known texts by A. Yariv are *Introduction to Optical Electronics* (Rinehart and Winston, 1971) and the more advanced *Quantum Electronics* (Wiley, 1975).

For full quantum-mechanical treatments of lasers, two good choices are M. Sargent III, M. O. Scully, and W. E. Lamb, Jr., *Laser Physics* (Addison-Wesley, 1977), and H. Haken, *Laser Theory* (Springer-Verlag, 1983).

A useful short bibliographic survey of laser references, aimed particularly at the college teacher, can be found in "Resource Letter L-1: Lasers," by D. C. O'Shea and D. C. Peckham, *Am. J. Phys.* **49**, 915–925 (October 1981).

For more advanced information on various laser topics, the four-volume *Laser Handbook*, edited by F. T. Arecchi and E. O. Schulz-Dubois (North-Holland, Amsterdam, 1972), provides an encyclopedic source with detailed articles on nearly every topic in laser physics, devices, and applications. If you'd like to look at some of the important original literature on lasers for yourself, well-chosen selections can be found in F. S. Barnes, ed., *Laser Theory* (IEEE Press Reprint Series, IEEE Press, 1972), or in D. O'Shea and D. C. Peckham, *Lasers: Selected Reprints* (American Association of Physics Teachers, Stony Brook, N. Y., 1982).

If you would like to do experiments with a home-made laser or just see how one might be constructed, a useful collection of articles from the "Amateur Scientist" section of *Scientific American* has been reprinted under the title *Light and Its Uses*, with introduction by Jearl Walker (W. H. Freeman and Company, 1980). Topics covered include simple helium-neon, argon-ion, carbon-dioxide, semiconductor, tunable dye, and nitrogen lasers, plus experiments on holography, interferometry, and spectroscopy.

---

Problems for 1.1

1. *Diagramming the electromagnetic spectrum.* On a large sheet of paper lay out a logarithmic frequency scale extending from the audio range (say, $f = 10$ Hz) to the far ultraviolet or soft X-ray region (say, $\lambda = 100$ Å). Mark both frequency and wavelength below the same scale in powers of 10 in appropriate units, e.g., Hz, kHz, MHz, and m, mm, $\mu$m. (You might also mark a "wavenumber" scale for $1/\lambda$ in units of cm$^{-1}$, and an energy scale for $\hbar\omega$ in units of eV.) Above the scale indicate the following landmarks (plus any other significant ones that occur to you):

   - Audio frequency range (human ear) (20–15000 Hz)
   - Standard AM and FM broadcast bands (535–1605 kHz, 88–108 MHz)
   - Television channels 2–6 (54–88 MHz) and 7–13 (174–216 MHz)
   - Microwave radar "S" and "X" bands (2–4 and 8–12 GHz)
   - Visible region (human eye)
   - Important laser wavelengths, including:

     HCN far-IR laser (311, 337, 545, 676, 744 $\mu$m)

     $H_2O$ far-IR laser (28, 48, 120 $\mu$m)

     $CO_2$ laser (9.6–10.6 $\mu$m)

     CO laser (5.1–6.5 $\mu$m)

     HF chemical laser (2.7–3.0 $\mu$m)

     Nd:YAG laser (1.06 $\mu$m)

     He-Ne lasers (1.15 $\mu$m, 633 nm)

     GaAs semiconductor laser (870 nm)

     Ruby laser (694 nm)

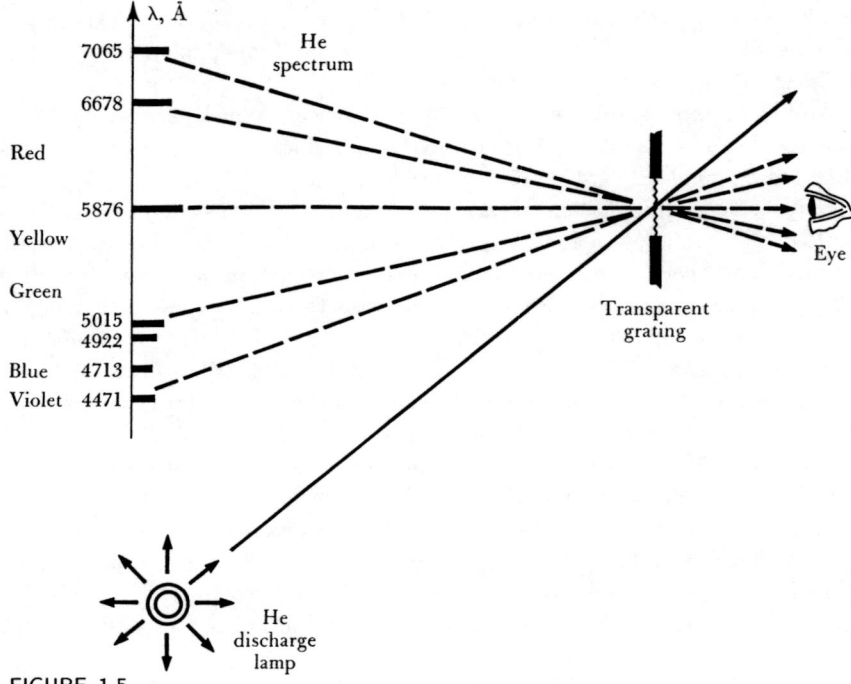

**FIGURE 1.5**
Helium discharge spectrum observed through an inexpensive replica transmission grating.

> Rhodamine 6G dye laser (560–640 nm)
> Argon-ion laser (488–515 nm)
> Pulsed $N_2$ discharge laser (337 nm)
> Pulsed $H_2$ discharge laser (160 nm)

## 1.2 ATOMIC ENERGY LEVELS AND SPONTANEOUS EMISSION

Our objective in this section is to give a very brief introduction to the concepts of atomic energy levels and of spontaneous emission between those levels. We attempt to demonstrate heuristically that atoms (or ions, or molecules) have quantum-mechanical energy levels; that atoms can be pumped or excited up into higher energy levels by various methods; and that these atoms then make spontaneous downward transitions to lower levels, emitting radiation at characteristic transition frequencies in the process. (Readers already familiar with these ideas may want to move on to Section 1.3.)

### The Helium Spectrum

Figure 1.5 illustrates a simple experiment in which a small helium discharge lamp (or lacking that, a neon sign) is viewed through an inexpensive transmission diffraction grating of the type available at scientific hobby stores. (If you have never done such an experiment, try to do this demonstration for yourself.)

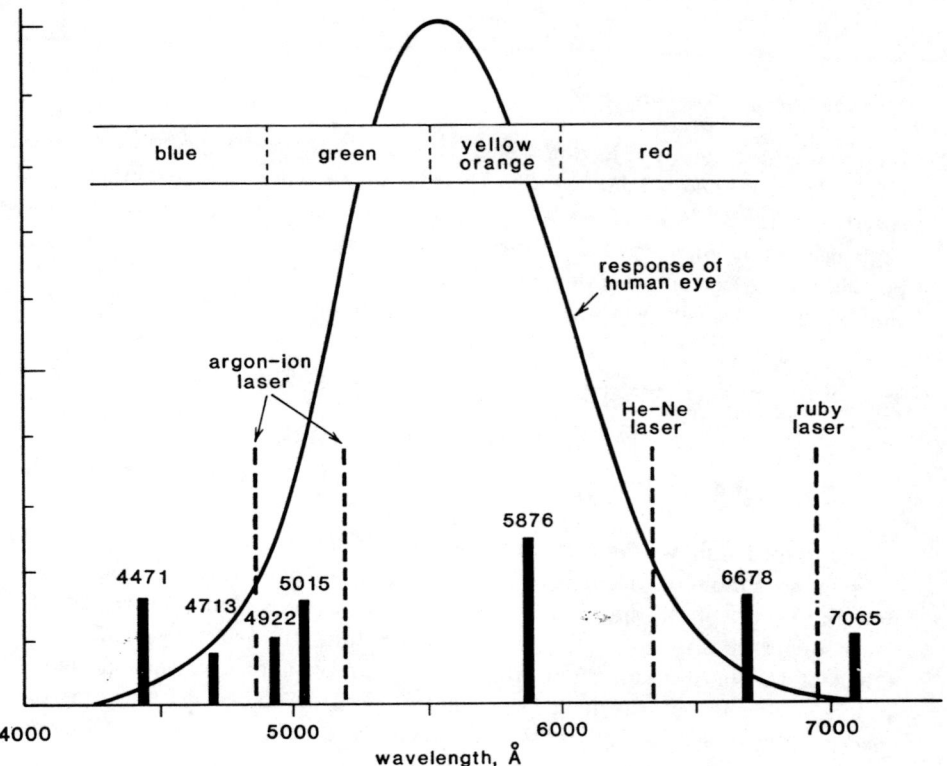

FIGURE 1.6
Helium spectral lines, four common laser lines, and human visual sensitivity.

When viewed directly the discharge helium lamp appears to emit pinkish-white light. When viewed through the diffraction grating, however, each wavelength in the light is diffracted at a different angle. Upon looking through the grating, you therefore observe multiple images of the lamp, each displaced to a different discrete angle, and each made up of a different discrete wavelength or color emitted by the helium discharge. A strong yellow line at 5876Å (or 588 nm) is particularly evident, but violet, green, blue, red, and deep red lines are also readily seen. These visible wavelengths are plotted in Figure 1.6, along with (as a matter of curiosity) the relative response of the human eye, and the wavelengths of four of the more common visible lasers.

These different wavelengths are, of course, only a few of the discrete components in the fluorescence spectrum of the helium atoms. In the helium discharge tube a large number of neutral helium atoms are present, along with a small number of free electrons and a matching number of ionized helium atoms to conduct electrical current. The free electrons are accelerated along the tube by the applied electric field, and collide after some distance with the neutral helium atoms. The helium atoms are thereby excited into various higher *quantum energy levels* characteristic of the helium atoms. A small fraction are also ionized by the electron collisions, thereby maintaining the electron and ion densities against recombination losses, which occur mostly at the tube walls.

After being excited into upper energy levels, the helium atoms soon give up their excess energy by dropping down to lower energy levels, emitting spontaneous electromagnetic radiation in the process. This *spontaneous emission* or *fluorescence* is the mechanism that produces the discrete spectral lines.

### The Discovery of Helium

Helium was first identified as a new element by its fluorescence spectrum in the solar corona. During the solar eclipse of 1868 a bright yellow line was observed in the emission spectrum of the Sun's prominences by at least six different observers. This line could be explained in relation to the known spectral lines of already identified elements only by postulating the existence of a new element, helium, named after the Greek word Helios, the Sun. This same element was later, of course, identified and isolated on Earth.

### Quantum Energy Levels

Figure 1.7 shows the rather complex set of quantum energy levels possessed by even so simple an atom as the He atom. The solid arrows in this diagram designate some of the spontaneous-emission transitions that are responsible for the stronger lines in the visible spectrum of helium. The dashed arrows indicate a few of the many additional transitions that produce spontaneous emission at longer or shorter wavelengths in the infrared or ultraviolet portions of the spectrum, lines which we can "see" only with the aid of suitable instruments.

Every atom in the periodic table, as well as every molecule or ion, has its own similar characteristic set of quantum energy levels, and its own characteristic spectrum of fluorescent emission lines, just as does the helium atom. Understanding and explaining the exact values of these quantum energy levels for different atoms and molecules, through experiment or through complex quantum analyses, is the task of the spectroscopist. The complex labels given to each energy level in Figure 1.7 are part of the working jargon of the spectroscopist or atomic physicist. In this text we will not be concerned with predicting the quantum energy levels of laser atoms, or even with understanding their complex labeling schemes, except in a few simple cases. Rather, we will accept the positions and properties of these levels as part of the data given us by spectroscopists, and will concentrate on understanding the dynamics and the interactions through which laser action is obtained on these transitions.

### Planck's Law

The relationship between the frequency $\omega_{21}$ emitted on any of these transitions and the energies $E_2$ and $E_1$ of the upper and lower atomic levels is given by Planck's Law

$$\omega_{21} = \frac{E_2 - E_1}{\hbar}, \qquad (1)$$

where $\hbar \equiv h/2\pi$, and Planck's constant $h = 6.626 \times 10^{-34}$ Joule-second.

In this text, as in real life, optical and infrared radiation will sometimes be characterized by its frequency $\omega$, and sometimes by its wavelength $\lambda_0$ expressed in units such as Ångstroms (Å), nanometers (nm), or microns (μm). Quantum transitions and the associated transition frequencies are also very often characterized by their transition energy or photon energy, measured in units of electron volts (eV), or their inverse wavelength $1/\lambda_0$ measured in units of "wavenumbers"

## 1.2 ATOMIC ENERGY LEVELS AND SPONTANEOUS EMISSION

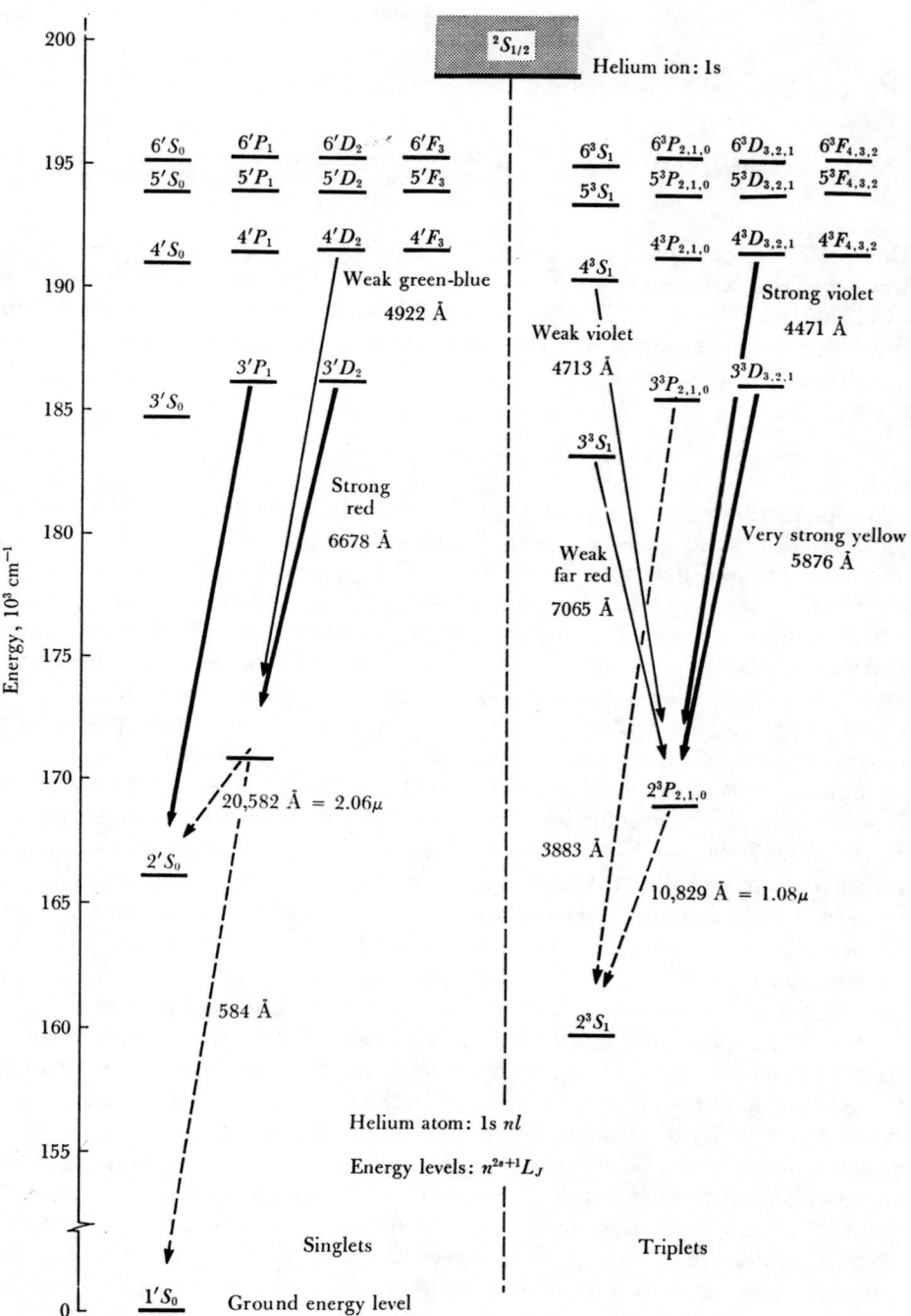

**FIGURE 1.7**
An energy-level diagram for the helium atom, showing the transitions responsible for the strong visible spectrum, as well as various ultraviolet and infrared transitions.

or cm$^{-1}$. Since we will be jumping back and forth between these units, it will be worthwhile to gain some familiarity with their magnitudes. Some useful rules of thumb to remember are that

$$1 \ \mu\text{m} \ (\text{``one micron''}) \equiv 1{,}000 \ \text{nm} \equiv 10000 \text{Å} \qquad (2)$$

and that, in suitable energy units,

$$[\text{transition energy } E_2 - E_1 \text{ in eV}] \approx \frac{1.24}{[\text{wavelength } \lambda_0 \text{ in microns}]}. \qquad (3)$$

Hence 10000Å or 1 $\mu$m matches up with 10,000 cm$^{-1}$ or $\sim$ 1.24 eV. A visible wavelength of 500 nm or 5000Å or 0.5 $\mu$m thus corresponds to a photon energy of 20,000 cm$^{-1}$ or $\sim$ 2.5 eV. Note that this also corresponds to a transition frequency of $\omega_{21}/2\pi = 6 \times 10^{14}$ Hz, expressed in the conventional units of cycles per second, or Hertz.

### Energy Levels in Solids: Ruby or Pink Sapphire

As another simple illustration of energy levels, try shining a small ultraviolet lamp (sometimes called a "mineral light") on any kind of fluorescent mineral, such as a piece of pink ruby or a sample of glass doped with a rare-earth ion, or on a fluorescent dye such as Rhodamine 6G. These and many other materials will then glow or fluoresce brightly at certain discrete wavelengths under such ultraviolet excitation. A sample of ruby, for example, will fluoresce very efficiently at $\lambda \approx 694$ nm in the deep red, a sample of crystal or glass doped with, say, the rare-earth ion terbium, $Tb^{3+}$, will fluoresce at $\lambda \approx 540$ nm (bright green), and a liquid sample of Rhodamine 6G dye will fluoresce bright orange.

Since ruby was the very first laser material, and is still a useful and instructive laser system, let us examine its fluorescence in more detail. Figure 1.8 shows a more sophisticated version of such an experiment, in which a scanning monochromator plus an optical detector are used to examine the ruby fluorescent emission in more detail. The lower trace shows the two very sharp (for a solid) and very closely spaced deep-red emission lines that will be observed from a good-quality ruby sample cooled to liquid-helium temperature. (At higher temperatures these lines will broaden and merge into what appears to be a single emission line.)

Figure 1.9 shows the crystal structure of ruby. Ruby consists essentially of lightly doped sapphire, $Al_2O_3$, with the darker spheres in the figure indicating the $Al^{3+}$ ions. (The lattice planes shown in the figure are $\sim$ 2.16Å apart.) Sapphire is a very hard, colorless (when pure), transparent crystal which can be grown in large and optically very good samples by flame-fusion techniques. The transparency of pure sapphire in the visible and infrared means that its $Al^{3+}$ and $O^{2-}$ atoms, when they are bound into the sapphire crystal lattice, have no absorption lines from their ground energy levels to levels anywhere in the infrared or visible regions. Indeed, no optical absorption appears in pure sapphire below the insulating band gap of the crystal in the ultraviolet.

We can, however, replace a significant fraction (several percent) of the $Al^{3+}$ ions in the lattice by chromium or $Cr^{3+}$ ions. The sapphire lattice as a result acquires a pink tint at low chromium concentrations, or a deeper red color at higher concentrations, and becomes what is called "pink ruby." The individual chromium ions, when they are bound into the sapphire lattice, have a set of quantum energy levels that are associated with partially filled inner electron shells in the $Cr^{3+}$ ion. These energy levels are located as shown in Figure 1.10.

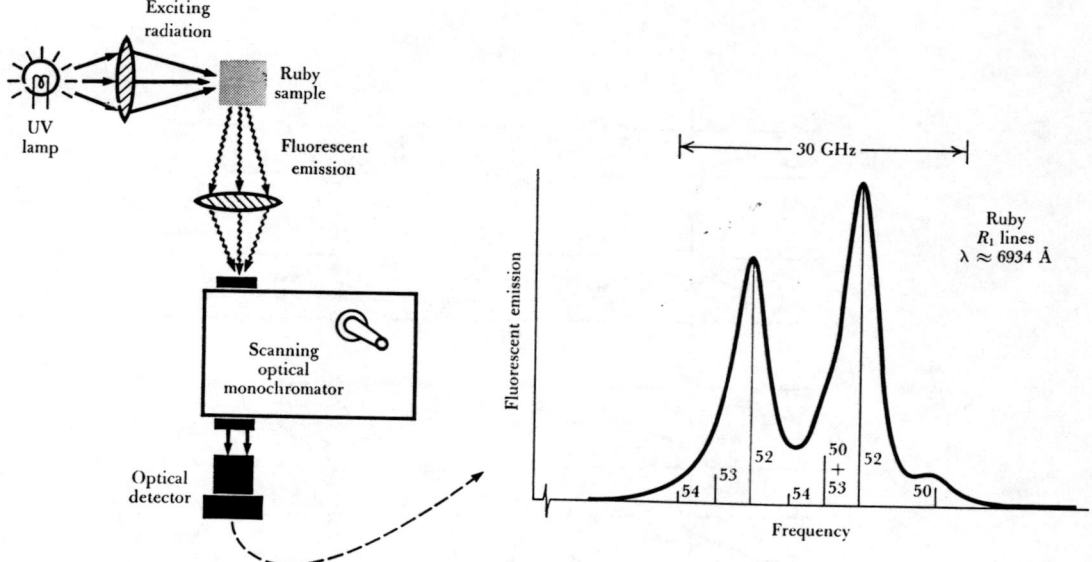

**FIGURE 1.8**
Fluorescent emission from a ruby crystal. The numbers under the spectrum indicate the slightly shifted transition frequencies corresponding to different isotopes of chromium.

The chromium ions can then absorb incident light in broad wavelength bands extending across much of the visible and near ultraviolet, by making transitions upward from the ground or $^4A_2$ $Cr^{3+}$ energy level to the series of broad bands or groups of levels labeled $^4F$ and $^2F$ in Figure 1.10. The chromium ions that are excited up into these levels then drop down by rapid nonradiative processes (which we will discuss shortly) to the two sharp $^2E$ levels shown in the figure. From there, these ions relax across the remaining energy gap down to the ground state by almost totally radiative relaxation, emitting the deep-red fluorescent emission characteristic of ruby. (The two sharp $^2E$ levels are often called the $R_1$ and $R_2$ levels, with most of the fluorescent emission coming from the lower or $R_1$ level. The two very sharp emission lines shown in Figure 1.10 then represent the separate transitions from the $R_1$ level down to the two closely spaced sublevels of the $^4A_2$ ground level.)

### Synthetic Sources of Pink Ruby

Sapphire, or rather pink ruby, was first grown in large amounts for use as jewels in the Swiss watch industry (it is said the pink color was added to make the tiny jewels easier to see and handle). Note that the energy levels of the $Cr^{3+}$ ion in ruby are very strongly shifted by Stark effects associated with the bonding of the $Cr^{3+}$ ion to the surrounding lattice ions. Hence these levels are very different from what would be the energy levels of an isolated $Cr^{3+}$ ion in free space. Many other colors of sapphire can also be created by adding other impurities, such as Fe, Mn, or Co, but only chromium-doped sapphire makes a good laser material.

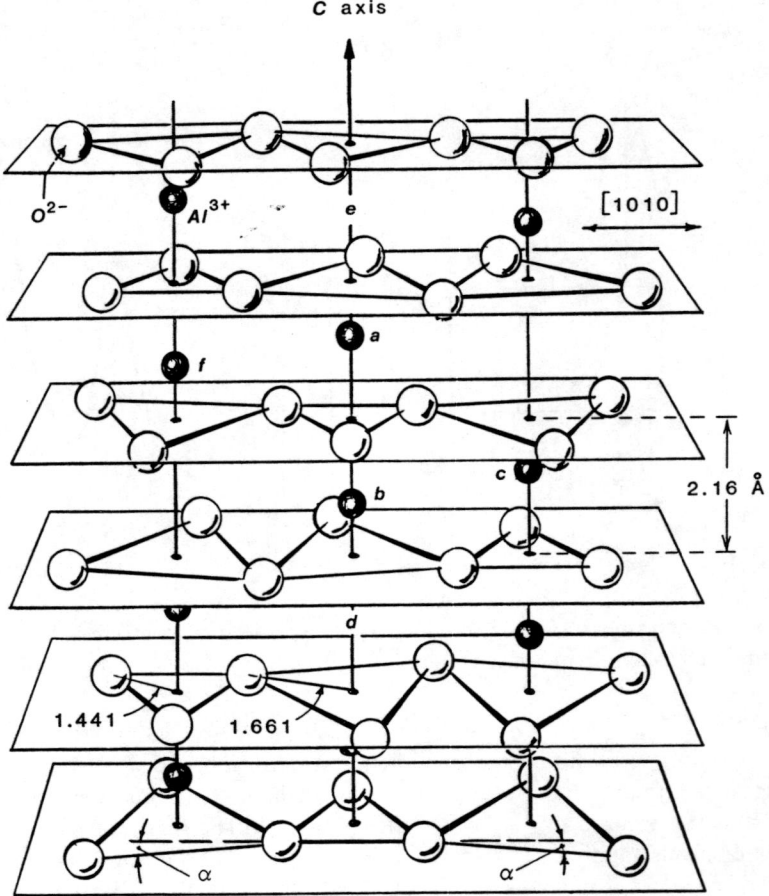

FIGURE 1.9
Sapphire crystal lattice.

### Energy Levels in Solids: Rare Earth Ions

Figures 1.11 and 1.12 show how a typical rare-earth ion such as $Nd^{3+}$ or $Tb^{3+}$ can be bonded into an irregular glassy lattice structure, together with the quantum energy levels associated with a trivalent terbium $Tb^{3+}$ ion when such an ion is dispersed at low concentration, either in a glass or in a crystal structure (for example, $CaF_2$).

Note that the energy levels of rare-earth ions such as $Tb^{3+}$ or $Nd^{3+}$ are associated with the electrons in the partially filled $4f$ inner shell of the rare-earth atom. In nearly all solid materials, these inner electrons are well shielded, by surrounding outer filled electron shells, from the crystalline Stark effects caused by the bonds to surrounding atoms in the crystal or glass material. Hence the quantum energy levels of such rare-earth ions are almost unchanged in many different crystalline or glass host materials.

Almost any material containing small amounts of $Tb^{3+}$, for example, will fluoresce with the same brilliant green color around 540 nm, and materials containing $Nd^{3+}$ all fluoresce strongly around 1.06 $\mu$m in the near infrared. There are also several other such rare-earth ions, including $Dy^{2+}$, $Tm^{2+}$, $Ho^{3+}$, $Eu^{3+}$, and $Er^{3+}$, that make good to excellent laser materials.

## 1.2 ATOMIC ENERGY LEVELS AND SPONTANEOUS EMISSION

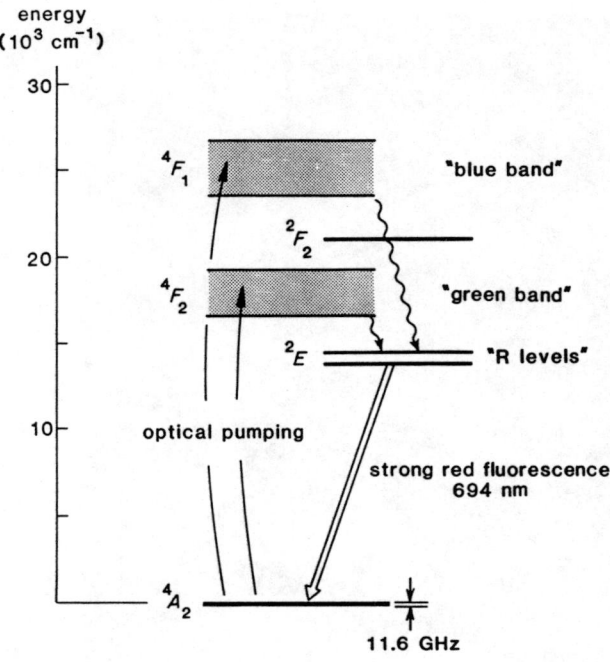

FIGURE 1.10
Quantum-mechanical energy levels of the $Cr^{3+}$ ions in a ruby crystal.

### Optical Pumping of Atoms

All of these minerals illustrate another basic method for pumping or exciting atoms into upper energy levels, that is, through the absorption of light at an appropriate pumping wavelength. The high-pressure mercury lamp used as the excitation source in a "mineral light" emits a broad continuum of visible and ultraviolet wavelengths. As shown in Figures 1.8 and 1.12, some of these wavelengths will coincide with the transition frequencies from the lowest or ground levels of the chromium or terbium ions (nearly all the ions are located at ground level when in thermal equilibrium) up to some of the higher energy levels of these ions.

These ions can thus absorb radiation ("absorb photons") from the UV light source at these particular frequencies, and as a result be lifted up to various of the upper levels. This excitation is enhanced by the fact that in solids the higher energy levels are often rather broad bands of levels. The absorption linewidths of the ruby and terbium absorption lines are thus relatively broad, permitting reasonably efficient absorption of the continuum radiation from the mercury lamp.

Once they are lifted upward by this so-called "optical pumping," the ions in each case then relax or fluoresce down to lower energy levels, as shown in Figure 1.12, emitting a relatively sharp fluorescence at two or three visible wavelengths as they drop from upper to lower levels.

### Spontaneous Energy Decay or Relaxation

Let us discuss a little more the spontaneous decay or relaxation process we have introduced here. Suppose that a certain number $N_2$ of such atoms have been pumped into some upper energy level $E_2$ of an atom or molecule, whether

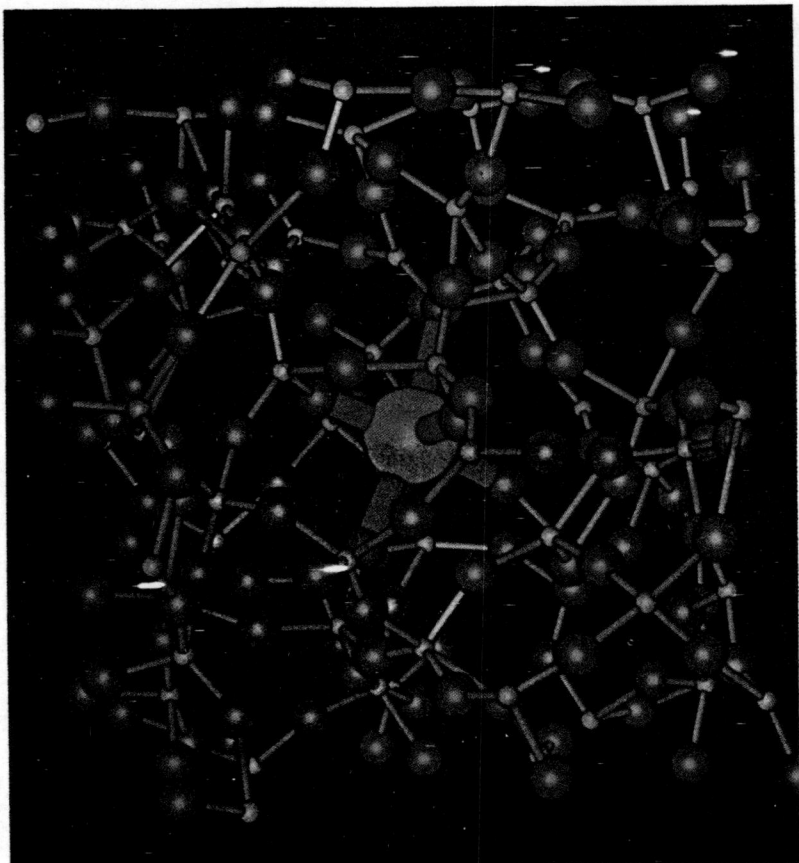

FIGURE 1.11
A single rare-earth ion (largest sphere) imbedded in a BaF$_2$ glass matrix. The larger spheres in the matrix represent barium, the smaller fluorine.

by electron collision in a gas like helium, or by optical pumping in a solid like ruby, or by some other mechanism. These atoms will then spontaneously drop down or relax to lower energy levels, giving up their excess internal energy in the process (Figure 1.13). (We will see where this energy goes in a moment.)

The rate at which atoms spontaneously decay or relax downward from any upper level $N_2$ is given by a spontaneous energy-decay rate, often called $\gamma_2$, times the instantaneous number of atoms in the level, or

$$\left.\frac{dN_2}{dt}\right|_{\text{spon}} = -\gamma_2 N_2 \equiv -N_2/\tau_2. \tag{4}$$

If an initial number of atoms $N_{20}$ are pumped into the level at $t = 0$, for example, by a short intense pumping pulse, and the pumping process is then turned off, the number of atoms in the upper level will decay exponentially in the form

$$N_2(t) = N_{20} e^{-\gamma_2 t} = N_{20} e^{-t/\tau_2}, \tag{5}$$

where $\tau_2 \equiv 1/\gamma_2$ is the lifetime of the upper level $E_2$ for energy decay to all lower levels.

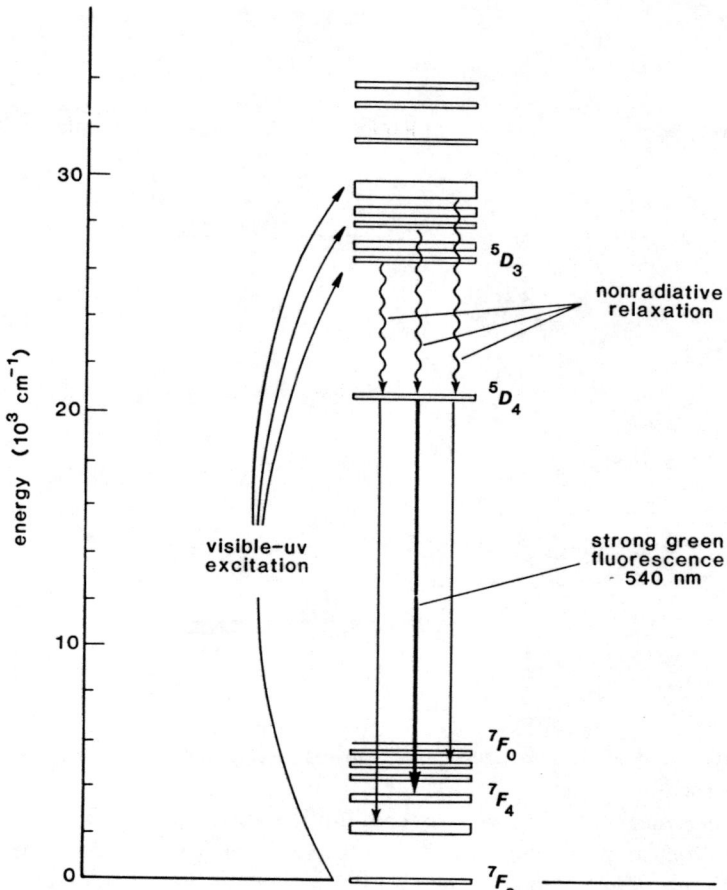

FIGURE 1.12
Optical pumping of the upper quantum-mechanical energy levels in the rare-earth ion terbium, $Tb^{3+}$.

The lifetime of the $R$ levels in the ruby crystal happens to be long enough (about 4 msec), and the visible fluorescence strong enough, that we can rather easily demonstrate this kind of exponential decay by using the simple apparatus shown in Figure 1.15. The pulsed stroboscopic light source emits a broadband flash of visible and ultraviolet light about 60 $\mu$sec long. This flash of light optically pumps the $Cr^{3+}$ ions in the ruby sample up to upper levels, from which they very rapidly decay to the metastable $R$ levels. These levels then decay to the ground level by emitting visible red fluorescence with a decay time $\tau \approx 4.3$ msec. (Similar fluorescence lifetime measurements can also be made for any of the other materials we have mentioned, but some of the lifetimes are much shorter, and the fluorescent intensities much smaller, making the experiment more difficult.)

### Radiative and Nonradiative Relaxation

There are actually two quite separate kinds of downward relaxation that occur in these solid-state materials, as well as in most other atomic systems. One mechanism is *radiative relaxation*, which is to say the spontaneous emission of electromagnetic or fluorescent radiation, as we have already discussed. We

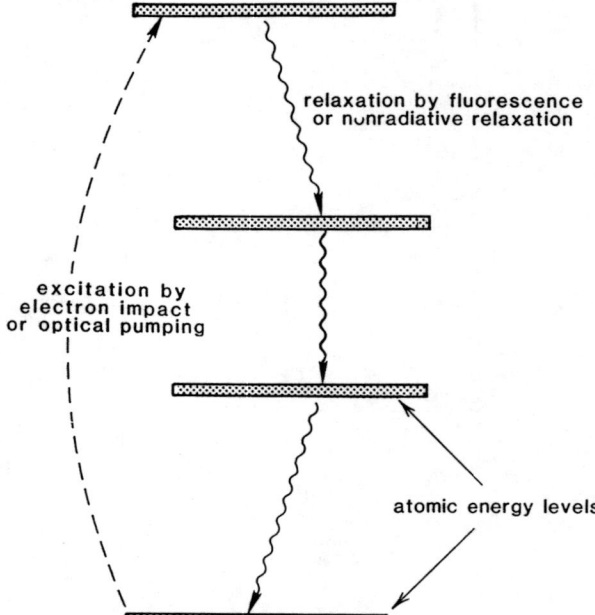

FIGURE 1.13
General concept of upper-level excitation by electron impact or optical pumping.

can usually measure this emitted radiation directly, with some suitable kind of photodetector.

The other mechanism is what is commonly called *nonradiative relaxation*. In terbium, for example, when the terbium ions relax from higher energy levels shown in Figure 1.12 down into the $^5D_4$ level, they get rid of the transition energy not by radiating electromagnetic radiation somewhere in the infrared, but by setting up mechanical vibrations of the surrounding crystal lattice. To put this in another way, the excess energy is emitted as *lattice phonons*, or as heating of the surrounding crystal lattice, rather than as electromagnetic radiation or *photons*—hence the term *nonradiative relaxation*. This kind of nonradiative emission is usually difficult to measure directly, since it mostly goes into a very small warming up of the surrounding medium. This same kind of nonradiative relaxation process also allows excited ruby atoms to relax down into the $^2E$ levels.

The total relaxation rate $\gamma$ on any given transition will thus be, in general, the sum of a *radiative* or *fluorescent* or *electromagnetic* part, described by a purely radiative decay rate that we often write as $\gamma_{rad}$; plus a *nonradiative* part, with a nonradiative decay rate that we often write as $\gamma_{nr}$. The total or measured decay rate for atoms out of the upper level will then be the sum of these, or $\gamma_{tot} \equiv \gamma_{rad} + \gamma_{nr}$. The actual numerical values for these rates, and the balance between radiative and nonradiative parts, will in general be different for every different atomic transition, and may depend greatly on the immediate surroundings of the atoms, as we will discuss in much more detail later. The one certain thing is that atoms placed in an upper level will decay downward, by some combination of radiative and/or nonradiative decay processes.

Nonradiative relaxation can be a particularly rapid process for relaxation across some of the smaller energy gaps for rare-earth ions and other absorbing ions in solids, as we will see in more detail later. For example, in terbium as in many other rare-earth ions, there may be many rather closely spaced levels or bands at higher energies; but then the energy gap down from the lowest of these

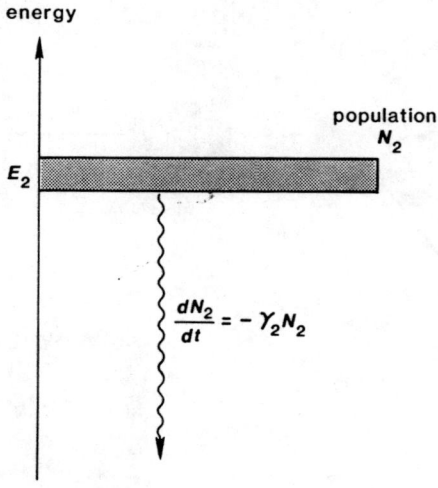
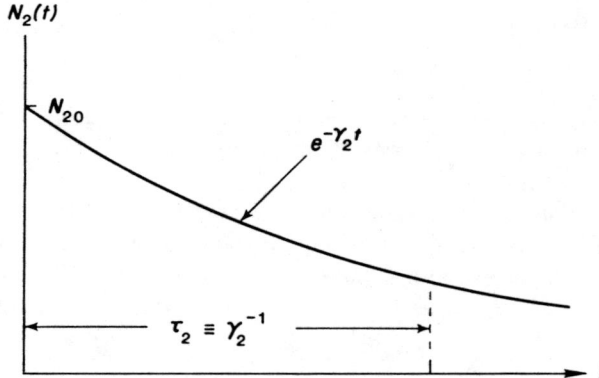

FIGURE 1.14
Spontaneous energy decay rate.

upper levels (the $^5D_4$ level in terbium) to the next lower group of levels may be larger than the frequency $\hbar\omega$ of the highest phonon mode that the crystal lattice can support.

As a result, the terbium ion cannot relax across this gap very readily by nonradiative processes, i.e., by emitting lattice phonons, since the lattice cannot accept or propagate phonons of this frequency. Instead the atoms relax across this gap almost entirely by radiative emission, i.e., by spontaneous emission of visible fluorescence. Across other, smaller gaps, however, the nonradiative relaxation rate is so fast that any radiative decay on these transitions is completely overshadowed by the nonradiative rate.

This behavior is typical for many other rare-earth ions in crystals and glasses. Following optical excitation to high-lying levels, the atoms relax by rapid nonradiative relaxation into some lower *metastable level*, from which further nonradiative relaxation is blocked by the size of the gap to the next lower level. Efficient fluorescent emission from here to the lower levels then occurs, followed by further fast nonradiative relaxation across any remaining energy gaps to the ground level. The nonradiative decay time of the atoms via phonon emission across the smaller energy gaps may be in the subnanosecond to picosecond range—too fast

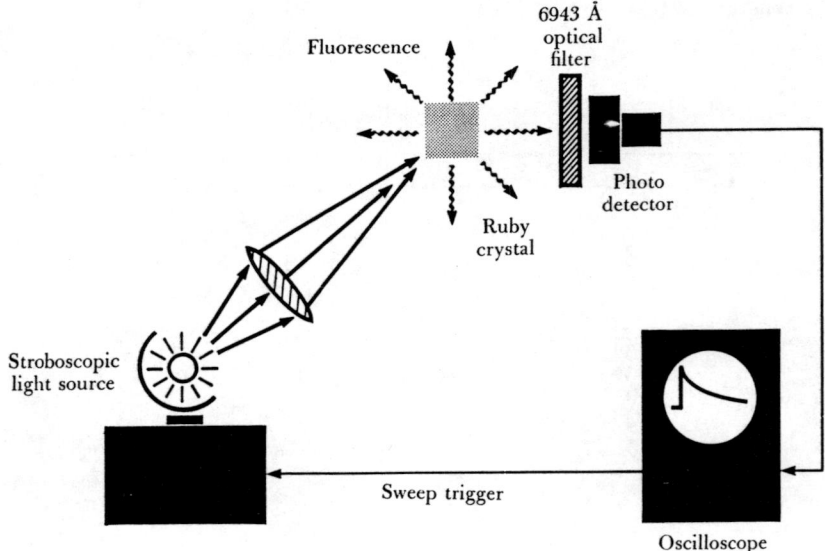

**FIGURE 1.15**
Measurement of ruby fluorescent lifetime.

to be easily measured—and the average lifetime of the same rare-earth ions in their metastable levels, before they radiate away their energy and drop down, is typically between a few hundred $\mu$sec and a few msec.

We will see later that in many rare-earth samples it is possible, by pumping hard enough, to actually build up enough of a population inversion between the metastable level and lower levels to permit laser action on these transitions. Several different rare-earth atoms can thus be used as good optically pumped solid-state lasers (though terbium itself is not among the best of these).

## REFERENCES

Brief but useful introductions to the whole range of spectroscopy, on many different kinds of atomic systems, in widely different frequency ranges and using widely different experimental techniques, can be found in D. H. Whiffen, *Spectroscopy* (John Wiley, 1966), or in Oliver Howarth, *Theory of Spectroscopy* (Halsted Press, John Wiley, 1973).

There exist innumerable books on the theory and practice of atomic and molecular spectroscopy, of which two recent examples are H. G. Kuhn, *Atomic Spectra* (Academic Press, 1969), and J. I. Steinfeld, *Molecules and Radiation: An Introduction to Modern Molecular Spectroscopy* (Harper and Row, 1974).

For tables of detailed data on energy levels of isolated atoms, the standard reference sources are the National Bureau of Standards *Tables of Atomic Energy Levels*, edited by Charlotte E. Moore (U.S. Government Printing Office, 1971).

## 1.3 STIMULATED ATOMIC TRANSITIONS

Having introduced *spontaneous* (downward) transitions, we will now look at the *stimulated* (upward and downward) transitions that are the essential processes in all kinds of laser and maser action.

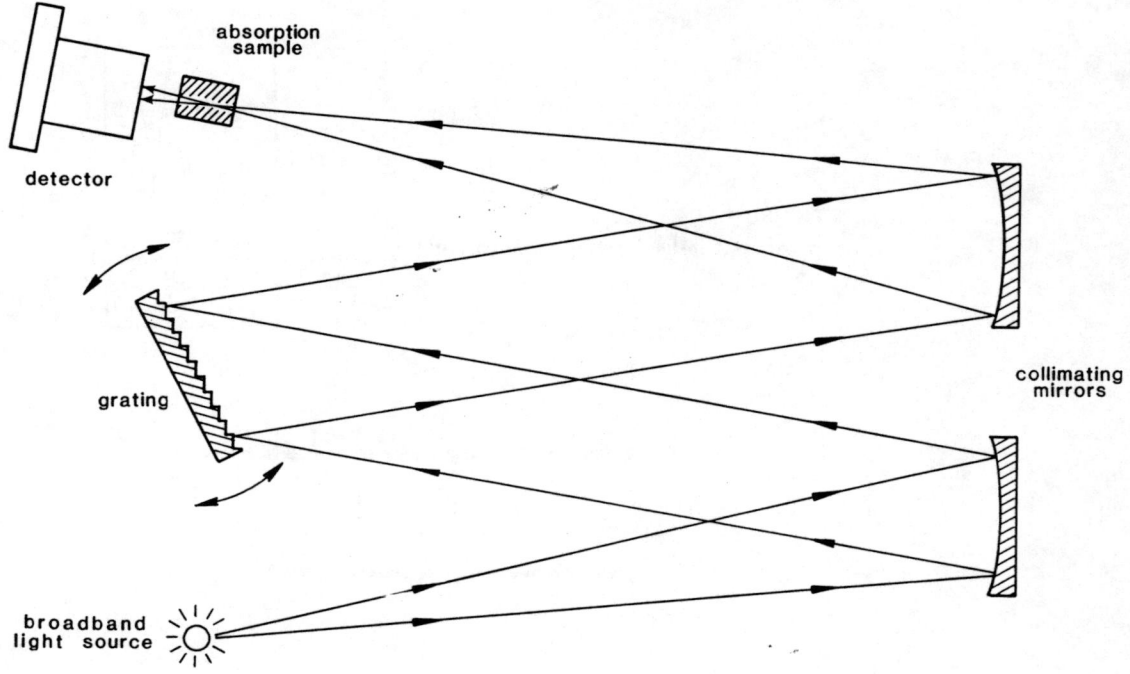

FIGURE 1.16
An elementary grating spectrometer.

### Atomic Absorption Lines

Suppose we now examine more carefully the absorption of radiation by a collection of atoms as a function of the wavelength of the incident radiation. Figure 1.16 shows a very elementary example of a grating spectrometer such as might be used for such measurements. (A tunable laser would be a very useful alternative, if one were conveniently available.)

In this spectrometer the radiation from a broadband continuum light source is collected into a roughly parallel beam by a collimating mirror, and is then reflected from a diffraction grating located on a rotatable mount. At any one orientation of the grating, only one wavelength (rather, a finite but narrow band of wavelengths) is reflected at the correct angle to be collected by another curved mirror, focused down through a narrow slit, and passed through the experimental sample onto a detector. By rotating the grating, we can tune the wavelength of the radiation that passes through the sample and thereby measure the transmission through the sample as a function of frequency or wavelength. (Figure 1.17 shows a more compact in-line version of such an instrument.)

The result of such an experiment will often appear as shown schematically in Figure 1.18. The atomic sample will have absorption transitions from the lowest energy level to higher energy levels; so it will exhibit discrete absorption lines—that is, narrow bands of frequency in which the sample exhibits more or less strong absorption—at exactly those wavelengths. These wavelengths will correspond through Planck's law to the energy gaps between the lowest and higher levels. If there happen to be some atoms already located in higher-lying levels, then absorption lines from those levels to still higher levels may also be seen, as illustrated by transition $C$ in the figure. These excited-state absorptions, how-

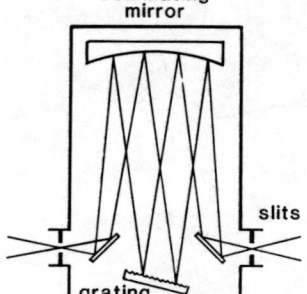

FIGURE 1.17
A compact in-line grating monochromator.

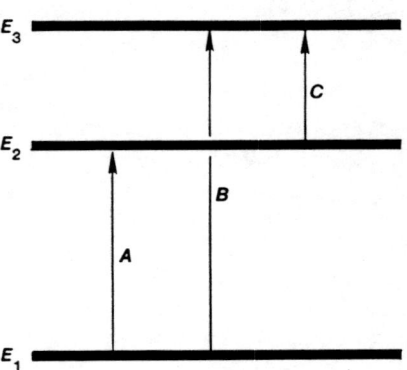

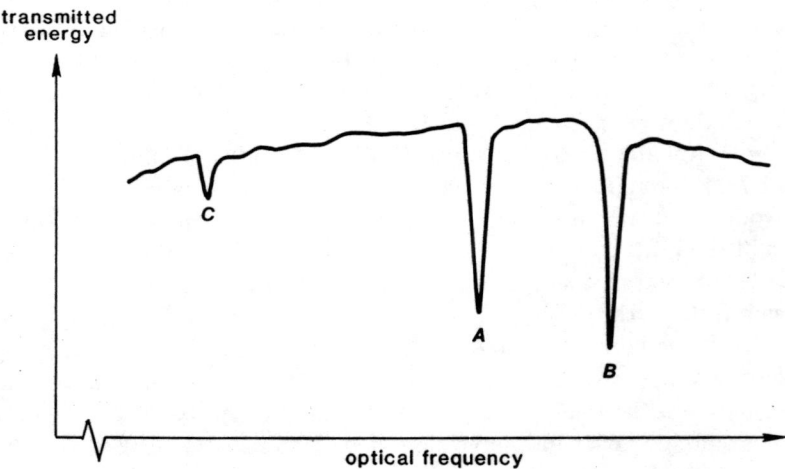

FIGURE 1.18
Absorption transitions (top) and absorption lines (bottom).

ever, will usually appear substantially weaker, simply because there will normally be many fewer atoms in the higher energy levels.

As a specific illustration of atomic absorption, Figure 1.19 shows some of the sharp absorption lines observed when radiation at wavelengths around 540 nm in the visible is transmitted through a crystal of lanthanum fluoride ($LaF_2$)

## 1.3 STIMULATED ATOMIC TRANSITIONS

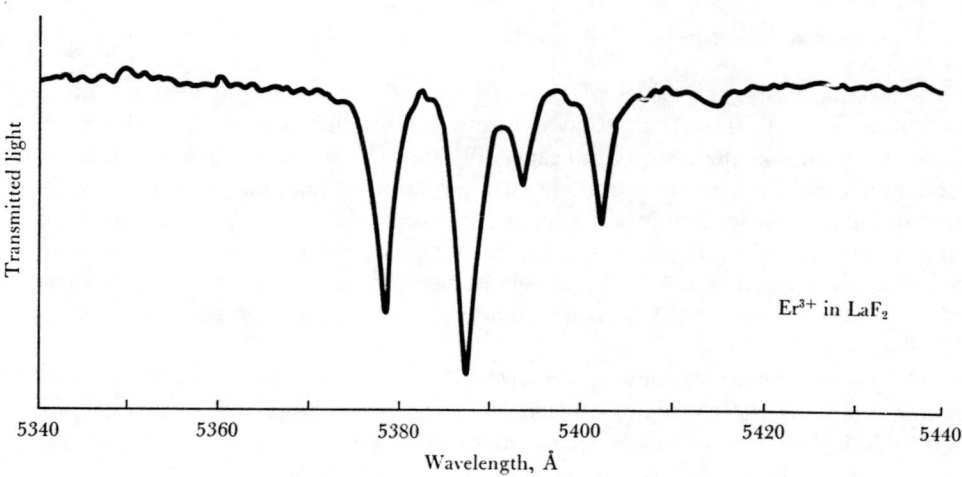

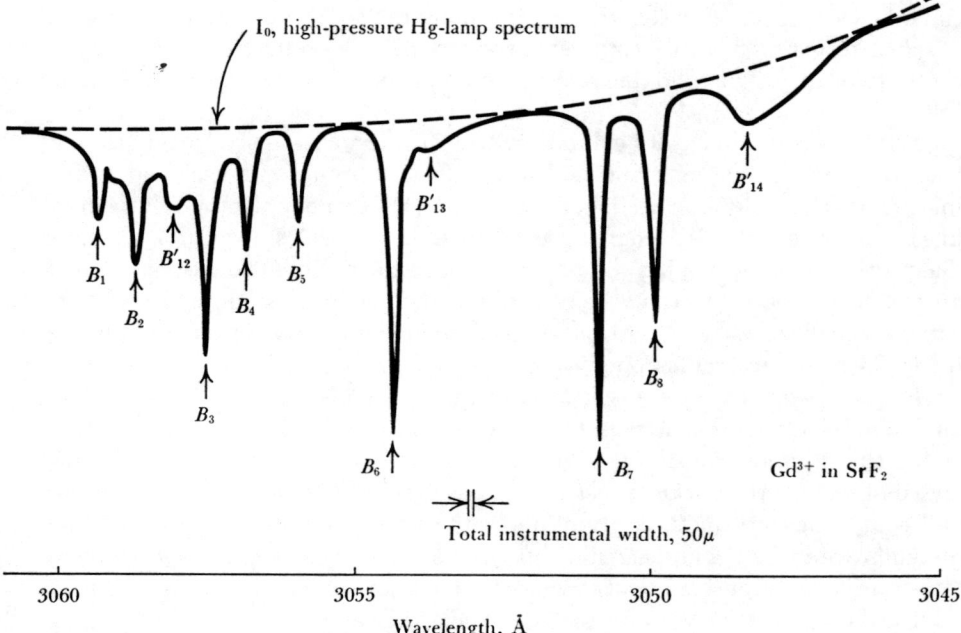

**FIGURE 1.19**
Light transmission versus wavelength through crystals of lanthanum fluoride (LaF$_2$) containing a small amount of the rare-earth ion erbium Er$^{3+}$ (upper trace), and strontium fluoride (SrF$_2$) containing a small amount of the rare-earth ion gadolinium Gd$^{3+}$ (lower trace).

containing a small percentage of the rare-earth ion erbium, or when radiation at wavelengths around 300 nm in the near ultraviolet is transmitted through a crystal of strontium fluoride (SrF$_2$) containing a small percentage of the rare-earth ion gadolinium. These absorption lines all represent different transitions from the lowest or ground levels of the Er$^{3+}$ or Gd$^{3+}$ ions to higher-lying levels, exactly analogous to the terbium levels shown in Figure 1.13. Of course, if a pure lanthanum or strontium fluoride crystal is grown without any erbium or gadolinium present, no such absorption lines are observed.

### Absorption Lines in Gases, and Molecular Spectroscopy

Absorption experiments of this sort are, of course, by no means limited to solids or to rare earths. Isolated atoms or ions in gases will exhibit such absorption lines in the visible, and especially the UV. Molecules in gases, liquids, and solids will exhibit an extremely rich spectrum of absorption lines, notably in the infrared as well as in the visible and ultraviolet. The absorption lines of atoms and molecules in gases are typically sharper or narrower than those in solids or liquids, since the energy levels in gases are not subject to some of the perturbing influences that tend to broaden or smear out the energy levels in liquids or solids.

As just one more example to illustrate absorption spectroscopy, Figure 1.20 shows a few of the sharp absorption lines characteristic of the formaldehyde molecule $H_2CO$ in a narrow range of wavelengths near 3.57 $\mu$m. This particular spectrum was taken by using a continuously tunable laser source (a cw injection diode laser using a lead/cadmium sulfide diode) rather than an incoherent spectrometer. The dashed envelope in Figure 1.20(a) is the power output of the tunable laser versus wavelength, over a tuning range that is extremely large in absolute terms ($\sim 3 \times 10^{10}$ Hz), yet extremely narrow ($\sim 0.04\%$) relative to the center frequency. The solid line is the power transmitted through the vapor-filled cell.

Many different molecules exhibit exactly such characteristic sharp lines, specific to the individual molecules, in rich profusion through the near and middle infrared regions. These sharp lines are extremely useful not only as potential laser lines, but as characteristic signatures of different molecules, for use in chemical diagnostics or in identifying the presence of specific pollutant molecules or hazardous chemicals. Note that the sensitivity and the laser scanning rate in the experiment allow a small portion of the formaldehyde absorption spectrum to be displayed on an oscilloscope in real time.

*Emission spectroscopy*, using the spontaneous emission lines radiated from an excited sample as in Figure 1.5, is thus one way of observing and learning about the discrete transitions and the quantum energy levels of atoms, ions, and molecules. *Absorption spectroscopy*, as briefly described here, is another and complementary method of obtaining the same kind of information. These methods are in fact complementary in their utility, since emission spectroscopy tends to give information about downward transitions emanating from high-lying levels, whereas absorption spectroscopy tends to give information about upward transitions from the ground level or low-lying atomic levels. The formaldehyde example illustrates the possibilities for applying tunable lasers to spectroscopy, to analytical chemistry, and to practical applications such as pollution detection.

### Stimulated versus Spontaneous Atomic Transitions

We have now seen that there are two basically different kinds of transition processes that can occur in atoms or molecules.

First, there are *spontaneous emission* or *relaxation transitions*, in which atoms spontaneously drop from an upper to a lower level while emitting electromagnetic and/or acoustic radiation at the transition frequency. *Fluorescence, energy decay*, and *energy relaxation* are other names for this process. When atoms emit this kind of fluorescence or spontaneous electromagnetic radiation, each individual atom acts almost exactly like a small randomly oscillating antenna—in

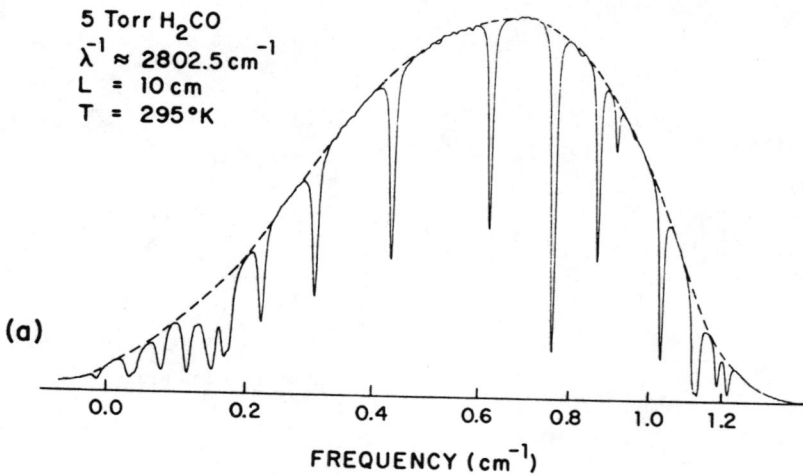

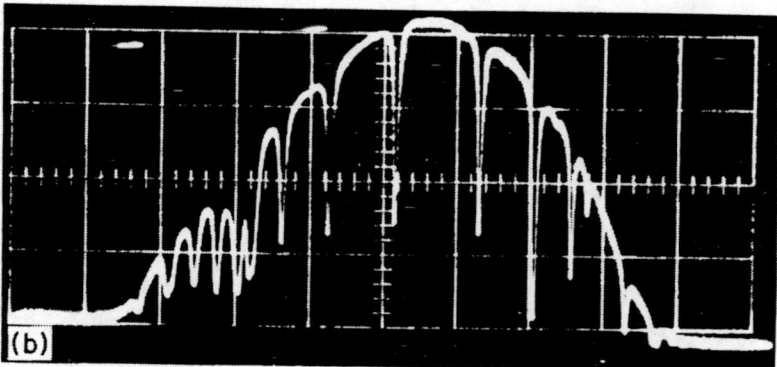

FIGURE 1.20
Absorption spectroscopy of formaldehyde using a tunable laser source near $\lambda = 3.57\mu m$.

most common cases, a small *electric dipole* antenna—internally driven at the transition frequency. Each individual atom radiates independently, with a temporal phase angle that is independent of all the other radiating atoms. Thus, the total fluorescent emission from a collection of spontaneously emitting atoms is noise-like in character (Figure 1.21), even though it will be limited in spectral width to the comparatively narrow linewidth of the atomic transition. Indeed, such spontaneously emitted radiation has all the statistical properties of narrowly bandlimited gaussian noise. We usually refer to it as *incoherent* emission.

Second, there are the *stimulated responses* or *stimulated transitions*—both stimulated absorption and stimulated emission—that occur when an **external** radiation signal is applied to an atom. In these transitions each individual **atom** acts like a miniature passive resonant antenna (again, usually an electric **dipole** antenna) *that is set oscillating by the applied signal itself.* That is, the internal motion or oscillation in the atom is not random, but is driven by and coherent with the applied signal.

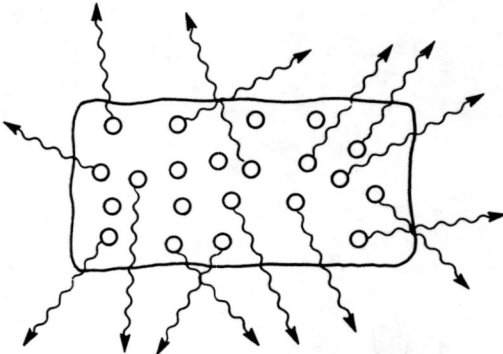

FIGURE 1.21
Spontaneous emission is incoherent or noise-like, emerging randomly in all directions.

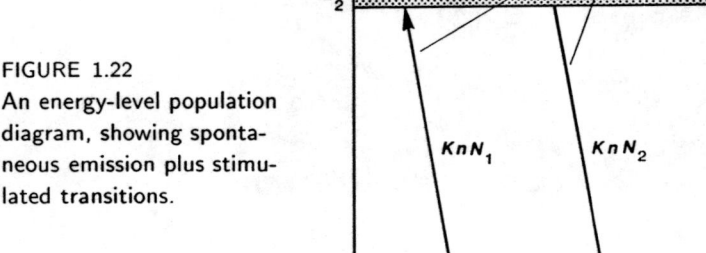

FIGURE 1.22
An energy-level population diagram, showing spontaneous emission plus stimulated transitions.

### Atomic Rate Equations

Suppose we have very many identical atoms, each of which has two just energy levels, $E_1$ and $E_2$. (Real atoms will undoubtedly have many other energy levels as well, but we will ignore other levels for the moment.) Suppose that $N_1(t)$ of the atoms present are in level $E_1$ and $N_2(t)$ atoms are in level $E_2$. This situation can be illustrated by an *energy-level population diagram*, as in Figure 1.22.

We have already stated that the spontaneous-relaxation rate down from level $E_2$ to level $E_1$ is directly proportional to the upper-level population $N_2(t)$ and is not influenced at all by the lower-level population $N_1(t)$. Hence the *spontaneous-emission rate* out of level 2 and into level 1 is given by

$$\left.\frac{dN_2(t)}{dt}\right|_{\text{spontaneous}} = -\left.\frac{dN_1(t)}{dt}\right|_{\text{spontaneous}} = -\gamma_{21} N_2(t), \qquad (6)$$

where $\gamma_{21}$ indicates the total spontaneous-transition rate or decay rate (radiative plus nonradiative) from level 2 to level 1.

Suppose now an optical signal is applied to these atoms to cause stimulated transitions, as in the optical pumping or absorption spectroscopy experiments we have just discussed. This signal must, of course, be tuned in frequency close to the transition frequency of interest, i.e., $\omega \approx \omega_{21} \pm \Delta\omega_a$, where $\Delta\omega_a$ is the linewidth of the atomic transition. We might then characterize the strength of this signal by its intensity $I$ (dimensions of power per unit area), or by the strength of its $E$ or $H$ fields. In discussions of stimulated transitions, however, the applied signal intensity or energy density is often expressed in units of the number of signal photons $n(t)$ per unit volume in the applied signal. This does not necessarily imply anything about photons as being billiard-ball-like point particles; it merely means that $n(t)$ is the electromagnetic energy density of the applied signal divided by the quantum energy unit $\hbar\omega$.

Such an applied signal will cause atoms initially in the lower energy level to begin making *stimulated transitions* or "jumps" upward to the upper energy level, at a rate proportional to the applied signal intensity (or power density) times the number of atoms in the starting level. The number of stimulated upward transitions per unit time caused by the applied signal can then be written as

$$\left. \frac{dN_2(t)}{dt} \right|_{\text{stimulated atop upward}} = Kn(t)N_1(t). \qquad (7)$$

That is, the stimulated upward transition rate is directly proportional to the photon density $n$ of the applied signal. Each such upward transition absorbs one quantum of energy from the applied signal and—at least in an elementary description—transfers it to one of the atoms which is lifted upward. This is the process of *stimulated absorption*.

But the essential point is that the same applied signal will also cause any atoms initially in the *upper energy level* to begin making similar stimulated transitions or jumps *downward* in energy, at a rate which is again proportional to the applied signal intensity times the number of atoms in the initial (i.e., upper) level. The number of stimulated downward transitions per unit time can thus similarly be written as

$$\left. \frac{dN_2(t)}{dt} \right|_{\text{stimulated atop downward}} = -Kn(t)N_2(t). \qquad (8)$$

This is the process of *stimulated emission*. The atoms in this case jump downward, giving up energy. This energy must go into the stimulating optical signal, which is therefore strengthened or amplified.

The constant $K$ in each of these equations is just a proportionality constant that measures the absolute strength of the stimulated response on the particular atomic transition. A fundamental and essential point, however, is that *this proportionality constant necessarily has exactly the same value for transitions in either direction*. This constant $K$ will also be largest for an applied signal tuned exactly to the atomic transition frequency, and will rapidly become small to negligible as the signal frequency $\omega$ is tuned away from the transition frequency $\omega_{21}$ by more than a linewidth or so.

The *total rate equation* for the atomic populations in this simple example, including stimulated plus spontaneous transitions, is thus given by

$$\left.\frac{dN_2(t)}{dt}\right|_{\text{total}} = \left.\frac{dN_2(t)}{dt}\right|_{\substack{\text{stimulated}\\\text{upward}}} + \left.\frac{dN_2(t)}{dt}\right|_{\substack{\text{stimulated}\\\text{downward}}} + \left.\frac{dN_2(t)}{dt}\right|_{\text{spontaneous}},$$

$$= Kn(t) \times [N_1(t) - N_2(t)] - \gamma_{21} N_2(t) = -\left.\frac{dN_1(t)}{dt}\right|_{\text{total}} \quad (9)$$

where $n(t)$ is directly proportional to the applied signal intensity or power density.

### Quantum Derivation of the Spontaneous Emission Process

In a quantum-mechanical analysis the constant $K$ in the stimulated-transition rate $Kn(t)N(t)$ is usually derived by using a *semiclassical quantum analysis*, in which the atoms are treated quantum-mechanically but the applied electromagnetic signal is treated classically. The spontaneous emission processes described by the spontaneous-relaxation probability $\gamma_{21} N_2(t)$ can, however, only be derived from a fully quantum electrodynamic analysis in which both the atoms and the electromagnetic field itself are treated quantum-mechanically.

Some people note that there is a correspondence in quantum theory between the spontaneous-emission rate, which corresponds to the downward stimulated-transition rate that would be caused by one extra photon, and the presence of zero-point fluctuations in the quantum electromagnetic fields, with a magnitude equivalent to an energy of one photon per mode; and deduce from this that zero-point fluctuations "cause" or "stimulate" the spontaneous emission. This can be a convenient way to calculate the spontaneous-emission rate or the quantum noise magnitude in a laser calculation, but attributing a causal relation to the zero-point fluctuations is a more dubious proposition. Zero-point fluctuations and spontaneous emission are both predicted, separately and independently, by quantum field theory, but nothing in the theory says that either one *causes* the other: they each arise independently of the other, from the commutation properties of the quantum field operators.

### Stimulated Transitions and Laser Amplification

The total rate at which atoms make signal-stimulated transitions between two energy levels (i.e., "up" minus "down") is thus given by $Kn(t) \times [N_1(t) - N_2(t)]$. Each upward transition transfers $\hbar\omega$ of energy from the signal to the atoms; each downward transition does the reverse.

But this implies that the net rate at which energy per unit volume is absorbed *from the signal by* the atoms is then given by this net flow rate times the energy $\hbar\omega$ per jump. That is, the net energy transfer rate to the atoms is

$$\frac{dU_a}{dt} = Kn(t) \times [N_1(t) - N_2(t)] \times \hbar\omega, \quad (10)$$

where $U_a$ is the energy density in the forced internal oscillation of the atoms.

This same energy must at the same time be coming out of the signal. Hence the energy density $U_{\text{sig}}(t) = n(t) \times \hbar\omega$ in the applied signal must be decreasing with time according to the reverse expression

$$\frac{dU_{\text{sig}}}{dt} = -K[N_1(t) - N_2(t)] \times n(t) \times \hbar\omega = -K[N_1(t) - N_2(t)] \times U_{\text{sig}}(t) \quad (11)$$

or, in terms of photon density,

$$\frac{dn}{dt} = -K[N_1(t) - N_2(t)]n(t). \quad (12)$$

The signal energy density $U_{\text{sig}}(t)$, or the photon density $n(t)$, may thus either decay or grow with time, depending on the sign of the population difference $\Delta N(t) = N_1(t) - N_2(t)$ in the square brackets.

The signal growth rate described by Equation 1.12 leads to the essential concept of laser amplification. This equation says that if an external signal is applied to a collection of atoms where there are more atoms in the lower energy level than in the upper, or where $N_1(t) > N_2(t)$, then the net transition rate or net flow of atoms between the levels will be upward. In this case net energy is being supplied to the atoms by the applied signal; so the applied signal must become absorbed or attenuated.

If, however, we can somehow produce a condition of *population inversion*, in which there are more atoms in the upper level than in the lower, or $N_2 > N_1$, then both the quantity $N_1 - N_2$ and hence the net energy flow between signal and atoms will change sign. The net stimulated-transition rate for the atoms will now be in the downward direction. Net energy will then be given up by the atoms, and taken up by the applied signal. This energy flow will in fact produce a *net amplification* of that signal, at a rate proportional to the population difference and to the strength of the external signal.

### Boltzmann's Principle

One of the fundamental laws of thermodynamics, Boltzmann's Principle, states that when a collection of atoms is in thermal equilibrium at a positive temperature $T$, the relative populations of any two energy levels $E_1$ and $E_2$ are given by

$$\frac{N_2}{N_1} = \exp\left(-\frac{E_2 - E_1}{kT}\right), \quad (13)$$

which of course means that

$$\Delta N \equiv N_1 - N_2 = \left(1 - e^{-\hbar\omega/kT}\right) N_1. \quad (14)$$

Thus for a collection of atoms in equilibrium at a normal positive temperature $T$, an upper-level population is always smaller than a lower-level population (much smaller if the energy gap $E_2 - E_1$ is an optical-frequency gap).

The total stimulated-transition rate on such an equilibrium transition is thus always absorptive or attenuating rather than amplifying. To create laser amplification, we must find some pumping process which will put more atoms into an upper level than into a lower level, and thus create a nonequilibrium condition

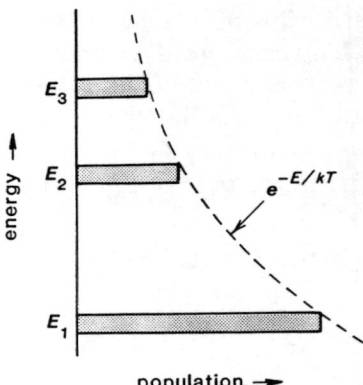

FIGURE 1.23
Boltzmann population factors.

of population inversion. In Section 1.5 we give some information on how this can be done in practice.

### Coherence in Stimulated Transitions

If we want, we can think of the basic stimulated transition process as the sum of two separate processes: in one, atoms initially in the lower energy level are stimulated by the applied signal to make transitions upward; in the other, atoms initially in the upper energy level are stimulated by the applied signal to make transitions downward. It is vital to understand, however, that the stimulated-transition probability produced by an applied signal (probability of transition per atom and per second) is always exactly the same in both directions. The net flow of atoms is thus always from whichever level has the larger population at the moment, to whichever level has the smaller population.

There is also no conceivable way to "turn off" one or the other of the stimulated absorption or emission processes separately. If the lower level is more heavily populated, the signal is attenuated. If the upper level is more heavily populated, the signal is amplified. This is the essential amplification process in all lasers and other stimulated-emission devices.

It is also essential to keep in mind that the stimulated transition process we have been introducing here results from a resonant response of the atomic wave function, or of the atomic charge cloud in each individual atom, to the applied signal. That is, the internal induced oscillation or dipole response that is produced in each atom is stimulated by and thus fully coherent with the applied signal.

The net amplification (or attenuation) process is thus a fully *coherent* one, in which the atomic oscillations follow the driving optical signal coherently in amplitude and phase. The output signal from an amplifying laser medium is a linear reproduction of the input signal, and of any amplitude modulation or phase modulation that may be on the input signal, except that (i) the output signal is amplified or increased in magnitude; (ii) the signal modulation may be decreased somewhat in bandwidth because of the finite bandwidth of the atomic response; and (iii) the signal in general has a small amount of spontaneous emission noise added to it.

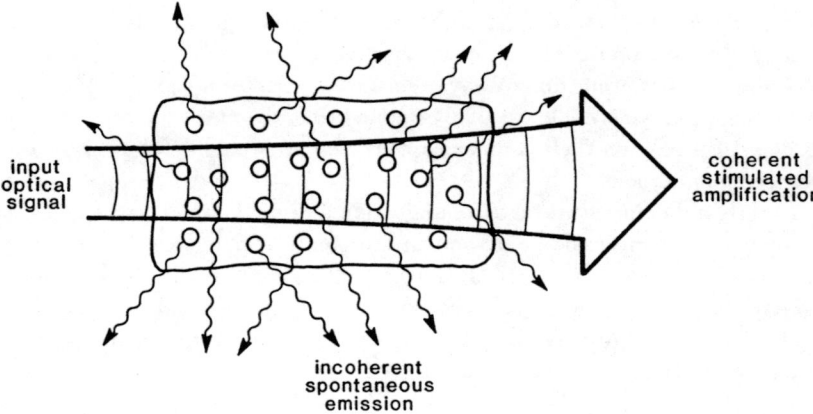

**FIGURE 1.24**
Incoherent spontaneous emission and coherent stimulated amplification occur simultaneously and in parallel in the laser medium.

### Spontaneous Versus Stimulated Transitions

Note also that in a collection of laser atoms with a population inversion, and with an applied signal present, both the spontaneous transitions and the stimulated transitions will occur simultaneously and essentially independently. The stimulated-transition rates and the spontaneous-relaxation rate can be simply added together. The spontaneous emission, however, will emerge in all directions, as in Figure 1.24, and will have the spectral and statistical character of narrowband random noise; whereas the stimulated emission (and absorption) will all be in the same direction and at the same frequency as the applied signal.

In a laser amplifier the input signal will thus be amplified by the stimulated transitions. At the same time, a small amount of the spontaneous emission (in essence, that portion traveling exactly parallel to the applied signal) will be added to the output signal by the spontaneous emission process. The spontaneous emission in this situation thus acts essentially like a small additive amplifier noise source insofar as the stimulated amplification process is concerned. Unless the applied signal is very small, approaching the noise limit of the laser amplifier, the added spontaneous-emission noise can normally be ignored in discussions of the basic stimulated amplification process.

## REFERENCES

The basic concepts of the stimulated emission process and the possibility of coherent "negative absorption" from atoms in the upper level of an atomic transition were clearly outlined by A. Einstein in "On the quantum theory of radiation," *Physikalische Zeitschrift* **18**, 121 (1917), and again by R. C. Tolman, "Duration of molecules in upper quantum states," *Rev. Mod. Phys.* **23**, 693-709 (June 1924).

An interesting and instructive early study on purely spontaneous emission from atoms is reported by E. Gaviola in "An experimental test of Schrödinger's theory," *Nature* **122**, 772 (1928). Gaviola observed the spontaneous emission lines from a mercury discharge at 435.8 nm and 404.6 nm from a common $2^3S_1$ upper level down to the $2^3P_1$ and $2^3P_0$ lower levels, under widely varying conditions of pressure and with various added buffer gases. The relative populations of these levels could then be expected to vary widely under these different conditions. Although Gaviola had no way to measure any of these populations, he could observe that the ratio of the intensities on the 435.8 nm and 404.6 nm lines always remained fixed, even though their absolute intensities changed widely. This strongly implied that the emission rates on these transitions depended only on their common upper-level population and not on either of the lower-level populations.

The first successful demonstration of amplification and oscillation using stimulated emission, employing an inverted population in ammonia at a microwave frequency, was accomplished by J. P. Gordon, H. J. Zeiger, and C. H. Townes, as reported in "The maser—new type of microwave amplifier, frequency standard, and spectrometer," *Phys. Rev.* **18**, 1264-1274 (August 15, 1955). Other references to the early history of stimulated emission devices are given in Section 1.10.

## 1.4 LASER AMPLIFICATION

Using the principles of stimulated emission outlined in the preceding section as a foundation, we next outline briefly how a laser material with an inverted atomic population produces useful laser amplification.

### Signal Absorption and Attenuation

Suppose first that we send a wave of tunable optical radiation through a collection of absorbing atoms, as illustrated in Figure 1.25, with this radiation tuned to a frequency $\omega$ near the transition frequency $\omega_{21}$ between two energy levels $E_1$ and $E_2$ of the atoms. Let the populations of these energy levels be $N_1$ and $N_2$ as shown earlier. (The symbols $N_1$ and $N_2$ nearly always in this book mean *population densities*; i.e., they have dimensions of atoms per unit volume inside the laser medium.)

For an absorbing population difference, we will find that this wave will be absorbed or attenuated with distance in passing through the atoms, in the form

$$\mathcal{E}(z) = \mathcal{E}_0 \times \exp[-\alpha(\omega)z]. \qquad (15)$$

For many atomic transitions the attenuation coefficient $\alpha(\omega)$ due to the atoms will be given (as we will derive later) by an expression of the general form

$$\alpha(\omega) = \frac{\lambda^2}{4\pi} \frac{\gamma_{\text{rad}}}{\Delta\omega_a} \frac{N_1 - N_2}{1 + [2(\omega - \omega_{21})/\Delta\omega_a]^2}. \qquad (16)$$

This expression contains factors such as the transition wavelength $\lambda$ (in the laser material); the radiative decay rate $\gamma_{\text{rad}}$ of the transition; and the transition linewidth $\Delta\omega_a$. Most important, it contains the population difference $N_1 - N_2$, and a lineshape factor (in the final term) giving the frequency lineshape of the transition. This lineshape will in general be a sharp resonance curve, as illustrated in Figure 1.25, with a finite linewidth or bandwidth $\Delta\omega_a$.

The particular lineshape given by Equation 1.16 is known as a *lorentzian lineshape*, and is characteristic of many real atomic transitions. Other transitions, for various reasons, may have somewhat different lineshapes, for example, a doppler-broadened or gaussian lineshape. The general dependence of the gain coefficient on the important atomic parameters for any real atomic transition

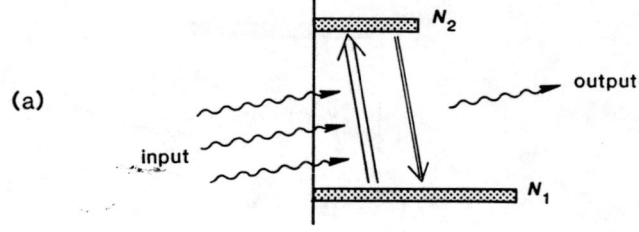

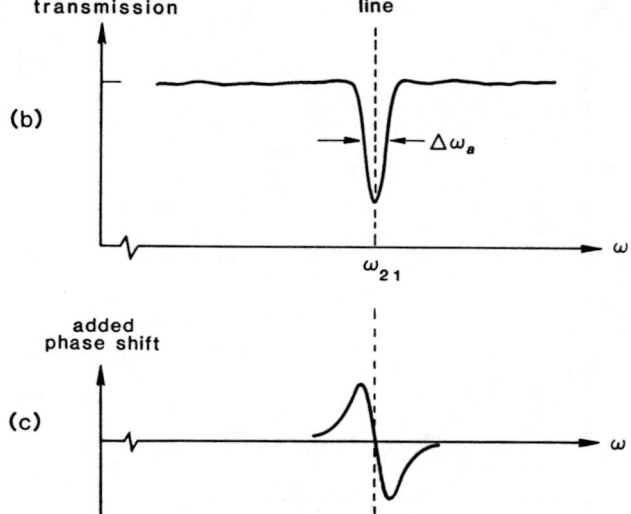

FIGURE 1.25
Stimulated absorption on an uninverted atomic transition.

will still be very much like Equation 1.16, even though the exact lineshape is somewhat different.

The signal wave passing through such an absorbing laser medium will also experience a small frequency-dependent *phase shift* due to the atoms, as shown by Figure 1.25(c). This atomic phase shift can have practical implications (such as laser frequency-pulling effects), which we will discuss in later chapters.

### Attenuation Coefficients

Note that the power flow carried by the wave passing through the atoms, or the wave intensity $I(z)$ (in units of power per unit area), is given by

$$I(z) = |\mathcal{E}(z)|^2 = I_0 \exp[-2\alpha(\omega)z]. \tag{17}$$

Hence the power or intensity attenuates with distance in the form $dI(z)/dz = -2\alpha(\omega)I(z)$. Thus in our notation the power-attenuation coefficient is given by $2\alpha(\omega)$. We will consistently use $\alpha$ in this text to represent an amplitude or "voltage" attenuation (or gain) coefficient, and $2\alpha$ to represent a power or intensity coefficient. In the journal literature, however, $\alpha$ by itself is often used to represent a power-attenuation or power-gain coefficient.

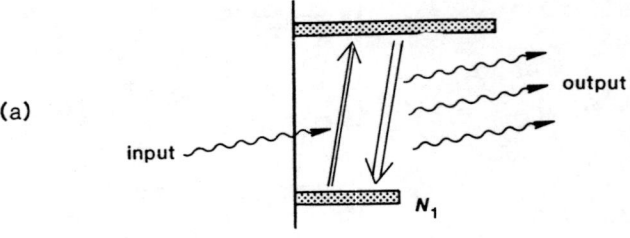

FIGURE 1.26
Stimulated amplification on an inverted atomic transition. Note that the phase shift versus frequency is also inverted relative to Figure 1.25.

### Laser Amplification

Suppose now the population difference on an atomic transition can, through some "pumping" process, be made to change sign, creating a *population inversion*. The same expression for the absorption coefficient $\alpha(\omega)$ as in Equation 1.16 then remains valid, *except that the population difference and absorption coefficient are both reversed in sign*. To emphasize this, let us rewrite Equation 1.16 in the form

$$-\alpha(\omega) \equiv \alpha_m(\omega) = \frac{\lambda^2 \gamma_{\text{rad}}}{4\pi \Delta\omega_a} \frac{N_2 - N_1}{1 + [2(\omega - \omega_{21})/\Delta\omega_a]^2}, \tag{18}$$

where $\alpha_m(\omega)$ means the "molecular" or "maser" or "laser" amplification coefficient. The wave amplitude and power will now *grow* or *amplify* with distance in the form

$$\mathcal{E}(z) = \mathcal{E}_0 \exp[+\alpha_m(\omega)z] \quad \text{and} \quad I(z) = I_0 \exp[+2\alpha_m(\omega)z] \tag{19}$$

as shown in Figure 1.26(b). The energy for this amplification comes, of course, from the inverted atoms—that is, the upper-level atoms supply energy to the wave, whereas the lower-level atoms still absorb energy. But since there are more upper-level atoms, the net effect is amplification rather than attenuation.

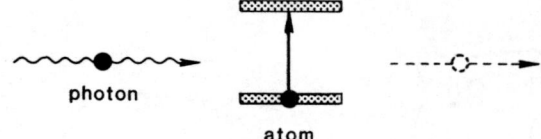

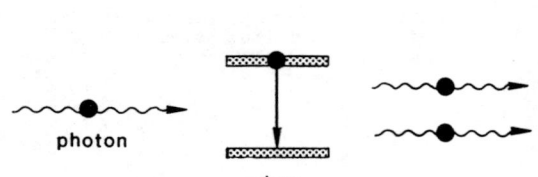

FIGURE 1.27
Photon description of stimulated absorption (top) and stimulated emission (bottom). (This viewpoint is not recommended!)

The laser amplification coefficient $\alpha_m(\omega)$ thus has exactly the same lineshape and all other properties as the absorption coefficient $\alpha(\omega)$ for the same transition without inversion. The only difference between stimulated absorption and stimulated emission is in the sign of the population difference. The net atomic phase shift, in fact, also changes sign as the population difference goes from absorbing to amplifying.

### Coherence and "Photons"

We have hardly mentioned *photons* yet in this book. Many descriptions of laser action use a photon picture like Figure 1.27, in which billiard-ball-like photons travel through the laser medium. Each photon, if it strikes a lower-level atom, is absorbed and causes the atom to make a "jump" upward. On the other hand, a photon, when it strikes an upper-level atom, causes that atom to drop down to the lower level, releasing another photon in the process. Laser amplification then appears as a kind of photon avalanche process.

Although this picture is not exactly incorrect, we will avoid using it to describe laser amplification and oscillation, in order to focus from the beginning on the *coherent* nature of stimulated transition processes. The problem with the simple photon description of Figure 1.27 is that it leaves out and even hides the important *wave* aspects of the laser interaction process. A photon description leads students to ask questions like, "How do we know that the photon emitted in the stimulated emission process is coherent with the stimulating photon?" The answer is that the whole stimulated transition process should be treated not as a "photon process" but as a coherent or wave process. These coherence effects are present, and must be considered, in at least two different ways.

First, when an electromagnetic signal wave passes through a collection of atoms, a much more accurate description of the stimulated transition process is that the electromagnetic fields in the wave cause the electronic charges inside the atoms to begin vibrating or oscillating in a coherent relationship to the driving signal fields. The atoms in fact both respond and reradiate like miniature atomic antennas. The fields reradiated by the individual atoms combine coherently with the incident signal fields to produce absorption or amplification (and also phase

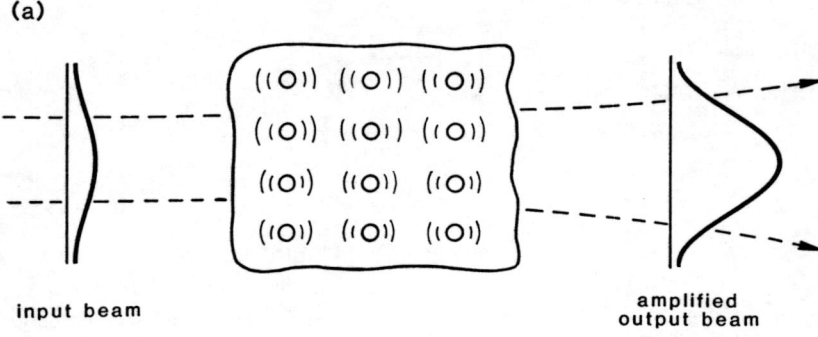

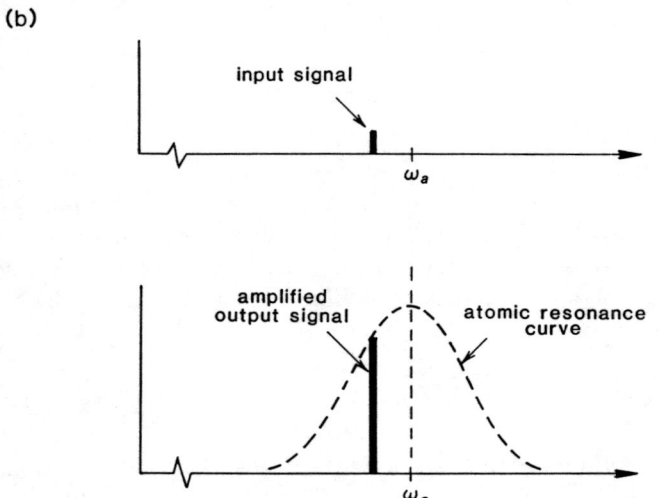

**FIGURE 1.28**
(a) The stimulated emission or reradiation from each laser atom is spatially coherent or spatially in phase with the incident signal radiation. (b) The stimulated emission is also temporally coherent with, and at the same frequency as, the incident signal radiation.

shift) in a manner that is both *spatially* and *spectrally* coherent, as illustrated in Figure 1.28.

Quantum mechanics tells us in fact that these atoms respond very much like little classical electronic dipole oscillators (as we will discuss in great detail in a later chapter), except that atoms initially in the lower energy level respond in a way that tends to cancel or absorb the incident signal, whereas atoms initially in the upper level respond in exactly opposite phase to the applied signal. The waves reradiated by the upper-level atoms thus tend to *add to* the driving signal wave, and amplify it, whereas the wavelets reradiated by lower-level atoms tend to add out of phase to the driving signal and thus attenuate it. Other than this phase difference, the stimulated absorption and emission processes are identical.

## Quantum Description of Stimulated Transitions

A second important aspect of stimulated transitions can also be obscured by the photon picture. In a fully correct quantum description, most atoms are not likely to be exactly "in" one quantum level or another at any given instant of time. Rather, the instantaneous quantum state of any one individual atom is usually a time-varying *mixture* of quantum states, for example, the upper and lower states of a laser transition. The populations $N_1$ and $N_2$ do not really represent discrete integer numbers of atoms in each level. Rather, each individual atom is partly in the lower level and partly in the upper level (that is, its quantum state is a mixture of the two eigenstates); and the numbers $N_1$ and $N_2$ represent averages over all the atoms of the fractional amount that each atom is in the lower or the upper quantum state in its individual state mixture.

Applying an external signal therefore does *not* cause an individual atom to make a sudden discrete "jump" from one level to the other. Rather, it really causes the quantum-state mixture of each atom to begin to evolve in a continuous fashion. Quantum theory says that an atom initially more in the lower level tends to evolve under the influence of an applied signal toward the upper level, and vice versa. This changes the state mixture or level occupancy for each atom, and hence the averaged values $N_1$ and $N_2$ over all the atoms. Individual atoms do not make sudden jumps; rather, the quantum states of all the atoms change somewhat, but each by a very small amount.

We should emphasize, finally, that laser materials nearly always contain a very large number of atoms per unit volume. Densities of atoms in laser materials typically range from $\sim 10^{12}$ to $\sim 10^{19}$ atoms/cm$^3$. This density is sufficiently high that laser amplification is an essentially smooth and continuous process, with very little "graininess" or "shot noise" associated with the discrete nature of the atoms involved.

---

Problems for 1.4

1. *Numerical values for the Boltzmann ratio.* The relative numbers of atoms $N_1$ and $N_2$ in two energy levels $E_1$ and $E_2$ separated by an energy gap $E_2 - E_1$ are given at thermal equilibrium by the Boltzmann ratio. To gain some feeling for real situations, evaluate the ratio $N_2/N_1$ for the following cases:

    (a) an optical transition, $\lambda = 500$ nm, at room temperature, 300 K;

    (b) a microwave transition, $f = 3$ GHz, at room temperature;

    (c) a 10 GHz transition at liquid-helium temperature, 4.2 K.

    For an optical transition at $\lambda = 500$ nm to have $N_2/N_1 = 0.1$, what temperature is required? What is the energy $kT$ corresponding to room temperature, expressed in wave numbers?

---

## 1.5 LASER PUMPING AND POPULATION INVERSION

Let us now examine in elementary terms the kind of pumping process that can produce the population inversion needed for laser amplification.

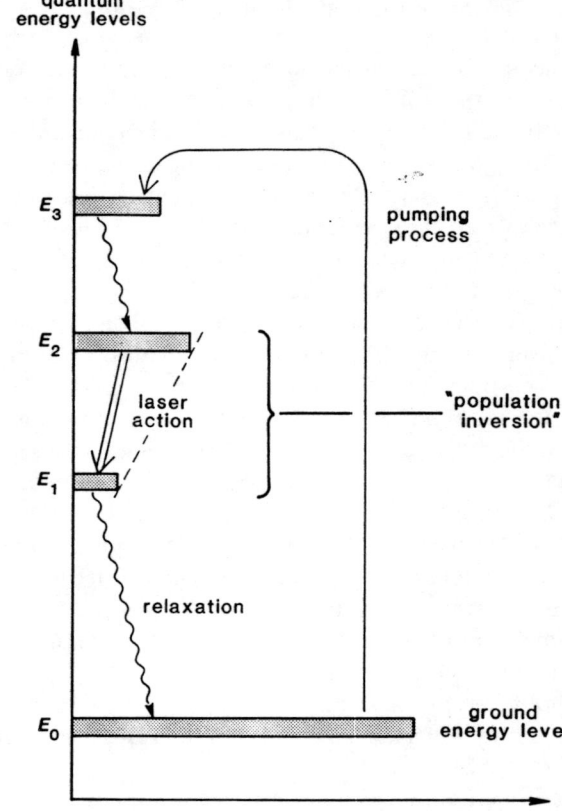

FIGURE 1.29
A four-level laser pumping system.

### Four-Level Pumping Model

As a simplified but still quite realistic model of many real laser systems, we can consider the four-level atomic energy system shown in Figure 1.29. We assume here that there is a lowest or ground energy level $E_0$ and two higher energy levels $E_1$ and $E_2$, between which laser action is intended to take place, plus a still higher level, or more often a group of higher levels, into which there is effective pumping from the ground level $E_0$. We can for simplicity group all these higher levels into a single upper pumping level $E_3$. At thermal equilibrium, under the Boltzmann relation, essentially all the atoms will be in the ground energy level $E_0$.

We then assume that there is a pumping rate $R_{p0}$ (atoms/second) from the ground level $E_0$ into the upper pumping level or levels $E_3$. This pumping rate may be produced by electron impact with the ground-level atoms in a gas discharge, as in many gas lasers; or by pumping with intense incoherent light from a pulsed flashlamp or a cw arc lamp, as in many optically pumped solid-state lasers; or by several other mechanisms we have not yet discussed. In any event, the properties of atoms do permit selective excitation from a lowest level primarily into certain selected upper levels, as assumed in this example.

It is then a realistic description of many practical lasers that a certain fraction $\eta_p$ of the atoms excited upward will relax down, perhaps through a series of cascaded steps, from the upper pumping level $E_3$ into the intended upper laser

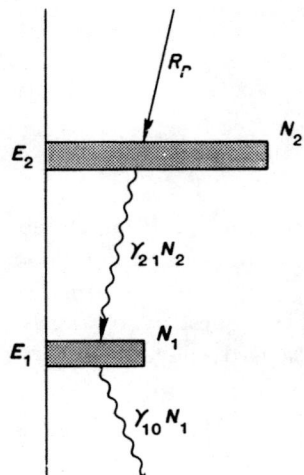

FIGURE 1.30
Rates of flow between atomic energy levels in an ideal four-level laser system.

level $E_2$. We might call $\eta_p$ the pumping efficiency for the laser system, since the effective pumping rate into the upper laser level (again in atoms/second) is $R_p = \eta_p R_{p0}$. This pumping efficiency can be close to unity in some solid-state and organic dye lasers, and only parts per thousand or less in many gas laser systems.

We can also assume in the simplest case that atoms relax from level $E_2$ down to level $E_1$ with a relaxation rate $\gamma_{21}$ and from level $E_1$ down to level $E_0$ with a relaxation rate $\gamma_{10}$. The relaxation processes between these levels may be a combination of the radiative and nonradiative processes we have described in preceding sections. In many practical lasers the fractional number of atoms lifted up out of the ground level $E_0$ into all the upper excited levels also remains small, so that the ground-level population remains essentially unchanged whether the pumping process is on or not.

The flow of atoms between energy levels under the influence of these pumping and relaxation processes (but not laser action for the minute) can then be described by *atomic rate equations* which we will discuss in much more detail in later chapters. For example, the rate equations describing the laser-level populations in the system shown in Figure 1.29 may be written as (Figure 1.30)

$$\frac{dN_2}{dt} \approx R_p - \gamma_{21} N_2 \tag{20}$$

and

$$\frac{dN_1}{dt} \approx \gamma_{21} N_2 - \gamma_{10} N_1. \tag{21}$$

These equations include the upward pumping rate and the downward relaxation rates into and out of levels $E_1$ and $E_2$.

If the pumping process is applied in a continuous fashion and the system comes to a steady-state equilibrium in which $dN_1/dt = dN_2/dt \equiv 0$, we can solve these equations for the steady-state populations and population difference on the laser transition, in the form

$$N_{2,ss} = R_p/\gamma_{21} \quad \text{and} \quad N_{1,ss} = (\gamma_{21}/\gamma_{10}) N_{2,ss}, \tag{22}$$

and hence

$$(N_2 - N_1)_{ss} = \frac{R_p(\gamma_{10} - \gamma_{21})}{\gamma_{10}\gamma_{21}} = R_p\tau_{21} \times (1 - \tau_{10}/\tau_{21}), \tag{23}$$

where $\tau_{21} \equiv 1/\gamma_{21}$ and $\tau_{10} \equiv 1/\gamma_{10}$.

This formula shows that if the lower-level decay rate $\gamma_{10}$ is fast compared to the upper-level decay rate $\gamma_{21}$, so that $\tau_{10} < \tau_{21}$, then there will inevitably be a population inversion on the $2 \rightarrow 1$ laser transition produced by the pumping process. Whether this inversion will be large enough to permit continuous laser amplification or oscillation on this transition is another question, obviously depending in part on the pumping efficiency and on how hard we can pump.

### Conditions for Population Inversion

The basic physical requirement to obtain continuous population inversion in this system is that atoms should relax out of the lower laser level $E_1$ down to still lower levels faster than atoms relax into this level from the upper laser level $E_2$. The absolute strength of the population inversion also depends on a strong pumping rate $R_p$ and a long upper-level lifetime $\tau_{21} \equiv 1/\gamma_{21}$; but the essential condition for population is still that the relative relaxation rates obey the condition that $\gamma_{10} > \gamma_{21}$.

The rate equations for real laser systems can become considerably more complicated, and involve more energy levels and relaxation rates than this simplest example; but the essential features will still be quite similar. The upper levels in many real lasers, for example, are more or less *metastable*—that is, they have comparatively long lifetimes. If we can pump efficiently into such a longer-lived upper level, and if there is a lower energy level with a short lifetime or rapid downward relaxation rate, then a population inversion is very likely to be established between these levels by the pumping process.

As we have mentioned, gas discharges and optical pumping are the two most widely used laser pumping processes. The gas discharges may be continuous (usually in lower-pressure gases) or pulsed (typically in higher-pressure gases). Direct electron impact with atoms or ions, and transfer of energy by collisions between different atoms, are the two main mechanisms involved in gas discharge pumping.

Optical pumping techniques may also be continuous or pulsed. The sources of the pumping light may be continuous-arc lamps, pulsed flashlamps, exploding wires, another laser, or even focused sunlight. Other more exotic pumping mechanisms include chemical reactions in gases, especially in expanding supersonic flows; high-voltage electron-beam pumping of gases or solids; and direct current injection across the junction region in a semiconductor laser.

### Problems for 1.5

1. *Slightly more complicated laser pumping system.* Add a third relaxation rate directly from level 2 to level 0 in Figure 1.29, with a downward rate on this transition given by $\gamma_{20}N_2$. Write out and solve the new rate equations including this term, and discuss the new conditions that are now necessary for population inversion.

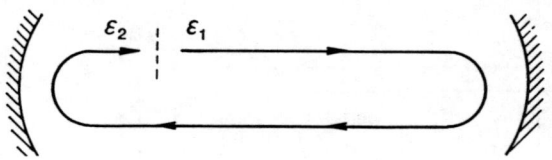

FIGURE 1.31
Round-trip amplification in a laser cavity.

## 1.6 LASER OSCILLATION AND LASER CAVITY MODES

Adding laser mirrors and hence signal feedback, as we will do in this section, is then the final step necessary to produce coherent laser oscillation and thus to obtain a working laser oscillator.

### Condition for Build-Up of Laser Oscillation

Suppose in fact that we have a laser rod or a laser tube containing atoms that are properly pumped so as to produce population inversion and amplification on a certain laser transition. To make a coherent oscillator using this medium, we must then add partially reflecting, carefully aligned end mirrors to the laser medium, as shown in Figures 1.1 or 1.4.

Suppose that we do this, and that a small amount of spontaneous emission at the laser transition frequency starts out along the axis of this device, being amplified as it goes. This radiation will reflect off one end mirror and then be reamplified as it passes back though the laser medium to the other end mirror, where it will of course again be sent back though the laser medium (Figure 1.31).

If the round-trip laser gain minus mirror losses is less than unity, this radiation will decrease in intensity on each pass, and will die away after a few bounces. But, if the total round-trip gain, including laser gain and mirror losses, is greater than unity, this noise radiation will build up in amplitude exponentially on each successive round trip; and will eventually grow into a coherent self-sustained oscillation inside the laser cavity formed by the two end mirrors. The *threshold condition* for the build-up of laser oscillation is thus that the total round-trip gain—that is, net laser gain minus net cavity and coupling losses—must have a magnitude greater than unity.

### Steady-State Oscillation Conditions

Net gain greater than net loss for a circulating wave thus leads to signal build-up at the transition frequency within the laser cavity. This exponential growth will continue until the signal amplitude becomes sufficiently large that it begins to "burn up" some of the population inversion, and partially saturate the laser gain.

Steady-state oscillation within a laser cavity, just as in any other steady-state oscillator, then requires that net gain just exactly equal net losses, or that the total round-trip gain exactly equal unity, so that the recirculating signal neither grows nor decays on each round trip, but stays constant in amplitude.

In mathematical terms, using the more detailed model shown in Figure 1.32, the steady-state oscillation condition for a linear laser cavity with spacing $L$ between the mirrors is that the total voltage gain and phase shift for a signal

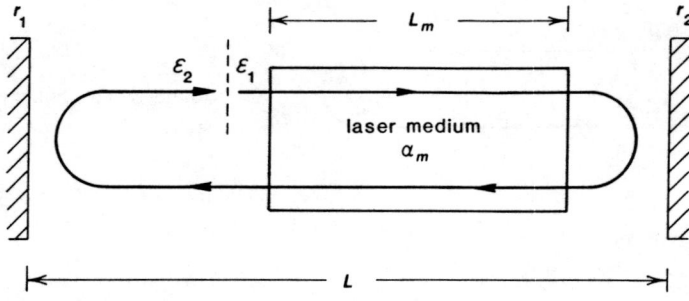

**FIGURE 1.32**
More detailed model of a linear or "standing-wave" laser oscillator.

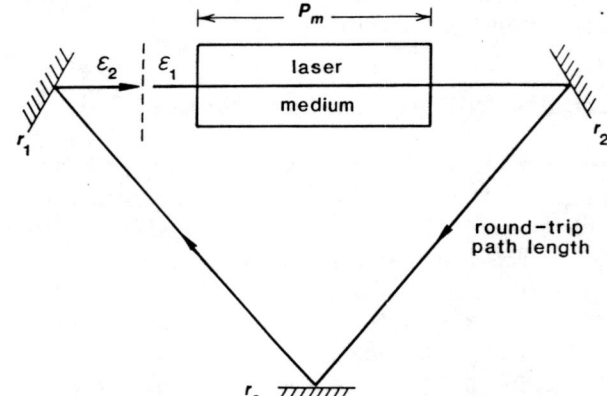

**FIGURE 1.33**
Analytical model of a ring laser oscillator.

wave at frequency $\omega$ in one complete round trip must satisfy the condition that

$$\mathcal{E}_2/\mathcal{E}_1 \equiv r_1 r_2 \exp\left(2\alpha_m L_m - j2\omega L/c\right) = 1 \qquad \text{at steady state.} \qquad (24)$$

where the coefficients $r_1$ and $r_2$ ($|r| \leq 1$) are the wave-amplitude or "voltage" reflection coefficients of the end mirrors; $\exp(2\alpha_m L_m)$ is the round-trip voltage amplification through the laser gain medium of length $L_m$; and $\exp(-2j\omega L/c)$ is the round-trip phase shift around the laser cavity of length $L$. (For simplicity we have left out here any internal losses inside the laser cavity, and also any small additional phase-shift effects caused by the laser atoms or the cavity mirrors.)

If the laser employs instead a ring cavity of the type shown in Figure 1.33—as is becoming more common in laser systems—then this condition becomes instead

$$\mathcal{E}_2/\mathcal{E}_1 = r_1 r_2 r_3 \exp\left(\alpha_m p_m - j\omega p/c\right) = 1 \qquad \text{at steady state,} \qquad (25)$$

where now $p$ is the perimeter or full distance around the ring, and $p_m$ is again the single-pass distance through the laser medium.

### Round-Trip Amplitude Condition

Either of these conditions on steady-state round-trip gain then leads to two separate conditions, one on the amplitude and the other on the phase shift of the round-trip signal transmission. For example, the magnitude part of the steady-state oscillation condition expressed by Equation 1.24 requires simply that

$$r_1 r_2 \exp(2\alpha_m L_m) = 1 \quad \text{or} \quad \alpha_m = \frac{1}{4L_m} \ln\left(\frac{1}{R_1 R_2}\right), \tag{26}$$

where $R_1 = |r_1|^2$ and $R_2 = |r_2|^2$ are the power reflectivities of the two end mirrors.

This condition determines the net gain coefficient or the minimum population inversion in the laser medium that is required to achieve oscillation in a given laser system. Using Equation 1.18 for the laser gain coefficient, we can convert this to the often-quoted *threshold inversion density*

$$\Delta N \equiv N_2 - N_1 \geq \Delta N_{\text{th}} \equiv \frac{\pi \Delta \omega_a}{\lambda^2 \gamma_{\text{rad}} L_m} \ln\left(\frac{1}{R_1 R_2}\right). \tag{27}$$

This expression on the one hand gives the minimum or threshold population inversion $\Delta N_{\text{th}}$ that must be created by the pumping process if oscillation build-up toward sustained coherent oscillation is to be achieved. On the other hand, Equations 1.26 and 1.27 also give the saturated gain coefficient $\alpha_m$ or the saturated inversion density $\Delta N$ (atoms per unit volume) that must just be maintained to have unity net gain at steady state.

A laser oscillator will always start out with inversion somewhat greater than threshold. It will then build up to an oscillation level that just saturates the net laser gain down to equal net loss. This saturation occurs (as we will show in more detail later) when the laser oscillation begins to use up atoms from the upper level at a rate which begins to match the net pumping rate into that level; and it is just this gain saturation process which stabilizes the amplitude of a laser oscillator at its steady-state oscillation level.

Equation 1.27 makes clear that reaching laser threshold will be easiest if the laser has a narrow transition linewidth $\Delta \omega_a$, and low cavity losses, including $R_1, R_2 \to 1$. Note also that laser action generally gets more difficult to achieve as the wavelength $\lambda$ gets shorter—infrared lasers are often easy, ultraviolet lasers are hard.

### Round-Trip Phase or Frequency Condition

Equations 1.24 or 1.25 also express a round-trip *phase shift condition* which says that the complex gain in these equations must actually be equal to unity modulo some large factor of $e^{-j2\pi}$; so for a linear cavity,

$$\exp(-j2\omega L/c) = \exp(-jq2\pi) \quad \text{or} \quad \frac{2\omega L}{c} = q2\pi, \quad q = \text{integer.} \tag{28}$$

In other words, the round-trip phase shift $2\omega L/c$ inside the cavity must be some (large) integer multiple of $2\pi$, or the round-trip path length must be an integer number of wavelengths at the oscillation frequency.

In the linear cavity case this phase condition is met at a set of discrete and equally spaced *axial-mode frequencies* given by

$$\omega = \omega_q \equiv q \times 2\pi \times \left(\frac{c}{2L}\right). \tag{29}$$

The phase shift condition thus leads to a *resonance frequency condition* for the laser cavity, or equivalently to an *oscillation frequency condition* for the laser oscillator. The set of frequencies $\omega_q$ are called *axial modes* because they represent

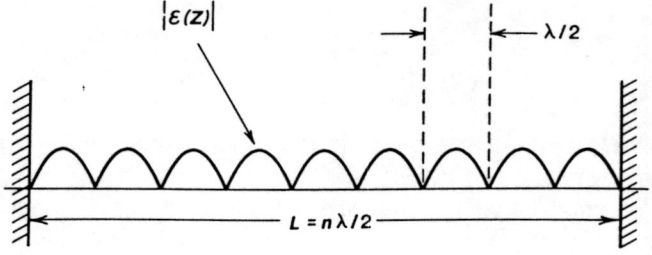

FIGURE 1.34
Axial-mode resonance condition in a standing-wave laser cavity.

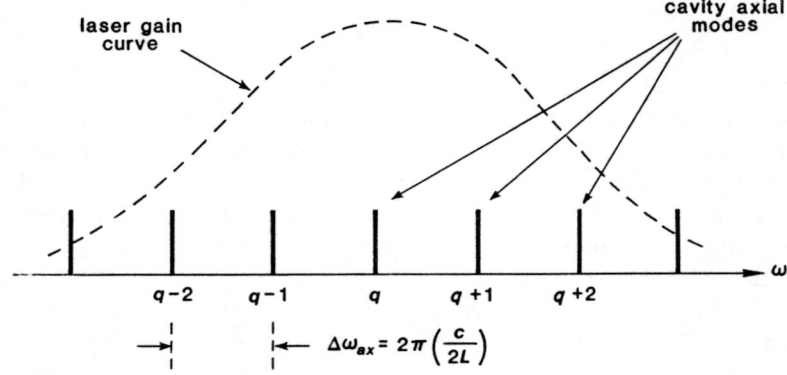

FIGURE 1.35
Multiple axial-mode frequencies under the atomic gain profile in a typical laser system.

the resonant frequencies at which there are exactly $q$ half-wavelengths along the resonator axis between the laser mirror in the linear or standing-wave case.

This same round-trip phase shift condition becomes $\omega p/c = q2\pi$ in the ring cavity case, and the resonant frequencies $\omega_q = q \times 2\pi \times (c/p)$ are then the frequencies at which the ring perimeter $p$ is an integer number of full wavelengths. The axial-mode integer $q$ is typically a very large number in any real laser; e.g., for the standing-wave case

$$q = \frac{\omega_q L}{\pi c} = \frac{2L}{\lambda_q} = \frac{p}{\lambda_q} \approx 10^5 - 10^6, \tag{30}$$

since $L$ (or $p$) is always $\gg \lambda$ for any except very unusual laser cavities.

The axial resonant modes of the laser cavity are thus equally spaced in frequency, with axial-mode separation $\Delta\omega_{ax}$ given by

$$\Delta\omega_{ax} \equiv \omega_{q+1} - \omega_q = 2\pi \times \frac{c}{2L} = 2\pi \times \frac{c}{p}$$

$$\approx 2\pi \times 300 \text{ MHz} \quad \text{for} \quad L = 50 \text{ cm}. \tag{31}$$

For many (though not all) practical lasers, this mode spacing is smaller than the atomic linewidth $\Delta\omega_a$; and hence there will be several axial-mode cavity resonances within the atomic gain curve, as shown in Figure 1.35. The laser may then oscillate, depending on more complex details, on just the centermost one of these axial modes, or on several (or even many) axial modes simultaneously.

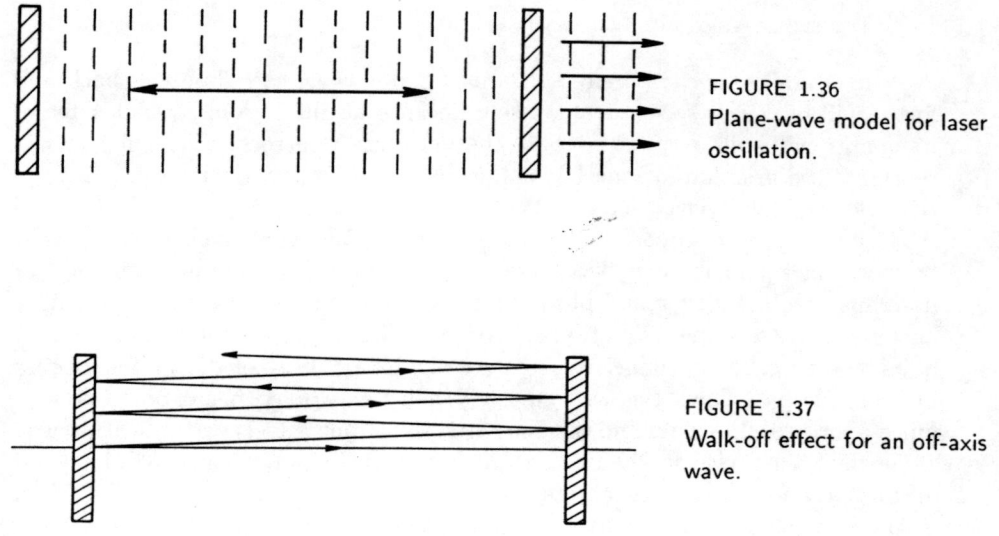

FIGURE 1.36
Plane-wave model for laser oscillation.

FIGURE 1.37
Walk-off effect for an off-axis wave.

### Transverse Spatial Properties: The Plane Mirror Approximation

We need to consider also the *transverse variation* of the optical fields in the laser cavity—that is, the variation over the cross-sectional planes perpendicular to the laser axis—since it is this variation that determines the *spatial coherence* or the *transverse-mode properties* of the laser oscillator.

In the simplest description, a laser will oscillate in the form of a more or less uniform, quasi-plane-wave optical beam bouncing back and forth between carefully aligned mirrors at the two ends of the laser resonator, as in Figure 1.36. The earliest successful lasers, and even some practical lasers today, in fact used flat or planar mirrors carefully aligned exactly parallel to each other and perpendicular to the axis of the laser.

If the optical wave in Figure 1.36 travels at even a slight angle to the resonator axis running perpendicular to the two mirrors, the radiation will walk out the open sides of the cavity, past the mirror edges, after some small number of bounces, as in Figure 1.37. This will represent a large "walk-off" loss from the laser cavity, so that only waves that are very accurately aligned with the resonator axis will remain within the cavity and be able to oscillate. Hence the beam direction for the oscillating waves will lie very accurately along the cavity axis. (This of course also requires strictly parallel alignment of the two mirrors.)

To the extent that the oscillating beam then approximates a finite diameter beam with a nearly planar (or possibly slightly spherical) wavefront, the phase of the emerging wavefront will be essentially uniform across the output mirror, a condition sometimes referred to as a "uniphase" wavefront. There will also then be a very high degree of coherence between the instantaneous phase of the wavefront emerging from widely separated points across the output mirror (but within the overall envelope of the laser beam); and so we can also say that there is a very high degree of "spatial coherence" to the laser output. The laser output beam coming through a partially transmitting end mirror, at least in this simplified description, will thus be a highly directional beam with a uniform phase across the mirror surface and hence essentially perfect spatial coherence in the output beam.

### Transverse Modes in Real Laser Cavities

In a real laser cavity, any such quasi-plane wave, as it bounces back and forth, will of course spread transversely because of diffraction, so that some of its energy will spill over the edges of the finite laser mirrors. This spillover will represent a diffraction loss mechanism, which becomes part of the overall round-trip losses of the laser cavity.

It is even more important to recognize, however, that such a wave, as it bounces back and forth between two mirrors, will also undergo distortion of its transverse amplitude and phase profile in each trip around the laser cavity because of these same diffraction effects. A uniform plane wave coming from a finite aperture, for example, will acquire significant Fresnel diffraction ripples in even one pass down the laser cavity. When this rippled beam bounces off a finite-aperture end mirror and the truncated wavefront travels back the other way along the laser cavity, it will acquire still further distortion because of additional diffraction and propagation effects.

The simple bouncing-plane-wave description of Figure 1.36 therefore cannot be fully correct, first because the uniform plane waves will spread and distort because of diffraction, and second because real laser cavities most often employ spherically curved mirrors, as in Figure 1.38, rather than flat or planar mirrors, for reasons we will soon consider. These mirrors have finite transverse widths or diameters, which effectively act as apertures for the circulating laser beam; and in addition there are often additional apertures elsewhere along the laser axis, either deliberately added or caused by the finite diameter of the laser tube or other intracavity elements.

To understand the transverse beam properties in real laser cavities, therefore, we must examine more carefully what happens to a propagating optical wave with a given transverse amplitude and phase pattern when it propagates through one complete round trip around a laser cavity, including all the focusing, aperturing, and diffraction effects in the round trip.

### Self-Reproducing Transverse Mode Patterns

The round-trip wave propagation in a real laser cavity can be studied by carrying out analytical or computer calculations of the manner in which the transverse field pattern of the optical beam changes on repeated round trips within a given resonator. Optical resonator mode calculations of this type were first pioneered in the early 1960s by A. G. Fox and T. Li at the Bell Telephone Laboratories, and are often referred to as "Fox and Li" calculations.

Such calculations are usually carried out with the laser gain omitted for simplicity. It then turns out that for any given laser cavity, employing either finite-diameter planar or (more usually) finite-diameter curved end mirrors, one will always find a certain discrete set of transverse eigenmodes, or distinct amplitude and phase patterns for the circulating beam in the cavity, *which will reproduce themselves in form, though slightly reduced in overall amplitude, after one round trip*. A typical example of such a self-reproducing transverse beam pattern is shown in Figure 1.38. These self-reproducing transverse field patterns represent the characteristic set of *lowest-order and higher-order transverse eigenmodes* or *transverse spatial modes* characteristic of that particular laser resonator.

These self-reproducing transverse eigenmodes, with amplitude and phase patterns that depend on the specific curvature and shape of the laser mirrors,

## 1.6 LASER OSCILLATION AND LASER CAVITY MODES

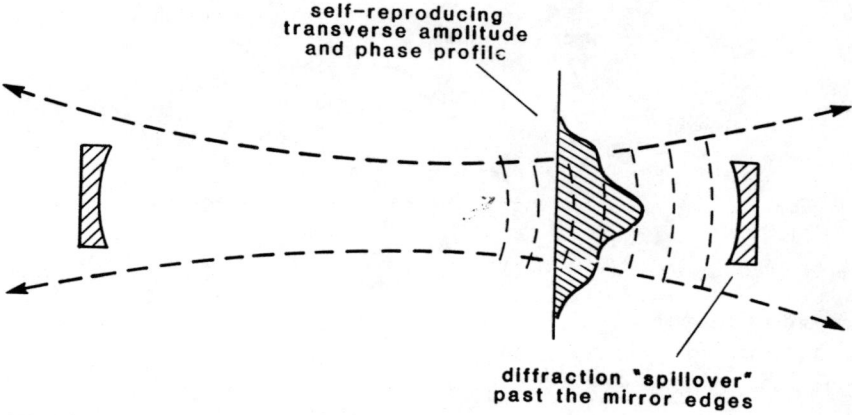

FIGURE 1.38
Example of a self-reproducing transverse mode pattern with finite diffraction losses in a typical real laser cavity.

are analogous to the transverse modes in a closed waveguide, or even more closely analogous to the lowest-order and higher-order propagation modes in a leaky optical lensguide. Indeed, we can view the repeated round trips in either a standing-wave or a ring laser resonator as essentially equivalent to passage through repeated sections of an iterated periodic lensguide, with reflection from the finite-aperture cavity mirrors being replaced by transmission through equivalent finite-aperture lenses having the same focal power.

These transverse eigenmodes can then provide self-consistent oscillation beam patterns for an oscillating laser. The amplitude reduction on each pass—which is generally different for each such transverse mode—simply represents the diffraction or spillover losses for that particular mode, caused by whatever finite apertures are present in the cavity. If the laser then begins oscillating in one of these patterns, and if the laser medium can maintain sufficient round-trip gain to overcome the diffraction losses of that particular transverse mode, along with all the other losses in the cavity, this will be one possible steady-state beam pattern or beam profile for the laser oscillation.

### Planar Resonator Modes

In any reasonably well-designed laser cavity with finite-width or finite-diameter end mirrors, we will normally find that there is one such lowest-order transverse mode pattern, which is usually reasonably smooth in its transverse amplitude and phase profile, and which has the lowest diffraction loss of all the self-reproducing transverse mode patterns in that particular resonator.

In a properly aligned planar resonator, for example, the lowest-order transverse mode will generally have an amplitude profile which looks something like the upper part of figure 1.39. That is, this mode will typically look something like the central lobe of a $J_0(r)$ Bessel function across the mirror for circular end mirrors, or like a single lobe of a cosine wave, that is $\mathcal{E}(x,y) \approx \cos(\pi x/a)\cos(\pi y/b)$ for rectangular mirrors of width $2a$ by $2b$. The exact amplitude pattern of this lowest-order mode will, however, also have diffraction ripples, as in the upper part of Figure 1.39, whose amplitude and spacing depend on the finite mirror size; and the quasi-Bessel function or cosine variation will not drop quite to zero

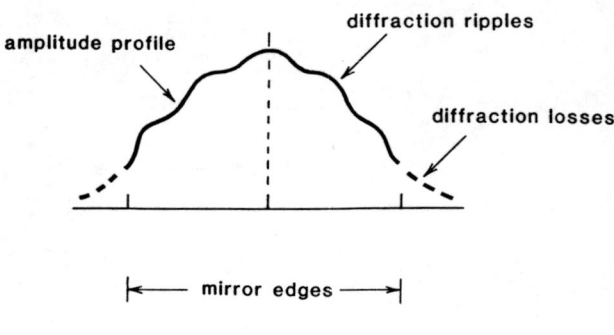
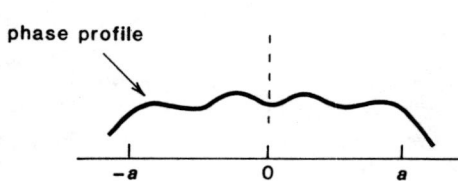

FIGURE 1.39
A typical lowest-order transverse mode profile in a planar-mirror cavity like Figure 1.36.

at the mirror edges, in agreement with the inevitable diffraction losses in such an open-sided resonator.

The phase variation of the lowest-order mode in a typical planar resonator will also exhibit some small Fresnel diffraction ripples, along with some small curvature of the wavefront along the outer edges of the resonator, as in the lower part of Figure 1.39; but over the major portion of the mode the wavefront will in fact be a very good approximation to a planar wavefront. A plane-mirror cavity oscillating in this lowest-order transverse mode will thus in fact have output beam properties very close to those of the simple plane wave described earlier.

The unwanted diffraction losses past the mirror edges for this lowest-order transverse mode will also typically be very small, unless the mirror sizes are made very small. The lower-order self-consistent transverse modes in almost any type of resonator in fact exhibit an uncanny ability to shape their amplitude and phase patterns in ways that minimize their diffraction losses on each round trip.

### Higher-Order Modes

This same laser cavity will generally also have many higher-order transverse modes. These will generally have larger diffraction losses and also more complex transverse amplitude and phase variations, like the higher-order transverse modes in waveguides. And they will generally have several transverse nulls and phase reversals, with either even or odd symmetry in simple cases. Their transverse spread inside the cavity is generally larger, which makes their diffraction losses larger than those of the lowest-order transverse mode; and their diffraction spread or beam spread outside the cavity is also generally larger than that for the lowest-order transverse mode. For these reasons, laser oscillation in these higher-order modes is generally considered undesirable. We will analyze the mode properties of these transverse modes in great detail in Chapters 14 through 23.

## 1.6 LASER OSCILLATION AND LASER CAVITY MODES    47

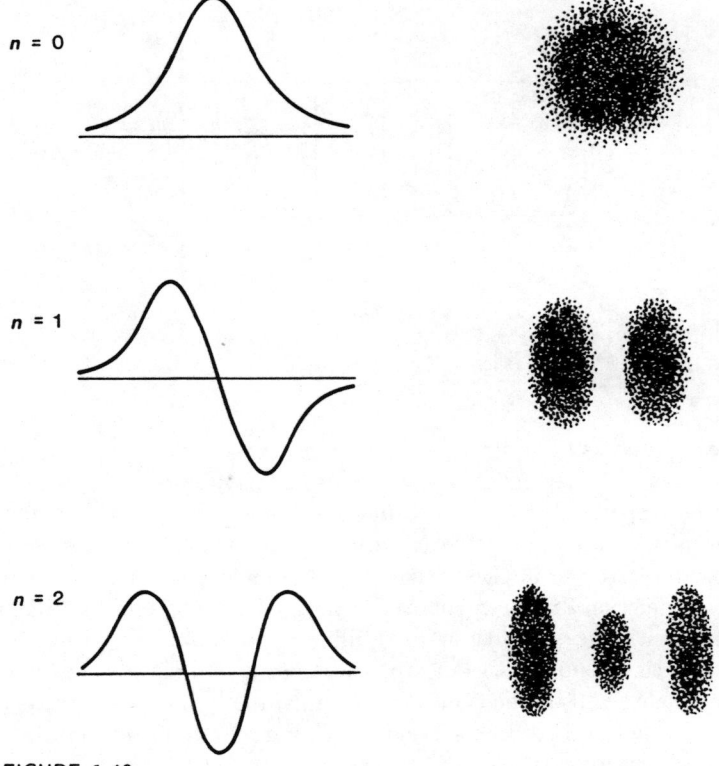

FIGURE 1.40
Hermite-gaussian transverse-mode patterns in a stable laser resonator.

### Stable and Unstable Laser Resonators

Practical laser cavities most often employ curved rather than planar end mirrors, in order to shape the transverse modes of the cavity and control the diffraction losses. There is one broad class of such curved-mirror resonator designs, the so-called *stable laser resonators*, in which the diffraction losses are generally very small, and the lowest-order and higher-order modes have the form (very nearly) of Hermite-gaussian functions, as in Figure 1.40, with the lowest-order mode having a gaussian transverse profile of the form $\mathcal{E}(r) = \exp(-r^2/w^2)$. Such gaussian modes and the resulting gaussian output beams are particularly easy to handle both analytically and in experiments, and practical lasers are very often designed in this fashion.

On the other hand, these Hermite-gaussian modes in realistic laser cavities do turn out to be very slender in diameter, so that they do not readily fill all the volume of larger-diameter laser tubes or rods. The laser must then oscillate in a mixture of lowest-order and higher-order modes (which tends to spoil the beam collimation properties) in order to fill and extract all the available power from the laser volume.

There is also a class of so-called *unstable optical resonators*, which make use of deliberately diverging laser wavefronts as shown in Figure 1.41. These resonators have transverse mode patterns that much more readily fill large laser volumes, but still suppress higher-order transverse modes. These unstable optical resonators necessarily have much larger output coupling or lower effective mirror reflectivity than stable resonators, since the diffraction spread past the output

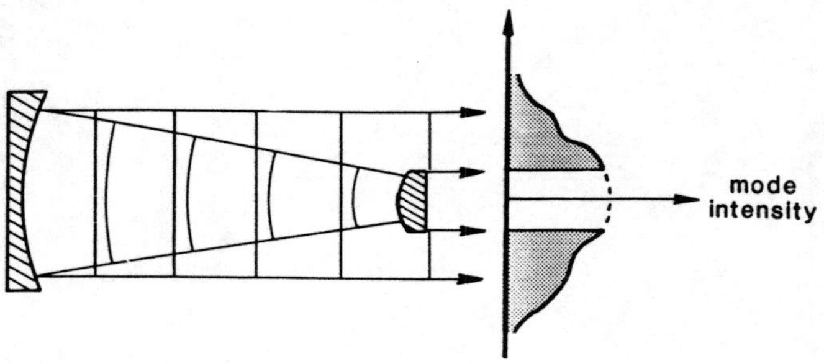

**FIGURE 1.41**
A typical unstable resonator transverse-mode profile.

mirror edges is used as the output coupling mechanism. This property limits the usefulness of unstable resonators for low-gain laser systems.

The mode properties of such unstable resonators are also rather more complex and esoteric than the simple Hermite-gaussian stable modes. (Note that the "stability" referred to in these resonator classifications is that of geometrical rays bouncing back and forth in the cavity designs in question, and has nothing directly to do with the stability or instability of the laser oscillation in the resulting transverse eigenmodes.) Perhaps the most useful class of laser resonator modes in the future will be the geometrically unstable but still Hermite-gaussian modes that can be obtained in so-called "complex paraxial" resonators by using variable reflectivity mirrors, as will be described in Chapter 23.

### General Transverse-Mode Oscillation Properties

Each different optical-resonator design, whether planar, stable, unstable, or still more complex, will thus possess some lowest-order transverse mode pattern which can circulate repeatedly around the laser cavity without changing its amplitude or phase profile. The phase profile of this lowest-order transverse mode will usually be comparatively smooth and regular across the output mirror of the laser cavity (as well as at any other transverse plane within the cavity). The phase front is often quasi spherical across the output plane of the laser, but this spherical curvature can be removed by a simple lens to convert the output beam into a fairly well-collimated plane wave.

A laser cavity which oscillates only in this lowest-order transverse mode will thus generally produce an output beam with good transverse characteristics and with a nearly uniphase character across the output mirror. If the laser oscillates simultaneously in several transverse modes, however, as can readily happen in real lasers, the output wavefront will no longer be "uniphase," and the collimation and focusing properties of the beam will generally deteriorate.

Forcing laser oscillation to occur only in the lowest-order transverse mode is thus a practical design objective, which is achieved in some though not all practical lasers. The primary obstacle to achieving single-transverse-mode oscillation in higher-power (or higher-gain) lasers is that the diffraction losses of the lowest- and higher-order modes in a large-diameter cavity are all small and nearly identical; so there is little or no loss discrimination between the different transverse modes. A designer must then add mode-control apertures, employ unstable resonator designs, or use other tricks to suppress the unwanted higher-order transverse modes.

Note also that the transverse mode properties we have just been discussing, and the axial mode or resonant frequency properties we discussed earlier, are almost independent of each other. There are some important secondary connections between these properties, and we will discuss them in detail in later chapters. In simplified terms, however, the round-trip propagation length determines the resonant axial-mode frequencies of the laser, whereas the focusing and diffraction effects associated with mirrors and apertures in the round-trip propagation determine the transverse mode patterns.

## REFERENCES

The original reference on the Fox and Li approach to optical resonator modes is A. G. Fox and T. Li, "Resonant modes in a maser interferometer," *Bell Sys. Tech. J.* **40**, 453–458 (March 1961); with later and more extensive results in "Modes in a maser interferometer with curved and tilted mirrors," *Proc. IEEE* **51**, 80–89 (January 1963).

A good review of standard stable resonator theory is given by H. Kogelnik and T. Li, "Laser beams and resonators," *Proc. IEEE* **54**, 1312 (October 1966) and *Appl. Optics* **5**, 1550–1567 (October 1966). Unstable resonators are reviewed in A. E. Siegman, "Unstable optical resonators," *Appl. Optics* **13**, 353 (February 1974).

## 1.7 LASER OUTPUT-BEAM PROPERTIES

The output beam from a laser oscillator thus basically consists of electromagnetic radiation, or light, that is not fundamentally different in kind from the radiation emitted by any other source of electromagnetic radiation. There are several important and fundamental differences in detail, however, between the "incoherent" light emitted by any thermal light source, such as the flashlight in Figure 1.42, and the "coherent" light emitted by a laser oscillator.

The output beams produced by laser oscillators in fact have much more in common with the outputs of conventional low-frequency electronic oscillators, such as transistors or vacuum tubes, than they do with any kind of thermal light sources. Laser beams are often described as being different from ordinary light sources in being both *spatially coherent* and *temporally or spectrally coherent*. These rather vague phrases refer to some characteristic laser output-beam properties that we will review briefly in this section.

An important point to keep in mind is that all these coherence properties arise primarily from the *classical resonant-cavity properties* of the laser resonator, as we described in the preceding section, rather than from any of the quantum transition properties of the laser atoms.

### Ideal Laser Monochromaticity and Frequency Stability

The flashlight shown in Figure 1.42, like any other thermal light source, emits a generally broadband continuum of light at many different wavelengths. There are light sources, such as discharge lamps, that emit only comparatively few spectral lines or narrow bands of wavelengths, but the spectral widths of the light emitted by even the best such sources are still limited by the linewidths of the atomic transitions in the discharge atoms.

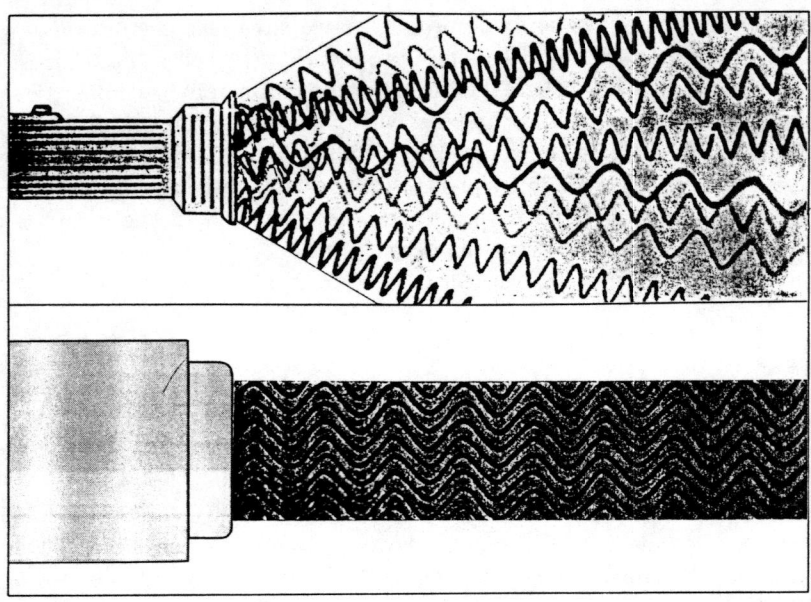

**FIGURE 1.42**
Incoherent light from a flashlight (top) and coherent light from a laser (bottom).

The output beams from most lasers can be, by contrast, *highly monochromatic*, and in ideal lasers can consist almost entirely of a *single frequency*. That is, the output signal from a near-ideal laser will be a nearly pure, constant-amplitude, highly stable, single-frequency sine wave, exactly like the signal generated by a highly stable electronic oscillator in any other frequency range.

Atomic transitions typically have fractional atomic linewidths $\Delta\omega_a/\omega$ ranging from 1 part in 100 (broadband dye or semiconductor materials) to narrower than 1 part in $10^6$ (narrow-line atomic transitions in gases); and it is this linewidth that characterizes the spontaneous or fluorescent emission from such atoms. In absolute terms such linewidths range from a few GHz (as in typical doppler-broadened gas lasers in the visible) to a few tens or hundreds of GHz (as in typical solid-state lasers). The short-term spectral purity of a good-quality single-frequency laser oscillator, by contrast, can range from a few tens of MHz (in a moderately well-stabilized gas laser) down to only a few Hz in a very highly stabilized system.

As we have said, it is the laser cavity and not the laser atomic transition that is primarily responsible for these spectral properties. Continuous oscillations can be sustained in a laser resonator only at those discrete axial-mode frequencies where the round-trip phase shift inside the laser cavity is an integer $q$ times $2\pi$. The laser atomic transition then serves primarily to provide gain at these cavity resonance frequencies, not to determine the oscillation frequency (except for small, second-order frequency-pulling effects that we have not yet discussed).

### Spectral Purity in Practical Lasers

Both the short-term frequency jitter and the long-term frequency drift of a laser oscillator usually result primarily from mechanical vibrations and noise, thermal expansion, and other effects that tend to change the length $L$ of the

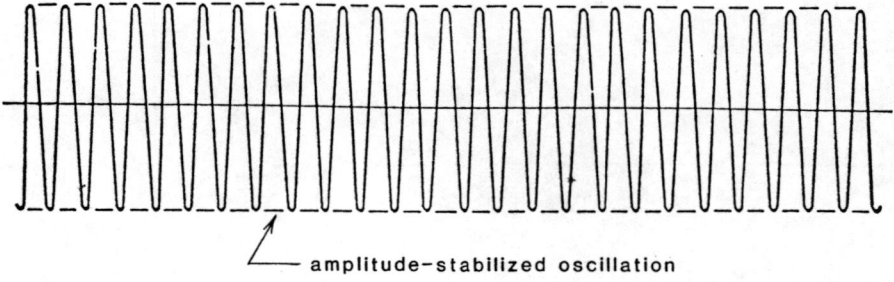

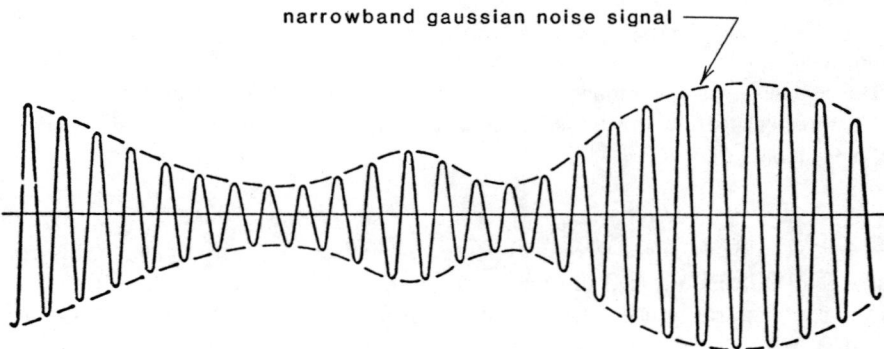

FIGURE 1.43
Sine wave from a coherent oscillator (top) and "noise wave" from a narrowband thermal source (bottom).

laser cavity. Very highly stabilized laser oscillators can nonetheless have long-term absolute frequency stabilities better than 1 part in $10^{10}$, and short-term spectral purities as high as 1 part in $10^{13}$, making them equal to or better than the best atomic clocks available in any frequency range.

The ultimate limit on laser spectral purity is finally set by quantum noise fluctuations caused by the spontaneous emission from the atoms inside the laser cavity. These quantum noise effects, which are described by the so-called "Schawlow-Townes formula," can be observed with great difficulty only on the very best and most highly stabilized laser oscillators.

### Laser Statistical Characteristics

In addition to being highly frequency-stable, a good-quality laser oscillator will generally have all the other statistical and amplitude-stabilization properties associated with a coherent electronic oscillator in any frequency range.

The most basic of these properties is that the instantaneous optical field in a single oscillating cavity mode will be essentially a pure optical-frequency sine wave, whose amplitude remains closely stabilized to the steady-state value at which the saturated laser gain just equals the net mode losses. This is usually a self-stabilizing situation: if the gain increases slightly above the loss because of some random fluctuation, the oscillation amplitude begins to grow slightly, and the slightly increased signal amplitude pulls the gain back down. Conversely, if the amplitude fluctuates slightly above its average value, this pushes the gain down below the loss, and pulls the oscillation amplitude back down.

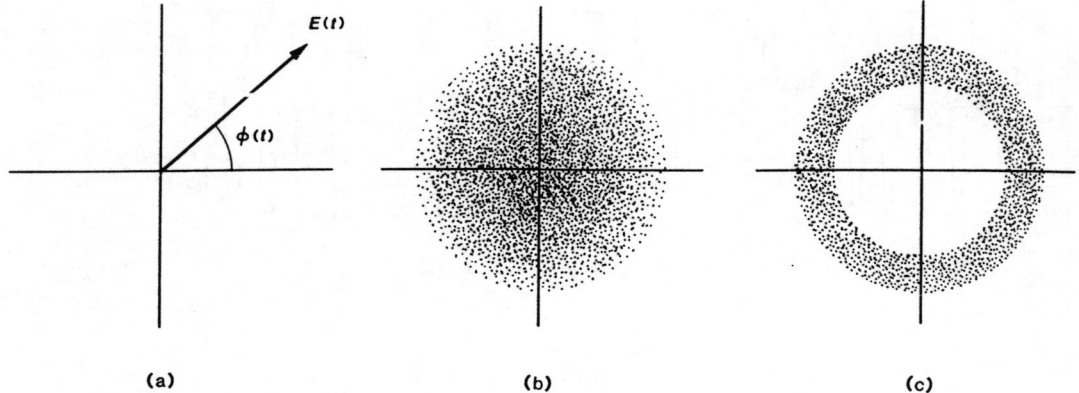

**FIGURE 1.44**
(a) The complex phasor amplitude of a sinusoidal signal at any one instant of time, and its statistical distributions for (b) a narrowband gaussian noise source, and (c) an amplitude-stabilized oscillator.

A well-stabilized single-frequency laser can in fact have almost negligible amplitude fluctuations, limited mostly by random fluctuations in the pumping rate and the cavity parameters. The output signal from a well-stabilized high-quality single-frequency laser can thus be best described as an optical sine wave with a highly stabilized amplitude and frequency, whose amplitude changes very little, but whose absolute phase drifts randomly and slowly through all possible values, because of small random environmental fluctuations and ultimately because of quantum noise.

### Laser Signals Versus Narrowband Incoherent (Thermal) Signals

The output signal from such a high-quality laser will also differ in another quite fundamental way from the spontaneous emission emitted by any thermal or "incoherent" light source. Suppose that the output signal from some very bright thermal light source could be first filtered through some extraordinarily narrowband optical filter, and then amplified through some very high-gain linear optical amplifier (perhaps a laser amplifier), so that the resulting signal was both as narrowband and as powerful as a typical high-quality laser beam. (Though this conceptual experiment would be extremely difficult in practice, there is no fundamental barrier to it in principle.) This output will then also look like an optical-frequency sine wave, but this sine wave will not have constant amplitude or phase, no matter how narrowly filtered it may be. Rather, it will always look something like the incoherent narrowband noise wave in the lower part of Figure 1.43.

Suppose that we write the instantaneous electric field for both the signals in Figure 1.43 in the form $\mathcal{E}(t) = E(t)\cos[\omega_0 t + \phi(t)]$, where $\omega_0$ is the midband or carrier frequency, and $E(t)$ and $\phi(t)$ are the slowly varying amplitude and phase of the signals. We can then represent each signal during any short interval of time by its instantaneous phasor amplitude $E(t)e^{j\phi(t)}$, where this phasor amplitude moves around in time in the complex plane as shown in Figure 1.44(a).

For a thermal noise source, the instantaneous phasor amplitude will then move slowly but randomly through many different phase angles and amplitudes, tracing out a two-dimensional random walk as shown in Figure 1.44(b). The

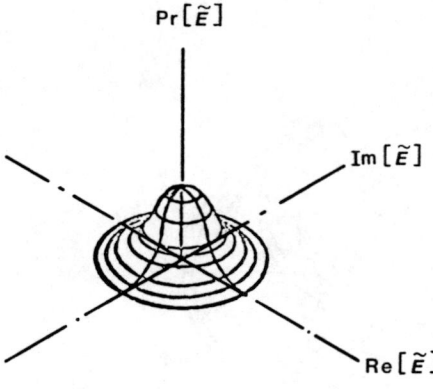

FIGURE 1.45
In three dimensions, the distribution in Figure 1.44(b) is a "gaussian molehill."

bandwidth of the noise signal, no matter how narrow, will determine only how rapidly the phasor moves around within this region—not how far it moves, or with what probability distribution. This noise signal, though having the same power and bandwidth (and hence the same power spectral density) as the laser, will still have the statistical character of *narrowband gaussian noise*. That is, both the phase and the amplitude of this thermal signal will fluctuate slowly with time, at a rate given essentially by the inverse bandwidth of the signal. The probability distribution for the instantaneous phasor amplitude of the thermal signal will be a "gaussian molehill" (Figure 1.45), with the $x$ and $y$ axes corresponding to the amplitudes of the $\sin(\omega_0 t)$ and $\cos(\omega_0 t)$ components of the signal, and the height of the molehill corresponding to the probability of the signal having these sin and cos components at any instant.

However, the laser oscillator signal, like any other conventional oscillator, will fluctuate primarily only in phase, with only small fluctuations in amplitude about its steady-state value. Its phase angle will wander slowly but randomly through all possible phase angles, in a manner corresponding to its small residual frequency uncertainty; but its amplitude will not. Its probability distribution will thus be a "gaussian molerun" (Figure 1.46) rather than a molehill.

### Amplitude Fluctuations in Semiconductor Diode Injection Lasers

The active volume in a semiconductor diode is very small; the passive cavity $Q$ is comparatively low; the atomic lifetimes are fairly short; and the atomic linewidth is very wide compared to most other lasers. As a result of all these characteristics, spontaneous emission effects or fundamental quantum noise fluctuations are generally more significant in semiconductor lasers than in many other types of lasers, and the resulting amplitude and phase fluctuations are larger and more easily observed than in most other lasers. A particularly clean illustration of amplitude-fluctuation effects in semiconductor injection lasers is given, for example, by P.-I. Liu, *et. al.*, in "Amplitude fluctuations and photon statistics of InGaAsP injection lasers," *IEEE J. Quantum Electron.* **QE–19**, 1348–1351 (September 1983). One of the conclusions of this study is that the output signal from a laser oscillator can be very accurately described as the combination of a coherent (highly stabilized) sinusoidal oscillation, plus an additive

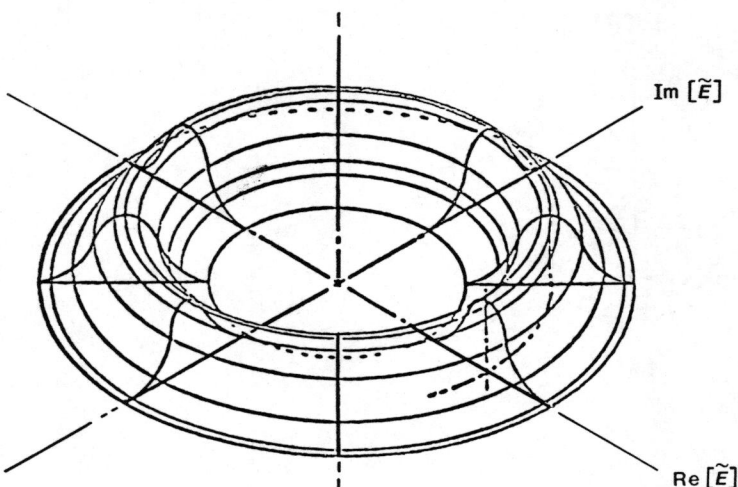

FIGURE 1.46
In three dimensions, the distribution in Figure 1.44(c) is a "gaussian molerun."

gaussian noise component which represents the net effect of spontaneous emission from the inverted laser medium inside the cavity.

### Laser Temporal Coherence

The preceding descriptions make more precise what is generally meant by the "temporal coherence" of a laser output signal. However, the term "coherence" is often used carelessly, both in discussions of lasers and in other situations, and this has led to some confusion. The term coherence necessarily refers not to one property of a signal at a single point in space and time, but to a *relationship*, or *a family of relationships*, between one signal at one point in space and time, and the same or another signal at other points in space and time.

There are, for example, certain precise mathematical definitions of coherence functions as used in coherence theory. These functions give the degree of correlation, described in a specific mathematical fashion, between two signals observed at different points in space and/or time. More colloquially, a signal is called "temporally coherent" if there is strong correlation in some sense between the amplitude and/or phase of the signal at any one time and at earlier or later times.

Both the amplitude and the phase of a good-quality laser oscillator will in fact change only slowly with time, so that the amplitude and phase of the output sine wave from the laser at any one time will be strongly correlated with the amplitudes and phases at considerably earlier or later times. A good laser beam might thus be said to be temporally coherent because of this strong correlation between the amplitudes and phases of the signal at not very different points in time. Much the same might be said, however, of the narrowband noise signal described earlier, since there is considerable coherence between the signals at any two times that are less than one reciprocal bandwidth apart. In fact, a high degree of coherence in the formal mathematical sense does not by itself imply that signals are the kind of "clean" and amplitude-stable sinusoidal oscillation signal generated by a good laser oscillator. Two highly disorderly or irregular signals can still have a very high degree of coherence *between* themselves.

Laser Spatial Coherence

We have already noted that a good-quality laser oscillator can also oscillate in a single transverse-mode pattern, which has a definite and specific amplitude and phase pattern across any transverse plane inside the laser, and particularly across the output mirror. In this situation there is a very high degree of correlation between the instanteous amplitudes. and especially between the instantaneous phase angles, of the wavefront at any two points across the output beam. We can then also say that the output beam possesses a very high degree of "spatial coherence" (in the transverse direction) as well as the temporal coherence discussed above.

Often this lowest-order output-beam pattern will vary reasonably smoothly in amplitude, and its phase variation will approximate reasonably closely either a plane wave or a spherical wave (which can be converted into a plane wave with a simple lens). In contrast, if there are badly distorted optical elements inside the laser cavity, the amplitude and especially the phase profile across the beam may be badly distorted. But if this pattern still represents a single transverse cavity mode, however badly distorted, then there will still be a high degree of coherence between the wavefront phasor at different transverse points; i.e., this beam will still be "spatially coherent" in some sense. In principle, we could therefore design a complex "deaberrating lens" or deaberrating spatial filter that can convert this distorted but stationary wavefront into a smooth and uniphase wavefront of the type that is desirable in a laser output beam.

Laser Beam Collimation

Thermal light sources not only usually emit many wavelengths, but also emit them quite randomly, in essentially all directions. Even if we capture some fraction of this radiation and collimate it with a lens or mirror, as in a searchlight or in the flashlight in Figure 1.42, the resulting degree of collimation, or the amount of radiation emitted per unit solid angle, is still much smaller than in even a very poor quality laser oscillator.

A single-transverse-mode laser oscillator can produce (usually in practice, and always in principle) an output beam that is more or less uniform in amplitude and constant in phase ("uniphase") across its full output aperture of width or diameter $d$. Such a beam can propagate for a sizable distance with very little diffraction spread; will have a very small far-field angle at still larger distances; and can be focused into a spot only a few wavelengths in diameter.

Elementary diffraction theory says, for example, that a uniphase plane wave coming from an aperture of diameter $d$ will have a minimum angular diffraction spread $\Delta\theta$ in the far field (Figure 1.47) given by

$$\Delta\theta \approx \frac{\lambda}{d} \quad \text{(in radians)}. \tag{32}$$

For a visible laser with $\lambda = 0.5$ μm and an output aperture of, say, $d = 0.5$ cm, this gives an angular spread of $\Delta\theta \approx 10^{-4}$ radians, which we might alternatively express as 0.1 milliradians or 100 μrad.

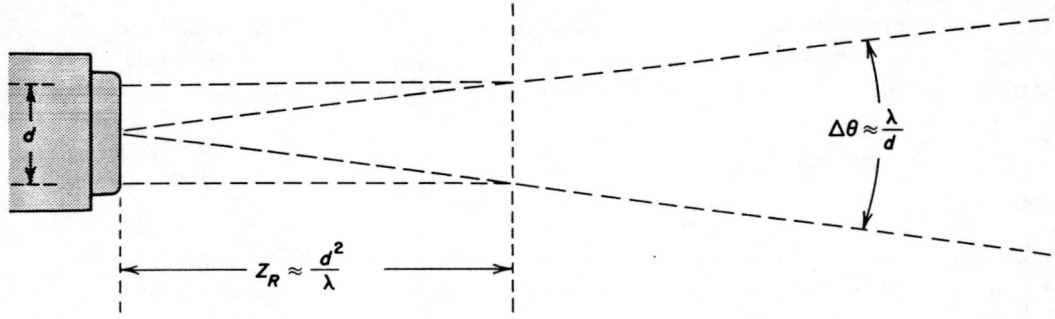

**FIGURE 1.47**
Laser beam collimation and diffraction spreading.

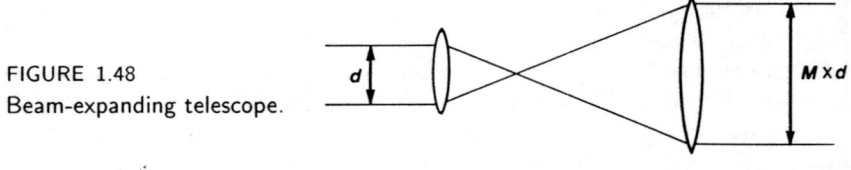

**FIGURE 1.48**
Beam-expanding telescope.

The axial distance over which this same beam will stay approximately parallel and collimated before diffraction spreading begins to significantly increase the beam size—sometimes called the Rayleigh range—is then given (see Figure 1.47) by $d/z_R \approx \lambda/d$, or

$$z_R \approx d^2/\lambda. \tag{33}$$

A visible beam with a diameter of 5 mm thus has a Rayleigh range of $z_R \approx 50$ meters.

Suppose this same uniphase beam is magnified by a 20-power telescope attached to the laser output and focused to infinity, as in Figure 1.48. Then the source aperture diameter is increased to $d = 10$ cm, and these results change to $\Delta\theta \approx 5$ $\mu$rad and $z_R \approx 20$ km. Uniphase laser beams can be propagated for very large distances with very small diffraction spreads.

### Laser Beam Focusing

Suppose this same uniphase laser beam with initial diameter $d$ is focused down to a spot of diameter $d_0$ by means of a simple lens of focal length $f$. The diameter $d_0$ of the focused spot can then be calculated by applying the same angular spread condition in reverse, to obtain

$$\Delta\theta \approx \frac{\lambda}{d_0} \approx \frac{d}{f} \tag{34}$$

or

$$d_0 d \approx f\lambda, \tag{35}$$

since the focal point will occur essentially one focal length $f$ beyond the lens.

Suppose we follow the common practice in optics of defining the "$f$-number" or "$f$-stop" of the focusing lens by $f^\# \equiv f/d$, i.e., focal length over diameter. (We

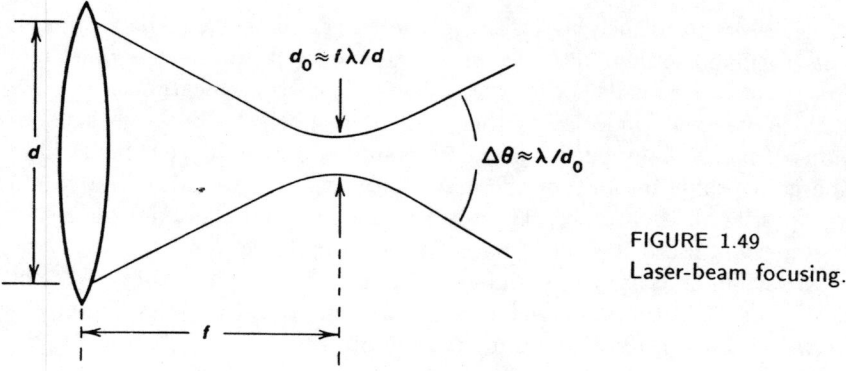

FIGURE 1.49
Laser-beam focusing.

are really defining this quantity in terms of the input beam diameter rather than of the lens diameter, but this of course determines the minimum lens diameter that can be employed.) The approximate diameter of the focused spot can then be written as simply

$$d_0 \approx f^{\#} \lambda. \tag{36}$$

Photography buffs will know that lenses with $f^{\#} \geq 10$ are fairly easy to obtain; lenses with $f^{\#}$ less than about 2 become expensive; and lenses with $f^{\#}$ approaching unity become very expensive.

All the power in a truly uniphase laser beam can thus be focused into a spot a few laser wavelengths in diameter, if we use a powerful lens. (Microscope objectives are usually used for this purpose, at least for laser beams that are not too high in power. A focusing lens for single-wavelength laser radiation of course requires no correction for chromatic aberration, which helps.)

### Nonideal Laser Oscillators: Multimode and Multifrequency Oscillation

Many real lasers can produce output beams which come very close to the ideal temporal and spatial behavior described in the preceding paragraphs. Other lasers, however—especially including some of the higher-power laser systems—are more likely to oscillate in both multiple axial and multiple transverse cavity modes. The coherence properties, both temporal and spatial, of such lasers then necessarily deteriorate relative to more ideal single-mode lasers; and the effort to obtain both single-axial-mode (or single frequency) oscillation, and single-transverse-mode (or "diffraction limited") beam quality, provides a continuing struggle for those who design and construct lasers.

Forcing a practical laser to oscillate in only a single centermost axial mode within the atomic linewidth is most easily accomplished if the laser cavity is made short in order to increase the $c/2L$ axial mode spacing, and if the atomic linewidth is narrow. The laser transition should also preferably be "homogeneously" rather than "inhomogeneously" broadened (we will define these specialized terms later). Special mode-selection techniques employing intracavity etalons and other special filters can also be used to reinforce one selected axial mode and suppress others.

Many practical lasers, however, actually oscillate in several axial modes simultaneously, usually in only a few, but perhaps in a few hundred in extreme cases. The outputs from such lasers, though no longer single-frequency, can still be quite narrowband compared to incoherent light sources; and multi-axial-mode oscillation is not a serious defect for many practical laser applications.

In such multi-axial-mode lasers there are more likely to be large random fluctuations of individual mode amplitudes, as individual mode frequencies drift across the gain profile because of thermal cavity expansion, and as individual modes compete with each other. The total intensity in all the axial modes is, however, somewhat more likely to remain constant. Real laser devices can also be operated in various internally modulated and pulsed forms, and may be subject to various kinds of instabilities and relaxation oscillations, such as "spiking," which we will discuss in more detail in later chapters.

The output signals from such less-than-perfect lasers may thus usually be described as the summation of several simultaneous and independent oscillation frequencies, and may have substantial random variations in amplitude and frequency for each separate oscillation. Such a rather random multifrequency output, though not really the same as a gaussian random noise signal, may appear much like random noise according to various statistical and spectral measures.

### Real Laser Oscillators: Multiple-Transverse-Mode Oscillation

Many real lasers produce output beams which also approach the desirable single-transverse-mode character. A laser beam having the necessary single-mode and uniphase character is often said to be "diffraction limited," since its far-field diffraction angle and focal spot size will approach the ideal limits given just above; whereas beams whose far-field angular spread or focused spot size are $k$ times larger than this are said to be $k$ times diffraction-limited in performance.

More detailed diffraction calculations show that the far-field beam spread of a nonideal beam from an aperture of diameter $d$ is not greatly affected by the exact amplitude pattern of the beam across the aperture; that is, it does not matter greatly whether the amplitude pattern is uniform, gaussian, cosine, or Bessel function, nor do moderate amplitude ripples on the beam lead to serious far-field beam spreading. However, phase variations across the beam wavefront, whether random or regular in character, do begin to substantially increase the far-field beam spread or the focal spot size as soon as they approach the order of 90° phase shift—a distortion of more than a quarter of an optical wavelength—anywhere across the beam width.

A rough argument for the deterioration in beam quality that results from multiple-transverse-mode operation can be developed as follows. Let us call the number of simultaneously oscillating transverse modes in some real laser $N_{tm}$. Then the far-field angular spread of the output beam from that laser will usually be $\sim N_{tm}^{1/2}$ times larger than the ideal value for a uniphase beam coming from an aperture of the same size, and the focused spot diameter will be $\sim N_{tm}^{1/2}$ times larger than for an ideal beam. (The spot area will, of course, be $\sim N_{tm}$ times larger.)

The ratio $N_{tm}^{1/2}$ is sometimes referred as the "times diffraction limited" or "TDL" ratio of the real laser oscillator. This TDL ratio may range from about 1 up to a few factors of ten in real lasers. (In practice, a designer can often insert some suitable aperture inside a real laser cavity to improve the transverse beam quality, at the price of a corresponding reduction in total output power.)

## REFERENCES

In using this text, you may wish to have an optics text handy as a reference. A list of good optics books includes:

C. L. Andrews, *Optics of the Electromagnetic Spectrum* (Prentice-Hall, 1960). Good especially for simple descriptions of diffraction, interference, and optical wave phenomena, using microwave demonstrations. Not very mathematical.

Max Born and Emil Wolf, *Principles of Optics* (Pergamon Press, 1959). *The* classic advanced-level optics text, found on every laser worker's bookshelf.

Earle B. Brown, *Modern Optics* (Reinhold, 1965). Largely nonanalytical, giving extensive detailed illustrations of practical optical instruments and devices as used in practice.

R. W. Ditchburn, *Light* (Wiley, 1952/1963). Another classic optics text, not as advanced as Born and Wolf, but very extensive and newly revised in 1962.

Grant R. Fowles, *Introduction to Modern Optics* (Holt, Rinehart, and Winston, 1968). Very good elementary optics text, modern and well illustrated.

Max Garbuny, *Optical Physics* (Academic Press, 1965). Not really an optics text; concerned rather with topics in physics that involve optical radiation, including thermal radiation, atomic spectra, and the interaction of optical radiation with matter.

Eugene Hecht and Alfred Zajac, *Optics* (Addison-Wesley, 1974). Another modern basic optics text, including elementary introductions to lasers and holography.

Miles V. Klein, *Optics* (Wiley, 1970). Modern introductory text focused primarily on interference and diffraction.

A. Nussbaum and R. A. Phillips, *Contemporary Optics for Scientists and Engineers* (Prentice-Hall, 1976). Modern coverage of geometrical and physical optics, emphasizing matrix optics and the Fourier analysis approach, plus holography, interferometry, and nonlinear optics.

John M. Stone, *Radiation and Optics* (McGraw-Hill, 1963). More analytical, detailed, and mathematically sophisticated, and with more emphasis on atomic phenomena than the other basic texts in this list.

Robert W. Wood, *Physical Optics* (Dover Publications, 1967). Though many parts of this classic book have become outdated by the passage of four decades since its last revision, this Dover reprint is still valuable for clear descripions, physical and historical insights, and ingeniously simple demonstrations of optical phenomena.

---

Problems for 1.7

1. *Fraunhofer (far field) aperture diffraction patterns.* From an optics text find the Fraunhofer diffraction patterns for (a) a square aperture of width $d$, or (b) a circular aperture of diameter $d$, when illuminated by a uniform plane wave. Let the beam width of either of these diffraction patterns be defined arbitrarily as the full width between the first nulls in each pattern. Determine the angular width (in radians) and the full solid angle (in steradians) of either far-field pattern as a function of wavelength and aperture area. Compare with the $\lambda/d$ rule of thumb developed in this chapter.

2. *Huygens' integral and the on-axis intensity in the far field.* Look up a mathematical statement of Huygens' principle in its simplest form. Then suppose a collimated plane wave (i.e., uniform intensity and phase) emerges with total power $P_0$ through a transmitting aperture of total area $A_0$. Using Huygen's integral, show that optical intensity or power density (Watts per unit area) on the beam

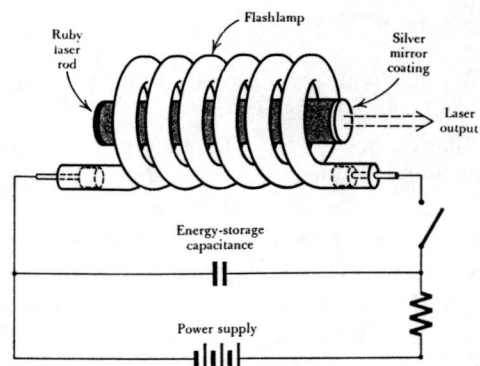

FIGURE 1.50
Design of the first pulsed ruby laser.

axis at large distance $z$ in the far field will be $I = A_0 P_0/(z\lambda)^2$ *independent of the shape of the transmitting aperture.* Verify that this is compatible with the far-field angular spread $\Delta\theta \approx \lambda/d$ asserted in this section.

## 1.8 A FEW PRACTICAL EXAMPLES

Let us look at just a few practical examples of real lasers that illustrate some of the points we have been discussing, notably the ruby solid-state laser, and the helium-neon gas laser.

### The Ruby Laser

The first laser of any type ever to be operated was in fact the flash-pumped ruby laser demonstrated by T. H. Maiman at the Hughes Research Laboratory in early 1960. We have already shown in Figure 1.10 the quantum energy levels associated with the unfilled $3d$ inner shell of a $Cr^{3+}$ ion when this ion replaces one of the $Al^{3+}$ ions in the sapphire or $Al_2O_3$ lattice. Up to $\sim 1\%$ of such replacements can be made in the sapphire lattice to create pink ruby.

By placing such a ruby rod shaped roughly like a slightly overweight cigarette inside a spiral flashlamp filled with a few hundred Torr of xenon (Figure 1.50), and then discharging a high-voltage capacitor bank through this lamp, Maiman was able to use the blue and green wavelengths from this lamp to optically pump atoms from the $^4A_2$ ground level of the $Cr^{3+}$ ions in the lattice into the broad $^4F_2$ and $^4F_1$ bands of excited levels. In ruby, atoms excited into these levels will relax very rapidly, and with close to 100% quantum efficiency, down into the comparatively very sharp $^2E$ levels, or $R_1$ and $R_2$ levels, lying $\sim$14,400 cm or 694 nm ($\sim$1.8 eV) above the ground level.

The ruby laser is, however, a three-level laser system, in which the lower laser level is also the ground energy level. By pumping hard enough, we can nonetheless cycle more than half of the $Cr^{3+}$ ions from the ground level up through the pumping bands and into the highly metastable upper laser level, with its fluorescent lifetime of $\tau \approx 4.3$ msec. Thus, even though ruby is a three-level system rather than a four-level system, which is usually very unfavorable,

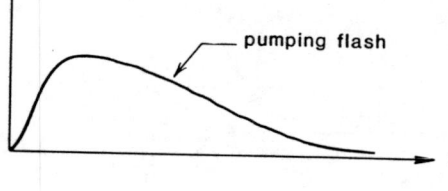
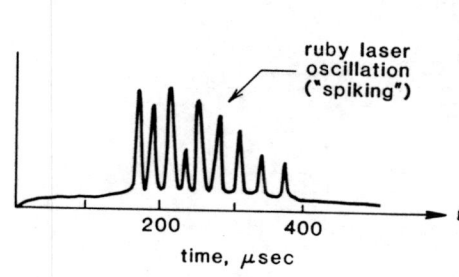

FIGURE 1.51
Output versus time from a typical "long-pulse" ruby laser oscillator.

with sufficiently hard pumping Maiman was able to produce a powerful burst of laser action from the ruby rod.

In a small flash-pumped laser such as ruby, or others, the flashlamp may be connected to a capacitor bank of perhaps 10 to 100 microfarads charged to a prebreakdown voltage of perhaps 1,000 to 1,500 volts, corresponding to ~5 to 50 J of stored energy. The lamp itself is then triggered or ionized by a high-voltage pulse, so that it becomes conducting. The capacitor energy then discharges through the lamp with a typical pulse length of perhaps 200 $\mu$sec, peak currents of up to a few hundred amperes, and peak electrical power input of 25 to 250 kW. The laser rod may convert the pump light in a typical solid-state laser into laser energy with $\sim 1\%$ efficiency, leading to laser output energies of 50 mJ to 0.5 J per shot, and average powers during the pulse of 2.5 to 25 kW. (We will discuss later the technique of "$Q$-switching," which can extract the same laser energy in a very much shorter pulse with very much higher peak power.)

The laser action in ruby actually occurs not as a clean and continuous laser action during the pulse, but as a series of short "spikes" or relaxation-oscillation bursts during the entire pumping time (see Figure 1.51). We will discuss this spiking behavior in more detail in a later chapter.

### Other Solid-State Lasers

There are many such solid-state lasers besides ruby (though unfortunately not many in the visible region). The most common of these are the rare-earth ions in crystals or glasses, with by far the most widely used examples being $Nd^{3+}$ lasers using Nd:YAG ($Nd^{3+}$ ions in yttrium aluminum garnet) and Nd:glass materials. The spiral flashlamp and diffusely reflecting pump enclosure used in Maiman's first ruby laser is now almost always replaced by one or more straight lamps placed parallel to the rod along the axes of an elliptical pump cavity (Figure 1.52).

In the first ruby lasers, partially transparent metallic silver mirrors were evaporated directly onto the polished ends of the laser rod (though such metallic mirrors are quite sensitive to optical damage at higher powers). Later solid-state lasers quickly shifted to the use of external dielectric-coated mirrors, just as in gas

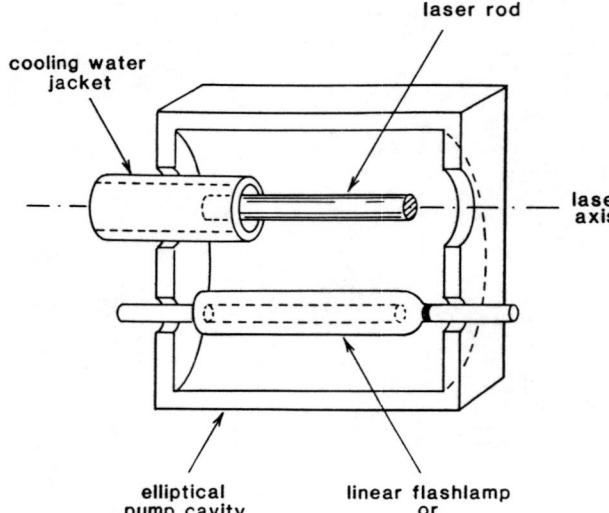

FIGURE 1.52
Elliptical pump cavity used in many optically pumped solid-state lasers.

lasers. The round-trip gains in ruby and other solid-state lasers are often much higher than in gas lasers—up to round-trip power gains of 10X and higher—so that mirrors with much lower reflectivity or higher transmission output can be employed.

Pulsed solid-state lasers are used for a variety of smaller-scale laser cutting, drilling, and marking applications; as military rangefinders and target designators; and in an enormous variety of scientific and technological experiments. By taking advantage of improved lamp efficiencies and laser materials, as well as the fact that most other materials are four-level lasers, we can also operate several solid-state lasers continuously at cw power outputs in the 1–100 W range with efficiencies of $\sim$ 1% or slightly higher, using electrical inputs of 100 W to 10 kW into xenon or krypton-filled arc lamps. (Both laser rod and lamps must, of course, be carefully water-cooled.) Even ruby can, with some difficulty, be made to oscillate on a cw basis. We will discuss the very useful $Nd^{3+}$ laser system in detail in later chapters.

### The Helium-Neon Laser

Another of the most common and familiar types of laser is the helium-neon gas laser developed at the Bell Telephone Laboratories in 1960 and 1961. The laser tube in a He-Ne laser consists of a few Torr of helium combined with approximately one-tenth that pressure of neon inside a quartz plasma discharge tube, which is usually provided with an aluminum cold cathode and an anode, as in Figure 1.53. This discharge tube may be 10 to 50 cm long and a few mm in diameter in a typical small laser. To avoid broadening of the laser transition by isotope shifts (and for other more complex reasons), a mixture of single-isotope $He^3$ and $Ne^{20}$ is usually employed; and it is found empirically that the optimum pressure-diameter product $pd$ in such a laser is a few Torr-mm and that the optimum gain per unit length varies inversely with tube diameter $d$.

This tube is then excited with a dc discharge voltage typically of order 1,000 to 1,500 vdc, producing a dc current typically of order $\sim$10 mA from a special

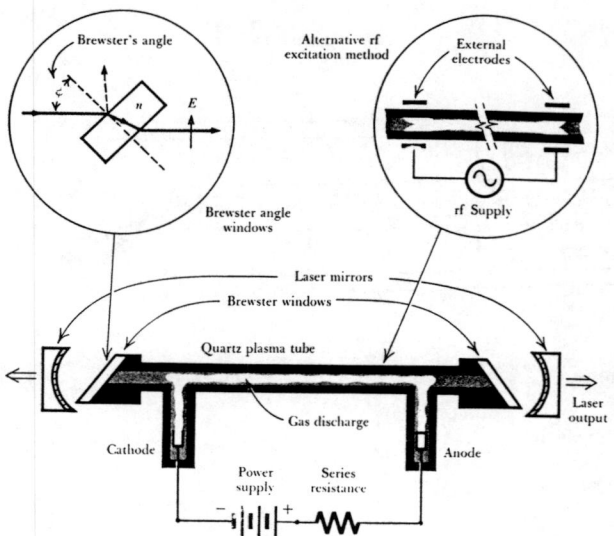

FIGURE 1.53
Elementary design for a helium-neon laser.

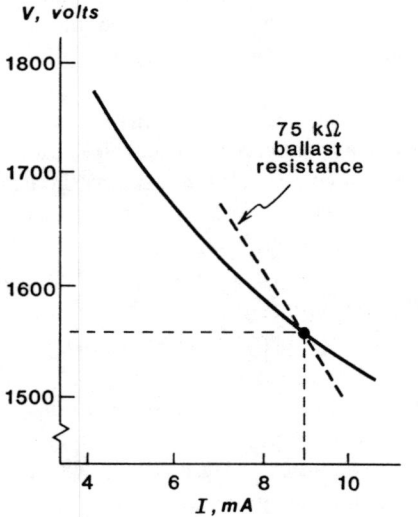

FIGURE 1.54
The glow discharge in a He-Ne laser tube has a negative-resistance $I$-$V$ curve.

high-purity aluminum cold cathode. (Radio-frequency excitation through external electrodes was also employed in many early lasers, but has been found to be generally less convenient.) Because a dc glow discharge in this pressure range has a negative-resistance I-V curve (Figure 1.54), a ballast resistance in series with the dc voltage supply is necessary to stabilize the discharge; and an initial higher-voltage spike must be supplied to ionize the gas and break down the gas discharge each time the tube is turned on.

The discharge tubes in many gas lasers (especially with longer lasers, or lasers for research purposes) may be provided with Brewster-angle end windows which transmit light of the proper linear polarization with essentially zero reflection loss at either face. (Because of the very low gain in the He-Ne system, reflection losses of several percent at each of the air-dielectric interfaces would be totally intolerable.) In many small inexpensive internal-mirror He-Ne lasers, however,

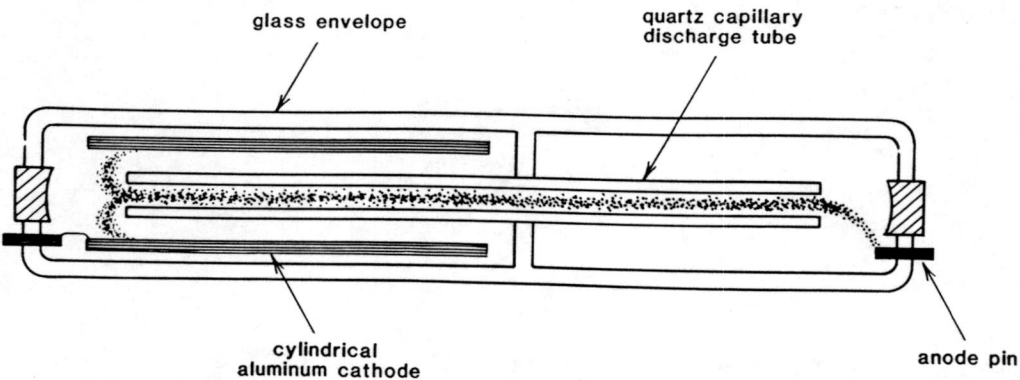

FIGURE 1.55
An internal-mirror He-Ne laser design.

the end mirrors are sealed directly onto the discharge tube, as part of the laser structure (Figure 1.55). Extreme cleanliness and purity of the laser gas fill is vital in the inherently low-gain He-Ne system; the tube envelope must be very carefully outgassed during fabrication, and a special aluminum cathode employed, at least in long-lived sealed-off lasers. The end mirrors themselves are carefully polished flat or curved mirrors with multilayer evaporated dielectric coatings, having as many as 21 carefully designed and evaporated layers to give power reflectivities in excess of 99.5% in some cases.

The pumping mechanism in the He-Ne laser is slightly more complex than those we have discussed so far. The helium gas, as the majority component, dominates the discharge properties of the He-Ne laser tube. Helium atoms have in fact two very long-lived or metastable energy levels, generally referred to as the $2^1S$ ("2-singlet-S") and $2^3S$ ("2-triplet-S") metastable levels, located $\sim 20$ eV above the helium ground level. Free electrons that are accelerated by the axial voltage in the laser tube and that collide with ground-state neutral helium atoms in the laser tube then can excite helium atoms up into these metastable levels, where they remain for long times.

There is then a fortuitous—and very fortunate—near coincidence in energy between each of these helium metastable levels and certain sublevels within the so-called $2s$ and $3s$ groups of excited levels of the neutral neon atoms, as shown in Figure 1.56. (The atomic energy levels in neon, as in other gases, are commonly labeled by means of several different forms of spectroscopic notation of various degrees of obscurity.)

When an excited He atom in one of the metastable levels collides with a ground-state Ne atom, the excited He atom may drop down and give up its energy, while the Ne atom simultaneously takes up almost exactly the same amount of energy and is thus excited upward to its near-coincident energy level. This important type of collision and energy-exchange process between the He and Ne atoms is commonly referred to as a "collision of the second kind." Any small energy defect in the process is taken up by small changes in the kinetic energy of motion of one or the other atom.

This process thus amounts to a selective pumping process, carried out via the helium atoms, which efficiently pumps neon atoms into certain specified excited energy levels. As Figure 1.56 shows, laser action is then potentially possible from these levels into various lower energy levels in the so-called $2p$ and $3p$ groups.

## 1.8 A FEW PRACTICAL EXAMPLES

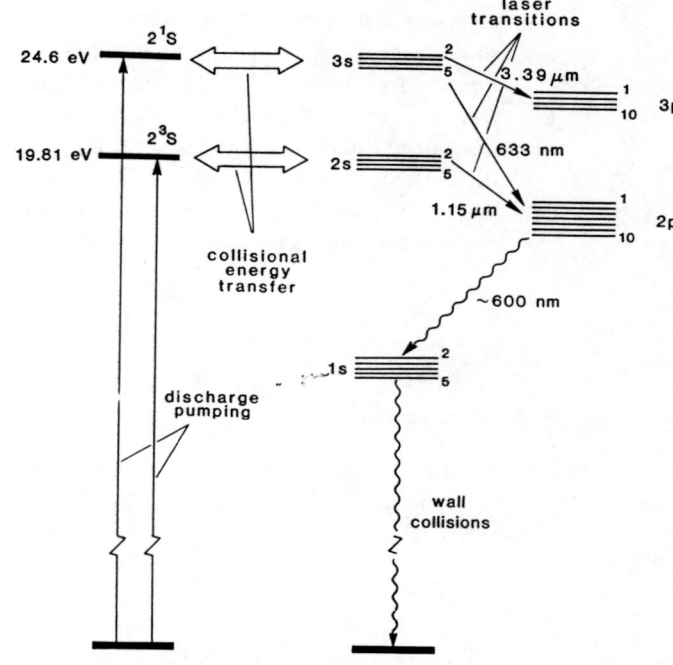

FIGURE 1.56
Energy levels in the He-Ne laser.

The first successful laser action in any gas laser was in fact accomplished by A. Javan and co-workers at Bell Labs in late 1960 on the $2s_2 \to 2p_4$ transition of helium-neon at 1.1523 microns in the near infrared. Shortly thereafter A. D. White and J. D. Rigden discovered that the same system would lase on the familiar and very useful $3s_2 \to 2p_4$ visible red transition at 633 nm (or 6328Å), as well as on a much stronger and quite high-gain set of $3s \to 3p$ transitions near 3.39 microns. (A half-dozen or so different nearby transitions within each of these groups can actually be made to lase. with the strongest transition in each group being determined in part by the relative pumping efficiencies into each sublevel and in part by the relative transition strengths of the different transitions.)

### Characteristics of Gas Lasers

The laser gain in the He-Ne 633 nm system is quite low, with perhaps $2\alpha_m \approx 0.02$ to $0.1$ cm$^{-1}$ (often expressed as "2% to 10% gain per meter"); and the typical power output from a small He-Ne laser may be 0.5 to 2.0 mW. With a dc power input of ~10 W, this corresponds to an efficiency of ~0.01%. Several manufacturers supply inexpensive self-contained laser tubes of this type for about $100 retail and considerably less in volume production. Such lasers are very useful as alignment tools in surveying, for industrial and scientific alignment purposes, supermarket scanners, video disk players, laser printers, and the like. (The dominance of the He-Ne laser in such applications may soon be ended by even cheaper and simpler semiconductor injection lasers.) Larger He-Ne lasers with lengths of 1 to 2 meters that can yield up to 100 mW output at comparable efficiencies are also available.

There are also scores of other gas lasers that are excited by using electrical glow discharges, higher-current arc discharges, hollow-cathode discharges, and transverse arc discharges. One notable family of such lasers are the rare-gas ion lasers, including argon, krypton, and xenon ion lasers, in which much larger electron discharge currents passing through, for example, a He-Ar mixture can directly excite very high-lying argon levels to produce laser action in both singly ionized $Ar^+$ and doubly ionized $Ar^{++}$ ions. Such ion lasers are generally larger than the He-Ne lasers, and even less efficient, but when heavily driven can produce from hundreds of milliwatts to watts of cw oscillation at various wavelengths in the near infrared, visible, and near ultraviolet. Longer-wavelength molecular lasers, such as the $CO_2$ laser, and shorter-wavelength excimer lasers are other examples of important gas laser systems.

## REFERENCES

For a recent summary of practical laser systems and many of their applications, see, for example W. W. Duley, *Laser Processing and Analysis of Materials* (Plenum Press, 1983).

## 1.9 OTHER PROPERTIES OF REAL LASERS

Practical lasers in fact come in a great variety of forms and types, using many different kinds of atoms, molecules, and ions, in the form of gases, liquids, crystals, glasses, plastics, and semiconductors. These systems oscillate at a great many different wavelengths, using many different pumping mechanisms. Nearly all real lasers have, however, certain useful properties in common.

### Temporal and Spatial Coherence

As we have discussed in some detail in earlier sections, nearly all lasers can be:

(a) *Very monochromatic.* Real laser oscillators can in certain near-ideal situations oscillate in a single, essentially discrete oscillation frequency, exactly like a coherent single-frequency electronic oscillator in more-familiar frequency ranges. This oscillation will, as with any other real oscillator, still have some very small residual frequency or phase modulation and drift, because of mechanical vibrations and thermal expansion of the laser structure and other noise effects, as well as small amplitude fluctuations due to power supply ripple and the like. Such a high-quality laser can still be, however, one of the most spectrally pure oscillators available in any frequency range.

More typically, a real laser device will oscillate in some number of discrete frequencies, ranging from perhaps 5 or 10 simultaneous discrete axial modes in narrower-line lasers up to a few thousand discrete and closely spaced frequencies in less well-behaved lasers with wider atomic linewidths.

Real lasers will also in many cases jump more or less randomly from one oscillation frequency to another, and the amplitudes and phases of individual modes will fluctuate randomly, because of mode competition combined with the kinds of unavoidable mechanical and electronic perturbations mentioned above. Nonetheless, the degree of temporal coherence in even a rather bad laser will generally be much higher than in any purely thermal or incoherent light source, and especially in any thermal source providing anywhere near the same power output as the laser's oscillation output power.

(b) *Very directional.* The output beam from a typical real laser will also be very directional and spatially coherent. This occurs because, with properly designed mirrors, many lasers can oscillate in a cavity resonance mode which is essentially a single transverse mode; and this mode can approximate a more or less ideal quasi-plane wave bouncing back and forth between carefully aligned end mirrors.

As we discussed in the preceding section, the resulting output beam from the laser can then be a highly collimated or highly directional beam, which can also be focused to a very tiny spot. Such a beam can be projected for long distances with the minimum amount of diffraction spreading allowed by electromagnetic theory. It can also be focused to a spot only a few wavelengths in diameter, permitting all the power in the laser beam to be focused onto an extremely small area.

Even lasers with nonideal spatial properties (perhaps because of distorted laser mirrors or, more commonly, because of optical aberrations and distortions in the laser medium or in other elements inside the laser cavity) will typically oscillate in only some moderate number of transverse modes, representing some lowest-order transverse mode and a number of more complicated higher-order transverse modes.

Note again that the longitudinal-mode or frequency properties and the transverse-mode or spatial properties of most laser oscillators are more or less independent, so that, for example, even wide-line or multifrequency lasers can very often have well-controlled transverse mode properties and can oscillate in a nearly ideal single transverse mode.

### Other Real Laser Properties

Besides these two basic properties, specific individual lasers can be:

(c) *Very powerful.* Continuous powers of kilowatts or even hundreds of kilowatts are obtained from some lasers, and peak pulse powers exceeding $10^{13}$ Watts are generated by other lasers. (It is interesting to note that this peak power is an order of magnitude more than the total electrical power-generating capacity of the United States—but of course for a very short time only.)

(d) *Very frequency-stable.* Both the spectral purity and the absolute frequency stability of certain lasers can equal or surpass that of any other electronic oscillator; so these lasers can provide an absolute wavelength standard with an accuracy exceeding that of any other presently known technique.

(e) *Very widely tunable.* Although most common lasers are limited to fairly sharply defined discrete frequencies, those of the spectral lines of the specific atoms employed in certain lasers (e.g., organic dye lasers and to a lesser extent semiconductor lasers) can be tuned over enormous wavelength ranges, and so are extremely useful for spectroscopic and chemical applications.

(f) *Very broadband.* Many laser transitions, though very narrowband in fractional terms, have extremely wide linewidths compared to those of conventional radio or microwave frequencies. Hence, such lasers can provide very broadband amplification and, more important, can generate and amplify extraordinarily short optical pulses. Laser pulses as short as a few picoseconds in duration are relatively commonplace, and some mode-locked lasers can generate light pulses as short as 30 femtoseconds ($30 \times 10^{-15}$ seconds) in length.

(g) *Very efficient.* The efficiency of most common lasers is smaller than designers would like, ranging from $\sim 0.001$ to $0.1\%$ in many gas lasers, up to typically 1 or 2% in optically pumped solid-state lasers. A few selected lasers, such as the $CO_2$ laser and the semiconductor injection lasers, can have efficiencies as high as 50% to 70% in converting electrical power directly into coherent radiation.

### Examples of Practical Laser Systems

It is impossible to catalog all the laser devices that have been demonstrated to date, especially since the variety of laser materials, laser pumping methods, and laser experimental techniques is almost endless. The *Laser Handbook* (see the References at the end of this section) gives a comprehensive list of most current laser systems. Some laser systems that are particularly well-known, particularly useful for practical applications, or particularly interesting for other reasons will be discussed in more detail later in this text.

The overall situation at present is that at least $10^5$, and up to $10^6$, distinct laser transitions that have been demonstrated, at wavelengths ranging from $\lambda \geq 600$ $\mu$m (0.6 mm) in the far infrared to the present short-wavelength record (1983) of $\lambda = 1160$Å from the pulsed-discharge $H_2$ laser in the near ultraviolet. At still longer wavelengths, besides more familiar vacuum tubes and semiconductor devices, there are several varieties of millimeter-wave and microwave masers, including molecular beam masers, solid-state electron paramagnetic resonance masers, and nuclear magnetic resonance (NMR) masers. These last devices in fact carry the stimulated-emission principle down to frequencies below 100 Hz. Lasers operating in the X-ray region do not yet seem to have been successfully demonstrated, though several candidates in the soft X-ray region (100–200Å) appear very promising.

A critical study by Bennett in 1979 identified a total of 1,329 distinct laser wavelengths coming from 51 different elements, considering only neutral atoms and ions in gases (no molecules). This data was culled from a source file of 30,000 (!) literature articles. There were, for example, 203 identified laser lines from neutral neon alone, grouped into the clusters of transitions shown in Figure 1.57. Another study identified 270 new lasing lines on various vibrational-rotational transitions in a limited wavelength range for the CO molecule alone. When we consider the enormous diversity of potential molecular species, and the very large number of distinct rotational-vibrational transitions in any one such molecule, it is not impossible that the number of potential distinct molecular laser lines could exceed one million.

Laser action has been obtained thus far in atoms, molecules, and ions in vapor (gas) phase; in atoms, ions, or molecules in crystals, glasses, and liquid solutions; in organic dye molecules in liquids, vapors, gels, and plastics; in semiconductors of several varieties; and in molecules and molecular radicals in planetary atmospheres and in interstellar space.

## 1.9 OTHER PROPERTIES OF REAL LASERS

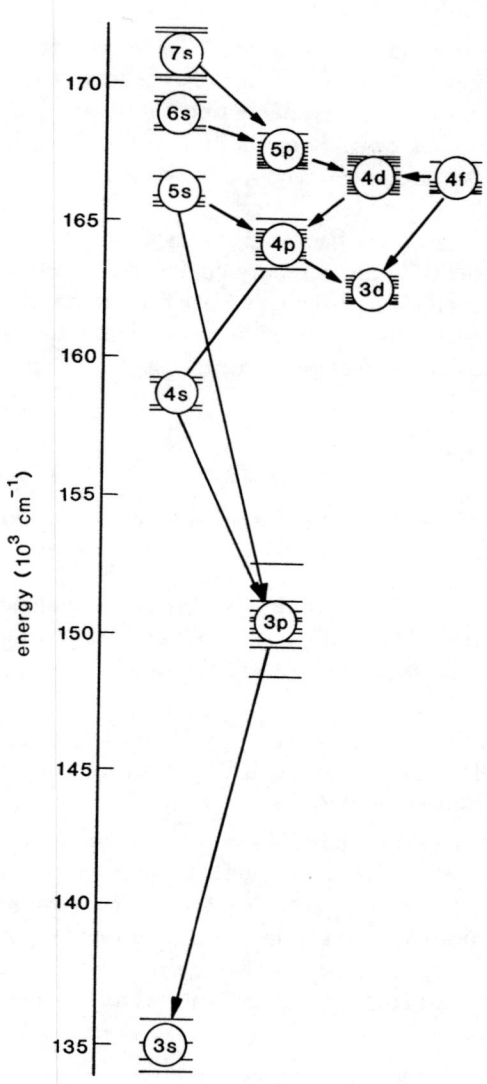

FIGURE 1.57
Groups of available laser transitions in the energy-level spectrum of atomic neon.

### Commercially Available Lasers

Of the laser systems mentioned so far, those that are now readily available in routine commercial production include the He-Ne 633 nm laser; many different sizes of both cw and TEA $CO_2$ lasers at 9 to 11 $\mu$m; various argon, krypton, and other noble-gas ion lasers in the visible and UV; the pulsed $N_2$ laser at 377 nm; the blue cadmium ion laser; the Nd:YAG laser (including many Q-switched, mode-locked, and wavelength-doubled versions); similar ruby, Nd:glass, and alexandrite solid-state lasers; various KrF and other excimer lasers; several varieties of flash-pumped, $N_2$-laser-pumped, YAG-laser-pumped, and cw-argon-laser-pumped tunable dye lasers; and of course many versions of the GaAs injection laser.

In addition there is much development work in government, industrial, and university laboratories on large Nd:glass laser systems and on the atomic iodine laser for laser fusion systems, and on various chemical lasers (HF, DF, CO, $CO_2$)

for military applications. There are also development efforts to a lesser extent on the copper vapor laser, various hollow-cathode visible gas lasers, and a few others. Most of the other known laser systems are available only as (expensive) custom prototypes, or by constructing one's own "home-built" version. (Many chemists, biologists, solid-state physicists, and spectroscopists have now become expert amateur laser builders.)

Commercial development of many other lasers has been rather slow, because the expensive engineering effort to develop a commercially engineered product cannot be justified until a market has been clearly identified. At the same time, commercially significant applications for certain lasers cannot be easily developed if the lasers are not available in commercially developed form.

### Laser-Pumping Methods

The list of successful laser-pumping methods that have been demonstrated to date includes the following.

- *Gas discharges*, both dc, rf, and pulsed, including glow discharges, hollow cathodes, arc discharges, and many kinds of pulsed axial and transverse discharges, and involving both direct electron excitation and two-stage collision pumping.
- *Optical pumping*, using flashlamps, arc lamps (pulsed or dc), tungsten lamps, semiconductor LEDs, explosions and exploding wires, other lasers, and even gas flames and direct sunlight.
- *Chemical reactions*, including chemical mixing, flash photolysis, and direct laser action in flames. It is instructive to realize that the combustion of one kg of fuel can produce enough excited molecules to yield several hundred kilojoules of laser output. A chemical laser burning one kg per second, especially if combined with a supersonic expansion nozzle, can thus provide several hundred kW of cw laser output from what becomes essentially a small "jet-engine laser."
- *Direct electrical pumping*, including high-voltage electron beams directed into high-pressure gas cells, and direct current injection into semiconductor injection lasers.
- *Nuclear pumping* of gases by nuclear-fission fragments, when a gas laser tube is placed in close proximity to a nuclear reactor.
- *Supersonic expansion of gases*, usually preheated by chemical reaction or electrical discharge, through supersonic expansion nozzles, to create the so-called *gasdynamic lasers*.
- *Plasma pumping in hot dense plasmas*, created by plasma pinches, focused high-power laser pulses, or electrical pulses. There are also widely believed rumors that X-ray laser action has in fact been demonstrated in a rod of some laser material pumped by the ultimate high-energy pump source, the explosion of a nuclear bomb.

In general, any nonequilibrium situation that involves intense enough energy deposition is reasonably likely to produce laser action, given the right conditions. Schawlow's Law (attributed to A. L. Schawlow, but apparently thus far unpublished) asserts in fact that anything will lase if you hit it hard enough. Schawlow himself has attempted to illustrate this by building, and then consuming, the first edible laser— a fluorescein dye in Knox gelatine, "prepared in accordance with the directions on the package" and then pumped with a pulsed $N_2$ laser. The fumes of Scotch whiskeys are also rumored to give molecular laser action in the far infrared when pumped with $CO_2$ radiation at 10.6 $\mu$m; and Israeli ingenuity has demonstrated a gasoline-fueled chemical laser which is ignited by an automobile spark plug (kilojoules per gallon and resulting pollution problems not identified).

### Lasers and Masers as Carnot-Cycle Heat Engines

A microwave laser or maser can be pumped in principle—and even in practice—by connecting a very hot, purely thermal source to the pumping transition, and connecting much colder thermal reservoirs to the other transitions (other than the laser transition) on which efficient downward relaxation is required. In practice, connecting a thermal source only to the pumping transition can mean either varying the emissivity versus wavelength of the pumping source, or putting appropriate wavelength filters between the pumping source and the laser medium, so that the laser medium "sees" the pumping source only within the desired pumping bands.

The maser or laser then functions as a heat engine, extracting energy from the hot pumping source, and converting it partly into coherent oscillation or work, and partly into waste heat delivered to the cold thermal reservoirs with which the other transitions must be in contact. The elementary thermodynamics of this have been discussed by H. E. D. Scovil and E. O. Schulz-DuBois, "Three-level masers as heat engines," *Phys. Rev. Lett.* **2**, 262–263 (March 15, 1959), who show that the limiting efficiency of these engines is exactly given by the Carnot-cycle efficiency between the hot pumping source and the cold reservoirs. For an experimental example, see J. M. Sirota and W. H. Christiansen. "Lasing in $N_2O$ and $CO_2$ isotope mixtures pumped by blackbody radiation," *IEEE J. Quantum Electron.* **QE-21**, 1777–1781 (November 1985),

Scovil and Schulz-DuBois also point out that a multilevel atomic system can be used as an atomic refrigerator, in which coherent radiation is applied to one of the transitions in order to reduce the Boltzmann temperature appropriate to some other transition in the same atomic system. Atomic refrigeration experiments of this sort have in fact been demonstrated, using another laser or coherent oscillator as the pump.

### Laser Performance Records

Much ingenuity as well as much sophisticated physics and engineering have thus far gone into laser research and development. As a result of this, plus the enormous flexibility of the stimulated-emission principle, in nearly every performance characteristic that we can define, the world record for *any* type of electronic device can be claimed by some laser device or laser system (generally a different laser for each characteristic). Different lasers can claim the current performance records in the following areas.

(a) *Instantaneous peak power.* A rather modest amplified mode-locked solid-state laser system can generate a peak instantaneous power in excess of

~ $10^{13}$ W—or several times the total installed-electrical generating capacity of the United States—though only for a few picoseconds.

(b) *Continuous average power.* The unclassified power outputs from certain infrared chemical lasers are in the range of several hundred kilowatts to one-half megawatt of continuous power output. The classified figures for cw power output are, at a guess, probably several megawatts cw or greater.

(c) *Absolute frequency stability.* The short-term spectral purity of a highly stabilized cw laser oscillator can be at least as good as $1:10^{13}$. The absolute reproducibility of, for example, a He-Ne 3.39 $\mu$m laser stabilized against a methane absorption line will exceed 1 part in $10^{10}$, and may become much better. The absolute standard of time at present is already an atomic stimulated absorption device, the cesium atomic clock. This may be replaced in the future as an absolute standard for both frequency and time by a very stable laser, stabilized against an IR or visible absorption line.

(d) *Short pulsewidth.* Mode-locked laser pulses shorter than 1 ps ($10^{-12}$ sec) in duration are now fairly routine. The current record is in fact a mode-locked and then compressed dye laser pulse with duration (full width at half maximum) of $\tau_p \approx 12$ femtoseconds, or $1.2 \times 10^{-14}$ seconds. Since this corresponds to a burst of light only ~6 optical cycles in duration, further sizeable improvements may be difficult.

(e) *Instantaneous bandwidth and tuning range.* Most common lasers are limited to sharply defined discrete frequencies of operation that depend on the transitions of the specific atoms employed in the laser, and to fairly narrow tuning ranges that depend on the linewidths of these atomic transitions. Both organic dye lasers in the visible and semiconductor lasers in the near infrared can offer, however, instantaneous amplification bandwidths of order $\Delta\lambda \approx 200$Å. This corresponds, for the former, to a frequency bandwidth $\Delta f \approx 24 \times 10^{12}$ Hz, or 24,000 GHz, or about one telephone channel for every person on Earth.

(f) *Antenna beamwidths.* The diffraction-limited beamwidth of a visible laser beam coming from a telescope 10 cm in diameter is considered easy to obtain. In order to obtain such a beamwidth at even a high microwave frequency of 30 GHz ($\lambda = 1$ cm), we would have to use diffraction-limited microwave antenna two kilometers in diameter.

(g) *Noise figure.* Laser amplifiers actually do not offer particularly good noise-figure performance in the usual sense of this term, because of the unavoidable added noise that comes from spontaneous emission in the laser medium. (It is simply not possible to have an inverted laser population without also having spontaneous emission from the upper level.)

This comparatively poor noise performance is, however, really an inherent limitation of the optical-frequency range rather than of the laser principle. That is, it can be shown that no coherent or linear phase-preserving amplifier of any kind can be a highly sensitive receiver or detector at optical frequencies, because "quantum noise" imposes a rather poor noise limitation, equivalent to an input noise of one photon per inverse amplifier bandwidth, on any such optical amplifier, no matter how it operates. Spontaneous emission is the putative source of this noise in a laser device, but any other conceivable optical amplifier with the same performance characteristics will have some equivalent noise source. (This noise limitation can be viewed as representing, if you like, the quantum uncertainty principle appearing in another guise.) Real lasers can, however, operate very close to this quantum noise limit.

Maser amplifiers can, in any case, provide noise figures in the microwave and radio-frequency ranges that are lower than those for any other electronic types of amplifiers at the same frequencies (though both cooled parametric amplifiers and even microwave traveling-wave tubes can come very close to the same values).

### Natural Masers and Lasers

It is also very challenging to realize that naturally occurring molecular masers and lasers with truly enormous power outputs have been oscillating for eons in interstellar space, on comets, and in planetary atmospheres in our own solar system.

Naturally occurring maser action was first identified from observations that certain discrete molecular lines in the radio emission coming from interstellar clouds had enormously large intensities (equivalent to blackbody radiation temperatures of $10^{12}$ to $10^{15}$ K), but at the same time had very narrow doppler linewidths, corresponding to kinetic temperatures below 100 K. The radiation was also found to be sometimes strongly polarized, and to occur only on a very few discrete lines in the complex spectra of these molecules.

The only reasonable explanation is that these emissions represent naturally occurring microwave maser action on these particular molecular transitions. Such astronomical maser amplification has been seen on certain discrete vibrational and rotational transitions of molecules, such as the hydroxyl radical ($OH^-$, 1,600 to 1,700 MHz), water vapor ($H_2O$, $\sim$ 22 GHz), silicon monoxide (SiO, mm wave region), and a few others. The pumping mechanism responsible for producing inversion is still uncertain, but may involve either radiative pumping by IR or UV radiation from nearby stellar sources or collision pumping by energetic particles. There is of course no feedback; so the observed radiation represents highly amplified spontaneous emission or "ASE" rather than true coherent oscillation.

More recently, amplified spontaneous-emission lines corresponding to population inversion on known $CO_2$ laser transitions near 10.4 and 9.4 $\mu$m have similarly been observed coming from the planetary atmospheres, or mesospheres, of the planets Mars and Venus. The pumping mechanism is believed to be absorption of sunlight by the $CO_2$ molecules. The net gains through the atmospheric layers are remarkably small ($\leq 10\%$) but the total powers involved quite large, because of the large volumes involved in these "natural lasers."

## REFERENCES

The edible laser medium is reported by T. W. Hänsch, M. Pernier, and A. L. Schawlow, "Laser action of dyes in gelatine," *IEEE J. Quantum Electron.* **QE-7**, 45–46 (January 1971).

The list of atomic laser lines comes from W. R. Bennett, Jr., *Atomic Gas Laser Transition Data: A Critical Evaluation* (Plenum Press, 1979). The reference to the carbon-monoxide laser lines is from D. W. Gregg and S. J. Thomas, "Analysis of the $CS^2-O^2$ chemical laser showing new lines and selective excitation," *J. Appl. Phys.* **39**, 4399 (August 1968).

For other extensive listings of most known laser transitions, see the Chemical Rubber Company *Handbook of Laser Science and Technology, Vol. I: Lasers and Masers* and *Vol. II: Gas Lasers*, ed. by M. J. Weber (CRC Press, 1982); or B. Beck, W. Englisch, and K. Gürs, *Table of (> 6000) Laser Lines in Gases and Vapors* (Springer-Verlag, 3d ed., 1980).

A good review of the important basic atomic and plasma processes in gas discharge lasers is given in C. S. Willett, *Introduction to Gas Lasers: Population and Inversion Mechanisms* (Pergamon Press, 1974). For an introduction to gasdynamic laser pumping see S. A. Losev, *Gasdynamic Laser* (Springer-Verlag, 1981).

Good reviews of naturally occurring masers are given by W. H. Kegel, "Natural masers: Maser emission from cosmic objects," *Appl. Phys.* **9**, 1–10 (1976); by M. J. Reid and J. M. Moran, "Masers," *Ann. Rev. Astron. Astrophys.* **19**, 231–276 (1981); and by M. Elitzur, "Physical characteristics of astronomical masers," *Rev. Mod. Phys.* **54**, 1225–1260 (October 1982).

Recent reports of natural laser action can be found in M. J. Mumma et. al., "Discovery of natural gain amplification in the 10 $\mu$m $CO_2$ laser bands on Mars: A natural laser," *Science* **212** 45–49 (1981); and in D. Deming et. al., "Observations of the 10-$\mu$m natural laser emission from the mesospheres of Mars and Venus," *Icarus* **55**, 347–355 (1983).

## 1.10 HISTORICAL BACKGROUND OF THE LASER

Readers of H. G. Wells' novel *The War of the Worlds* might quite reasonably conclude that the first laser device to be operated on Earth was in fact brought here by Martian invaders a century ago, at least according to the description that:

> "In some way they (the Martians) are able to generate an intense heat in a chamber of practically absolute nonconductivity.... This intense heat they project in a parallel beam against any object they choose, by means of a polished parabolic mirror of unknown composition.... However it is done, it is certain that a beam of heat is the essence of the matter. What is combustible flashes into flame at its touch, lead runs like water, it softens iron, cracks and melts glass, and when it falls upon water, that explodes into steam."

(From *Pearson's Magazine*, 1897.)

Those who have seen the effects produced by the beam from a modern multikilowatt $CO_2$ laser will not be surprised at the recent discovery that the atmosphere of Mars consists primarily of carbon dioxide, and that natural laser action occurs in it!

Whether or not Martians operated $CO_2$ lasers in 1897, the first man-made stimulated-emission device on Earth came in early 1954, when Charles H. Townes at Columbia University, assisted by J. P. Gordon and H. Zeiger, operated an ammonia beam maser, a microwave-frequency device that oscillated (very weakly) at approximately 24 GHz. This was closely followed by a similar development by N. G. Basov and A. M. Prokhorov in the Soviet Union. The Columbia group coined the name *maser* to represent *microwave amplification by stimulated emission of radiation.*

There was then much discussion and some experimental work in subsequent years on radio and microwave-frequency maser devices, using both molecular beams and magnetic resonance in solids, and also on theoretical developments toward an optical-frequency maser or laser. Perhaps the most important of these developments was when Nicolaas Bloembergen of Harvard University in 1956 suggested a continuous three-level pumping scheme for obtaining a continuous population inversion on one microwave resonance transition, by pumping with continuous microwave radiation on another transition.

Bloembergen's ideas were quickly verified in other laboratories, leading to a series of microwave paramagnetic solid-state masers. These microwave masers were useful primarily as exceedingly low noise but rather complex and narrow-band microwave amplifiers. They are now largely obsolescent, except for a few highly specialized radio-astronomy experiments or deep-space communications receivers.

The extension of microwave maser concepts to obtain maser or laser action at optical wavelengths was being considered by many scientific workers in the late 1950s. A widely cited and influential paper on the possibility of optical masers was published by Charles Townes and A. L. Schawlow in 1958. Much recent attention has been given to a series of patent claims based on notebook entries recorded at about the same time by Gordon Gould, then a graduate student at Columbia.

The first experimentally successful optical maser or laser device of any kind, however, was the flashlamp-pumped ruby laser at 694 nm in the deep red operated by Theodore H. Maiman at the Hughes Research Laboratories in 1960. The very important helium-neon gas discharge laser was also successfully operated later in the same year by Ali Javan and co-workers at the Bell Telephone Laboratories. This laser operated initially at 1.15 $\mu$m in the near infrared, but was extended a year later to the familiar helium-neon laser transition oscillating at 633 nm in the red.

An enormous number of other laser devices have of course since emerged, not only in the first few years following the initial demonstration of laser action, but steadily during the more than two decades since that time. The variety of different types of lasers now available is enormous, with several hundred thousand different discrete wavelengths available, from perhaps close to a thousand different laser systems. Commercially important and widely used practical lasers are very much fewer, of course, but still numerous. Some of the more interesting and/or useful laser systems have been described earlier in this chapter.


## REFERENCES

A useful summary of the early history of masers and lasers is given by B. A. Lengyel, "Evolution of lasers and masers," *Am. J. Phys.* **34**, 903-913 (October 1966). Additional details are in an excellent survey on laser work at IBM by P. P. Sorokin, "Contributions of IBM to laser science—1960 to present," *IBM J. Res. Develop.* **23**, 476-489 (September 1979); and a more personal account of work at General Electric by R. N. Hall, "Injection lasers," *IEEE Trans. Electron Devices* **ED-23**, 700-704 (July 1976).

The widely cited early article by A. L. Schawlow and C. H. Townes setting forth some of the fundamental considerations for laser action is "Infrared and optical masers," *Phys. Rev.* **112**, 1940-1949 (December 15, 1958). Related personal details are given in A. L. Schawlow, "Masers and lasers," *IEEE Trans. Electron Devices* **ED-23**, 773-779 (July 1976); and in R. Kompfner, "Optics at Bell Laboratories—optical communications," *Appl. Optics* **11**, 2412-2425 (November 1972).

The first successful laser operation was published (after a rather ludicrous series of publication misadventures) by T. H. Maiman in *Nature* **187**, 493 (August 6, 1960).

Two news stories on the emergence of Gordon Gould's early patent claims are by Nicholas Wade, "Forgotten inventor emerges from epic patent battle with claim to laser," *Science* **198**, 379–381 (October 28, 1977), with a reply by A. J. Torsiglieri and W. O. Baker, "The origins of the laser," *Science* **199**, 1022-1026 (March 10, 1978), and by Eliot Marshall, "Gould advances inventor's claim on the laser," *Science* **216**, 392–395 (April 23, 1982).

W. E. Lamb, Jr., has written a scholarly and detailed study of some of the ideas behind the laser in "Physical Concepts in the Development of the Maser and Laser," in *Impact of Basic Research on Technology*, edited by B. Kursunoglu and A. Perlmutter (Plenum Publishing Corporation, 1972), pp. 59–111. Another book covering some of the same material is M. Bertolotti's *Masers and Lasers: An Historical Approach* (Adam Hilger, 1983).

Soviet views on the early laser contributions of V. A. Fabrikant are in an encomium published on his 70th birthday by the editors of *Optics and Spectroscopy (USSR)* **43**, 708 (December 1977).

Finally, an excellent series of historical reminiscences by pioneers in the laser and maser field will be found in the "Centennial Papers" in a Centennial Issue of the *IEEE J. Quantum Electron.* **QE–20**, 545–615 (June 1984).

## 1.11 Additional Problems for Chapter 1

1. *Energy storage and Q-switching in a solid-state laser.* Solid-state lasers (and some gas lasers) can be operated in a useful fashion known as "Q-switching," in which laser oscillation is prevented by blocking (or misaligning) one of the cavity end mirrors, and building up a very large population inversion in the laser medium using a long pump pulse. At the end of this pumping pulse, the mirrors are suddenly unblocked, and the laser then oscillates in a short but very intense burst that "dumps" most of the energy available in the inverted atomic population.

    Pink ruby of the type used in ruby lasers contains $\sim 2 \times 10^{19}$ chromium $Cr^{3+}$ ions/cm$^3$. In a typical Q-switched ruby laser, almost all the ions in the laser rod can be pumped into the upper laser level while the mirrors are blocked, by a flashlamp pump pulse lasting $\sim 1$ ms. Since the resulting Q-switched pulse when the mirrors are unblocked typically lasts only $\sim 50$ ns, there will be no further pumping or repumping once the Q-switched pulse begins. What will be the maximum possible energy output in such a single-shot Q-switched burst from a cylindrical ruby rod 7.5 cm long by 1 cm diameter? What will be the peak laser power output (approximately)?

2. *Optical intensity in a focused laser-beam spot.* If the laser pulse in the preceding problem is focused onto a circular spot 1 mm in diameter, what will be the peak power density (in W/cm$^2$) in the spot? What will be the optical $E$ field strength in the spot?

3. *Stimulated transition rate for molecules in a $CO_2$ laser.* A typical low-pressure glow-discharge-pumped $CO_2$ laser uses a mixture of He, $N_2$, and $CO_2$ with an 8:1:1 ratio of partial pressures for the three gases and a total gas pressure at room temperature of 20 Torr (though this may vary somewhat depending on tube diameter). The cw laser power output at $\lambda = 10.6$ $\mu$m from an optimized $CO_2$ laser tube 1 cm in diameter by 1 meter long might be 50 W. At this power output, how many times per second is an individual $CO_2$ molecule being pumped upward to the upper laser level and then stimulated downward to the lower laser level by stimulated emission? Note that the relation between pressure $p$ and density $N$ in a gas is $N$(molecules/cm$^3$) $= 9.65 \times 10^{18} p$(Torr)$/T$(K).

4. *Stored energy and energy output in a TEA $CO_2$ laser.* A $CO_2$ laser at $10.6\mu$m can be operated at low gas pressures, in the range of 20 to 50 Torr, as a low- to medium-power cw gas laser pumped by a cw glow discharge. It can also be operated at much higher gas pressures, in the range of 1 to 10 atmospheres (1 atmosphere = 760 Torr) as a pulsed laser with much higher peak power output. Since it is impossible to maintain a stable glow discharge at such high gas pressures, and since the discharge voltage per unit length goes up rapidly with increasing gas pressure, such a laser must be pumped with a very short high-voltage discharge, lasting perhaps a few microseconds, which is usually applied transversely across the laser tube rather than along the tube. A laser of this type is thus referred to as a *Transverse Electric Atmospheric*, or TEA, type of laser.

Suppose every $CO_2$ molecule in such a laser is lifted up to the upper laser level and then drops down by laser action just once during a single laser pulse. Calculate the resulting pulse energy output in Joules per 1,000 cm$^3$ of gas volume per Torr of $CO_2$ gas pressure. Calculate also the total energy output per pulse from a laser 1 meter long by 2 cm diameter operating with 760 Torr partial pressure of $CO_2$.

Real TEA $CO_2$ lasers more typically yield $\sim$ 40 Joules of output per liter-atmosphere of gas volume during a laser oscillation pulse lasting from a few hundred nanoseconds to perhaps half a microsecond. How many times on average does each $CO_2$ molecule circulate up through the upper laser level during the pulse?

5. *Heating effects due to focused laser beams.* We wish to gain some feeling for the heating effects of focused laser beams, by calculating these effects for some highly idealized (and hence not fully realistic) examples, as follows.

(a) A 1-Joule, 100-nanosecond pulse from a $Q$-switched Nd:glass laser is focused onto a metallic surface and totally absorbed in a volume of material 20 microns in diameter by 10 nm (100Å) deep. Neglecting surface losses and heat conduction into the material, what will be the initial rate of rise of the temperature in the absorbing volume?

(b) A 1-Watt laser beam (perhaps from a 1-Watt cw Nd:YAG laser) is focused by a good-quality lens into the same spot. If both heat conduction and vaporization of the material are ignored (which is clearly *not* realistic), what will be the predicted steady-state temperature of the surface in the focused spot?

(c) Suppose a 100-Watt cw beam is used, and all the laser power goes into vaporizing material in and near the spot, so that the laser beam tunnels a hole with a constant 50 $\mu$m diameter into the medium. What is the drilling rate in meters/second?

In each of (a) to (c), assume for simplicity a material density of 2 gms/cm$^3$, a material specific heat of 1 cal/gm-deg K, and in (c) a vaporization temperature of 1,800 K.

6. *Laser fusion: laser design and fundamental economics.* Fusion researchers hope it may be possible in the future to heat and compress tiny nuclear-fuel pellets with short, intense laser pulses until nuclear fusion occurs inside the compressed pellet. Such a process would release useful energy in the form of neutrons emitted from a nuclear micro-explosion. This potentially unlimited energy source faces many practical difficulties, however. Some estimates say laser pulses of $\sim 10^5$ Joules in $\sim$ 100 psec may be needed even to reach "scientific break-even," i.e., the point where nuclear energy released just equals laser energy incident.

Preliminary laser fusion experiments use a small mode-locked neodymium-YAG laser oscillator to generate a 10 mJ input pulse at $\lambda = 1.06$ $\mu$m, followed by a chain of successively larger Nd:glass amplifiers to amplify the pulse to the required final energy. The amplifier material consists of a special glass doped with $\sim 5\%$ by weight of $Nd_3O_5$ to give $\sim 4.6 \times 10^{20}$ $Nd^{3+}$ ions/cm$^3$. The laser transition is between two excited energy levels of the Nd ions. Some of the design considerations for a Nd:glass fusion laser are as follows.

(a) As a reasonable estimate, perhaps 10% of the available Nd ions can be pumped into the upper laser energy level, and then 1% of those excited ions can be stimulated to make downward transitions by the ultrashort laser pulse as it passes down the amplifier chain. What minimum total volume of laser glass will be required in the amplifier chain?

(b) The laser glass when fully pumped has a power gain coefficient $2\alpha_m \approx 0.1$ cm$^{-1}$. What overall length of glass will be required in the amplifier chain?

(c) Laser glass may be permanently damaged if the optical power density in a short optical pulse exceeds $\sim 10^{10}$ W/cm$^2$. What aperture size will be required at the output of the final amplifier stage?

(d) The energy efficiency of this type of laser, from electrical energy initially stored in the power supply to potential laser energy stored in the upper energy level, is about 1%; and then only $\sim 1\%$ of this is usefully extracted by a short pulse. If energy-storage capacitors for laser power supplies cost about 10 cents per Joule of energy stored, what will the capacitor bank for this system cost?

(e) Suppose things go well, and each pellet releases $10^6$ Joules (1 MJ) of energy when it is "zapped" by the laser. If the price of electricity is currently 10 cents per kilowatt-hour, what is the retail value of the fusion energy produced per shot?

7. *Thermal light sources versus coherent light sources.* To gain some appreciation for the differences between a thermal and a laser light source, we can compare the visual brightness of a weak laser beam and of a powerful searchlight beam as seen by a distant observer standing in the center of each beam and looking back toward the source.

(a) Consider first a 10-mW 6328Å Ne-Ne laser with a beam expansion telescope attached. The beam from such a laser usually has a gaussian transverse intensity profile; but let's assume for simplicity that the output beam has a uniform plane-wave distribution across an output aperture 1 cm in diameter. What will be the power density (W/m$^2$) at the center of this beam as a function of distance in the far field, i.e., at large distances from the source? Note that such a laser will require an electrical power input of perhaps 100 Watts.

A human eye can, under optimum conditions, detect as little as 100 photons per second entering a fully dark-adapted eye with an entrance pupil diameter of $\sim 8$ mm. From how far away could this laser be seen (assuming you are standing in the center of the beam in the far field)?

(b) Consider next a simple searchlight consisting of a spherical hot spot (for example, an electric arc) located at the focal point of a large lens or, more likely, a large spherical mirror. Assume the hot spot can be modeled as a thermally emitting ball-shaped blackbody radiator 1 cm in diameter with a temperature of 6,000 K. (Note: melting point of tungsten $\sim 3,700$ K and of carbon $\sim 3,850$ K.) What is the total thermal radiation in Watts from the surface area of this ball at this temperature (which is also the minimum electrical input power required to drive the searchlight)?

Assume that roughly 15% of this total radiated power falls within the 100-nm-wide band of visible wavelengths. Let both the searchlight mirror diameter and its focal length be 1 meter. What fraction of the total visible emitted radiation is then collimated by the searchlight mirror, and what is the far-field beamspread of this collimated radiation? When all factors are included, what is the far-field visible power density as a function of distance in the searchlight beam? How do the small milliwatt laser and the large kilowatt searchlight compare in far-field brightness?

(c) How do these comparisons change (i) if another 10-power telescope expands the laser beam initially to 10 cm in diameter? (ii) If the diameter of the searchlight hot spot is increased to 2 cm? (iii) If the searchlight mirror is changed to 2 meters diameter with the same focal length for the mirror? (iv) If the searchlight mirror diameter and its focal length are both doubled, so that the $f$-number stays the same?

8. *Legal and illegal laser applications.* How many methods (legal or illegal) can you think of to measure the height of a tall building using a laser—*without turning on the laser?*

# 2
# STIMULATED TRANSITIONS: THE CLASSICAL OSCILLATOR MODEL

Our first major objective in this text is to understand how optical signals act on atoms (or ions, or molecules) to excite resonance responses and to cause transitions between the atomic energy levels. In later chapters we will examine how the excited atoms or molecules react back on the optical signals to produce gain and phase shift. Eventually we will combine these two parts of the problem into a complete, self-consistent description of laser action. For the minute, however, all we want to consider is what optical fields do to atoms.

The effect of a near-resonant applied signal on a collection of atoms can be divided into two parts. First, there is a *resonance excitation of some individual transition in the atoms*. This can be modeled by a resonant oscillator model, which leads to a resonant atomic susceptibility, among other things. In this chapter we will develop the classical electron oscillator model for an atomic resonance, and show how this model can lead to equations that describe all the essential features of a single atomic transition. In Chapter 3 we will show in more detail how this purely classical model can in fact describe and explain even the most complex quantum-mechanical aspects of real atomic transitions.

The second aspect of the atomic response in real atoms is that, under the influence of an applied signal, atoms begin to make stimulated transitions between the upper and lower levels involved in the transition, so that the atomic level populations begin to change. These stimulated transition rates are described by the *atomic rate equations* that we introduced in the opening chapter of this book. We will discuss these rate equations in more detail in several later chapters.

## 2.1 THE CLASSICAL ELECTRON OSCILLATOR

Let us first review some of the important physical properties of real atoms. Note that throughout this text, we will speak of "atoms" as a shorthand for simple free atoms, ions, or molecules in gases; or for individual laser atoms, ions, or molecules in solids or in liquids (such as the $Cr^{3+}$ ions in ruby, or the Rhodamine 6G dye molecules in a laser dye); or even for the valence and

## 2.1 THE CLASSICAL ELECTRON OSCILLATOR

conduction electrons responsible for optical transitions in semiconductors. Some of the important background facts about real atoms are as follows.

- Atoms consist in simplified terms of a massive fixed nucleus plus a surrounding *electron-charge distribution*, whether we think of this distribution as a fuzzy charge cloud, or as a set of electronic orbits, as shown in Figure 2.1(a), or as a quantum wavefunction.
- Atoms exhibit *sharp resonances* both in their spontaneous-radiation wavelengths and in their stimulated response to applied signals.
- These resonances are usually *simple harmonic resonances*—that is, there are usually no additional responses at exactly integer multiples of these sharp resonant frequencies.
- Most (though not all) atoms respond to the *electric field* of an applied signal rather than the magnetic field. In more technical terms, the strongest atomic transitions, and those most important for laser action, are usually of the type known as *electric dipole transitions*. (There do exist other types of atomic transitions, including some laser transitions, that are classified as magnetic dipole, electric quadrupole, or even higher order. Magnetic dipole transitions are described, using a different classical model, in a later chapter of this text.)

All these properties lead us to use the *classical electron oscillator (CEO) model* shown in Figure 2.1 as a classical model to represent a single electric-dipole transition in a single atom. With some simple extensions, which we will describe later, this CEO model will give a complete and accurate description of every significant feature of a real atomic quantum transition.

### Analysis of the Classical Electron Oscillator Model

The CEO model envisions that the electronic charge cloud in a real atom may be displaced from its equilibrium position with an instantaneous displacement $x(t)$, as shown in Figure 2.1(b). Because of the positive charge on the nucleus, this displacement causes the electronic charge cloud to experience a linear restoring force $-Kx(t)$. The electronic charge cloud is thus in many ways similar to a point electron with mass $m$ and charge $-e$ that is located in a quadratic potential well, with potential $V = Kx^2(t)$, or that is attached to a spring with spring constant $K$. An externally applied signal with an electric field $\mathcal{E}_x(t)$ may also be applied to this charge cloud.

The classical equation of motion for an electron trapped in such a potential well, or suspended on such a spring, and subjected to an applied electric field $\mathcal{E}_x(t)$, is then

$$m\frac{d^2x(t)}{dt^2} = -Kx(t) - e\mathcal{E}_x(t), \qquad (1)$$

which we may write in more abstract form as

$$\frac{d^2x(t)}{dt^2} + \omega_a^2 x(t) = -(e/m)\mathcal{E}_x(t), \qquad (2)$$

The frequency $\omega_a$ is then the classical oscillator's resonance frequency, given by $\omega_a^2 \equiv K/m$. We will equate this resonance frequency for the CEO model with the

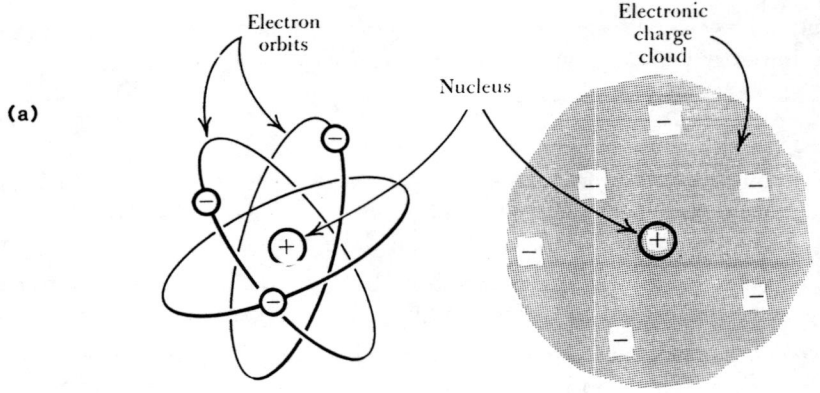

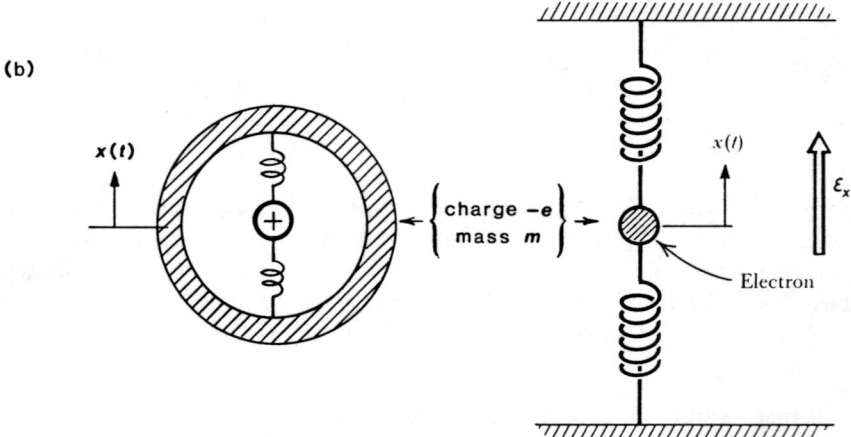

**FIGURE 2.1**
(a) Electronic models for a real atom. (b) The classical electron oscillator model.

transition frequency $\omega_{21} \equiv (E_2 - E_1)/\hbar$ of a real atomic transition in a real atom. More generally, we will identify any one single transition in an individual atom with a corresponding classical electron oscillator, so that from here on we will refer to real atoms or to individual classical oscillators almost interchangeably.

### Damping and Oscillation Energy Decay

The oscillatory motion of the electron in the CEO model, or of the charge cloud in a real atom, must be damped in some fashion, however, since it will surely lose energy with time. Hence we must add a damping term to the equation of motion in the form

$$\frac{d^2x(t)}{dt^2} + \gamma \frac{dx(t)}{dt} + \omega_a^2 x(t) = -\frac{e}{m}\mathcal{E}_x(t), \tag{3}$$

where $\gamma$ is a damping rate or damping coefficient for the oscillator. The electronic motion $x(t)$ without any applied signal will then oscillate and decay in the fashion

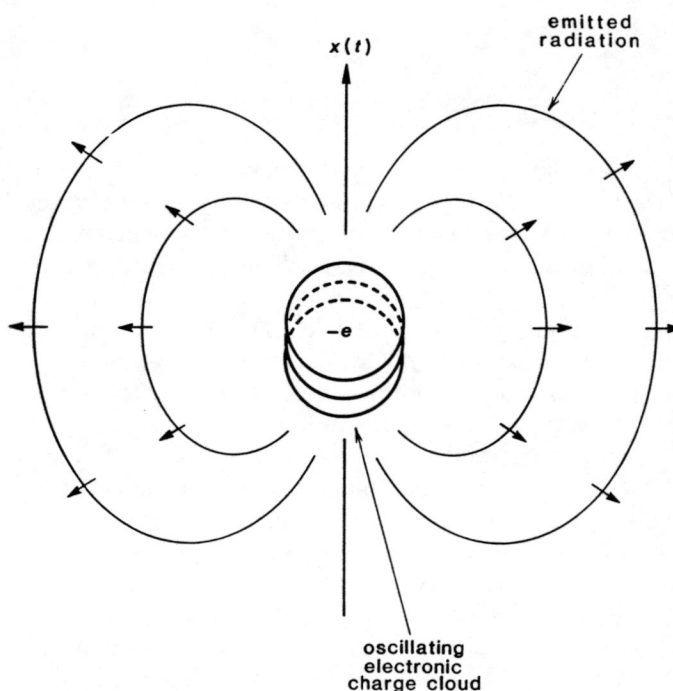

FIGURE 2.2
Emission of electromagnetic radiation by a sinusoidally oscillating electronic charge.

$$x(t) = x(t_0)\exp[-(\gamma/2)(t-t_0) + j\omega'_a(t-t_0)], \quad (4)$$

where $\omega'_a$ is the exact resonance frequency given by

$$\omega'_a \equiv \sqrt{\omega_a^2 - (\gamma/2)^2}. \quad (5)$$

The $Q$ of an optical frequency transition in an atom will always be high enough to allow us to simplify life from now on by ignoring the difference between $\omega_a$ and $\omega'_a$. The energy associated with the internal oscillation in the CEO model, which we will write as $U_a(t)$, thus decays as

$$U_a(t) = \frac{1}{2}Kx^2(t) + \frac{1}{2}mv_x^2(t) = U_a(t_0)e^{-\gamma(t-t_0)} \equiv U_a(t_0)e^{-(t-t_0)/\tau}. \quad (6)$$

The decay rate $\gamma$ is thus the *energy decay rate*, and the lifetime $\tau \equiv \gamma^{-1}$ is the *energy decay time* for the oscillator model.

Both classical electron oscillators and real atomic transitions will always lose energy in part by radiating away electromagnetic radiation, in what we call *spontaneous emission* or *fluorescence*, at the transition frequency $\omega_a$. This radiation of electromagnetic energy from the oscillating charge cloud, as shown in Figure 2.2, leads to a *purely radiative* part of the decay rate $\gamma$, which we will call $\gamma_{\rm rad}$.

Real atomic transitions in many cases, however, also lose additional oscillation energy by other "nonradiative" mechanisms, such as collisions with other atoms, or the emission of heat vibrations into a surrounding crystal lattice. This additional energy loss leads to an additional *nonradiative* part of the total decay rate, which we will denote by $\gamma_{\rm nr}$. The total energy decay rate is then generally given by

$$\gamma \equiv \frac{1}{U_a}\frac{dU_a}{dt} = \gamma_{\text{rad}} + \gamma_{\text{nr}}. \tag{7}$$

Note that the energy $U_a$ we are talking about here is the energy associated with the *internal charge cloud oscillation within the atom*. This energy is quite distinct from other kinds of energy the atom may also possess, such as the kinetic energy of motion the same atom may possess if the atom as a whole is moving rapidly in a gas.

The energy decay rate for an atomic transition may thus include both radiative and nonradiative parts. Radiative decay, which is exactly the same thing as spontaneous electromagnetic emission or fluorescent emission from the atom, is always present, though sometimes very weak. Nonradiative decay can also be present, sometimes much more strongly and sometimes much less strongly than the radiative part of the total decay, depending on individual circumstances. The causes of nonradiative decay can include inelastic collisions of atoms with each other, or with the walls of a laser tube, so that the internal oscillating energy of the atoms gets converted into kinetic energy of the gas atoms, or goes into heating up the tube walls. Nonradiative decay in solids or liquids can also involve the loss of energy from the electronic oscillation of the atoms into lattice vibrations and hence into heat in the surrounding crystal lattice in a solid. *The general property of all nonradiative atomic relaxation or decay mechanisms is that energy is lost from the internal oscillatory motion of the individual atomic charge clouds, and that this energy goes into simple heating up of surrounding gas atoms or tube walls or crystal lattices.*

### Radiative Decay Rates

The purely radiative decay rate or spontaneous emission rate for a classical electron oscillator can be calculated from classical electromagnetic theory (see Problems). The sinusoidally oscillating electron radiates energy outward exactly like an oscillating dipole antenna or an oscillating current source; and this energy is the spontaneous emission. The resulting decay rate for a classical electron oscillator imbedded in an infinite medium of dielectric permittivity $\epsilon$ is given by

$$\gamma_{\text{rad,ceo}} = \frac{e^2 \omega_a^2}{6\pi \epsilon m c^3}. \tag{8}$$

Note that according to the conventions used in this text, $\epsilon$ and $c$ are the dielectric permeability and the velocity of light *in any surrounding dielectric medium*, and not necessarily the free-space values $\epsilon_0$ and $c_0$. (You might now review the discussion of units and notation for this text given in the Introduction.) This classical oscillator radiative decay rate has a value $\gamma_{\text{rad,ceo}} \approx 10^8$ sec$^{-1}$ for a visible frequency oscillator, compared to an oscillation frequency of $\omega_a \approx 4 \times 10^{15}$ sec$^{-1}$. Hence, the decay rate is very small compared to the oscillation frequency.

Real atomic transitions have radiative decay rates that are determined by quantum considerations. These rates for real atoms are different from the classical expression just given, and are different for each different atomic transition. For many transitions, however, the real atomic decay rates for so-called *strongly allowed* transitions are of the same order of magnitude as the purely classical radiative decay rate for a CEO with the same resonance frequency.

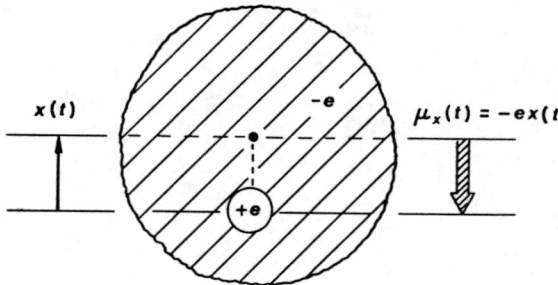

FIGURE 2.3
Microscopic electric-dipole moment in an atom with a displaced electron-charge cloud.

---

**More On Radiative Damping.**

The radiative damping process for a classical oscillating electron has other, more complex aspects that we have avoided discussing here. These properties are discussed, for example, in Chapter 25 of W. K. H. Panofsky and M. Phillips, *Classical Electricity and Magnetism* (Addison-Wesley, 1955), or in Chapter 12 of J. M. Stone, *Radiation and Optics* (McGraw-Hill, 1963). An advanced discussion is given by F. Rohrlich, *Classical Charged Particles* (Addison-Wesley, 1965). Other interesting discussions can be found in R. G. Newburgh, "Radiation and the classical electron," *Am. J. Phys.* **36**, 399 (May 1968); in W. L. Burke, "Runaway solutions: Remarks on the asymptotic theory of radiation damping," *Phys. Rev.* **A2**, 1501 (October 1, 1970); and in G. N. Plass, "Classical electrodynamic equations of motion with radiative reaction," *Rev. Mod. Phys.* **33**, 37 (January 1961). A short summary can also be found on pp. 70–71 of A. E. Siegman, *An Introduction to Lasers and Masers* (McGraw-Hill, 1971).

---

### Microscopic Dipole Moments and Macroscopic Polarization

The next important step we must take is to go from microscopic individual atoms, represented by individual electron oscillators, to macroscopic electromagnetic effects in real laser materials. We do this by adding up the *microscopic electric dipole moments* from many individual atoms or classical oscillators to produce a *macroscopic electromagnetic polarization* in the laser material.

We first note that displacement of the electronic charge cloud of an atom away from its equilibrium position around the nucleus by an effective distance $x(t)$ means that there is a displacement of the center of the negative electronic charge, with value $-e$, away from the matching positive charge $+e$ of the heavy and nearly immobile nucleus. This displacement creates a microscopic electric dipole moment $\mu_x(t)$ associated with that individual oscillator or atom, which is given by

$$\mu_x(t) = [\text{charge}] \times [\text{displacement}] = -ex(t) \qquad (9)$$

as shown in Figure 2.3.

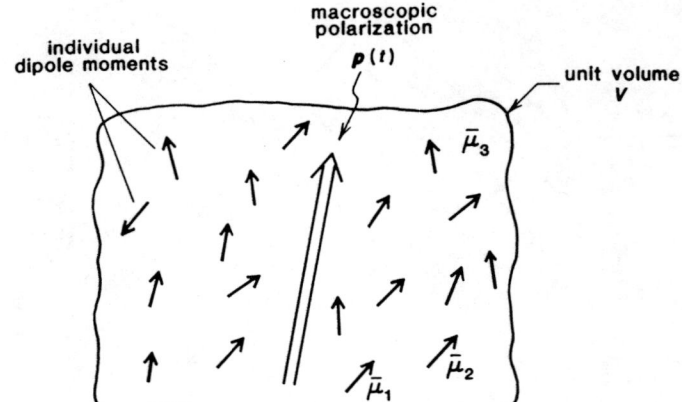

FIGURE 2.4
Macroscopic electric polarization produced by a collection of individual dipole moments.

Let us then recall that in electromagnetic theory Maxwell's equations are written in the form

$$\nabla \times \mathcal{E}(r,t) = -\frac{\partial b(r,t)}{\partial t},$$

$$\nabla \times h(r,t) = j(r,t) + \frac{\partial d(r,t)}{\partial t}, \tag{10}$$

together with the definitions

$$d(r,t) = \epsilon_0 \mathcal{E}(r,t) + p(r,t),$$

$$b(r,t) = \mu_0 b(r,t) + m(r,t), \tag{11}$$

in which $p(r,t)$ and $m(r,t)$ are the *electric and magnetic polarizations*, or *dipole moments per unit volume*, at point $r$ and time $t$.

The electric polarization $p(r,t)$ at any point in an atomic medium is thus, by definition, the net *electric dipole moment per unit volume* in a small differential volume surrounding that point. In a laser medium in particular, this polarization $p$ must be calculated by *adding up the vector sum of the individual dipole moments $\mu_x$ of all the atoms in that unit volume*.

Consider, for example, a tiny volume of a laser medium containing a very large number of microscopic atoms or classical oscillators, as shown schematically in Figure 2.4. (Note that in a typical laser medium the density of atoms may be anywhere from $10^{12}$ to $10^{19}$ laser atoms/cm$^3$; so there may be anywhere from $10^3$ to $10^{10}$ atoms even in a tiny cube only 10 optical wavelengths on a side.) Let each atom in this volume be labeled by an index $i$, and let each atom have an instantaneous electric dipole moment $\mu_{xi}(t) = -ex_i(t)$.

This medium will then have a macroscopic electric polarization $p$ around that point $r$ in the medium whose $x$ component is given by

$$p_x(r,t) \equiv V^{-1} \sum_{i=1}^{NV} \mu_{xi}(t). \tag{12}$$

The volume $V$ here can represent any small unit volume (but still containing many dipoles) surrounding the point $r$, and $N$ is the density of individual dipoles in that volume, so that $NV$ is the total number of dipoles.

We could, to be more general, write both the microscopic dipole moments and the macroscopic polarization in this formula as vector quantities, in which case the macroscopic polarization $p$ would be the vector sum over all individual dipoles $\mu_i$ within that volume. However, for now we are focusing only on the linearly polarized $x$ components of $p(r,t)$ and $\mu_i(t)$. Also, in real materials both the applied field $\mathcal{E}(r,t)$ and the polarization $p(r,t)$ will in general be functions of position $r$, though the changes in value will be very small compared to interatomic spacings. We will not be worrying about the spatial variation of this macroscopic polarization until later, however.

The step we have just taken, of going from individual microscopic atomic dipole moments $\mu_{xi}$ to a macroscopic electric polarization $p_x$, is a crucial step in the theoretical analysis of laser action. To analyze the response of a laser material, we use quantum theory—or as a substitute we use the CEO model—to calculate the *microscopic* dipole moments of individual laser atoms. These responses are then summed over large numbers of such atoms per unit volume in a real laser medium to produce the *macroscopic* polarization. This polarization then goes into Maxwell's equations to produce laser absorption, gain, and/or phase shift (as we will see later). We measure in the laboratory, or employ in laser devices, only the *macroscopic* effects of this atomic polarization. We seldom if ever observe the minute *microscopic* effects produced by one tiny single atom acting alone.

### Discussion

The primary concept introduced in this section is that we can use the classical electron oscillator model, with resonance frequency $\omega_a$, as a substitute for a single atomic transition with transition frequency $\omega_{21}$ in a single real quantum atom. The very great utility of the CEO model for this purpose will become apparent in following sections. The essential accuracy of this simple classical model can, however, be further illustrated by the following point.

Suppose a classical oscillating electric dipole antenna is placed close to a reflecting metallic surface, or close to one or more dielectric layers or surfaces. The spatial radiation pattern, the radiative decay rate, and even the resonance frequency of the classical dipole will then all be changed by significant amounts. This occurs, in classical terms, because the radiating dipole is influenced by its own radiated fields reflected back from the nearby surfaces. These effects are strongest, of course, when the oscillator is close to the surface, within one wavelength or less.

Experimental studies of exactly these same effects have also been carried out on real atomic transition dipoles, using real radiating atoms placed very close to dielectric or metal surfaces, with exactly the same results being obtained for the real atoms. Such experiments have been carried out, for example, by using thin monomolecular layers of radiating dye molecules adsorbed onto dielectric films one wavelength or less thick attached to a reflecting silver surface or to another dielectric surface or layer. The observed changes in the radiative behavior of these real atomic (or molecular) dipoles have been found to agree completely with theoretical calculations using purely classical models for both the radiating atoms and the electromagnetic fields.

## REFERENCES

The CEO model, often referred to as the Lorentz model of an atom, has a long history and has been widely used. An early reference, still worth reading, is the second edition of H. A. Lorentz, *The Theory of Electrons*, (reprinted in paperback by Dover Publications, 1952), especially Chapter III, Sections 77–81. Other discussions can be found in, for example, M. Garbuny, *Optical Physics* (Academic Press, 1965); in B. Rossi, *Optics* (Addison-Wesley, 1957); and in J. M. Stone, *Radiation and Optics* (McGraw-Hill, 1963), where a particularly extensive discussion is given. An interesting short discussion of some fine points of the CEO model is given in L. Mandel, "Energy flow from an atomic dipole in classical electrodynamics," *J. Opt. Soc. Am.* **62**, 1011–1012 (August 1972).

For further information on microscopic dipole moments and macroscopic polarization, consult any good text on electromagnetic theory, such as W. K. H. Panofsky and M. Phillips, *Classical Electricity and Magnetism* (Addison-Wesley, 1955), pp. 20–35 and 117–118; D. T. Paris and F. K. Hurd, *Basic Electromagnetic Theory* (McGraw-Hill, 1969), pp. 65–70; R. S. Elliott, *Electromagnetics* (McGraw-Hill, 1966), Chapter 6; or S. Ramo, J. Whinnery, and T. H. Van Duzer, *Fields and Waves in Communication Electronics* (Wiley, 1965), pp. 63–64 and 131–149. An extensive review of much the same elementary ideas as in this chapter is given in R. W. Christy, "Classical theory of optical dispersion," *Am. J. Phys.* **40**, 1403 (October 1972).

For interesting references on oscillating atomic dipoles close to surfaces, see K. H. Drexhage in *Scientific American* **222**, March 1970, p. 108; and also in *Progress in Optics, Vol. XII*, edited by E. Wolf (North Holland, Amsterdam, 1974). Other clever experiments done by W. Lukosz and R. E. Kunz are described in "Changes in fluorescence lifetimes induced by variations of the radiating molecules' optical environment," *Optics Commun.* **31**, 42–46 (October 1979). See also R. R. Chance, A. Prock, and R. Silbey, *Phys. Rev. A* **12**, 1448 (1975); J. P. Wittke, "Spontaneous-emission-rate alteration by dielectric and other waveguiding structures," *RCA Rev.* **36**, 655–665 (December 1975); and P. W. Milonni and P. L. Knight, "Spontaneous emission between mirrors," *Optics Commun.* **9**, 119–122 (October 1973).

---

Problems for 2.1

1. *More detailed classical electron oscillator model.* A more detailed semiclassical model of an atom might picture the electronic charge cloud as a rigid, uniform, spherical distribution of negative charge with total charge $-Ze$, total mass $Zm$, and diameter $2a$, surrounding a point nucleus of mass $ZM$ and charge $+Ze$, where $Z$ is the atomic number, $m$ the electron mass, $M$ the proton mass, and $-e$ the charge on an electron. Suppose this rigid electronic charge cloud is displaced slightly from a concentric position about the nucleus (the charge cloud is assumed to be "transparent" to the nucleus, so that they can easily move with respect to each other).

   Find the net restoring force on the displaced charge cloud (or, alternatively, find the resulting change in total potential energy of the system) for small displacements of the charge cloud with respect to the nucleus; and then find the classical resonance frequency at which the charge cloud will oscillate about the nucleus. (It may be assumed that only the electronic charge cloud will move appreciably, since $M \gg m$.)

   In the simplified Bohr model of the hydrogen atom, the radius of the first electron orbit is $a_0 = 0.53$Å ($1$Å $= 10^{-10}$ m or 0.1 nm). Using twice this value as a first guess for the outside radius of the charge cloud in a typical atom, compute a numerical value for the resonance frequency derived above. To what wavelength does this correspond?

2. *Q-value for a classical electron oscillator.* One way (though not the most general way) of defining the $Q$ or "quality factor" of any resonant system is as the ratio of its resonant frequency to its energy decay rate. At what frequency and what wavelength will the $Q$ of a classical electron oscillator be reduced to unity, provided that purely radiative decay is the only energy decay mechanism that is operative?

3. *Classical derivation of the radiative decay rate.* The time-averaged rate (averaged over a few cycles) at which power is radiated into the far field in all directions by a dipole antenna or by a sinusoidally oscillating charge with an electric dipole moment $\mu_x(t) = \mu_1 \cos \omega t$ is, from classical electromagnetic theory, $P_{\text{av}} = \omega^4 \mu_1^2 / 12\pi\epsilon c^3$. Use this formula to verify Equation 2.8 for the radiative decay rate $\gamma_{\text{rad}}$ of a classical electron oscillator.

---

## 2.2 COLLISIONS AND DEPHASING PROCESSES

The next important concept that we have to introduce—a particularly fundamental and important concept—is the effect of *dephasing events*, such as atomic collisions, on the oscillation behavior of classical oscillators or of real atoms.

### Coherent Dipole Oscillations

Any single microscopic electric dipole oscillator, when left by itself, obeys the equation of motion

$$\frac{d^2 \mu_x(t)}{dt^2} + \gamma \frac{d\mu_x(t)}{dt} + \omega_a^2 \mu_x(t) = (e^2/m)\mathcal{E}_x(t) \tag{13}$$

which is obtained by multiplying $-e$ into both sides of Equation 2.3 and using Equation 2.9. Hence the oscillating moment of a single atom with no applied field $\mathcal{E}_x$ present has the exponentially decaying sinusoidal form

$$\mu_x(t) = \mu_{x0} \exp\left[-(\gamma/2)(t - t_0) + j\omega_a(t - t_0) + j\phi_0\right], \tag{14}$$

where $\mu_{x0}$ is the magnitude and $\phi_0$ the phase (at time $t_0$) of the initial oscillation that has been set up in the dipole oscillator, perhaps by some pulsed applied signal.

We have already pointed out that even a small volume of laser material may contain a large number of laser atoms, or tiny oscillating dipoles. We might therefore label each individual atom or dipole by an index $i$, and write the oscillating dipole moment of the $i$-th atom as

$$\mu_{x,i}(t) = |\mu_{x0,i}(t_0)| \exp\left[-(\gamma/2)(t - t_0) + j\omega_a(t - t_0) + j\phi_i\right], \tag{15}$$

where $\phi_i$ is the phase angle of the $i$-th dipole oscillator at the starting time $t_0$.

Now suppose first that these dipoles are all oscillating together, all at the same frequency, and more importantly all initially in phase—that is, all with the same value of $\phi_i$ at the same reference time $t_0$. Then the total dipole moment due to the vector sum of all these moments in some small volume will be

$$\mu_{x,\text{tot}}(t) = \sum_{i=1}^{NV} \mu_{x,i}(t) = NV\mu_x(t) \quad \begin{cases} \text{all dipoles} \\ \text{oscillating} \\ \text{in phase,} \end{cases} \quad (16)$$

where $\mu_x(t)$ is the moment of any one dipole; $N$ is the density of dipoles (i.e., the number per unit volume); and $NV$ is the total number of dipoles in a small volume $V$. The macroscopic polarization, or the dipole moment per unit volume, will then be given by $p_x(t) = \mu_{x,\text{tot}}(t)/V$, or

$$p_x(t) = N\mu_{x0} \exp\left[[-(\gamma/2) + j\omega_a](t-t_0) + j\phi_0\right] \quad \begin{cases} \text{all dipoles} \\ \text{oscillating} \\ \text{in phase.} \end{cases} \quad (17)$$

The macroscopic polarization $p_x(t)$ in the atomic medium will thus have the same natural oscillation frequency $\omega_a$ and the same energy decay rate $\gamma/2$ as the individual dipoles. In this example its magnitude will also be $N$ times as large as any one individual dipole—but if (and only if) the individual dipoles all keep oscillating unperturbed and with the same phases.

This macroscopic polarization when all the dipoles are oscillating in time-phase with each other may be rather large in real situations. The dipoles are then said to be oscillating *coherently*, or *fully aligned* with each other.

### Dephasing Effects: Random Collisions

This is not the usual situation with real atoms, however. There are almost always perturbation effects, or *dephasing effects*, which scramble or randomize the time-phases $\phi_i$ of individual dipole oscillators, and which thereby cause the macroscopic polarization $p_x(t)$ to become much smaller than the result given by the two preceding equations. To understand this, let us look first at a very simple example of how a particular type of dephasing process, namely, randomly occurring and instantaneous dephasing events or "collisions," might operate to destroy the macroscopic polarization or coherent dipolar oscillation in a collection of atoms.

Figure 2.5 shows three assumed dipole moments, which we label $\mu_{x,1}(t)$, $\mu_{x,2}(t)$, and $\mu_{x,3}(t)$, all oscillating initially in phase and at the same oscillation frequency. The total moment $\mu_{x,\text{tot}}(t)$, as shown at the bottom of the figure, is then initially three times as large as the moment of any one dipole. Suppose, however, that after random time intervals first one and then another of the dipoles suffers an instantaneous dephasing event or "collision," which does not reduce the amplitude of the oscillating moment, but does shift it to a new phase angle in time.

After each such collision the amplitude of the total moment is reduced, because the individual moments no longer add in phase. (The individual dipole oscillations will also slowly decay in amplitude themselves because of energy decay, as discussed in the previous section; but we have not illustrated this point here.) Random collisions thus gradually destroy the macroscopic polarization, even without any energy decay.

Figure 2.6 illustrates in another manner this difference between dipole oscillators that are "aligned" or oscillating coherently in phase, and randomly phased

## 2.2 COLLISIONS AND DEPHASING PROCESSES 91

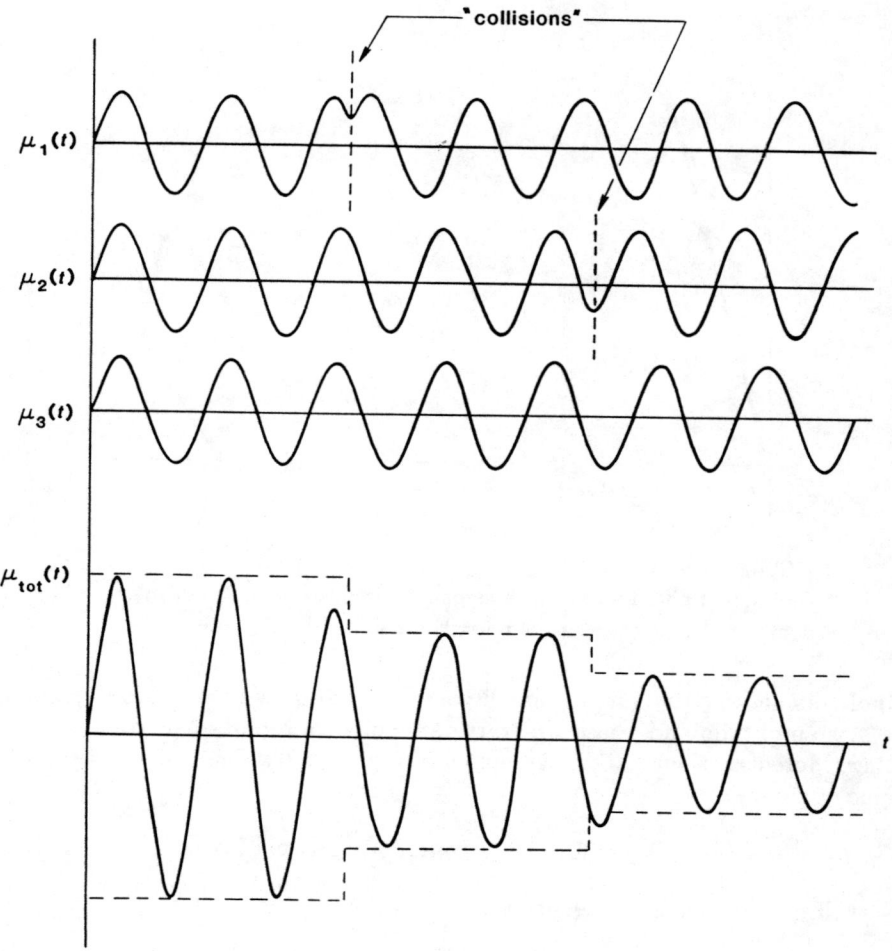

**FIGURE 2.5**
Decay of the total dipole moment resulting from random dephasing collisions in a collection of oscillating dipoles.

dipole oscillators, by showing the results of adding up three phasors that represent the amplitude and instantaneous time-phase of the three separate individual dipole oscillators. In (a) the three phasors are fully aligned; in (b) and (c) they are gradually shifted in phase or "dephased" to produce a smaller and smaller resultant sum. Note that these are *phasor* diagrams, in which the horizontal and vertical axes for each vector are the real and imaginary parts of the phasor amplitudes of the oscillating moments, or the cosine and sine parts of the sinusoidal oscillations. These axes do not represent the vector coordinates of the dipoles in space, since we are talking here for the moment only about the $x$ component of the dipole oscillations $\mu_x(t)$.

### Large Numbers of Dipoles

Suppose that we add up the phasor amplitudes of a large number $NV$ of dipoles, but with the phase angles $\phi_i$ randomly distributed over all values between 0 and $2\pi$. This then becomes the standard statistical problem of adding up many randomly phased sine waves, and we can find that the resulting total

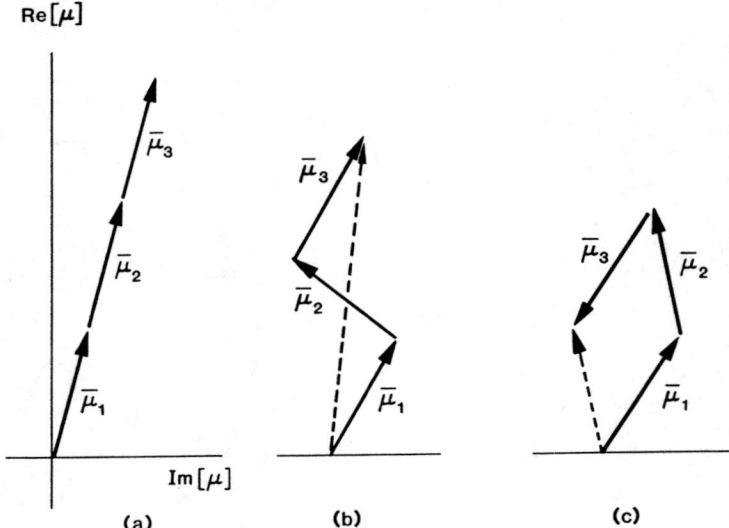

FIGURE 2.6
Addition of phasors. (a) All three phasors in phase. (b) Partially randomized phases. (c) More completely randomized.

dipole moment $\mu_{x,\text{tot}}(t)$ in any small volume $V$ will now be a random quantity; i.e., its amplitude and phase will vary randomly from one small volume to another. Moreover, the total dipole moment in any small volume will have a mean value of zero, i.e.,

$$\langle \mu_{x,\text{tot}}(t) \rangle = 0 \qquad \text{(randomly phased dipoles)}, \tag{18}$$

but will have a root-mean-square value given by

$$\langle \mu_{x,\text{tot}}^2(t) \rangle^{1/2} = (NV)^{1/2} |\mu_x(t)| \qquad \text{(randomly phased dipoles)}, \tag{19}$$

where $|\mu_x|$ refers to the value for any one single dipole by itself.

The quantity $NV$ will be a very large number even for very small volumes $V$. Hence the rms moment, or rms macroscopic polarization, for the randomly phased case, which is proportional to $(NV)^{1/2}|\mu_x|$, will be very much smaller than the possible coherent polarization of order $NV|\mu_x|$ that could be produced by the same number of dipoles oscillating in phase. (The rms polarization in the randomly phased case is in fact essentially random noise; and this noise is essentially the same thing as the spontaneous emission from a collection of quantum atoms, although we will not discuss this topic here.)

### Dephasing Mechanisms

The crucial point, then, is that any effect which tends to randomize the oscillation phases $\phi_i$ in a large collection of individual dipoles (such as are present in even a small volume of laser material) will act to destroy any coherent macroscopic polarization that may be present in this collection of dipole oscillators. Such dephasing effects do exist in real atomic systems, and understanding these additional dephasing effects is our primary task in this section.

These dephasing effects that cause the oscillation phases of individual atomic oscillators to become randomized, even though each dipole continues to oscillate with the same decaying amplitude and average frequency, are often referred to for simplicity as *collisions*. However, the actual physical processes that can cause dephasing of the individual internal atomic oscillations in a collection of atoms can include the following.

- Atoms (or ions or molecules) in gases, moving with their normal thermal or Brownian motion, can in fact make *random physical collisions* with each other, or with other gas atoms, or with the walls of the laser tube. Even if these collisions are elastic—that is, even if they do not take any energy away from the internal electronic oscillation energy of the atoms—in general such collisions between atoms will scramble and randomize the phases of the electronic oscillations inside the colliding atoms.
- For laser atoms in solids, the quantum energy-level spacings and hence the exact transition frequencies $\omega_a$ of the laser atoms are affected by nearby host atoms, and hence depend on the exact distances to nearby atoms in the host crystal lattice. *Thermal vibrations of the crystal lattice* will modulate these distances slightly, and thus modulate the atomic transition frequencies $\omega_a$ by small but random amounts with time. This is called *phonon broadening*, and it produces in turn a random phase modulation and hence a "phase smearing" of the dipole oscillations in the laser atoms.
- In materials where the laser atoms are sufficiently dense, the local time-varying electric (or magnetic) fields produced by any one oscillating dipole may spread out to, and be felt by, other neighboring laser atoms. The individual oscillating dipoles are then no longer totally independent, but become weakly coupled to each other through what is called *dipolar coupling*. This kind of weak coupling between individual resonant systems, even if they are all identical, always tends to randomize and to broaden the overall response of the collection. (This is true of weakly coupled resonant electric circuits, as well as weakly coupled resonant atoms.) This process of dipolar coupling is thus still another mechanism for producing random phase smearing in the atomic dipoles. Moreover, this dipolar coupling itself will be randomly modulated in atoms by the thermal motion of the atoms, whether by gas kinetics in gases or lattice vibrations in solids.

Whatever may be the physical cause, the net result of each of these physical processes is to randomize or "dephase" the phases of individual atomic oscillators with respect to each other. The coherent dipole oscillations get converted eventually into *incoherent* oscillations.

---

### More on Collision Broadening.

In a more sophisticated description of collision broadening in gases, the discrete collision between two atoms in the gas is not really instantaneous. Rather, when two quantum atoms in a gas come very near each other, their quantum wavefunctions overlap. The presence of each atom then causes a small but not insignificant shift in the quantum energy levels of the other atom. (More precisely, we must calculate the shifted quantum levels of the two atoms taken together as a single combined quantum system.) The resonance frequency of each atom is thus shifted by a small, time-varying amount as the atoms pass near each other. The collision interval during which the atoms are close enough to influence each other is short enough ($\approx 10^{-13}$ sec) to be "instantaneous" on a practical time-scale; yet it is long enough for the accumulated shifts in optical frequency cycles to leave the final phases of the oscillators essentially randomized relative to their initial phases. This brief interaction acts for all practical purposes, then, like an instantaneous randomizing collision.

---

### Exponential Decay: The Dephasing Time $T_2$

A simple formula for the rate at which a macroscopic polarization $p_x(t)$ will be destroyed by random dephasing events can be developed as follows. Suppose that not just a few dipoles, as in the preceding examples, but a very large number of individual (but identical) oscillators are involved. Suppose also that the dephasing events for individual dipoles happen randomly both in their times of occurrence and in the phase changes they produce. Let there then be some large number $N_0$ of dipoles in a unit volume, all initially oscillating in phase, so that the magnitude of the initial polarization at a starting time $t_0$ is

$$p_x(t_0) = N_0 \mu_{x0}. \tag{20}$$

At any later time $t > t_0$, we can then divide these $N_0$ dipoles into (a) a decreasing number of dipoles $N(t)$ that have not yet suffered any collisions at all; and (b) an increasing number $N_0 - N(t)$ of dipoles that have suffered at least one collision, and perhaps more. The $N(t)$ dipoles that have not yet undergone any collisions or dephasing events will then continue to oscillate in phase and to produce a macroscopic polarization

$$p_x(t) = N(t)\mu_x(t) = N(t)\mu_{x0} \cos \omega_{\text{at}} t. \tag{21}$$

Those dipoles that have suffered even one collision, however, will have phases that are entirely random (assuming, as is normally done, that the phase of a dipole oscillation is entirely randomized after each collision). Hence those dipoles will add up to produce no coherent polarization at all, on the average.

The coherent polarization after any time $t > t_0$ thus comes entirely from the remaining uncollided dipoles. [A more precise statement is that the $N$ dipoles oscillating coherently in phase will add up to produce a macroscopic polarization proportional to $N\mu_{x0}$, whereas the $N_0 - N$ dipoles oscillating with random phases will add up to produce a macroscopic sum with a mean value of zero and an rms value proportional to $(N_0 - N)^{1/2}\mu_{x0}$. Since the number of atoms involved in any atomic system is always very large, the latter quantity is negligible compared to the coherent part of the oscillation; and we can neglect the contribution from the randomly phased dipoles in the latter group.]

The number of uncollided dipoles $N(t)$ will of course decrease steadily with time. How can we calculate the rate at which the number of uncollided atoms $N(t)$ decreases? Let us suppose that collisions occur at a random rate of $1/T_2$ collisions per atom per second. Then, the total number of collisions $dN$ that

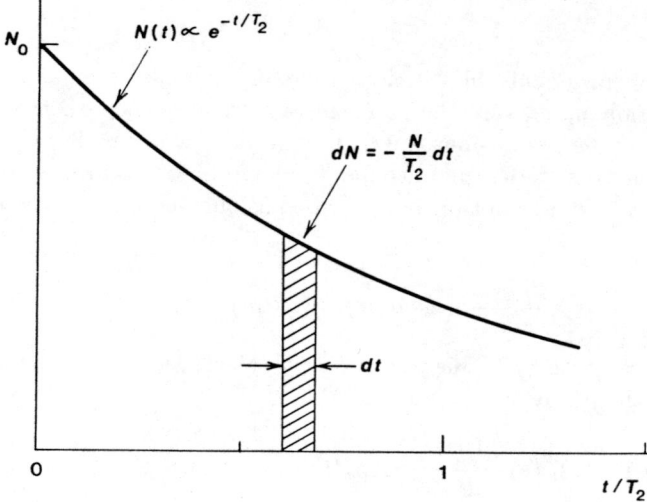

FIGURE 2.7
Decay of the number of uncollided dipoles.

members of this uncollided group will undergo in a little time interval $dt$ about time $t$, or the loss rate from the uncollided group $N(t)$ in time $dt$, will be given by

$$dN(t) = -\frac{N(t)}{T_2} dt. \tag{22}$$

The size of the uncollided group will thus decay as

$$N(t) = N_0 e^{-(t-t_0)/T_2}, \qquad t > t_0. \tag{23}$$

The coherent macroscopic polarization produced by these still uncollided oscillators will therefore also decay as

$$\begin{aligned} p_x(t) &= N(t)\mu_x(t) \\ &= N_0 e^{-(t-t_0)/T_2} \times \mu_{x0} \exp\left[-(\gamma/2)(t-t_0) + j\omega_a(t-t_0) + j\phi_0\right] \quad (24) \\ &= p_{x0} \exp\left[-(\gamma/2 + 1/T_2)(t-t_0) + j\omega_a(t-t_0) + j\phi_0\right]. \end{aligned}$$

In other words, *the amplitude decay rate $\gamma/2$ appropriate for the individual dipoles must be replaced by $\frac{1}{2}\gamma + T_2^{-1}$ as the effective amplitude decay rate for the coherent polarization $p_x(t)$*, or

$$\left(\frac{\gamma}{2}\right) \begin{pmatrix} \text{single-dipole} \\ \text{decay rate} \end{pmatrix} \Rightarrow \left(\frac{\gamma}{2} + \frac{1}{T_2}\right) \begin{pmatrix} \text{macroscopic} \\ \text{polarization} \\ \text{decay rate} \end{pmatrix}. \tag{25}$$

It may seem slightly odd that in this substitution the dephasing rate $1/T_2$ gets added to the quantity $\gamma/2 \equiv 1/2\tau$, which is *half* the energy decay rate. The reason for this difference of a factor of two—which will continue to be with us—is essentially that $1/T_2$ and $1/2\tau$ are both decay rates for sinusoidal *amplitudes* like $\mu_x(t)$ or $p_x(t)$; whereas $\gamma$ itself is an *energy* decay rate for the quantity $|\mu_x(t)|^2$.

### Summary

The primary conclusion of this section, therefore, is that although the macroscopic polarization $p_x(t)$, has the same resonance frequency as the individual microscopic dipole oscillations $\mu_x(t)$, it may have a faster decay rate because of dephasing effects. Individual atomic dipole oscillations, in the intervals between dephasing events, can thus be described as obeying the equation of motion

$$\frac{d^2\mu_x(t)}{dt^2} + \gamma\frac{d\mu_x(t)}{dt} + \omega_a^2\mu_x(t) = (e^2/m)\mathcal{E}_x(t) \tag{26}$$

with an amplitude decay rate $\gamma/2$. But the coherent polarization $p_x(t)$ must be described as obeying the equation

$$\frac{d^2p_x(t)}{dt^2} + (\gamma + 2/T_2)\frac{dp_x(t)}{dt} + \omega_a^2 p_x(t) = (Ne^2/m)\mathcal{E}_x(t), \tag{27}$$

where there is an additional factor of $1/T_2$ in the amplitude decay rate because the dephasing processes cause the oscillations of individual atoms to become randomized in phase at a rate $1/T_2$. (Note again the difference of a factor of 2 between the $\gamma$ and $1/T_2$ terms.)

In the analysis we have presented here—which is in fact very similar to the approach in much more sophisticated quantum treatments—the time constant $T_2$ thus has the physical significance of the mean time between dephasing events or collisions for any one individual atom, so that $1/T_2$ is the *collision frequency* for any one individual atom. The time constant $T_2$ is thus often called the *collision time*. This same time constant is often referred to more broadly as the *dephasing time*, or even the *dipolar interaction time* for $p_x(t)$. In quantum analyses or in the Bloch equations for magnetic resonance, $T_2$ is also called the *off-diagonal* or *transverse relaxation time*.

### REFERENCES

The concepts of collisions and dephasing as presented here were apparently first introduced by H. A. Lorentz in the *Proceedings of the Amsterdam Academy of Science* **8**, 591 (1906). The number of discussions of collisions and line broadening in the literature is now extremely large. We will give many references to these in Section 3.2, where we discuss collision broadening and line broadening in real atoms in more detail.

---

### Problems for 2.2

1. *Collision broadening with a different kind of collision statistics.* Consider a somewhat unusual collection of atoms, in which each individual atom suffers its dephasing collisions at absolutely regular intervals, spaced by an intercollision time $T_c$, so that the collisions for a given atom occur at instants $t = t_0, t_0 + T_c, t_0 + 2T_c$, and so forth, with the intercollision time $T_c$ being the same for all the atoms. (You might think of each individual atom as bouncing back and forth between two walls at the same constant velocity, not hitting any other atoms, but being dephased each time it hits the walls.) The reference time or "first-collision time" $t_0$ is different for each atom, however, with values uniformly distributed between

0 and $T_c$ (i.e., the atoms at any instant are uniformly distributed in the space between the walls).

Suppose that a group of atoms with these oddball collision properties are all set oscillating internally, with their internal oscillations initially in phase at $t = 0$. Describe how the macroscopic polarization $p_x(t)$ will decrease with time, including both energy decay and this oddball dephasing; and plot $p_x(t)$ versus $t/T_c$ for values of the parameter $\gamma T_c = 0.1$, 1, and 10.

---

## 2.3 MORE ON ATOMIC DYNAMICS AND DEPHASING

Since dephasing effects are a particularly important aspect of atomic dynamics, let us look a bit further at some of the additional consequences and varieties of dephasing effects in real atomic systems.

### Dephasing Effects Plus Applied Signals

One basic assumption in discussions of dephasing is that dephasing effects and applied signal fields such as $\mathcal{E}_x(t)$ will act on individual dipoles simultaneously and independently—that is, we can simply add up their effects in computing the total internal motion of individual atoms. What, then, are the relative strengths of these two effects? To explore this, let us consider, for example, the response of a single dipole oscillator subjected to an on-resonance sinusoidal applied field $\mathcal{E}_x(t) = E_1 \cos \omega_a t$ during the period just after this atom has suffered a randomizing collision at $t = t_0$. How rapidly can the applied signal $\mathcal{E}_x(t)$ "pull" the individual dipole moment $\mu_x(t)$ back into a coherent phase relationship with the applied signal, following a dephasing collision?

The problem here is clearly to solve the single-dipole equation of motion (Equation 2.26) with the specified applied signal and with an arbitrary initial condition on the phase (i.e., the position and velocity) of the oscillator at time $t = 0$. This can be done straightforwardly, although the exact solution is a bit messy. We know, however, that an equation of this type has a transient or homogeneous solution, independent of $\mathcal{E}_x(t)$, of the form

$$\mu_x(t) = \mu_{x0} \exp\left[-(\gamma/2)(t - t_0) + j\omega_a(t - t_0) + j\phi_0\right]. \tag{28}$$

(There is a minor approximation in this expression, namely, the replacing of $\omega_a'$ by $\omega_a$.) We will also show in Section 2.4 that an on-resonance applied signal will produce a steady-state or forced sinusoidal solution of the form

$$\mu_x(t) = \text{Re}\left[-j \frac{e^2 E_1}{m \omega_a \gamma} e^{j\omega_a t}\right] = \text{Re}\left[\tilde{\mu}_{ss} e^{j\omega_a t}\right], \tag{29}$$

where $\tilde{\mu}_{ss} \equiv j(e/m\omega_a\gamma)E_1$ is the steady-state phasor amplitude of the motion produced by the field $E_1$. Suppose we also define $\tilde{\mu}_0 \equiv \mu_{x0} \exp j\phi(0)$ as the complex phasor amplitude (magnitude and phase) of the sinusoidal motion of $\mu_x(t)$ immediately after the collision. (The phase $\phi(0)$ will take on random values for different dipoles after different collisions.)

The total solution for $\mu_x(t)$ following any given collision will then be a linear combination of the forced plus transient solutions, with just enough transient solution included to meet the initial boundary condition at $t_0$, or

$$\mu_x(t) = \text{Re}\left[\tilde{\mu}_{ss} + (\tilde{\mu}_0 - \tilde{\mu}_{ss})e^{-(\gamma/2)(t-t_0)}\right]e^{j\omega_a(t-t_0)}. \tag{30}$$

This says, in effect, that we can write $\mu_x(t)$ in the form

$$\mu_x(t) = \text{Re}\left[\tilde{\mu}(t)e^{j\omega_a(t-t_0)}\right], \tag{31}$$

where $\tilde{\mu}(t)$ is a slowly changing complex phasor amplitude given by

$$\tilde{\mu}(t) = \tilde{\mu}_{ss} + [\tilde{\mu}_0 - \tilde{\mu}_{ss}]e^{-(t-t_0)/2\tau}. \tag{32}$$

In other words, *following a collision the phasor amplitude $\tilde{\mu}(t)$ of the sinusoidal motion "pulls in" from the initial random post-collision value $\tilde{\mu}_0$, toward the forced or steady-state value $\tilde{\mu}_{ss}$, with an exponential time constant $2\tau$*. Note that this pull-in time constant does not depend at all on the strength of the applied field.

Now, we will see later that for most real laser transitions the dephasing time $T_2$ is usually much shorter than the energy decay time $\tau$; so the dephasing time constant $T_2$ is also much shorter than the pull-in time constant $\tau$ (or $2\tau$). Any individual atom is, therefore, very likely to be dephased again by another collision, after a short time $\approx T_2$, well before it gets pulled completely into phase by the applied signal $\mathcal{E}_x(t)$. *In real laser systems, therefore, even with applied signals present, the motions of the individual dipoles are mostly dephased, or randomly phased, by the dephasing processes.* A coherent applied signal $\mathcal{E}_x(t)$ can usually only struggle to impose a small amount of phase ordering on this unruly bunch of oscillators.

Exceptions to this usual situation occur only for applied signals that are strong enough to produce the kind of Rabi flopping behavior that we will discuss in Chapter 5. Most signals in common lasers are "weak signals" which do not produce this kind of behavior; and the dipole motion in these system will be mostly random, with a small fractional amount of signal-imposed coherent ordering.

### Dephasing by Random Frequency Modulation

Let us also look in a bit more detail at another type of dephasing that occurs in many solid-state laser materials.

The most graphic way of picturing dephasing effects in any collection of atoms is probably the kind of sudden, sharp, discrete, randomly occurring dephasing events or "collisions" that we have described above. An important alternative dephasing process for atoms, however, especially in crystals and other solids, is *phonon broadening*, or phonon frequency modulation of the atomic transitions, rather than genuine collisions between different atoms as in a gas.

In systems with phonon broadening (or with dipolar coupling as well) the dephasing process results not from sudden collisions, but from a more continuous but still random *frequency modulation* of each individual dipole's oscillation frequency. The net result, however, is essentially the same: dipoles that begin oscillating in phase gradually end up, after a time on the order of $T_2$, with their phases completely randomized.

## 2.3 MORE ON ATOMIC DYNAMICS AND DEPHASING

Consider, for example, the chromium $Cr^{3+}$ ions in a ruby crystal or the neodymium $Nd^{3+}$ ions in a Nd:YAG crystal or glass lattice, such as we showed in Chapter 1; and suppose the internal electronic charges of several such ions have been set oscillating in an internal dipole oscillation with the same initial phase. Now, the surrounding lattice itself will also be vibrating slightly at any finite temperature, because of thermal agitation; so the spacing between each ion and its nearest neighbors in the lattice will be changing slightly in a random way that is different for each ion. But for ions in solids, small changes in the lattice spacing will cause very small but finite shifts in the exact resonance frequency $\omega_a$ of the transition in each ion. The sinusoidal dipole oscillations of the various ions, as a result, will proceed at slightly different and randomly changing frequencies; and the dipole oscillations will thus drift slowly and randomly out of phase with each other.

This same argument can hold for dipole oscillations in a crystal lattice, in a glassy solid, in a liquid, or in any condensed atomic medium. Dipoles initially oscillating in phase will eventually be converted to random phases. It is not so evident here that this will lead to an exponential decrease in the coherent polarization component. The fact is, however, that the same assumption of exponential decay that we made for the macroscopic polarization is just as good an approximation for these situations also.

### Note on Phonon Broadening.

The shifts in resonance frequencies of ionic transitions in solids due to changes in the local lattice spacing can be demonstrated experimentally simply by squeezing the crystals to compress the lattice spacing slightly, and noting that there are small but finite *pressure shifts* for the transition frequencies of the ions in the crystals. These pressure shifts may be viewed as small changes in the *Stark shifts* of the atomic energy levels which are produced by the electric fields associated with the bonds between atoms in the crystal lattice.

### Some Typical Numbers for Dephasing Effects

The magnitudes of the dephasing effects and the values of the dephasing time $T_2$ exhibit very large variations in different kinds of atomic media. Recall first that visible transitions have oscillation frequencies on the order of $6 \times 10^{14}$ Hz and thus oscillation periods of the order of $10^{-15}$ sec.

The collision frequencies for atoms in real gases can vary over a wide range, depending on gas pressure; but values in the range of $1/T_2 \approx 10^8$ to $10^9$ $\text{sec}^{-1}$, or dephasing times of $T_2 \approx 10^{-8}$ to $10^{-9}$ sec at low pressures, are not uncommon. Energy decay times in the range of $\tau \equiv \gamma^{-1} \approx 10^{-5}$ to $10^{-7}$ sec for transitions of interest are also reasonable. The general conclusion is thus that there are always an enormous number of optical cycles between each collision or dephasing event. The collision rate is usually an order of magnitude or more higher than the energy decay rate; so the $1/T_2$ term in the polarization decay often dominates over the $\gamma/2$ part of the decay.

For atoms in solids, the lattice vibrational frequencies that are excited by thermal agitation, and that cause the thermal frequency dephasing, range from zero up to $\approx 10^{13}$ Hz. The lattice modulation of the atomic transition frequencies is thus in general very fast compared to any measurements we might try to make on the atoms, but still slow compared to the actual transition frequencies $\omega_a$. We must then ask not only how rapidly the lattice atoms vibrate, but also how strongly they modulate or shift the transition frequencies $\omega_a$. The answer in typical lasers (e.g., ruby or Nd:YAG) is that these frequencies are shifted randomly by amounts on the order of $10^{11}$ to $10^{12}$ Hz.

Now, two dipoles having a random frequency difference of $\omega_{a2} - \omega_{a1}$ will get $2\pi$ out of time-phase with each other after an interval of $2\pi/(\omega_{a2} - \omega_{a1})$ seconds. The effective $T_2$ dephasing times for ionic transitions in solids are thus often of the order of $10^{-11}$ to $10^{-12}$ sec. The energy decay times in solids on good laser transitions are sometimes as slow as $10^{-3}$ to $10^{-4}$ sec. Again, the $1/T_2$ dephasing component dominates, generally by a very large amount, over the $\gamma/2$ energy decay rate.

### Coherent Versus Incoherent Decay

Suppose, as a final mental exercise, that a large number of atoms $N_0$ are initially all oscillating and radiating together in phase, as we described earlier. (Preparing a group of atoms in this coherently phased initial condition is not always easy to accomplish, as we will see later. It generally requires very strong applied signals, applied in very short pulses.)

Given this initial preparation, all these atoms or oscillators will then radiate together as one giant coherent dipole. The initial value of this dipole will be $N_0 \mu_{x0}$, where $\mu_{x0}$ is the dipole moment of one individual atom; and the rate at which this collection of coherently oscillating atoms radiates energy will be proportional to $N_0^2 |\mu_{x0}|^2$. The essential point is that all the dipoles are radiating *coherently*, that is, in time-phase with each other.

This coherence will, however, be destroyed by dephasing processes in a time of order $T_2$, which for real atomic transitions is often very short (from nanoseconds down to less than picoseconds). Once the coherent oscillation is destroyed, after a few dephasing times $T_2$, the individual dipoles will in general still be oscillating and radiating energy, since their energy decay time $\tau \equiv \gamma^{-1}$ is generally longer (sometimes much longer) than the dephasing time $T_2$. The individual dipoles will continue, in fact, to radiate energy through the $\gamma_{\rm rad}$ and $\gamma_{\rm nr}$ processes, but they now radiate individually and *incoherently*, with random phase relationships between the dipoles. The radiation that now comes out from the sample is essentially narrowband noise, or spontaneous emission, or fluorescence centered at the atomic transition frequency $\omega_a$. It comes out in all directions, and with a narrowband but essentially noise-like spectrum. The power radiated is simply the sum of the individual powers radiated by the $N_0$ individual dipoles, and hence is now proportional to $N_0 |\mu_{x0}|^2$ rather than to $N_0^2 |\mu_{x0}|^2$.

Suppose we perform an experiment in which we set a collection of atomic dipoles oscillating coherently, perhaps using some kind of pulsed applied signal, and then observe how the atoms radiate afterward. We can expect to see two transients: first the coherent transient radiation, which may be strong but very fast (time constant $\approx T_2$); and then the incoherent transient radiation, generally much weaker but longer-lived, corresponding to normal spontaneous emission or fluorescence (time constant $\approx T_1$). So-called *coherent pulse* or *coherent free-*

*induction decay experiments* displaying the first type of behavior can be performed on atoms at optical frequencies. These experiments are generally rather difficult, however, requiring short but intense coherent laser pulses for excitation, together with high-speed detectors for the coherently radiated signals.

Much more common are ordinary *fluorescent lifetime experiments*, such as we will illustrate later. In these, a group of atoms are again set oscillating, but the excitation mechanism is some form of incoherent excitation, such as a pulse of broadband light from an incoherent flashlamp, or a short burst of current through a collection of gas atoms. There is no initial phase coherence to the excitation in these cases, and hence no coherent initial polarization to either radiate coherently or decay at the $T_2$ rate. The radiation in this case comes entirely from incoherent spontaneous emission or fluorescence, and the measured decay rate will be simply the energy decay rate $\gamma \equiv \gamma_{\text{rad}} + \gamma_{\text{nr}}$. Understanding the distinction between these coherent and incoherent types of processes is extremely important in understanding the atomic phenomena involved in lasers.

Problems for 2.3

1. *Dephasing by random frequency modulation.* It is claimed in this section that a continuous but random modulation of the resonance frequencies of a system can lead to a net exponential result for dephasing behavior. To examine this basic idea in more detail, consider a large number of individual dipole oscillators all having the same natural oscillation frequency $\omega_a$, and all oscillating initially in phase with each other in the general form $\cos \omega_a t$. Suppose, however, that each oscillator is randomly phase-modulated (or frequency-modulated) by some external perturbation, so that after a time $t$ any individual oscillator oscillates as $\cos[\omega_a t + \phi_i(t)]$, where $\phi_i(t)$ is a random phase angle for the $i$-th oscillator after time $t$.

   If the individual oscillators randomly diffuse in phase angle with increasing time, the probability density distribution of oscillation phases for different oscillators after a time $t$ may take on the gaussian form $Pr[\phi_i] = \exp[-\phi_i^2/2\sigma^2(t)]/\sqrt{2\pi\sigma^2(t)}$. If, for example, the phases of different oscillators diffuse as a random-walk process, standard statistical arguments say that the random phases will have just this kind of gaussian probability distribution, and that the variance $\sigma^2$ of this distribution will increase linearly in time in the form $\sigma^2(t) = 2Dt$, where $D$ is a diffusion coefficient for the phases.

   Using this probabilistic model, evaluate how the macroscopic polarization $p(t)$ obtained by summing over a large number of microscopic dipoles $\mu_i(t)$ per unit volume will decay in time, assuming all the dipoles start out oscillating in phase (i.e., with $\phi_i(0) = 0$). Hints: The expected value or average value for some function $f(y)$, where $y$ is a random variable with probability distribution $\Pr[y]$, is given by $<f> \equiv \int f(y) \Pr[y] \, dy$. A useful formula to remember is that

   $$\int_{-\infty}^{\infty} e^{-Ay^2 - 2By} \, dy = \sqrt{\pi/A} \, e^{B^2/A}.$$

   This formula holds for $A$ and $B$ complex, provided only that $\text{Re}[A] > 0$.

## 2.4 STEADY-STATE RESPONSE: THE ATOMIC SUSCEPTIBILITY

Our next task is to compute the steady-state response of a collection of oscillators or atoms to a sinusoidal applied signal, and to express this response as a linear resonant electric susceptibility.

### Phasor Analysis

Suppose that the electric field $\mathcal{E}_x(t)$ applied to a collection of classical oscillators, or electric dipole atoms, is a sinusoidal signal with frequency $\omega$, which we write in the form

$$\mathcal{E}_x(t) = \text{Re}[\tilde{E}_x e^{j\omega t}] = \frac{1}{2}[\tilde{E}_x e^{j\omega t} + \tilde{E}_x^* e^{-j\omega t}]. \tag{33}$$

In electrical engineering jargon the complex quantity $\tilde{E}_x$ is a "phasor" whose magnitude and phase angle give the amplitude and phase of the real quantity $\mathcal{E}_x(t)$. Suppose, for example, that the complex phasor $\tilde{E}_x$ has the magnitude and phase angle $\tilde{E}_x \equiv |\tilde{E}_x| e^{j\phi}$. Then the real field $\mathcal{E}_x(t)$ will be given by $\mathcal{E}_x(t) = \text{Re}[|\tilde{E}_x| e^{j(\omega t + \phi)}] = |\tilde{E}_x| \cos(\omega t + \phi)$, so that obviously $|\tilde{E}_x|$ is the magnitude and $\phi$ the phase angle (in time) of the cosinusoidal signal.

The steady-state response from a linear atomic system will then have the same sinusoidal form, i.e.,

$$p_x(t) = \text{Re}[\tilde{P}_x e^{j\omega t}] = \frac{1}{2}[\tilde{P}_x e^{j\omega t} + \tilde{P}_x^* e^{-j\omega t}], \tag{34}$$

so that a similar description will obviously prevail for the magnitude and phase angle of the real polarization $p_x(t)$ and its complex phasor amplitude $\tilde{P}_x$.

Both the $e^{j\omega t}$ and the $e^{-j\omega t}$ terms in these phasor expansions are needed to give the complete real fields; but in any *linear* system with a linear differential equation, such as we are considering here, the $\tilde{E}_x e^{j\omega t}$ part of the applied field will be connected only to the $\tilde{P}_x e^{j\omega t}$ part of the induced polarization, and similarly for the $\tilde{E}_x^* e^{-j\omega t}$ and $\tilde{P}_x^* e^{-j\omega t}$ parts of these quantities. Moreover, these separate responses in any real physical system will be simply the complex conjugates of each other, so that the complex-conjugate or $e^{-j\omega t}$ terms really contain no additional information over and above the $e^{j\omega t}$ terms.

Following the usual practice in phasor analyses, therefore, we will focus only on the $e^{j\omega t}$ terms from now on. Moreover, for simplicity we will generally leave off the "Re" notation from now on and write the real fields in the form $\mathcal{E}_x(t) = \tilde{E}_x e^{j\omega t}$, with the operation of taking the real part being understood.

If we put these sinusoidal phasor expansions into the equation of motion for $p_x(t)$, Equation 2.27, and separate out the $e^{j\omega t}$ terms, we obtain a relation between the complex phasor amplitudes:

$$\left[-\omega^2 + j\omega(\gamma + 2/T_2) + \omega_a^2\right]\tilde{P}_x = \frac{Ne^2}{m}\tilde{E}_x, \tag{35}$$

which we will rearrange into the form

$$\frac{\tilde{P}_x}{\tilde{E}_x} = \frac{Ne^2}{m}\frac{1}{\omega_a^2 - \omega^2 + j\omega(\gamma + 2/T_2)}. \tag{36}$$

## 2.4 STEADY-STATE RESPONSE: THE ATOMIC SUSCEPTIBILITY

This is the linear steady-state relationship between the phasor polarization $\tilde{P}_x$ induced in the collection of oscillators or atoms and the field $\tilde{E}_x$ applied to them. In linear-system terms, it is the *transfer function* for the response of the atomic medium.

### Electric Polarization and Susceptibility: Standard Definitions

This transfer function is more commonly known as the *electric susceptibility of the atomic medium*, as produced by the polarization response of the atoms or oscillators. We can recall that the electric field $\tilde{E}$, the electric polarization $\tilde{P}$, and the electric displacement $\tilde{D}$ in any arbitrary dielectric medium are related under all circumstances by the basic definition from electromagnetic theory

$$\tilde{D} = \epsilon_0 \tilde{E} + \tilde{P}. \tag{37}$$

In the more restrictive case of a linear and isotropic dielectric medium, the polarization $\tilde{P}$ and the electric field $\tilde{E}$ will also be related, by an expression which is conventionally written in the form

$$\tilde{P}(\omega) = \tilde{\chi}(\omega) \epsilon_0 \tilde{E}(\omega), \tag{38}$$

so that the quantity $\tilde{\chi}(\omega)$ defined by

$$\tilde{\chi}(\omega) \equiv \frac{\tilde{P}(\omega)}{\epsilon_0 \tilde{E}(\omega)} \tag{39}$$

is the *electric susceptibility* of the medium, with $\epsilon_0$ being the dielectric permeability of free space. We will adopt a slightly modified version of this definition a few paragraphs further on.

The relationship between the electric displacement $\tilde{D}$ and the electric field $\tilde{E}$ in a linear medium can then be written, using the standard definition of Equation 2.39, as

$$\tilde{D} = \epsilon_0 \left[1 + \tilde{\chi}\right] \tilde{E} = \tilde{\epsilon} \tilde{E}, \tag{40}$$

which means that the complex dielectric constant $\tilde{\epsilon}(\omega)$ is given by

$$\tilde{\epsilon}(\omega) \equiv \epsilon_0 (1 + \tilde{\chi}). \tag{41}$$

For a completely general description, the field quantities $\tilde{D}$, $\tilde{E}$, and $\tilde{P}$ really should be treated as vector quantities in these relations; and in the more general linear but anisotropic case the susceptibility $\tilde{\chi}$ then becomes a tensor quantity. For simplicity, however, let us stick with scalar notation at this point.

The electric susceptibility relating the applied signal $\tilde{E}_x$ and the atomic polarization $\tilde{P}_x$ in an atomic medium is very important in calculating laser gain, phase shift, and many other properties, as we will see shortly. Before going further with this discussion, however, we must introduce a slightly nonstandard definition of the electric susceptibility $\tilde{\chi}$, which is peculiar to this book, but which will turn out to be very useful in simplying later formulas.

### Atomic Susceptibility: A Modified Definition

In a sizable fraction of the laser materials of interest to us, the resonant oscillators or the laser atoms that produce the resonant polarization $\tilde{P}_x$ are not located in free space. Rather, these atoms are imbedded in a laser crystal, or perhaps in a glass or a liquid host material. In the laser material ruby, for example, the $Cr^{3+}$ laser atoms that are responsible for the laser behavior are dispersed (at $\approx 1\%$ density) in a host lattice of colorless $Al_2O_3$, or sapphire. In dye laser solutions the dye molecules, for example, Rhodamine 6G, are dissolved at perhaps $10^{-3}$ molar concentration in a liquid solvent such as water or ethanol.

In all these devices, the host materials in the absence of the laser atoms are transparent dielectric materials that are nearly lossless at the laser wavelength, but have a relative dielectric constant $\epsilon/\epsilon_0$ or an index of refraction $n$ that is significantly greater than unity. These materials will possess, therefore, a large nonresonant linear electric polarization $P_{host}$ that is associated with the host material by itself, and that has no direct connection with the generally much weaker resonant polarization $P_{at}$ that comes from the resonant response of the classical oscillators or from the resonant transitions in the laser atoms.

We can therefore write the total displacement vector in such a material in more detail as

$$\tilde{D} = \epsilon_0 \tilde{E} + \tilde{P}_{host} + \tilde{P}_{at}. \tag{42}$$

In this equation $\tilde{P}_{host}$ refers to the large, broadband, linear nonresonant polarization associated with the host material by itself; whereas $\tilde{P}_{at}$ refers to the weak, narrowband, linear resonant polarization produced by the classical oscillators or atoms imbedded in the host material. Following conventional electromagnetic notation, we can then define a nonresonant susceptibility $\tilde{\chi}_{host}$ and a dielectric constant $\epsilon_{host}$ for the host material according to the usual definitions, in the form

$$\tilde{P}_{host} = \tilde{\chi}_{host} \epsilon_0 \tilde{E} \quad \text{and} \quad \epsilon_{host} = \epsilon_0 (1 + \tilde{\chi}_{host}). \tag{43}$$

The total polarization can therefore be written as

$$\tilde{D} = \epsilon_0 [1 + \tilde{\chi}_{host}] \tilde{E} + \tilde{P}_{at} = \epsilon_{host} \tilde{E} + \tilde{P}_{at}. \tag{44}$$

Note that in typical laser crystals or liquids the host dielectric constant (at optical frequencies) will have magnitude $\epsilon_{host}/\epsilon_0 \approx 2$ to 3, so that the dimensionless host susceptibility will have magnitude $\tilde{\chi}_{host} \approx 1$ to 2. To put this in another way, the index of refraction of typical laser host materials, given by $n_{host} \equiv \sqrt{\epsilon_{host}/\epsilon_0}$, will have values of $n_{host} \approx 1.5$ to 2.0 for typical liquids or crystals.

Suppose now that we were also to define a separate susceptibility $\tilde{\chi}_{at}$ for the atomic or resonant oscillator part of the response in the laser medium, using the same conventional definition as given earlier, namely,

$$\tilde{\chi}_{at} = \tilde{P}_{at}/\epsilon_0 \tilde{E} \quad \begin{pmatrix} \text{conventional} \\ \text{definition} \end{pmatrix}. \tag{45}$$

Then we would end up with a result of the form

$$\tilde{D} = \epsilon_{host}\tilde{E} + \tilde{\chi}_{at}\epsilon_0 \tilde{E} = \epsilon_{host}[1 + (\epsilon_0/\epsilon_{host})\tilde{\chi}_{at}]\tilde{E} \quad \begin{pmatrix} \text{conventional} \\ \text{definition} \end{pmatrix}. \tag{46}$$

Now, there will be many times in later chapters when we will want to expand the bracketed factor involving $\tilde{\chi}_{at}$ to various orders in $(\epsilon_0/\epsilon_{host})\tilde{\chi}_{at}$, since this quantity is always small compared to unity. If we follow this conventional definition for $\tilde{\chi}_{at}$, using $\epsilon_0$, we will end up carrying along perpetual factors of $\epsilon_0/\epsilon_{host}$ to various powers in all these expressions.

To avoid this, *we will instead consistently use in this book an alternative nonstandard definition* for $\tilde{\chi}_{at}$ which we obtain by writing

$$\tilde{P}_{at} \equiv \tilde{\chi}_{at}\epsilon_{host}\tilde{E} \qquad \begin{pmatrix}\text{this book's}\\\text{definition}\end{pmatrix}, \qquad (47)$$

so that the *atomic resonance* part of the susceptibility is defined by the expression

$$\tilde{\chi}_{at} \equiv \frac{\tilde{P}_{at}(\omega)}{\epsilon_{host}\tilde{E}(\omega)} \qquad \begin{pmatrix}\text{this book's}\\\text{definition}\end{pmatrix}. \qquad (48)$$

Note that we are not rewriting any laws of electromagnetic theory by doing this—we are merely introducing a slightly unconventional way of defining $\tilde{\chi}_{at}$ for an atomic transition, in which $\epsilon_{host}$ is used as a normalizing constant in the denominator, rather than $\epsilon_0$ as in the standard definition. If we use this definition, as we will from here on, the total electric displacement in any laser material is then given by the simpler form

$$\tilde{D} = \epsilon_{host}\tilde{E} + \tilde{P}_{at} = \epsilon_{host}[1 + \tilde{\chi}_{at}]E \qquad \begin{pmatrix}\text{this book's}\\\text{definition}\end{pmatrix}. \qquad (49)$$

This alternative form avoids the factor of $\epsilon_0/\epsilon_{host}$ in Equation 2.46. For simplicity, from here on we will also drop all the "host" subscripts and simply write $\epsilon_{host}$ as $\epsilon$.

To keep all this straight, just remember: from here on $\epsilon$ is the dielectric constant of the host lattice or dielectric material *without* the laser atoms; whereas $\tilde{\chi}_{at}$, defined according to the alternative definition of Equation 2.48, is the additional (weak) contribution due to the resonant atomic transition in the laser atoms. Of course, if the laser material is a dilute gas with $\epsilon_{host} = \epsilon_0$, there is no difference anyway.

### Resonant Susceptibility: The Resonance Approximation

With this definition we can write the general susceptibility $\tilde{\chi}_{at}$ for the resonant response in a collection of resonant oscillators or atoms by combining Equations 2.36 and 2.48 to obtain

$$\tilde{\chi}_{at}(\omega) \equiv \frac{\tilde{P}_x}{\epsilon \tilde{E}_x} = \frac{Ne^2}{m\epsilon}\frac{1}{\omega_a^2 - \omega^2 + j\omega\Delta\omega_a}. \qquad (50)$$

We have introduced here the important quantity

$$\Delta\omega_a \equiv \gamma + 2/T_2, \qquad (51)$$

which we will shortly identify as the *atomic linewidth* (FWHM) of the atomic resonance. Since both $\gamma$ and $2/T_2$ are always small compared to optical frequencies, this linewidth $\Delta\omega_a$ is very small compared to the center frequency $\omega_a$ for essentially all transitions of interest in lasers—never more than 10% at absolute most, and usually much, much narrower. (In fact, fractional linewidths greater than a fraction of a percent occur in practice only in semiconductor injection lasers and organic dye lasers.)

We are most often interested only in the response of the atoms to signal frequencies $\omega$ that lie within a few linewidths of either side of the resonant frequency $\omega_a$. Within this region we can make what is called the *resonance approximation* by writing

$$\omega^2 - \omega_a^2 = (\omega + \omega_a)(\omega - \omega_a) \approx 2\omega_a(\omega - \omega_a) \approx 2\omega(\omega - \omega_a), \quad (52)$$

so that the frequency-dependent part of the susceptibility expression becomes

$$\frac{1}{\omega_a^2 - \omega^2 + j\omega\Delta\omega_a} \approx \frac{1}{2\omega(\omega_a - \omega) + j\omega\Delta\omega} \approx \frac{1}{2\omega_a(\omega_a - \omega) + j\omega\Delta\omega}. \quad (53)$$

By using this we can then convert Equation 2.50 into the simpler resonant form

$$\tilde{\chi}_{at}(\omega) = \frac{-jNe^2}{m\omega_a\epsilon\Delta\omega_a}\frac{1}{1 + 2j(\omega - \omega_a)/\Delta\omega_a}. \quad (54)$$

It is evident that this response will decrease rapidly compared to its midband value as soon as the frequency detuning $\omega - \omega_a$ becomes more than a few times the linewidth $\pm\Delta\omega_a$; and hence it really does not matter at all whether we use $\omega$ or $\omega_a$ in the denominator in the first part of this expression, so long as we do not tune away from $\omega_a$ by more than, say, $\pm 10\%$.

### The Lorentzian Lineshape

The right-hand part of Equation 2.54 exhibits a very common frequency dependence known as a *complex lorentzian lineshape*. Since we will be seeing this frequency dependence over and over in the remainder of this text, let us gain a little familiarity with its properties.

Suppose that, for simplicity, we define a normalized frequency shift away from line center by

$$\Delta x \equiv 2\frac{\omega - \omega_a}{\Delta\omega_a}, \quad (55)$$

so that $\Delta x = 0$ corresponds to midband and $\Delta x = \pm 1$ corresponds to a frequency shift of half a linewidth away from line center on either side. Then the complex lorentzian lineshape is given by

$$\tilde{\chi}_{at}(\omega) = -j\chi_0''\frac{1}{1 + 2j(\omega - \omega_a)/\Delta\omega_a} = -j\chi_0''\frac{1}{1 + j\Delta x}, \quad (56)$$

where

$$\chi_0'' \equiv \frac{Ne^2}{m\omega_a\epsilon\Delta\omega_a} \quad (57)$$

is the magnitude of the negative-imaginary value at midband.

Readers familiar with Fourier transforms will recognize that this complex lorentzian lineshape is simply the Fourier transform in frequency space of the exponential time decay of the polarization $p_x(t)$. (Whether the $-j$ factor in front of the $1/(1 + j\Delta x)$ frequency dependence is to be considered part of the complex lorentzian lineshape or not is entirely a matter of style.) Note once again that in examining the frequency dependence of lorentzian transitions—for example, in solving some of the problems at the end of this section—the frequency dependence of the constant $\chi_0''$ can be entirely ignored; i.e., it makes no practical difference whether we use $\omega$ or $\omega_a$ in the denominator of Equation 2.57. This constant in front can be treated as entirely independent of frequency within the resonance approximation.

The real and imaginary parts of this complex lorentzian lineshape then have the forms

$$\tilde{\chi}_{at}(\omega) \equiv \chi'(\omega) + j\chi''(\omega) = -\chi_0'' \left[ \frac{\Delta x}{1 + \Delta x^2} + j \frac{1}{1 + \Delta x^2} \right], \tag{58}$$

where $\tilde{\chi}'(\omega)$ and $\tilde{\chi}''(\omega)$ are the real and imaginary parts of this function, as plotted in Figure 2.8. The imaginary part of this response, or $\chi''(\omega)$, has a resonant response curve of the form

$$\chi''(\omega) = -\chi_0'' \frac{1}{1 + [2(\omega - \omega_a)/\Delta\omega_a]^2} = -\chi_0'' \frac{1}{1 + \Delta x^2}. \tag{59}$$

This lineshape is conventionally called the *real lorentzian lineshape*, with a response centered at $\Delta x = 0$ or $\omega = \omega_a$, and with a full width between the half-power points $\omega - \omega_a = \pm\Delta\omega_a$ given by

$$\Delta\omega_a = \gamma + 2/T_2. \tag{60}$$

The linewidth $\Delta\omega_a$ is thus the *full width at half maximum (FWHM) linewidth* of the atomic transition. We will shortly identify $\chi''(\omega)$ as the absorbing (or amplifying) part of the atomic response.

The real part of the lorentzian susceptibility, or $\chi'(\omega)$, has the frequency dependence

$$\chi'(\omega) = -\chi_0'' \frac{2(\omega - \omega_a)/\Delta\omega_a}{1 + [2(\omega - \omega_a)/\Delta\omega_a]^2} = -\chi_0'' \frac{\Delta x}{1 + \Delta x^2}, \tag{61}$$

which has the antisymmetric or roughly first-derivative form shown in Figure 2.8. We will shortly identify this $\chi'(\omega)$ part as the reactive, or phase-shift, or dispersive part of the atomic response.

Note that the literature on atomic transitions and lasers uses many different linewidth definitions for $\Delta\omega$, $\Delta f$, $\Delta\lambda$, etc., which in different publications are sometimes defined as the half widths of resonance lines; sometimes as the full widths, as here; and sometimes even as rms linewidths, or $1/e^2$ linewidths, or other exotic widths. We will be consistent in this text in always using a FWHM definition for any linewidth $\Delta\omega$, $\Delta f$, or $\Delta\lambda$, unless we explicitly say otherwise.

### Magnitude of the Steady-State Atomic Response

Let us emphasize once more that the atomic response of a collection of atoms to an applied signal is *coherent*, in the sense that the steady-state induced polarization $\tilde{P}(\omega)$ follows, in amplitude and time-phase, the driving signal field $\tilde{E}(\omega)$ in the manner described by the complex susceptibility or transfer function $\tilde{\chi}(\omega)$. The susceptibility $\tilde{\chi}(\omega)$ given by Equations 2.50, 2.54 or 2.56 is a dimensionless quantity. We will see later that in essentially every case of interest to us, the numerical value of this quantity is very small compared to unity. Only for very large atomic densities, very strongly allowed transitions, and very narrow linewidths does the numerical magnitude of $\tilde{\chi}$ approach unity; and these conditions are not normally all present at once in laser materials.

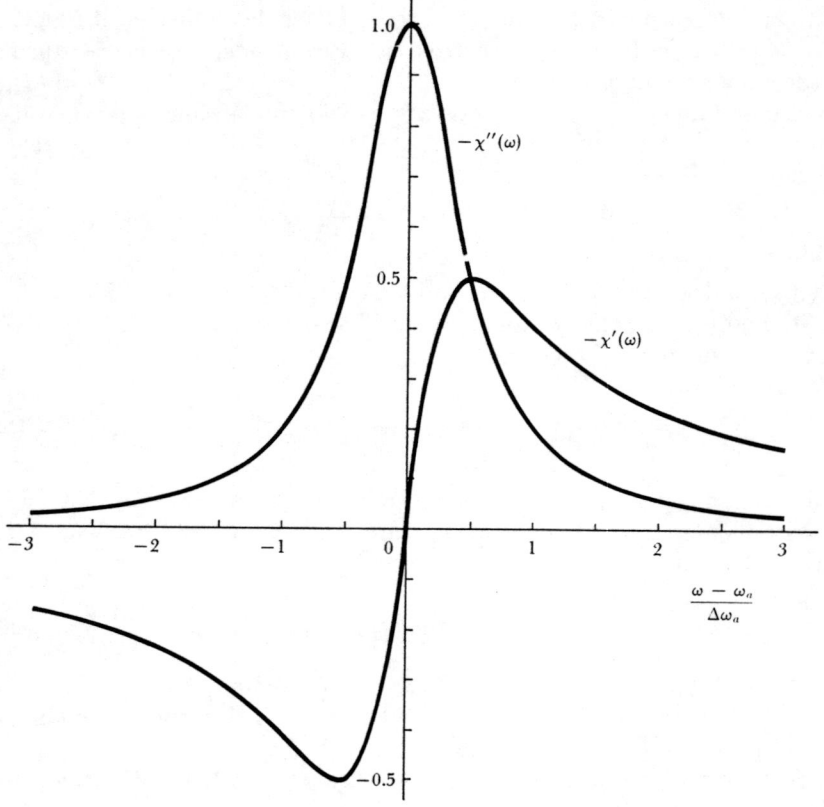

FIGURE 2.8
Real and imaginary parts of the complex lorentzian lineshape.

We might also note that the magnitude of the atomic susceptibility at midband is proportional to the inverse linewidth $1/\Delta\omega_a$, where the linewidth $\Delta\omega_a$ for classical oscillators is given by

$$\Delta\omega_a = \gamma_{\rm rad} + \gamma_{\rm nr} + 2/T_2. \tag{62}$$

As the dephasing time $T_2$ becomes smaller, this linewidth becomes larger, and hence the relative strength of the induced atomic response decreases in direct proportion to the dephasing processes as measured by $2/T_2$.

If there were no dephasing processes, so that $T_2 \to \infty$, then the applied signal field $\tilde{E}$ could drive all the individual atomic dipoles to oscillate completely in phase, and would produce the largest possible induced response, limited only by the dipole energy decay rate. The dephasing processes associated with any finite $T_2$ value, however, operate to "fight" the coherent phasing effects of the applied signal, and to reduce the coherent polarization that can be developed. The usual situation in most (though not all) laser materials is that the $2/T_2$ dephasing term dominates over the energy decay rate $\gamma$; as a result, the applied signal can produce only a small fractional coherent ordering of the dipole oscillation's steady state, working against the randomizing effects of the dephasing processes.

## Problems for 2.4

1. *Radiative decay rate at the He-Ne laser transition frequency.* Evaluate the radiative decay rate and the radiative lifetime for a classical electron oscillator with the same resonance frequency as the He-Ne 6328Å laser line. Also evaluate the atomic linewidth $\Delta\omega_a$ that would apply if this were the only damping or line-broadening effect present. (Note: The actual 6328Å transition in neon is much weaker, i.e., has a much slower radiative decay rate; and other broadening mechanisms, including both collision and doppler-broadening, are present in the real He-Ne laser.)

2. *Classical-mechanics description of power transfer from field to atoms.* Describe, using suitable plots, the magnitude and phase of the induced dipole response $\mu_x(t)$ of a classical electron oscillator to a sinusoidal applied field $\mathcal{E}_x(t)$ as a function of the driving frequency $\omega$, from well below to well above resonance. Using the fact that the work done by a force $f_x(t)$ pushing on an object moving with velocity $v_x(t)$ is given by $dW/dt = f_x(t)v_x(t)$, explain in mechanical terms the steady-state power transfer from the applied signal to the oscillating dipole, both for frequencies near resonance and in the reactive regimes well away from resonance.

3. *Lorentzian lineshape for a resonant electrical circuit.* Show that the electrical impedance $Z(\omega)$ as a function of frequency for a resistance $R$, an inductance $L$, and a capacitance $C$ connected in parallel can also be approximated by a complex lorentzian lineshape.

4. *Lorentzian lineshape locus in the complex plane.* The complex lorentzian susceptibility $\tilde{\chi}(\omega)$ can be plotted as a contour in the complex plane with $\chi'$ and $\chi''$ as the horizontal and vertical axes, respectively, and $\omega$ as a parameter along this contour. Plot a few points to trace the geometric form of this contour. Can you give a simple analytic form for the contour (in the resonance approximation)?

5. *Range of validity for the resonance approximation.* The resonance approximation leading to the lorentzian lineshape for a high-$Q$ classical oscillator is said to be valid "near resonance." How far from resonance on either side can you in fact vary the frequency tuning $\omega - \omega_a$ before the magnitude of the difference between the exact form and the approximate lorentzian form for the complex susceptibility of a classical electron oscillator becomes as large as 10 percent?

6. *Derivative spectroscopy.* Some types of spectrometers used to study atomic resonances give an output signal proportional not to the atomic absorption line $\chi''(\omega)$ itself as a function of $\omega$, but rather to its first derivative $d\chi''(\omega)/d\omega$. This first-derivative curve has two peaks of opposite sign centered about $\omega_a$. Find the spacing $\Delta\omega_{pk}$ between these two peaks in terms of the usual FWHM atomic linewidth $\Delta\omega_a$ for a high-$Q$ lorentzian line.

7. *Overlapping lineshapes: maximally flat condition.* Suppose an atomic medium contains *two* groups of resonant oscillators with the same linewidth, density, and other parameters, but with slightly different resonant frequencies $\omega_{a1}$ and $\omega_{a2}$. Using the resonance approximation for each line, plot the total susceptibilities $\chi'(\omega)$ and $\chi''(\omega)$ versus $\omega$ due to both groups of atoms combined, for frequency separations $\omega_{a2} - \omega_{a1}$ of 0.2, 0.5, 1, 2, and 5 times the linewidth $\Delta\omega_a$. What

frequency separation will cause the first derivative $d\chi'(\omega)/d\omega$ to be exactly zero at the midpoint between the two lines?

---

## 2.5 CONVERSION TO REAL ATOMIC TRANSITIONS

The classical electron oscillator results derived in the preceding sections can be converted into *quantum-mechanically correct* formulas for real atomic transitions in real atoms by making a few simple and almost obvious substitutions. These substitutions are briefly introduced in this section, and then discussed in more detail in the following chapter.

### Substitution of Radiative Decay Rate

The first step in converting from the CEO model to real atomic transitions is to notice a similarity between the constant in front of the classical oscillator susceptibility expression of Equation 2.57 in the preceding section, which has the form

$$\chi_0'' \equiv \frac{Ne^2}{m\omega_a \epsilon \Delta\omega_a}, \tag{63}$$

and the classical oscillator radiative decay rate that we introduced in Equation 2.8, which is given by

$$\gamma_{\text{rad,ceo}} = \frac{e^2 \omega_a^2}{6\pi\epsilon m c^3}. \tag{64}$$

In fact, if we substitute the second of these into the first, we can write the amplitude of the classical oscillator susceptibility at midband in the form

$$\chi_0'' = \frac{3}{4\pi^2} \frac{N\lambda^3 \gamma_{\text{rad,ceo}}}{\Delta\omega_a}. \tag{65}$$

In this form all the atomic and electromagnetic constants appearing in the classical oscillator model (charge $e$, mass $m$, and the dielectric constant $\epsilon$) drop out; and the resulting expression depends only on directly measurable properties of the classical oscillator, namely, the transition wavelength $\lambda$, the density of oscillators $N$, the radiative decay rate $\gamma_{\text{rad,ceo}}$, and the linewidth $\Delta\omega_a$ of the transition itself. This expression is a more fundamental and useful way of writing the susceptibility, since it is now equally valid for either classical oscillators or real atoms, provided only that we use the appropriate values of $\lambda$, $\gamma_{\text{rad}}$, and $\Delta\omega_a$ in each case.

### Introduction of Population Difference

The second and more fundamental step in converting from classical oscillators to real atoms is to notice that the classical electron oscillator response we have derived here is proportional to the number density $N$ of the classical oscillators. But we learned in Chapter 1 that the response on real quantum transitions is proportional to the *population difference density* $\Delta N_{12} = N_1 - N_2$ between the populations (atoms per unit volume) in the lower and upper levels of the atomic transition.

That is, a collection of classical oscillator "atoms" can only *absorb energy*, at least in steady state. Both quantum theory and experiments show, however, that when a signal is applied to a collection of real quantum atoms, the steady-state response is always such that the lower-level atoms *absorb energy* through upward transitions, but the upper-level atoms *emit energy* through downward transitions. The lower-level atoms thus act essentially like conventional classical oscillators, but the upper-level atoms act somehow like "inverted" classical oscillators.

The single most crucial step in converting our classical oscillator results to accurate quantum formulas for real atomic transitions is thus to replace the classical oscillator density $N$ by a quantum population difference $\Delta N_{12} \equiv N_1 - N_2$, where $N_1$ and $N_2$ are the number of atoms per unit volume, or the "level populations," in the lower and upper energy levels. This substitution is the primary point where quantum theory enters the classical oscillator model.

### Quantum Susceptibility Result

If we make both of these substitutions, and also for simplicity leave off all the classical oscillator labels, then the resonant susceptibility expression for either a collection of classical oscillators or a real atomic transition is given by the same expression, namely,

$$\tilde{\chi}_{\text{at}}(\omega) = -j\frac{3}{4\pi^2}\frac{\Delta N \lambda^3 \gamma_{\text{rad}}}{\Delta \omega_a}\frac{1}{1 + 2j(\omega - \omega_a)/\Delta \omega_a}. \tag{66}$$

It will often be convenient to write this expression for the complex lorentzian susceptibility in the form

$$\begin{aligned}\tilde{\chi}_{\text{at}}(\omega) &= -j\chi_0'' \times \frac{1}{1 + 2j(\omega - \omega_a)/\Delta \omega_a} \\ &= -\chi_0'' \left[\frac{2(\omega - \omega_a)/\Delta \omega_a}{1 + [2(\omega - \omega_a)/\Delta \omega_a]^2} + j\frac{1}{1 + [2(\omega - \omega_a)/\Delta \omega_a]^2}\right],\end{aligned} \tag{67}$$

where $-j\chi_0''$ is the given midband susceptibility, with magnitude given by

$$\chi_0'' = \frac{3}{4\pi^2}\frac{\Delta N \lambda^3 \gamma_{\text{rad}}}{\Delta \omega_a}. \tag{68}$$

*This expression then becomes an essentially quantum-mechanically correct expressions for the resonant susceptibility of any real electric-dipole atomic transition, provided simply that we use in these formulas the real (i.e., measured) values of the parameters $\lambda$, $\gamma_{\text{rad}}$, $\Delta \omega_a$, and $\Delta N$ for that particular atomic transition.*

### Discussion

The preceding results thus say that the linear response to an applied signal, as expressed by $\tilde{\chi}(\omega)$, for a collection of classical oscillators or for a real atomic transition, depends only on the following.

- For the classical case the *number of oscillators* $N\lambda^3$, or for the quantum case the *net population difference* $\Delta N \lambda^3$, contained in a volume of one wavelength cubed, where $\lambda \equiv \lambda_0/n$ is the wavelength in the host crystal medium.

- The *radiative decay rate* $\gamma_{\text{rad}}$ characteristic of that particular oscillator, or of that particular atomic transition. This is a very fundamental and important point: different transitions in real atoms will have very different strengths, as measured by their radiative decay rates. We see here that the *induced* or *stimulated* response on each such transition will be directly proportional to the *spontaneous* emission rate on that same transition. Oscillators that radiate strongly also respond strongly.

- The *inverse linewidth* $1/\Delta\omega_a$ of that transition. This says in effect that there is a characteristic area under each such resonance (with a magnitude proportional to $\Delta N \lambda^3 \gamma_{\text{rad}}$). Transitions that are broadened or smeared out by dephasing effects, or by other line-broadening mechanism, then have proportionately less response at line center.

- And finally, there is the *complex lorentzian lineshape* that gives the frequency variation of the atomic response as a system is tuned on either side of the resonance frequency.

Each of these points is fundamental, and will reoccur many times in discussions of real atomic responses later on.

### The Quantum Polarization Equation of Motion

We can also make the same substitutions in the differential equation of motion for $p(t)$ in the time domain, and rewrite Equation 2.27 in the form

$$\frac{d^2 p_x(t)}{dt^2} + \Delta\omega_a \frac{dp_x(t)}{dt} + \omega_a^2 p_x(t) = \frac{3\omega_a \epsilon \lambda^3 \gamma_{\text{rad}}}{4\pi^2} \Delta N(t) \mathcal{E}_x(t). \tag{69}$$

after which *this also becomes an essentially quantum-mechanically correct equation for the induced polarization response $p(t)$, or at least for its quantum expectation value, on a real atomic transition.* We will make further use of this quantum equation in later chapters.

Notice that after making this conversion to the quantum case, we now have a situation in which the population difference $\Delta N(t)$ on the right-hand side of the equation may itself be an explicit function of time, as a result of stimulated transitions, pumping effects, and/or relaxation processes, rather than being a constant value $N$ as in the classical case. This makes the quantum equation essentially nonlinear, as contrasted with the essentially linear character of the classical oscillator model.

We will see later that in most cases of interest in lasers, the rate of change of the population difference $\Delta N(t)$ is slowl compared to the inverse of the atomic linewidth $\Delta\omega_a$. This represents the so-called *rate-equation limit*, in which we can validly solve the polarization differential equation of motion in a linear fashion, just as we did in this chapter, and thereby obtain the linear resonant sinusoidal susceptibility given above.

There are other situations, however, in which the applied signal becomes strong enough (or the transition linewidth is narrow enough) that we move into a large-signal regime where the time-variation of $\Delta N(t)$ does become important.

In this large-signal regime it is no longer possible to solve Equation 2.69 as a simple linear differential equation; and hence the linear susceptibility $\tilde{\chi}(\omega)$ is no longer an adequate description of the atomic response. We must instead solve the nonlinear polarization equation for $p(t)$, Equation 2.69, together with a separate rate equation for the time variation of $\Delta N(t)$ that we will derive in a later chapter, in order to get the full large-signal atomic behavior. The result in the large-signal limit is a more complex form of behavior, commonly referred to as *Rabi flopping behavior*, which we will describe in more detail in a later chapter.

The polarization equation of motion is thus more general than the sinusoidal susceptibility results, which are valid only within the so-called "rate-equation limits." Most laser devices in fact operate in the rate-equation regime; but there are also more complex large-signal phenomena, often referred to as "coherent pulse phenomena," which occur only in the Rabi-frequency regime. Such coherent pulse effects can be demonstrated experimentally using appropriate high-power laser beams and narrow-line atomic transitions.

### Additional Substitutions

Let us finally give a brief but complete list of all the other steps that are necessary to convert the classical oscillator results derived in this chapter into the completely correct quantum results for any real electric-dipole atomic transition. Converting the classical oscillator formulas to apply to a real atomic transition requires the following steps.

1. *Transition frequency.* Any single kind of atom will of course have numerous resonant transitions among its large number of quantum energy levels $E_i$. The classical electron oscillator can model only one such transition between two selected levels, say $E_i$ and $E_j > E_i$, at a time. To treat several different signals applied to different transitions at different frequencies simultaneously, we must in essence employ multiple CEO models, one for each transition. The level populations $N_i(t)$ in the different levels involved are then interconnected by rate equations, as we will discuss in a later section.

The classical resonance frequency $\omega_a$ must be replaced by the actual transition frequency $\omega_{ji}$ in the real atom, i.e.,

$$\omega_a \Rightarrow \omega_{ji} = \frac{E_j - E_i}{\hbar}. \tag{70}$$

The actual transition wavelength $\lambda$, measured in the laser host medium, must of course also be used.

2. *Atomic population difference.* The population difference must be replaced by the population difference on that particular transition, i.e.,

$$\Delta N \Rightarrow \Delta N_{ij} = N_i - N_j, \tag{71}$$

where $N_i$ is the lower-level and $N_j$ the upper-level population density.

3. *Radiative decay rate.* The radiative decay rate $\gamma_{\text{rad}}$ must be replaced by a quantum radiative decay rate appropriate to the specific $i \to j$ transition under consideration, i.e.,

$$\gamma_{\text{rad}} \Rightarrow \gamma_{\text{rad},ji}. \tag{72}$$

Every real atomic transition between two energy levels $E_i$ and $E_j$ will have such a characteristic spontaneous-emission rate, which is the same thing as the Einstein A coefficient on that transition, i.e., $\gamma_{\text{rad},ji} \equiv A_{ji}$.

4. *Transition linewidth.* The linewidth $\Delta\omega_a$ must be replaced by a linewidth $\Delta\omega_{a,ij}$ characteristic of the real transition in the real atoms, i.e.,

$$\Delta\omega_a \Rightarrow \Delta\omega_{a,ij}. \tag{73}$$

This involves using real-atom values for the linewidth contributions of both the energy decay rate, i.e., $\gamma_{ij}$, and the dephasing time $T_{2,ij}$ on that particular transition, as well as any other broadening mechanisms that may be present. We will say more later about what these real-atom values mean and how they are obtained. Note also that different $i \to j$ transitions in a given atom may have quite different linewidths $\Delta\omega_{a,ij}$.

5. *Transition lineshape.* More generally, the complex lineshape of $\tilde{\chi}_{\text{at}}(\omega)$ for a real atomic transition may not be exactly lorentzian, although many real atomic transitions are. It may be necessary for some transitions to replace the lorentzian frequency dependence with some alternative lineshape or frequency dependence for $\tilde{\chi}(\omega)$. Whatever this lineshape may be, however, the real and imaginary parts $\chi'(\omega)$ and $\chi''(\omega)$ near resonance will almost always have lineshapes much like those in Figure 2.8.

6. *Tensor properties.* We assumed in previous sections a classical oscillator model that was linearly polarized along the $x$ direction. We have thus derived essentially only one tensor component of the linear susceptibility, that is, the component defined by

$$\tilde{P}_x(\omega) = \tilde{\chi}_{xx}(\omega)\epsilon\tilde{E}_x(\omega). \tag{74}$$

The response of a real atomic transition may involve a more complicated and anisotropic (though still linear) response of all three vector components $\boldsymbol{P}(\omega) = [\tilde{P}_x, \tilde{P}_y, \tilde{P}_z]$ to the vector field components $\boldsymbol{E}(\omega) = [\tilde{E}_x, \tilde{E}_y, \tilde{E}_z]$. The susceptibility $\tilde{\chi}(\omega)$ must then be replaced by a *tensor susceptibility* $\boldsymbol{\chi}(\omega)$, i.e.,

$$\tilde{\chi}(\omega) \Rightarrow \boldsymbol{\chi}(\omega). \tag{75}$$

where $\boldsymbol{\chi}(\omega)$ is a $3 \times 3$ susceptibility tensor defined by

$$\boldsymbol{P}(\omega) = \boldsymbol{\chi}(\omega)\epsilon\boldsymbol{E}(\omega). \tag{76}$$

We discuss the resulting tensor properties of real transitions in more detail later.

7. *Polarization properties.* The magnitude of the response of an atomic transition to an applied signal in the tensor case will also depend on how well the applied field polarization $\boldsymbol{E}$ lines up or overlaps with the tensor polarization needed for optimum response from the atoms. If the applied field is not properly polarized or oriented with respect to the atoms, the observed response will be reduced. We can account for this by replacing the numerical factor of 3 that appears in the susceptibility expression with a factor we call "three star," i.e.,

$$\frac{3}{4\pi^2} \Rightarrow \frac{3^*}{4\pi^2}, \tag{77}$$

where the numerical value of this $3^*$ factor (to be explained in more detail in the following chapter) is $0 \leq 3^* \leq 3$.

8. *Degeneracy effects.* What appears to be a single quantum energy level $E_i$ may be in many real atomic systems some number $g_i$ of *degenerate energy levels*, i.e., separate and quantum-mechanically distinct energy states all with the same or very nearly the same energy eigenvalue $E_i$. To express the net small-signal response summed over all the distinct but overlapping transitions between these degenerate sublevels, the population-difference term $N_i - N_j$ for systems with degeneracy must be replaced by

$$\Delta N_{ij} = (N_i - N_j) \Rightarrow \Delta N_{ij} = (g_j/g_i) N_i - N_j, \qquad (78)$$

where $E_i$ is the lower and $E_j$ the upper group of levels; $g_i$ and $g_j$ are the statistical weights or degeneracy factors of these lower and upper groups of levels; and $N_i$ and $N_j$ are the *total* population densities in the degenerate groups of lower and upper levels.

9. *Inhomogeneous broadening.* Finally, additional line-broadening and line-shifting mechanisms, the so-called "inhomogeneous" broadening mechanisms, will often broaden and change the lineshapes of real atomic resonances, over and above the broadening due to energy decay and dephasing as expressed in the linewidth formula $\Delta\omega_a = \gamma + 2/T_2$. The *homogeneous linewidth* $\Delta\omega_a$ then gets replaced (at least for certain purposes) by an *inhomogeneous linewidth* $\Delta\omega_d$, i.e.,

$$\Delta\omega_a \Rightarrow \Delta\omega_d. \qquad (79)$$

When this happens, the lineshape often gets changed also, from lorentzian to something more like gaussian in shape; and the $3^*/4\pi^2$ numerical factor in front of the susceptibility expression may be increased by $\approx 50\%$. What is meant by inhomogeneous broadening, and how these additional broadening mechanisms affect real atomic resonances, is described in the final section of the following chapter.

Further details on all the topics introduced in this section are given in the next chapter. With these conversion factors included, the basic polarization equation of motion and the resulting linear susceptibility formula for a real homogeneously broadened atomic transition become quantum-mechanically and quantitatively correct for real quantum atomic transitions.

## REFERENCES

An interesting historical review of early attempts to develop purely classical theories of atomic and molecular absorption of radiation, and of their connections to quantum mechanics, including dipole moments, collision broadening, and sum rules, is given by J. H. Van Vleck and D. L. Huber, "Absorption, emission, and linebreadths: A semihistorical perspective," *Rev. Mod. Phys.* **49**, 939–959 (October 1977).

---

Problems for 2.5

1. *Classical oscillator model for the index of refraction in gases.* Can the classical oscillator model be used to explain not only the resonance behavior of atomic

transitions, but also the low-frequency dielectric properties of gases and solids? For example, the *CRC Handbook of Chemistry and Physics* gives the values shown in the following table for the low-frequency relative dielectric constant $\epsilon/\epsilon_0$ for some simple gases at 0°C and atmospheric pressure (760 torr), measured from dc through radio and microwave frequencies, as well as the optical index of refraction $n$ measured across the visible region.

| Gas | $(\epsilon/\epsilon_0 - 1) \times 10^4$ | $(n - 1) \times 10^4$ |
| --- | --- | --- |
| He | 0.6 | 0.36 |
| Ar | 5.1–5.4 | 2.8 |
| $H_2$ | 2.5 | 1.3–1.4 |
| $CO_2$ | 9.2 | 4.5 |
| Air | 5.3 | 2.9 |

Can these values be explained, at least as to order of magnitude, using a simple CEO model for the atoms involved?

To answer this question, we must realize that in simple gases such as He, the strongest upward electric-dipole transition from the atomic ground state is usually to some first excited level that is located well into the ultraviolet. Such an atom can then be modeled for many purposes by a classical electron oscillator with a resonance frequency $\omega_a$ located somewhere in the ultraviolet. The low-frequency dielectric constant, as well as the index of refraction through the visible region, are then both caused primarily by the low-frequency "tail" of this first strong ultraviolet transition, with both of these quantities being only very slightly larger than unity in numerical value.

To demonstrate this analytically, suppose a dc electric field $E_0$ is applied to a collection of such classical oscillators with the same density $N$ as a standard gas at room temperature and atmospheric pressure (which implies a density $N \approx 2.5 \times 10^{19}$ atoms/cm$^3$). Assume these oscillators have a resonance frequency $\omega_a$ corresponding to $\lambda_a = 100$ nm, which is in the vacuum ultraviolet. Using the CEO model, what will be the induced dc polarization $P_0/E_0$ in this gas? What will be its dc dielectric constant $\epsilon$ compared to the value of $\epsilon_0$ for a vacuum, and by how much will its index of refraction $n = \sqrt{\epsilon/\epsilon_0}$ differ from unity? Why does argon have a larger value than helium, and why do the molecular gases also have significantly larger values?

2. *Classical oscillator model for the index of refraction in solids.* Let us now apply the same argument as in Problem 2 to a solid material. The host crystal in a typical solid-state laser material, for example, itself consists of atoms, and these atoms (in the crystalline form) usually have their lowest atomic resonance frequency in the near ultraviolet. Can the CEO model also give a reasonable explanation of the dielectric polarization properties of the host laser material itself, independent of any laser atoms that may be present in the material?

To test this, evaluate the dielectric polarization $P$ and the relative dielectric constant $\epsilon_{\text{host}}/\epsilon_0$ at visible and near-infrared wavelengths for a medium consisting of a collection of classical oscillators, if the classical oscillators have a resonance frequency $\omega_a$ in the near ultraviolet, say, at $\lambda = 300$ nm (which is not an unrealistic value for the band edge or ultraviolet edge in typical solid materials). Find the numerical value of this relative dielectric constant, assuming the oscillators have a density $N$ comparable to typical solid densities, e.g., $10^{22}$ atoms/cm$^3$.

# 3
# ELECTRIC-DIPOLE TRANSITIONS IN REAL ATOMS

In the previous chapter we developed the classical electron oscillator model for an atomic transition, and showed how it could lead to quantum-mechanically correct expressions for the equation of motion and for the resonant susceptibility on a single atomic transition in a real quantum atom. In this chapter we continue this discussion to show how, with some simple extensions, this same purely classical model can explain even the most complex quantum-mechanical aspects of real atomic transitions. We also give some typical numerical values and experimental examples of these properties in real laser transitions.

## 3.1 DECAY RATES AND TRANSITION STRENGTHS IN REAL ATOMS

This section discusses in more detail the energy decay rates and the transition strengths of real atomic transitions, and their relationship to the purely classical oscillator model introduced in Chapter 2.

### Energy Decay Processes in Real Atoms

Real atoms of course have a large number of quantum energy levels, with many transitions and decay rates among these levels. The atoms in an upper energy level $E_j$ in a collection of real atoms will relax to many different lower levels $E_i$ via both radiative and nonradiative decay mechanisms, as illustrated in Figure 3.1. The total rate at which atoms will decay from an upper energy level $E_j$ through all downward relaxation paths may be expressed by a "rate equation" of the form

$$\frac{dN_j}{dt} = -\sum_{E_i < E_j} \gamma_{ji} N_j = -\gamma_j N_j = -N_j/\tau_j, \qquad (1)$$

where $\tau_j$ is the total lifetime of the excited state $E_j$, and $\gamma_j$ is its total decay rate. The total decay rate $\gamma_j$ is given by the sum over all the downward decay

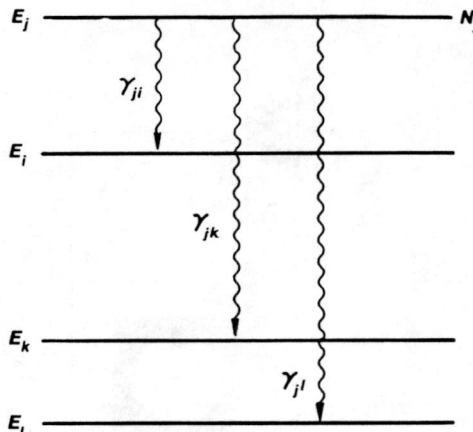

FIGURE 3.1
Downward relaxation of atoms from an upper energy level.

paths, i.e.,

$$\gamma_j \equiv \frac{1}{\tau_j} = \sum_{E_i < E_j} \gamma_{ji} = \sum_{E_i < E_j} [\gamma_{\text{rad},ji} + \gamma_{\text{nr},ji}], \quad (2)$$

so that this sum includes both radiative and nonradiative rates to all lower levels $E_i < E_j$.

In the absence of any applied signals, the population $N_j(t)$ of the upper level will thus decay with time in the exponential form

$$N_j(t) = N_j(t_0)e^{-\gamma_j(t-t_0)} = N_j(t_0)e^{-(t-t_0)/\tau_j}. \quad (3)$$

The decay rate $\gamma_j$ given by these equations is the quantum analog for level $E_j$ of the energy decay rate $\gamma$ in the classical oscillator model.

### Fluorescent Lifetime Measurements

The lifetime $\tau_j$ of an upper energy level can be measured by observing the fluorescent emission from the upper level $E_j$ to any other lower level $E_i$ immediately after a short pulse of pumping light applied to a solid-state laser material, or a short current pulse sent through a gaseous atomic system, has lifted an initial number of atoms up into the upper level. Figure 3.2 illustrates this kind of fluorescent lifetime measurement on a ruby sample, using a stroboscopic light source that produces repeated pumping pulses a few tens of $\mu$s long, and an optical filter that blocks most of the excitation light, so that only the exponentially decaying ruby fluorescence ($\tau \approx 4.3$ ms) reaches the detector.

The measured intensity of the fluorescent emission on some specific $j \rightarrow i$ transition will be proportional to the radiative decay rate $\gamma_{\text{rad},ji}$ on that transition and to the upper-level population as a function of time, i.e.,

$$I_{\text{fl}}(t) = \text{const} \times \gamma_{\text{rad},ji} N_j(t). \quad (4)$$

Since the upper-level population $N_j(t)$ decays with an exponential decay rate equal to the total decay rate $\gamma_j$, the measured exponential behavior for the fluorescent emission will be like

$$I_{\text{fl}}(t) = \text{const} \times N_j(t) = \text{const} \times e^{-t/\tau_j}. \quad (5)$$

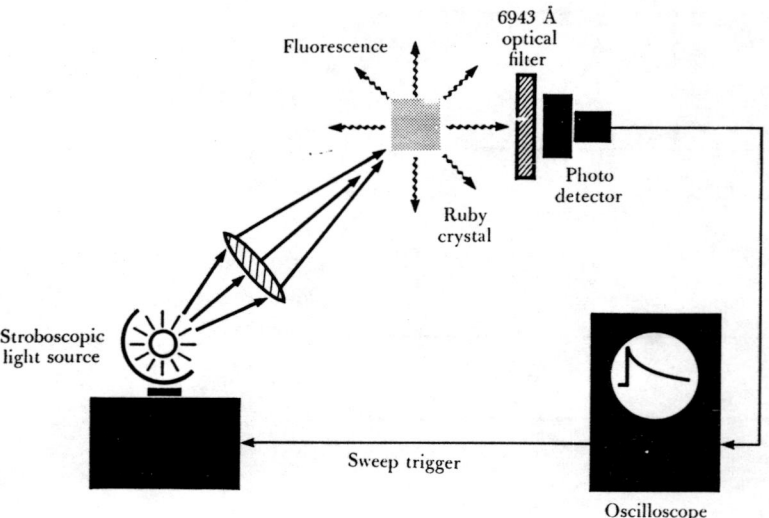

**FIGURE 3.2**
Fluorescent lifetime measurement using a pulsed light source.

A fluorescence decay measurement thus measures only the *total lifetime* $\tau_j$ of the upper level $E_j$, not the radiative decay rate $\gamma_{\text{rad},ji}$ on any individual transition (even though it is this radiative decay rate that produces the observed fluorescent emission).

### Nonradiative Decay Rates

It is important to distinguish between the radiative and the nonradiative parts of the total energy decay rate $\gamma_{ji}$ on each downward transition. The total decay rate, just as for a classical oscillator, is the sum of both mechanisms; so the total decay rate on the $j \rightarrow i$ transition is

$$\gamma_{ji} = \gamma_{\text{rad},ji} + \gamma_{\text{nr},ji}. \tag{6}$$

The radiative part of this decay represents spontaneous emission of electromagnetic radiation, which is physically the same thing as fluorescence. Radiative decay is always present (although sometimes very weak) on any real atomic transition.

The nonradiative part of the decay represents loss of energy from the atomic oscillations into heating up the immediate surroundings in all other possible ways, such as into inelastic collisions, collisions with the laser tube walls, lattice vibrations, and so forth. Nonradiative decay may or may not be significant on any specific transition, and its magnitude may change greatly for different local surroundings of the atoms (e.g., the nonradiative decay rate can be quite different for the same solid-state laser ion in different host lattices).

### Purely Radiative Decay in Real Atoms

The radiative part of the total decay rate on a real atomic transition has a very close analogy to the radiative decay rate of a classical oscillator. An oscillating atom, like an oscillating classical dipole, will radiate electromagnetic energy (photons) at its discrete oscillation frequencies; this radiation will decrease or decay with time; and many real atomic transitions will radiate with an electric-dipole-like radiation pattern.

The radiative decay rate for a real atomic transition is exactly the same thing as the *Einstein A coefficient* for that transition, i.e., $\gamma_{\text{rad},ji} \equiv A_{ji}$, where $E_j$ is the upper and $E_i$ the lower energy level involved. The numerical value of $A_{ji}$ or $\gamma_{\text{rad},ji}$ on any real atomic transition is given by the quantum-mechanical integral

$$A_{ji} = \frac{8\pi^2}{\epsilon\hbar\lambda^3} \left| \iiint \psi_j^*(r)\, er\, \psi_i(r)\, dr \right|^2, \tag{7}$$

which involves the dipole moment operator $er$ and the product of the quantum wave functions of the two quantum states involved. Hence the quantum radiative decay rate for a real atomic transition can in principle always be calculated (though not always easily) if the quantum wave functions $\psi_i$ and $\psi_j$ of the two levels are known. Calculated values for simpler atoms and molecules can also be found in handbooks and in the literature.

We pointed out earlier that the classical electron oscillator has a radiative decay rate given by

$$\gamma_{\text{rad,ceo}} = \frac{e^2 \omega_a^2}{6\pi\epsilon mc^3} \approx 2.47 \times 10^{-22} \times n f_a^2, \tag{8}$$

where $n$ is the index of refraction of the medium in which the oscillator is imbedded, and the oscillation frequency is measured in Hz ($\equiv$cycles/second). A useful rule of thumb is that the purely radiative lifetime for a classical oscillator is approximately given by

$$\tau_{\text{rad,ceo}}(\text{ns}) \approx \frac{45 \times [\lambda_0(\text{microns})]^2}{n}, \tag{9}$$

where $n$ is again the index of refraction and $\lambda_0$ is the free-space wavelength in $\mu$m. (These two equations are among the few formulas in this book where an index of refraction term is explicitly needed.) For example, at the visible wavelength $\lambda_0 = 500$ nm or 0.5 $\mu$m, the classical oscillator lifetime is $\tau_{\text{rad,ceo}} \approx 11$ ns, or $\gamma_{\text{rad,ceo}} \approx 10^8$ sec$^{-1}$. Note also the wavelength-squared dependence of this lifetime: infrared oscillators will have substantially longer lifetimes than UV or especially X-ray oscillators.

### Oscillator Strength

It is then a general rule that the radiative decay rate for any real atomic or molecular transition will always be slower than, or at best comparable to, the radiative decay rate for a classical oscillator at the same frequency, so that $\gamma_{\text{rad},ji} \leq \gamma_{\text{rad,ceo}}$, or $\tau_{\text{rad},ji} \geq \tau_{\text{rad,ceo}}$. We can also recall that the induced response to an applied signal of either a real atomic transition or a classical electron oscillator will be directly proportional to the radiative decay rate $\gamma_{\text{rad}}$, with essentially the same proportionality constant in each case.

Because of this, it has become conventional to define a dimensionless *oscillator strength* as a measure of the strength of the response on a real atomic transition relative to the response of a classical electron oscillator at the same frequency.

## CHAPTER 3: ELECTRIC-DIPOLE TRANSITIONS IN REAL ATOMS

This oscillator strength is defined formally for a transition from level $j$ down to level $i$ by

$$\mathcal{F}_{ji} \equiv \frac{\gamma_{\text{rad},ji}}{3\gamma_{\text{rad,ceo}}} = \frac{\tau_{\text{rad,ceo}}}{3\tau_{\text{rad},ji}}. \tag{10}$$

A factor of 3 appears in this definition because of the polarization properties of real atoms compared to classical oscillators, in a fashion which will emerge later.

TABLE 3.1
Typical Radiative Decay Rates

| Transition | Wavelength | Radiative decay rate | Oscillator strength | Comments |
|---|---|---|---|---|
| *Atomic sodium resonance lines:* | | | | |
| $3s \to 3p$ | 589 nm | $6.3 \times 10^7$ s$^{-1}$ (1.6 ns) | 0.33 | Strong sodium $D$ line |
| $3s \to 4p$ | 330 nm | $2.9 \times 10^6$ s$^{-1}$ (350 ns) | 0.0047 | Weaker UV transition |
| *He-Ne laser transitions:* | | | | |
| $3s_2 \to 2p_4$ | 633 nm | $1.4 \times 10^6$ s$^{-1}$ (0.7 µs) | 0.0084 | Red laser line |
| $2s_2 \to 2p_4$ | 1.153 µm | $4.4 \times 10^6$ s$^{-1}$ (0.23 µs) | 0.09 | Near IR laser |
| $3s_2 \to 3p_4$ | 3.392 µm | $9.6 \times 10^5$ s$^{-1}$ (1.04 µs) | 0.17 | Middle IR laser |
| *Selenium quasi forbidden laser lines:* | | | | |
| $^1S_0 \to {}^3P_1$ | 489 nm | $7.7$ s$^{-1}$ (130 ms) | $3 \times 10^{-8}$ | Magnetic-dipole transition |
| $^1S_0 \to {}^1D_2$ | 777 nm | $2.3$ s$^{-1}$ (430 ms) | $2 \times 10^{-9}$ | Electric-quadrupole |
| *Neodymium YAG laser transition:* | | | | |
| $^4F_{3/2} \to {}^4I_{3/2}$ | 1.064 µm | $820$ s$^{-1}$ (1.22 ms) | $\approx 8 \times 10^{-6}$ | Measured $\tau_2$ is 230 µs |
| *Ruby laser transition:* | | | | |
| $^2E \to {}^4A_2$ | 694 nm | $230$ s$^{-1}$ (4.3 ms) | $\approx 10^{-6}$ | Decay is almost purely radiative |
| *Rhodamine 6G dye laser transition:* | | | | |
| $S_1 \to S_0$ | 620 nm | $3 \times 10^8$ s$^{-1}$ (3.3 ns) | $\approx 1.1$ | Decay is almost purely radiative |

Some typical oscillator strengths for real atomic transitions are given in Table 3.1. Note that strongly allowed transitions starting from the ground level of a simple atom in a gas to the first excited level of opposite parity—for example, the $3s \rightarrow 3p$ transition in Na, or the $2s \rightarrow 2p$ transition in a Li atom—have oscillator strengths very close to unity, and hence radiative decay rates close to the classical oscillator values. These transitions are sometimes called the *resonance lines* of the atoms, since they show up very strongly in both the spontaneous emission and the absorption spectra of these atoms. Other allowed electric-dipole transitions in the same atoms may be from $10^{-2}$ to $10^{-5}$ times weaker, and magnetic-dipole and electric-quadrupole transitions may have oscillator strengths of $\mathcal{F} \approx 10^{-7}$ or smaller. Laser transitions in solids or in gaseous molecules typically have similarly weak oscillator strengths, whereas the strong visible singlet-to-singlet transitions in organic dye molecules, such as the Rhodamine 6G dye laser molecule, may have oscillator strengths near unity, and hence radiative decay rates close to the classical oscillator value (e.g., radiative decay times of several nanoseconds).

A strongly allowed atomic transition with oscillator strength of the order of unity will thus have a stimulated response to an applied signal of the same magnitude as a classical electron oscillator at the same frequency. Very weakly allowed atomic transitions, on the other hand, may have an oscillator strength or response ratio as small as $\mathcal{F} \approx 10^{-6}$ to $10^{-7}$ times weaker. So-called "forbidden transitions," or atomic transitions on which virtually no response can be obtained, will have $\gamma_{\text{rad},ji} \ll \gamma_{\text{rad,ceo}}$ and hence $\mathcal{F}_{ji} \rightarrow 0$ in principle, although in fact the decay rate is never absolutely zero.

### Sum Rules, and Oscillator Strengths for Degenerate Transitions

When the upper and lower energy levels are degenerate, with degeneracy factors $g_j$ and $g_i$ (to be explained in a later section), the upward and downward oscillator strengths for a given transition are usually defined more precisely by

$$\mathcal{F}_{ji}\big|_{\text{down}} = -\frac{\gamma_{\text{rad},ji}}{3\gamma_{\text{rad,ceo}}} \quad \text{and} \quad \mathcal{F}_{ij}\big|_{\text{up}} = +\frac{g_j}{g_i}\frac{\gamma_{\text{rad},ji}}{3\gamma_{\text{rad,ceo}}}. \tag{11}$$

With these more precise definitions also go *quantum-mechanical sum rules*, which say that the numerical sum of the oscillator strengths $\sum_{j \neq i} \mathcal{F}_{ji}$ (including sign) from a given level $E_j$ to all other levels above and below it in the same atom has some simple value, which is usually close to unity.

### Example: The Nd:YAG Laser Transition

The 1.06 $\mu$m transition in the Nd:YAG laser is not only of great practical importance, but can provide a good illustration of many of the practical factors that determine the radiative decay rate and the oscillator strength for a real atomic transition.

The solid arrow in Figure 3.3 shows the strong laser transition at $\lambda_0 = 1.0642$ $\mu$m on the $^4F_{3/2}$ to $^4I_{11/2}$ group of transitions in Nd:YAG. (The dashed lines on the left in this figure indicate other transitions near 1.35 $\mu$m and 880 nm on which useful laser oscillation can also be obtained; the transitions from the $^4F_{3/2}$ to $^4I_{15/2}$ levels, with wavelengths near 1.8 $\mu$m, are very weak and oscillate only with difficulty if at all.)

The measured fluorescent lifetime of the $^4F_{3/2}$ upper energy level (call this level $E_2$) in this material is $\tau_2 \approx 230$ $\mu$s; so the total decay rate for this compound

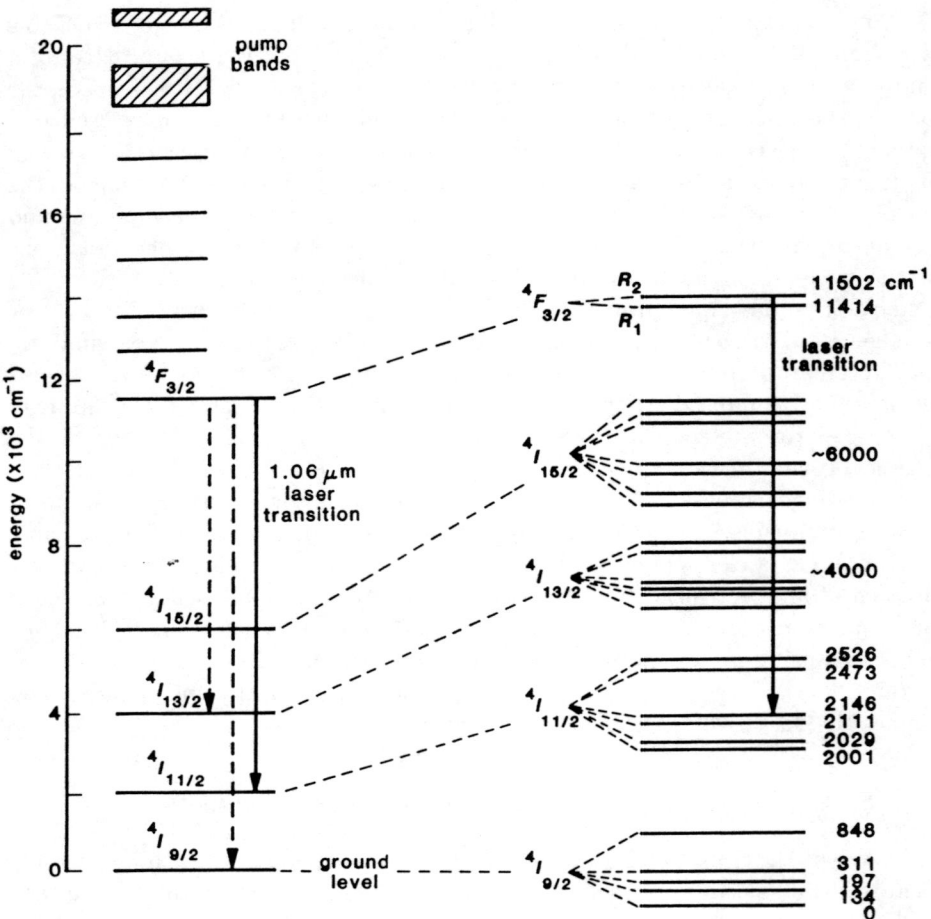

FIGURE 3.3
Quantum-mechanical energy levels of the $Nd^{3+}$ ion in a Nd:YAG laser crystal.

level or group of levels is $\gamma_2 = \gamma_{\text{rad}} + \gamma_{\text{nr}} \approx 4350$ s$^{-1}$. The measured quantum efficiency for this level, however, defined as the ratio of radiative decay (photons emitted) to total decay (i.e., total atoms relaxing down) turns out to have an experimental value

$$\frac{\text{radiative decay rate}}{\text{total decay rate}} \equiv \frac{\gamma_{\text{rad}}}{\gamma_{\text{rad}} + \gamma_{\text{nr}}} \approx 0.56; \tag{12}$$

so the purely radiative decay rate is $\gamma_{\text{rad}} \approx 0.56 \times 4350 \approx 2435$ s$^{-1}$. (The quantum efficiency is measured by shining a calibrated light source onto the crystal, and making a difficult measurement of the total number of input photons absorbed compared to total fluorescent photons emitted.)

The upper level $E_2$ in Nd:YAG really consists, however, of two distinct but closely spaced and partially overlapping levels (call them $E_{2a}$ and $E_{2b}$), which are sometimes called the $R_1$ and $R_2$ levels, and which have an energy spacing of $\approx 80$ cm$^{-1}$. The upper level $E_{2b}$ is the actual upper laser level. These two levels at room temperature will have Boltzmann population ratios $N_{2b}/N_2 \approx 0.4$ and $N_{2a}/N_2 \approx 0.6$, and will be held to these ratios by fast relaxation processes between the two levels. Both of these levels will then radiate spontaneously with different strengths to six different lower levels; so there are actually 12 closely spaced fluorescent lines from the two upper levels to the six lower levels in the 1.06 μm group, with the relative strengths of these lines varying by more than an order of magnitude.

The branching ratio, or the amount of spontaneous radiation on the actual 1.0642 μm laser transition, relative to the total radiative emission from both $^4F_{3/2}$ levels to all lower levels, has been measured to be

$$\frac{\gamma_{\text{rad}}(1.0642 \text{ μm laser line}) \times N_{2b}}{\gamma_{\text{rad}}(\text{all } 1.06 \text{ μm lines}) \times N_2} \approx 0.135. \tag{13}$$

Hence we can finally deduce that the purely radiative decay rate for the isolated YAG laser transition by itself is

$$\gamma_{\text{rad}}(1.0642 \text{ μm}) \approx (0.135/0.40) \times 2.435 \times 10^3 \approx 820 \text{ sec}^{-1}. \tag{14}$$

This corresponds to a purely radiative lifetime of $1/820$ sec $\approx 1.22$ ms (to be compared to the measured fluorescent lifetime of 230 μs).

The numbers quoted here represent a current best estimate, at the time of writing, for the value of $\gamma_{\text{rad},ji}$ that should be used in formulas for the response on this particular Nd:YAG laser transition. However, even in a system as heavily studied as Nd:YAG, these numbers are uncertain, largely because of the experimental difficulties of measuring accurately such quantities as the branching ratio and the absolute fluorescent quantum efficiency. There is no observable physical quantity anywhere in this system that actually decays with this radiative lifetime of 1.22 ms.

## REFERENCES

The concept of an effective oscillator strength for each real atomic transition traces back at least to R. Ladenburg in *Zeitschrift für Physik* **4**, 451 (1921). For early but still interesting reviews of this subject, see S. A. Korff and G. Breit, "Optical dispersion," *Rev. Mod. Phys.* **4**, 471–502 (July 1932); or R. Ladenburg, "Dispersion in electrically excited gases," *Rev. Mod. Phys.* **5**, 243–256 (October 1933).

Calculations of oscillator strengths or decay rates for real atomic transitions in real environments are very complex. You may have to search a wide variety of scientific literature to find whatever numbers are available for any specific laser system. In practice, these rates are either measured or simply estimated. Tabulations of oscillator strengths and radiative decay rates for individual transitions in isolated atoms (at least the simpler ones) are published in such references as the National Bureau of Standards' volumes on *Atomic Transition Probabilities, Volume I: Hydrogen Through Neon*, by W. L. Wiese, M. W. Smith, and B. M. Glennon, and *Volume II: Sodium Through Calcium*, by Wiese, Smith, and B. M. Miles, available from the U. S. Government Printing Office.

Atomic transitions in isolated atoms are also of great interest to plasma physicists and astrophysicists, and the literature in these fields includes many useful references. Two outstanding examples are H. R. Griem, *Plasma Spectroscopy* (McGraw-Hill, 1964); and the extensive compilation of formulas and data in C. W. Allen, *Astrophysical Quantities* (London: Athlone Press, 1973).

A good example of how oscillator strengths and transition probabilities are calculated for a rare-earth ion in various crystals is M. J. Weber *et al.*, "Optical transition probabilities for trivalent holmium in $LaF_2$ and $YAlO_3$," *J. Chem. Phys.* **57**, 11–16

(July 1, 1972). The oscillator strengths for the various transitions of this ion all turn out to have values of $10^{-6}$ to $10^{-7}$, typical of such rare-earth ions.

As we said, the numbers given in the literature for the basic properties of the Nd:YAG laser transition are by no means all in agreement, despite the widespread use of this laser material. The most extensive and detailed review of all aspects of the Nd:YAG laser is probably the chapter on "Progress in Nd:YAG Lasers" by H. G. Danielmeyer in Vol. 4 of *Lasers: A Series of Advances*, edited by A. K. Levine and A. J. DeMaria (Marcel Dekker, 1976), pp. 1–71. The numbers given in this section come primarily from the careful measurements and analysis by S. Singh, R. G. Smith, and L. G. Van Uitert, "Stimulated-emission cross section and fluorescent quantum efficiency of $Nd^{3+}$ in yttrium aluminum garnet at room temperature," *Phys. Rev. B* **10**, 2566–2572 (September 15, 1974).

As if to illustrate the difficulty of accurate optical measurements, a more recent publication by M. Birnbaum, A. W. Tucker, and C. L. Fincher, "Laser emission cross section of Nd:YAG at 1064 nm," *J. Appl. Phys.* **52**, 1212–1215 (March 1981), argues that the stimulated transition cross section and the quantum efficiency for Nd:YAG are both about twice the values previously given by Singh *et al*.

Problems for 3.1

1. *Quantum calculation: Hydrogen-atom oscillator strengths.* The energy eigenstates for the hydrogen atom, and the formula for calculating the transition strength or the Einstein $A$ coefficient of a transition given the upper and lower quantum wave functions, can be found in any standard quantum-theory text. Using these, calculate the oscillator strengths for the three allowed transitions from the $n = 1$, $l = 0$, $m = 0$ ground state of the hydrogen atom to the $n = 2$, $l = 1$, $m = -1$, 0, and +1 levels (taken together, these transitions form the 1216Å Lyman $\alpha$ transition). Note: This calculation is straightforward, but becomes a bit messy.

## 3.2 LINE-BROADENING MECHANISMS IN REAL ATOMS

Let us now consider a few of the more important line-broadening mechanisms responsible for the atomic linewidths $\Delta\omega_a$ in real atoms. All of these mechanisms are, as we will see, basically extensions of those derived for the classical electron oscillator. In this section we give more information on homogeneous line-broadening mechanisms in real atoms, and on how these relate to the CEO model.

### Homogeneous Broadening

All the line-broadening mechanisms we have considered thus far produce what is called *homogeneous broadening*. This means simply that all the energy-decay and dephasing mechanisms we have discussed thus far act on all the dipoles in a collection in the same way, so that the response of each individual oscillator or atom in the collection is broadened in the same fashion. The homogeneous lorentzian linewidth (FWHM) that we derived for the stimulated response of a collection of classical oscillators is then

$$\Delta\omega_a = \gamma + 2/T_2, \qquad (15)$$

where $\gamma$ is the energy decay rate and $1/T_2$ the rate at which "dephasing events" occur, whatever may be the cause of these dephasing events. There do exist additional and basically different types of broadening effects called *inhomogeneous broadening effects*, which we will introduce in the last section of this chapter. Doppler broadening is one primary example of such an inhomogeneous broadening mechanism.

### Lifetime Broadening in Real Atomic Transitions

That part of the homogeneous linewidth $\Delta\omega_a$ caused by the total energy decay rate $\gamma \equiv \gamma_{\text{rad}} + \gamma_{\text{nr}}$ is called *lifetime broadening*. Lifetime broadening is basically a Fourier-transform effect. An exponentially decaying signal of the form $\mathcal{E}(t) = \exp[-(\gamma/2 + j\omega_a)t]$ for $t > 0$ has a complex lorentzian Fourier transform of the form $\tilde{E}(\omega) = 1/[1 + 2j(\omega - \omega_a)/\gamma]$, which has a FWHM linewidth $\Delta\omega_a = \gamma$.

If dephasing effects are absent, only this lifetime broadening will remain. If in addition all nonradiative mechanisms are turned off, then only radiative decay will be left, and the linewidth will take on its minimum possible value $\Delta\omega_a = \gamma_{\text{rad}}$. This is called *purely radiative lifetime broadening*. This purely radiatively broadened condition may sometimes occur for real atoms in very low-pressure gases, where the atoms are highly isolated, and where no collisions or nonradiative effects can occur (although doppler broadening, to be discussed later, will also be present and of great importance in such a gas).

In a collection of real atoms, the transition at frequency $\omega_{ji}$ between two energy levels $E_j$ and $E_i$ with total decay rates $\gamma_j$ and $\gamma_i$, respectively, will generally have a lifetime-broadening contribution that is given in a more exact analysis by

$$\Delta\omega_a = \gamma_i + \gamma_j + 2/T_{2,ij}, \qquad (16)$$

where $2/T_{2,ij}$ is the dephasing rate appropriate to that particular transition. The main point here is that in most cases the $\gamma$ term in the classical oscillator linewidth is replaced by the sum of the upper-state and lower-state energy decay rates $\gamma_i + \gamma_j$, so far as lifetime-broadening effects are concerned.

We have noted previously that energy decay rates $\gamma_j$ for real atomic transitions take on widely different values, depending on both radiative and nonradiative processes. For strong visible-wavelength atomic transitions in gases, $\gamma_{\text{rad}}$ may become as large as $\approx 10^7$ to $10^8$ s$^{-1}$, leading to a lifetime-broadening contribution $\Delta\omega_a/2\pi$ ranging from a few MHz to a few tens of MHz. This can be a significant source of homogeneous line broadening for a transition in a low-pressure gas.

For the Nd:YAG laser on the other hand, the upper-level energy decay time is $\tau_j \approx 230$ $\mu$s. This gives a lifetime-broadening contribution of only 700 Hz, which is absolutely insignificant compared to the enormously larger phonon-broadening dephasing contribution of $\Delta\omega_a/2\pi \approx 120$ GHz.

### Dephasing Collisions and Pressure Broadening in Gases

The primary dephasing events for atoms or molecules in gases are real collisions between the radiating atoms or molecules and various collision partners. In

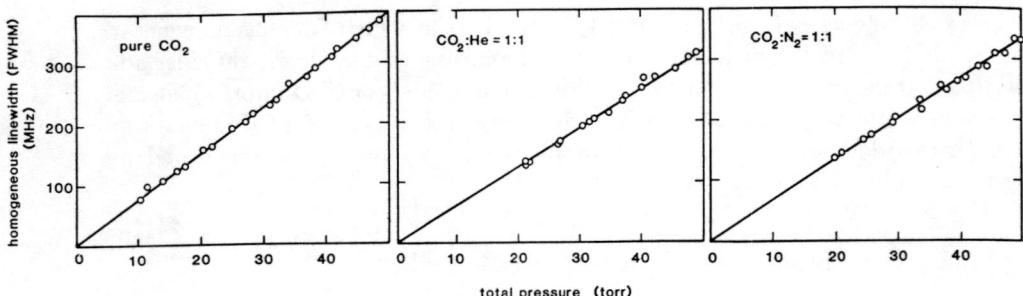

FIGURE 3.4
Pressure broadening of the $CO_2$ laser transition in various gas mixtures. (Adapted from R. L. Abrams, *Appl. Phys. Lett.* **25**, 609–611, November 15, 1974.)

a typical gas mixture atoms may collide with other atoms of the same kind (called "self-broadening"); with atoms of different kinds (called "foreign-gas broadening"); or with the tube walls (generally not of importance at optical frequencies). The total collision-broadening contribution to the homogeneous linewidth of a given atomic transition will then be directly proportional to the density, or to the partial pressure, of each species that is present. The homogeneous linewidth will therefore increase linearly with total gas pressure (assuming a constant gas mixture) in the general form

$$\Delta\omega_a = A + BP, \qquad (17)$$

where $A$ and $B$ are constants that are different for different atomic transitions and gas mixtures. This behavior is naturally referred to as *pressure broadening*, and Equation 3.17 is sometimes referred to as the *Stern-Vollmer equation*. (The coefficients $A$ and $B$ used here have nothing at all to do with the Einsten $A$ and $B$ coefficients).

Figure 3.4 illustrates some measured homogeneous pressure-broadening results for the 10.6 $\mu$m laser transition in $CO_2$ caused by $CO_2$ molecules colliding with other $CO_2$ molecules and also with He atoms or $N_2$ molecules in various gas mixtures. Note that here (as in many other common gases) a few tens of torr of total pressure gives a few hundreds of MHz of pressure broadening. Note also that the lifetime-broadening contribution in these mixtures is apparently negligible, as indicated by the essentially zero intercept of the pressure-broadening curves at zero pressure.

### Typical Numerical Values

The amount of dephasing and line broadening that actually occurs in a real collision between two atoms (or molecules, or ions) depends on how close the two partners come to each other; how their quantum wave functions overlap and interact with each other during the collision; and (to a slight extent) how fast the atoms are traveling. The atomic wave functions that are involved are, of course, different for different energy states $E_i$ or $E_j$ of the colliding partners. Therefore the amount of pressure broadening, or the constant factor $B$ in the

Stern-Vollmer formula, can often be different for different transitions even in the same atom.

TABLE 3.2
Typical pressure-broadening coefficients

| Wavelength | Collision partners | Pressure broadening |
|---|---|---|
| *Mercury resonance line:* | | |
| 2537Å | Hg + Ar,$N_2$,$CO_2$ | 10–20 MHz/torr |
| *Sodium resonance line:* | | |
| 589 nm | Na + Na | $\approx$ 2000 MHz/torr |
| *He-Ne laser transitions:* | | |
| 633 nm | He+Ne | $\approx$ 70 MHz/torr |
| 3.39 $\mu$m | He+Ne | 50–80 MHz/torr |
| *$CO_2$ laser transition:* | | |
| 10.6 $\mu$m | $CO_2$ + $CO_2$ | 7.6 MHz/torr (5.8 GHz/atm) |
| 10.6 $\mu$m | $CO_2$ + $N_2$ | 5.5 MHz/torr (4.2 GHz/atm) |
| 10.6 $\mu$m | $CO_2$ + He | 4.5 MHz/torr) (3.5 GHz/atm) |
| 10.6 $\mu$m | $CO_2$ + $H_2O$ | 2.9 MHz/torr (2.2 GHz/atm) |

Pressure-broadening coefficients are often expressed in practice in units of MHz/torr or, in some cases, GHz/atmosphere, as in Table 3.2. Collision-broadening coefficients are also sometimes given in the literature as frequency broadening (in various units) versus gas density $N$ rather than gas pressure $P$. It is then convenient to remember that

$$N(\text{atoms/cm}^3) = 9.65 \times 10^{18} \frac{P(\text{torr})}{T(K)} \qquad (18)$$

for the relation between partial pressure and density of each species in a gas mixture.

The results for the $CO_2$ laser transition in Table 3.2 and in Figure 3.4 also illustrate how the pressure-broadening coefficient, or the effective cross section of a gas molecule for dephasing collisions, can be different for different collision partners. In a typical He:$N_2$:$CO_2$ laser gas mixture, the total pressure broadening of the 10.6 micron $CO_2$ laser transition must be written as an expression like

$$\Delta\omega_a(CO_2) = A + B_{He}P_{He} + B_{N_2}P_{N_2} + B_{CO_2}P_{CO_2}, \tag{19}$$

where each $P_x$ is the partial pressure of a different gas, and the pressure-broadening coefficients $B_x$ have different values for each different collision partner.

### Nonlorentzian Lineshapes in Collision Broadening.

It can be shown from various statistical arguments that dephasing collisions that have zero duration and that completely randomize the oscillation phases after each collision should in theory produce an exponential polarization decay, and hence an associated exact lorentzian lineshape. It can also be shown that zero-duration collisions that shift the oscillation phases by very small but randomly distributed amounts ($\delta\phi \ll 2\pi$ after each collision) should also produce a lorentzian lineshape. Collisions that last for a short but finite duration, however, may lead to small but observable deviations from the ideal lorentzian lineshape.

The simplest form of extended theory for finite-duration collisions predicts a modified lorentzian lineshape, in which the linewidth $\Delta\omega_a$ itself becomes frequency-dependent, with a midband value $\Delta\omega_{a0}$ plus an added term of the form $-C \times (\omega - \omega_a)$ at frequencies away from line center. Hence the lineshape deviates increasingly from an exact lorentzian with increasing detuning from line center, with this deviation becoming most observable in the outer wings of the atomic line, many linewidths from line center.

Clearcut measurements of this small deviation in the wings of the sodium $D_1$ and $D_2$ lines at ≈589 nm, caused by collisions with He, Ne, Ar, Kr, and Xe atoms, have recently been made by observing the scattering of a tunable single-frequency laser beam from a sodium cell. Results in good agreement with an extended theory are reported by R. E. Walkup, A. Spielfiedel, and D. E. Pritchard, "Observation of non-lorentzian spectral lineshapes in Na-noble-gas systems." *Phys. Rev. Lett.* **45**, 986–989 (September 22, 1980).

### Phonon Broadening (FM Broadening) of Real Atoms in Solids

Another kind of homogeneous line broadening that is important for many solid-state laser transitions is *phonon broadening*. Phonon broadening refers to a rapid and random frequency modulation of the instantaneous atomic-transition frequency for an atom in a solid (or liquid) caused by high-frequency lattice vibrations in the surrounding crystal lattice. This process is physically quite different from a discrete collision-type process having a mean time $T_2$ between collisions, but the net result in terms of randomizing the phases and broadening the response of a collection of oscillators is very much the same, and can in fact be described by an effective dephasing time $T_2$.

Phonon broadening does not depend directly on atomic density $N$ as does pressure broadening. It does, however, depend strongly on lattice temperature,

## 3.2 LINE-BROADENING MECHANISMS IN REAL ATOMS

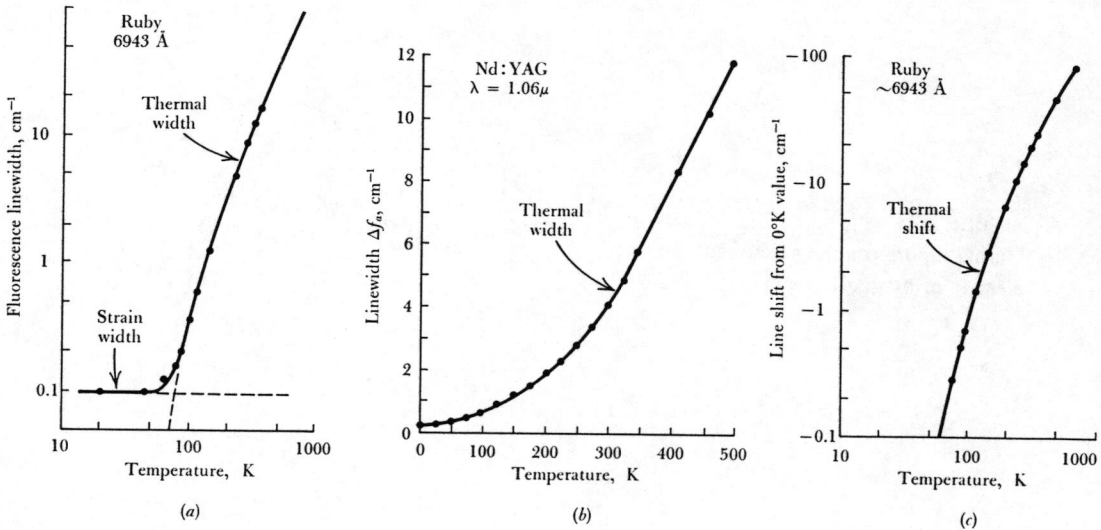

**FIGURE 3.5**
Phonon broadening and resonance frequency shifting versus temperature in two solid-state laser materials. (From D. E. McCumber and M. D. Sturge, *J. Appl. Phys.* **34**, 1682, June 1963.)

since the lattice vibrations result from thermal excitation of the lattice modes. Figure 3.5 shows, for example, the linewidths of two common solid-state laser transitions plotted versus temperature. The 694 nm laser transition in ruby shows a residual inhomogeneous strain broadening at lower temperature, changing over to thermal FM or phonon broadening at higher temperatures, whereas the linewidth of the 1.06 $\mu$m laser transition in Nd:YAG shows strongly temperature-dependent thermal phonon broadening over essentially the entire range plotted.

The phonon-broadening contribution will become very small for temperatures below a few tens of degrees Kelvin. There may then be a residual linewidth contribution of inhomogeneous type, which arises from residual static strains and imperfections in the solid-state material. This residual strain broadening may be quite different from sample to sample, depending on the perfection of individual crystal samples.

Note also that besides phonon broadening in these solids, there may also be a significant *thermal shift* of the exact center frequencies of the transitions, which can sometimes be useful (and sometimes not so useful).

### Dipolar Broadening

A third important mechanism that produces homogeneous dephasing and line broadening in certain materials at higher densities is *dipolar broadening*. Dipolar broadening results from the random interaction and coupling between nearby atoms through their overlapping dipolar electric or magnetic fields (Figure 3.6). The random perturbation of each dipole oscillator by the random fields from its neighbors can cause a time-varying frequency shift in the exact resonance frequency of each such dipole; and this in turn leads to an effective dephasing and line broadening in a fashion somewhat similar to phonon broadening.

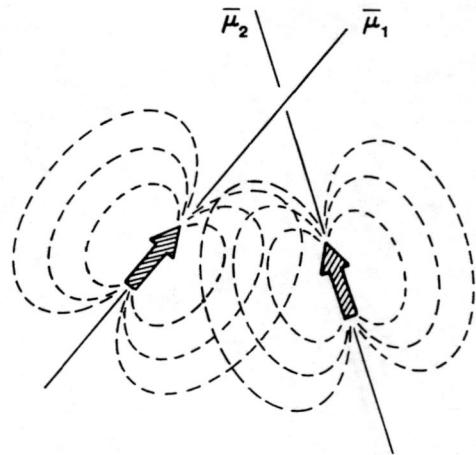

FIGURE 3.6
Dipole-dipole coupling between two nearby electric or magnetic dipoles.

Dipolar broadening is not commonly of great importance in laser materials, since they do not usually have the combination of high atomic density and strong atomic dipoles needed to make dipolar broadening predominate over the collision or thermal phonon-broadening mechanisms. However, dipolar broadening can be observed, for example, in rare-earth pentaphosphates and certain other solid-state materials that can have a high intrinsic density of rare-earth atoms, and that are sometimes used for miniature optically pumped solid-state lasers. If these materials are carefully prepared and cooled to liquid-helium temperatures, where thermal phonon broadening becomes negligible, dipolar-broadened linewidths of a few kHz can be observed by various sophisticated experiments.

### Transit-Time Broadening

There can be some experiments in which the atoms in a gas move across the full width of the optical beam with which they are interacting in a transit time $T_{tr}$ which is small compared with either the energy decay lifetime $\tau = 1/\gamma$ or the dephasing time $T_2$ of the atoms. In such a situation, it is this transit time which limits the duration of the coherent interaction between the atoms and the applied signals, and which thus determines a kind of effective lifetime broadening. This is generally referred to as *transit-time broadening*, with an effective linewidth contribution on the order of $\Delta\omega_a \approx 1/\mathrm{T}_{tr}$.

Since the thermal velocity of an atom or molecule in a gas is typically on the order of $\approx 10^5$ cm/s, transit-time broadening will produce only a few hundred kHz of broadening for a beam width or interaction length even as small as a few mm. Transit-time broadening thus becomes significant only in special situations, for example, very high-resolution molecular-beam experiments involving tightly focused optical beams and high-speed molecules. Transit-time broadening must also sometimes be considered with larger gas cells in experiments using extraordinarily high-resolution laser frequency standards, very low-pressure gases, and very long-lived molecular absorption lines.

### Coherent Pulse Experiments: Dephasing Versus Energy Decay

As we have noted in earlier discussions, it is important, and somewhat subtle, to distinguish clearly between those effects involved in *energy decay* and those involved in *line broadening* and *dephasing* of real atoms.

We described earlier, for example, an excited-state lifetime measurement in which atoms were excited into an upper energy level $E_j$, and the spontaneous emission or fluorescence on a downward transition $E_j \to E_i$ was then observed. This fluorescent emission is purely spontaneous emission, that is, incoherent random noise with a narrow spectrum (of width $\Delta\omega_a$) centered at the transition frequency $\omega_{ji}$. The excitation mechanism (pumping light or electric current) excites the atoms into level $E_j$ in an incoherent fashion. The atoms then oscillate spontaneously at frequencies like $\omega_{ji}$, but with no phase coherence between individual atoms. We add the radiated powers from each atom (not the voltages) to get the total spontaneous emission. This emission comes out randomly in all directions, and has the statistical and spectral characteristics of narrowband random noise.

It is also possible, though usually much more difficult, to perform a more complicated experiment to demonstrate *coherent* atomic emission and the effects of dephasing on this coherent emission. Suppose some incoherent excitation mechanism, such as a flash of light or a current pulse in a gas, excites some of the atoms in an atomic medium up into some excited level $E_j$ or $E_i$, or maybe even into a mixture of both. Spontaneous emission will then start. But before the populations $N_j$ or $N_i$ have decayed away, let us send a strong but short coherent signal pulse at the transition frequency $\omega_{ji}$ through the atoms. This pulse will then excite a coherent response $p(t)$ in the atoms on the $j \to i$ transition.

This induced polarization $p(t)$ will be given by the transient solution of the polarization equation of motion (Equation 2.69), taking into account the applied signal pulse. The applied signal pulse may be too short for the steady-state solution $\tilde{P}(\omega)$ given by the linear susceptibility to be reached. But nonetheless, after the signal pulse passes through the collection of atoms, they will be left with a *coherently oscillating macroscopic polarization* $p(t)$ in the medium. The atoms have all been driven in phase by the same applied signal; and after it passes they will continue to oscillate coherently and in phase at least for a brief while.

In the jargon of quantum electronics, we say that the atoms have been "coherently prepared" or "transversely aligned" by the strong signal pulse. They will then continue to radiate coherently and in the same direction as the applied signal pulse. This radiation, like the applied signal, will be spatially and temporally coherent radiation, not noise. The atoms will have some memory of how they were coherently excited by the signal pulse; and we must add vectorially the radiated voltage, not power, from each oscillating atomic dipole.

The amplitude of this coherent oscillation and radiation will, however, decay away at a total rate $(\gamma/2 + 1/T_2)$ because of the dephasing plus lifetime processes. This decay will be faster—often very much faster—than the energy decay rates $\gamma_i$ or $\gamma_j$ of the level populations. If the dephasing rate $1/T_2$ is rapid compared to $\gamma_i$ and $\gamma_j$, the coherent radiation will rapidly disappear, leaving behind the much weaker but longer-lasting incoherent spontaneous emission.

This kind of more sophisticated experiment is referred to generally as a "coherent pulse" experiment. The presence of a coherent initial signal pulse to set up the transient coherent polarization $p(t)$ is essential. The exponentially decaying coherent radiation after the coherent signal pulse is turned off is often called "free induction decay." Note that a very narrow atomic transition in a gas might have a linewidth $\Delta\omega_a/2\pi \approx 1$ MHz, so that $T_2 \approx 300$ ns. Optical signal pulses shorter than this can be generated, and lifetimes this short can be measured with fast photodetectors; hence coherent-pulse measurements on such a transition are feasible.

In Nd:YAG, the 1.06 $\mu$m laser transition has an upper-level energy-decay lifetime of $\tau_2 \approx 230$ $\mu$s. The transverse dephasing time of this transition (its inverse phonon-broadened linewidth) is, however, more like $T_2 \approx 1$ psec at room temperature. This is simply too fast to be either excited or observed with conveniently available optical tools.

## REFERENCES

The descriptions of collisions, dephasing processes, and energy decay that we have presented have been largely "phenomenological"—that is, we have added reasonable terms to the equations of motion for atomic phenomena in order to make theoretical predictions that agree reasonably well with the phenomena observed in real atoms. Essentially the same phenomenological approach is also used, however, even in more sophisticated and detailed quantum analyses of atomic behavior and atomic transitions. See, for example, R. G. Breene, Jr., *The Shift and Shape of Spectral Lines* (Pergamon Press, 1970), which uses this same approach to the theory of collision broadening. A newer and more advanced book on this topic by the same author is *Theories of Spectral Line Shape* (Wiley, 1981).

An early but very clear review of the same semiclassical theory of collision broadening may be found in H. Margenau and W. W. Watson, "Pressure effects on spectral lines," *Rev. Mod. Phys.* **8**, 22 (January 1936). A more recent review is given by A. Ben-Reuven, "The meaning of collision broadening of spectral lines: the classical analog," in *Advances in Atomic and Molecular Physics*, Vol. 5, ed. by D. R. Bates and I. Estermann (Academic Press, 1969).

In real gases, collisions between atoms can lead not only to broadening of the transition, but also to a somewhat smaller shifting of the atomic transition frequency, usually to a lower frequency (a "red shift"). These pressure shifts often amount to as much as $\approx 30\%$ to $\approx 50\%$ of the line broadening. A good discussion of the physics underlying both collision broadening and shifting of atomic lines in gases is given in Chapter 4 of A. C. G. Mitchell and M. W. Zemansky, *Resonance Radiation and Excited Atoms* (Cambridge University Press, 1961). Various theories and confirming experimental data for gaseous transitions are reviewed in S. Y. Chen and M. Takeo, "Broadening and shifting of spectral lines due to the presence of foreign gases," *Rev. Mod. Phys.* **29**, 20 (January 1957).

More advanced reviews of collision broadening and shifting include an excellent review by H. M. Foley, "The pressure broadening of spectral lines," *Phys. Rev.* **69**, 616 (1946); as well as W. R. Hindmarsh and Judith M. Farr, "Collision broadening of spectral lines by neutral atoms," in *Progress in Quantum Electronics*, Vol. 2, ed. by J. H. Sanders and S. Stenholm (Pergamon Press, 1974), pp. 141–214. A similar reference is H. van Regemorter, "Spectral line broadening," in *Atoms and Molecules in Astrophysics*, ed. by T. R. Carson and M. J. Roberts (Academic Press, 1972), pp. 85–119. The most recent and extensive review is perhaps that by N. Allard and J. Kielkopf, "The effect of neutral nonresonant collisions on atomic spectral lines," *Rev. Mod. Phys.* **54**, 1103–1182 (October 1982).

## Problems for 3.2

1. *Derivative spectroscopy on a variable-pressure gas sample.* A spectrometer of the type that measures $d\chi''(\omega)/d\omega$ versus $\omega$ is used to study an atomic transition in a gas for different gas pressures in the sample cell of the spectrometer. The atomic transition exhibits lifetime broadening plus pressure broadening of the Stern-Vollmer type. When all pressure-dependent factors are included, what is the optimum pressure for obtaining the strongest peak-derivative signal in the spectrometer? Explain physically.

## 3.3 POLARIZATION PROPERTIES OF ATOMIC TRANSITIONS

The transitions between quantum energy levels in real atoms exhibit anisotropic vector characteristics, or tensor characteristics, in both their spontaneous emission behavior and their stimulated response; and we need to understand the tensor nature of this behavior in order to fully understand real atomic transitions. In the simplest case, the response of a real atomic transition may be either *linearly polarized* or *circularly polarized* on different transitions. In the most general case, any single transition in an atom or molecule may have an *elliptically polarized response* relative to some specific set of $(x, y, z)$ axes. The induced response in all these situations must then be described by a *tensor susceptibility* connecting the vector signal field and the vector atomic polarization.

We can gain a great deal of insight into these tensor properties by examining the transitions in a collection of single free atoms (not molecules) when these atoms are placed in a dc magnetic field. The dc field then both provides a reference axis and also Zeeman-splits the energy levels to eliminate all degeneracy in the system. In this section we will examine the behavior of such Zeeman-split transitions; in the next section we will introduce the general tensor-analytical method.

### Zeeman-Split Atomic Transitions

The simplest example of a real atomic transition is probably the transition between a single lower energy level $E_1$ that is an $S$ state, having quantized angular momentum $J = O$, and an upper level $E_2$ that is a $P$ state, having quantized total angular momentum $J = 1$. (Such states are characteristic of isolated single atoms in gases.) An angular-momentum value greater than 0 means that the upper level really consists of $2J + 1 = 3$ distinct quantum levels, which are degenerate in energy in zero magnetic field. These levels will, however, be split apart by a dc magnetic field $B_0$ into 3 distinct energy levels labeled by $M_J = 1, 0$, and $-1$, as illustrated in Figure 3.7. (This splitting into separate energy levels is, of course, known as Zeeman splitting.) There are then three separate and distinct transitions from the upper levels to the lower level, at three slightly different transition frequencies as illustrated in Figure 3.7. Figure 3.8 shows some real spectral lines recorded on photographic plates in a high-resolution spectrometer for various spontaneous emission lines from excited zinc or sodium atoms,

# CHAPTER 3: ELECTRIC-DIPOLE TRANSITIONS IN REAL ATOMS

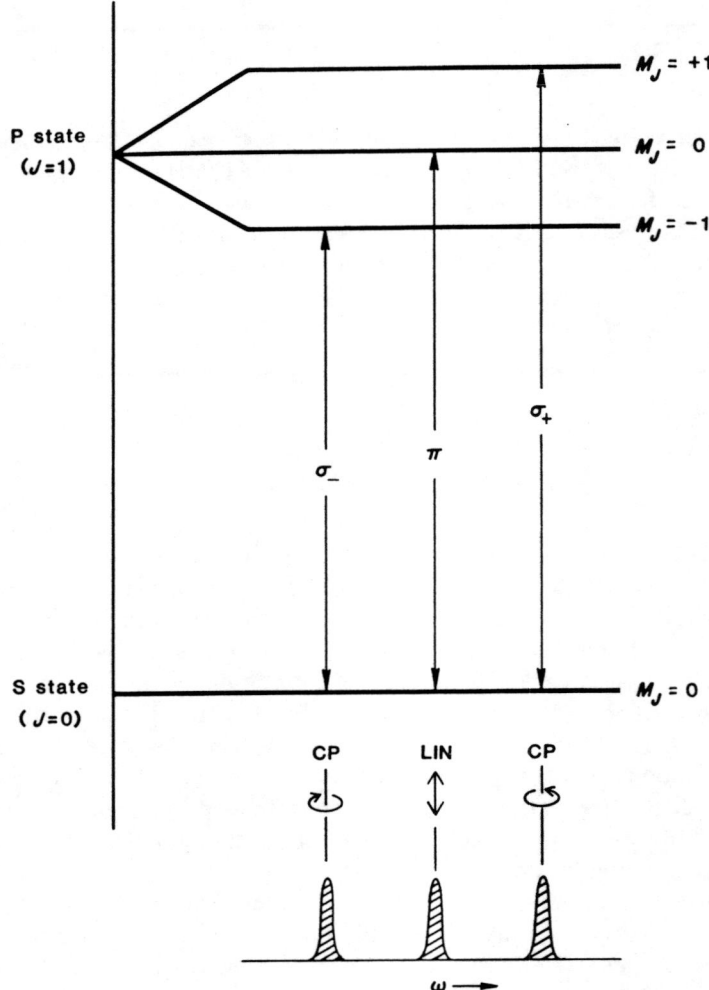

FIGURE 3.7
Zeeman splitting of atomic energy levels in a simple case.

with and without dc magnetic fields, illustrating both the simplest and more complicated types of Zeeman splitting.

### Pi and Sigma Transitions

If we study the polarization behavior of the central transition (from $M_J = 0$ to $M_J = 0$) in the example shown in Figure 3.7, we will find that this transition behaves exactly like a dipole oscillator that is linearly polarized along the direction of the dc magnetic field, both in its spontaneous radiation and in its stimulated response to an applied signal. That is, on this particular transition the atoms act just like our linearly polarized CEO model, with their linear axis along the dc field. No spontaneous emission comes out in the direction directly along the dc field axis, for example, since a linearly oscillating dipole does not radiate along its polarization axis; and there will be no stimulated response to applied $E$ fields perpendicular to that direction. Such a linearly polarized $\Delta M_J = 0$ transition is often called a $\pi$ transition.

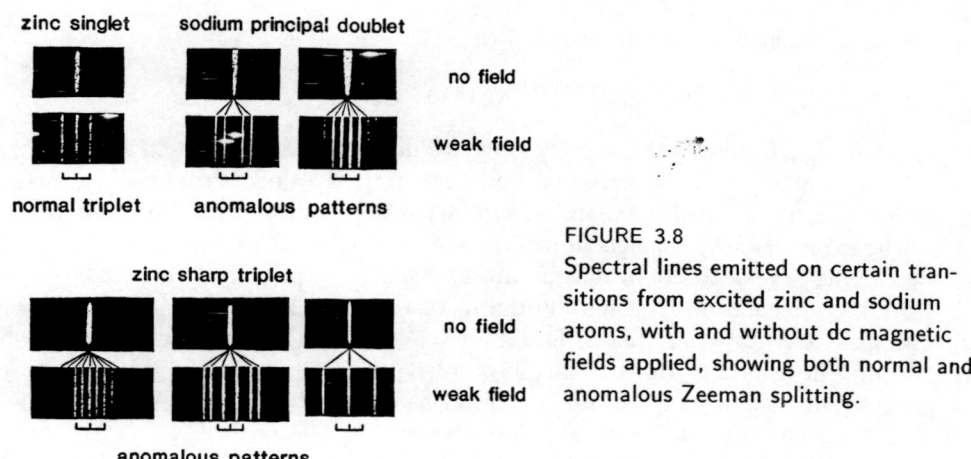

FIGURE 3.8
Spectral lines emitted on certain transitions from excited zinc and sodium atoms, with and without dc magnetic fields applied, showing both normal and anomalous Zeeman splitting.

The outer two lines in Figure 3.7 (connected to the $M_J = +1$ and $-1$ levels) are then found to be circularly polarized with respect to the magnetic field axis, with opposite senses of circularity in both their spontaneous emission and their stimulated responses. These circularly polarized lines are called $\sigma_+$ and $\sigma_-$ transitions.

We will need to use tensors to describe the susceptibility properties of these transitions. Before we discuss this, however, a brief summary of some of the quantum properties of these atomic transitions may be very useful in understanding both their polarization properties and the relationship between the quantum theory and the classical models of these transitions. Readers with limited backgrounds in quantum theory should skim the next few paragraphs, and not be concerned if all the details are not clear to them.

### Quantum Description of Atomic Transitions

In quantum theory, the quantum state of any real atom at time $t$ is completely specified by a quantum wave function $\psi(r, t)$, where $r$ indicates a general position in space. The evolution of this wave function in space and time is governed, according to quantum theory, by Schrödinger's equation of motion. We can, at least in principle, solve Schrödinger's equation to find $\psi(r, t)$ for a given atom with given initial conditions and a given applied signal; and we will then know everything there is to know physically about that atom.

Any isolated quantum system such as a single atom will also have a special set of quantum energy eigenstates or "stationary states" with associated quantum wave functions $\psi_j(r)$. These wave functions $\psi_j(r)$ are time-independent solutions of Schrödinger's equation with no applied signal present. Each such eigenstate corresponds to one of the energy levels and energy eigenvalues $E_j$ of the atom. These stationary eigenstates then provide a basis set, or a set of normal modes, for expanding any quantum state of the atom at any time.

A real atom at any instant of time will in general not be in a single energy eigenstate or energy level. Rather, it will be in a time-varying *quantum state mixture* of two or more such eigenstates. The wave function for a single atom at any instant of time may then be written in general as

$$\psi(r,t) = \tilde{a}_1(t)e^{-iE_1 t/\hbar}\psi_1(r) + \tilde{a}_2(t)e^{-iE_2 t/\hbar}\psi_2(r) + \cdots, \qquad (20)$$

where $E_1$, $E_2$, etc., are the energy eigenvalues. In the absence of an applied signal or any other external perturbation, the complex-valued expansion coefficients $\tilde{a}_1(t)$, $\tilde{a}_2(t)$, ... in this expansion will be constant in time, and there will be only the $\exp(-iE_j t/\hbar)$ frequency factor associated with each eigenstate.

One key idea here is that an atom is generally *not* in just one energy level. Rather, each atom is generally in a mixture of levels. An individual atom with a quantum state like that in Equation 3.20 then has a probability $|\tilde{a}_1|^2$ of being found in level $E_1$; a probability $|\tilde{a}_2|^2$ of being found in level $E_2$; and so forth. Averaging these probabilities over many atoms gives the same net effect as if $N_1$ atoms were in level $E_1$, $N_2$ atoms in level $E_2$, and so on.

A second key point is that these state mixtures are "stationary," in the sense that the $\tilde{a}_j$'s do not change with time unless there is an external signal or external perturbation applied to the atom. The time-varying phase rotation factor $\exp(-jE_j t/\hbar)$ associated with each term in the expansion is necessary to make $\psi(r,t)$ satisfy the Schrödinger equation in the absence of an applied signal; but these phase factors do not, of course, change the magnitudes of the coefficients.

### Physical Interpretation of the Quantum State

One physical interpretation for the wave function $\psi(r,t)$ of an electron charge cloud surrounding a fixed nucleus is that $|\psi(r,t)|^2$ gives the probability density for finding an orbital electron at point $r$ at time $t$. More generally, we can say that $\rho(r,t) \equiv -e|\psi(r,t)|^2$ gives the value (more precisely, the "quantum expectation value") of the local charge density in the electron charge cloud around the atom. If the wave function $\psi(r,t)$ is a mixture of, say, two energy states, the charge density in the atom has the form

$$\begin{aligned}\rho(r,t) &= \left|\tilde{a}_1(t)e^{-iE_1 t/\hbar}\psi_1(r) + \tilde{a}_2(t)e^{-iE_2 t/\hbar}\psi_2(r)\right|^2 \\ &= |\tilde{a}_1(t)|^2\,|\psi_1(r)|^2 + |\tilde{a}_2(t)|^2\,|\psi_2(r)|^2 \\ &\quad + \tilde{a}_1(t)\tilde{a}_2^*(t)\psi_1(r)\psi_2^*(r)\exp[i(E_2-E_1)t/\hbar] \\ &\quad + \tilde{a}_1^*(t)\tilde{a}_2(t)\psi_1^*(r)\psi_2(r)\exp[-i(E_2-E_1)t/\hbar] \\ &= \rho_{dc}(r) + \rho_{ac}(r,t).\end{aligned} \qquad (21)$$

The key observation here is that the atomic charge density contains both two *static parts*, proportional to the individual level occupancies $|\tilde{a}_j(t)|^2\,|\psi_j(r)|^2$, and a *sinusoidally oscillating component* given by the **mixed term** or **cross term**

$$\rho_{ac}(r,t) = \mathrm{Re}\left[\tilde{a}_1(t)\tilde{a}_2^*(t)\psi_1(r)\psi_2^*(r)e^{i\omega_{21}t}\right]. \qquad (22)$$

This oscillating component inherently oscillates at the **transition frequency** $\omega_{21} = (E_2-E_1)/\hbar$ between the two levels involved. *There is in effect a natural quantum oscillating dipole moment in the real quantum atom*, which can be compared with the oscillating moment $\mu_x(t)$ of the CEO model. This is **a quantum-mechanically predicted oscillation** in the atom, at the transition frequency between any two occupied levels.

## 3.3 POLARIZATION PROPERTIES OF ATOMIC TRANSITIONS

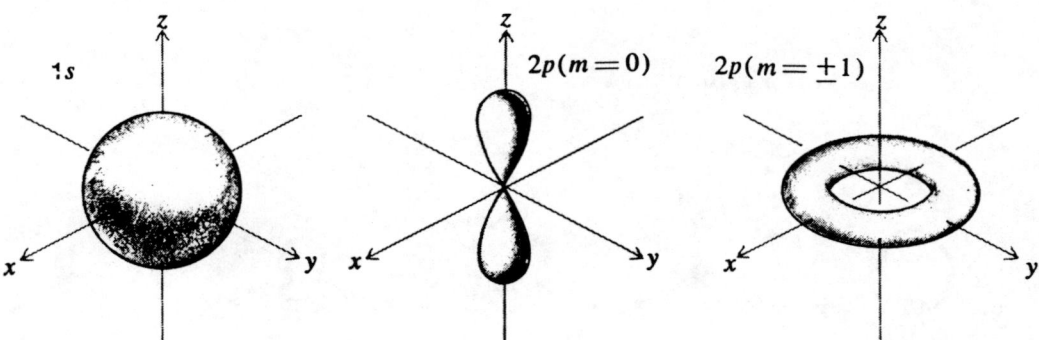

FIGURE 3.9
Schematic representations of the electronic charge distributions for Zeeman-split quantum eigenstates.

The magnitude of this oscillating component is proportional to the cross term $\tilde{a}_1(t)\tilde{a}_2^*(t)$ between the two level occupancies, and it obviously decays away as the level occupancy coefficients $\tilde{a}_1(t)$ and $\tilde{a}_2(t)$ decay away, just as $\mu_x(t)$ decays at rate $\gamma$ in the classical oscillator. The phase $\phi_i$ of the atomic oscillation in the $i$-th atom depends on the *phase-angle difference* of the complex coefficients $\tilde{a}_1 = |\tilde{a}_1|e^{-i\phi_1}$ and $\tilde{a}_2 = |\tilde{a}_2|e^{-i\phi_2}$ in the combination $\tilde{a}_1\tilde{a}_2^* = |\tilde{a}_1\tilde{a}_2|e^{i(\phi_2-\phi_1)}$. This phase can be randomized by dephasing processes that randomize the individual phases of $\tilde{a}_1$ and $\tilde{a}_2$, without necessarily changing the occupancies $|\tilde{a}_1|^2$ or $|\tilde{a}_2|^2$ of either level.

### Zeeman Transitions: Linear and Circular Dipoles

As a specific example of such an oscillating charge pattern and oscillating dipole moment in a real atom, let us examine the simple but realistic Zeeman-split example described earlier. We will look in the following paragraphs at simplified three-dimensional representations of the volume charge distributions $\psi(r, t)$ that correspond to various eigenstates and state mixtures, keeping in mind that $\psi(r, t)$ itself is a complex function with a sign or phase angle as well as a magnitude at each point in space.

Figure 3.9 shows in schematic form, for example, the wave functions $|\psi(r, t)|^2$ for a $J = O$ eigenstate or $S$ state (a spherically symmetric charge cloud); for a $J = 1$, $M_J = 0$ or $P_0$ eigenstate (dumbbell shape); and for a $J = 1$, $M_J = \pm 1$ or $P_{\pm 1}$ eigenstate (toroidal ring). Note that in the dumbbell the wave function $\psi(r)$ has opposite sign in the upper and lower lobes, whereas in the $M_J = \pm 1$ states the wave function has an $\exp(\pm j\theta)$ phase variation around the torus.

### Linearly Polarized ($\pi$) Transition

Suppose, then, that the quantum state $\psi(r, t)$ of an atom is a mixture of, say, the $1S$ and the $2P_0$ states of the hydrogen atom (the ball and the dumbbell in Figure 3.9; the transition between these two levels in the hydrogen atom is, in fact, the Lyman $\alpha$ line at 1216Å). When the phases of the complex coefficients $\tilde{a}_1(t)$ and $\tilde{a}_2(t)$ are included, the complex-valued wave functions $\tilde{a}_1\psi_1(r)$ and $\tilde{a}_2\psi_2(r)$ associated with these states may interfere constructively and/or

**linear dipole:** $S(M=0) + P(M=0)$ states

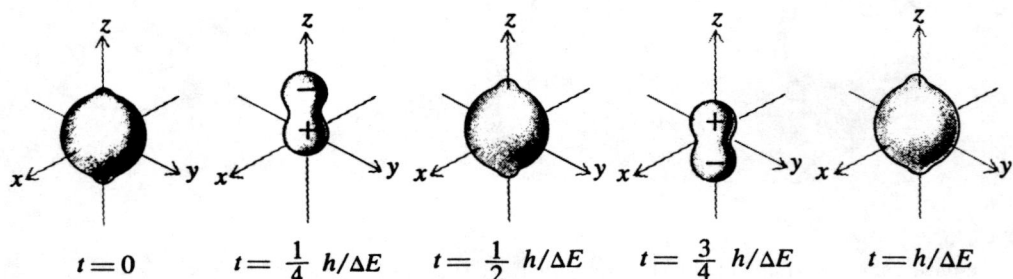

**circular dipole:** $S(M=0) + P(M=\pm 1)$ states

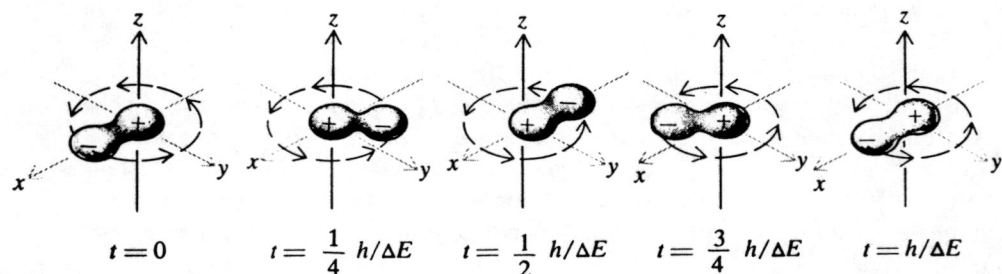

**FIGURE 3.10**
Upper part: Oscillating charge distribution in the coherent state mixture $1S + 2P_0$ as a function of time for a quantum atom. The atom acts as a linearly oscillating dipole. Lower part: Corresponding probability density, or quantum charge distribution, for a quantum state mixture of $1S + 2P_1$ states. This quantum state mixture acts like a rotating, circularly polarized electric dipole. (Adapted from G. R. Fowles, *Introduction to Modern Optics*, Holt, Rinehart, and Winston, 1968.)

destructively at different points to create the total wavefunction $\psi(r,t)$; and this interference will, moreover, rotate through all possible phases at the transition frequency $\omega_{21}$ because of the $\exp(-iE_j t/\hbar)$ terms.

The upper part of Figure 3.10 shows what the total wave function $|\psi(r,t)|^2$ produced by summing and squaring the $1S$ and the $2P_0$ states will look like at successive times during one oscillation cycle of the $\exp(j\omega_{21}t)$ variation. The center of charge of the total atomic charge cloud clearly oscillates back and forth linearly along what is here labeled the $z$ axis. The quantum atom with this particular mixture of $1S + 2P_0$ states acts exactly like a linearly oscillating dipole.

### Circularly Polarized ($\sigma$) Transitions

The lower part of Figure 3.10 shows the same type of result when the lower state $E_1$ is again a $1S$ state, but the upper level $E_2$ is now a $2P_1$ state with $M_J = +1$. Because of the $\exp(+j\theta)$ variation of the $P_1$ state wave function

π transitions

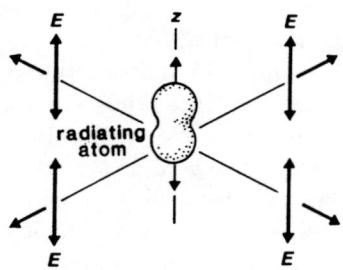

σ transitions

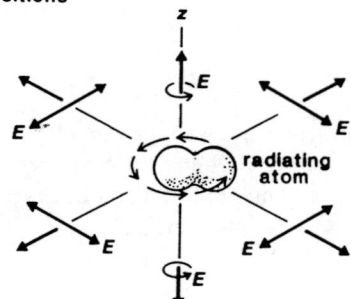

FIGURE 3.11
Polarization properties and oscillation characteristics of simple Zeeman-split atomic transitions.

around the equatorial plane, the wave functions $\psi_1$ and $\psi_2$ corresponding to the "ball" and the "torus" interfere constructively on one side and destructively on the other side of the rotational axis, producing a cancellation on one side and a "bulge" on the other side.

As the coefficients $\tilde{a}_1 e^{-iE_1 t/\hbar}$ and $\tilde{a}_2 e^{-iE_2 t/\hbar}$ rotate in time-phase, however, the resulting bulge in the quantity $|\psi_1(r)+\psi_2(r)|^2$ rotates about the $z$ axis at the transition frequency $\omega_{21}$. The atom radiates like an oscillator that is circularly polarized in the $x, y$ plane. The polarization of this rotation will be of opposite sense (opposite circularity) depending on whether the magnetic quantum number $M_j = +1$ or $-1$ in the upper level.

Figure 3.11 summarizes the polarization properties and the radiation characteristics into various directions of these simple Zeeman-split oscillating-electric-dipole charge distributions. These results represent the quantum-mechanical polarization properties of real atomic transitions. They obviously can be very well represented, however, by the kinds of purely classical electron oscillator models we have been developing.

These polarization properties of the quantum oscillations in the atomic wave functions determine both the spontaneous and the stimulated properties of the real atoms. That is, an atom whose charge distribution can oscillate only in a certain direction on a given transition will obviously respond only to applied fields that have the same direction or sense of polarization. Hence Figure 3.11 illustrates equally well both the stimulated response and the spontaneous emission properties of these transitions.

### Elliptically Polarized Transitions

Many real atomic transitions, particularly the transitions of isolated single atoms in gases, as well as many molecular transitions, will have either pure linear or pure circular polarization properties exactly like those illustrated in Figures 3.10 and 3.11. Atomic transitions in crystals or in complex molecules may, however, have more complex polarization properties. It turns out (though we will not attempt to illustrate this in detail here) that the most general possible polarization for either an electric-dipole or a magnetic-dipole type of atomic transition is an *elliptical polarization* in an arbitrarily oriented plane of polarization, with arbitrary ellipticity and arbitrary orientation of the elliptical axes in that plane. Linear and circular polarization are then elementary limiting cases of this general form.

## REFERENCES

The sketches of oscillating atomic dipoles in this section are adapted from the excellent text by G. R. Fowles, *Introduction to Modern Optics* (Holt, Rinehart and Winston, 1968), especially Chap. 7. See also G. R. Fowles, "Quantum dynamical description of atoms and radiative processes," *Am. J. Phys.* **31**, 407–409 (June 1963).

---

### Problems for 3.3

1. *Two-dimensional Zeeman-split classical oscillator model.* Let us see if it is possible to develop a purely classical oscillator model that will reproduce in more detail some of the circular polarization and Zeeman-splitting properties of the real atomic transitions described in this section.

   To do this, consider a *two-dimensional* classical electron oscillator, consisting of a point electron that is free to move in two dimensions on an $x, y$ plane. Assume that there is a central restoring force such that the restoring force terms in the two transverse directions are $f_x = -Kx$ and $f_y = -Ky$. Assume that there is also a dc magnetic field $B_0$ normal to the $x, y$ plane. Such a field will cause forces $-eB_0(dy/dt)$ in the $x$ direction and $+eB_0(dx/dt)$ in the $y$ direction. Assume also that there are equal damping factors $\gamma$ in both directions, in exact analogy to the one-dimensional case.

   Assuming then a sinusoidal $\tilde{E}_x$ field applied (for simplicity) only in the $x$ direction, find the resulting steady-state displacements $\tilde{X}(\omega)$ and $\tilde{Y}(\omega)$ of the oscillator as a function of the applied frequency. Discuss the resonance behavior of the response, and identify the resonance frequencies of the oscillator. (Note that you will need to solve two coupled equations of motion, and that the resulting equation for the resonance frequencies will be quartic rather than quadratic as for the simple one-dimensional classical electron oscillator.)

   Discuss also, with appropriate sketches, the nature of the induced steady-state electron motion $x(t)$ and $y(t)$ for signals tuned to one or the other of the Zeeman-split resonance peaks. The behavior calculated should be similar to the Zeeman splitting of real atomic resonances when a dc magnetic field is applied. Discuss also the induced electron motion at $\omega_a$, exactly halfway between the peaks.

Hints: It will be convenient to define a cyclotron frequency $\omega_c \equiv eB_0/m$, and to make the assumptions that $\gamma \ll \omega_c \ll \omega_a$. (These assumptions imply that the magnetic field splitting, or Zeeman splitting, of the resonance at $\omega_a$ will be large compared to the damping $\gamma$ but still small compared to the unperturbed center frequency $\omega_a$.) The algebra involved in this problem will also be easier if you use a resonance approximation, as well as the other approximations noted above, as early as possible in the calculations.

2. *Computer plots of oscillating atomic charge distributions (research problem)*. Using whatever computer graphics facilities may be available to you, carry out further computer investigations of the oscillating charge density distributions for quantum state mixtures like those shown in this section. Try making, for example, contour plots or three-dimensional display plots at different phases in the oscillation cycle, to illustrate the dynamic motion of the charge density—and please send me copies of any particularly good results! You might also investigate such plots for simpler one-dimensional cases, such as an electron in a one-dimensional quadratic or square well potential.

---

## 3.4 TENSOR SUSCEPTIBILITIES

Real atomic transitions thus have a tensor character that must be taken into account to give a complete and accurate description of the stimulated response on these transitions. In this section we summarize these tensor aspects of electric (or for that matter magnetic) dipole transitions in real atoms.

### Tensor Susceptibility: Linear Dipole Oscillators

Suppose that a sinusoidal signal with frequency $\omega$ on or near a single atomic transition is applied to a collection of real electric-dipole atoms. Then the steady-state vector polarization $\boldsymbol{P}(\omega)$ induced in the collection of atoms must be related to the vector field $\boldsymbol{E}(\omega)$ by a tensor equation of the form

$$\boldsymbol{P}(\omega) = \boldsymbol{\chi}(\omega)\epsilon\boldsymbol{E}(\omega), \tag{23}$$

where $\boldsymbol{\chi}(\omega)$ is a $3 \times 3$ tensor form of the susceptibility $\tilde{\chi}(\omega)$, with components $\tilde{\chi}_{xx}(\omega)$, $\tilde{\chi}_{xy}(\omega)$, and so forth. Let us first examine the tensor character of this susceptibility for some simple examples, to get a feeling for the nature of these tensor responses.

The most elementary example is the linear classical electron oscillator. For the classical oscillator we calculated the $x$ component of polarization $\tilde{P}_x$ induced by an $x$-polarized field component $\tilde{E}_x$. In tensor notation this gives us only the $xx$ tensor component of $\boldsymbol{\chi}$, or

$$\tilde{P}_x(\omega) = \tilde{\chi}_{xx}(\omega)\epsilon\tilde{E}_x(\omega). \tag{24}$$

It is physically evident that no $\tilde{P}_y$ or $\tilde{P}_z$ polarization components will occur in the linear oscillator model (since the electron is by definition not free to move along those coordinates in the linear model); and also that no response will be induced in the linear model by field components $\tilde{E}_y$ or $\tilde{E}_z$. Hence we can write

144     CHAPTER 3: ELECTRIC-DIPOLE TRANSITIONS IN REAL ATOMS

this response in expanded tensor or matrix form as

$$\boldsymbol{P}(\omega) = \begin{bmatrix} \tilde{P}_x(\omega) \\ \tilde{P}_y(\omega) \\ \tilde{P}_z(\omega) \end{bmatrix} = \tilde{\chi}(\omega)\epsilon \begin{bmatrix} 3 & 0 & 0 \\ 0 & 0 & 0 \\ 0 & 0 & 0 \end{bmatrix} \begin{bmatrix} \tilde{E}_x(\omega) \\ \tilde{E}_y(\omega) \\ \tilde{E}_z(\omega) \end{bmatrix}. \tag{25}$$

Following a pattern that we will use repeatedly in this section, we have separated the right-hand side of this equation into a dimensionless tensor part with a trace of magnitude 3, plus a purely scalar (but still complex) susceptibility $\tilde{\chi}(\omega)$.

The scalar susceptibility part of this expression for a homogeneously broadened lorentzian transition will then have the usual form

$$\tilde{\chi}(\omega) = -j\frac{1}{4\pi^2} \frac{\Delta N \lambda^3 \gamma_{\text{rad}}}{\Delta \omega_a} \frac{1}{1 + 2j(\omega - \omega_a)/\Delta \omega_a} \tag{26}$$

in which the factor of 3 has been left with the dimensionless tensor for reasons that will become apparent later. Subscripts $ij$ might also be attached to each factor in Equations 3.25 and 3.26 if necessary to identify the specific transition in a real atom that is involved.

Note that the choice of the $x$ axis for the direction of the linear response here is entirely arbitrary. We might choose to label the linear response as being along the $y$ or the $z$ axes, or along some more arbitrary linear axis. If we made this last choice, the tensor would become more complicated in form, corresponding to an arbitrary rotation of the coordinate axes with respect to the $x, y, z$ axes. It would still be, however, a purely real tensor.

### Circularly Polarized (Gyrotropic) Responses

Let us next consider circularly polarized transitions, such as the $\sigma_\pm$ transitions we saw in the previous section. For a transition that is circularly polarized in the $x, y$ plane (which is true of many simple transitions in free atoms), the tensor susceptibility becomes

$$\boldsymbol{P} = \begin{bmatrix} \tilde{P}_x \\ \tilde{P}_y \\ \tilde{P}_z \end{bmatrix} = \tilde{\chi}(\omega)\epsilon \times \frac{3}{2} \begin{bmatrix} 1 & \mp j & 0 \\ \pm j & 1 & 0 \\ 0 & 0 & 0 \end{bmatrix} \begin{bmatrix} \tilde{E}_x \\ \tilde{E}_y \\ \tilde{E}_z \end{bmatrix}. \tag{27}$$

where $\tilde{\chi}(\omega)$ is exactly the same as in Equation 3.26, and the factor of 3/2 is attached to the tensor part of this circularly polarized expression, in order to make its trace (i.e., its diagonal sum) have the same value of 3 as for the linearly polarized expression.

Suppose that the applied signal in this case is linearly polarized along the $x$ axis, i.e.,

$$\tilde{E}_x = \tilde{E}_0 \quad \text{and} \quad \tilde{E}_y = \tilde{E}_z = 0. \tag{28}$$

Then the induced polarization components will be

$$\tilde{P}_x = (3\tilde{\chi}\epsilon/2)\tilde{E}_0 \quad \text{and} \quad \tilde{P}_y = \mp j(3\tilde{\chi}\epsilon/2)\tilde{E}_0, \tag{29}$$

where $(3/2)\tilde{\chi}\epsilon$ is in general a complex-valued quantity. Hence the real polarization terms will be of the form

$$p_x(t) = \text{Re}\left[\tilde{P}_x e^{j\omega t}\right] = |(3\chi\epsilon/2)E_0|\cos(\omega t + \theta),$$
$$p_y(t) = \text{Re}\left[\tilde{P}_y e^{j\omega t}\right] = \pm|(3\chi\epsilon/2)E_0|\sin(\omega t + \theta),$$
(30)

where $\theta$ is the net phase angle of $(3\tilde{\chi}\epsilon/2)\tilde{E}_0$. Although the applied signal field is linearly polarized, the induced polarization $\boldsymbol{p}(t)$ is *circularly polarized* in the $x, y$ plane, rotating from $x$ to $y$ for the $+$ sign or from $x$ to $-y$ for the $-$ sign.

The circularly polarized tensor form given in Equation 3.27 inherently leads to circularly polarized behavior of the induced polarization. This form is characteristic of $\sigma$-type electric-dipole transitions and many simple magnetic-dipole transitions, and is often referred to as a *gyrotropic tensor response*. As before, rotation to a different coordinate orientation will make the tensor appear more complicated, but the essential character will remain the same.

### Elliptically Polarized Responses

Suppose a sinusoidal electric field $\tilde{\boldsymbol{E}}$ is applied to an arbitrary two-level nondegenerate electric-dipole transition in a real atom. Such a transition will have a *quantum dipole matrix element* $\tilde{\boldsymbol{\mu}}_{21}$ given by the integral

$$\tilde{\boldsymbol{\mu}}_{21} = -e \iiint \psi_2^*(\boldsymbol{r}) \times \boldsymbol{r} \times \psi_1(\boldsymbol{r})\, d\boldsymbol{r} \equiv \begin{bmatrix} \tilde{\mu}_x \\ \tilde{\mu}_y \\ \tilde{\mu}_z \end{bmatrix} \quad (31)$$

between the upper and lower levels of the transition. That is, $\tilde{\boldsymbol{\mu}}_{21}$ may be interpreted as a column vector with elements given by the $x, y, z$ vector components of the integral. The hermitian conjugate $\tilde{\boldsymbol{\mu}}_{21}^\dagger$ of this column vector is then a row vector whose elements $[\tilde{\mu}_x^*, \tilde{\mu}_y^*, \tilde{\mu}_z^*]$ are the complex conjugates of the elements in the column vector.

An exact quantum analysis then says that the expectation value for the phasor amplitude $\tilde{\boldsymbol{\mu}}$ of the dipole moment induced in the atom by the applied field will be given by

$$\tilde{\boldsymbol{\mu}} = \text{const} \times \left(\tilde{\boldsymbol{\mu}}_{21}^\dagger \cdot \tilde{\boldsymbol{E}}\right) \times \tilde{\boldsymbol{\mu}}_{21}$$
$$= \text{const} \times \begin{bmatrix} \tilde{\mu}_x^* & \tilde{\mu}_y^* & \tilde{\mu}_z^* \end{bmatrix} \cdot \begin{bmatrix} \tilde{E}_x \\ \tilde{E}_y \\ \tilde{E}_z \end{bmatrix} \times \begin{bmatrix} \tilde{\mu}_x \\ \tilde{\mu}_y \\ \tilde{\mu}_z \end{bmatrix},$$
(32)

where the dot product is taken in the usual matrix-multiplication fashion between the row vector $\tilde{\boldsymbol{\mu}}_{21}^\dagger$ and the column vector $\tilde{\boldsymbol{E}}$ with elements $[\tilde{E}_x, \tilde{E}_y, \tilde{E}_z]$.

The induced macroscopic polarization $\tilde{\boldsymbol{p}}(\omega)$ in a collection of atoms will then be just the microscopic dipole moment $\tilde{\boldsymbol{\mu}}$ in each individual atom, as given by Equation 3.32, summed over all the atoms in any small unit volume. Equation 3.32 contains a scalar constant, times a scalar dot product, times the column vector $\tilde{\boldsymbol{\mu}}_{21}$, which is the net vector quantity on the right-hand side of the equation. Equation 3.32 says, therefore, that *the induced response $\tilde{\boldsymbol{\mu}}$ or $\tilde{\boldsymbol{p}}$ of the atoms will always have exactly the same polarization properties as the transition's dipole matrix element $\tilde{\boldsymbol{\mu}}_{21}$, regardless of the polarization properties of the applied signal*

$\tilde{E}$. That is, you can drive the atoms with any polarization $\tilde{E}$ you want; but they will always respond with their own fixed characteristic form of polarization, as given by $\tilde{\mu}_{21}$.

The *magnitude* of this induced response, however, will depend on the dot product between the applied field $E$ and the hermitian conjugate of the moment $\mu_{21}$; and this dot product is mathematically the same thing as matrix multiplication between these two quantities. By invoking the associative properties of matrix and vector multiplication, therefore, we can reorder Equation 3.32 into the alternative form

$$\tilde{\mu} = \text{const} \times \tilde{\mu}_{21} \times \tilde{\mu}_{21}^\dagger \times \tilde{E} = \text{const} \times \begin{bmatrix} \tilde{\mu}_x \\ \tilde{\mu}_y \\ \tilde{\mu}_z \end{bmatrix} \times \begin{bmatrix} \tilde{\mu}_x^* & \tilde{\mu}_y^* & \tilde{\mu}_z^* \end{bmatrix} \times \begin{bmatrix} \tilde{E}_x \\ \tilde{E}_y \\ \tilde{E}_z \end{bmatrix}. \quad (33)$$

In this reorganized form, the middle product $\tilde{\mu}_{21} \times \tilde{\mu}_{21}^\dagger$ can now be interpreted as the matrix product, computed according to the usual rules, of the two vector (or matrix) quantities $\tilde{\mu}_{21}$ and its hermitian conjugate. But the result of this multiplication will be a $3 \times 3$ matrix or tensor $T$, often called a *dyadic product*, which we will write as

$$T \equiv \text{const} \times \tilde{\mu}_{21} \times \tilde{\mu}_{21}^\dagger = \text{const} \times \begin{bmatrix} \tilde{\mu}_x^* \\ \tilde{\mu}_y^* \\ \tilde{\mu}_z^* \end{bmatrix} \times \begin{bmatrix} \tilde{\mu}_x & \tilde{\mu}_y & \tilde{\mu}_z \end{bmatrix} = \begin{bmatrix} \tilde{t}_{xx} & \tilde{t}_{xy} & \tilde{t}_{xz} \\ \tilde{t}_{yx} & \tilde{t}_{yy} & \tilde{t}_{yz} \\ \tilde{t}_{zx} & \tilde{t}_{zy} & \tilde{t}_{zz} \end{bmatrix}, \quad (34)$$

where the constant is some suitable normalization constant. Note that the $nm$-th element of the $T$ matrix is obtained in the usual matrix-multiplication way, by multiplying the $n$-th row of the $\tilde{\mu}_{21}^\dagger$ column vector (just one element) times the $m$-th column of the $\tilde{\mu}_{21}$ row vector (also just one element).

Hence we can write the macroscopic polarization in a general tensor form as

$$\tilde{p}(\omega) = \text{const} \times \tilde{\mu}_{21} \tilde{\mu}_{21}^\dagger \times \tilde{E}(\omega) = \tilde{\chi}(\omega) \epsilon \times T \times \tilde{E}(\omega), \quad (35)$$

where the most general form of the susceptibility tensor $T$ for a dipole transition is given by the dyadic product

$$T = \text{const} \times \tilde{\mu}_{21} \tilde{\mu}_{21}^\dagger. \quad (36)$$

Suppose the transition matrix element $\tilde{\mu}_{21}$ is a column vector with elements $[1, -j, 0]$ corresponding to RHCP motion in the $x, y$ plane. The hermitian conjugate $\tilde{\mu}_{21}^\dagger$ is then a row vector with elements $[1, +j, 0]$, and the tensor susceptibility has the form

$$T = \frac{3}{2} \times [1 \ -j \ 0] \times \begin{bmatrix} 1 \\ j \\ 0 \end{bmatrix} = \frac{3}{2} \times \begin{bmatrix} 1 & j & 0 \\ -j & 1 & 0 \\ 0 & 0 & 0 \end{bmatrix}. \quad (37)$$

This is, of course, just the RHCP gyrotropic result given in Equation 3.27.

### Most General Tensor Form

Simple linearly polarized and circularly polarized responses are the most common and elementary forms for the tensor responses of electric-dipole and magnetic-dipole atomic transitions. To obtain the most general possible form for a dipole susceptibility tensor, we can note that the quantum transition moment

$\tilde{\mu}_{21}$ can have at most three complex-valued vector components, namely, $\tilde{\mu}_x$, $\tilde{\mu}_y$, and $\tilde{\mu}_z$, or six independent real numbers. Using these values, we can then carry out the matrix multiplication of the dyadic product as defined in Equation 3.34 to obtain the most general tensor form $T$.

It can then be shown that for any such dipole transition the most general allowed form of this dyadic-product response will be an *elliptically polarized tensor response*, with the resulting induced polarization $\tilde{P}(\omega)$ having some arbitrary (but fixed) degree of ellipticity and arbitrary orientation of the elliptical axes in some reference plane which is itself arbitrarily oriented with respect to the $x, y, z$ axes. This behavior is inherent in the mathematical form itself, independent of physical properties of the transitions.

There seems to be little point in writing out this general elliptical tensor form in more detail here. If you wish to know what the resulting tensor looks like, first add together the tensor responses for two independent linear responses along the $x$ and $y$ axes, but with an arbitrary amplitude ratio and arbitrary phase angle between them. This will produce the tensor form for an arbitrary elliptical response in the $x, y$ plane. Performing a conventional coordinate rotation from the $x, y, z$ axes to an arbitrarily oriented set of new $x', y', z'$ axes will then generate the most general possible form for the susceptibility tensor.

Note that the degree of ellipticity of the original ellipse, plus the orientation of this ellipse in space, accounts for four real parameters. The normalization condition that the trace of the resulting tensor should be normalized to three, i.e., $\tilde{t}_{xx} + \tilde{t}_{yy} + \tilde{t}_{zz} = 3$, then accounts for the remaining two of the six real numbers mentioned above. (Alternatively, we could require only that the magnitude of the trace be unity, leaving an arbitrary overall phase shift in all the tensor elements.) There are thus really only four adjustable real parameters among the nine complex elements of the normalized susceptibility tensor.

### Tensor Axes

But what determines the direction of the relevant axes of polarization and the degree of ellipticity for a real transition in a real atom? A simplified answer is as follows.

Single atoms floating freely in a gas always have degenerate electronic energy levels, for example, the Zeeman levels described earlier (except, of course, for $J = 0$ or $S$ states, which are not degenerate). In this situation we must apply some static perturbation, such as a dc magnetic field (Zeeman splitting) or a dc electric field (Stark splitting), to "break" this degeneracy and to separate the individual transitions into distinct transition frequencies. Each of these separate Zeeman-split transitions will then have a distinct type and direction of tensor polarization.

The direction of the static perturbation in this situation will determine one of the reference axes for the tensor susceptibility; this direction is often chosen to be the $z$ direction. The dc field direction will thus serve as the reference axis for the tensor responses on these transitions. For free atoms in such a static field, the response is then always either linear along this $z$ axis ($\pi$ transitions) or else circularly polarized about it ($\sigma$ transitions), so that no unique choice for the $x$ and $y$ axes is either necessary or possible.

Atoms in a crystal will have a more complex environment, with more clearly determined reference axes, but often with a lower order of symmetry. In a crystal, each individual atom will be imbedded in some surrounding lattice structure with a distinctive orientation in space. The orientation of this lattice structure gives the reference axes against which the polarization-tensor properties of the atomic transitions can be uniquely evaluated. The most general possible result, as already noted, is an elliptically polarized tensor response with respect to these axes.

Finally, in molecules the structural axes of the molecular structure itself give reference axes for the electronic transitions of the electron charge cloud of the molecule. As the molecule rotates, these axes rotate with it. If a simple molecule has only a single axis of symmetry (e.g., a diatomic molecule like $N_2$), all its electronic transitions are either linear along this axis or circular about it.

### Isotropic Responses?

An important observation is that it is *not* physically possible for the tensor response of a single, nondegenerate atomic or molecular transition to be isotropic (that is, to be linear and equal in all directions). That is, a single nondegenerate transition cannot have a tensor response of the form

$$\begin{bmatrix} \tilde{P}_x \\ \tilde{P}_y \\ \tilde{P}_z \end{bmatrix} = \tilde{\chi}(\omega)\epsilon \times \begin{bmatrix} 1 & 0 & 0 \\ 0 & 1 & 0 \\ 0 & 0 & 1 \end{bmatrix} \times \begin{bmatrix} \tilde{E}_x \\ \tilde{E}_y \\ \tilde{E}_z \end{bmatrix}. \tag{38}$$

A response that is effectively isotropic in this fashion can, however, be obtained by averaging over a collection of atoms. There are two different ways in which this can occur, as follows.

- The response of each individual atom in a collection may be anisotropic, with one of the nondegenerate tensor forms given earlier; but the atoms may have their reference axes randomly oriented in all directions. This would be expected in a randomly oriented collection of gas molecules, for example, or in a noncrystalline material, such as a liquid, a powder, or a glass, in which the local surroundings for different atoms may be randomly oriented. Averaging over all directions of the atomic axes leads to an isotropic overall response as in Equation 3.38.

- The observed response may be the summation over a complete set of degenerate atomic transitions that are not resolved in frequency, because no external perturbation has been applied to break the degeneracy. These degenerate transitions all coincide in frequency, and hence cannot be separately excited. Adding up the small-signal tensor responses of such a complete set of overlapping degenerate transitions then leads to an isotropic response here also. (Or we could say that there is no way to define any unique reference axes in the atoms; so the atoms are in effect randomly oriented.)

In either situation the tensor response will have the apparently isotropic form given in Equation 3.38, where the scalar $\tilde{\chi}(\omega)$ is again the same as in Equation 3.26, that is, without the factor of 3 in front.

To look at this in another way, suppose we have a collection of randomly oriented atoms, so that $N/3$ of them will be in effect oriented along each axis. The linear response due to these atoms will then be the value of $\tilde{\chi}(\omega)$ derived in the previous chapter, including the initial factor of 3, but with a population (or population difference) of only $N/3$ instead of $N$. Therefore, the response along each axis will be given by the scalar $\tilde{\chi}(\omega)$ formula in the form we have used in this section *without the initial factor of 3 that appeared in $\tilde{\chi}(\omega)$ in earlier chapters.*

The isotropic tensor form of Equation 3.38 is thus both mathematically simple and characteristic of certain common physical situations. It also has a trace equal to three—at least if we give the diagonal elements the simplest value of unity. It is largely for this reason that we have adopted the convention that $\text{Tr}[T] \equiv 3$ in writing all of the preceding normalized tensor susceptibilities. The significance of this factor of 3 is discussed in more detail in the following section.

---

## Problems for 3.4

1. *Negative circular polarization response of a gyrotropic tensor.* Using the gyrotropic form of tensor response, verify that if you apply a circularly polarized $\tilde{E}$ field which has one sense of circular polarization to an atom whose natural response is the opposite sense of circular polarization, then this applied $\tilde{E}$ field will produce no atomic response at all.

2. *Tensor response of an anisotropic two-dimensional classical oscillator.* Suppose that you have a two-dimensional classical electron oscillator in which the electron moves in an anisotropic potential well in such a way that the restoring force in the $x$ direction is $-K_x x$, but in the $y$ direction is $K_y y$, where $K_x$ and $K_y$ differ by an amount that is small compared to their average value, but large compared to the fractional linewidth of the atomic transitions. The damping and collision-broadening rates for motion along both axes are the same, and no magnetic field is present. Write the classical electron oscillator susceptibility, including the tensor form, for a macroscopic collection of such oscillators.

3. *Tensor response of a three-dimensional Zeeman-split classical oscillator.* Consider as a classical model for an electric-dipole atom an electron that can move in the $x$, $y$, and $z$ directions about a nucleus with a linear central restoring force, where a dc magnetic field $B_0$ is also present in the $z$ direction as in Problem 1 of Section 3.3. Using the same notation and results as in that problem (except that the electron is now also free to move in the $z$ direction), derive a general expression for the tensor electric susceptibility of a collection of these classical atoms. As in the earlier problem, work out the three separate resonance frequencies of this system (corresponding to separate atomic transitions). Then, making the reasonable assumptions that $\omega_0 \gg \omega_c \gg \Delta\omega_a$ (i.e., small Zeeman splitting, but even smaller atomic linewidths), discuss the tensor character of $\tilde{\chi}(\omega)$.

4. *Field patterns in a "twisted-mode" laser cavity.* Circularly polarized optical signals can be confusing but interesting. Consider, for example, a uniform optical wave that is right-hand circularly polarized looking along its direction of propagation (that is, at any single transverse plane, the field $\mathcal{E}$ of this particular wave rotates from $x$ into $y$ as time $t$ increases). Suppose this wave passes through a quarter-wave plate (QWP); bounces off a mirror at normal incidence; and passes back out through the QWP along the same optical axis, but propagating in the opposite direction.

    [A quarter-wave plate is an optical element made of an anistropic or birefringent material, e.g., crystal quartz, that has two transverse axes; call them the $x$ and $y$

or "fast" and "slow" axes. When an optical wave passes through such an element in the $z$ direction, the $\mathcal{E}_y$ component of the optical field vector $\mathcal{E}$ sees a slightly higher index of refraction $n_y$ than the value $n_x$ seen by the $\mathcal{E}_x$ field component. A quarter-wave plate has a thickness $d$ such that $(n_y - n_x)\omega d/c_0 = (\pi/2)$, i.e., the optical path length through the QWP is one quarter-wavelength longer for one polarization component than the other.]

(a) Develop a "snapshot" of the vector field pattern $\mathcal{E}(z,t)$ in the standing-wave region on the side of the QWP away from the mirror, showing how the $\mathcal{E}$ fields appear at any single instant of time $t$, with whatever amount of analysis or explanation is needed to support this "snapshot." How does this resulting field pattern differ from a RHCP propagating wave?

(b) In an ordinary standing-wave laser cavity with linearly polarized $\mathcal{E}$ fields, there are nulls in the standing-wave field pattern every half-wavelength along the cavity. Laser atoms located at or near these nulls are thus essentially unaffected by the optical signal, and in particular deliver no power or gain to the optical signal, leading to a phenomenon referred to as "spatial hole burning." Would the vector field pattern analyzed above eliminate the problems caused by spatial hole burning?

5. *More on the twisted-mode cavity.* A laser cavity with no Brewster angle surfaces, with a quarter-wave plate at each end of the cavity, and with the principal axes of the two quarter-wave plates rotated by 45° with respect to each other, was invented once as a means of eliminating spatial inhomogeneity effects in lasers. Analyze the axial modes in this cavity, and explain why it may be useful for this purpose. (See my article, "Historical note on spatial hole burning and twisted-mode laser resonators," *Opt. Commun.* **24**, 365, March 1978).

---

## 3.5 THE "FACTOR OF THREE"

One of the more confusing and often-argued aspects of atomic transitions is the "factor of three" that appeared in Equations 3.10 and 3.11, in the definition of oscillator strength, as well as in the trace of the tensor susceptibility in the preceding section. This section gives a brief but accurate explanation both of how this factor arises and of how it must be included in the appropriate theoretical formulas.

### Tensor Power Transfer Rates

We showed in Section 3.4 that the tensor susceptibility for a real atomic transition can be written in the form

$$\chi(\omega) = \tilde{\chi}(\omega)\boldsymbol{T} = -j\chi_0''\boldsymbol{T} \quad \text{(at midband)}, \tag{39}$$

where $\tilde{\chi}(\omega)$, or its midband value $-j\chi_0''$, is a scalar susceptililiy formula (*without* the numerical factor of 3); and $\boldsymbol{T}$ is a dimensionless tensor that we will always normalize to make

$$\operatorname{Tr}[\boldsymbol{T}] = 3. \tag{40}$$

## 3.5 THE "FACTOR OF THREE"

Let us now do some energy-storage and power-transfer calculations. For example, the time-averaged rate of energy transfer per unit volume from an applied field $\mathcal{E}(t)$ to a collection of atoms through the induced polarization $p(t)$ may be written as

$$\frac{dU_a}{dt} = \left\langle \mathcal{E}(t) \cdot \frac{dp(t)}{dt} \right\rangle. \tag{41}$$

For a steady-state sinusoidal response given by $\boldsymbol{P}(\omega) = \boldsymbol{\chi}(\omega)\epsilon \boldsymbol{E}(\omega)$, this leads at midband, $\omega = \omega_a$, to the time-averaged result

$$\frac{dU_a}{dt} = -\frac{\omega_a \chi_0'' \epsilon \left[\boldsymbol{E}^* \cdot \boldsymbol{T}\boldsymbol{E} + \boldsymbol{E} \cdot \boldsymbol{T}^* \boldsymbol{E}^*\right]}{4}. \tag{42}$$

The multiplications on the right-hand side of this equation must be carried out using the standard rules for matrix multiplication, with vectors to the right of the dots considered as column vectors and quantities to the left of the dots considered as row vectors.

The time-averaged stored energy per unit volume in the same signal fields is, however,

$$U_{\text{sig}} = \left\langle \frac{1}{2}\epsilon|\mathcal{E}(t)|^2 \right\rangle = \frac{\epsilon\left[\boldsymbol{E}^* \cdot \boldsymbol{E}\right]}{2}. \tag{43}$$

Hence a ratio of energy transfer rate (to the atoms) over energy stored (in the signal fields) may be written as

$$\frac{1}{U_{\text{sig}}} \frac{dU_a}{dt} = \omega_a \chi_0'' \times \left[\frac{\boldsymbol{E}^* \cdot \boldsymbol{T}\boldsymbol{E} + \boldsymbol{E} \cdot \boldsymbol{T}^* \boldsymbol{E}^*}{2\boldsymbol{E}^* \cdot \boldsymbol{E}}\right]. \tag{44}$$

When the dimensionless ratio in the brackets on the right-hand side of this equation is calculated for various forms of the susceptibility tensor $\boldsymbol{T}$, as given in the previous section, and for various signal field polarizations $\boldsymbol{E}$, its value always turns out to be somewhere between a maximum of 3 and a minimum of 0. (Some examples are shown in Table 3.3. Work a few of these out for practice, using the tensor forms from Section 3.4.)

**TABLE 3.3**
Normalized tensor responses

| Saturated Tensor Form | Gain Applied Field Polarization | Normalized Response |
|---|---|---|
| Circular, $x \to \pm y$ | Circular, $x \to \pm y$ | 3 |
| Circular, $x \to \pm y$ | Circular, $x \to \mp y$ | 0 |
| Circular, $x \to \pm y$ | Linear ($x$ or $y$) | 1.5 |
| Circular, $x \to \pm y$ | Linear ($x$) | 0 |
| Circular, $x \to \pm y$ | Random | 1 |
| Linear ($x$) | Linear ($x$) | 3 |
| Linear ($x$) | Linear (angle $\theta$ from $x$) | $3\cos^2\theta$ |
| Linear ($x$) | Circular, $x \to \pm y$ | 1.5 |
| Linear | Random | 1 |
| Linear ($x$) | Linear ($y$ or $x$) | 0 |
| Isotropic | Arbitrary | 1 |

### The Factor of "Three-Star"

The dimensionless factor that multiplies $\omega_a \chi_0''$ in Equation 3.44 thus always ranges between 0 and 3, depending on the nature of the signal polarization and the normalized tensor response. In fact, for different situations this dimensionless factor takes on values as follows.

- For *aligned* atoms—that is, for any collection of atoms that have a nondegenerate transition, and all of whose atomic axes are aligned in parallel to give an identical tensor response—there is always some optimum signal field polarization that will give this dimensionless factor its maximum value of 3, and thus make $(1/U_{\text{sig}})(dU_a/dt) = 3 \times \omega_a \chi_0''$.
- For such aligned atoms there is always also an "anti-optimum" signal polarization, for which the corresponding value is identically zero. (Linear dipole transitions have, in fact, an entire plane in which the induced response is identically zero.)
- Combining aligned atoms with any other signal polarization between the optimum and anti-optimum forms gives a value for the dimensionless factor somewhere between $3 \times \omega_a \chi_0''$ and $0 \times \omega_a \chi_0''$.

- For *nonaligned* (which is to say, randomly aligned) atoms, and hence an isotropic tensor response, the dimensionless response always has the value of unity, so that $(1/U_{\text{sig}})(dU_a/dt) = 1 \times \omega \chi_0''$.
- In a similar manner, for *randomly polarized* signal fields combined with any atomic alignment, the dimensionless response is also always unity.

In our discussions from here on, it would be nice if we did not have to keep track of the explicit vector nature of the signals or the tensor nature of the atomic responses. In order to do this, while allowing for the tensor nature of the atomic response, we will give this dimensionless factor in Equation 3.44 a name, and include it in the atomic susceptibility expression from now on. That is, we will from now on in this book often write the susceptibility expression for a homogeneous lorentzian atomic transition in the form

$$\tilde{\chi}(\omega) = -j \frac{3^*}{4\pi^2} \frac{\Delta N \lambda^3 \gamma_{\text{rad}}}{\Delta \omega_a} \frac{1}{1 + 2j(\omega - \omega_a)/\Delta \omega_a}, \qquad (45)$$

where the parameter 3* ("three-star") indicates what we will from now on call the "factor of three." This parameter, depending on circumstances, may have the numerical values:

- $3^* = 3$ for fully aligned atoms plus optimally polarized fields; or
- $3^* = 1$ either for randomly aligned atoms with arbitrarily polarized fields, or for randomly polarized fields with any atomic alignment; or
- $3^* = 0$ for fully aligned atoms and "anti-optimum" fields; or
- $0 \leq 3^* \leq 3$ for any intermediate case.

This notation will prove very convenient, especially since, as we will see, this same factor of 3* carries over into many other stimulated-transition and gain formulas as well.

---

Problems for 3.5

1. *Averaging* $\cos^2 \theta$ *over* $4\pi$ *steradians.* Show by direct integration that the average value of $\cos^2 \theta$ averaged over all directions—that is, the value of $(4\pi)^{-1} \int \int \cos^2 \theta \, d\Omega$, where $d\Omega$ is the integral over all solid angles—is 1/3.

---

## 3.6 DEGENERATE ENERGY LEVELS AND DEGENERACY FACTORS

In many real atomic systems, what appears to be a single atomic resonance in a collection of atoms, with a single transition frequency $\omega_{21}$ between upper and lower energy levels $E_2$ and $E_1$, may in fact be the summation of a number of overlapping transitions, with different strengths and polarization properties, between distinct but degenerate sublevels of the upper and lower levels. It is still possible in discussing the small-signal response of such a system to treat such a set of degenerate transitions as a single transition with an isotropic susceptibility. This section shows, however, that we must modify the definition of population inversion on such a degenerate transition by adding certain lower-level and upper-level *degeneracy factors*, in order to take into account the unresolved degeneracies of both the upper and lower levels.

### Degeneracy Factors

Suppose, to be specific, that two apparently discrete energy levels $E_1$ and $E_2$ really each consist of $g_1$ and $g_2$ quantum-mechanically distinct sublevels, respectively, as shown in Figure 3.12. The integers $g_1$ and $g_2$ are then called the *statistical weights* or *degeneracy factors* of the levels. Let $N_1$ and $N_2$ be the *total* populations in levels $E_1$ and $E_2$. At thermal equilibrium the atoms in each level will then be divided equally among the sublevels, with populations $N_1/g_1$ or $N_2/g_2$ in each of the respective sublevels. (There are, moreover, very rapid relaxation processes that usually act to rapidly equalize the populations of degenerate sublevels, even if they are somehow perturbated from equal populations, for example, by a strong applied signal.)

Boltzmann's Law, which relates the relative populations of an upper and lower energy level at thermal equilibrium, then applies rigorously to each distinct energy sublevel. In other words, it says that for any pair of such sublevels the population ratio at thermal equilibrium must be

$$\frac{N_2/g_2}{N_1/g_1} = \exp\left(-\frac{E_2 - E_1}{kT}\right). \tag{46}$$

Hence for the *total* level populations the Boltzmann ratio really must be written in the form

$$\frac{N_2}{N_1} = \frac{g_2}{g_1} \exp\left(-\frac{E_2 - E_1)}{kT}\right). \tag{47}$$

This is a more precise generalization of the Boltzmann Law. Note that as a consequence of this, a highly degenerate upper level might possibly have, at thermal equilibrium, a larger total population than a lower level that is less degenerate (that is, if $g_2/g_1 > \exp[(E_2 - E_1)/kT]$). This is not a population inversion in any sense, however—for example, it does not lead to net stimulated emission or gain, as we will now show.

### Net Susceptibility of a Degenerate Transition

To evaluate the overall stimulated response on a degenerate transition, we must sum over all the individual subtransitions, as shown in Figure 3.12. Let us label all the upper sublevels by an index $m$ that runs from $m = 1$ to $m = g_2$, and all the lower sublevels by a similar index $n$. The total response on the transition is then the sum over $n$ and $m$ of all transitions between all the sublevels $E_{1n}$ and $E_{2m}$.

The tensor susceptibility on any one such transition between level $E_{1n}$ and level $E_{2m}$ may then be written in the form

$$\chi_{1n,2m}(\omega) = \tilde{g}(\omega) \times \gamma_{\text{rad},2m\to 1n} \times \left(\frac{N_1}{g_1} - \frac{N_2}{g_2}\right) \times \boldsymbol{T}_{1n,2m}, \tag{48}$$

## 3.6 DEGENERATE ENERGY LEVELS AND DEGENERACY FACTORS

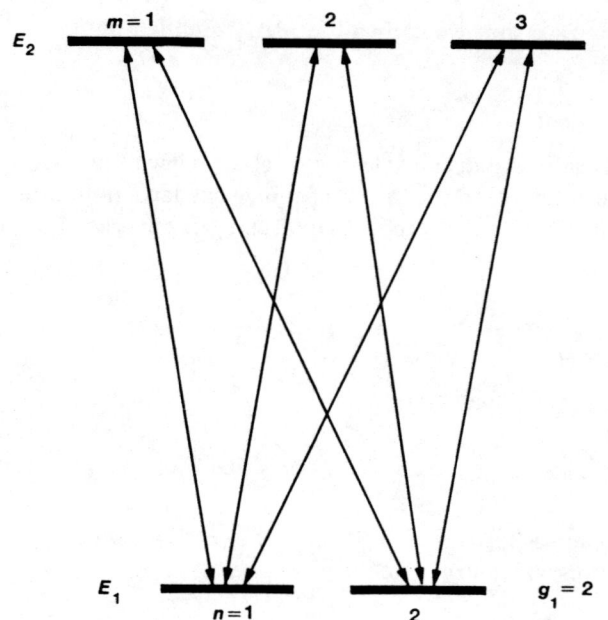

FIGURE 3.12
Degenerate sublevels of two quantum energy levels $E_1$ and $E_2$. Each sublevel is a separate and distinct quantum energy eigenstate, but the degenerate sublevels all have the same energy eigenvalue.

where $\boldsymbol{T}_{1n,2m}$ is the tensor response on that particular transition; $N_1/g_1 - N_2/g_2$ gives the population difference on that particular transition; $\gamma_{\text{rad},2m\to 1n}$ gives the strength of that particular transition; and the lineshape function $\tilde{g}(\omega)$ can usually be assumed to be the same for all transitions, namely,

$$\tilde{g}(\omega) = -j\frac{1}{4\pi^2}\frac{\lambda^3}{\Delta\omega_a}\frac{1}{1+2j(\omega-\omega_a)/\Delta\omega_a}. \tag{49}$$

(If different subtransitions have different linewidths, this will complicate the following analysis, but probably not change the results.)

The total response on all the $1n \to 2m$ transitions can then be written as

$$\chi_{\text{tot}}(\omega) = \sum_{n=1}^{g_1}\sum_{m=1}^{g_2} \chi_{1n,2m}(\omega)$$

$$= \tilde{g}(\omega) \times \sum_{n=1}^{g_1}\sum_{m=1}^{g_2} \gamma_{\text{rad},2m\to 1n}\left(\frac{N_1}{g_1}-\frac{N_2}{g_2}\right) \times \boldsymbol{T}_{\text{av}}, \tag{50}$$

where $\boldsymbol{T}_{\text{av}}$ is some averaged tensor susceptibility over all the transitions involved. This tensor will simply be isotropic if the average is over a complete set of degenerate transitions.

At the same time, the *total* radiative decay rate downward out of the upper level $E_2$ will be given by

$$\frac{dN_2}{dt} = -\sum_{n=1}^{g_1}\sum_{m=1}^{g_2} \gamma_{\text{rad},2m\to 1n}\left(\frac{N_2}{g_2}\right). \tag{51}$$

That is, we must sum over all radiative decay rates from all upper sublevels to all lower sublevels. This total downward rate may then be equated to an averaged or measured radiative decay rate that we will call $\gamma_{\text{rad},2\to1}$, defined by

$$\frac{dN_2}{dt} = -\gamma_{\text{rad},2\to1} N_2. \tag{52}$$

This will be the measured radiative decay rate for level $E_2$ viewed as a single effective level without degeneracy taken into account. Since the level populations can be taken outside the sums in all the preceding equations, this averaged decay rate is given by

$$\gamma_{\text{rad},2\to1} \equiv \frac{1}{g_2} \sum_n \sum_m \gamma_{\text{rad},2m\to1n} \tag{53}$$

Combining Equations 3.48 to 3.53 then gives

$$\chi_{\text{tot}}(\omega) = \tilde{g}(\omega) \times \gamma_{\text{rad},2\to1} \times \left(\frac{g_2}{g_1} N_1 - N_2\right) \times T_{\text{av}}. \tag{54}$$

If we absorb the tensor properties into a factor 3* as in the previous section, this may be written as a scalar susceptibility

$$\tilde{\chi}_{\text{tot}}(\omega) = -j \frac{3^*}{4\pi^2} \frac{\lambda^3 \gamma_{\text{rad},2\to1}}{\Delta\omega_a} \left(\frac{g_2}{g_1} N_1 - N_2\right) \frac{1}{1 + 2j(\omega - \omega_a)/\Delta\omega_a}, \tag{55}$$

where $T_{\text{av}}$ will normally be isotropic and 3* will be equal to unity.

This final result now looks exactly like the nondegenerate susceptibility expression in earlier sections, except that the population difference $\Delta N$ is replaced by

$$\Delta N \Rightarrow \left(\frac{g_2}{g_1} N_1 - N_2\right). \tag{56}$$

A more precise condition for population inversion and gain on an atomic transition is thus

$$\frac{N_2}{g_2} > \frac{N_1}{g_1} \tag{57}$$

and not just $N_2 > N_1$. In physical terms, there must be true population inversion on the individual sublevels, and not simply $N_2 > N_1$.

For degenerate transitions in gases, with randomly aligned atoms, the averaged tensor susceptibility $T_{\text{av}}$ will in fact always be isotropic, leading to 3* = 1 in Equation 3.55. For degenerate transitions in solids the situation may be somewhat more complex, and a degenerate transition may still have some anisotropic character to its tensor response $T_{\text{av}}$.

### Discussion

The main result of this section, then, is that the small-signal steady-state response on a degenerate transition is exactly the same as for a nondegenerate transition, except that the effective value of $\gamma_{\text{rad},2\to1}$ must be employed, and the effective population difference becomes $\Delta N \equiv (g_2/g_1)N_1 - N_2$ rather than just $N_1 - N_2$. We will use this result where appropriate in future sections.

This result does assume that the various sublevels of each main level remain equally populated, so that we can assign an equal fraction $N_j/g_j$ of the level population to each one of them. For very strong signals, and perhaps also for very short pulses (short compared to $T_2$), some of the transitions between sublevels $E_{1n}$ and $E_{2m}$ will respond more strongly to an applied signal than will others, because of substantially different values of $\gamma_{\text{rad},2m\to 1n}$, as well as different polarization properties. A strong applied signal will then cause atoms to flow from certain sublevels $E_{1n}$ to other sublevels $E_{2m}$ at quite different rates; and this difference will tend to unbalance the otherwise equal sublevel populations, especially if for some reason the relaxation between the sublevels is slowed down. This kind of selective pumping between lower and upper sublevels, especially when the degeneracy has been slightly broken, is in fact an essential element of a spectroscopic technique referred to as *optical pumping*.

Unless the degeneracy between sublevels is at least partially broken, however, there will usually also be relaxation processes between sublevels that will tend to rapidly return the sublevel populations to equality. These so-called "cross-relaxation" processes can be especially fast, because no energy change is required to relax an atom from one sublevel to another sublevel within the same degenerate main level. Strong applied signals can thus override these relaxation processes, but only temporarily.

The general warning to be taken is the following: In considering the effects of very strong (or very short-pulse) signals, for example, in so-called "coherent pulse" experiments, a degenerate transition can no longer be treated as a slightly modified single transition. It must instead be treated in detail as a set of multiple, independent, though still closely coupled transitions all at the same frequency.


## REFERENCES

A good readable discussion of some of the limitations warned about in the last paragraph is A. Dienes, "On the physical meaning of the 'two nondegenerate levels' atomic model in nonlinear calculations," *IEEE J. Quantum Electr.* **QE-4**, 260–263 (May 1968). See also B. W. Shore, "Effects of magnetic sublevel degeneracy on Rabi oscillations," *Phys. Rev. A* **17**, 1739–1746 (May 1978).

Self-induced transparency is one of the large-signal situations in which the behavior with degeneracy present is considerably different from that in the simple nondegenerate case. This is discussed in detail by C. K. Rhodes, A. Szöke, and A. Javan, "The influence of level degeneracy on the self-induced transparency effect," *Phys. Rev. Lett.* **21**, 1151–1155 (October 14, 1978).

Rather complex equations result if we take full account of the level degeneracies in a gas laser, especially if we include the small Zeeman splitting of the nearly degenerate transitions that results when either a longitudinal or a transverse dc magnetic field is applied. One example from among the extensive literature on this topic is M. Sargent III and W. E. Lamb, Jr., "Theory of a Zeeman laser. I and II," *Phys. Rev.* **164**, 436–465 (December 10, 1967).


## 3.7 INHOMOGENEOUS LINE BROADENING

As the final step in describing the resonant response of real atomic transitions, we must introduce an additional and important type of line broadening known as *inhomogeneous broadening*, of which doppler broadening is the premier example.

### Homogeneous Broadening

The steady-state response of a homogeneously broadened transition in a collection of oscillators or atoms is given by the complex lorentzian formula

$$\tilde{\chi}_h(\omega;\omega_a) = -j\frac{3^*}{4\pi^2}\frac{\Delta N \lambda^3 \gamma_{\rm rad}}{\Delta\omega_a}\frac{1}{1+2j(\omega-\omega_a)/\Delta\omega_a}. \qquad (58)$$

We have attached a subscript $h$ to indicate that this is the usual form for a *homogeneous* transition; and we have added the second argument to $\tilde{\chi}_h(\omega;\omega_a)$ to indicate the explicit dependence on the resonance frequency $\omega_a$ along with the applied frequency or signal frequency $\omega$. This kind of broadening is called *homogeneous broadening* because the response of each individual atom in the collection is equally and homogeneously broadened. Many real atomic transitions, under appropriate conditions, exhibit exactly this lineshape.

### Inhomogeneous Broadening

In many other real atomic situations, however, different atoms in a collection of nominally identical atoms may, for various reasons, have slightly different resonant frequencies $\omega_a$, such that the $\omega_a$ values for different atoms are randomly distributed about some central value $\omega_{a0}$. We must then think of the resonance frequencies $\omega_a$ for different atoms as being randomly shifted by small but different amounts for each atom in the collection.

An applied signal passing through such a collection of atoms will then see only a total response due to all the atoms—it will have no way to pick out only those atoms with certain specific frequency shifts. If the random shifting of the individual center frequencies is sizable compared to the linewidth $\Delta\omega_a$ of each individual response, any measurement of the overall response from all the atoms in the collection will then give a smeared-out or broadened summation of the randomly shifted responses of all the individual atoms (see Figure 3.13). The overall response of the collection of atoms will be substantially broadened, and the response at line center will be substantially reduced in amplitude. This general type of behavior is referred to as *inhomogeneous broadening*.

### Spectral Packets

That subgroup of atoms whose resonant frequencies $\omega_a$ all fall within a range of roughly one homogeneous linewidth $\Delta\omega_a$ about a given value of $\omega_a$ is often referred to as a single *spectral packet* (or *spin packet* in magnetic-resonance jargon). All the atoms in a single packet have essentially the same (homogeneous) response to an applied signal. The total response of an inhomogeneously broadened line is then the sum of the individual responses of all the spectral packets, each at a different resonance frequency.

If the individual packets are spread out in frequency about $\omega_{a0}$ by an amount large compared to their individual homogeneous widths $\Delta\omega_a$, as in Figure 3.13, the line is said to be *strongly inhomogeneous*. If the inhomogeneous shifting is small compared to the homogeneous packet widths, the line will remain essentially homogeneous, and the amount of inhomogeneous broadening that does exist will be of little importance.

## 3.7 INHOMOGENEOUS LINE BROADENING

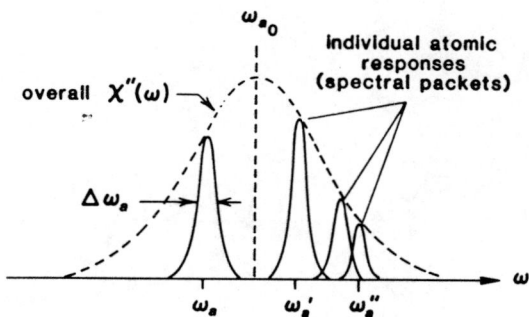

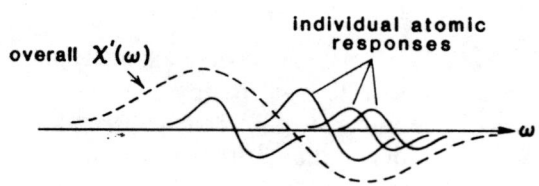

FIGURE 3.13
Individual atomic responses, or "spectral packets," within an inhomogeneously broadened atomic transition.

### Causes of Inhomogeneous Broadening

There are several possible causes of random resonance-frequency shifting and thus of inhomogeneous broadening in typical atomic systems.

- In gases, different atoms will have different kinetic velocities through space. This kinetic motion produces a doppler shift in the frequency of an applied signal as seen by the atom, or alternatively a doppler shift in the apparent resonance frequency $\omega_a$ of the atom as seen by the applied signal. This so-called *doppler broadening* is an important and widespread source of inhomogeneous broadening for optical-frequency transitions in atomic and molecular gases.

- In solids, laser atoms at different sites in a crystal may see slightly different local surroundings, or different local crystal structures, because of defects, dislocations, or lattice impurities. This produces slightly different values for the exact energy levels of the atoms, and thus slight shifts in transition frequencies. To the extent that the local lattice surroundings are similar for every atom but vibrate rapidly and randomly in time, they produce a dynamic *homogeneous phonon broadening*. To the extent that the surroundings are different from site to site but static in time, they produce a static *inhomogeneous lattice broadening* or *strain broadening*.

Other types of inhomogeneous broadening also exist (for example, inhomogeneous dc magnetic fields in magnetic resonance experiments), but these are two of the most important for optical transitions.

**160**   CHAPTER 3: ELECTRIC-DIPOLE TRANSITIONS IN REAL ATOMS

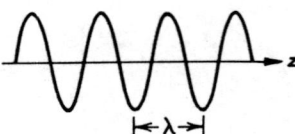

**FIGURE 3.14**
Doppler shift for an atom moving through an electromagnetic wave.

### Doppler Broadening

One of the most common examples of inhomogeneous broadening is *doppler broadening* of the resonance transitions in gases. The atoms in an atomic or molecular gas will have, in addition to their internal oscillations, thermal or Brownian kinetic motion through space, with a maxwellian distribution of kinetic velocities. When an atom moving with velocity $v_z$ as in Figure 3.14 interacts with a wave of signal frequency $\omega$ traveling at velocity $c$ along the $z$ direction (for example, a wave traveling down the axis of a laser tube), the frequency of the wave as seen by the atom will be doppler-shifted to a new value $\omega'$ given by

$$\omega' = (1 - v_z/c)\,\omega. \tag{59}$$

Resonance of the applied signal with the atomic transition in that particular atom will then occur when the doppler-shifted signal frequency $\omega' = \omega(1 - v_z/c)$ seen by the moving atom equals the atom's internal resonance frequency $\omega_{a0}$.

From an alternative viewpoint, resonance will occur when the signal frequency $\omega$ measured in the laboratory frame equals the shifted resonance value $\omega_{a0}(1 + v_z/c)$. In other words, as seen from the lab the resonance frequency of the atom appears to be doppler-shifted to a new value,

$$\omega_a = (1 + v_z/c)\,\omega_{a0}. \tag{60}$$

For an atom or molecule of mass $M$ in a gas at temperature $T$, the kinetic velocity $v_z$ has a mean-square value given by $M\langle v_z^2 \rangle \approx kT$. Hence the average doppler shift for a moving gas atom will be of order

$$\frac{\omega_a - \omega_{a0}}{\omega_{a0}} \approx \sqrt{\frac{kT}{Mc^2}} \approx 10^{-6} \quad \left(\begin{array}{c}\text{for typical atomic}\\ \text{masses and temperatures}\end{array}\right). \tag{61}$$

The amount of doppler broadening in a real gas thus depends (but only rather slowly) on the kinetic temperature $T$ of the gas and on the molecular weight of the atom or molecule involved.

### Doppler Lineshape

To be more precise, the distribution of axial velocities $v_z$ in a gas at thermal equilibrium will be a maxwellian, or gaussian, probability distribution given by

$$g(v_z) = \left(\frac{1}{2\pi\sigma_v^2}\right)^{1/2} \exp\left(-\frac{v_z^2}{2\sigma_v^2}\right) \tag{62}$$

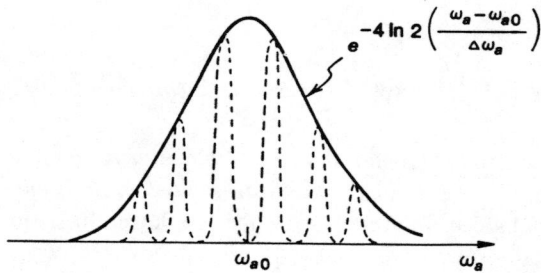

FIGURE 3.15
A gaussian inhomogeneous atomic lineshape, such as is produced by doppler broadening in atoms.

with an rms spread given by $\sigma_v^2 = kT/M$. The inhomogeneous distribution of shifted resonant frequencies, call it $g(\omega_a)$, for a doppler-broadened atomic transition will then similarly have a gaussian form that can be written as

$$g(\omega_a) = \left(\frac{4\ln 2}{\pi \Delta\omega_d^2}\right)^{1/2} \exp\left[-(4\ln 2)\left(\frac{\omega_a - \omega_{a0}}{\Delta\omega_d}\right)^2\right] \qquad (63)$$

as illustrated in Figure 3.15. This expression has been written so that $\omega_{a0}$ is the center frequency; and following our standard convention, the linewidth $\Delta\omega_d$ has been defined to be the FWHM linewidth of the gaussian distribution, which means it must take on the form

$$\Delta\omega_d = \sqrt{\frac{(8\ln 2)kT}{Mc^2}}\,\omega_{a0}. \qquad (64)$$

It is useful to remember that in units of electron volts, $kT$ at room temperature is 1/40 of an electron volt or 25 meV; and $Mc^2$ is the rest-mass energy of the atom, which for a single proton is $\approx 10^9$ eV. For an atom or molecule with an atomic number of 20, the fractional doppler broadening is thus

$$\frac{\Delta\omega_d}{\omega_{a0}} = \sqrt{\frac{(8\ln 2)kT}{Mc^2}} \approx \sqrt{\frac{5.5 \times 25 \times 10^{-3}}{20 \times 10^9}} \approx 2.6 \times 10^{-6} \qquad (65)$$

or typically a few parts per million. A visible laser transition will have a center frequency on the order of $\omega_a/2\pi \approx 6 \times 10^{14}$ Hz, and a doppler broadening on the order of $\Delta\omega_d/2\pi \approx 2 \times 10^9$ Hz $\approx 2$ GHz. The room-temperature doppler broadening of the He-Ne laser transition at 633 nm, in fact, is just about $\Delta\omega_d/2\pi \approx 1{,}500$ MHz.

### General Analysis of Inhomogeneous Broadening

As a more general approach to inhomogeneous broadening, suppose we consider a large collection of nominally identical atoms, with the fractional number of atoms whose exact resonant frequency is between some value $\omega_a$ and $\omega_a + d\omega_a$ being given by

$$dN(\omega_a) = Ng(\omega_a)\,d\omega_a, \qquad (66)$$

where $N$ is the total number of atoms. (We really should use the population difference $\Delta N$ here, but let's write $N$ instead for simplicity.) The function $g(\omega_a)$ is thus the probability density distribution over the resonant frequencies $\omega_a$, with the normalization that

$$N^{-1}\int_0^\infty dN(\omega_a) = \int_{-\infty}^\infty g(\omega_a)\,d\omega_a = 1. \tag{67}$$

The function $g(\omega_a)$ is always very narrowly clustered about the center frequency $\omega_{a0}$, so any portion of the analytic function $g(\omega_a)$ extending below $\omega_a = 0$ can be ignored. It therefore makes negligible difference whether the lower limit in this normalization integral is actually 0 or $-\infty$.

To calculate the overall complex susceptibility of any such collection of atoms, we must then multiply the homogeneous response $\tilde{\chi}_h(\omega;\omega_a)$ produced by any one atom whose resonance frequency is $\omega_a$ by the fractional number of atoms $g(\omega_a)d\omega_a$ that have the same resonance frequency $\omega_a$, as illustrated in Figure 3.13, and then integrate that response over all values of $\omega_a$ in the form

$$\tilde{\chi}(\omega) = \int_{-\infty}^\infty \tilde{\chi}_h(\omega;\omega_a)\,g(\omega_a)\,d\omega_a. \tag{68}$$

Suppose the distribution $g(\omega_a)$ is gaussian, as it often is. The full-blown equation for the complex small-signal susceptibility of an inhomogeneously broadened transition thus becomes a gaussian distribution of frequency-shifted lorentzian lines. If we write this out in full, it takes the general form

$$\tilde{\chi}(\omega) = -j\frac{3^*}{4\pi^2}\sqrt{\frac{4\ln 2}{\pi}}\,\frac{N\lambda^3\gamma_{\mathrm{rad}}}{\Delta\omega_a\Delta\omega_d}\int_{-\infty}^\infty \frac{1}{1+2j(\omega-\omega_a)/\Delta\omega_a}$$
$$\times \exp\left[-(4\ln 2)\left(\frac{\omega_a-\omega_{a0}}{\Delta\omega_d}\right)^2\right]d\omega_a. \tag{69}$$

This rather messy integral must be evaluated each time an accurate calculation is needed of the susceptibility of an atomic transition in which doppler broadening is important. (Even this integral still ignores certain large-signal saturation and "hole-burning" effects that we will discuss in a later chapter.)

Inhomogeneous broadening in general, whether due to doppler broadening or to other mechanisms, is usually caused by some kind of random distribution of velocities, or defects, or whatever; and random distributions, whatever their cause, are very often gaussian in form (as sometimes expressed in the Central Limit Theorem). We will therefore interpret the gaussian expression for doppler broadening in Equation 3.63 somewhat more broadly, and use it as a general expression for $g(\omega_a)$ in any kind of inhomogeneous broadening. Similarly, we will use $\Delta\omega_d$ as a general notation for the inhomogeneous linewidth of an inhomogeneously broadened distribution, whether this is due to doppler broadening or to some other cause.

### Strongly Homogeneous Limit

The integral in Equation 3.69 cannot be evaluated analytically, at least not for arbitrary ratios of inhomogeneous broadening $\Delta\omega_d$ to homogeneous broadening $\Delta\omega_a$. The limiting cases of strongly homogeneous broadening and strongly inhomogeneous broadening can, however, be handled, at least approximately, as follows.

Let us suppose first that the inhomogeneous broadening effects are small, which means either that the resonance frequencies of individual packets are shifted by very little compared to the homogeneous linewidth $\Delta\omega_a$, or alternatively that the individual packets have a wide homogeneous linewidth $\Delta\omega_a$ compared to the inhomogeneous linewidth $\Delta\omega_d$. The inhomogeneous distribution, whether gaussian or otherwise, is then essentially a delta function, i.e.,

$$g(\omega_a) \approx \delta(\omega_a - \omega_{a0}) \qquad \text{if } \Delta\omega_d \ll \Delta\omega_a. \tag{70}$$

The integral in Equation 3.69 is now trivial, and physically obvious: the overall response is simply the unperturbed homogeneous form $\tilde{\chi}_h(\omega;\omega_{a0})$. In effect there is no inhomogeneous or doppler broadening. This is commonly known as the *strongly homogeneous* limit.

As one practical example of this, consider the 10.6 $\mu$m TEA $CO_2$ laser operating at atmospheric pressure. The inhomogeneous doppler broadening for this long-wavelength transition is $\Delta\omega_d/2\pi \approx 60$ MHz, whereas the homogeneous pressure broadening at one atmosphere is $\Delta\omega_a/2\pi \approx 6$ GHz or 6,000 MHz. The individual packets are thus $\approx 100$ times wider than the doppler broadening, and the line is essentially homogeneous.

### Strongly Inhomogeneous Limit

Now suppose instead that the inhomogeneous linewidth $\Delta\omega_d$ is large enough to shift the spectral packets widely in frequency compared to their homogeneous linewidth $\Delta\omega_a$, so that there are many packets within the overall linewidth. It is then possible in this limit to obtain an analytic approximation to Equation 3.69 that is reasonably accurate for the imaginary part $\chi''(\omega)$ of the overall inhomogeneous susceptibility, though not for the $\chi'(\omega)$ part.

The approximation for the absorptive part of the overall susceptibility is obtained by expanding the complex lorentzian $\tilde{\chi}_h(\omega;\omega_a)$ inside the general integral into its real and imaginary parts. In the limit as $\Delta\omega_a$ becomes small, the $\chi''_h$ part of the homogeneous function becomes roughly like a delta function, i.e.,

$$\frac{2}{\pi\Delta\omega_a} \frac{1}{1 + [2(\omega - \omega_a)/\Delta\omega_a]^2} \approx \delta(\omega - \omega_a) \qquad \text{if } \Delta\omega_a \ll \Delta\omega_d. \tag{71}$$

This lorentzian curve is not a very good delta function, since its wings fall off only as $1/(\omega - \omega_a)^2$ far from line center, but it is adequate here. Putting this into the general equation and integrating over the delta function then gives for the $\chi''(\omega)$ part of the susceptibility

$$\chi''(\omega) \approx -\sqrt{\pi \ln 2}\, \frac{3^*}{4\pi^2}\, \frac{\Delta N \lambda^3 \gamma_{\text{rad}}}{\Delta\omega_d}\, \exp\left[-(4\ln 2)\left(\frac{\omega - \omega_{a0}}{\Delta\omega_d}\right)^2\right] \tag{72}$$

in the strongly inhomogeneous limit where $\Delta\omega_d \gg \Delta\omega_a$.

This expression for a strongly inhomogeneous absorption line has the following interesting features in comparison with the usual homogeneous lorentzian absorption line.

- It has a gaussian, not a lorentzian, lineshape for the absorption profile of $\chi''(\omega)$, with a FWHM linewidth of $\Delta\omega_d$, not $\Delta\omega_a$.

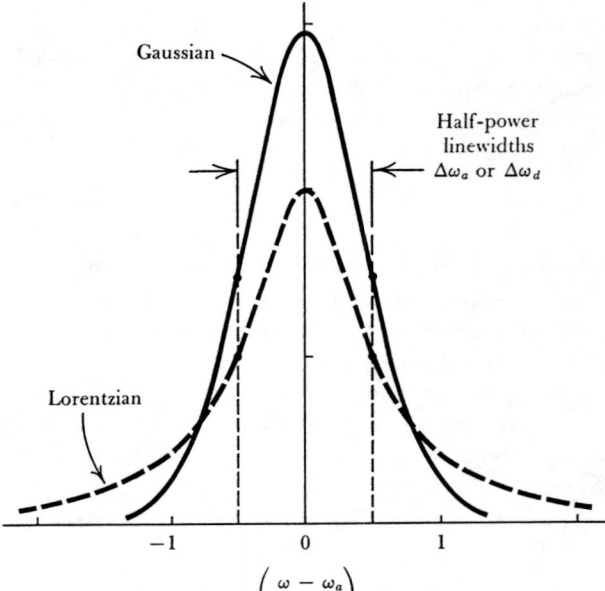

FIGURE 3.16
Comparison of gaussian and lorentzian lineshapes having the same half-power linewidth and the same total area.

- It has essentially the same constant factors in front as does the homogeneous lorentzian response for $\chi''(\omega)$, except that the response now varies as $1/\Delta\omega_d$, instead of $1/\Delta\omega_a$.
- But it has an extra numerical factor of $\sqrt{\pi \ln 2} \approx 1.48$ in front of the other factors that appear in the lorentzian expression.

In fact, *these three simple modifications convert the $\chi''(\omega)$ susceptibility expression for a homogeneous lorentzian transition into the corresponding expression for a strongly inhomogeneous gaussian or doppler-broadened transition.*

Figure 3.16 shows lorentzian and gaussian (i.e., strongly homogeneous and strongly inhomogeneous) susceptibilities $\chi''(\omega)$ normalized to the same FWHM linewidth and the same area. Note that the gaussian absorption curve $\chi''(\omega)$ has a peak value that is $\approx 50\%$ higher, but that it drops off much faster in the wings than does the lorentzian. The integrated area under each curve is the same, since the smaller area in the wings of the gaussian profile is balanced by the 50% larger peak intensity at the center.

It is also interesting to note that the homogeneous packet linewidth $\Delta\omega_a$ actually does not appear at all in the strongly inhomogeneous expression given in Equation 3.72. Measuring the $\chi''(\omega)$ response of a strongly inhomogeneous line tells you $\Delta\omega_d$, but it does not give any information about the homogeneous linewidth $\Delta\omega_a$ of the packets buried within the line—at least not to first order.

### Complex Susceptibility in the Strongly Inhomogeneous Limit

It is not possible to develop a similar approximation for the reactive part of the susceptibility, $\chi'(\omega)$, in the strongly inhomogeneous limit. The reason is essentially that although $\chi''(\omega)$ varies like $1/(1+\omega^2)$ in frequency, which is a weak delta function, the real part $\chi'(\omega)$ varies like $\omega/(1+\omega^2)$, which is not a delta

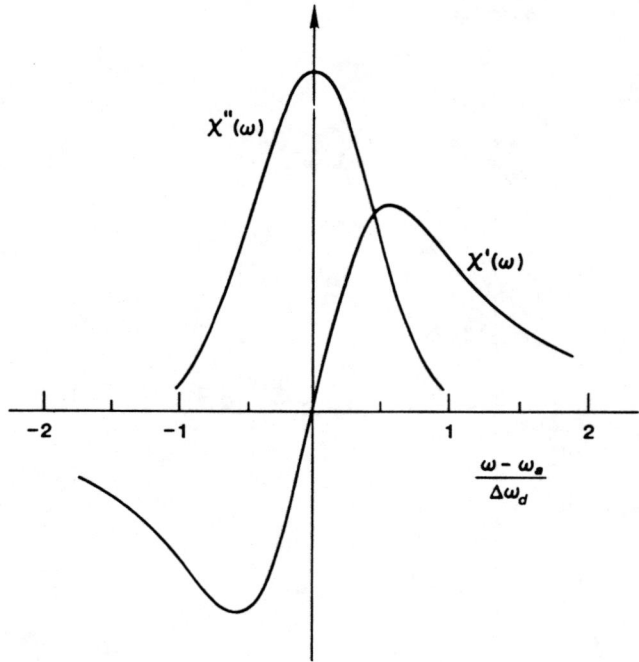

FIGURE 3.17
Exact plots of the real and imaginary parts of the complex susceptibility for a strongly inhomogeneous gaussian transition.

function at all. There is thus no analytic approximation to the exact integral of Equa;tion 3.69 for $\chi'(\omega)$ even in the strongly inhomogeneous limit.

Figure 3.17 does show numerically computed plots of both $\chi'(\omega)$ and $\chi''(\omega)$ for the strongly inhomogeneous gaussian limit, $\Delta\omega_d \gg \Delta\omega_a$. The inhomogeneous susceptibility $\chi'(\omega)$, though it cannot be analytically approximated, looks in general very much like the lorentzian case; i.e., it is antisymmetric and looks generally (though not exactly) like the first derivative of the $\chi''(\omega)$ curve.

### Intermediate Region: Voight Profiles and Their Uses

In the intermediate region where $\Delta\omega_a \approx \Delta\omega_d$ and neither of the limiting approximations is valid, the general expression for $\tilde{\chi}(\omega)$ in Equation 3.69 can only be integrated numerically and then plotted for different ratios of $\Delta\omega_a/\Delta\omega_d$. The lineshapes for the $\chi''(\omega)$ curves that are obtained in this region are obviously intermediate between lorentzian and gaussian lineshapes, and are generally referred to as *Voight profiles*. The exact shape of the Voight profile depends on both the homogeneous and the inhomogeneous linewidths, or, more precisely, on the ratio of these two linewidths.

Figure 3.18 shows, for example, the measured absorption profile for a molecular transition in carbon monoxide (the $v'' = 0, J'' = 11$ to $v' = 1, J' = 10$ transition) at a wavelength of $\lambda = 4.76$ μm or $1/\lambda = 2,099$ cm$^{-1}$, as measured with a tunable laser in a 10:2:88 mixture of $CO_2$:$H_2$:Ar at a temperature of 3,340 K and a pressure of 0.195 atm. (These rather unusual conditions were obtained in a special shock-tube measuring apparatus.) This absorption profile is clearly best matched by a Voight profile somewhere intermediate between a gaussian and a lorentzian.

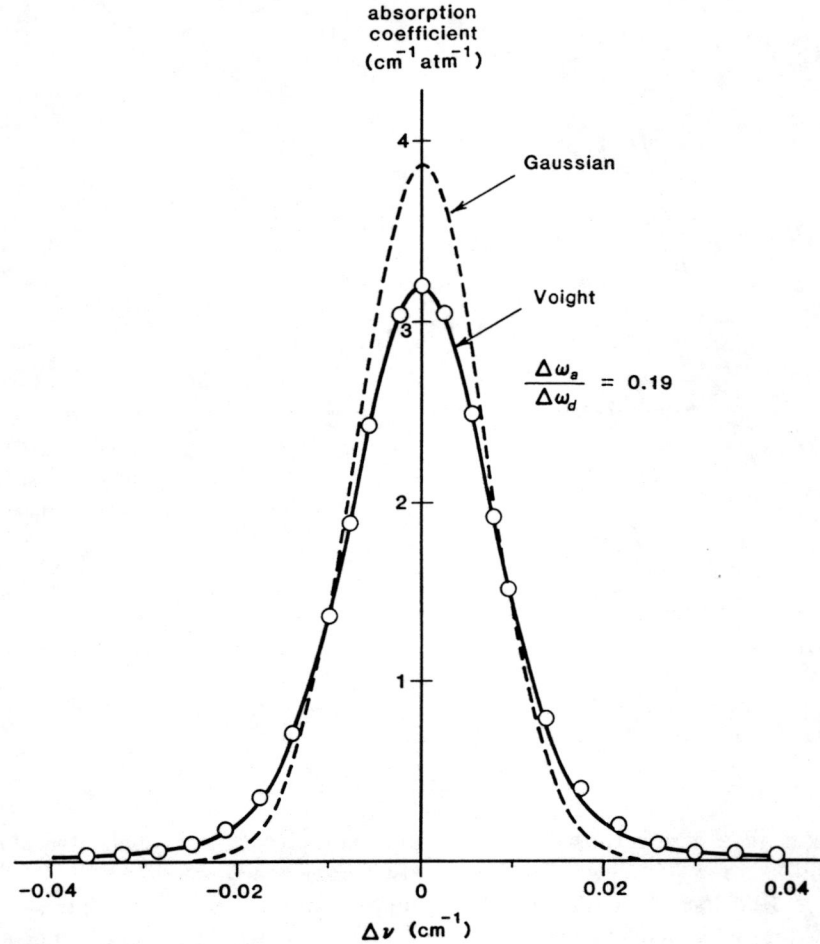

FIGURE 3.18
Measured values of the absorption profile for carbon monoxide (circles) compared with gaussian and Voight functions having the same half-power values.

If we have an experimental plot of $\chi''(\omega)$ for a transition in this intermediate region that has been measured with sufficient accuracy, we can in fact deconvolve the lorentzian and gaussian contributions by fitting the measured curve to numerically a computed Voight profile with the proper ratio of $\Delta\omega_a/\Delta\omega_d$. Since we can predict the doppler linewidth for a given transition in a gas fairly accurately from the theoretical expression in Equation 3.64, we can then use the ratio of $\Delta\omega_a/\Delta\omega_d$ from this kind of Voight profile determinations to derive the homogeneous linewidth $\Delta\omega_a$, provided it is not too small compared to $\Delta\omega_d$. Figure 3.19 shows, for example, absorption data for various pressures of pure $CO_2$ taken with a tunable $CO_2$ laser and fitted to Voight profiles. (The absorption measurement technique used here was actually a more effective way of measuring weak absorptions, called photoacoustic spectroscopy.) The top trace shows the laser tuning curve, and the middle traces show raw data, with frequency markers every 30 MHz of frequency tuning. The lower plot shows this data normalized and fitted to a series of Voight profiles with increasing amounts of lorentzian pressure broadening.

## 3.7 INHOMOGENEOUS LINE BROADENING

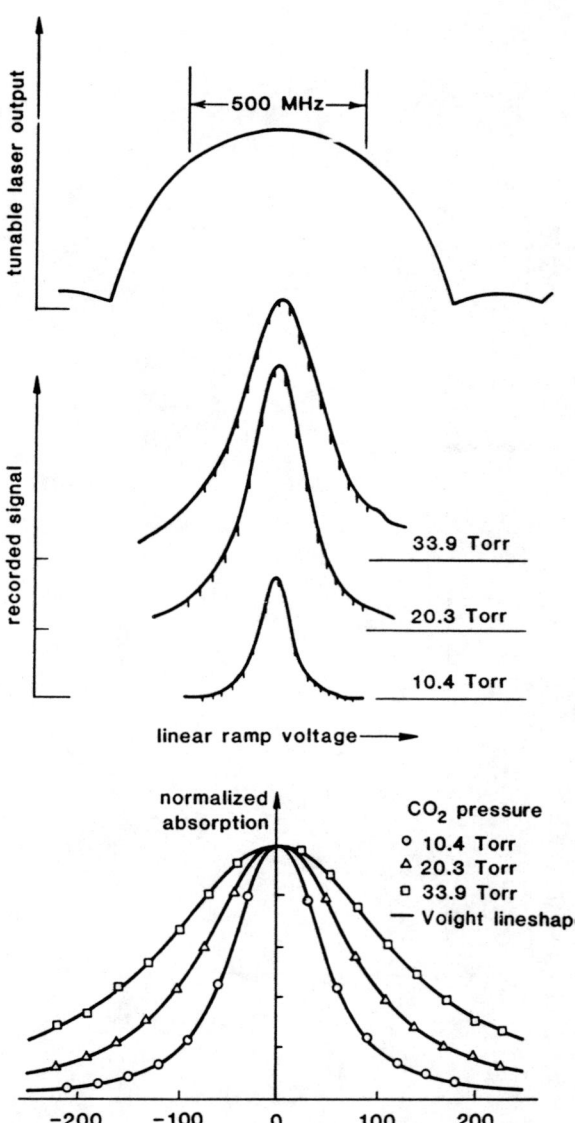

FIGURE 3.19
Top: The power output from a tunable $CO_2$ laser versus frequency tuning. Middle: Three absorption profiles for pure $CO_2$ at different pressures measured using this laser. Bottom: This same data fitted to Voight profiles with different degrees of inhomogeneous broadening.

### The Transition From Doppler to Pressure Broadening

If we gradually increase the gas pressure in an absorption cell, the measured absorption profile of a transition in the gas atoms will change over from being doppler-broadened at low pressures ($\Delta\omega_a \ll \Delta\omega_d$) to being pressure broadened at high pressures ($\Delta\omega_a \gg \Delta\omega_d$).

Figure 3.20(a) shows, as one example, an apparatus for making accurate measurements of the absorption profiles of various $CO_2$ gas mixtures at different pressures using a tunable $CO_2$ laser. In (b) we see direct midband absorption data versus total gas pressure measured on a typical $He:N_2:CO_2$ gas mixture, and (c) shows the atomic linewidth deduced from this data. Both curves illustrate the changeover from inhomogeneous doppler broadening at low pressures to homogenous pressure (collision) broadening at high pressures. Figure 3.21 which

168     CHAPTER 3: ELECTRIC-DIPOLE TRANSITIONS IN REAL ATOMS

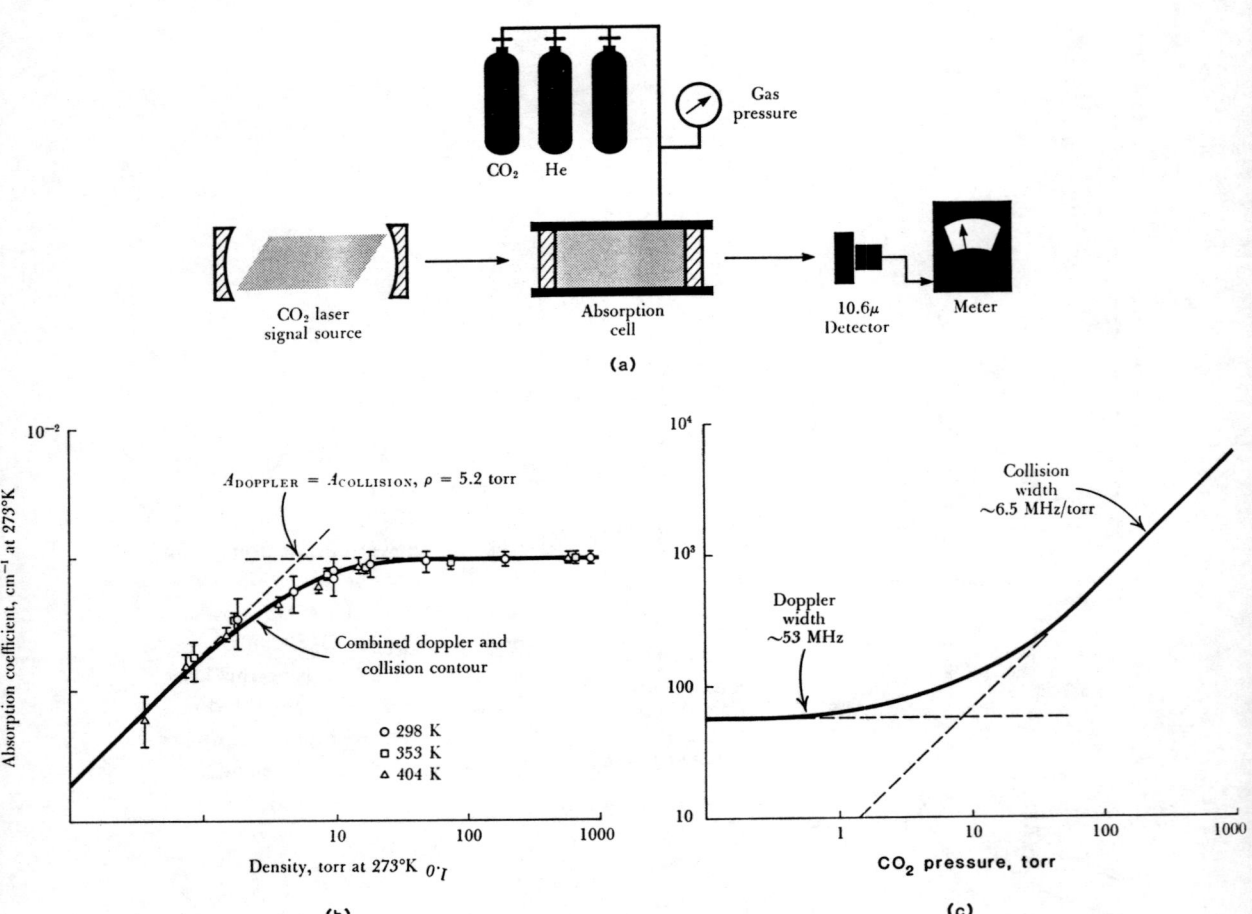

FIGURE 3.20
(a) Apparatus using a tunable $CO_2$ laser to measure absorption versus frequency in a variable-pressure $CO_2$ cell. (b) Midband (peak) absorption versus gas pressure; and (c) linewidth versus pressure in a typical He:Ne:$CO_2$ mixture, showing changeover from doppler broadening at low pressures to pressure (collision) broadening above about 10 torr total pressure. (Data from E. T. Gerry and D. A. Leonard, *Appl. Phys. Lett.* **8**, 227, May 1, 1966.)

shows a very similar variation with pressure of the absorption coefficient on a certain chemical laser transition in the mid-IR using deuterium fluoride (DF) molecules (see Problems).

### An Alternative Notation: $T_2$ and $T_2^*$

The lorentzian and gaussian lineshapes that we have developed in this section are often expressed in an alternative notation, which we can briefly summarize as follows.

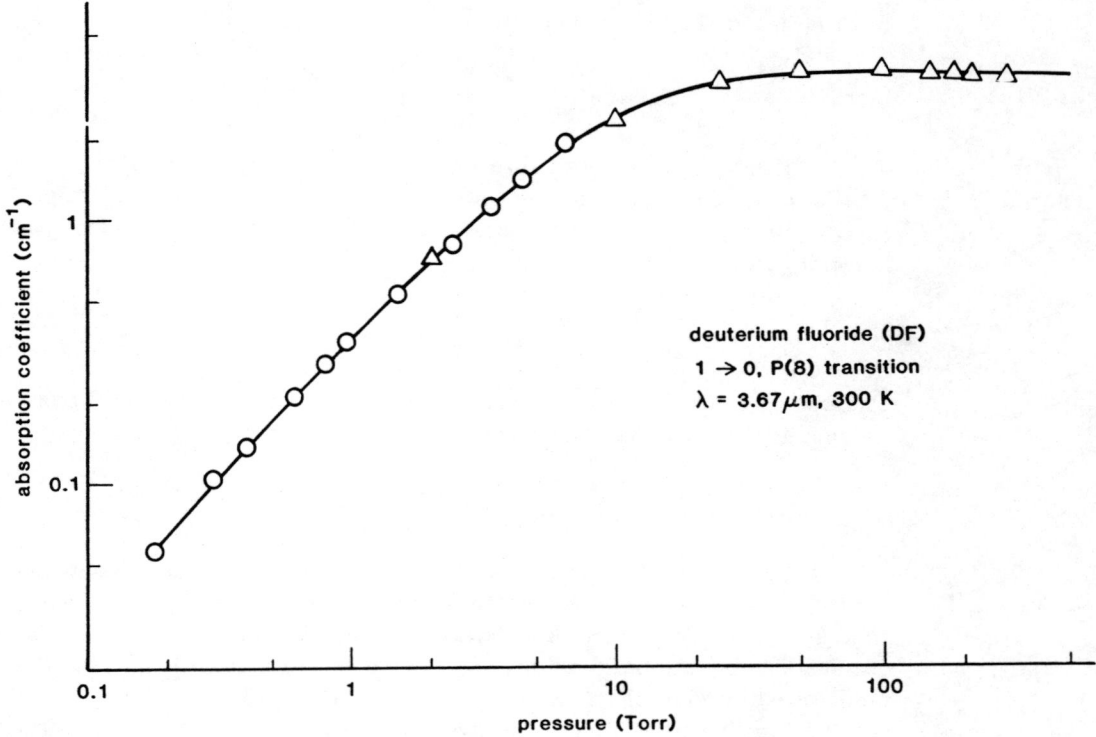

**FIGURE 3.21**
Midband absorption coefficient versus pressure on a deuterium fluoride (DF) transition, showing the transition from doppler broadening at low pressures to pressure broadening at higher pressures.

In the scientific literature on magnetic resonance, where inhomogeneous broadening was first studied, as well as in other areas of resonance physics, the complex homogeneous lorentzian lineshape is often written in the alternative notation

$$\tilde{\chi}_{\text{lor}}(\omega) = -j\chi_0'' \frac{1}{1 + jT_2(\omega - \omega_a)}, \qquad (73)$$

and the real lorentzian lineshape for a homogeneous absorption line is then commonly written in normalized form as

$$g_{\text{lor}}(\omega) = \frac{2}{\pi \Delta\omega_a} \frac{1}{1 + [2(\omega - \omega_a)/\Delta\omega_a]^2} \equiv \frac{T_2}{\pi} \frac{1}{1 + T_2^2(\omega - \omega_a)^2} \qquad (74)$$

with the same normalization that $\int g_{\text{lor}}(\omega)\,d\omega = 1$. In this notation the FWHM homogeneous linewidth is usually written in the simpler form

$$\Delta\omega_a \equiv 2/T_2 \qquad (75)$$

rather than as $\Delta\omega_a = \gamma + 2/T_2$. In essence, the $\gamma$ contribution to the homogeneous linewidth has been absorbed into an expanded definition of $2/T_2$ that includes both the dephasing and lifetime-broadening contributions. (We will occasionally use this expanded definition of $T_2$ later in this book.)

Then, in order to make the gaussian lineshape function $g_{\text{gauss}}(\omega)$ have the same algebraic constants in front for the same normalization, the inhomogeneous function is written in the analogous form

$$g_{\text{gauss}}(\omega) \equiv \left(\frac{T_2^*}{\pi}\right) \exp\left[-\frac{T_2^{*2}(\omega - \omega_a)^2}{\pi}\right], \qquad (76)$$

which satisfies the same normalization that $\int g_{\text{gauss}}(\omega)\, d\omega = 1$. The parameter $T_2^*$ that has been introduced here is the inhomogeneous analog to the dephasing time $T_2$ in the homogeneous case. It is related to the gaussian inhomogeneous linewidth $\Delta\omega_d$ by

$$\Delta\omega_d \equiv \frac{\sqrt{4\pi \ln 2}}{T_2^*} \approx \frac{3}{T_2^*}. \qquad (77)$$

The quantity $T_2^* \approx 3/\Delta\omega_d$ is thus the inhomogeneous (or gaussian) analog to the quantity $T_2 = 2/\Delta\omega_a$ for the homogeneous (or lorentzian) lineshape.

Physical Significance of $T_2$ and $T_2^*$

If we leave out the complications involving the additional $\gamma$ contribution, the time constant $T_2$ is what we identified earlier as the *homogeneous dephasing time*. It defines the average time duration within which the coherent oscillations of two different atomic dipoles are likely to be permanently and irreversibly randomized by collisions, or by other homogeneous dephasing events.

The time constant $T_2^*$ can be given an analogous interpretation as the *inhomogeneous dephasing time* due to inhomogeneous broadening mechanisms for a group of oscillating atoms. Consider, for example, two atoms located in different spectral packets within a gaussian inhomogeneous line. The natural oscillation frequencies $\omega_{a1}$ and $\omega_{a2}$ of these two atoms will differ by an amount $\omega_{a1} - \omega_{a2}$ that will typically be of order $\approx \Delta\omega_d$. Even without any homogeneous dephasing events, therefore, these two oscillating dipoles will get out of phase by one half-cycle after a length of time $\delta t$ given by $(\omega_{a1} - \omega_{a2})\, \delta t = \pi$, or $\delta t = \pi/(\omega_{a1} - \omega_{a2}) \approx \pi/\Delta\omega_d \approx T_2^*$.

The time constant $T_2^*$ is thus the time duration after which different packets within an inhomogeneous line are likely to have become dephased because of their different oscillation frequencies, even without any collisions or similar dephasing events. The condition for a strongly inhomogeneous atomic line can be written in either of the alternative forms

$$\text{strongly inhomogeneous line:} \quad \Delta\omega_d \gg \Delta\omega_a \quad \text{or} \quad T_2^* \ll T_2. \qquad (78)$$

Thus in a strongly inhomogeneous line the $T_2^*$ dephasing of different packets because of different oscillation frequencies will happen much more rapidly than the homogeneous dephasing within each packet that is caused by $T_2$.

In practical experiments, therefore, if all the different atoms or packets within a strongly inhomogeneous line are initially set oscillating coherently and in phase by means of some suitable initial preparation pulse, the coherent macroscopic polarization $p(t)$ in the collection will disappear after the shorter time $T_2^*$, not the longer homogeneous dephasing time $T_2$, because of the inhomogeneous frequency-difference effects. In an inhomogeneous line under small-signal conditions, $T_2^*$ and not $T_2$ is the significant dephasing time.

For example, in the He-Ne 633 nm laser transition the doppler linewidth is $\Delta f_d \approx 1{,}500$ MHz, and the inhomogeneous dephasing time is thus $T_2^* \approx 3/(2\pi \Delta f_d) \approx 320$ psec. This must be compared with a homogeneous linewidth for individual packets of more like $\Delta f_a \approx 100$ MHz, and hence a homogeneous dephasing time of $T_2' \approx 1/\Delta\omega_a \approx 3.2$ ns.

One vitally important difference between strongly homogeneous and strongly inhomogeneous systems, however, is that the inhomogeneous dephasing after the time $T_2^*$ is fundamentally reversible: the different oscillation phases $\omega_{a1}t$, $\omega_{a2}t$, and so forth, that develop for different atoms after a time $t$ can in principle be "unwound" by certain sophisticated large-signal or coherent-pulse techniques. We will discuss these later, in connection with coherent photon echo experiments.

### Inhomogeneous Strain Broadening: Glass Laser Materials

Inhomogeneous broadening is also of considerable importance in certain solid-state laser transitions. Random strains, defects, and other site-to-site variations in solid-state laser materials can significantly change the local crystal fields seen by laser ions that are imbedded in these materials, and this in turn can randomly shift the exact resonance frequencies of laser atoms in those materials, sometimes by quite large amounts.

This type of inhomogeneous broadening predominates in inhomogeneous materials such as laser glasses at room temperature, or in more organized crystalline laser materials at very low temperatures (approaching liquid-helium temperatures), where it is no longer masked by the much larger phonon-broadening effects. Since this kind of broadening, often called *strain broadening*, is caused basically by random defects in the laser material, its magnitude may depend strongly on material growth and perfection, impurities, and annealing. It is thus not possible to give any general formulas, since the amount of strain broadening may vary from sample to sample of the same material.

If these random strains and defects have a gaussian distribution, the resulting inhomogeneous broadening effects can look and act much like doppler broadening, even though the underlying physical mechanism is totally different. The ratio of homogeneous linewidth $\Delta\omega_a$ to inhomogeneous linewidth $\Delta\omega_d$ will still be the crucial parameter in determining whether the transition will be strongly homogeneous, strongly inhomogeneous, or somewhere in between.

For example, the widely used yttrium aluminum garnet (YAG) crystal can be grown with high crystal quality. The linewidth of the $Nd^{3+}$ ion in Nd:YAG laser crystals therefore exhibits only a small amount of inhomogeneous strain broadening. The laser transition is primarily phonon broadened and thus homogeneous at room temperature. Reducing the temperature to below liquid-nitrogen temperature (77 K) greatly reduces the phonon broadening and makes the residual strain broadening observable.

On the other hand, the same $Nd^{3+}$ ion placed in a Nd:glass laser material, with its much larger amount of structural randomness, has a much larger inhomogeneous strain-broadening component, which is significant even at room temperature. This broadening is due to variations in the local crystal fields seen by the laser ions at different sites within the glassy material. The ratio of inhomogeneous to homogeneous broadening in Nd:glass laser materials is not fully understood and varies considerably (by at least a factor of three) from one glass composition to another.

The inhomogeneous linewidths in different glasses at room temperature, for example, vary over a linewidth range of from at least 40 to 120 $cm^{-1}$. (Linewidths this wide are more often expressed in units of $cm^{-1}$ or wavenumbers than in more conventional units; remember that 1 cm$^{-1}$ ≡ 30 GHz.) The homogeneous linewidth in the same materials varies over a range from 20 to 75 cm$^{-1}$, and is strongly correlated (for reasons that are not well understood) with the velocity of sound in the glass. This homogeneous linewidth reduces to ≪ 1 cm$^{-1}$ at 4.2 K, where the lattice vibrations and hence the homogeneous phonon broadening are reduced to nearly zero.

As a general rule, therefore, Nd:glasses are found to fall somewhere in the intermediate or mixed category between homogeneous and inhomogeneous broadening, with ratios of $\Delta\omega_a/\Delta\omega_d$ ranging from 0.16 to 1.9 in different glasses.

### Far Outside the Resonance Linewidth: All Lines Become Homogeneous

Suppose we go out into the far wings of any atomic resonance transition, homogeneous or inhomogeneous, and measure the atomic response at 5 or 10 linewidths out from the line center. (Note that in any usual atomic transition we can do this and still be well within the "resonance approximation" we introduced earlier, so that the lorentzian and gaussian lineshapes will still apply.)

The gaussian response characteristic of an inhomogeneous transition—for example, a doppler-broadened transition—will then fall off as $\approx \exp\left[-(\omega - \omega_a)^2\right]$, whereas the lorentzian response characteristic of a homogeneous transition—or of a homogeneous packet within an inhomogeneous transition—will fall off only at the much slower rate of $\approx 1/(\omega - \omega_a)^2$ for the $\chi''$ part of the susceptibility, or the even slower rate of $\approx 1/(\omega - \omega_a)$ for the $\chi'$ part of the susceptibility. Figure 3.22 shows, for example, the $\chi''(\omega)$ parts of the susceptibility plotted on the same frequency scale for a gaussian transition with a given linewidth $\Delta\omega_d$ and for a lorentzian line—or a lorentzian packet within the gaussian line—whose linewidth $\Delta\omega_a$ is only 1/5 as large as the gaussian linewidth $\Delta\omega_d$.

This example makes it clear that if we go far enough out from line center, the lorentzian response, though it may be 20 or 30 dB down from the midband value, will clearly dominate over the gaussian response. In other words, *far enough out in the wings, all transitions—even strongly inhomogeneous transitions—once again appear to be homogeneous in character*. If we tune away from an inhomogeneous transition by a sufficient number of inhomogeneous lineshapes, the atomic response will be very weak, though possibly still measurable; and the lineshape of that response will look like a homogeneous lineshape characterized by the $\Delta\omega_a$ of the individual spectral packets, rather than the $\Delta\omega_d$ of the inhomogeneous frequency spreading. For sufficiently strong transitions, this homogeneous response far out in the wings of an inhomogeneous transition can still be of interest, as we will see later on.

### Summary

The differences between homogeneous and inhomogeneous broadening in the central part of the atomic line play a very significant role in the performance of a laser material, especially when saturation effects are taken into account. Many practical laser materials, particularly gases, are strongly inhomogeneous, but others are strongly homogeneous. We will return to the detailed "hole burning" properties of inhomogeneous laser systems in a later chapter.

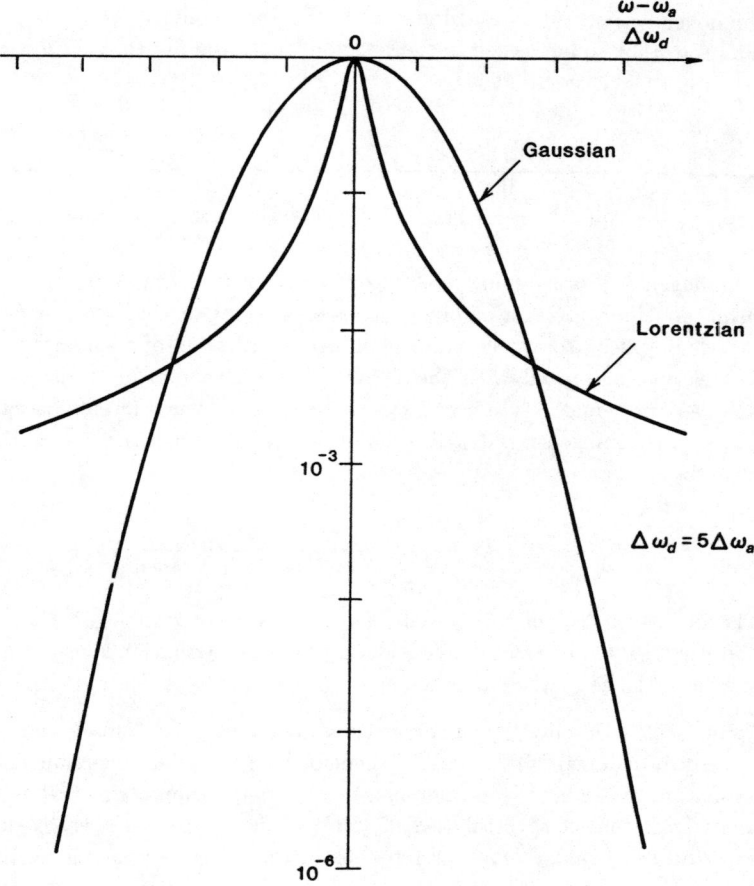

**FIGURE 3.22**
Comparison of lorentzian and gaussian lineshapes in the wings far from line center.

## REFERENCES

The concepts of inhomogeneous broadening and of spectral packets (originally called "spin packets") were first originated in magnetic resonance, notably in the pioneering paper by A. M. Portis, "Electronic structure of F centers: saturation of the electron spin resonance," *Phys. Rev.* **91**, 1071 (1953).

The exact integral for the inhomogeneous susceptibility in the Voight region, with mixed homogeneous and inhomogeneous broadening, can be transformed mathematically into a closed expression involving an error function of complex argument (which doesn't help much). One of many references on this is W. J. Surtrees, "Calculation of combined doppler and collision broadening," *J. Opt. Soc. Am.* **55**, 893 (1965).

The experimental results in Figure 3.20 and nearby illustrations are from E. T. Gerry and D. A. Leonard, "Measurement of 10.6 $\mu$m $CO_2$ laser transition probability and optical broadening cross sections," *Appl. Phys. Lett.* **8**, 227 (May 1, 1966); from R. K. Hanson, "Shock tube spectroscopy: advanced instrumentation with a tunable diode laser," *Appl. Optics* **16**, 1479 (1977); and from R. L. Abrams, "Broadening coefficients for the P(20) $CO_2$ laser transition," *Appl. Phys. Lett.* **25**, 609–611 (November 15, 1974).

For an example of the use of Voight profiles in a solid—specifically in ruby at low temperature, when the linewidth is a combination of homogeneous phonon broadening and inhomogeneous strain broadening—see D. F. Nelson and M. D. Sturge. "Relation between absorption and emission in the region of the $R$ lines of ruby," *Phys. Rev.* **137**, A1117–A1130 (February 15, 1965).

---

Problems for 3.7

1. *Inhomogeneous broadening with a lorentzian (rather than gaussian) inhomogeneous distribution.* Most inhomogeneous broadening mechanisms, such as doppler broadening, lead to a gaussian probability distribution of resonance frequencies, for reasons associated with the Central Limit Theorem of statistics. Suppose, however, we could create a collection of atoms having a lorentzian rather than gaussian inhomogeneous distribution of resonance frequencies, i.e., a distribution given by

$$g(\omega_a) = \frac{2}{\pi \Delta\omega_d} \frac{1}{1 + [2(\omega_a - \omega_{a0})/\Delta\omega_d]^2}.$$

The linewidth $\Delta\omega_d$ of this distribution is then exactly analogous to $\Delta\omega_d$ in the doppler case. (The reason for considering such a distribution is primarily because it will make the mathematics easier.)

Using such a lorentzian inhomogeneous distribution, calculate the inhomogeneously broadened small-signal susceptibility $\tilde{\chi}(\omega)$ of a collection of atoms or oscillators, assuming the usual homogeneous complex lorentzian response for each individual atom or spectral packet. (Hint: This calculation is easily done if you know how to evaluate contour integrals in the complex plane using the residue method; if you're not familiar with this, ask an acquaintance.) Discuss the resulting general inhomogeneous lineshape, its real and imaginary parts, and their overall linewidth for various values of the inhomogeneity parameter $\Delta\omega_a/\Delta\omega_d$.

2. *Inhomogeneous broadening with a uniform inhomogeneous distribution.* There might exist an oddball laser crystal with a distribution of defects such that the atomic frequency shifts produced by these defects were *uniformly* distributed between some maximum positive and negative shift values about the unshifted center frequency $\omega_{a0}$. (Or there might be an even more remarkable gas in which the axial velocities, instead of having a maxwellian distribution, were uniformly distributed between a minimum and a maximum value.) Either of these unusual situations would lead to an inhomogeneously broadened transition with a *rectangular* inhomogeneous lineshape $g(\omega_a)$, rather than the much more common gaussian lineshape discussed in the text.

   Let $\Delta\omega_d$ here mean the full width of this rectangular distribution. Find an exact analytic expression for the complex susceptibility $\tilde{\chi}(\omega)$, and plot $\chi'(\omega)$ and $\chi''(\omega)$ versus $(\omega - \omega_a)/\Delta\omega_a$ for different degrees of homogeneous versus inhomogeneous broadening, for example, for $\Delta\omega_d/\Delta\omega_a = 1/20$, 1, and 20. Find also the midband value $\tilde{\chi}(\omega_{a0})$ in the limits of very large and very small inhomogeneous broadening. (Hint: You may have to do some thinking about how to interpret the natural logarithm of a complex argument.)

3. *Ditto with a triangular distribution.* Repeat the previous problem with a triangular inhomogeneous lineshape having FWHM linewidth $\Delta\omega_d$ and base width $2\Delta\omega_d$.

4. *Midband absorption versus pressure in a gas.* Why does the midband absorption value shown in Figure 3.20 at first increase with increasing pressure and then saturate in the form shown?

5. *Chemical lasers, and absorption versus pressure in a deuterium fluoride gas cell.* By burning deuterium with fluorine to get chemically excited molecules of deuterium fluoride (DF), and then letting the resulting molecules expand through a supersonic nozzle, we can make a very powerful chemical laser (hundreds of kilowatts cw) at wavelengths around $\lambda = 3.6$ to $4.1$ microns in the near infrared. Such lasers have been considered as military weapons (if the laser beam doesn't get you, the toxic chemicals will). The quantum transitions in deuterium fluoride molecules are thus of some interest.

Figure 3.21 shows the measured signal-absorption coefficient $2\alpha$ at midband ($\omega = \omega_a$) versus pressure on a certain DF transition at $\lambda = 3.67$ microns starting from the ground state (lowest energy level) of the DF molecule. The power absorption coefficient $2\alpha$ is related to the transition susceptibility $\chi''$ by $2\alpha(\omega) = (2\pi/\lambda)\chi''(\omega)$ (as we will learn later). The midband absorption is plotted against gas pressure in a cell containing unexcited DF molecules at room temperature. The DF transition is presumably pressure-broadened, lorentzian, and homogeneous at high gas pressures; but doppler-broadened, gaussian, and inhomogeneous at low gas pressures (pure lifetime broadening will be negligible in all cases).

Explain the shape of this experimental curve, and use it to deduce as much as you can about the properties and numerical coefficients of this particular DF transition. Some useful numbers: molecular weight of a DF molecule = 21; mass of a proton, $M = 1.67 \times 10^{-27}$ kg; Boltzmann constant $k = 1.38 \times 10^{-23}$ in mks units; gas density $N(\text{molecules/cm}^3) = 9.65 \times 10^{18} P(\text{torr})/T(\text{K})$; room temperature $\approx 300$ K; and 1 atmosphere = 760 torr.

6. *Inhomogeneous Voight profiles far out in the wings.* Using any suitable numerical procedure, calculate the Voight profile for $\chi''(\omega)$ versus $\omega$ for $\Delta\omega_d/\Delta\omega_a = 10$, extending the calculations out to several inhomogeneous linewidths from line center. Plot the results on a log amplitude scale, and compare the exact Voight profile to a gaussian curve that matches the Voight profile near line center. Are there significant differences in the outer wings? Explain.

# 4
## ATOMIC RATE EQUATIONS

Applying a sinusoidal signal to a collection of atoms, with the frequency $\omega$ tuned near one of the atomic transition frequencies $\omega_a$, will produce a coherent induced polarization $p(t)$ or $\tilde{P}(\omega)$ in the collection of atoms, as we have described in the preceding two chapters. The strength of this induced response will be proportional to the instantaneous population difference $\Delta N$ on that particular transition.

At the same time, however, this applied signal field will also cause the populations $N_1(t)$ and $N_2(t)$ in the collection of atoms to begin changing slowly because of *stimulated transitions between the two levels* $E_1$ and $E_2$, as we will discuss in this chapter. The rates of change of the populations are given by *atomic rate equations*, which contain both stimulated terms and relaxation or energy-decay terms (and possibly also other kinds of pumping terms). Deriving the quantum form for these stimulated and relaxation terms is the primary objective of this chapter.

These atomic rate equations are of great value in analyzing pumping and population inversion in laser systems. Solutions of the rate equations for strong applied signals also lead to population saturation effects, which are of very great importance in understanding the large-signal saturation behavior of laser amplifiers and the power output of laser oscillators. Solving the atomic rate equations and understanding these solutions for some simple cases will therefore be the principal objective of Chapter 6.

### 4.1 POWER TRANSFER FROM SIGNALS TO ATOMS

We will derive the stimulated transition rates for an atomic transition in this chapter by examining the power flow or the energy transfer between an applied optical signal and an atomic transition. To get started on this, let us learn something about the rate at which power is transferred from an applied signal to any material medium, including a collection of resonant oscillators or atoms.

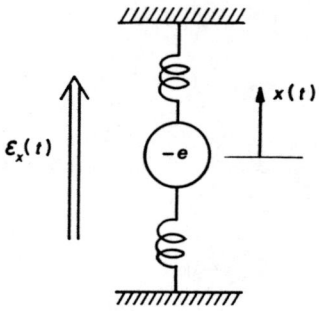

FIGURE 4.1
Mechanical model for a classical electron oscillator with an applied $\mathcal{E}$ field.

### Power Transfer to a Collection of Oscillators: Mechanical Derivation

When an electric field $\mathcal{E}_x(t)$ acts on a moving charge, it delivers power to (or perhaps receives power from) that moving charge. In a single classical oscillator (as in Figure 4.1), a purely mechanical argument says that the amount of work $dU$ done by a force $f_x$ acting on the electron, when the electron moves through a distance $dx$ is

$$dU = f_x \, dx = -e\mathcal{E}_x \, dx. \tag{1}$$

The instantaneous rate at which power is delivered by the field to the classical oscillator is then

$$\frac{dU(t)}{dt} = -e\mathcal{E}_x(t)\frac{dx(t)}{dt} = \mathcal{E}_x(t)\frac{d\mu_x(t)}{dt}, \tag{2}$$

where $\mu_x(t)$ is, of course, the dipole moment of the oscillator.

If we sum this power flow over all the oscillators or atoms in a small unit volume $V$, this result says that the average power per unit volume, $dU_a/dt$, delivered by the field to the atoms or oscillators is

$$\frac{dU_a}{dt} = V^{-1}\mathcal{E}_x(t)\sum_{i=1}^{NV}\frac{d\mu_{xi}(t)}{dt} = \mathcal{E}_x(t)\frac{dp_x(t)}{dt}. \tag{3}$$

This equation, although derived from a mechanical argument, is a very general electromagnetic or even quantum-mechanical result. That is, this equation still holds true whether $p_x(t)$ represents the sum of a large number of classical oscillator dipoles with a number density $N$, or whether $p_x(t)$ represents the effect of a large number of quantum dipole expectation values proportional to a population difference $\Delta N$.

### Time-Averaged Power Flow

To obtain the time-averaged power delivered to a collection of atoms by a sinusoidal signal field, we can write the applied signal and the resulting polarization in phasor form as

$$\mathcal{E}_x(t) = \frac{1}{2}[\tilde{E}(\omega)e^{j\omega t} + \tilde{E}^*(\omega)e^{-j\omega t}] \tag{4}$$

and

$$p_x(t) = \frac{1}{2}[\tilde{P}(\omega)e^{j\omega t} + \tilde{P}^*(\omega)e^{-j\omega t}]. \tag{5}$$

The steady-state sinusoidal polarization $\tilde{P}(\omega)$ on an atomic transition will then be related to the applied field by

$$\tilde{P}(\omega) = \tilde{\chi}(\omega)\epsilon\tilde{E}(\omega) = [\chi'(\omega) + j\chi''(\omega)]\,\epsilon\tilde{E}(\omega). \tag{6}$$

(Remember that we use the host dielectric constant $\epsilon$ and not $\epsilon_0$ in this relation, for the reasons explained in the previous chapter.)

If we substitute these phasor forms into Equation 4.3 and take the time average (by dropping the $e^{\pm 2j\omega t}$ terms) we obtain a useful result for the average power absorbed from the fields, by the atoms, per unit volume, namely,

$$\left.\frac{dU_a}{dt}\right|_{\text{av}} = \frac{j\omega}{4}\left(\tilde{E}^*\tilde{P} - \tilde{E}\tilde{P}^*\right) = -\frac{1}{2}\epsilon\omega\chi''(\omega)|\tilde{E}(\omega)|^2. \tag{7}$$

The most important point here is that the power absorption (or emission) by the atoms depends only on the $\chi''(\omega)$ part of the complex susceptibility $\tilde{\chi}(\omega)$. This is the "resistive" or lossy part of $\tilde{\chi}(\omega)$, whereas $\chi'(\omega)$ is the purely reactive part.

The minus sign in the final term of Equation 4.7 merely means that if we use the definition $\tilde{\chi} \equiv \chi' + j\chi''$, then the quantity $\chi''$ for an absorbing medium will turn out to be a negative number, as indeed we have already found for the classical electron oscillator. (Some authors, attempting to avoid this minus sign, use instead the definition that $\tilde{\chi} \equiv \chi' - j\chi''$.)

### Poynting Derivation of Energy Transfer

The results for power transfer obtained above are in fact general electromagnetic results, having nothing directly to do with the particular atomic or quantum process that creates the polarization $p_x(t)$. To verify this, let us carry through a standard electromagnetic derivation of this same result, starting by writing Maxwell's equations

$$\nabla \times \mathcal{E} = -\partial b/\partial t, \quad \nabla \times h = j + \partial d/\partial t,$$
$$d = \epsilon_0\mathcal{E} + p, \quad b = \mu_0(h + m), \tag{8}$$

and then substituting them into the vector identity

$$h \cdot (\nabla \times \mathcal{E}) - \mathcal{E} \cdot (\nabla \times h) \equiv \nabla \cdot (\mathcal{E} \times h). \tag{9}$$

Note that all the vector quantities here, for example, $\mathcal{E}(r, t)$, are general vector functions of space and time at this point.

Equation 4.9 can then be integrated over an arbitrary volume $V$, bounded by a closed surface $S$ as in Figure 4.2, using the additional vector identity that

$$\int_V \nabla \cdot (\mathcal{E} \times h)\,dV = -\int_S (\mathcal{E} \times h) \cdot d\mathbf{S}, \tag{10}$$

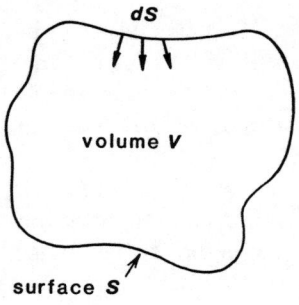

FIGURE 4.2
Volume $V$ with a surface $S$ for the evaluation of electromagnetic power flow.

where $d\mathbf{S}$ is an inward unit vector normal to the surface $S$. Rearranging terms then gives as a general formula

$$\int_S (\mathcal{E} \times \mathbf{h}) \cdot d\mathbf{S} = \frac{d}{dt} \int_V \left( \frac{1}{2}\epsilon_0 |\mathcal{E}|^2 + \frac{1}{2}\mu_0 |\mathbf{h}|^2 \right) dV$$
$$+ \int_V (\mathcal{E} \cdot \mathbf{j})\, dV \qquad (11)$$
$$+ \int_V \left( \mathcal{E} \cdot \frac{d\mathbf{p}}{dt} + \mu_0 \mathbf{h} \cdot \frac{d\mathbf{m}}{dt} \right) dV.$$

We can give a physical interpretation to each term in this equation.

The surface integral on the left-hand side of this equation gives the integral over the closed surface $S$ of the inwardly directed instantaneous Poynting vector $\mathcal{E} \times \mathbf{h}$. According to the standard interpretation of electromagnetic theory, this Poynting integral gives the total electromagnetic power being carried by fields $\mathcal{E}$ and $\mathbf{h}$ and flowing into the volume $V$ at any instant.

The terms on the right-hand side of Equation 4.11 tell where this power is going. The volume integral on the right-hand side of the first line is a purely reactive or energy-storage term. It gives the instantaneous rate of increase or decrease in the stored electromagnetic field energy terms $\frac{1}{2}\epsilon_0 \mathcal{E}^2$ and $\frac{1}{2}\mu_0 h^2$ in the volume $V$. (These are vacuum energy density terms—that is, they do not include any energy going into atomic polarizations $\mathbf{p}(t)$ or $\mathbf{m}(t)$ in the volume $V$.)

The integral on the right-hand side of the second line gives the instantaneous power per unit volume being delivered by the $\mathcal{E}$ field to any currents $\mathbf{j}$, whether these currents come from ohmic losses ($\mathbf{j} = \sigma\mathcal{E}$) or any other real currents $\mathbf{j}(\mathbf{r},t)$ that may be present in the volume.

The integral on the right-hand side of the final line of Equation 4.11 then accounts for the instantaneous powers per unit volume $\mathcal{E} \cdot (d\mathbf{p}/dt)$ and $\mu_0 \mathbf{h} \cdot (d\mathbf{m}/dt)$ that are being delivered by these fields to any electric and magnetic polarizations $\mathbf{p}(\mathbf{r},t)$ and $\mathbf{m}(\mathbf{r},t)$ that may be present, as a consequence of any kind of atomic or material medium. The $\mathcal{E} \cdot (d\mathbf{p}/dt)$ term represents in particular the vector generalization of the simple mechanical derivation we gave at the beginning of this section.

### Reactive Versus Resistive Power Flow

Note that power transfer from the signal fields to these atomic polarization terms does not necessarily mean this power is being *dissipated* in the atoms. If the medium has a purely reactive susceptibility, with $\chi'' = 0$, then there can be no time-averaged power transfer, because $\mathcal{E}(t)$ and $p(t)$ will be 90° out of time-phase, and the time-averaged value of the $\mathcal{E} \cdot dp/dt$ term will be zero: energy will flow from the signal into the medium during one quarter cycle, and back out during the following quarter cycle.

The energy transfer into the polarization in this situation is basically reactive stored energy, which flows into the atoms during one half-cycle and back out during the following half-cycle. This reactive energy flow could be combined with the first integral on the right-hand side of Equation 4.11. If this were done, the expanded first term would become the time derivative of the more familiar expressions $\frac{1}{2}\epsilon|\mathcal{E}|^2$ and $\frac{1}{2}\mu|h|^2$, which give the total electromagnetic energy stored in a medium rather than in vacuum.

We can also see that the rate of change of polarization $dp/dt$ in the $\mathcal{E} \cdot dp/dt$ term plays the same role as the current density $j$ in the $\mathcal{E} \cdot j$ term. It is sometimes convenient to define a "polarization current density" $j_p$ by

$$j_p \equiv dp/dt, \qquad (12)$$

which can be added to the real current density $j$ in Maxwell's equations. From Chapter 3 we realize that this polarization current simply represents the sloshing back and forth of the bound but oscillating atomic charge clouds that lead to the oscillating dipole moments $\mu(t)$ in each atom and to the macroscopic polarization $p(t)$ in the collection of atoms. To the extent that this current is in phase with the $\mathcal{E}(t)$ term [that is, comes from the $\chi''(\omega)$ part of the susceptibility], it represents additional resistive loss or dissipation in the medium; to the extent that it is 90° out of phase [the $\chi'(\omega)$ part], it represents reactive energy storage.

### Quality Factor

The absorptive susceptibility $\chi''$ in an atomic medium can be interpreted as a kind of inverse $Q$ or quality factor for the ratio of signal energy stored in a volume to signal power dissipated in that volume, in just the same fashion as the $Q$ factor is defined for a mechanical system or an electrical circuit. That is, the time-averaged stored signal energy per unit volume associated with a sinusoidal signal field in a host medium of dielectric constant $\epsilon$ can be written as

$$U_{\text{sig}} = \frac{1}{2}\epsilon|E|^2. \qquad (13)$$

The inverse $Q$ factor for this little volume can then be defined as

$$\frac{1}{Q} \equiv \frac{\text{energy dissipated}}{\omega \times \text{energy stored}} = \frac{1}{\omega U_{\text{sig}}} \frac{dU_a}{dt} = -\chi''. \qquad (14)$$

The dimensionless atomic susceptibility $\chi''$, as we have defined it in this text, is thus essentially an inverse $Q$ factor describing the average power absorption per unit volume, by the atoms, from the signal. Of course for an amplifying transition, this $Q$ becomes a negative number.

For real laser transitions this $Q$ is always very high, since in all practical laser situations $|\chi''| \ll 1$. We usually think of a high $Q$ value in a system as being in some sense "good". Here, however, a high $Q$ means a *weak* susceptibility, and hence a small gain in an amplifying laser medium, which is generally *not* what we would like to have.

### Tensor Formulation of Power Flow

Real laser transitions may have a linear but anisotropic response, in which the induced polarization must be described by a tensor susceptibility. To describe the power transfer properly in this case we must employ a more sophisticated form for the analysis in terms of the *hermitian* and *antihermitian* parts of this susceptibility tensor.

To do this, we note that the full vector formula for instantaneous power delivered per unit volume is

$$\frac{dU_a(t)}{dt} = \boldsymbol{\mathcal{E}} \cdot d\boldsymbol{p}/dt. \tag{15}$$

The time-averaged power flow in an atomic medium with a tensor atomic susceptibility $\chi$ is then given by

$$\left.\frac{dU_a}{dt}\right|_{\text{av}} = \frac{j\omega\epsilon}{4}\left[\boldsymbol{E}^* \cdot \boldsymbol{\chi} \boldsymbol{E} - \boldsymbol{E} \cdot \boldsymbol{\chi}^* \boldsymbol{E}^*\right] = \frac{j\omega\epsilon}{4} \sum_{i=1}^{3}\sum_{j=1}^{3} \tilde{E}_i^* \left(\tilde{\chi}_{ij} - \tilde{\chi}_{ji}^*\right) \tilde{E}_j, \tag{16}$$

where $i$ and $j$ are both summed over the three directions $x, y, z$. If $\chi$ happens to be an isotropic or even a diagonal tensor, then these sums reduce directly to our previous scalar results. For a general anisotropic tensor susceptibility, however, we must separate the complex tensor $\chi$ not into its real and imaginary parts, but into its *hermitian* and *antihermitian* parts, as given by

$$\chi \equiv \chi_h + j\chi_{ah}, \tag{17}$$

where $\chi_h$ and $\chi_{ah}$ are defined by

$$\chi_h \equiv (1/2)\left(\chi^\dagger + \chi\right) \quad \text{and} \quad \chi_{ah} \equiv (j/2)\left(\chi^\dagger - \chi\right), \tag{18}$$

with $\chi^\dagger$ being the hermitian conjugate of $\chi$. Note that $\chi_h$ and $\chi_{ah}$ are not necessarily the same as the real and imaginary parts of $\chi$, since $\chi^\dagger$ and $\chi$ are in general not simply the complex conjugates of each other.

It can then be shown that the time-averaged power transfer is given by

$$\left.\frac{dU_a}{dt}\right|_{\text{av}} = -\frac{1}{2}\omega\epsilon \boldsymbol{E}^*(\omega)\chi_{ah}(\omega)\boldsymbol{E}(\omega). \tag{19}$$

In the general tensor case it is the *antihermitian* part $j\chi_{ah}$ of the susceptibility tensor, and not just the imaginary part $\chi''$, that is the resistive or power-absorbing part.

## 4.2 STIMULATED-TRANSITION PROBABILITY

We will next use these results to derive a stimulated-transition probability, which gives the stimulated-transition rate at which atoms make transitions back and forth between quantum energy levels under the influence of an applied signal. We will do this by considering the rate at which an applied signal will deliver energy to a collection of real quantum atoms, and the manner in which these atoms can accept this energy.

**CHAPTER 4: ATOMIC RATE EQUATIONS**

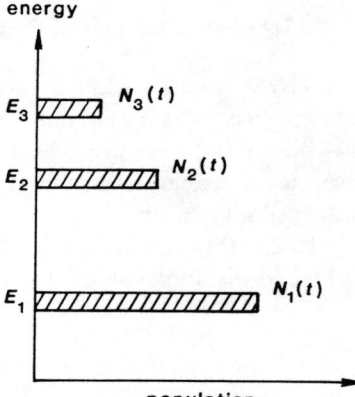

FIGURE 4.3
Energy levels and level populations in an atomic system.

### Energy Transfer From Signal To Atoms

We have learned that when a sinusoidal signal field $\tilde{E}(\omega)$ produces a steady-state polarization $\tilde{P}(\omega)$ on a transition in a collection of atoms, the time-averaged power transfer per unit volume from the signal to the atoms must be given by

$$\frac{dU_a}{dt} = -\frac{1}{2}\omega\chi''(\omega)\epsilon|\tilde{E}|^2, \tag{20}$$

where the susceptibility for a homogeneous lorentzian atomic transition is given from the previous chapter by

$$\tilde{\chi}''(\omega) = -\frac{3^*}{4\pi^2}\frac{\Delta N \gamma_{\rm rad}\lambda^3}{\Delta\omega_a}\frac{1}{1+[2(\omega-\omega_a)/\Delta\omega_a]^2}. \tag{21}$$

(For an inhomogeneous gaussian transition exactly the same expression as Equation 4.21 would apply, except that the homogeneous linewidth $\Delta\omega_a$ would be replaced by the inhomogeneous linewidth $\Delta\omega_d$: the lorentzian frequency dependence would be replaced by a gaussian; and an additional factor of $\sqrt{\pi\ln 2}\approx 1.48$ would appear in front.)

The rate of energy transfer from the signal to the atoms can thus be written in the general form

$$\frac{dU_a}{dt} = \left[\frac{3^*}{8\pi^2}\frac{\gamma_{\rm rad}}{\Delta\omega_a}\frac{\omega\epsilon|\tilde{E}|^2\lambda^3}{1+[2(\omega-\omega_a)/\Delta\omega_a]^2}\right](N_1 - N_2). \tag{22}$$

Note that this energy absorption is directly proportional to the population difference $\Delta N \equiv N_1 - N_2$.

### Energy Storage by the Atoms

Now, what will the atoms do with this energy, or how can they accept this energy from the applied fields? From a quantum viewpoint, the total oscillation energy stored in a collection of atoms is given by the number of atoms $N_j$ in each quantum energy level, times the energy eigenvalue $E_j$ associated with that level, summed over all the energy levels $E_j$, as in Figure 4.3. In a collection of identical two-level atoms, for example. the total oscillation energy density $U_a$ (energy per unit volume) is given by

$$U_a(t) = N_1(t)E_1 + N_2(t)E_2, \tag{23}$$

where $N_1$ and $N_2$ are the populations in levels $E_1$ and $E_2$. More generally, the total energy in a collection of multilevel atoms is the sum over all levels

$$U_a(t) = \sum_{j=1}^{M} N_j(t)E_j. \tag{24}$$

Since the energy eigenvalues $E_j$ are fixed quantities, if the collection of atoms is to accept energy the level populations $N_j$ must change, with atoms flowing from a lower energy level to a higher energy level.

In the classical oscillator model the energy of each oscillator was associated with the internal oscillatory motion $|x(t)|^2$. In a quantum description, however, the internal energy of each atom must be calculated from the level populations and energy eigenvalues as above. These two descriptions are not unrelated—for example, we noted earlier that an atom in a mixture of populations at levels $E_1$ and $E_2$ has an internal oscillating dipole moment $\mu(t)$ at the transition frequency $\omega_{21}$ that is associated with that mixture of populations.

In any event, if energy is to be delivered from a signal to a collection of atoms, the only way in which the atoms can accept this energy and change their total internal energy $U_a(t)$ is by changing the populations $N_j(t)$ in the collection of atoms. The signal field, as we have seen in the previous chapter, induces a dipole moment in each atom, and thus produces a coherent macroscopic polarization proportional to the population difference $\Delta N \equiv N_1 - N_2$. But, it must also cause the quantum state of each individual atom to begin to change in such a way that the populations $N_1(t)$ and $N_2(t)$ in the collection of atoms also begin to change.

### Stimulated Transition Probabilities

We can emphasize this point by rewriting Equation 4.22 in the alternative form

$$\frac{dU_a}{dt} = W_{12}N_1\hbar\omega_a - W_{21}N_2\hbar\omega_a, \tag{25}$$

where either of the quantities $W_{12}\hbar\omega_a = W_{21}\hbar\omega_a$ corresponds to the collection of factors inside the set of square brackets in Equation 4.22. By rewriting Equation 4.22 in this alternative form, however, we make the energy flow from the signal to the atoms seem to be produced by two flows of atoms, one *upward* from level 1 to level 2 at an upward stimulated-transition rate given by $W_{12}N_1$ (units of atoms per second), as shown by the upward arrow in Figure 4.4; plus an opposite flow of atoms *downward* from level 2 to level 1 at a downward stimulated-transition rate given by $W_{21}N_2$.

In other words, the energy transfer from the signal to the atoms, as given by Equation 4.25, can be accounted for by a net flow rate of atoms across the gap, upward minus downward, given by

$$\left.\frac{dN_2}{dt}\right|_{\text{stim}} = -\left.\frac{dN_1}{dt}\right|_{\text{stim}} = W_{12}N_1 - W_{21}N_2, \tag{26}$$

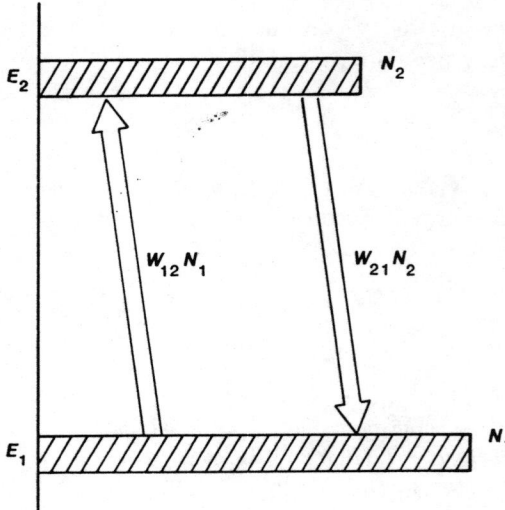

FIGURE 4.4
Upward and downward stimulated transitions between two energy levels.

as illustrated in Figure 4.4. Both of these flow rates are expressed in units of atoms per second. To get the net energy flow, each of these rates must be multiplied by the transition energy or photon energy $\hbar\omega_a$, since each transit of one atom across the energy gap represents a net absorption or emission of one quantum of energy by the atoms.

The quantities $W_{12}$ and $W_{21}$ are then referred to as the *upward and downward stimulated-transition probabilities*, per atom and per unit time, produced by the applied signal acting on the lower-level and upper-level atoms, respectively. By equating Equations 4.22 and 4.25, we can see that these stimulated-transition probabilities are given by

$$W_{12} \equiv W_{21} = \frac{3^*}{8\pi^2} \frac{\gamma_{\text{rad}}}{\hbar \Delta \omega_a} \frac{\epsilon |\tilde{E}|^2 \lambda^3}{1 + [2(\omega - \omega_a/\Delta\omega_a]^2}. \qquad (27)$$

With this interpretation we may say that *the applied signal gives each atom in the lower level $E_1$ a probability $W_{12}$ per unit time of making a stimulated transition to the upper level, absorbing a quantum of energy in the process; similarly, the applied signal gives each atom in the upper level an equal probability $W_{21}$ per unit time of making a transition downward to the lower level, giving up one quantum of energy to the signal in the process.*

Equations 4.25 through 4.27 provide an important and general result, sometimes referred to as "Fermi's Golden Rule", which is usually derived only with the aid of quantum theory. We have obtained the correct quantum answer, however, from a simple energy argument, thus further illustrating the power of the classical oscillator arguments used in this text. The reader might reasonably ask, however, since $W_{12} = W_{21}$, why didn't we describe them both by a single symbol? The answer is, first of all, that later on in writing multilevel rate equations it may help to keep various terms straight if we use $W_{ij}$ to mean a transition probability from level $i$ to level $j$, and $W_{ji}$ to mean the same transition probability in the reverse direction. In addition, there are some slight additional complications for the rate equations between degenerate energy levels, as we will see in a moment.

### Quantum Description of Stimulated Transitions.

These signal-stimulated transition rates for atoms between two energy levels are often described in simplified fashion as a process in which the applied signal causes individual atoms to make discrete jumps back and forth between the two levels under the influence of the applied signal, exchanging one photon with each jump. This is the "billiard-ball" or photon model of laser amplification.

A much more correct description, however, *even in quantum theory*, is to say that each individual atom in the collection of atoms, rather than making a discrete "jump" or transition from one level to the other, in fact really only changes its quantum state by a small amount in response to the applied signal. We have pointed out earlier that the quantum state of an atom involved in a transition between two levels $E_1$ and $E_2$ can be written in the general form

$$\psi(r,t) = \tilde{a}_1(t) e^{-iE_1 t/\hbar} \psi_1(r) + \tilde{a}_2(t) e^{-iE_2 t/\hbar} \psi_2(r). \tag{28}$$

where the expansion coefficients $\tilde{a}_1(t)$ and $\tilde{a}_2(t)$ are constant (stationary) in the absence of an applied signal. The time evolution of this quantum state in each individual atom in the presence of an applied signal must then be calculated in a proper quantum analysis by solving the Schrödinger equation of motion for the atom in the presence of the signal field.

The net result of such a calculation will be that, under the influence of an applied signal, the expansion coefficients $\tilde{a}_1(t)$ and $\tilde{a}_2(t)$ of each individual atom will begin to change *slowly but continuously* with time. In other words, the quantum state makeup of each individual atom will begin to shift by a small but continuous amount from quantum state $\psi_1$ towards quantum state $\psi_2$ or vice versa. The probability of finding each atom in one level or the other begins to change by a small amount; and when these probabilities for individual atoms are averaged over the entire collection, it appears as if the population in one level has decreased and in the other has increased.

For many purposes, however, it is acceptable to summarize the results of this calculation by simply saying that, in the presence of an applied signal, atoms begin to make stimulated transitions or jumps back and forth between the two levels $E_1$ and $E_2$, thus changing $N_1(t)$ and $N_2(t)$. The final result averaged over the collection of atoms is basically the same whether we think of each individual atom changing its quantum state by a small amount (which is what really happens), or whether we think of a small fraction of the atoms making discrete "jumps" from one level to the other (which is how the situation is often described).

### General Atomic Transition With Degeneracy

To take care of the more general case in which we have transition rates between two arbitrary energy levels $E_i$ and $E_j > E_i$, where these levels may have degeneracy factors $g_i$ and $g_j$, we must note that the complex susceptibility on such a transition is given by

$$\tilde{\chi}''_{ij}(\omega) = -\frac{3^*}{4\pi^2} \frac{\gamma_{\text{rad},ji} \lambda_{ij}^3}{\Delta\omega_a} \frac{[(g_j/g_i)N_i - N_j]}{1 + [2(\omega - \omega_{ji})/\Delta\omega_{a,ij}]^2}, \tag{29}$$

Hence, the power transfer from signal to atoms can be written, first in electromagnetic form, and then in rate-equation form, as

$$\frac{dU_a}{dt} = \left[\frac{3^*}{8\pi^2}\frac{\gamma_{\text{rad},ji}}{\Delta\omega_a}\frac{\omega\epsilon|\tilde{E}_{ij}|^2\lambda_{ij}^3}{1+[2(\omega-\omega_a)/\Delta\omega_a]^2}\right]\left(\frac{g_j}{g_i}N_i - N_j\right) \quad (30)$$

$$= W_{ij}N_i\hbar\omega_{ji} - W_{ji}N_j\hbar\omega_{ji}.$$

By equating these two forms, we see that the general expression for the stimulated transition probabilities in this case becomes

$$W_{ji} \equiv \frac{g_i}{g_j}W_{ij} = \frac{3^*}{8\pi^2}\frac{\gamma_{\text{rad},ji}}{\hbar\Delta\omega_{a,ij}}\frac{\epsilon|\tilde{E}_{ij}|^2\lambda_{ij}^3}{1+[2(\omega-\omega_{ji})/\Delta\omega_{a,ij}]^2}, \quad (31)$$

where $\tilde{E}_{ij}$ is the electric field of the applied signal on the $i-j$ transition.

Again the flow of atoms upward from level $E_i$ to level $E_j$ is given by the number of atoms in the lower level $N_i$ times an upward stimulated-transition probability $W_{ij}$, and the quantity $W_{ji}$ is similarly the probability of an upper level atom being stimulated to make a downward transition. The stimulated-transition probabilities $W_{ij}$ and $W_{ji}$ in the two directions are, however, related in general by

$$g_iW_{ij} = g_jW_{ji}. \quad (32)$$

A very fundamental point is that the stimulated-transition probabilities in the two directions are still identical, except for the minor complication of the degeneracy factors $g_i$ and $g_j$.

### Fundamental Properties of the Stimulated-Transition Probabilities

From Equations 4.27 or 4.31, the important physical parameters involved in these signal-stimulated transition probabilities $W_{ij}$ and $W_{ji}$ are evidently

- The applied signal strength, or the signal energy per wavelength cubed, as measured by $\epsilon|\tilde{E}|^2\lambda^3$.
- The relative strength of the atomic transition, measured by its radiative decay rate or Einstein $A$ coefficient, $\gamma_{\text{rad}}$.
- The inverse atomic linewidth, $1/\Delta\omega_a$.
- The frequency of the applied signal $\omega$ relative to the atomic transition frequency $\omega_a$, as measured by the atomic lineshape. Applied signals tuned away from line center are less effective in producing stimulated transitions.
- And, finally, the tensor alignment between the applied field and the atoms, as measured by the factor $0 \leq 3^* \leq 3$.

For an inhomogeneous gaussian transition the formulas in Equations 4.27 or 4.31 must be modified by replacing $1/\Delta\omega_a$ by $1/\Delta\omega_d$; replacing the lorentzian frequency dependence by a gaussian; and adding a factor of $\sqrt{\pi\ln 2} \approx 1.48$ in front.

Note also that in the preceding analysis we speak of the upward and downward stimulated rates $W_{12}N_1$ and $W_{21}N_2$ as if they were separate and distinct processes. It is, however, only the net transition rate between levels, $W_{12}N_1 - W_{21}N_2$, that really counts. There is no way to "turn off" one of these rates and produce only the other one. They are physically identical or at least physically inseparable.

The transition rates discussed in this section are called *stimulated transition rates* because they are caused by applied signals producing changes in the populations $N_1(t)$ and $N_2(t)$. Populations of atomic levels also change with time because of pumping effects, and because of energy decay or relaxation transitions between the levels. These relaxation processes produce additional terms in the rate equations, which we must describe in subsequent sections. The stimulated and relaxation terms must be added together in the total rate equations to describe how the populations change with time.

## REFERENCES

Nearly all the discussions in this book will speak of atoms being acted upon by optical fields that are part of some traveling wave or beam of light. The atoms really respond, however, to the local $E$ field strength of the optical signal (at least in an electric-dipole transition), without caring whether these fields are part of a propagating wave or perhaps of an evanescent field distribution, as in frustrated total internal reflection, or in the evanescent fields outside the core of an optical fiber. Experiments to show that the stimulated-transition probability is in fact exactly the same either for propagating photons or for evanescent fields have been carried out by C. K. Carniglia, L. Mandel, and K. H. Drexhage, "Absorption and emission of evanescent photons," *J. Opt. Soc. Am.* **62**, 479–486 (April 1972).

## 4.3 BLACKBODY RADIATION AND RADIATIVE RELAXATION

The next objective in this chapter must be to understand how thermal fluctuations, or blackbody radiation fields, can also cause stimulated transitions between atomic energy levels. We will then go on to show how these "noise-stimulated" transitions are related to the spontaneous emission or radiative decay processes we have discussed earlier, and how they provide a very important part of the relaxation processes in an atomic system.

### Blackbody Radiation Density

One of the most basic conclusions of thermodynamics is that any volume of space that is in thermal equilibrium with its surroundings must contain a *blackbody radiation energy density*, made up of noise-like *blackbody radiation fields*. Furthermore, the magnitude of these fields depends only on the temperature of the region and of its immediate surroundings and not at all on the shape or construction of the volume (provided only that the volume is large compared to a wavelength of the radiation involved). The electromagnetic fields that make up this blackbody radiation energy are real, measurable, broadband, noise-like $E$ and $H$ fields, with random amplitudes, phases, and polarization, which are present everywhere in the volume.

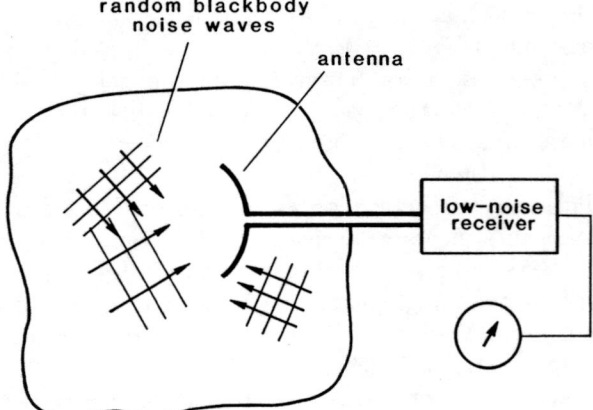

FIGURE 4.5
Measurement of the blackbody radiation fields inside an arbitrary enclosure.

The amount of blackbody radiation energy per unit volume that is present within a region at temperature $T_{\text{rad}}$, at frequencies within a narrow frequency range $d\omega$ about $\omega$, is given in fact by the *blackbody radiation density*

$$dU_{\text{bbr}} = \frac{8\pi}{\lambda^3} \frac{\hbar d\omega}{\exp(\hbar\omega/kT_{\text{rad}}) - 1}. \tag{33}$$

In more precise terms, $dU_{\text{bbr}}/d\omega$ is the spectral density of the blackbody radiation energy, i.e., the amount of energy per unit volume and per unit frequency range centered at frequency $\omega$. We write the temperature as $T_{\text{rad}}$ in this expression to indicate that it is the temperature of the "radiative surroundings" of this region—that is, the temperature of the nearest electromagnetically absorbing walls or boundaries—that determines the blackbody radiation energy density in the region.

The energy density $dU_{\text{bbr}}$ in any narrow frequency range $d\omega$ will have associated with it a mean-square electric field intensity $d|E_{\text{bbr}}|^2$ given by

$$dU_{\text{bbr}} = \frac{\epsilon}{2} d|\tilde{E}_{\text{bbr}}|^2. \tag{34}$$

That is, there will be real measurable electric fields of noise-like character associated with the blackbody energy within the frequency range $d\omega$, and these fields will have a root-mean-square phasor amplitude $\tilde{E}_{\text{bbr}}$ given by

$$d|\tilde{E}_{\text{bbr}}|^2 = \frac{16\pi}{\lambda^3} \frac{\hbar d\omega}{\exp(\hbar\omega/kT_{\text{rad}}) - 1}. \tag{35}$$

With a sensitive enough antenna or probe and a receiver with a low enough noise figure (Figure 4.5), these noise-like fields can be detected and measured as a function of center frequency $\omega$ and temperature $T_{\text{rad}}$ inside the enclosure.

### Blackbody-Stimulated Atomic Transitions

Any atoms that may be present in the region under consideration are then exposed to these entirely real though noise-like $\tilde{E}_{\text{bbr}}$ fields. These $E$ fields will in fact act on the atoms just like signal fields, and will cause stimulated transitions and power absorption at exactly the same rate as would be caused by any other

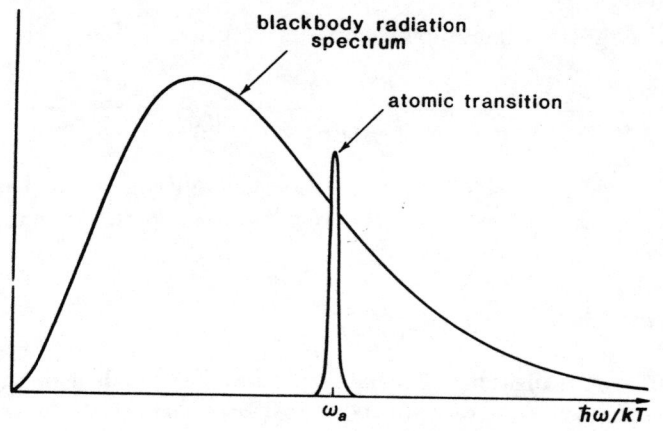

FIGURE 4.6
Blackbody radiation spectrum, and the absorption spectrum of a single narrow atomic transition.

applied signal fields with the same mean-square amplitude. We can calculate the stimulated-transition rates that will be caused by these blackbody radiation fields by means of the following argument.

Figure 4.6 illustrates how the broadband continuum spectral distribution of the blackbody noise fields will overlap with the narrow atomic lineshape of a typical atomic transition. The amount of stimulated-transition probability $dW_{12}$ that will be caused in a two-level atomic system by those blackbody radiation components lying within a small frequency bandwidth $d\omega$ centered at $\omega$ within the atomic linewidth will then be given (for a lorentzian transition) by exactly the same stimulated-transition probability expression as was derived in the previous section, namely,

$$dW_{12,\text{bbr}} = dW_{21,\text{bbr}} = \frac{3^*}{8\pi^2}\frac{\gamma_{\text{rad}}}{\hbar\Delta\omega_a}\frac{\epsilon\, d|\tilde{E}_{\text{bbr}}|^2\lambda^3}{1+[2(\omega-\omega_a)/\Delta\omega_a]^2}, \quad (36)$$

with $d|\tilde{E}_{\text{bbr}}|^2$ given by Equation 4.35. Because these blackbody $\tilde{E}$ fields will be randomly polarized, the 3* factor will have an averaged value of unity; so we will drop it from here on.

The total transition rate on the $1 \to 2$ transition due to blackbody radiation fields at all frequencies is then easily calculated by integrating the contribution from each narrow range $d\omega$, as given by Equation 4.36, over all the blackbody signals that are present at all frequencies, in the form

$$W_{12,\text{bbr}} = W_{21,\text{bbr}} = \int dW_{12,\text{bbr}}$$
$$= \int_{-\infty}^{\infty} \begin{bmatrix}\text{radiation density}\\ \text{at frequency }\omega\end{bmatrix} \times \begin{bmatrix}\text{transition response}\\ \text{at frequency }\omega\end{bmatrix} d\omega. \quad (37)$$

For any reasonable atomic transition, the atomic linewidth will always be very much narrower than the blackbody spectral distribution, as in Figure 4.6. It is then an entirely valid approximation to give the blackbody distribution function its value at the line center, $\omega = \omega_a$, and take it outside the integral over $d\omega$. The integral of Equation 4.36 over the lineshape then reduces to the simple form

$$W_{12,\text{bbr}} = W_{21,\text{bbr}} = \frac{\gamma_{\text{rad}}}{\exp(\hbar\omega_a/kT_{\text{rad}}) - 1} \int_{-\infty}^{\infty} \frac{2}{\pi\Delta\omega_a} \frac{d\omega}{1 + [2(\omega - \omega_a)/\Delta\omega_a]^2}. \tag{38}$$

But the integral on the right-hand side of this equation has unity area independent of its linewidth $\Delta\omega_a$; hence we obtain the very simple and fundamental result that

$$W_{12,\text{bbr}} = W_{21,\text{bbr}} = \frac{\gamma_{\text{rad}}}{\exp(\hbar\omega_a/kT_{\text{rad}}) - 1}. \tag{39}$$

These blackbody-stimulated transition rates turn out to be independent of any properties of the atomic transition except its radiative decay rate $\gamma_{\text{rad}}$.

The very basic result that we obtain here is thus that *the stimulated transition rate between any two atomic levels caused by blackbody fields depends only on the radiative decay rate for that transition, and on the Boltzmann factor at the temperature of the radiation, and on nothing else.* In particular this result does not depend at all on the linewidth, or even the lineshape, of the transition.

### Power Absorption from the Surroundings?

This argument says that even without any externally applied signals, thermal-noise-stimulated transitions or "jumps" will be continually taking place in both directions between any two energy levels $E_1$ and $E_2$, with stimulated-transition probabilities $W_{12,\text{bbr}}$ and $W_{21,\text{bbr}}$ given by Equation 4.39. More precisely, for two energy levels $E_i$ and $E_j > E_i$ having level degeneracies $g_i$ and $g_j$, respectively, these thermally stimulated transition rates will be given by

$$W_{ji,\text{bbr}} = \frac{g_i}{g_j} W_{ij,\text{bbr}} = \frac{\gamma_{\text{rad},ji}}{\exp(\hbar\omega_{ji}/kT_{\text{rad}}) - 1} \tag{40}$$

These transitions are caused entirely by the unavoidable blackbody radiation fields in which the atoms are always immersed (unless the electromagnetic surroundings can be cooled all the way to absolute zero).

But this in turn implies that there will necessarily be net power absorption, proportional to the atomic population difference $\Delta N = N_1 - N_2$, from the blackbody fields to the atoms. In other words, the blackbody fields will be continuously delivering energy, or heat, to the atoms through these stimulated transitions. But this in turn raises serious questions about thermal equilibrium between the atoms and the surroundings. How can a collection of atoms, which are nominally in thermal equilibrium, remain in equilibrium if they are continually absorbing energy from their surroundings? Even more serious, how can a collection of atoms which are supposedly at an atomic temperature $T_a$ (defined by the Boltzmann ratio) continually absorb energy from surroundings that might be at a different thermodynamic temperature $T_{\text{rad}}$—especially if the surrounding temperature $T_{\text{rad}}$ might in some cases be colder than the atomic temperature $T_a$?

### Power Emission to the Surroundings

The answer to these questions comes in remembering that there will also be in any atomic system purely spontaneous and entirely downward transitions, due to the spontaneous emission or radiative decay from the upper-level atoms;

and these spontaneous transitions or purely radiative decays will transfer power from the atoms back to the electromagnetic surroundings, with a spontaneous decay rate given by $\gamma_{\text{rad}}$.

These spontaneous downward transitions in the atoms are to be viewed as genuinely "spontaneous" and not as "noise-induced" transitions, at least in the approach we are taking here, since they simply occur spontaneously in a manner explainable only by quantum theory. However, we will see that these spontaneous-emission transitions from the atoms to the surroundings can and do exactly balance the noise-stimulated absorption from the surroundings to the atoms, when the atoms are in thermal equilibrium with their electromagnetic surroundings. (Some people find it helpful to describe the spontaneous downward transitions as being "one-way stimulated transitions" which are stimulated by quantum zero-point fluctuations in the electromagnetic field; but we will not get involved in that argument here.)

### Thermal Balance with the Electromagnetic Surroundings

Figure 4.7 shows schematically the overall transfer of energy that takes place in both directions between a collection of atoms and their "electromagnetic surroundings," through stimulated absorption and emission of blackbody radiation, plus spontaneous emission of energy from the atoms to the surroundings.

Each arrow in Figure 4.7 indicates the direction and magnitude of an energy flow. The ratio of energy flow from the atoms into the surroundings caused by blackbody-stimulated plus spontaneous emission, compared to energy flow in the reverse direction due to blackbody-stimulated absorption, is given by

$$\frac{\text{energy flow out of atoms}}{\text{energy flow into atoms}} = \frac{(W_{21,\text{bbr}} + \gamma_{\text{rad}}) N_2}{W_{12,\text{bbr}} N_1}$$

$$= \frac{W_{21,\text{bbr}} + \gamma_{\text{rad}}}{W_{12,\text{bbr}}} \times \frac{N_2}{N_1}. \tag{41}$$

Now, the population ratio in a collection of two-level atoms can be described at any instant by an "atomic temperature" $T_a$, in the sense that the Boltzmann ratio between the energy-level populations is given by

$$\frac{N_2}{N_1} = \exp\left(-\frac{\hbar \omega_a}{k T_a}\right). \tag{42}$$

At the same time, by using Equation 4.39 the ratio of spontaneous and noise-stimulated emission rates to noise-stimulated absorption rates is related to the temperature $T_{\text{rad}}$ of the electromagnetic surroundings by

$$\frac{W_{21,\text{bbr}} + \gamma_{\text{rad}}}{W_{12,\text{bbr}}} = \exp\left(\frac{\hbar \omega_a}{k T_{\text{rad}}}\right). \tag{43}$$

The ratio of the energy flow rates in the two directions is thus given, in terms of the temperatures of the atoms and the surroundings, by

$$\frac{\text{energy flow out of atoms}}{\text{energy flow into atoms}} = \exp\left(\frac{\hbar \omega_a}{k T_{\text{rad}}} - \frac{\hbar \omega_a}{k T_a}\right). \tag{44}$$

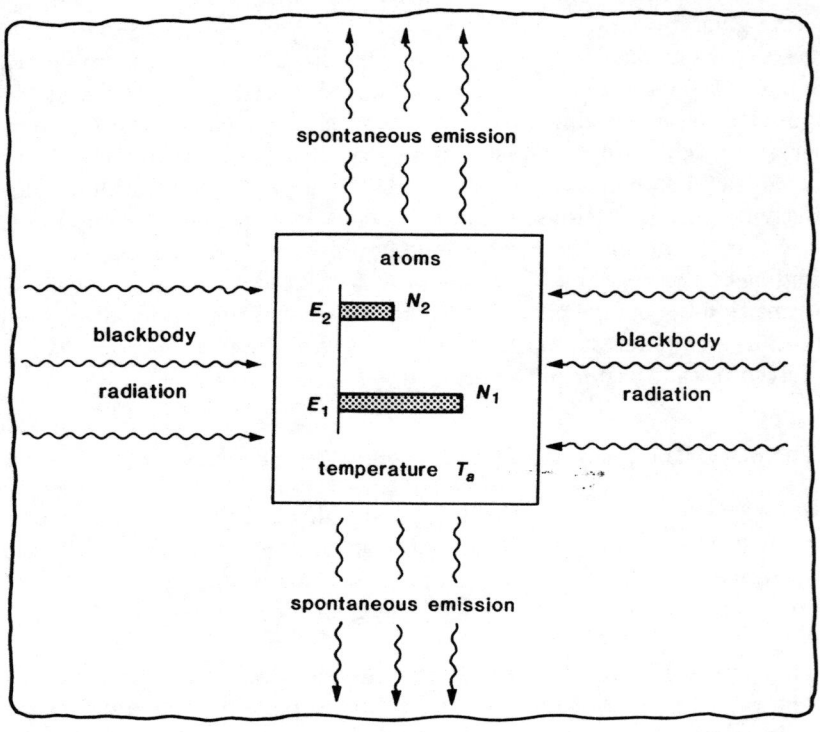

**FIGURE 4.7**
The blackbody radiation fields inside any volume whose surroundings are at temperature $T_{\text{rad}}$ will produce stimulated transitions, and thus power absorption, in a collection of atoms; the atoms in turn will radiate power back to the surroundings through spontaneous emission.

These rates will be equal and opposite *if and only if the atomic temperature $T_a$ exactly equals the surrounding electromagnetic temperature $T_{\text{rad}}$*.

The net energy received by the atoms from the blackbody fields will thus, at thermal equilibrium, exactly equal the energy radiated back to the surroundings by the atoms. There will be no net flow of atoms between levels $E_1$ and $E_2$, and no net power transfer between atoms and surroundings—as should certainly be the case at thermal equilibrium.

### Discussion: Thermal Equilibrium

There are several very fundamental conclusions that can be drawn from the preceding analytical results. First, the existence of a spontaneous, purely downward emission in any collection of atoms appears to be essential, if for no other reason than to maintain energy balance with the atomic surroundings at thermal equilibrium. A collection of atoms in thermal equilibrium at any finite temperature will always have a net power absorption on its atomic transitions; and the volume containing these atoms will always have finite blackbody signals to be absorbed by the atoms (unless the surroundings are at absolute zero). The atoms will therefore always absorb energy from the blackbody fields, producing a net flow of atoms into the upper energy levels.

These upper-level atoms must then spontaneously drop down and radiate away energy at a rate given by $\gamma_{\rm rad}$ times the number of atoms in the upper level. This energy reradiation will exactly equal the energy that the same atoms inevitably absorb from the blackbody radiation fields in which they are immersed, if the atoms and the surroundings are at the same temperature.

In the more general situation, the atomic temperature $T_a$ of a collection of atoms and the electromagnetic temperature $T_{\rm rad}$ of their surroundings might be different, at least on a temporary basis. That is, the atoms might be in internal thermal equilibrium at a temperature $T_a$, in the sense that all the phases of individual atomic oscillations are fully dephased or randomized, and all level populations satisfy the Boltzmann ratios with this temperature value. This temperature might, for example, be relatively hot because the atoms have been immersed in a hot environment. These atoms might then be suddenly moved into an enclosure which has walls at a substantially colder (or hotter) temperature $T_{\rm rad}$.

The atoms will now form one thermal reservoir at temperature $T_a$, and the walls and the blackbody radiation will form another reservoir at $T_{\rm rad} \neq T_a$. Whichever is hotter, energy will flow from the hotter system to the colder. The total system will eventually come to a thermal equilibrium at some temperature in between the initial temperatures, depending on the relative heat capacities of the two systems. This kind of "atomic transition calorimetry" can in fact be carried out experimentally, on nuclear magnetic transitions, for example.

### Detailed Balance

Overall thermal equilibrium requires, in fact, that the blackbody absorption and spontaneous emission rates be in exact equilibrium *transition by transition*, for each one of the $E_i \rightarrow E_j$ pairs in a collection of multilevel atoms. This necessity for the net absorption and spontaneous emission to be in balance on each individual transition at thermal equilibrium is sometimes referred to as "detailed balance." Detailed balance applies, in fact, not just transition by transition, but also frequency component by frequency component within any single transition: the net absorption rate by the atoms at any frequency $\omega$ and the spontaneous emission in a very narrow range $d\omega$ about that same $\omega$ must also balance. An atomic transition must, therefore, by fundamental thermodynamic arguments, have exactly the same atomic lineshape for spontaneous emission as it does for stimulated absorption, whether this lineshape be lorentzian, gaussian, or whatever.

The simple relationship derived in Equation 4.39 between $W_{ij,\rm bbr}$ and $\gamma_{\rm rad,\it ji}$ is therefore hardly accidental. This relation is rather a basic and necessary condition for thermal equilibrium to ensue. The same relation between $W_{ij,\rm bbr}$ and $\gamma_{\rm rad,\it ji}$ must hold generally for *any* kind of stimulated transition, with any lineshape or form of tensor response, and any order of electric or magnetic dipole or multipole character. The direct proportionality we noted earlier between the stimulated response $\tilde{\chi}(\omega)$ and the spontaneous emission rate $\gamma_{\rm rad}$ for an atomic transition is also a necessary consequence of the balance between net blackbody absorption and spontaneous emission that is required to reach thermal equilibrium.

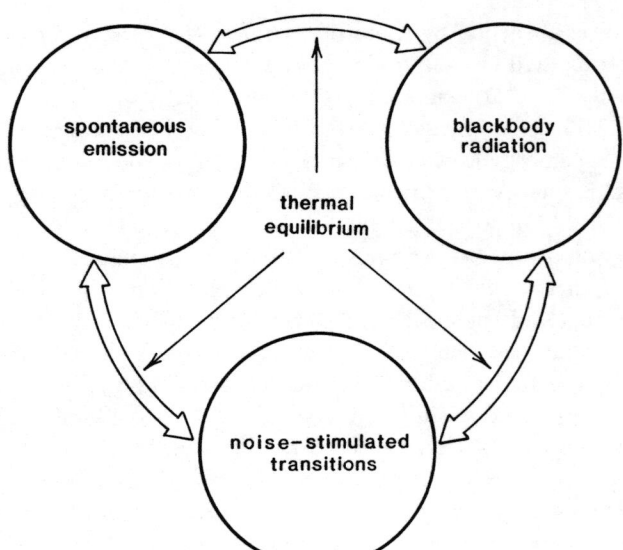

FIGURE 4.8
Stimulated atomic transitions, spontaneous emission, and blackbody radiation are connected by the right sort of circular argument: given any two of them we can calculate the third.

The logical arguments we have developed here might be represented by Figure 4.8. The stimulated-transition circle indicates the processes of noise-stimulated absorption and emission in a collection of atoms. These noise-stimulated processes can be derived by a *semiclassical derivation*—that is, a derivation in which the atoms are quantized but the electromagnetic fields are not. The blackbody-radiation and spontaneous-emission circles then indicate the existence of these two phenomena, either of which can be derived independently of the other, but only by employing a full quantum electrodynamic calculation in which the electromagnetic field itself is quantized.

The connecting arrows then indicate that we can use the existence of any two of these phenomena, plus the criterion of thermal equilibrium, to derive the existence and magnitude of the third. It is a matter of choice, for example, whether we begin with the existence of blackbody radiation, and then use this to imply the necessity for spontaneous emission; or whether we take some other direction around the circle. Any two of these processes imply the third.

## REFERENCES

An early paper which develops almost exactly the same argument as in this section is R. C. Tolman, "Duration of molecules in upper quantum states," *Rev. Mod. Phys.* **23**, 693–709 (June 1924). This paper is interesting and instructive to read even now because of how clearly Tolman understands (and presents) the fundamental ideas, despite the confusion over the quantum theory which still prevailed in 1924; and also because he mentions experimental evidence which confirms the theory. Tolman also clearly foresees the possibility of coherent "negative absorption" and hence laser amplification.

Problems for 4.3

1. *Thermal equilibration in a two-level atomic system: purely radiative case.*
   Suppose a collection of two-level atoms has a specified radiative decay rate $\gamma_{\text{rad}}$,

and no nonradiative decay, $\gamma_{nr} = 0$. The atoms are pre-cooled to absolute zero for long enough to come into equilibrium with $N_2 = 0$, and then are suddenly moved at $t = 0$ into an enclosure with walls held at a finite temperature $T_{\rm rad}$. Find formulas for the populations $N_1(t)$ and $N_2(t)$, and for the temperature $T_a(t)$ of the collection of atoms for $t > 0$.

## 4.4 NONRADIATIVE RELAXATION

The total energy-decay rates for quantum energy levels in atoms can involve both radiative and nonradiative transfer of energy from atoms to their surroundings. In a broader viewpoint, therefore, we must really be concerned with the *total atomic relaxation processes* that result from interactions between the atoms and their thermal surroundings, both through electromagnetic or "radiative" interactions and through nonelectromagnetic or "nonradiative" interactions. In this section we will try to make clear how an atomic transition interacts with both its electromagnetic and its nonelectromagnetic surroundings; how these interactions lead to both radiative and nonradiative decay; and how these in turn lead to two different but similar kinds of relaxation transitions associated with these two mechanisms.

### Radiative Relaxation Rates and Transition Probabilities

In the preceding section we obtained the remarkable and very fundamental result that blackbody radiation from the "electromagnetic surroundings" of a nondegenerate two-level atom will cause "blackbody stimulated transitions" with upward and downward transition probabilities given by

$$W_{12,\rm bbr} = W_{21,\rm bbr} = \frac{\gamma_{\rm rad}}{\exp(\hbar\omega_a/kT_{\rm rad}) - 1}, \qquad (45)$$

where $T_{\rm rad}$ is the temperature of the electromagnetic surroundings. This is a very fundamental relationship. We can view it as being imposed by the necessity for thermal equilibrium between the rate at which an atom spontaneously radiates energy and the rate at which it absorbs energy from blackbody fields.

The transition rates $W_{12,\rm bbr}$ and $W_{21,\rm bbr}$ are thus from one viewpoint stimulated transitions caused by the real (if weak), random, omnipresent blackbody radiation fields. The existence of these fields depends only on the temperature of the surroundings, however, and on nothing else. There is nothing we can do in practice to control or modify these blackbody fields (short of cooling everything in the vicinity down toward absolute zero). Hence we may just as well think of the blackbody-stimulated transition rates as being part of the *relaxation mechanisms* which are always present among the atomic-level populations, independent of anything that we ourselves do.

In earlier chapters we spoke for simplicity only of energy decay, i.e., only of spontaneous downward relaxation from upper levels to lower levels. The possibility of "upward relaxation," caused by energy coming back from the thermal surroundings to the atoms, was not mentioned. We are now seeing that, in a complete and accurate description, when an atom is coupled to external surroundings it can do more than just relax downward and give energy to those surroundings, as we said earlier. It can also (but with inherently lower probability) *receive energy from its thermal surroundings* and be lifted or relaxed *upward in energy*. This is directly related to the fact that in thermal equilibrium there are always some numbers of atoms, given by the Boltzmann ratios, in upper energy levels (though these may be very small numbers). At any temperature greater than absolute zero, the atoms never relax completely into the lowest energy level, as would always happen if only downward relaxation occurred.

Of course, for optical-frequency transitions at room temperature, the Boltzmann ratio is enormously small ($\approx 10^{-36}$). Both the upper-level populations and the upward relaxation rates are truly negligible, and only downward relaxation need be considered. For lower frequencies and more closely spaced levels, however, Boltzmann ratios and upward relaxation rates do need to be taken into account, and therefore we do need to understand the full situation described here. For microwave and lower-frequency transitions, in fact, the Boltzmann ratio becomes nearly unity, and upward and downward relaxation rates become very nearly equal.

### Nonradiative Relaxation Rates and Transition Probabilities

Blackbody relaxation and energy-exchange mechanisms represent, however, only the interactions of the atoms with their *electromagnetic* surroundings, acting through the blackbody radiation and the radiative decay rate. These interactions are shown in a schematic form in the top part of Figure 4.9.

We must recognize, however, that real atoms will usually also be in thermal contact with what we will refer to, in general terms, as "other surroundings" or "nonradiative surroundings," as shown schematically in Figure 4.9(b). These nonradiative surroundings, to which the atoms can also be coupled, can include a crystal lattice in which the laser atoms are imbedded; or a surrounding liquid medium in which the laser molecules are dissolved; or other atoms or walls with which the atoms of interest are colliding in a gas.

The atoms may then exchange energy with these "nonradiative surroundings" by means of the nonradiative decay processes that are included in the nonradiative decay rate $\gamma_{nr}$, in essentially the same way as the atoms exchange energy with the "electromagnetic surroundings" through the purely radiative processes that are involved in $\gamma_{rad}$. But this necessarily implies, from the same kind of thermodynamic reasoning we employed earlier, that these "nonradiative surroundings" must also be able to cause "nonradiatively stimulated transitions" between the atomic levels, with stimulated-transition probabilities $W_{12,nr}$ and $W_{21,nr}$, in a manner exactly analogous to the blackbody transitions $W_{12,bbr}$ and $W_{21,bbr}$ described earlier.

These additional transitions we will refer to generally as *nonradiative relaxation transitions*. The basic physics involved in the nonradiative interaction of a collection of atoms with their "nonradiative surroundings" will then be the same in every important aspect as that of the radiative interaction of these same atoms with their electromagnetic surroundings.

### Example: Phonon Interactions in Crystal Lattices

As a specific example of this, let us consider the interaction between a collection of laser or maser atoms and the lattice vibrations in a surrounding host crystal lattice, since this is one important type of "nonradiative surroundings."

## 4.4 NONRADIATIVE RELAXATION

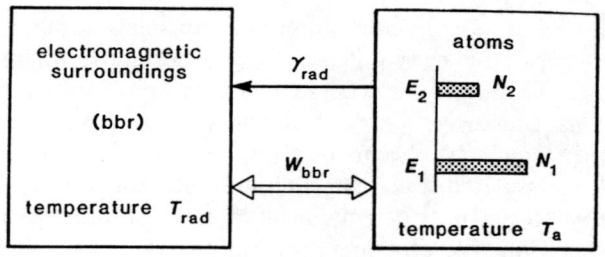

(a)

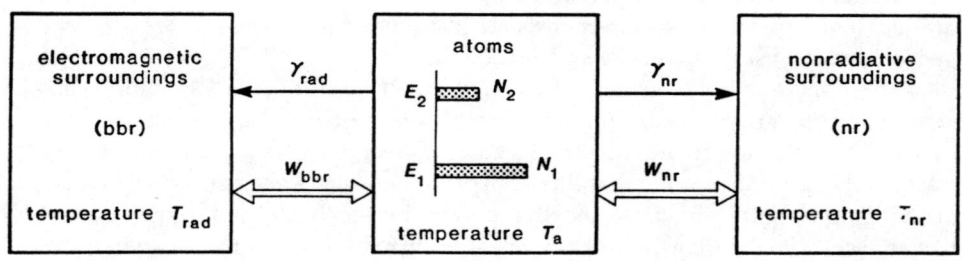

(b)

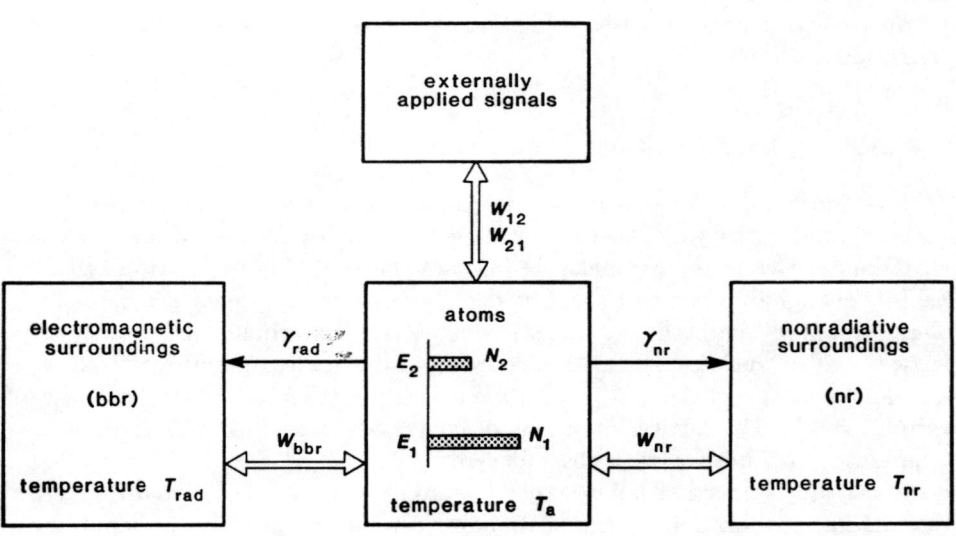

(c)

**FIGURE 4.9**
(a) Interaction of a collection of atoms with the "electromagnetic surroundings" only.
(b) Interaction with both "electromagnetic" and "nonradiative" surroundings (which may in general be at different temperatures). (c) Interaction with both of these types of surroundings, and with an externally applied signal.

A crystal lattice containing laser atoms can propagate acoustic waves, often referred to as *phonons*, at many different frequencies and in many different directions, just as a vacuum or dielectric medium can propagate electromagnetic waves, or *photons*. Moreover, like electromagnetic waves, these acoustic waves can interact with atomic transitions of atoms contained in the crystal lattice, and can produce stimulated transitions and induced atomic responses.

That is, there will generally be some weak coupling or interaction between the quantum wave function of an atom imbedded in a crystal lattice and the acoustic vibrations in the surrounding crystal lattice. This coupling is very analogous to the weak electric-dipole coupling between the atomic wave functions and the electromagnetic vibrations (fields) in the surrounding electromagnetic "ether." The basic physical principles that apply to *electromagnetic interactions* with atoms therefore apply in almost exactly the same way to what we may call generalized *acoustic interactions* with the atoms.

For example, a coherent acoustic signal in the form of a lattice vibrational wave at a frequency $\omega$ near an atomic transition frequency $\omega_a$ can be absorbed or amplified through its interaction with the atoms, just like an electromagnetic wave; and this absorption or amplification of the acoustic wave will depend on the atomic population difference (and the atomic linewidth and lineshape) exactly like an electromagnetic wave interaction. It is entirely possible to use an inverted atomic population to amplify acoustic waves and to produce acoustic-wave oscillation in a crystal at the atomic transition frequency. Such "acoustic lasers" or "acoustic masers" have been experimentally demonstrated at microwave frequencies, using some of the same pumping methods and maser materials used to produce electromagnetic maser oscillation at the same frequencies on the same transitions.

### Acoustic Transition Rates

Of more importance to us here is the fact that at any finite temperature such a crystal lattice will have thermal lattice vibrations, or "blackbody acoustic radiation," which is exactly analogous in character to blackbody electromagnetic fields (although the appropriate energy density formulas are somewhat different). These thermally induced vibrations represent the heat content of the crystal lattice, and as such can be characterized by a lattice temperature which we will label more generally as $T_{nr}$, with the subscripts standing for "nonradiative surroundings". The lattice vibrations of course go to zero only if the lattice temperature $T_{nr}$ itself goes to absolute zero.

A critically important point is that the atoms will then be affected by these thermal lattice vibrations in the surrounding crystal, in basically the same way that they are affected by the blackbody radiation in the electromagnetic surroundings. To describe this interaction we must use exactly the same arguments as for the electromagnetic surroundings, but now we refer to interactions with the "nonradiative" or lattice surroundings rather than with the "electromagnetic surroundings."

### Generalized Nonradiative Interaction Processes

In fact, by invoking the necessity for detailed thermal balance in the energy transfer processes between the atoms and the lattice acoustic modes, we can argue that these acoustically stimulated transition rates $W_{12,nr}$ and $W_{21,nr}$ must be related to the nonradiative decay rate $\gamma_{nr}$ by exactly the same fundamental relationship as Equation 4.45 for the radiative case, namely,

$$W_{12,nr} = W_{21,nr} = \frac{\gamma_{nr}}{\exp(\hbar\omega_a/kT_{nr}) - 1}. \tag{46}$$

Only if these relations hold will the power delivered to the atoms by the surrounding lattice through the $W_{12,nr}$ and $W_{21,nr}$ transitions always be exactly balanced, under thermal equilibrium conditions, by the power delivered from the atoms back to the nonradiative lattice surroundings through $\gamma_{nr}$. This equation applies, in fact, in a completely general fashion, not just to the interaction of atoms with acoustic lattice surroundings in crystals, but also to the nonradiative interactions of a collection of quantum atoms with *any kind of nonradiative thermal surroundings*.

That is, suppose the upper-level atoms in a collection of atoms do in fact lose some of their excitation energy by transferring energy into any kind of "nonradiative surroundings," whether to a surrounding crystal lattice or cell walls, or by collisions with other atoms in a gas mixture. Suppose this energy loss rate is described by a nonradiative decay rate $\gamma_{nr}$ times the upper-level population, and that the surroundings which receive this energy are describable by a temperature $T_{nr}$.

These other surroundings must then necessarily produce upward and downward thermally stimulated transitions on the same transition in the collection of atoms, with thermally stimulated transition probabilities $W_{12,nr}$ and $W_{21,nr}$ exactly as given by Equation 4.46. We use the notations $W_{nr}$ and $T_{nr}$ in this equation, and in Figure 4.9(b), to emphasize that the net interaction with any "nonradiative thermal surroundings" is completely analogous to the interaction of the same atoms with the blackbody radiation surroundings, even though electromagnetic radiation and blackbody radiation fields in the usual sense are not involved.

### Radiative Plus Nonradiative Surroundings

The nonradiative decay rate $\gamma_{nr}$ thus plays the same role in interacting with any kind of "nonradiative surroundings" as the radiative decay rate $\gamma_{rad}$ plays in interacting with the radiative or electromagnetic surroundings. The generalization of Equation 4.46 to degenerate transitions is also the same as for the electromagnetic case, namely,

$$W_{ji,nr} = \frac{g_i}{g_j} W_{ij,nr} = \frac{\gamma_{nr,ji}}{\exp(\hbar\omega_{ji}/kT_{nr}) - 1}. \tag{47}$$

The combined influence of radiative and nonradiative interactions for any collection of atoms (actually for any single transition in a collection of atoms) can then be illustrated by an expanded diagram like Figure 4.9(b), in which we indicate separately the interactions and the relaxation transition rates for the radiative and the nonradiative surroundings. The only significant parameters in these interactions are the two relaxation rates $\gamma_{rad}$ and $\gamma_{nr}$, and the associated temperatures of the surroundings $T_{rad}$ and $T_{nr}$, respectively.

These two temperatures $T_{rad}$ and $T_{nr}$ will usually have the same value; but in special cases the temperature $T_{nr}$ of the "nonradiative surroundings" could be different from the temperature $T_{rad}$ of the electromagnetic surroundings. Suppose the crystal lattice of an atomic medium is essentially lossless and transparent to electromagnetic radiation at all frequencies of interest, so that the lattice itself is not part of the electromagnetic surroundings. The temperature $T_{nr}$ characteristic of the lattice vibrations when the crystal is cooled, for example, in a liquid helium bath, may be much colder than the temperature $T_{rad}$ of the warmer electromagnetic surroundings seen by the atoms through the windows of the helium dewar.

### Another Nonradiative Example: Inelastic Collisions in Gases

As another example of nonradiative interactions, suppose that excited atoms of type $A$ in a mixture of two different gases can lose some of their excitation energy through inelastic collisions with atoms of type $B$, with this energy going into heating up the kinetic motion of the type $B$ atoms. This is a form of nonradiative decay for the excited atoms of type $A$, which can be accounted for by a nonradiative decay rate $\gamma_{nr}$ (which will probably be directly proportional to the pressure or density of the atoms of type $B$).

From the same arguments as before, these same collisions must then also produced collision-stimulated transitions in both directions between the levels of the type $A$ atoms, with transition rates $W_{12,nr}$ and $W_{21,nr}$ given by Equation 4.47, and with $T_{nr}$ given by the kinetic temperature of the type $B$ atoms. The physical details of how the kinetic motion of the type $B$ atoms can react back to produce collision-stimulated upward and downward transitions in the type $A$ atoms may not be particularly obvious; and it is certainly not at all clear how we might use a population inversion in the type $A$ atoms to "amplify" the type $B$ kinetic motion.

The general rule is, however, that if a collection of excited atoms can deliver energy in any fashion to some part of their nonradiative surroundings, then they are in some way coupled to those surroundings. As a result, these "other surroundings" are necessarily coupled back to the atoms, and thermal fluctuations in these "other surroundings" can cause upward and downward thermally stimulated transition rates in the atomic system by acting through the same nonradiative interaction mechanisms.

Note in this instance that collisions between atoms in a gas may contribute to the homogeneous line broadening of transitions in these atoms in either of two distinct ways. *Elastic collisions* between atoms cause dephasing effects, and thus give a homogeneous line-broadening contribution $2/T_2$ which is directly proportional to the collision frequency and thus to the gas pressure. *Inelastic collisions* may cause both additional dephasing and an additional nonradiative energy decay term $\gamma_{nr}$, which will in turn give an additional pressure-dependent lifetime broadening contribution.

### Total Relaxation Transition Rates

It is important to understand how there can be separate but essentially similar relaxation effects produced by both the radiative and the nonradiative surroundings, as illustrated in Figure 4.9(b). Once we understand the underlying physics, however, it is then much simpler to combine these two effects (including the spontaneous relaxation effects) into a single pair of *thermally stimulated relaxation transition probabilities*, which we will henceforth denote by $w_{12}$ and $w_{21}$, and which are defined as follows.

Let the transition rate or flow rate (in atoms/second) in the downward direction due to all these interactions be written in the form

$$\left|\frac{dN_2}{dt}\right|_{\substack{\text{downward} \\ \text{relaxation}}} = (W_{21,\text{bbr}} + \gamma_{\text{rad}} + W_{21,\text{nr}} + \gamma_{\text{nr}}) N_2 \qquad (48)$$

$$\equiv w_{21} N_2,$$

and let the corresponding flow rate in the upward direction be written as

$$\left|\frac{dN_1}{dt}\right|_{\substack{\text{upward} \\ \text{relaxation}}} = (W_{12,\text{bbr}} + W_{12,\text{nr}}) N_1 \qquad (49)$$

$$\equiv w_{12} N_1.$$

Obviously we then have

$$w_{21} \equiv W_{21,\text{bbr}} + W_{21,\text{nr}} + \gamma_{\text{rad}} + \gamma_{\text{nr}} \qquad (50)$$

in the downward direction, and

$$w_{12} \equiv W_{12,\text{bbr}} + W_{12,\text{nr}} \qquad (51)$$

in the upward direction. The downward relaxation transition probability $w_{21}$ includes both the thermally stimulated downward transitions and the spontaneous emission transitions from both radiative and nonradiative mechanisms, whereas the upward transition probability $w_{12}$ represents the thermally stimulated upward transitions due to both mechanisms.

Figure 4.10 illustrates these net relaxation rates between any pair of atomic levels. For an arbitrary pair of levels $E_i$ and $E_j > E_i$, the downward relaxation probability must be written as

$$w_{ji} \equiv W_{ji,\text{bbr}} + W_{ji,\text{nr}} + \gamma_{\text{rad},ji} + \gamma_{\text{nr},ji} \qquad (52)$$

and the upward relaxation probability on the same transition is written as

$$w_{ij} \equiv W_{ij,\text{bbr}} + W_{ij,\text{nr}}. \qquad (53)$$

We will from here on use these lowercase notations $w_{12}$ and $w_{21}$, or more generally $w_{ij}$ and $w_{ji}$, as defined above, to indicate the *total relaxation transition probabilities (per atom and per unit time)* in the upward and downward directions between any two levels $i$ and $j$, due to all the purely thermal interactions plus energy decay processes connecting the atoms to their surroundings.

Also, from now on we will restrict the uppercase symbols $W_{12}$ and $W_{21}$, or more generally $W_{ij}$ and $W_{ji}$, to indicate *signal-stimulated transition probabilities* that are produced by external signals or pumping mechanisms that we either deliberately apply to the atoms, or that we allow to build up in a laser cavity, as shown schematically in Figure 4.9(c). That is, from here on the uppercase $W_{ij}$'s signify deliberately induced transition probabilities that we can turn off or suppress; the lowercase $w_{ij}$'s are relaxation transition probabilities that we can in essence do nothing about (except possibly by cooling the surroundings).

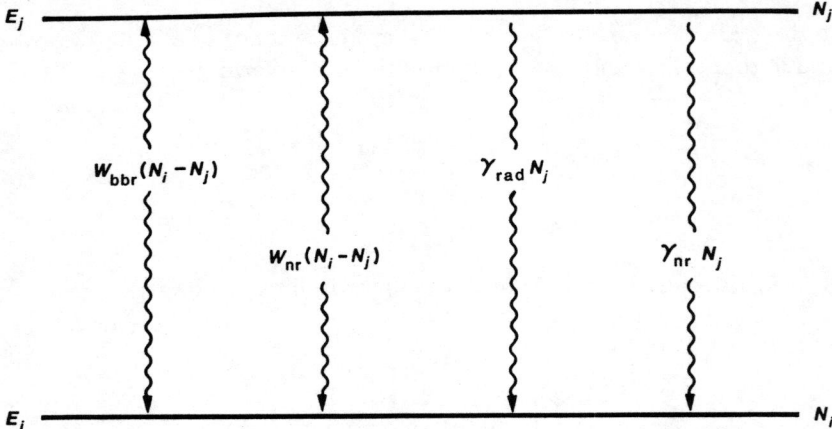

**FIGURE 4.10**
Total thermally stimulated plus spontaneous-emission transition rates between two energy levels, including both blackbody or radiative relaxation rates and nonradiative relaxation rates.

### Boltzmann Relaxation Ratios

Note that if the surroundings of an atom, radiative and nonradiative, are both at the same equilibrium temperature $T_{nr} = T_{rad} = T$, then the preceding expressions show that the ratio between upward and downward relaxation probabilities is always given by the Boltzmann ratio

$$\frac{w_{12}(\uparrow)}{w_{21}(\downarrow)} = e^{-\hbar\omega_a/kT} \tag{54}$$

or, more generally,

$$\frac{w_{ij}}{w_{ji}} = \frac{g_j}{g_i} \exp\left(-\frac{E_j - E_i}{kT}\right), \tag{55}$$

where $T$ is the temperature of the thermal surroundings. The upward thermally induced relaxation rate is always smaller (and on optical-frequency transitions usually much smaller) than the combination of downward thermally induced relaxation plus energy decay.

This Boltzmann relation does **not** depend on the nature or the strength of the radiative and/or nonradiative **relaxation** mechanisms that may be present; it will hold if they are all at the **same** temperature $T$. If the radiative and nonradiative surroundings are somehow at different temperatures, however, each interaction must be considered separately, and this ratio becomes somewhat more complicated.

### Optical Frequency Approximation

A convenient rule of thumb for visible frequencies is that the equivalent temperature corresponding to $\hbar\omega_a/k$ is $\approx 25{,}000$ K. For any reasonable temperature $T$ of the surroundings, therefore, the Boltzmann ratio at optical frequencies is always very small, on the order of

$$\exp(-\hbar\omega_a/kT) \approx \exp(-25{,}000/300) \approx 10^{-36}. \tag{56}$$

The thermally stimulated terms in the relaxation rates, either upward or downward, are then totally negligible compared to the spontaneous emission rates, and the relaxation transition probabilities in the two directions can be approximated by

$$w_{ij}(\uparrow) \approx 0 \quad \text{(upward direction)} \tag{57}$$

and

$$w_{ji}(\downarrow) \approx \gamma_{ji} \equiv \gamma_{\text{rad},ji} + \gamma_{\text{nr},ji} \quad \text{(downward direction)}. \tag{58}$$

When we write out the rate equations for lower-frequency transitions, such as for magnetic resonance or microwave maser experiments, where the photon energy $\hbar\omega$ is $\ll kT$, then the relaxation terms in both upward and downward directions must be included; and we must use the more complete formulation involving the relaxation probabilities $w_{ij}$ and $w_{ji}$ in both upward and downward directions. The simplified notation using only $\gamma_{ji}$ terms and including relaxation or energy decay in the downward direction only is more commonly employed in optical-frequency and laser analyses, where the optical-frequency approximation is almost always valid. Infrared and submillimeter laser transitions fall somewhere in between, and may require use of the more complete formulation on at least some of the transitions.

## REFERENCES

We refer in this section to the possibility of maser amplification of coherent acoustic signals rather than electromagnetic signals. Coherent amplification of microwave phonons using an inverted atomic transition was first demonstrated by E. B. Tucker. "Amplification of 9.3-kMc/s ultrasonic pulses by maser action in ruby," *Phys. Rev. Lett.* **6**, 547 (1961). A more lengthy review of these kinds of experiments is given in E. B. Tucker, "Interactions of phonons with iron-group ions," *Proc. IEEE* **53**, 1547 (October 1965).

---

Problems for 4.4

1. *Thermal equilibration: radiative and nonradiative contributions.* A collection of two-level atoms in a crystal is coupled both to the electromagnetic surroundings with radiative decay rate $\gamma_{\text{rad}}$ and to the crystal-lattice surroundings with decay rate $\gamma_{\text{nr}}$. Suppose the electromagnetic surroundings are somehow held at a fixed temperature $T_{\text{rad}}$ which is different from the fixed temperature $T_{\text{nr}}$ of the crystal lattice. Derive a formula for the steady-state equilibrium value of the Boltzmann temperature $T_a$ for the level populations of the two-level atoms in this case, as a function of the two surrounding temperatures $T_{\text{rad}}$ and $T_{\text{nr}}$, the normalized energy gap $\hbar\omega/k$, and the ratio $\gamma_{\text{rad}}/\gamma_{\text{nr}}$.

---

## 4.5 TWO-LEVEL RATE EQUATIONS AND SATURATION

The stimulated transition probabilities and relaxation transition probabilities derived in the preceding sections of this chapter can now be used to write the

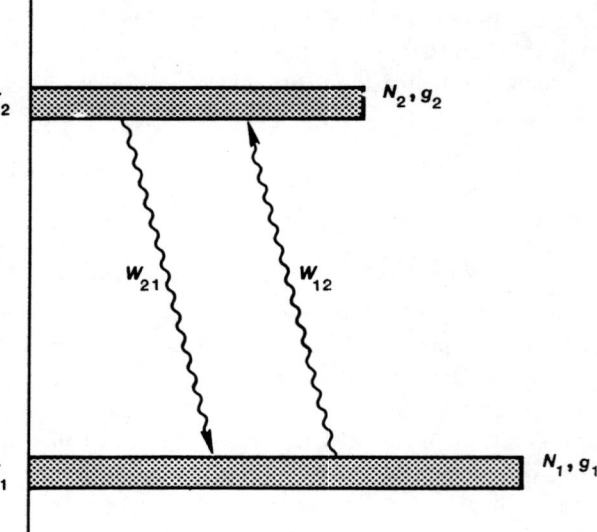

FIGURE 4.11
Total relaxation rates plus signal-stimulated transition rates between two energy levels.

general rate equations for any atomic system, taking into account both applied signals and relaxation processes. In this section we will explore the rate-equation solutions for an ideal two-level system. This will allow us to introduce a number of useful concepts, particularly the idea of *saturation* of the population difference $\Delta N$ at high enough applied signal levels.

Two-Level Rate Equation

In a simple two-level atomic system with an applied signal present, atoms flow from level 1 to level 2 at a rate $(W_{12} + w_{12}) N_1$, and from level 2 to level 1 at a rate $(W_{21} + w_{21}) N_2$, as illustrated in Figure 4.11. The total rate equation for the level populations $N_1$ and $N_2$ in this system is

$$\frac{dN_1(t)}{dt} = -\frac{dN_2(t)}{dt} = -[W_{12} + w_{12}]N_1(t) + [W_{21} + w_{21}]N_2(t). \qquad (59)$$

If the energy levels have no degeneracy, the stimulated-transition rates are related by $W_{12} = W_{21}$, and the relaxation rates are related by $w_{12}/w_{21} = \exp(-\hbar\omega_a/kT)$, where $T$ is the temperature of the surroundings of the atoms.

For a two-level system, however, it is usually more convenient to work with the total number of atoms $N_1(t) + N_2(t) = N$ and the population difference $N_1(t) - N_2(t) = \Delta N(t)$. Since the thermal equilibrium populations $N_{10}$ and $N_{20}$ with no signal present are related by the Boltzmann ratio $N_{20}/N_{10} = \exp(-\hbar\omega_a/kT)$, the population difference $\Delta N_0$ on a nondegenerate two-level transition at thermal equilibrium, with no applied signal, can be written as

$$\Delta N_0 \equiv N_{10} - N_{20} = \frac{w_{21} - w_{12}}{w_{12} + w_{21}} N = N \tanh\left(\hbar\omega_a/2kT\right). \qquad (60)$$

For a simple system with just two levels and a fixed total population, only one rate equation for the population difference $\Delta N(t)$ is then really needed. The equations for $dN_1(t)/dt$ and $dN_2(t)/dt$ can be combined into a single rate equation in the form

$$\frac{d}{dt}\Delta N(t) = -(W_{12} + W_{21})\Delta N(t) - (w_{12} + w_{21})\left(\Delta N(t) - \frac{w_{21} - w_{12}}{w_{12} + w_{21}}N\right). \quad (61)$$

We can make this equation appear even simpler by using the fact that $W_{12} = W_{21}$ for the signal-stimulated transition probability, and by defining a two-level energy relaxation time or population recovery time $T_1$ by

$$w_{12} + w_{21} \equiv 1/T_1. \quad (62)$$

If we also recognize that the final term in Equation 4.61 is just the thermal-equilibrium population difference $\Delta N_0$ for the atoms in equilibrium with the surroundings at temperature $T_{\text{rad}}$, then this two-level rate equation takes on the particularly simple and yet very general form

$$\frac{d}{dt}\Delta N(t) = -2W_{12}\Delta N(t) - \frac{\Delta N(t) - \Delta N_0}{T_1}. \quad (63)$$

This particularly simple form for the ideal two-level case with fixed total population turns out to be very useful and important for describing a great variety of laser and maser phenomena.

### Physical Interpretation: The Population Recovery Time $T_1$

Understanding this two-level rate equation is important for understanding many subsequent aspects of laser behavior. For example, the relaxation term on the right-hand side of Equation 4.63, namely, $-[\Delta N(t) - \Delta N_0]/T_1$, obviously causes the population difference $\Delta N(t)$ to relax toward its *thermal equilibrium value* $\Delta N_0$ in the absence of an applied signal, with an exponential time constant $T_1$. This time constant $T_1$ is therefore often called the *population recovery time* or the *energy relaxation time* of the system.

Suppose the two-level transition is an optical-frequency transition with $\hbar\omega_a \gg kT$. The upward relaxation probability $w_{12}$ is then essentially zero, whereas the downward relaxation probability $w_{21}$ is essentially the upper-level energy decay rate $\gamma_{21}$ as, we discussed earlier. The definition of $T_1$ therefore becomes

$$1/T_1 \equiv w_{12} + w_{21} \approx \gamma_{21} \equiv 1/\tau_{21}. \quad (64)$$

In the optical-frequency limit, the time constant $T_1$ is thus the same thing as the total lifetime or energy decay time $\tau_{21}$ of the upper energy level.

### Steady-State Atomic Response: Saturation

In contrast, the stimulated signal term $-2W_{12}\Delta N(t)$ on the right-hand side of Equation 4.63 obviously acts to drive the population difference $\Delta N(t)$ toward zero, that is, to *saturate the population difference*. The stimulated-transition probability $W_{12}$ is, of course, proportional to the strength of the applied signal, and so the rate at which $\Delta N(t)$ is driven toward zero is proportional to the applied signal intensity. Note that the factor of 2 appears in front of this stimulated term because the transition of a single atom from level 1 to level 2 both reduces

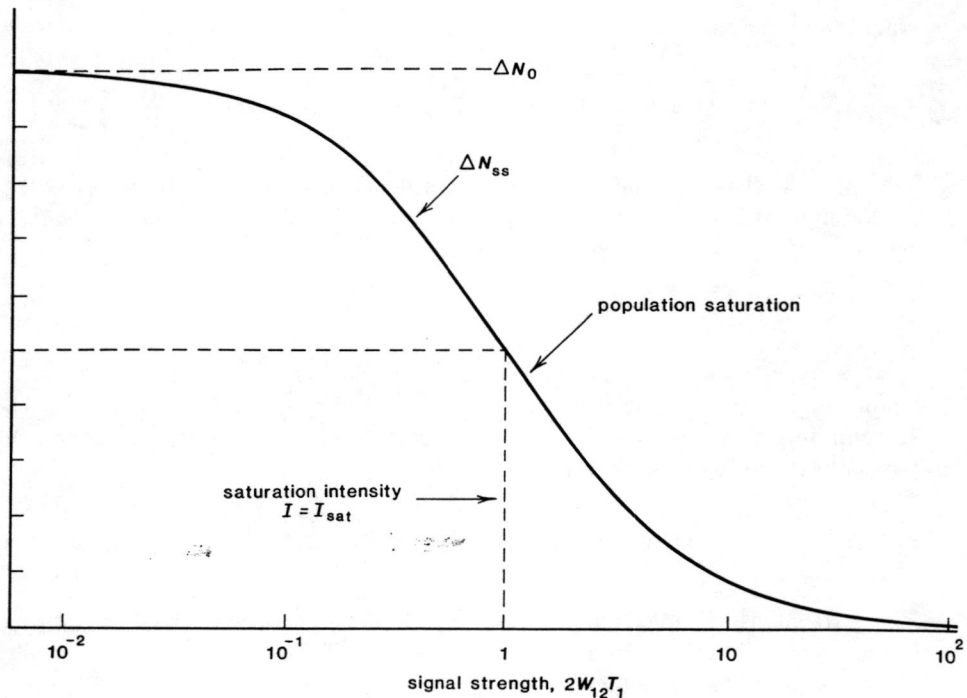

**FIGURE 4.12**
Saturation of the population difference $\Delta N$ with increasing applied signal strength in a two-level atomic transition.

$N_1(t)$ by one and increases $N_2(t)$ by one, and thus changes $\Delta N(t)$ by twice that much.

The steady-state behavior of the population difference $\Delta N$ in the presence of an applied signal $W_{12}$ must be a balance between these competing population-recovery and population-saturation effects. To obtain the steady-state solution, we can set the total time derivative in the rate equation equal to zero, i.e.,

$$\frac{d}{dt}\Delta N = 0 = -2W_{12}\Delta N - \frac{\Delta N - \Delta N_0}{T_1} \tag{65}$$

and obtain from this the steady-state population difference

$$\Delta N = \Delta N_{\text{ss}} \equiv \Delta N_0 \times \frac{1}{1 + 2W_{12}T_1}. \tag{66}$$

The ratio of the steady-state value $\Delta N_{\text{ss}}$ with signal present to the thermal-equilibrium value $\Delta N_0$ with no applied signal is plotted versus applied signal strength $W_{12}$ in Figure 4.12.

We see that as the applied signal strength or stimulated-transition rate $W_{12}$ increases, the steady-state population difference $\Delta N_{\text{ss}}$ is driven below the small-signal or thermal-equilibrium value $\Delta N_0$, and eventually is driven toward zero at large enough applied signal levels. This steady-state value of the population difference results from a balance between the stimulated-transition term, which acts to transfer atoms from the more heavily populated level $N_1$ toward the less heavily populated level $N_2$, and thus tends to equalize the populations, and the relaxation term, which tends to pull $\Delta N$ back toward its thermal-equilibrium value $\Delta N_0$.

This reduction in the steady-state population difference with increasing signal strength has the general form

$$\frac{\Delta N_{ss}}{\Delta N_0} = \frac{1}{1 + W_{12}/W_{sat}} = \frac{1}{1 + \text{const} \times \text{signal power}}, \quad (67)$$

where $W_{sat} \equiv 1/2T_1$ is the value of the stimulated-transition probability at which the population difference is driven down to exactly half its initial or small-signal value. This form of reduction in population difference with increasing signal strength is generally referred to as *homogeneous saturation of the population difference* on the two-level transition.

### Saturation in Real Laser Systems

This general type of saturation behavior is extremely important in laser theory. Gain coefficients and loss coefficients in laser materials are directly proportional to the population difference on the laser transition. We will see later on that in a great many atomic systems the population difference on the atomic transition will very often saturate with increasing signal strength in the form given by Equation 4.67, even for initially inverted population differences produced by laser pumping.

As a result, either the attenuation coefficient or the gain coefficient $\alpha_m$ in an atomic medium will very often saturate with increasing signal intensity $I$ in the general fashion given by

$$\alpha_m = \alpha_m(I) = \alpha_{m0} \times \frac{1}{1 + I/I_{sat}} = \alpha_{m0} \times \frac{1}{1 + \text{const} \times \text{signal power}}, \quad (68)$$

where $\alpha_{m0}$ is the small-signal (unsaturated) attenuation or gain coefficient; $I$ is the applied signal intensity (usually expressed as power per unit area); and $I_{sat}$ is a saturation intensity at which the gain or loss coefficient is saturated down to half its initial value $\alpha_{m0}$.

This form of saturation behavior is often referred to as *homogeneous saturation*, since it is characteristic of homogeneously broadened transitions. Inhomogeneously broadened transitions, such as doppler-broadened lines, exhibit a more complex saturation behavior, including "hole burning" effects, which we will describe in a later chapter.

### Saturable Absorption and Saturable Gain

Materials specially chosen to operate as saturable absorbers are often used in laser experiments for $Q$-switching, mode-locking, and isolation from low-level leakage signals. On the other hand, saturation of the inverted population difference and hence the gain in an amplifying laser medium is what determines a laser's power output. When a laser oscillator begins to oscillate, the oscillation amplitude grows at first until the intensity inside the cavity is sufficient to saturate down the laser gain exactly as we have described. Steady-state oscillation then occurs when the saturated laser gain becomes just equal to the total cavity losses, so that the net round-trip gain is exactly unity. Gain saturation is thus the primary mechanism that determines the power level at which a laser will oscillate.

Note that the reactive susceptibility $\chi'(\omega)$, and hence the phase shift on an atomic transition, is also directly proportional to the population difference $\Delta N$. An atomic transition will thus exhibit both *saturable absorption or gain* and *saturable phase shift* as the applied signal strength is increased.

### Transient Two-Level Solutions

Let us also look at the transient response of a two-level atomic system to an applied signal. Suppose a two-level system has some initial population difference $\Delta N(t_0)$ at time $t_0$ (where this initial value may or may not be the same as the thermal-equilibrium value $\Delta N_0$); and assume that an applied signal with constant amplitude $W_{12}$ is then turned on at $t = t_0$. The transient solution to the rate equation for $t > t_0$ is then

$$\Delta N(t) = \Delta N_{\text{ss}} + [\Delta N(t_0) - \Delta N_{\text{ss}}] \exp\left[-(2W_{12} + 1/T_1)(t - t_0)\right], \quad (69)$$

where $\Delta N_{\text{ss}}$ is the steady-state or saturated value of $\Delta N$ given earlier. This transient response is plotted for a few typical cases in Figure 4.13.

With no applied signal present, so that $2W_{12}T_1 = 0$, the population $\Delta N(t)$ relaxes from the initial value $\Delta N(t_0)$ toward the thermal-equilibrium value $\Delta N_0$ with exponential time constant $T_1$. When a constant applied signal is present, however, the population difference $\Delta N(t)$ relaxes—more accurately, is driven— toward the saturated steady-state value $\Delta N_{\text{ss}} < \Delta N_0$. Increasing the signal strength also speeds up the rate $(2W_{12} + 1/T_1)$ at which the population difference approaches this saturated condition.

### Two-Level Systems With Degeneracy

The same simple results derived above can also be obtained, though with slightly more algebraic complexity, even if the two-level system has degeneracies $g_1$ and $g_2$ in its lower and upper energy levels. To verify this we can recall that if degeneracy factors $g_1$ and $g_2$ are present, the stimulated transition rates are related by $g_1 W_{12} = g_2 W_{21}$, and the relaxation rates are related by $w_{12}/w_{21} = (g_2/g_1) \exp[-(E_2 - E_1)/kT]$. For the degenerate case, it also makes the most sense to define the **population difference** $\Delta N$ on the two-level transition in the form

$$\Delta N(t) \equiv (g_2/g_1)N_1(t) - N_2(t), \quad (70)$$

since this is the population difference that appears in the complex susceptibility $\chi(\omega)$, and hence in any absorption or gain expressions.

The **population difference** $\Delta N_0$ at thermal equilibrium must also now be written, using these definitions, in the slightly more complicated form

$$\Delta N_0 \equiv (g_2/g_1)N_{10} - N_{20} = N \frac{1 - \exp[-\hbar\omega_a/kT]}{1 + (g_1/g_2)\exp[-\hbar\omega_a/kT]}. \quad (71)$$

Here we must also define the effective signal-stimulated transition probability $W_{\text{eff}}$ by

$$W_{\text{eff}} \equiv \frac{1}{2}(W_{12} + W_{21}) \quad (72)$$

## 4.5 TWO-LEVEL RATE EQUATIONS AND SATURATION

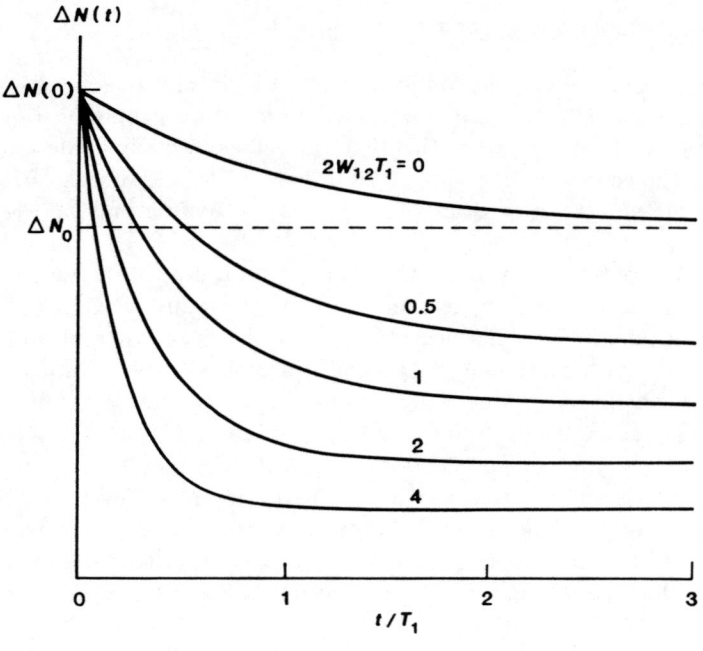

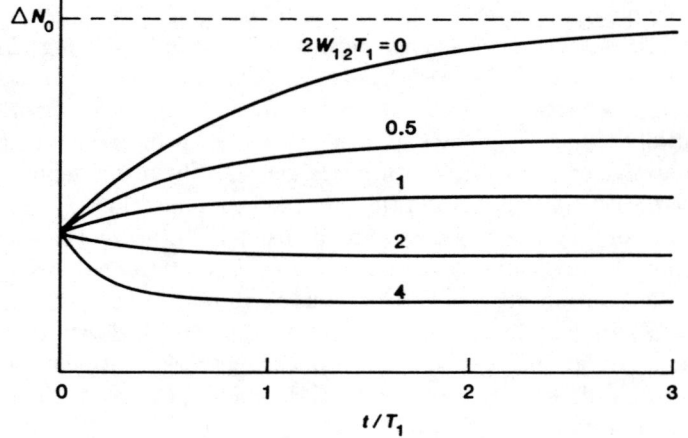

FIGURE 4.13
Transient saturation behavior following sudden turn-on of an applied signal.

and the energy relaxation time $T_1$ in the same fashion as above, namely, $w_{12} + w_{21} \equiv 1/T_1$. The two-level rate equation with degeneracy then takes on exactly the same simple form as in Equation 4.63, namely,

$$\frac{d}{dt}\Delta N(t) = -2W_{\text{eff}}\Delta N(t) - \frac{\Delta N(t) - \Delta N_0}{T_1}. \tag{73}$$

A two-level system even with degeneracy thus behaves exactly like an ideal nondegenerate two-level system, and all the results we have just derived remain valid, provided that we use the special definitions of $\Delta N(t)$ and $W_{\text{eff}}$ that we have just introduced.

### Atomic Time Constants: $T_1$, $T_2$, and $\tau$

The notation $T_1$ that we have introduced for the two-level rate equation in this section is one example of several different notations for atomic time constants that will appear frequently in the rest of this text, as well as in many analyses of atomic behavior in the scientific literature. It is important to keep track of the physical meanings of these time constants, as well as the distinctions between them.

The symbol $T_1$ is used rather widely in the scientific literature as we have used it here, namely, to indicate in general the time constant with which a population $N(t)$ or a population difference $\Delta N(t)$ will return to its equilibrium value or—what is essentially the same thing—the time constant with which an atomic system will exchange energy with its surroundings. The time constant $T_1$ is thus generally equivalent to the population recovery or energy decay times $\tau$ or $\gamma^{-1}$ often used in other analyses. For a two-level optical-frequency transition in particular, this time constant is essentially the same as the upper-level lifetime or energy-decay lifetime $\tau_{21}$. This same time constant $T_1$ is also, for reasons that we will learn later, sometimes referred to as the *longitudinal relaxation time*, especially in Bloch equation analyses, or the *on-diagonal relaxation time* in quantum analyses of atomic systems.

This time constant $T_1$ stands in contrast to the quite different time constant $T_2$ we introduced in an earlier chapter to describe the elastic dephasing of the coherent macroscopic polarization $p(t)$. The time constant $T_2$ is also widely used in the scientific literature, and is sometimes called the *atomic dephasing time*, the *transverse relaxation time*, or the *off-diagonal relaxation time* of the same atomic transition. In most situations the energy decay or population recovery time $T_1$ is substantially longer than the dephasing time $T_2$ (although for highly isolated individual atoms, as in a very low-pressure gas cell, the usual dephasing mechanisms may be nearly eliminated, and then, in the usual notation, $T_2 \approx T_1$).

The notations $T_1$ and $T_2$ are most commonly used to indicate these two different time constants in magnetic resonance and Bloch equation analyses, and for analyses on two-level atomic systems; the alternative notations $\tau$ or $\gamma$ and sometimes $\Delta \omega_a = 2/T_2$ are also commonly used, especially in optical-wavelength and multilevel laser calculations. We will jump back and forth between these alternative notations in different parts of this text, depending on what seems to match up best with the usual scientific literature.

---

### Problems for 4.5

1. *Signal-power absorption by a collection of atoms: Where does the absorbed power go?* Consider the net steady-state power per unit volume that is absorbed by the atoms, from the applied signal, in a simple two-level atomic system as a function of the applied signal strength $W_{12} = W_{21}$ (ignore degeneracy effects for simplicity). Plot how this absorbed power varies with signal strength $W_{12}$, and discuss the behavior especially at very small and very large $W_{12}$. Where does the absorbed power go?—and how does it get there?

2. *Effect of a sinusoidally modulated saturating signal: linearized analysis.* Suppose a time-varying signal is applied to a collection of nondegenerate absorbing two-level atoms. Assume this signal is tuned exactly to the transition frequency $\omega_a$, but is weakly amplitude-modulated in time at a low frequency $\omega_m$, so that

$W_{12}(t) = W_{21}(t) = W_a + W_b \cos \omega_m t$. Assume the modulation depth is small, $W_b \ll W_a$, and the modulation frequency is low, $\omega_m \ll \omega_{21}$ or $\Delta \omega_a$. The modulation frequency $\omega_m$ may, however, be of the same order as the inverse relaxation rate $1/T_1 = w_{12} + w_{21}$.

Try putting this time-modulated transition probability into the two-level rate equation and solving for the time-varying population difference $\Delta N(t)$, including saturation effects. Hints: Assume the population difference will vary like $\Delta N(t) = \Delta N_a + \Delta N_b(t)$, with $\Delta N_b \ll \Delta N_a$, and then linearize the equation by neglecting cross-products of the small terms $W_b$ and $\Delta N_b$. Find out in particular how the phase lag $\phi$ between the signal modulation $W_{12}(t)$ and the population modulation $\Delta N(t)$ will change with the modulation frequency $\omega_m$.

The instantaneous population of the upper level $N_2(t)$ in this experiment might be monitored as a function of time by observing the spontaneous emission $\gamma_{\mathrm{rad}} N_2(t)$ from the upper level with a suitable detector. By using a variable modulation frequency $\omega_m$ and a phase meter to measure the phase lag $\phi$ versus $\omega_m$, could someone use this technique to measure the lifetime $T_1$ (or $\tau$ ) of the two-level system?

3. *Effects of a square-wave modulated saturating signal.* The power level of the signal applied to a two-level atomic transition is modulated back and forth between two steady levels, say, $W(t) = W_a$ and $W_b$, in square-wave fashion, spending a length of time $T$ at each level before switching to the other level. Carry out an analysis to find the population difference $\Delta N(t)$ as a function of time through one complete cycle of this process, after many cycles have taken place. With the aid of a calculator if necessary, calculate and plot the peak-to-peak variation of $\Delta N(t)$ versus the quantity $T/T_1$ for various values of the quantities $2W_a T_1$ and $2W_b T_1$.

Verify that your answers go to the correct limiting values in the limits of $T/T_1 \ll 1$ and $T/T_1 \gg 1$.

---

## 4.6 MULTILEVEL RATE EQUATIONS

A real atomic system will, of course, have a very large number of energy levels $E_i$, with different degeneracies $g_i$ and time-varying populations $N_i(t)$. Signals may then be applied to this atomic system simultaneously at frequencies near several different transition frequencies $\omega_{ji} = (E_j - E_i)/\hbar$; and relaxation transitions will occur in general between all possible pairs of levels in the system. We will now show how to write the complete rate equations applicable to such a multilevel, multisignal, multifrequency case.

### Multilevel Atomic Systems

Figure 4.14 shows a typical multienergy-level atomic system to which several different signals tuned near different transition frequencies may be simultaneously applied. We assume for simplicity that all the transitions to which signals are applied have resonance frequencies $\omega_{ji}$ that differ from each other by at least a few atomic linewidths. This ensures that each applied signal is in

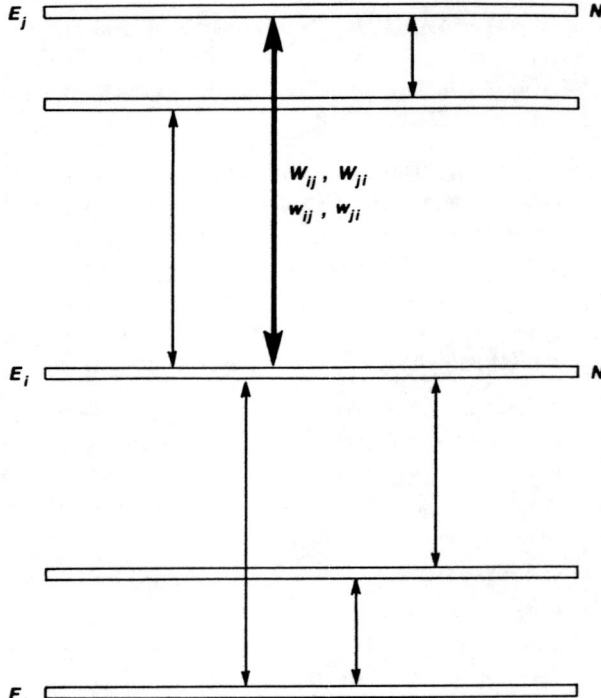

FIGURE 4.14
Multienergy-level rate equations.

resonance with (and thus affects) only the one transition to which it is tuned. We also assume that all the applied signals will be weak enough that a rate equation approach is valid. This is a point that will be discussed in more detail in a later chapter.

Consider first just the flow of atoms between some given level $E_i$ and some other higher-lying level $E_j$. If a signal is applied to this particular transition, the flow rate in the upward direction out of level $E_i$ will be $(W_{ij} + w_{ij})N_i$, and the flow rate in the downward direction into level $E_i$ will be $(W_{ji} + w_{ji})N_j$. The net flow rate between these two levels will thus be expressed by the rate equation terms

$$\frac{dN_i}{dt} = -\frac{dN_j}{dt} = -W_{ij}N_i + W_{ji}N_j - w_{ij}N_i + w_{ji}N_j \qquad (74)$$

for any pair of $i$ and $j$ levels.

The stimulated transition rate produced by the applied signal on this particular transition will have the same general form as Equation 4.31, namely,

$$W_{ji} = \frac{g_i}{g_j}W_{ij} = \frac{3^*}{8\pi^2}\frac{\gamma_{\text{rad},ji}}{\hbar\Delta\omega_{a,ij}}\frac{\epsilon|\tilde{E}_{ij}|^2\lambda_{ji}^3}{1+[2(\omega-\omega_{ji})/\Delta\omega_{a,ij}]^2}, \qquad (75)$$

where all the quantities have values appropriate to that particular $i \to j$ transition. This expression assumes a lorentzian homogeneous transition. An appropriately modified version must be substituted if the transition is a gaussian inhomogeneous transition. All atomic transition parameters such as $\gamma_{\text{rad},ji}$ and $\Delta\omega_{a,ij}$, as well as the applied signal field $\tilde{E}_{ij}$, will of course have different values for each transition in the system.

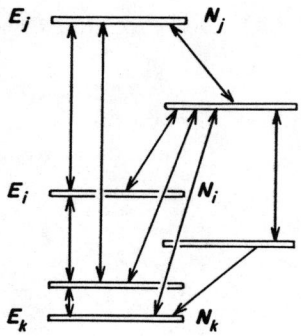

FIGURE 4.15
Example of multiple stimulated and relaxation transitions in a multilevel atomic system.

The relaxation transition probabilities $w_{ij}$ and $w_{ji}$ between an arbitrary pair of levels usually cannot be calculated in any simple fashion, and their numerical values are most often either measured or just guessed at. These numerical values may differ (widely!) for different transitions in any given system, and may depend strongly on gas pressure, crystal-lattice temperature, or other properties of the atomic surroundings. These probabilities on any given transition will, however, be related, as always, by the Boltzmann ratios

$$\frac{w_{ij}}{w_{ji}} = \frac{g_j}{g_i} \exp\left(-\frac{E_j - E_i}{kT}\right) \tag{76}$$

for any pair of levels $E_i$ and $E_j$.

### Multilevel Rate Equations

When multiple signals are present simultaneously on several different transitions, each applied signal will produce stimulated transitions that affect only the populations $N_i$ and $N_j$ of the two levels involved in that particular transition. The population changes produced by multiple signals are then taken into account by simply summing the stimulated transition rates produced by each applied signal on its particular transition, plus the relevant relaxation rates, as given by the preceding equations. The signals on different transitions do not (to first order) interfere with each other, even if they happen to terminate on the same energy level. The relaxation transition rates between all the levels are similarly taken into account simply by adding their independent effects on each energy-level population.

Suppose as a general example that energy level $E_i$ is acted on by several applied signals each close to a different transition frequency $|\omega_{ji}| = |E_j - E_i|/\hbar$ as illustrated in Figure 4.15. (The other levels $E_j$ may be below or above level $E_i$.) The general rate equation for the population $N_i(t)$ on this particular level is then

$$\frac{dN_i}{dt} = \sum_{j \neq i}(-W_{ij}N_i + W_{ji}N_j) + \sum_{j \neq i}(-w_{ij}N_i + w_{ji}N_j). \tag{77}$$

The first sum gives the stimulated transition rates to all other levels $E_j$ for which appropriate signals tuned at or near $|\omega_{ji}|$ are present. Each such signal will produce appropriate stimulated-transition probabilities related by $g_i W_{ij} = g_j W_{ji}$. The second sum gives the relaxation rates to and from all the other levels

$E_j$ of the atom, or at least all other levels $E_j$ for which the relaxation terms $w_{ij}N_i$ and $w_{ji}N_j$ have any appreciable magnitude.

In general, then, for an $M$-level atomic system we can write $M$ separate rate equations of the form in Equation 4.77, one for each level population $N_i(t)$ for $i = 1$ to $i = M$. The three rate equations for a 3-level atomic system, for example, have the form

$$\begin{aligned} dN_1/dt = &- (W_{12} + W_{13} + w_{12} + w_{13})\, N_1 \\ &+ (W_{21} + w_{21})\, N_2 + (W_{31} + w_{31})\, N_3, \\ dN_2/dt = &- (W_{21} + W_{23} + w_{21} + w_{23})\, N_2 \\ &+ (W_{12} + w_{12})\, N_1 + (W_{32} + w_{32})\, N_3, \\ dN_3/dt = &- (W_{31} + W_{32} + w_{31} + w_{32})\, N_3 \\ &+ (W_{13} + w_{13})\, N_1 + (W_{23} + w_{23})\, N_2, \end{aligned} \qquad (78)$$

if we assume that applied signals may be present on all three possible transitions.

Note that we organized these equations by systematically writing the stimulated plus relaxation terms in each equation, first for the level $N_i$ itself, and then for all the other levels $N_j$ connected to this level. It would be equally possible to organize the terms in each equation in a pair-wise fashion, for example, for level $N_2$,

$$\begin{aligned} dN_2/dt = &- (W_{21}N_2 - W_{12}N_1) - (W_{23}N_2 - W_{32}N_3) \\ &- (w_{21}N_2 - w_{12}N_1) - (w_{23}N_2 - w_{32}N_3). \end{aligned} \qquad (79)$$

where we first write all the stimulated terms and then all the relaxation terms, for each other level to which level $E_2$ is connected. The important point is obviously to include all the (necessary) terms, and then to organize them in a fashion which makes their solution the easiest.

### Conservation of Atoms

If the total number of atoms in all the energy levels is constant, however, there will also be a "conservation of atoms" equation, namely,

$$\sum_{i=1}^{M} N_i = N_1 + N_2 + \cdots + N_M = N, \qquad (80)$$

where $M$ is the total number of energy levels in the system. But if this condition applies, then only $M - 1$ of the $M$ rate equations will be linearly independent, since any one of the $M$ rate equations can be obtained as the negative sum of the other $M - 1$ equations.

We will really have, therefore, $M - 1$ rate equations for individual level populations, plus the conservation of atoms equation, to give a total of $M$ independent equations in the $M$ unknown populations $N_i(t)$, $i = 1$ to $M$.

## Multilevel Systems: Steady-State Behavior

We will employ general multilevel rate equations of the type described here to analyze several different laser pumping and signal saturation processes in future chapters. In the remainder of this chapter, however, let us look at some general characteristics of these multilevel equations and their solutions, without going into the details of any specific problems or specific examples.

Let us consider first the steady-state behavior of such a multilevel system when one or more signals of constant amplitude are applied to different transitions in the system. In particular, how will the populations and population differences in a multilevel system saturate or move away from their thermal-equilibrium values in the presence of one or more strong applied signals, and how will this compare to the simple two-level saturation result?

To find out how the steady-state level populations $N_{i,\text{ss}}$ will vary in a multi-level system with a set of constant-intensity applied signals $W_{ij}$, we must solve the appropriate set of $M - 1$ rate equations like those just described, with the time derivatives set equal to zero, plus the supplementary condition given by conservation of the total number of atoms. This gives a set of $M$ coupled linear algebraic equations of the general form

$$\begin{aligned}
\hat{W}_{11}N_{1,\text{ss}} + \hat{W}_{12}N_{2,\text{ss}} + \ldots + \hat{W}_{1M}N_{M,\text{ss}} &= 0, \\
\hat{W}_{21}N_{1,\text{ss}} + \hat{W}_{22}N_{2,\text{ss}} + \ldots + \hat{W}_{2M}N_{M,\text{ss}} &= 0, \\
\ldots + \ldots + \ldots + \ldots &= 0, \\
N_{1,\text{ss}} + N_{2,\text{ss}} + \ldots + N_{M,\text{ss}} &= N
\end{aligned} \quad (81)$$

where each of the $\hat{W}_{ij}$ elements is a linear combination of (constant) $W_{ij}$ and $w_{ij}$ factors. The first $M-1$ of these equations come from the $M-1$ rate equations, and the final one comes from the conservation of atoms. (These equations could, of course, be rearranged into any arbitrary order.)

The available methods for solving such a set of $M$ coupled algebraic equations in all their glory, in order to find the steady-state populations in an $M$-level system, are merely those used for solving any set of $M$ linear algebraic equations—and the calculations that are required, especially for $M \geq 3$, become just as messy. If you have not done this kind of calculation recently, try carrying out an explicit solution of the full-blown coupled equations for the 3-level system (Equations 4.78). It will rapidly become obvious how intractably messy the algebra becomes even for just three energy levels, let alone any case with $M > 3$.

Real laser problems do sometimes involve, however, atomic or molecular systems having anywhere from four to several dozen simultaneous coupled rate equations of this type. Possible methods for attacking these $M$-level steady-state problems then include the following.

- Adopt some standard algebraic algorithm, such as Cramer's Rule, and keep tirelessly turning the crank until algebraic solutions emerge. (However, Cramer's Rule is well known to be a poor procedure for numerical computer calculations, because of round-off error in the repeated additions and subtractions that are involved.)

- Use a computer with a good packaged linear-equation routine other than Cramer's rule (such as gaussian elimination).

- Eliminate as many terms from the equations as possible by means of physical arguments (for example, eliminate levels you know have negligible populations or terms corresponding to negligible relaxation rates); and then be clever in substituting the remaining equations into each other.

- Give up and substitute some alternative attack.

The second of these methods is the only feasible one if you have a really big multilevel problem and it really has to be solved. The third approach is the only useful one for most other simple cases.

### Saturation in Multilevel Systems

If we do solve one of these multilevel systems for the steady-state populations as a function of relaxation rates and applied signal strengths, what will the saturation behavior on any particular transition look like?

We found in the previous section that the steady-state population difference $\Delta N_{12}$ in a two-level system saturates with increasing signal intensity $W$ in a simple homogeneous fashion. We will also find quite generally in any multilevel system that increasing the signal strength $W_{ij}$ applied to any $i \to j$ transition will cause the population difference on that transition to saturate in essentially the same fashion, that is, in the form

$$\Delta N_{\text{ss},ij} = \Delta N_{0,ij} \frac{1}{1 + W_{ij}/W_{\text{sat},ij}}, \qquad (82)$$

where $W_{\text{sat},ij}$ is a saturation value or saturation intensity for that particular transition under those particular circumstances.

The initial inversion $\Delta N_{0,ij}$ on this transition might be, for example, the small-signal population inversion on the lasing transition in a system which is being pumped on some other transition. Both the initial inversion $\Delta N_{0,ij}$ and the saturation intensity $W_{1j,\text{sat}}$ will then depend in a complicated way on the relaxation rates of the system and the signals applied to any other transitions in the system. So long as these are fixed in amplitude, however, turning on and increasing the strength of a signal $W_{ij}$ applied to the $j \to i$ transition will cause the population difference $\Delta N_{\text{ss},ij}$ on that particular transition to saturate in exactly the same homogeneous fashion as for the two-level system. The saturation behavior on the inverted $i \to j$ transition will be formally identical to the saturation behavior of a two-level system, even though many other relaxation rates or other applied (fixed-intensity) signals may be present in the system.

### Proof of Saturation Behavior

The point that we have just made concerning saturation in a multilevel system is best illustrated by solving the multilevel rate equations for a few simple but still realistic practical cases, and examining the solutions, as we will do in later chapters. This point can also be proven in a slightly messy but quite general way, which we will outline here. Readers willing to accept this point may want to move on to the next subheading.

We first note that if a signal $W_{ij}$ is applied to a certain $i \to j$ transition, the rate equation for either of the two levels $N_i$ or $N_j$ involved in that transition may be rewritten at steady-state ($d/dt \equiv 0$) in the form

$$\frac{dN_i}{dt} = -W_{ij}N_i + W_{ji}N_j + f_i(N_k, w_{ik}, W_{ik}) = 0, \tag{83}$$

where $f_i(N_k, w_{ik}, W_{ik})$ is a complicated function of all the other level populations $N_k$, relaxation rates $w_{ik}$, and applied signals $W_{ik}$ on all the other levels in the system (but not $W_{ij}$ or $W_{ji}$).

Now, if the signal intensity on this one transition is turned up to a very large value, so that $W_{ij}$ and $W_{ji}$ approach $\infty$, the function $f_i$ must approach a finite limiting value, since all the factors in the $f_i$ expression remain finite. Hence, to keep the first pair of terms in Equation 4.83 finite as $W_{ij}$ and $W_{ji} \to \infty$, the population difference on the transition must decrease in the limiting form

$$\Delta N_{ij} \equiv \left(\frac{g_j}{g_i}N_i - N_j\right) \approx \frac{(g_j/g_i)f_i}{W_{ij}} \qquad (W_{ij}, W_{ji} \to \infty). \tag{84}$$

To examine this in more detail, we can note that since the rate equations are a linear coupled set, their steady-state solutions (for $dN_i/dt = 0$) for any of the steady-state level populations $N_k$, expressed in terms of any one particular transition probability $W_{ij} \equiv (g_j/g_i)W_{ji}$, must be of the general form

$$N_{\text{ss},k} = \frac{a_{kij} + b_{kij}W_{ij}}{c_{ij} + d_{ij}W_{ij}}, \tag{85}$$

where the constants $a_{kij}$, $b_{kij}$, etc., will in general be complicated mixtures of all the other relaxation probabilities $w_{pq}$ and transition probabilities $W_{rs}$ that are present in the system for all the $pq$ and $rs$ transitions other than $W_{ij}$ and $W_{ji}$. (These coefficients might be found, for example, by expanding the steady-state solution of the rate equations using Cramer's rule; and noting that all the population expressions $N_k$ will have the same denominator $c_{ij} + d_{ij}W_{ij}$.) As $W_{ij}$ and $W_{ji}$ become arbitrarily large, each of the level populations will approach some saturated steady-state value given by $N_k \to b_{kij}/d_{ij}$ as $W_{ij} \to \infty$.

Consider in particular the limit of the $i \to j$ population difference $\Delta N_{ij}$ as $W_{ij} \to \infty$. This can be gotten by subtracting two expressions like Equation 4.85 with $k = i$ and $k = j$. But then the infinite signal limit implies that the factors $(g_j/g_i)b_{iij}W_{ij}$ and $b_{jij}W_{ij}$ involved in these two expressions must exactly cancel. Hence $\Delta N_{ij}$ must have the form (for all values of $W_{ij}$)

$$\Delta N_{\text{ss},ij} = \frac{(g_j/g_i)a_{iij} - a_{jij}}{c_{ij} + d_{ij}W_{ij}}. \tag{86}$$

Note again that all the constants in this expression, like $a_{iij}$ and $c_{ij}$, depend on all the other $w_{pq}$'s and $W_{rs}$'s, but not on $W_{ij}$ or $W_{ji}$.

The conclusion of the derivation just given is that the saturated population difference for a strong signal applied to any one single transition in a multilevel system can be written in the general form

$$\Delta N_{\text{ss},ij} = \frac{\Delta N_{0,ij}}{1 + (d_{ij}/c{ij})W_{ij}}. \tag{87}$$

This is exactly like the saturation of a two-level system as derived in the preceding section.

### Transient Response of Multilevel Systems

We can also say some general things about the transient reponse of a multilevel atomic system, for example when an applied signal is suddenly turned on or turned off. The $M-1$ coupled rate equations given earlier, plus the conservation of atoms equation, form a set of $M$ coupled linear differential equations for the level populations $N_i(t)$ versus $t$ (or at least these equations are linear so long as the applied signals $W_{ij}$ have either zero or constant values). Linear coupled differential equations lead in general either to exponentially decaying or possibly to oscillating transient solutions. A two-level system exhibits, for example, a single exponential recovery with decay rate $1/T_1$ for no applied signal or $(W_{12} + W_{21} + 1/T_1)$ for a constant applied signal (cf. Equation 4.69).

The transient solutions for an $M$-level atomic system will similarly exhibit $M-1$ transient terms with $M-1$ exponential time constants, very much like the transient response of a multiloop RC electrical circuit containing $M-1$ independent capacitances. Each time constant of the multilevel system will in general be some complicated combination of all the $w_{ij}$'s and $W_{ij}$'s in the system. These time constants, or rather the corresponding decay rates, will be the multiple roots of a polynomial equation formed from the secular determinant for the coupled set of linear equations. Standard techniques such as Laplace transforms can be used to find these decay rates and transient solutions.

As a practical matter, however, in most multilevel systems one or two relaxation rates dominate in determining the transient behavior of any given level. Experimental results for the time behavior of any one level population $N_i(t)$ usually show either just one predominant time constant or perhaps in more complex cases a double-exponential type of transient behavior.

### Optical-Frequency Approximation

As we noted earlier, most laser systems at optical frequencies have $\hbar\omega/kT \gg 1$ for all the transitions involved. To express this in still another way, the energy gap corresponding to visible-frequency radiation is $\hbar\omega \approx 2$ eV, as compared to $kT \approx 25$ meV at room temperature, so that $\hbar\omega/kT \approx 40$.

In this limit, all upward relaxation probabilities $w_{ij}$ can be ignored, and all downward relaxation probabilities can be written as energy decay rates in the form $w_{ji} \approx \gamma_{ji}$. We can therefore use $\gamma_{ji}$ as an alternative notation for the downward relaxation probability from any level $E_j$ to any lower level $E_i$ in the rate equations, and there are no upward relaxation processes.

The relaxation terms for any given level $E_i$ in the general rate equations can then be simplified to the form

$$\left.\frac{dN_i}{dt}\right| = -\sum_{k<i} \gamma_{ik} N_i + \sum_{k>i} \gamma_{ki} N_k, \qquad (88)$$

where the first sum represents relaxation out of level $E_i$ into all lower levels $E_k$, and the second sum represents relaxation down into level $E_i$ from all higher levels $E_k$. If we consider only the first term, the net energy-decay rate from level $E_i$ to all lower levels is

$$\left.\frac{dN_i}{dt}\right| = -\sum_{k<i} \gamma_{ik} N_i = -\gamma_i N_i. \qquad (89)$$

The total decay rate $\gamma_i$ and the net lifetime $\tau_i$ for level $E_i$ are thus given by summing over all the radiative and nonradiative decay rates from level $E_i$ to all lower levels $E_k$, i.e.,

$$\gamma_i \equiv \frac{1}{\tau_i} = \sum_{k<i} \gamma_{ik}. \tag{90}$$

In the absence of pumping effects or relaxation from upper levels, an initial population $N_i(t_0)$ in level $E_i$ will decay as

$$N_i(t) = N_i(t_0) e^{-\gamma_i(t-t_0)} = N_i(0) e^{-(t-t_0)/\tau_i}. \tag{91}$$

Note again that multiple decay processes acting in parallel combine by summing the decay rates, or summing the *inverse lifetimes* associated with each process.

## REFERENCES

An early paper which discusses multilevel rate equations and the saturation behavior of multilevel systems, without any optical-frequency approximations, is J. P. Lloyd and G. E. Pake, "Spin relaxation in free radical solutions exhibiting hyperfine structure," *Phys. Rev.* **94**, 579 (May 1, 1954). A systematic procedure for solving rate equations and finding the transient populations $N_j(t)$ in a sequence of cascading energy levels, using the optical-frequency approximation, is given by L. J. Curtis, "A diagrammatic mnemonic for calculation of cascading level populations," *Am. J. Phys.* **36**, 1123 (December 1968).

A general proof that the populations of an $M$-level system will return to equilibrium in the form of a sum of $M-1$ decaying exponentials is given in M. W. P. Strandberg and J. R. Shaw, "General properties of thermal-relaxation rate equations," *Phys. Rev.* B **7**, 4809 (1973).

---

Problems for 4.6

1. *Saturation of the lower transition in a general three-level atomic system.* Suppose a signal that produces a stimulated transition probability $W_{12} = W_{21}$ (no degeneracy) is applied to the $1 \to 2$ transition of a three-level system. Solve the complete steady-state rate equations to obtain the saturation behavior of the population difference $\Delta N_{12}$ without making any optical-frequency approximation. In general terms, how does this saturation resemble or differ from the saturation of a simple two-level system?

2. *Ditto for saturation of the upper transition in a three-level system.* Repeat the previous problem for a signal $W_{23}$ applied only to the $2 \to 3$ transition. Compare the saturation behavior in this case to the ideal two-level case.

3. *Saturation of the 1-3 transition in a three-level system: no optical approximation.* Suppose a signal producing a stimulated transition probability $W_{13} = W_{31}$ (no degeneracy) is applied to the $1 \to 3$ transition of a 3-level system. Using the full rate equations (no optical-frequency approximation), calculate how the population difference $\Delta N_{13} \equiv N_1 - N_3$ on the signal transition will saturate with increasing signal intensity $W_{13}$. What does the general answer reduce to if all

relaxation rates have very nearly the same value $w_{ij} \approx w_{ji} \approx 1/2T$ for all $i$ and $j$?

4. *Ditto, using the optical approximation.* Repeat the previous problem, assuming that the optical approximation applies; i.e., all downward rates $w_{ji}$ are finite and all upward rates $w_{ij} \approx 0$.

5. *Rate-equation analysis of a thermally pumped laser (research problem).* We mentioned in Chapter 1 that a laser or maser can be pumped by a purely thermal or blackbody source, and that a laser oscillator of this sort is really a kind of heat engine with a limiting efficiency that should be equal to the Carnot-cycle efficiency between the pumping and relaxation temperature. Let us try to develop this analysis further.

As an idealized model for this analysis, let us consider a collection of identical atoms having just three energy levels $E_0$, $E_1$, and $E_2$; and suppose that this collection of atoms has very strong and primarily *radiative* decay on the 0–2 transition, and has very strong and primarily *nonradiative* decay on the 0–1 and 1–2 transitions. Suppose also that the effective temperature $T_{\rm rad}$ of the electromagnetic surroundings for the 0–2 radiative transition is very hot, but the effective temperature $T_{\rm nr}$ for the nonradiative surroundings of the 0–1 and 1–2 transitions is much colder. It may then be possible to achieve a thermally pumped laser inversion on the 2–1 transition; in fact, this situation sounds very much like a description of an idealized sun-pumped laser.

Suppose a laser signal that produces a stimulated transition probability $W_s$ on the 2–1 transition is also present. Solve the rate equations for this system to evaluate the steady-state populations $N_0$, $N_1$, and $N_2$, and evaluate under what conditions a population inversion can in fact be produced on the 2–1 transition. Then evaluate both the net rate at which energy is extracted from the laser medium by the laser signal, as given by $W_s(N_2 - N_1)\hbar\omega_{21}$, and also the net rate at which the thermal radiative pumping source delivers energy to the laser medium on the 0–2 transition.

Considering this device as a kind of heat engine, which converts thermal energy from the hot source at $T_{\rm rad}$ into work in the form of coherent radiation, evaluate the conversion efficiency in this device as a function of the applied signal level $W_s$. Can the resulting efficiency be made to look anything like a Carnot-cycle efficiency in any reasonable limit?

# 5
# THE RABI FREQUENCY

Both the linear susceptibility approach and the rate equation analysis we have developed in the past several chapters are approximations—though usually very good approximations—to the exact dynamics of an atomic system with an external signal applied.

If a very strong (or very fast) signal is applied to an atomic transition, however, the exact nonlinear behavior of the atomic response becomes more complicated, and the rate-equation approximation is no longer adequate to describe the atomic response. In this chapter, therefore, we explore the conditions under which the rate-equation approximation will remain valid, and some of the interesting new effects—particularly the very important Rabi flopping behavior—that a resonant atomic transition will display in response to a strong enough applied signal.

The material discussed in this chapter, though essential for understanding large signals and so-called "coherent transient" effects, is not essential for most straightforward laser amplification and oscillation effects. Readers who are primarily interested in the latter may therefore want to skip over this chapter.

## 5.1 VALIDITY OF THE RATE-EQUATION MODEL

Let us first do a quick review of the basic equations which lead to rate equations, and of the approximations involved in writing the rate equation for a simple two-level atomic system.

### The Resonant-Dipole Equation

The classical electron oscillator model, with suitable quantum extensions, led us in Chapter 2 to the "resonant-dipole equation"

$$\frac{d^2 p(t)}{dt^2} + \Delta\omega_a \frac{dp(t)}{dt} + \omega_a^2 p(t) = K\Delta N(t)\mathcal{E}(t), \tag{1}$$

where the constant $K$ is given by

$$K \equiv \frac{3^* \omega_a \epsilon \lambda^3 \gamma_{\rm rad}}{4\pi^2}. \tag{2}$$

We emphasize once more that this equation is a *quantum-mechanically correct* equation for the expectation value of the polarization $\langle p(t) \rangle$ in a two-level quantum system.

To derive the linear susceptibility $\tilde{\chi}(\omega)$ for the atomic transition, we solved this equation for sinusoidal steady-state signals, treating the population difference $\Delta N(t)$ as a constant. This susceptibility turns out to be, of course, directly proportional to the population difference $\Delta N$.

In Chapter 4 we then used these linearized sinusoidal results, based on assuming that the level populations are *constant*, to derive the rate equations that predict a *time rate of change* for the populations $N_1(t)$, $N_2(t)$, and $\Delta N(t)$. To do this, we made the assumption that both the linear susceptibility description (based on constant $\Delta N$) and the rate-equation results (which describe a time-varying $\Delta N(t)$) will remain valid provided that the time rate of change of the population difference $\Delta N(t)$ is "slow" in some meaningful sense. The conditions for this approximation to be valid are essentially the following.

### Transient Response of the Resonant Dipole Equation

The resonant-dipole equation 5.1 is basically a linear second-order resonant equation, with a linewidth $\Delta\omega_a$ or response time $2/\Delta\omega_a$ (often written as $T_2$). To examine its transient behavior, let us suppose that the population difference $\Delta N(t)$ is indeed constant, or nearly so, and that the applied signal $\mathcal{E}(t)$ is a cosinusoidal signal at $\omega = \omega_a$ that is turned on at $t = 0$ in the form

$$\mathcal{E}(t) = E_1 \sin \omega_a t, \qquad t \geq 0. \tag{3}$$

The induced response $p(t)$ will then be given, very nearly, by

$$p(t) \approx -K \frac{\Delta N E_1}{\omega_a \Delta \omega_a} \left[1 - e^{-\Delta\omega_a t/2}\right] \cos \omega_a t \tag{4}$$

as illustrated in Figure 5.1.

If the $2/T_2$ term dominates in the linewidth expression $\Delta\omega_a = \gamma + 2/T_2$, as is often the case, or if we simply absorb the $\gamma$ contribution into a broadened definition of $2/T_2$, this may be written as

$$p(t) \approx -K \frac{\Delta N E_1}{\omega_a \Delta \omega_a} \left[1 - e^{-t/T_2}\right] \cos \omega_a t. \tag{5}$$

The transient response is thus a forced cosinusoidal oscillation that builds up to a steady-state value with a time constant $2/\Delta\omega_a$, often written as just $T_2$ for simplicity.

The important conclusion to be drawn here is the following. We already know that the forced response of the atomic polarization $p(t)$ to a sinudoidal driving signal $\mathcal{E}(t)$ will in general be very small unless the driving signal frequency $\omega$ is at or close to the resonance frequency $\omega_a$ of the system. We now see that this forced sinusoidal response of $p(t)$ will follow any amplitude (or phase) variations in the envelope of the sinusoidal driving term $\mathcal{E}(t)$ with a transient time delay that is approximately $2/\Delta\omega_a \approx T_2$.

## 5 1 VALIDITY OF THE RATE-EQUATION MODEL

**sinusoidal signal input:**

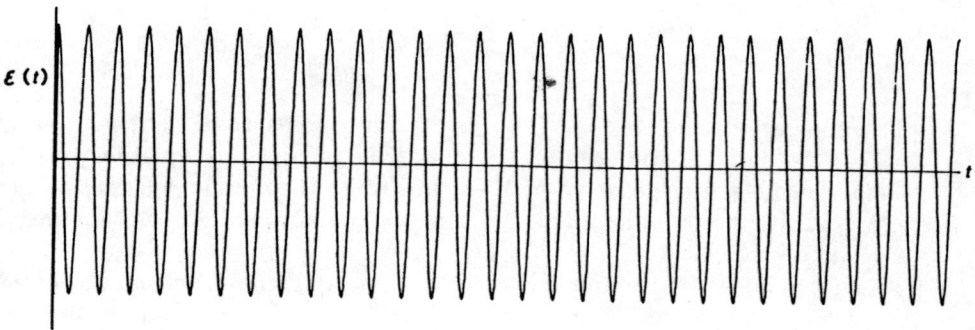

**induced polarization response:**

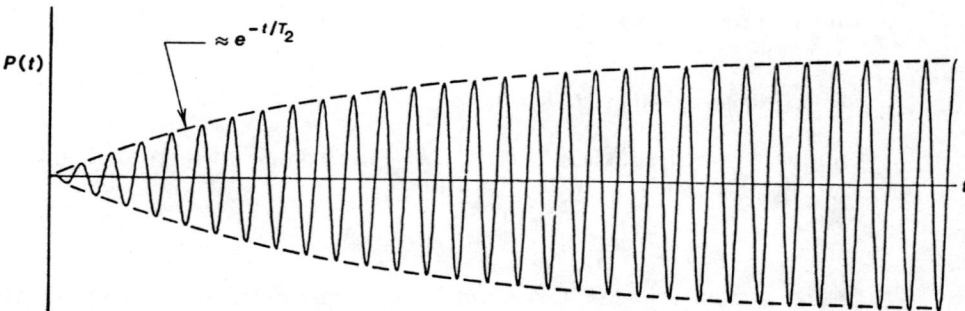

FIGURE 5.1
Transient build-up of induced polarization in response to a weak sinusoidal signal input which is suddenly turned on at $t = 0$.

It will therefore be a valid approximation to solve the resonant-dipole equation 5.1 and use it to find both the steady-state and transient responses of $p(t)$, making the approximation that $\Delta N(t)$ is a constant, *provided that the transient rate of change of $\Delta N(t)$ itself is slow compared to the time constant $2/\Delta\omega_a \approx T_2$*. It is the slow variation of $\Delta N(t)$ that is essential—not necessarily the slow variation (or weak amplitude) of $\mathcal{E}(t)$.

Note also that within this approximation, variations in either the phase or the amplitude of a signal field $\mathcal{E}(t)$ that are rapid compared to $T_2$ will simply not be "seen" or responded to by the atomic system. To put this in another way, the atomic transition has a finite bandwidth $\Delta\omega_a$, and rapid variations in phase or amplitude of $\mathcal{E}(t)$ represent frequency sidebands that are outside the linewidth $\Delta\omega_a$ of the atomic response. Hence these sidebands will induce little or no response (in the small signal limit).

### The Population Difference Equation

Let us now see what determines the time-variation of $\Delta N(t)$ itself, especially under the influence of stronger applied signals. From an energy-conservation argument, we showed in Chapter 4 that the population equation for a simple two-level quantum system may be written in the form

$$\frac{d\Delta N(t)}{dt} + \frac{\Delta N(t) - \Delta N_0}{T_1} = -\left(\frac{2}{\hbar\omega}\right)\mathcal{E}(t)\cdot\frac{dp(t)}{dt}. \tag{6}$$

The population difference $\Delta N(t)$ is essentially a measure of the energy in the atomic system; and the term on the right-hand side of this equation gives the instantaneous power delivered by the field $\mathcal{E}(t)$ to the atomic polarization $p(t)$, expressed in photon units. (The factor of 2 is present here for the same reason given in Section 4.5.)

This equation, written in this form, is also a *quantum-mechanically correct* equation for the expectation value of the energy, or the population difference $\Delta N$, in a two-level quantum system. To make this be generally true even at large signals, however, the right-hand side of this equation must be kept in the more general and fundamental form given here, rather than in the stimulated transition form $-2W_{12}\Delta N$, because we are now *not* limiting ourselves to the usual steady-state amplitude and phase relationship between $\mathcal{E}(t)$ and $p(t)$ that leads to the rate equations.

### Quasi-sinusoidal Applied Signals

Suppose we write the applied signal $\mathcal{E}(t)$ in a somewhat more general form, namely

$$\mathcal{E}(t) = \mathrm{Re}\,\tilde{E}(t)e^{j\omega_a t}. \tag{7}$$

That is, we assume that $\mathcal{E}(t)$ is basically a sinusoidal signal somewhere near the resonance frequency $\omega_a$, but with a possible amplitude or phase modulation that is contained in the time-varying complex amplitude $\tilde{E}(t)$. Similarly, we can write the resulting polarization in the same general form

$$p(t) = \mathrm{Re}\,\tilde{P}(t)e^{j\omega_a t}. \tag{8}$$

Let us put these signals into the right-hand side of the population difference equation 5.6. This right-hand side then becomes

$$\begin{aligned}-\frac{2}{\hbar\omega}\mathcal{E}(t)\frac{dp(t)}{dt} &= -\frac{j}{2\hbar}\left(\tilde{E}e^{j\omega t} + \tilde{E}^*e^{-j\omega t}\right)\times\left(\tilde{P}e^{j\omega t} - \tilde{P}^*e^{-j\omega t}\right)\\ &= \frac{j}{2\hbar}\left(\tilde{E}\tilde{P}^* - \tilde{E}^*\tilde{P}\right) - \frac{j}{2\hbar}\left(\tilde{E}\tilde{P}e^{2j\omega t} - \tilde{E}^*\tilde{P}^*e^{-2j\omega t}\right).\end{aligned} \tag{9}$$

The driving term on the right-hand side of the population equation will thus contain both *quasi constant or dc terms proportional to the imaginary part of* $\tilde{E}\tilde{P}^*$ and *second harmonic or $\pm 2\omega$ terms proportional to the imaginary part of* $\tilde{E}\tilde{P}$.

### Harmonic-Generation Terms

Let us first consider these $\pm 2\omega$ terms, which are essentially harmonic-generation terms. The response of the population difference $\Delta N(t)$ in Equation 5.6 is fundamentally a sluggish response, because of the normally very long relaxation time $T_1$ that appears on the left-hand side. We can expect therefore that the response of this equation to the second harmonic or $\pm 2\omega$ terms will be very small, compared to the response to the quasi-dc terms on the right-hand side.

To put this in another way, changes in $\Delta N(t)$ will result from the integration over time of the terms on the right-hand side of Equation 5.6. But those terms with a time-variation of the form $e^{\pm 2j\omega t}$ will tend to integrate to zero within a few optical cycles, whereas quasi-dc terms will tend to integrate into a significant change with time. The $2\omega$ terms on the right-hand side of Equation 5.6 can therefore normally be dropped.

At high enough signal levels, the harmonic terms appearing in Equation 5.6, which we are now discarding, will produce some small but nonzero modulation at $2\omega$ of the population difference $\Delta N(t)$. These small second-harmonic terms in $\Delta N(t)$ will then carry back into the right-hand side of the resonant-dipole equation 5.1 for $p(t)$, where they will mix with the $\pm \omega$ terms in $\mathcal{E}(t)$ to produce both $\pm 3\omega$ driving terms, and small additional $2\omega - \omega = \omega$ terms. The $\pm 3\omega$ driving terms will then produce third harmonic terms in the polarization $p(t)$, and these terms may in turn radiate and generate third-harmonic optical signals. This chain of harmonic effects can continue to higher orders as well, though the effects grow rapidly weaker with increasing order.

Large enough driving signals will, therefore, potentially produce even-order harmonic responses in $\Delta N(t)$, which in turn will feed back to produce odd-order harmonic responses in $p(t)$, and vice versa. These various higher-order harmonic responses can be observed as *harmonic generation and intermodulation or mixing effects* that occur at large signal intensities in atomic systems. These harmonic-generation effects are one part of the rich repertoire of large-signal nonlinear effects that can be observed in atomic systems, and that form the basis of the very useful field of nonlinear optics.

### Conventional Rate-Equation Approximation

If we ignore these weak harmonic terms, however, the population difference equation 5.6 simplifies to

$$\frac{d\Delta N(t)}{dt} + \frac{\Delta N(t) - \Delta N_0}{T_1} = -\frac{j}{2\hbar}\left[\tilde{E}(t)\tilde{P}^*(t) - \tilde{E}(t)^*\tilde{P}(t)\right]. \tag{10}$$

Now, if the linear-susceptibility or rate-equation condition holds, we can relate $\tilde{P}$ and $\tilde{E}$ to a good approximation by

$$\tilde{P} \approx \tilde{\chi}\epsilon\tilde{E} \approx (\chi' + j\chi'')\epsilon\tilde{E}, \tag{11}$$

where $\tilde{\chi}$ itself is directly proportional to $\Delta N$. The right-hand side of this equation then simplifies still further to become

$$\frac{d\Delta N(t)}{dt} + \frac{\Delta N(t) - \Delta N_0}{T_1} \approx (\epsilon/\hbar)\chi''|\tilde{E}|^2 \approx -2W_{12}\Delta N(t). \tag{12}$$

But this is, of course, simply the standard two-level rate equation.

### Conditions for Rate-Equation Validity

Having derived this rate equation for $\Delta N(t)$ on the assumption that any changes in $\Delta N(t)$ will be slow, we can then solve it for the predicted variation of $\Delta N(t)$ and see if the rate of change will in fact be slow. As we have seen in earlier chapters, a typical transient solution to the rate equation, assuming a constant-amplitude signal $W_{12}$ suddenly turned on at $t = t_0$, is

$$\Delta N(t) = \Delta N_{ss} + [\Delta N(t_0) - \Delta N_{ss}] \times \exp[-(2W_{12} + 1/T_1)(t - t_0)], \qquad (13)$$

where $\Delta N_{ss}$ is the partially saturated, steady-state value of $\Delta N(t)$ as $t \to \infty$.

The condition that the time rate of change of the population, or $(d/dt)\Delta N(t)$, be slow compared to the time constant $2/\Delta\omega_a$ in the resonant-dipole equation reduces to the condition that

$$[2W_{12} + 1/T_1] \ll [\Delta\omega_a \equiv 1/T_1 + 2/T_2]. \qquad (14)$$

One general condition for this to be satisfied, and for a rate equation to be valid, is that $1/T_1 \ll 1/T_2$, or that the energy relaxation time $T_1$ be long compared to the dephasing time $T_2$. To put this another way, the atomic transition should have a significant amount of broadening due to dephasing, as compared to the purely lifetime broadening in the system. This condition is generally true for most laser transitions.

Of more significance, once this condition is met, is that the signal strength must be weak enough so that

$$W_{12} \ll \Delta\omega_a. \qquad (15)$$

In other words, the stimulated-transition rate $W_{12}$ must be small compared to the transition linewidth $\Delta\omega_a$. For electric dipole transitions this can be converted into a condition on the applied signal strength given by

$$|\tilde{E}|^2 \ll \frac{(\hbar\Delta\omega_a)^2}{\epsilon\hbar\gamma_{\text{rad}}\lambda^3}. \qquad (16)$$

In a quantum analysis we can show that this condition is equivalent to requiring that *the quantum-mechanical perturbation of the energy of the atom caused by the applied field strength $\tilde{E}$ must be small compared to the homogeneous linewidth $\hbar\Delta\omega_a$ of the atom expressed in energy units.* We will express this condition in another and more meaningful way in the following section.

### Rate-Equation Validity in Typical Laser Systems

The great majority of signals present in even high-power laser systems will in fact satisfy the criteria expressed by Equations 5.14 through 5.16, and the rate-equation approximation will thus be valid. Higher-power laser systems, in fact, commonly use materials that have wider atomic linewidths, which helps to preserve this condition.

For example, the atomic linewidths in gas lasers may range from a few hundred Mhz up to a few Ghz, and solid-state linewidths are typically $10^{11}$ to $10^{12}$ Hz. The corresponding transient response times $T_2$ for the polarization equation are in the range from $10^{-8}$ to $10^{-12}$ sec. The stimulated-transition rates in these same lasers can be estimated by equating the actual laser power density extracted per unit volume from the laser medium to the inverted population density $\Delta N$ per unit volume, times $\hbar\omega$, times a signal-stimulated-transition rate $W_{ij}$. The resulting stimulated-transition rates are typically in the range from $10^3$ to $10^7$ sec$^{-1}$, and thus readily meet the criterion just given.

The level populations $N_j(t)$ in an atomic system also change with time as a result of relaxation and laser pumping. In practice, in useful laser materials population changes due to either relaxation or pumping are slow—in fact, most often very slow—compared to the inverse linewidth. As an elementary example, the relaxation and pumping time constants $w_{ij}^{-1}$ and $W_{ij}^{-1}$ in solid-state laser materials commonly range from milliseconds (e.g., ruby) to a few hundred microseconds (e.g., Nd:YAG or Nd:glass); whereas inverse linewidths in these materials are in the range $1/\Delta\omega_a \approx 10^{-11}$ seconds. In organic dye lasers the relaxation and stimulated transition times can be much faster, e.g., a few nanoseconds ($10^{-9}$ sec) down to even a few picoseconds ($10^{-12}$ sec). However, the inverse linewidths for these materials are even shorter, e.g., typically $1/\Delta\omega_a \approx 10^{-13}$ sec.

### Saturation Condition

We might also ask if we can obtain the saturation condition for a population difference $\Delta N$, as discussed in Chapter 4, while still remaining within the range of validity of the rate equation approach. The signal intensity required to achieve saturation in, for example, a simple two-level atomic system is given by $2W_{12}T_1. \geq 1$ or $W_{12} \geq 1/2T_1$, and the condition for remaining within the rate-equation regime is given in Equation 5.14. Combining these two equations then yields the double condition

$$[1/2T_1] \leq W_{12} \ll [\Delta\omega_a \equiv 1/T_1 + 2/T_2]. \tag{17}$$

This says that a population difference can be saturated without violating the rate-equation limitation only if the time constants $T_1$ and $T_2$ satisfy the condition, $1/T_1 \ll \Delta\omega_a$ or $T_2 \ll T_1$. This condition is met in virtually all useful laser materials: the energy relaxation rates are nearly always slow compared to the atomic linewidth, since the latter is determined primarily by dephasing (or even inhomogeneous) mechanisms that are substantially larger than pure lifetime broadening.

### Large-Signal Effects in Multilevel Systems

The classical oscillator or resonant-dipole model and the associated linear susceptibility, on the one hand, and the multilevel rate equations, on the other hand, provide two complementary sets of equations for analyzing the complete response not merely of a two-level system, but of a multilevel atomic system as well.

In a multienergy-level system, *a separate resonant-dipole model must be applied to each individual atomic transition*, with the associated populations $N_j(t)$ and $N_i(t)$ taken as quasi constants. The resulting polarization and susceptibility on each transition can then be used directly in Maxwell's equations, and can describe accurately the amplitude, phase, polarization, and even tensor characteristics of the atomic response on that particular transition. Note that the resonant-dipole equation for each transition is a second-order equation, as well as potentially a vector equation. Hence it can give both amplitude and phase, as well as tensor properties, of the response on that transition.

The rate equations, by contrast, treat only the energy flow, or the intensity part of the atomic response, with no phase information being available. However, they do provide a set of simple coupled first-order equations that can tie together the populations of all the energy levels in an atomic system, including relaxation and pumping mechanisms, as well as all the simultaneous signals applied to the system.

As a general approach, then, given a multilevel system with multiple signals applied, we can use the susceptibility and polarization results on each individual transition to find phase shifts, gains, and reactions back on the applied signals, and the rate equations to find the resulting populations on those transitions. Combining these two approaches provides a more or less complete, accurate, and self-consistent description of the atomic response.

There are, of course, situations in which applied signals may violate the rate-equation conditions. We can then expect to find nonlinear effects, including nonlinear mixing, intermodulation, harmonic signals, and the Rabi flopping behavior we will describe in the following section. Detailed analysis of these effects in a multilevel system requires more general analytical methods, of which the "density matrix" approach of quantum theory is among the most useful and widely employed. The Bloch equations of magnetic-resonance theory are also very useful for analyzing large-signal and nonlinear effects in simple two-level atomic systems.

## REFERENCES

An interesting discussion of when simple rate equations do and do not apply, including the reaction of the atoms back on the signal field, is given by C. L. Tang, "On maser rate equations and transient oscillations," *J. Appl. Phys.* **34**, 2935 (October 1963).

A few representative references on large-signal harmonic-generation and intermodulation effects in atomic systems include J. R. Fontana, R. H. Pantell, and R. G. Smith, "Parametric effects in a two-level electric dipole system," *J. Appl. Phys.* **33**, 2085 (June 1962); C. L. Tang and H. Statz, "Nonlinear effects in the resonant absorption of several oscillating fields by a gas," *Phys. Rev.* **128**, 1013 (1962); W. J. Tabor, F. S. Chen, and E. O. Schulz-DuBois, "Measurement of intermodulation and a discussion of dynamic range in a ruby traveling-wave maser," *Proc. IEEE* **52**, 656 (June 1964); F. Bosch, H. Rothe, and E. O. Schulz-DuBois, "Direct observation of difference frequency signal in a traveling-wave maser," *Proc. IEEE* **54**, 1243 (October 1964); A. Javan and A. Szoke, "Theory of optical frequency mixing using resonant phenomena," *Phys. Rev.* **137**, A536 (1965); and D. H. Close, "Strong-field saturation effects in laser media," *Phys. Rev.* **153**, 360 (January 10, 1967).

## Problems for 5.1

1. *Harmonic response of a two-level atomic system.* To gain some feel for the nonlinear harmonic response of an atomic system, retain the second-harmonic terms on the right-hand side of the population equation for a two-level electric dipole system, and evaluate the steady-state $\pm 2\omega$ component $\Delta N_2$ of the population difference $\Delta N(t)$ that will be produced by these second-harmonic driving terms, using the steady-state fundamental-frequency solutions for $\tilde{E}$ and $\tilde{P}$. (Assume for simplicity that the fundamental-frequency applied signal $\tilde{E}$ is exactly on resonance.)

   Then put this second-harmonic component of $\Delta N(t)$ into the resonant-dipole equation on the right-hand side, and evaluate how large a $\pm 3\omega$ component of polarization $\tilde{P}_3$ will be produced at steady state by mixing between the $\pm 2\omega$ components of $\Delta N(t)$ and the $\pm\omega$ components of the applied signal $\tilde{E}$. (This analytical process of computing and matching up successively higher powers of $\omega$ on both sides of a set of equations is sometimes referred to as "harmonic balancing.")

Verify that at low signal levels the third-harmonic component of the polarization will be weaker than the fundamental component by a ratio with a functional dependence like $\tilde{P}_3/\tilde{P}_1 \propto (\omega_R/\omega_a)^2$, where $\omega_R \ll \omega_a$ is the Rabi frequency defined in the following section.

## 5.2 STRONG-SIGNAL BEHAVIOR: THE RABI FREQUENCY

What happens to the atomic behavior when an applied signal is strong enough that the rate-equation approximation is no longer valid? Much additional insight into the range of validity of the rate equations, and into the quantum behavior of an atomic transition outside this range, can be obtained from a simplified analysis we will present in this section to describe the large-signal response of an elementary two-level electric dipole system. This analysis will introduce an important new concept, the *Rabi frequency* for a stimulated atomic transition.

Students who read this section on the large-signal behavior of electric dipole transitions may also want to read Chapter 31 on magnetic dipole transitions, especially the final section of that chapter, which shows how these same large-signal and Rabi-frequency effects can alternatively be described in magnetic dipole terms.

### Simplified Large-Signal Analysis: The Polarization Equation

To carry out a large signal analysis in a simplified and yet meaningful way, we will make three simplifying, though really not very limiting, assumptions. First, we will assume an on-resonance applied signal, which we will write as $\mathcal{E}(t) = E_1(t)\exp(j\omega_a t)$, where $E_1(t)$ is the slowly varying amplitude of this applied signal. Later on we will assume that this amplitude is constant, although it may be very strong, and may be turned on suddenly at $t = 0$.

Second, we will allow for possible large-signal and transient effects in the atomic response by writing the polarization $p(t)$ in the form

$$p(t) = \text{Re}\left[\tilde{P}_1(t)e^{j\omega_a t}\right] = \text{Re}\left[-jP_1(t)e^{j\omega_a t}\right]. \tag{18}$$

That is, the polarization amplitude $\tilde{P}_1(t)$ is itself assumed to be a time-varying quantity, to account for the transient dynamics of the atomic response. Because we know from experience that in the limiting case of an on-resonance applied signal $p(t)$ will turn out to be $-90°$ out of time-phase with $\mathcal{E}(t)$, we also write this phasor quantity as a real (but time-varying) amplitude $P_1(t)$ with a constant factor of $-j$ in front, corresponding to a fixed $90°$ phase shift,

Substituting Equation 5.18 into the resonant-dipole equation 5.1, and separating the $e^{+j\omega_a t}$ and $e^{-j\omega_a t}$ terms leads us to an equation of motion for the phasor amplitude $P_1(t)$, namely,

$$\frac{d^2 P_1(t)}{dt^2} + (2j\omega_a + \Delta\omega_a)\frac{dP_1(t)}{dt} + j\omega_a \Delta\omega_a P_1(t) = jK\, E_1(t)\, \Delta N(t). \quad (19)$$

Now, it is certainly true that $\Delta\omega_a \ll \omega_a$; so we can probably drop the $\Delta\omega_a$ factor in front of the $dP_1(t)/dt$ term. In addition, we can reasonably assume that the time-variation of $P_1(t)$ itself, though it may approach in magnitude the quantity $\Delta\omega_a P_1(t)$, will surely be slow compared to $\omega_a P_1(t)$. In simple terms, the phasor amplitude $P_1(t)$ may change significantly within a time of the order of one reciprocal linewidth, or $1/\Delta\omega_a$, but not in one optical cycle, or $1/\omega_a$.

As a result of this, we can drop the second-derivative term $d^2 P_1(t)/dt^2$ relative to the $2\omega_a dP_1(t)/dt$ term, and simplify the transient equation for $P_1(t)$ to

$$\frac{dP_1(t)}{dt} + \frac{\Delta\omega_a}{2} P_1(t) \approx \frac{K}{2\omega_a} E_1(t)\, \Delta N(t). \quad (20)$$

This approximation is commonly referred to as the *slowly varying envelope approximation* (SVEA). Note that it is a much less restrictive approximation than the rate-equation approximation—that it, it allows much faster time-variations and much stronger signals than in the rate-equation limit.

### The Population Difference Equation

Along with this slowly varying envelope approximation for the resonant-dipole equation, we must also use the population equation of motion (Equation 5.6). We have already noted that the transient response of that equation will be governed by the generally very slow relaxation time $T_1$. Therefore, as a third approximation we will use on the right-hand side of Equation 5.6 the time-averaged value of $\mathcal{E} \cdot dp/dt$, with the time average being taken over at least a few cycles of the sinusoidal quantities $\mathcal{E}(t)$ and $p(t)$. This approximation then takes out the second-harmonic factors, but still allows for relatively rapid envelope variations in either the signal $\mathcal{E}(t)$ or the polarization $p(t)$.

With this further approximation the population equation becomes

$$\frac{d\Delta N(t)}{dt} + \frac{\Delta N(t) - \Delta N_0}{T_1} \approx -\frac{1}{\hbar} E_1(t)\, P_1(t). \quad (21)$$

All complex conjugates have been dropped, since we will find that $P_1(t)$ always turns out to be purely real for the on-resonance case, $\omega = \omega_a$, which is all we are considering here.

### Large-Signal Solutions: The Rabi Frequency

These last two equations are the basis for our large-signal atomic analysis. Suppose we now assume a constant signal amplitude $E_1$ which is turned on suddenly at $t = 0$. The large-signal polarization and population equations 5.20 and 5.21 with $E_1$ constant form a simple pair of *linear coupled first-order differential equations* for the quantities $P_1(t)$ and $\Delta N(t)$ under the influence of the constant signal field $E_1$. By substituting one of these equations into the other, we can combine the two first-order equations to obtain a single second-order equation for $\Delta N(t)$, namely,

$$\frac{d^2 \Delta N(t)}{dt^2} + \left(\frac{\Delta\omega_a}{2} + \frac{1}{T_1}\right)\frac{d\Delta N(t)}{dt} + \left(\frac{\Delta\omega_a}{2T_1} + \frac{KE_1^2}{2\hbar\omega_a}\right)\Delta N(t) = \frac{\Delta\omega_a}{2T_1}\Delta N_0. \quad (22)$$

Now, the quantity $KE_1^2/2\hbar\omega$ appearing in the second set of brackets in this equation has the dimensions of a frequency squared. Suppose we define this frequency to be the *Rabi frequency* $\omega_R$, given by

$$\frac{KE_1^2}{2\hbar\omega_a} = \frac{3^*}{8\pi^2}\frac{\gamma_{\rm rad}\epsilon\lambda^3}{\hbar}E_1^2 \equiv \omega_R^2 \quad (23)$$

This Rabi frequency $\omega_R$ is proportional to the applied signal field strength $E_1$, and also depends on the transition strength as measured by $\gamma_{\rm rad}$. It has a very important physical significance, which we will develop in the following paragraphs.

Using this notation, we can rewrite the population difference equation in the form

$$\left[\frac{d^2}{dt^2} + \left(\frac{\Delta\omega_a}{2} + \frac{1}{T_1}\right)\frac{d}{dt} + \left(\frac{\Delta\omega_a}{2T_1} + \omega_R^2\right)\right]\Delta N(t) = \frac{\Delta\omega_a}{2T_1}\Delta N_0. \quad (24)$$

We can also write the $P_1(t)$ equation in exactly the same form

$$\left[\frac{d^2}{dt^2} + \left(\frac{\Delta\omega_a}{2} + \frac{1}{T_1}\right)\frac{d}{dt} + \left(\frac{\Delta\omega_a}{2T_1} + \omega_R^2\right)\right]P_1(t) = \frac{KE_1}{2\omega_a T_1}\Delta N_0, \quad (25)$$

which has exactly the same form as the $\Delta N(t)$ equation, except for a constant on the right-hand side.

### Large-Signal Limit: Rabi-Frequency Oscillations

Let us consider first the limiting case in which either the applied signal amplitude $E_1$ is extremely strong or the relaxation times $T_1$ and $T_2$ are very long and the linewidth $\Delta\omega_a$ is very narrow. We can then make the large-signal assumption that the Rabi frequency is large compared to all of these other rates, i.e., $\omega_R \gg \Delta\omega_a$ and $\omega_R \gg 1/T_1$. The differential equations 5.24 and 5.25 for the population difference $\Delta N(t)$ and the polarization amplitude $P_1(t)$ then reduce to the very much simplified forms

$$\frac{d^2\Delta N}{dt^2} + \omega_R^2 \Delta N \approx 0 \quad (26)$$

and similarly

$$\frac{d^2 P_1}{dt^2} + \omega_R^2 P_1 \approx 0. \quad (27)$$

The first of these equations has an elementary solution of the form

$$\Delta N(t) = \Delta N_0 \cos\omega_R t, \quad (28)$$

and the polarization amplitude $P_1(t)$ then has a matching solution of the form

$$P_1(t) = \sqrt{K\hbar/2\omega_a}\,\Delta N_0 \sin\omega_R t = P_m \sin\omega_R t. \quad (29)$$

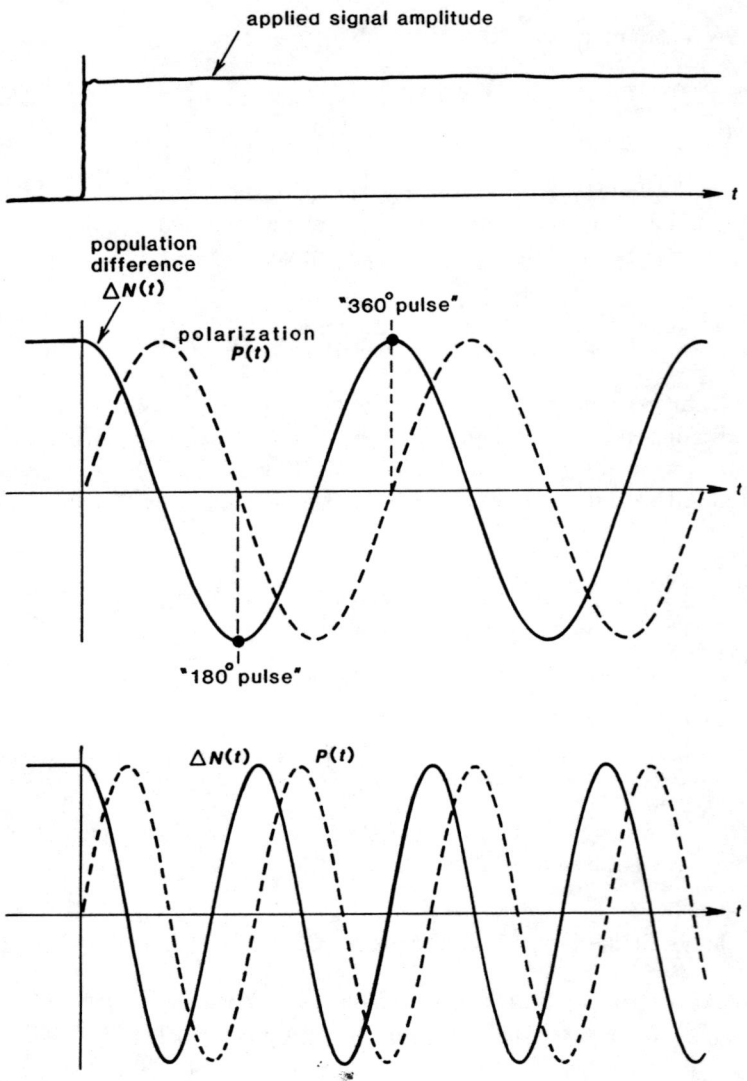

**FIGURE 5.2**
Rabi flopping behavior in response to a sudden and very strong sinusoidal signal input.

where $P_m$ is the maximum value of the oscillating polarization. Figure 5.2 shows the population difference $\Delta N(t)$ and the envelope of the polarization $P_1(t)$ as given by these very-large-signal solutions.

It is apparent that in this very-strong-signal limit, the atomic behavior is very different from the rate-equation limit. The population difference $\Delta N(t)$, rasther than going exponentially toward a saturated valued $\Delta N_{ss}$ as it does in the rate-equation limit, instead continually oscillates back and forth between its initial value and the opposite of that value, at the Rabi frequency $\omega_R$. At the same time, the induced polarization amplitude $P_1(t)$, instead of catching up to the applied signal with a time constant $\approx T_2$, continually chases but never catches up with the sinusoidal ringing of the population difference $\Delta N(t)$, so that the magnitude of $P_1(t)$ also oscillates sinusoidally, but lags behind $\Delta N(t)$ by 1/4 of a Rabi cycle.

### Discussion of the Rabi Flopping Behavior

This overall behavior of $\Delta N(t)$ and $P_1(t)$ is generally referred to as *Rabi flopping behavior*. It is a common result of quantum as well as classical analyses of large-signal atomic response.

The Rabi frequency $\omega_R$ in this very-large-signal limit is, by assumption, large compared to either $\Delta\omega_a \approx 2/T_2$ or to $1/T_1$. An essential feature of this regime is therefore that *the population difference $\Delta N(t)$ oscillates through many Rabi cycles in a time interval short compared to either $T_1$ or (more important) $T_2$.* This Rabi oscillation frequency will still, however, be very small compared to the optical carrier frequency $\omega_a$, so that there will still be very many optical cycles within each Rabi cycle. The slowly varying envelope approximation for $\Delta N(t)$ and $P_1(t)$ compared to $\omega_a$ is therefore still entirely valid.

It is also important to note that *the Rabi frequency $\omega_R$ at which these oscillations occur depends directly on the applied signal amplitude $E_1$, and on the square root of the transition strength as determined by the $\gamma_{\text{rad}}$ value.* Turning up the applied signal intensity will therefore give an even more rapid oscillation of the atomic population. The two lower plots in Figure 5.2 show two different applied signal strengths, with the stronger applied signal leading to a larger Rabi frequency. We emphasize again that all this oscillatory behavior occurs during a time interval short compared to either $T_1$ or $T_2$, and is entirely different from the rate-equation behavior at much lower applied signal levels.

Note also that the polarization amplitude oscillates with a 90° time lag relative to $\Delta N(t)$, so that the maximum value of $|P_1|$ occurs when $|\Delta N| = 0$, in sharp contrast to the usual rate-equation behavior, in which $|\tilde{P}(\omega)|$ is directly proportional to $\Delta N(t)$. What this means physically is that the oscillating dipoles are all fully aligned in phase in this situation (since the $T_2$ dephasing mechanisms are weak compared to the applied signal); in addition, the oscillatory quantum component $\tilde{a}_1 \tilde{a}_2^*$ that we discussed earlier has its maximum value, subject to the constraint that $|\tilde{a}_1|^2 + |\tilde{a}_2|^2 = 1$, at the midway point when $|\tilde{a}_1| = |\tilde{a}_2| = \sqrt{1/2}$.

This Rabi flopping behavior is thus, once again, *totally different from the usual rate-equation behavior.* It represents the most fundamental type of transient large-signal behavior that can be produced in either an isolated atom or a collection of atoms in which the applied signal is strong enough to override all the relaxation and dephasing processes.

### Large-Signal Case: Limiting Behavior at Long Times

To gain further insight into the distinction between small-signal and large-signal atomic behavior, we can write down the exact solutions to the full differential equations for $\Delta N(t)$ and $P_1(t)$, assuming a signal $E_1$ of arbitrary (but constant) amplitude that is again turned on at $t = 0$.

Let us first look, however, at the long-term steady-state solution to these equations. If we set all time derivatives to zero in the full differential equations 5.24 or 5.25, the eventual steady-state value of $\Delta N(t)$, when $d/dt = 0$, is given by

$$\lim_{t \to \infty} \Delta N(t) = \Delta N_{ss} \equiv \frac{\Delta N_0}{1 + 2(\omega_R^2/\Delta\omega_a)T_1}. \tag{30}$$

But this is exactly the same as the rate-equation saturation result for a two-level system, namely,

$$\Delta N_{ss} = \frac{\Delta N_0}{1 + 2W_{12}T_1}. \tag{31}$$

Comparing these two equations that the stimulated-transition probability $W_{12}$ which is valid in the small-signal or rate-equation regime can be related to the Rabi frequency $\omega_R$ and the transition linewidth $\Delta\omega_a$ by the very simple and useful form

$$\text{stimulated transition probability,} \quad W_{12} \equiv \frac{\omega_R^2}{\Delta\omega_a}. \tag{32}$$

(The reader can verify this formula by making use of the exact formulas for $\omega_R$ and $W_{12}$ given earlier.) We will make more use of this interesting result a little later.

Even in the very-large-signal or Rabi-flopping regime, so long as there is any finite $T_1$ and $T_2$, no matter how small, the Rabi flopping behavior will eventually die out. The population difference in our simple two-level mode will eventually saturate (possibly after many Rabi cycles) to a steady-state (and highly saturated) value given by

$$\lim_{t \to \infty} \Delta N(t) = \Delta N_0 \frac{1}{1 + 2W_{12}T_1} = \Delta N_0 \frac{1}{1 + S}, \tag{33}$$

where $S$ is the "saturation factor" given by

$$S \equiv 2W_{12}T_1 = I/I_{\text{sat}} = \frac{2T_1}{\Delta\omega_a} \times \omega_R^2. \tag{34}$$

This factor is, of course, directly proportional to the applied signal power, and will be very much greater than one for any signal falling in the very-large-signal or Rabi-flopping regime.

### Exact Solutions: Transient Response

The differential equations 5.20 and 5.21 (or 5.24 and 5.25) for $\Delta N(t)$ and $P_1(t)$ can, of course, be solved exactly, without approximations, for any level of signal strength. As a practical hint, the algebra involved in doing this becomes much easier if you convert the equations to a suitable set of normalized variables. A convenient choice is to normalize the time scale to the dephasing time $T_2$ by writing $t' = t/T_2$; to define a normalized signal amplitude by $R = \omega_R T_2 = 2\omega_R/\Delta\omega_a$, and a time-constant ratio by $D = T_2/T_1$ (note that $D$ will normally be a small number); and then to use normalized quantities $\hat{n} = \Delta N/\Delta N_0$ and $\hat{p} = P_1/P_0$, where $P_0 = \sqrt{\hbar K/2\omega_a}\,\Delta N_0$. The two coupled equations then become

$$\frac{d\hat{n}}{dt} + D(\hat{n} - 1) = -R\hat{p},$$

$$\frac{d\hat{p}}{dt} + \hat{p} = R\hat{n}.$$

Since these are linear coupled equations, the exact solutions will have a transient behavior that will take on either overdamped or oscillatory forms, depending on the ratio of $R$ to $(1-D)/2$, which in real terms corresponds to the ratio of $\omega_R^2$ to the quantity $(\Delta\omega_a/4 - 1/2T_1)^2$. Let us examine each of these limits in turn.

1. *The overdamped or weak-signal regime.* In the weak-signal regime the applied signal strength is small enough that $\omega_R < (\Delta\omega_a/4 - 1/2T_1)$, which means that the Rabi frequency is small compared to the atomic linewidth $\Delta\omega_a$ (and so the stimulated-transition probability $W_{12}$ is small compared to $\Delta\omega_a$ also). The exact solution is then overdamped, and has two exponential decay components given by

$$-\alpha \pm \beta = -\left(\frac{\Delta\omega_a}{4} + \frac{1}{T_1}\right) \pm \sqrt{\left(\frac{\Delta\omega_a}{4} - \frac{1}{2T_1}\right)^2 - \omega_R^2}, \qquad (36)$$

so that $\beta < \alpha$. This condition corresponds to the usual rate-equation limit, as we will now see.

If we assume for simplicity that the atomic system is initially at rest, so that $P_1(0) = 0$ and $\Delta N(0) = \Delta N_0$ when the signal is first turned on, then the solution in this limit may written as

$$\Delta N(t) = \Delta N_{\text{sat}} \left[1 + Se^{-\alpha t}(\cosh\beta t + (\alpha/\beta)\sinh\beta t)\right], \qquad (37)$$

where $\Delta N_{\text{sat}}$ and the saturation factor $S$ are as defined earlier. For the limiting case of a very weak signal, $\omega_R \ll \Delta\omega_a$, and also slow energy decay, $1/T_1 \ll \Delta\omega_a$, the two time constants approach the limits

$$\alpha + \beta \approx \Delta\omega_a/2 \quad \text{and} \quad \alpha - \beta \approx (2W_{12} + 1/T_1); \qquad (38)$$

so Equation 5.37 can be approximated by

$$\Delta N(t) \approx \Delta N_{\text{sat}}\left\{1 + S\exp[-(2W_{12} + 1/T_1)t]\right\}. \qquad (39)$$

But this is exactly the same as the transient two-level behavior developed in Section 4.5 using rate equations. This result thus verifies that the Rabi-frequency behavior blends smoothly into rate-equation behavior in the appropriate weak-signal limit.

2. *The oscillatory or strong-signal regime.* For signals strong enough that $\omega_R > (\Delta\omega_a/4 - 1/2T_1)$, the equations become underdamped, and we must use instead a pair of complex conjugate time constants given by

$$-\alpha \pm j\beta = -\left(\frac{\Delta\omega_a}{4} + \frac{1}{T_1}\right) \pm j\sqrt{\omega_R^2 - \left(\frac{\Delta\omega_a}{4} - \frac{1}{2T_1}\right)^2}. \qquad (40)$$

The imaginary part $\beta$ in particular now corresponds to a kind of modified Rabi frequency $\omega_R'$ given by

$$\beta = \omega_R' \equiv \sqrt{\omega_R^2 - \left(\frac{\Delta\omega_a}{4} - \frac{1}{2T_1}\right)^2} \qquad (41)$$

when the effects of damping and dephasing are included.

In terms of these quantities, the exact solution for the same initial conditions then becomes

$$\Delta N(t) = \Delta N_{\text{sat}}\left\{1 + Se^{-\alpha t}[\cos\beta t + (\alpha/\beta)\sin\beta t]\right\}. \qquad (42)$$

This result is a more exact form of the large-signal Rabi flopping limit given earlier, with the effects of weak relaxation terms $\Delta\omega_a$ and $T_1$ included.

To illustrate how the transient response of the atomic system changes as the applied signal amplitude increases from the weak-signal or rate-equation regime to the large-signal or Rabi-flopping regime, Figure 5.3 shows the calculated behavior of $\Delta N(t)$ and $P_1(t)$ from these exact solutions plotted versus $t/T_1$ in a two-level system, assuming that the dephasing time $T_2$ is $1/5$ of the energy decay rate $1/T_1$ and that the Rabi frequency ranges from 0.15 to 2.2 times the atomic linewidth $\Delta\omega_a$. This obviously covers a range from the weak-signal regime, exhibiting essentially rate-equation behavior, into the lower end of the strong-signal regime, exhibiting a significant amount of Rabi flopping behavior.

In the intermediate regime between weak and very strong applied signals, the population clearly oscillates back and forth at a modified Rabi frequency $\beta \equiv \omega'_R$ that is somewhat lower than $\omega_R$. This Rabi flopping behavior eventually dies out, however, as the dephasing effects described by $\Delta\omega_a$ gradually destroy the coherently driven transient behavior.

### Summary

There are two points concerning the results derived in this chapter that we should especially emphasize here.

- All the results we have developed in this section are *quantum-mechanically correct* (at least for an ideal two-level quantum system), since the initial polarization and population difference equations from which we started were quantum-mechanically correct. The Rabi flopping behavior is a very general and characteristic quantum phenomenon, readily predicted from Schrödinger's equation for any strongly perturbed two-level system.
- Even in the weak-signal regime where no Rabi flopping behavior is occurring, the Rabi frequency $\omega_R$ still provides a natural measure of the strength of the applied signal field, relative to the transition frequency $\omega_a$. In quantum-mechanical terms, $\hbar\omega_R$ is a measure of the *perturbation hamiltonian* caused by the applied field acting on the atom, just as $\hbar\Delta\omega_a$ is a measure of the random perturbation hamiltonian caused by the relaxation mechanisms and the dephasing or phonon-broadening mechanisms acting on the atoms, and $\hbar\omega_a$ is a measure of the static or unperturbed hamiltonian of the atom.

This point is especially illustrated by the fact that the stimulated-transition probability $W_{12}$ in any two-level system (electric dipole or any other kind) can always be written in terms of the Rabi frequency $\omega_R$ for that transition, in the form

$$W_{12} = \frac{\omega_R^2}{\Delta\omega_a}, \qquad (43)$$

where $\Delta\omega_a$ is the homogeneous linewidth for that transition. The condition for rate-equation behavior, which we said earlier was $W_{12} \ll \Delta\omega_a$, translates into the condition that

$$\omega_R \ll \Delta\omega_a. \qquad (44)$$

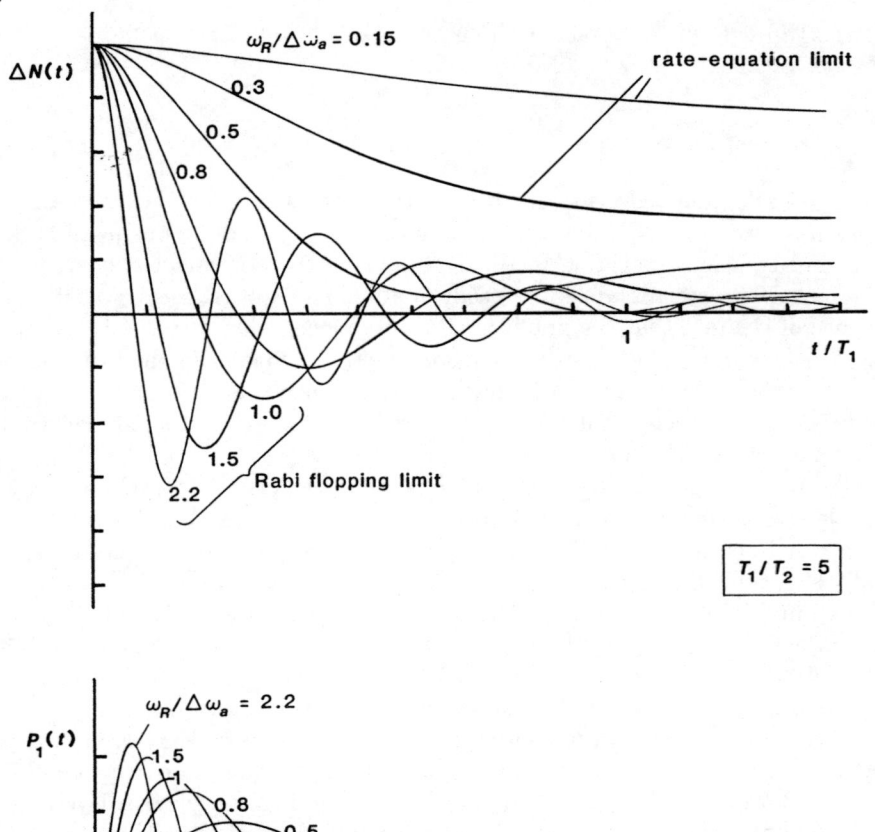

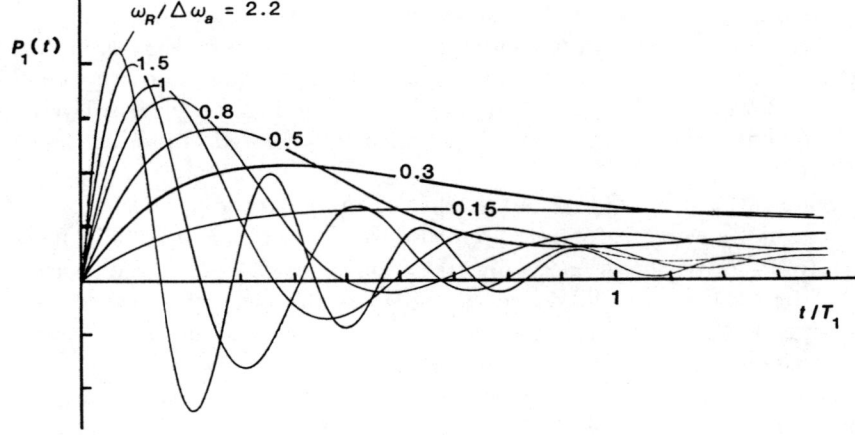

FIGURE 5.3
Exact solutions for the induced polarization $P_1(t)$ and the population difference $N(t)$ in a two-level system produced by sinusoidal applied signals of varying strengths that are suddenly turned on at $t = 0$. The parameters for the various curves are $\Delta\omega_a T_1 = 10$ (or $T_1 = 5T_2$) and $\omega_R/\Delta\omega_a$ ranging from 0.15 up to 2.2. The behavior changes over from weak-signal or rate-equation behavior for $\omega_R \ll \Delta\omega_a$ to strong-signal or Rabi flopping behavior for $\omega_R > \Delta\omega_a$.

In other words, in order to be in the rate-equation regime, the Rabi flopping frequency $\omega_R$ itself must be much less than the linewidth $\Delta\omega_a$.

In physical terms, rate-equation behavior results when the signal strength and hence the Rabi frequency are small enough that a dephasing event or a relaxation event is sure to occur, and to break up the Rabi flopping behavior, before even a fraction of a Rabi cycle is completed. So-called coherent or large-signal Rabi-flopping effects occur, on the other hand, when the atoms can be driven through one or several Rabi cycles in a time short compared to either of the relaxation times $T_1$ or $T_2$.

### Coherent Pulse Effects

Rabi flopping behavior, and other strong-signal effects and departures from elementary rate-equation behavior, are most easily observed by using pulsed signals and transient detection methods. This is both a practical matter, in that strong applied signals are more easily obtained in pulsed form, and a consequence of the fact that the nonlinear Rabi-frequency kind of behavior shows up most clearly in transient rather than steady-state behavior of the atoms. Hence a number of different pulsed large-signal experiments have been developed to demonstrate such coherent transient behavior; these are commonly referred to as "coherent pulse" experiments.

As one example, the Rabi flopping behavior predicts that if we apply a sufficiently strong signal pulse with a duration $T_p$ such that $\omega_R T_p \equiv \pi$ to an initially uninverted and absorbing two-level atomic transition, this pulse can flip the initially absorbing population difference $\Delta N_0$ over into a completely inverted and hence amplifying condition $-\Delta N_0$ at the end of the pulse. We simply turn off the applied signal in the Rabi flopping behavior at the point where the initial population inversion has been completely inverted, and then let this inverted population slowly decay back to equilibrium with time constant $T_1$.

This is commonly known as a "$\pi$ pulse" or "180° pulse" experiment. It provides one way (though in practice not a very useful way) to obtain pulsed inversion in a two-level system. Note that there is an inverse relationship between the signal amplitude $E_1$ and the pulse duration $T_p$ needed to produce an exactly 180° pulse.

Similarly, a pulse with an "area" (that is, an $\omega_R T_p$ product) such that $\omega_R T_p = 2\pi$ will first invert the atomic population difference, and then flip it exactly back to its initial condition, as illustrated in Figure 5.4. As a result such a "$2\pi$ pulse" or "360° pulse" will deliver no energy at all to the atoms (at least, not to first order). This means that a sufficiently strong pulse with this area can travel essentially unattenuated through an absorbing atomic medium which is otherwise opaque for lower-intensity signals. This phenomenon is known as "self-induced transparency" and has been demonstrated experimentally.

### Self-Consistent Coherent Pulse Analyses

More rigorous analysis of these coherent pulse experiments requires us to take into consideration not only the effect of the signal on the atoms, but also the reaction of the resulting induced atomic polarization $p(t)$ back on the signal. For instance, in self-induced transparency the first half of the signal pulse delivers energy from the signal to the atoms, but in the second half of the pulse the atoms radiate energy back to the signal. As a consequence the signal pulse soon distorts from a square pulse, or whatever its initial shape may be, into a unique self-consistent pulse shape. Also, the pulse velocity is reduced much below the free propagation velocity of the electromagnetic wave, in essence because the pulse energy spends a significant fraction of the time stored in the atoms rather than in the wave.

A detailed calculation of pulse propagation through a simple two-level absorbing medium has been carried out by Davis and Lin, using essentially the

## 5.2 STRONG-SIGNAL BEHAVIOR: THE RABI FREQUENCY

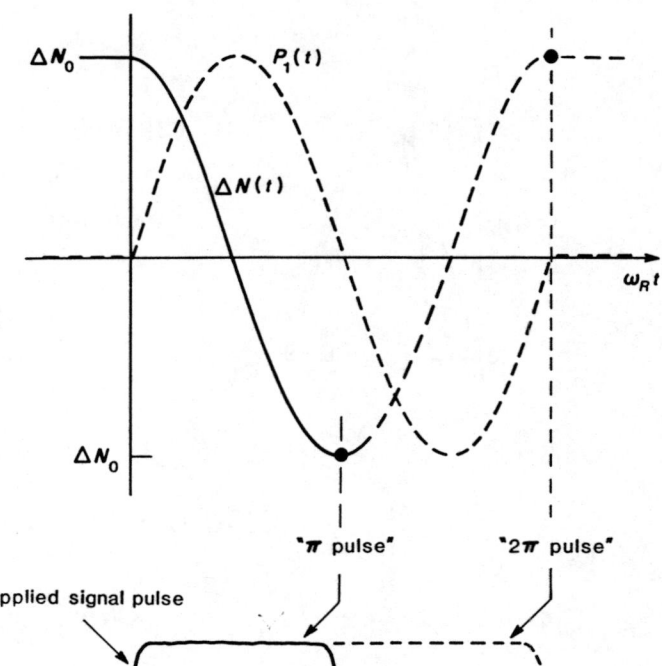

FIGURE 5.4
Large-signal "$\pi$" and "$2\pi$" pulses applied to an atomic system.

atomic equations presented in this section, combined with Maxwell's equations for the propagation of the signal pulse itself. Figure 5.5 illustrates some typical results from their calculations.

The left-hand plots show the initial smooth pulse sent into the absorbing medium (plotted as $E$ field amplitude, not intensity), and also the resulting modified pulseshapes at two different distances into the absorbing medium. The time scale for the modified pulses has been delayed in each case by the propagation time from the input plane to the observation plane, so that the two pulses will line up. The input pulse duration ($\approx 5$ ps) is much shorter than the assumed value of $T_2$ for the atomic medium, and the pulse intensity is large enough that the Rabi frequency at the peak is large compared to both the atomic linewidth and the inverse pulse duration. The right-hand plots show the time-variation of the population difference $\Delta N(t)$ at these same two observation planes, on the same delayed time scale, as the pulse sweeps past each plane.

In the top pair of plots, corresponding to the first observation plane, the early portion of the pulse (up to about 1.5 ps) has been strongly absorbed by the medium; but beyond that time the accumulated pulse energy has been enough to strongly saturate the absorber, so that the trailing edge of the pulse is nearly unattenuated. The right-hand plot also shows that the pulse intensity near the peak is more than adequate to produce significant Rabi flopping behavior in the atomic system. Note also that the population difference begins to recover toward its unsaturated value (plotted downward) with time constant $T_1$ as the pulse intensity dies away.

By the time the pulse reaches the second observation plane, which is five times further into the absorbing medium, the oscillatory Rabi behavior of the

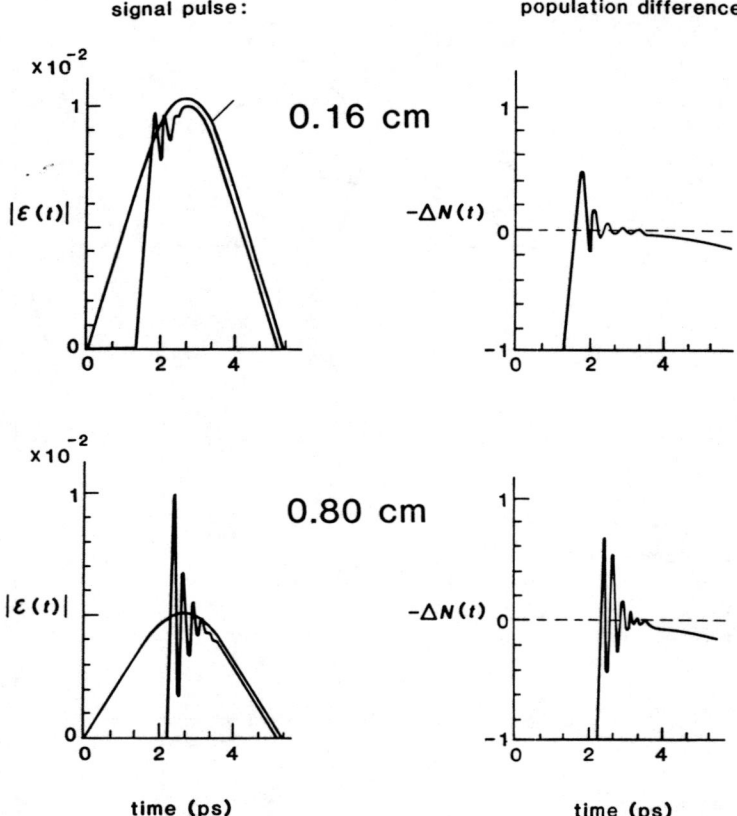

**FIGURE 5.5**
Results of numerical calculations for propagation of an intense optical pulse through a two-level absorbing medium, including both the nonlinear response of the atoms, and the reaction of the medium back on the intensity and phase of the pulse. The left-hand plots show the input pulseshape, and the partially absorbed and reshaped pulses at two different distances into the medium. The right-hand curves show the time-variation of the population difference at the same reference planes inside the medium as the pulse passes by in each case. From L. W. Davis and Y. S. Lin, *IEEE J. Quant. Electr.* **QE–9**, 1135–1138 (December 1973).

atomic polarization has begun to react back on the propagating pulse; and the pulse itself has acquired a strong oscillatory behavior as well. The first full cycle in the pulse oscillation has become, in fact, almost a full $2\pi$ pulse, sufficient to flip the population difference more than 60% of the way to complete inversion to the opposite sign. If this pulse were to propagate further, it would it fact break up into one or several such $2\pi$ pulses.

There exist a great many such transient, large-signal, coherent-pulse effects which can be demonstrated on atomic transitions using appropriate pulsed signals. These transient responses can be described analytically using either an electric-dipole model for the atomic transition, or a magnetic-dipole model, which often provides more insight into the transient behavior even for what are really electric-dipole transitions. We will therefore consider these transient responses in more extensive detail in Chapter 31, after we have introduced the magnetic-dipole model for atomic transitions.

### Multilevel Systems: Mixing and Intermodulation

Strong cw or long-pulse signals can also produce significant harmonic generation and intermodulation effects in a real atomic system, as we mentioned in the previous section. Suppose, for example, that two cw signals are simultaneously applied to the same atomic transition, with at least one of the signals being strong enough to violate the rate approximation conditions and produce significant Rabi flopping effects. Alternatively, suppose that this strong signal is applied to one transition, say, the $i \to j$ transition, and a weak signal is simultaneously applied to another transition which shares a common energy level, say, the $j \to k$ transition.

Then in either case, to give a somewhat simplified description, the strong signal will modulate the populations $N_i(t)$ or $N_j(t)$ in time according to the modified Rabi frequency. This modulation of the populations will then modulate the net absorption or emission seen by other, weaker signals on the same or on connecting transitions. In general, we can expect to see intermodulation and distortion products appearing in any strong-signal multiple-frequency experiment, with the weaker signals being modulated at something like the Rabi frequency produced by the stronger signal.

These mixing and intermodulation effects rapidly become very complicated when several energy levels or several applied frequencies are involved. Proper analysis of these effects usually requires carrying out what is called a multilevel quantum-mechanical density-matrix analysis. Fortunately, intermodulation effects of this type are usually small in most practical laser systems, although they can sometimes be observed. The general criterion for observing them is applying unusually strong signals to an atomic system that has unusually narrow linewidths and strong transitions, so that the Rabi frequencies involved can become larger than the linewidths. Some references on effects of this type are given below.

### REFERENCES

Discussions of the Rabi frequency will be found in many quantum-mechanics texts and books on atomic transitions and resonance physics. One example (although the discussion happens to be framed in magnetic-dipole or Bloch-vector terminology) is L. Allen and J. H. Eberly, *Optical Resonance and Two-Level Atoms* (Wiley-Interscience, 1975), pp. 52–60. Another is M. Sargent III, M. O. Scully, and W. E. Lamb, Jr. *Laser Physics* (Addison-Wesley, 1974).

---

### Problems for 5.2

1. *The slowly varying envelope approximation.* When you apply the slowly varying envelope approximation to the second-order resonance equation, why isn't it possible to simply drop the $d^2p/dt^2$ derivative term right from the beginning?

2. *Analysis of off-resonance Rabi flopping behavior.* Carry through the same derivation of large-signal atomic response as in the text for a constant but large-amplitude $E_1$ field applied to a two-level system, but assume that the applied signal frequency $\omega$ may be tuned well off the resonance frequency $\omega_a$, by an amount $(\omega - \omega_a) \approx \omega_R$ or greater. Describe in particular how the effective Rabi

flopping frequency changes as the signal frequency is tuned away from $\omega_a$. [Hint: You will need to treat $\tilde{P}(t)$ as a complex quantity, and to make a resonance approximation along with the slowly varying envelope approximation.]

3. *Coherent transients: The 90° pulse.* In addition to 180° and 360° pulses, coherent transient experimenters often make use of 90° pulses. What sort of atomic behavior would occur in an atomic system following a 90° applied signal pulse, and what sort of experimental uses might this behavior have?

4. *Large-signal atomic response: Two-frequency mixing and intermodulation effects.* Suppose that an applied signal with two steady-state sinusoidal frequency components, $\mathcal{E}(t) = \tilde{E}_1 e^{j\omega_1 t} + \tilde{E}_2 e^{j\omega_2 t}$, where $\omega_1$ and $\omega_2$ are both near $\omega_a$, is applied to a two-level atomic system. Try expanding both $\Delta N(t)$ and $p(t)$ in the form, for example,

$$\Delta N(t) = \sum_{n,m} \Delta N_{nm} \exp j(n\omega_1 + m\omega_2)t,$$

and then substituting these into the exact atomic equations and applying harmonic balance to find the coefficients $\Delta N_{nm}$ and $P_{nm}$.

Find in particular the magnitudes of the $n = 1$, $m = 1$ (sum frequency) and $n = -1$, $m = -1$ (difference frequency) components in $\Delta N(t)$, and the $n + m = \pm 3$ components in $p(t)$, as nonlinear functions of the signal amplitudes $E_1$ and $E_2$.

5. *Quantum transition matrix element for an electric-dipole atom.* In quantum theory, applying an electric field $\mathcal{E}$ to an atom produces a perturbation hamiltonian $\mathcal{H}' = -\mu_{op}\mathcal{E}$ where $\mu_{op}$ is a quantum-mechanical electric-dipole operator. In terms of this operator, an electric-dipole transition matrix element $\mu_{12}$ between any two atomic levels can then be calculated from an overlap integral $\mu_{12} \equiv \int \psi_1^*(\mathbf{r})\mu_{op}\psi_2(\mathbf{r})\,d\mathbf{r}$, where $\psi_1$ and $\psi_2$ are the quantum eigenstates for the two energy levels involved. This transition matrix element gives a quantum measure of the strength of the electric dipole response on that transition. One can show, for example, that the Rabi frequency produced by a sinusoidal field $E_1$ acting on that transition is given by

$$\omega_R \equiv \frac{\mu_{12} E_1}{\hbar}$$

This is the form in which the Rabi frequency is most often written in the scientific literature.

(a) By equating this form to the expressions for $\omega_R$ derived in this section, show that under strong Rabi-flopping conditions the maximum induced polarization $P_m$ of Equation 5.29 has the value $P_m = \Delta N_0 \mu_{12}$. In physical terms, this means that at the maximum points of $P_1(t)$, the applied field $E_1$ has set the quantum state of every single atom in the population difference $\Delta N_0$ oscillating in exactly the same phase, with an induced dipole moment per atom just equal to the quantum transition dipole matrix element $\mu_{12}$.

(b) Find the analytical connection between the **radiative decay rate** $\gamma_{\text{rad}}$ and the electric-dipole matrix element $\mu_{12}$ as two alternative ways of expressing the strength of any allowed electric-dipole transition; and again, if possible, give a physical interpretation of this.

# 6

# LASER PUMPING AND POPULATION INVERSION

The atomic rate equations introduced in the two previous chapters are of great value in analyzing laser pumping, population inversion, and gain saturation in laser systems. The primary objective of this chapter is to illustrate this point by solving the atomic rate equations and examining these solutions for some simple but important atomic systems.

## 6.1 STEADY-STATE LASER PUMPING AND POPULATION INVERSION

One of the most common applications for rate equations is in analyzing laser pumping. In this section, therefore, we will develop and solve the rate equations to analyze steady-state laser pumping in simplified four-level and three-level laser systems.

### Elementary Four-Level Laser System

Figure 6.1 shows a complicated multienergy-level system typical of many real laser systems: this one is for a solid-state laser system using the $Nd^{3+}$ ion in Nd:YAG or Nd:glass. The upward arrows indicate upward pumping rates to various higher levels produced by flashlamp pumping on strong absorption lines from the ground state: the downward arrow indicates the widely used 1.064 $\mu$m laser transition from the $^4F_{3/2}$ level to the $^4I_{11/2}$ level. (There are actually at least eight different laser transitions of widely varying strengths between these two clusters of closely spaced levels, with wavelengths extending from 1.0520 to 1.1226 $\mu$m. In addition, generally weaker laser action is possible from the same $^4F_{3/2}$ levels to the cluster of $^4I_{9/2}$ levels at four wavelengths between 0.89 and 0.9462 $\mu$m; to the $^4I_{13/2}$ levels at four wavelengths between 1.319 and 1.358 $\mu$m; and—very weakly—at 1.833 $\mu$m to one of the $^4I_{15/2}$ levels slightly above the lowest or ground level in this cluster.)

Figure 6.1 does not show the numerous downward radiative and nonradiative relaxation paths among all these levels. As in many other rare-earth and other solid-state laser systems, however, atoms excited into the higher excited levels in this material will nearly all relax, primarily by fast nonradiative relaxation, into the sharp and long-lived $^4F_{3/2}$ *metastable level* that provides the upper laser

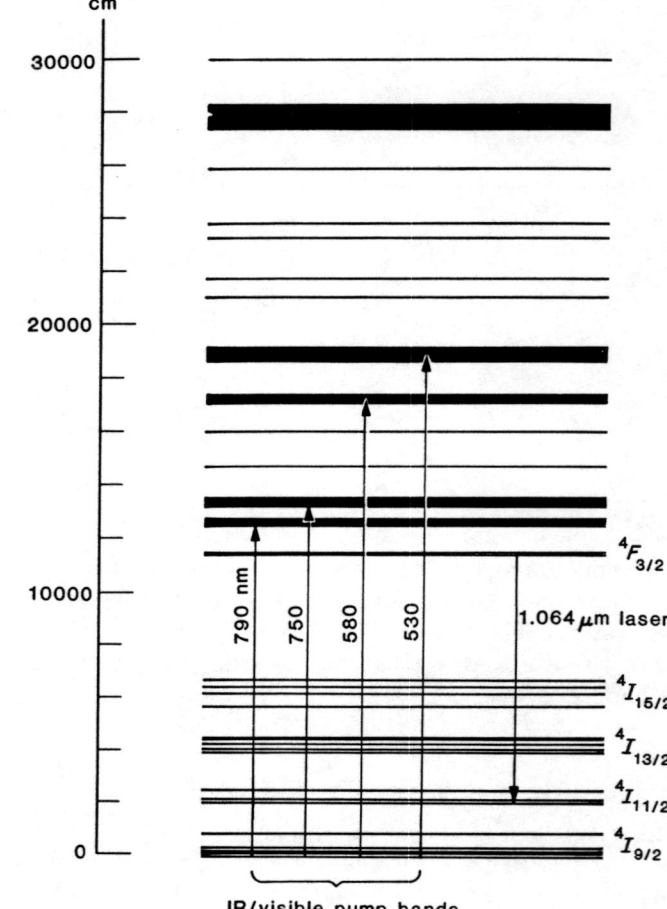

FIGURE 6.1
Important pump bands and the dominant 1.06 μm laser transition in the Nd:YAG or Nd:glass laser medium.

level in this system. A sizable population density can thus build up in this upper laser level.

For purposes of analysis this complicated set of levels can then be simplified into the idealized four-level laser system shown in Figure 6.2. This four-level model will in fact provide a simple but surprisingly accurate analytical model for many real laser systems. In this model level 4 represents the combination of all the levels lying above the upper laser level in the real atomic system. It is desirable that many of these levels be in fact broad absorption bands, so that the optical pumping into these levels by a broadband pump lamp can be very efficient.

Level 3 represents the upper laser level, usually a fairly sharp and long-lived level, with a large gap below it. Level 2 then represents the lower laser level, and level 1 the lowest or ground level. Other low-lying levels that may be present in the material, both above and below the lower laser level, are ignored in the model because they play no real role in the laser action. They act only as temporary way stations through which atoms may pass in relaxing from the other levels to the ground level $E_1$.

## 6.1 STEADY-STATE LASER PUMPING AND POPULATION INVERSION

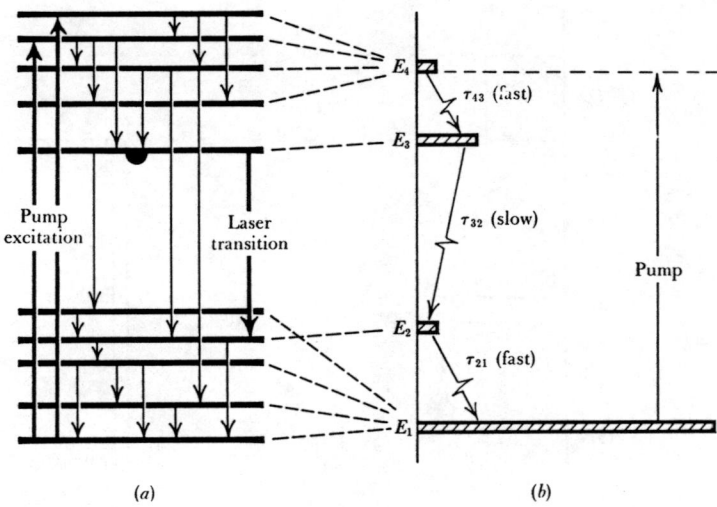

**FIGURE 6.2**
An idealized four-level pumping system which replaces the more complex scheme of Figure 6.1.

### Four-Level Pumping Analysis

To analyze this system we will write down the relevant rate equations one after another, using the model shown in Figure 6.3, and then solve for their steady-state solutions, making reasonable approximations as we go along. Since the condition $\hbar\omega/kT \gg 1$ is usually very well satisfied for all transitions in a visible laser system, we will write all of the following rate equations using the "optical approximation" introduced in Section 4.4.

We begin by assuming that the laser pumping process, whatever its physical cause, produces a stimulated pump transition probability $W_{14} = W_{41} = W_p$ between levels 1 and 4. The rate equation for level 4 in the optical approximation is then

$$\frac{dN_4}{dt} = W_p(N_1 - N_4) - (\gamma_{43} + \gamma_{42} + \gamma_{41})N_4 \quad (1)$$
$$= W_p(N_1 - N_4) - N_4/\tau_4,$$

where the lifetime $\tau_4$ given by

$$\frac{1}{\tau_4} \equiv \gamma_4 = \gamma_{43} + \gamma_{42} + \gamma_{41} \quad (2)$$

is the total lifetime for decay of level 4 to all lower levels. The steady-state population of level 4, when $dN_4/dt = 0$, is then given by

$$N_4 = \frac{W_p\tau_4}{1 + W_p\tau_4}N_1 \approx W_p\tau_4 N_1 \quad \text{if} \quad W_p\tau_4 \ll 1. \quad (3)$$

The normalized pumping rate $W_p\tau_4$, which will appear in many of the following expressions, will in fact have a value much less than unity in many (though not all) practical laser systems.

Direct pumping up from the ground level into the upper laser level 3 in this model can very often be assumed negligible, either because the $1 \to 3$ transition

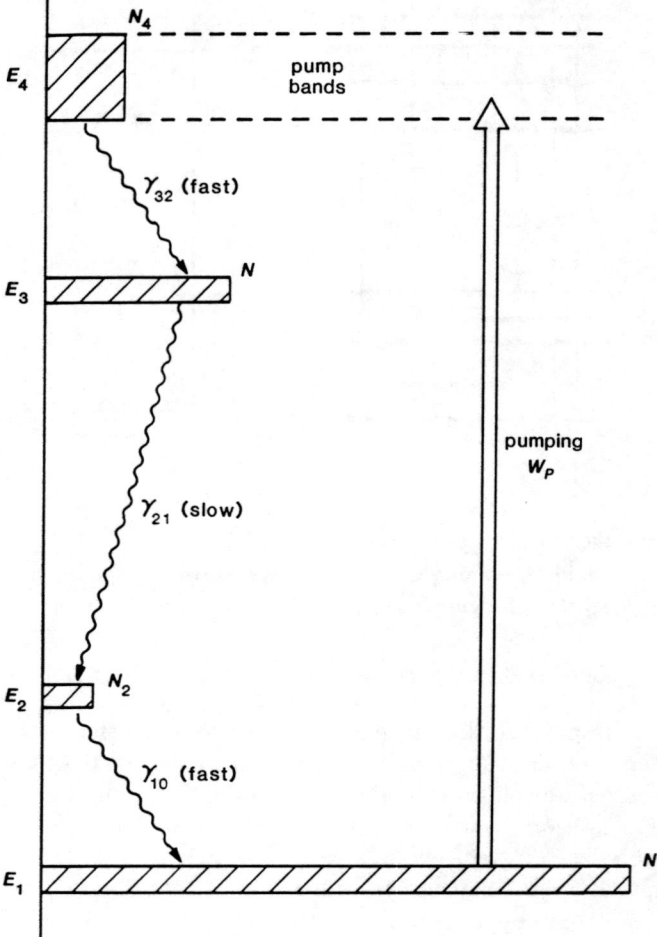

FIGURE 6.3
The idealized four-level pumping scheme of Figure 6.2 shown in more detail.

will have a weaker absorption cross section than the collected $1 \rightarrow 4$ transitions, or because this transition will be much narrower than the strong absorption bands from the ground level to the groups of levels that make up level 4. The rate equations for the two levels $N_3$ and $N_2$ are then

$$\frac{dN_3}{dt} = \gamma_{43} N_4 - (\gamma_{32} + \gamma_{31}) N_3 = \frac{N_4}{\tau_{43}} - \frac{N_3}{\tau_3} \qquad (4)$$

and

$$\frac{dN_2}{dt} = \gamma_{42} N_4 + \gamma_{32} N_3 - \gamma_{21} N_2 = \frac{N_4}{\tau_{42}} + \frac{N_3}{\tau_{32}} - \frac{N_2}{\tau_{21}}. \qquad (5)$$

The first of these then gives at steady state $(d/dt = 0)$

$$N_3 = \frac{\tau_3}{\tau_{43}} N_4. \qquad (6)$$

In a good laser system the $4 \rightarrow 3$ relaxation rate will be very fast, but the upper laser level 3 will have a long lifetime by comparison, so that $\tau_3 \gg \tau_{43}$ and hence $N_3 \gg N_4$.

Combining Equation 6.5 and 6.6 then gives the result

$$N_2 = \left(\frac{\tau_{21}}{\tau_{32}} + \frac{\tau_{43}\tau_{21}}{\tau_{42}\tau_3}\right) N_3 = \beta N_3, \qquad (7)$$

where the parameter $\beta$ is defined to be

$$\beta \equiv \frac{\tau_{21}}{\tau_{32}} + \frac{\tau_{43}\tau_{21}}{\tau_{42}\tau_3}. \qquad (8)$$

This parameter $\beta$ thus depends only on relaxation-time ratios, not absolute values. If this quantity is less than unity, the steady-state result will be $N_2 < N_3$, which means there will be the desired population inversion on the $3 \to 2$ transition.

In a good laser system the upper levels $E_4$ will relax primarily into the upper laser level $E_3$, so that $\gamma_{42} \approx 0$ or $\tau_{42} \approx \infty$. In this case $\beta \approx \tau_{21}/\tau_{32}$, and the condition for population inversion becomes simply

$$\beta \equiv \frac{N_2}{N_3} \approx \frac{\tau_{21}}{\tau_{32}} \ll 1. \qquad (9)$$

In other words, to have good inversion on the $3 \to 2$ transition, atoms should relax out of the lower laser level $E_2$ down into lower levels much faster than atoms relax into $E_2$ from above. Even if level 4 does not relax only into level 3, if the upper laser level has a long lifetime (both $\tau_{32}$ and $\tau_3$ long) and the lower laser level has a short lifetime ($\tau_{21}$ short), then population inversion on the $3 \to 2$ transition is virtually certain.

Whether this population inversion will be large enough to give sufficient gain to achieve laser action in a practical cavity is another matter. Nonetheless, these conditions are met and laser action can be produced on many transitions in many real atomic systems.

### Fluorescent Quantum Efficiency

Another dimensionless parameter often used in evaluating laser materials is the *fluorescent quantum efficiency* $\eta$, defined as the number of fluorescent photons spontaneously emitted on the laser transition divided by the number of pump photons absorbed on the pump transition(s) when the laser material is below threshold. For the four-level system this quantum efficiency is given by

$$\eta = \frac{\gamma_{43}}{\gamma_4} \times \frac{\gamma_{\text{rad}}}{\gamma_3} = \frac{\tau_4}{\tau_{43}} \times \frac{\tau_3}{\tau_{\text{rad}}}, \qquad (10)$$

where $\gamma_{\text{rad}} \equiv \gamma_{\text{rad}}(3 \to 2)$ is the radiative decay rate on the $3 \to 2$ transition. The first ratio in this expression tells what fraction of the total atoms excited to level 4 relax directly into the upper laser level 3, rather than bypassing 3 and dropping to lower levels, and the second ratio tells what fraction of the total decay out of level 3 is purely radiative decay to level 2.

### Four-Level Population Inversion

With the aid of the parameters $\beta$ and $\eta$, plus the conservation of atoms condition that $N_1 + N_2 + N_3 + N_4 = N$, we can solve for the population inversion

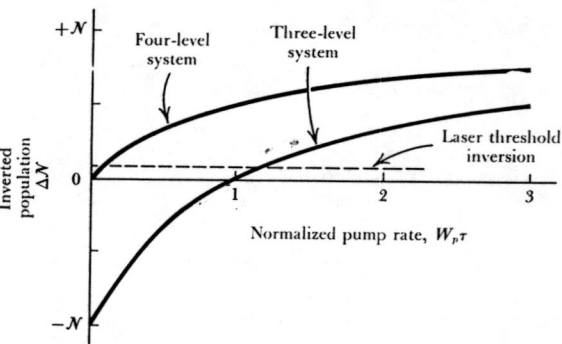

FIGURE 6.4
Laser population inversion versus normalized pumping rate for idealized four-level and three-level laser systems.

$N_3 - N_2$ versus pumping strength on the four-level system. After some algebra we can obtain

$$\frac{N_3 - N_2}{N} = \frac{(1-\beta)\eta W_p \tau_{\rm rad}}{1 + [1 + \beta + 2\tau_{43}/\tau_{\rm rad}]\eta W_p \tau_{\rm rad}}, \qquad (11)$$

where $\tau_{\rm rad} \equiv \tau_{\rm rad}(3 \to 2)$ is the radiative decay rate on the laser transition itself. In a good laser material the lifetime $\tau_{43}$ from the upper pump level into the upper laser level will be short compared to this radiative decay time, and this expression can then be simplified into

$$\frac{N_3 - N_2}{N} \approx \frac{(1-\beta)\eta W_p \tau_{\rm rad}}{1 + (1+\beta)\eta W_p \tau_{\rm rad}} \approx \frac{W_p \tau_{\rm rad}}{1 + W_p \tau_{\rm rad}} \quad \text{if} \quad \beta \to 0. \qquad (12)$$

The optimum situation is obviously $\beta \approx \tau_{21}/\tau_{32} \to 0$.

Figure 6.4 shows a plot of the inversion $N_3 - N_2$ on the four-level laser transition versus the normalized pumping rate $W_p \tau$, assuming $\beta = 0$. For a four-level system, the population inversion on the $3 \to 2$ transition increases linearly with the pumping intensity $W_p$ at lower pump levels, but then approaches a limiting value for $W_p \tau \gg 1$ as the ground state $E_1$ is depleted and a large fraction of the atoms are lifted into the upper laser level.

This four-level pumping model provides a surprisingly good analytical model for understanding the behavior of a large number of real laser systems, as we will show in later sections.

### Three-Level Laser System

Figure 6.5 illustrates how a three-level laser system can be similarly employed as a model for the real energy levels of the familiar ruby laser, just as the four-level system provided a model for the Nd:YAG and many similar solid-state and dye lasers.

A three-level laser differs from the four-level system in that the lower laser level is the ground level $E_1$. This is a serious disadvantage, since more than half the atoms initially in the ground state must be pumped through the upper pumping level $E_3$ into the upper laser level $E_2$ before any inversion at all is obtained on the $2 \to 1$ transition. Three-level lasers are, therefore, usually not as efficient as four-level lasers. One reason for analyzing the three-level system,

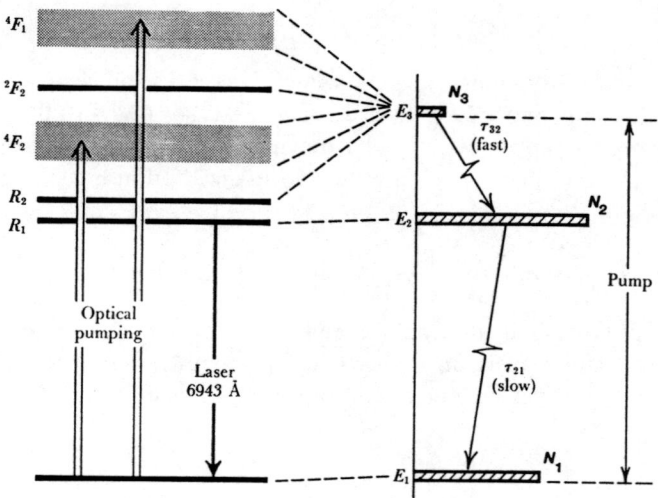

**FIGURE 6.5**
The relevant quantum energy levels in the laser material ruby, and an idealized three-level model for this system.

nonetheless, in addition to general background knowledge, is that the 694 nm ruby laser—the first laser ever to be operated and still a useful solid-state laser—is a nearly ideal three-level laser system.

Suppose the pumping process in a three-level system produces a stimulated transition probability $W_{13} = W_{31} = W_p$. Then the rate equations for the two upper levels are

$$\frac{dN_3}{dt} = W_p(N_1 - N_3) - \frac{N_3}{\tau_3} \tag{13}$$

and

$$\frac{dN_2}{dt} = \frac{N_3}{\tau_{32}} - \frac{N_2}{\tau_{21}}. \tag{14}$$

There is also the usual conservation equation $N_1 + N_2 + N_3 = N$, and it is again useful to define a fluorescence quantum efficiency given by

$$\eta = \frac{\tau_3}{\tau_{32}} \times \frac{\tau_{21}}{\tau_{\text{rad}}(2 \to 1)}. \tag{15}$$

The important relaxation-time ratio in this model is given by

$$\beta \equiv \frac{N_3}{N_2} = \frac{\tau_{32}}{\tau_{21}}. \tag{16}$$

The steady-state population difference on the $2 \to 1$ transition can then be found to be

$$\frac{N_2 - N_1}{N} = \frac{(1 - \beta)\eta W_p \tau_{\text{rad}} - 1}{(1 + 2\beta)\eta W_p \tau_{\text{rad}} + 1}. \tag{17}$$

Inversion in the three-level system can be obtained only if $\beta < 1$, and even then inversion can occur only when the pumping rate exceeds a threshold value given by

$$W_p \tau_{\text{rad}} \geq \frac{1}{\eta(1-\beta)}. \qquad (18)$$

The optimum situation obviously occurs when the relaxation from the upper pumping level 3 into the upper laser level 2 is very fast, so that $\beta \to 0$, and when the relaxation from the upper laser level 1 down to the ground level 1 is purely radiative, so that $\eta \to 1$. The inversion versus normalized pumping strength then reduces to

$$\frac{N_2 - N_1}{N} \approx \frac{W_p \tau_{\text{rad}} - 1}{W_p \tau_{\text{rad}} + 1} \qquad \text{if } \eta \to 1 \text{ and } \beta \to 0. \qquad (19)$$

The significant differences in inversion versus pumping for a three-level and a four-level system are illustrated in Figure 6.4. Other things being equal, a four-level laser system should have a much lower pumping threshold than a three-level laser system.

### More on the Ruby Laser

The well-known ruby laser is a special case that overcomes the inherent disadvantages of the three-level laser, since the ruby laser can in fact have a moderately low pulsed laser threshold, and can even (with some difficulty) be operated as a cw laser. The ruby laser works as well as it does because of an unusually favorable combination of other factors, including:

- Unusually broad and well-located pump absorption bands that make very efficient use of the broadband radiation from standard flashlamps.
- A fluorescent quantum efficiency $\eta$ very close to unity.
- An unusually narrow atomic linewidth for the laser transition.
- An unusually long and almost purely radiative lifetime $\tau_{21} \approx 4.3$ ms for the upper laser level.
- The availability of ruby synthetic crystals with very good optical quality, good thermal conductivity, and high optical power handling capability.

The ruby laser has been particularly useful because it has an oscillation wavelength located in (or at least on the edge of) the visible region, where more sensitive photodetectors are generally available, in contrast to the infrared wavelengths of most of the solid-state rare-earth lasers. On the other hand, modern flashlamp-pumped dye lasers also give good laser action all across the visible region.

---

### Problems for 6.1

1. *Three-level system with two pumping signals applied.* Consider a three-level atomic system with no degeneracies ($g_1 = g_2 = g_3 = 1$) and with two separate pumping signals applied to give $W_{12} = W_{21} = W_a$ and $W_{13} = W_{31} = W_b$. Using the optical-frequency approximation, find the population difference $N_2 - N_3$ as a function of pumping and relaxation rates. Suppose that $W_a$ and $W_b$ can have variable amplitudes but always with a fixed ratio of $W_a/W_b$. Describe and plot the variation of $(N_2 - N_3)$ with pumping strength for some illustrative cases.

## 6.2 LASER GAIN SATURATION

2. *Population inversion versus pumping in an "upper-level" three-level laser system.* It is possible (though not likely in practice) to have a three-level laser system which is pumped on the $1 \to 3$ transition and in which cw laser action takes place on the $3 \to 2$ rather than the $2 \to 1$ transition (no such real system is known). Suppose that level 3 in such a system is long-lived with lifetime $\tau_3$; level 2 has a short relaxation time to the ground state; and the system is pumped with transition probability $W_p$ on the $1 \to 3$ transition.

   Carry through the rate-equation analysis (in the optical approximation) necessary to find the population inversion on the $3 \to 2$ transition as a function of pumping power $W_p$. Compare the $\Delta N$ versus $W_p$ curve for this system to to those for the four-level and three-level models shown in Figure 6.4.

3. *Cascade pumping of a four-level laser system.* Suppose a four-level system is "cascade pumped" with two separate pumping transition probabilities $W_{13} \equiv W_A$ and $W_{34} \equiv W_B$. The optical approximation $\hbar\omega \gg kT$ applies for all transitions, and there are significant downward relaxation rates $\gamma_{ji}$ between all levels.

   Solve for the population difference $\Delta N_{42} \equiv N_4 - N_2$ in this system as a function of the two pumping powers $W_A$ and $W_B$, using the $\gamma_j$ and $\gamma_{ji}$ notations for the downward relaxation rates. Discuss what conditions are needed for an inversion on the $4 \to 2$ transition, and how this inversion depends on the two pumping powers.

4. *Analysis of a five-level laser system.* Consider a five-level atomic system in which the optical approximation applies between all levels. Assume that this system is pumped on the $1 \to 5$ transition with a pumping transition probability $W_{15} = W_{51} + W_p$, and that each upper level in the system relaxes only into the level immediately beneath it. Evaluate the population difference on the $3 \to 2$ transition as a function of $W_p$, and discuss the dependence of this population difference on various interlevel relaxation rates.

5. *Laser refrigeration?* In an earlier chapter we considered an analysis of the population inversion and energy transfer in a thermally pumped laser or maser system. The reverse of this, which is also physically possible, is an optically pumped "laser refrigerator."

   To see how such a refrigerator might operate, consider applying a strong coherent signal to the $1 \to 2$ transition in a three-level system. With the right ratios of relaxation times, we can then achieve a steady-state population difference on the $2 \to 3$ transition which is considerably "colder" than the Boltzmann ratio on the same transition before the signal was turned on. The atoms now look significantly colder to the thermal surroundings, at least in the frequency band at and around the $2 \to 3$ transition frequency. In other words, this becomes a *refrigerator*, which uses coherent work (the applied signal on the $1 \to 2$ transition) to achieve cooling of the atoms, and maybe even the atomic surroundings, at and near the $2 \to 3$ transition frequency.

   Assume again that, with the use of suitable filters, the atoms can see quite different physical surroundings at different frequency bands (this is a perfectly reasonable assumption). What sort of refrigeration efficiencies might we achieve? What relaxation time ratios and other conditions do we need to achieve high efficiency? Can we ever approach the thermodynamic limit?

## CHAPTER 6: LASER PUMPING AND POPULATION INVERSION

## 6.2 LASER GAIN SATURATION

In many real laser systems laser action takes place between two excited levels that are located high above the ground level, and the population density in these excited laser levels always remains small compared to the total density of atoms in the lowest energy level $E_0$ (as we will label the ground level in this section). This is particularly true in gas lasers, where linewidths are narrow, transitions are relatively strong, and only small inversion densities are necessary to give significant gain. It may be less true in solid-state lasers, such as the ruby example of the preceding section, where large fractions of the total atomic density may sometimes be pumped into the upper laser levels.

In any event, in this section we will use this as a simplified model to develop some further rate equation analyses, showing in particular how the laser gain itself saturates with increasing signal power in typical laser systems.

### Laser Gain Saturation Analysis

Figure 6.6 gives a simplified but yet realistic model for many laser systems of this type. Atoms are pumped by some pumping mechanism from the ground level $E_0$ into some upper level $E_3$. They then relax down (perhaps by cascade processes) into the upper laser level $E_2$, from where they relax or make stimulated laser transitions down to the lower laser level $E_1$, and thence back to ground. Note that we have specifically included a laser signal, corresponding to laser amplification or laser oscillation, and represented by the stimulated transition probability $W_{\text{sig}}$ in this diagram.

Suppose the upper-level populations all remain small compared to the initial ground-state population. Then the pumping rate from the ground level $E_0$ into the upper atomic level $E_3$ caused by a pumping transition probability $W_{03} = W_{30} = W_p$ may be written as

$$\left.\frac{dN_3}{dt}\right|_{\text{pump}} = W_p(N_0 - N_3) \approx W_p N_0, \tag{20}$$

where $N_0 \approx N$ is very nearly the total density of laser atoms in the system.

In this situation there is essentially no "back-pumping" from $E_3$ to $E_0$, since very few atoms accumulate in the upper levels and hence $N_3 \ll N_0$. It is then common and convenient practice to speak not of a pumping transition probability $W_p$ (probability per atom per second), but of a *net pumping rate* (atoms per second, per unit volume) being lifted up out of the ground level, as given by $W_p N_0 \approx W_p N$.

This pumping rate $W_p N_0$ in a real laser system will be more or less directly proportional to the pump light intensity (in an optically pumped laser), or to the discharge current density (in a discharge-pumped gas laser), or to a chemical reaction rate (in a chemically pumped laser). Moreover, in many real lasers some fixed fraction $\eta_p$ of the atoms pumped into an upper energy level will decay, often through some cascade process, down into the longer-lived upper laser level $E_2$. The number of atoms per second reaching the upper laser level is then given by an effective pumping rate $R_p = \eta_p W_p N_0$, where $\eta_p$ represents the quantum efficiency for pump excitation into this upper laser level. (This pumping efficiency may be quite high, even approaching unity, for many solid-state and organic dye lasers, and may be very small for many typical discharge-pumped gas lasers.)

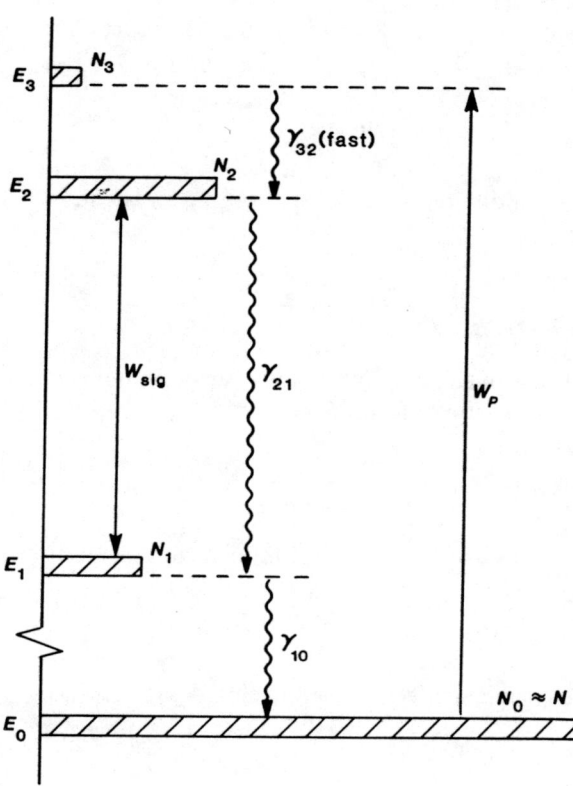

FIGURE 6.6
Simplified model for laser pumping and gain saturation between two high-lying energy levels $E_1$ and $E_2$.

With these generally valid assumptions, the rate equations for the excited laser levels $E_2$ and $E_1$, including a laser signal with stimulated transition probability $W_{12} = W_{21} = W_{\text{sig}}$ on the laser transition, may be written as

$$\frac{dN_2}{dt} = R_p - W_{\text{sig}}(N_2 - N_1) - \gamma_2 N_2 \qquad (21)$$

and

$$\frac{dN_1}{dt} = W_{\text{sig}}(N_2 - N_1) + \gamma_{21} N_2 - \gamma_1 N_1, \qquad (22)$$

where $\gamma_2 = \gamma_{21} + \gamma_{20}$ is the total decay rate downward from the upper laser level $E_2$ to all lower levels.

The steady-state solutions to these equations are then given by

$$N_1 = \frac{W_{\text{sig}} + \gamma_{21}}{W_{\text{sig}}(\gamma_1 + \gamma_{20}) + \gamma_1 \gamma_2} R_p,$$

$$N_2 = \frac{W_{\text{sig}} + \gamma_1}{W_{\text{sig}}(\gamma_1 + \gamma_{20}) + \gamma_1 \gamma_2} R_p. \qquad (23)$$

Note particularly that only the two rate equations 6.21 and 6.22 were used in obtaining these results. No "conservation of atoms" condition stating that $N_1 + N_2$ remains constant was necessary or even possible in this case, since in fact $N_1 + N_2$ does not remain constant in this system when either the pump rate $R_p$ or the signal strength $W_{\text{sig}}$ changes. Two rate equations for the populations at

## CHAPTER 6: LASER PUMPING AND POPULATION INVERSION

the two levels are thus both necessary and also sufficient in this particular type of rate-equation calculation.

### Gain Saturation Behavior

The steady-state population difference $\Delta N_{21} \equiv N_2 - N_1$ on the laser transition is then given by

$$\Delta N_{21} \equiv N_2 - N_1 = \left(\frac{\gamma_1 - \gamma_{21}}{\gamma_1 \gamma_2}\right) \times \frac{R_p}{1 + [(\gamma_1 + \gamma_{20})/\gamma_1 \gamma_2] W_{\text{sig}}}. \tag{24}$$

The inverted population difference in this simple example varies with both pumping rate and signal intensity in the simple form

$$\Delta N_{21} = \Delta N_0 \frac{1}{1 + W_{\text{sig}} \tau_{\text{eff}}}. \tag{25}$$

where $\Delta N_0$ is a small-signal or unsaturated population inversion given by

$$\Delta N_0 = \frac{\gamma_1 - \gamma_{21}}{\gamma_1 \gamma_2} R_p = (1 - \tau_1/\tau_{21}) \times R_p \tau_2 \tag{26}$$

and $\tau_{\text{eff}}$ is an effective recovery time or lifetime for the signal gain given by

$$\frac{1}{\tau_{\text{eff}}} = \frac{\gamma_1 \gamma_2}{\gamma_1 + \gamma_{20}} \quad \text{or} \quad \tau_{\text{eff}} = \tau_2 \left(1 + \tau_1/\tau_{20}\right), \tag{27}$$

If the upper laser level $E_2$ relaxes primarily into the lower laser level $E_1$ and not directly down to any lower levels $E_0$, then the expression for the population inversion reduces to simply

$$\Delta N_{21} \approx R_p(\tau_2 - \tau_1) \times \frac{1}{1 + W_{\text{sig}} \tau_2} \tag{28}$$

where we have used the approximation that $\gamma_2 \approx \gamma_{21}$ and $\gamma_{20} \approx 0$.

### Discussion

The analytical results in Equations 6.24 through 6.28 illustrate several typical aspects of laser behavior, including:

- The crucial requirement for obtaining inversion on this transition is that the time-constant ratio $\tau_{21}/\tau_1$ should be $> 1$, or that the condition $\gamma_1 > \gamma_{21}$ be satisfied. In physical terms, this means that inversion is obtained only if atoms relax downward out of the lower level $E_1$ at rate $\gamma_1$ faster than they relax in at rate $\gamma_{21}$ from the upper level.
- If this condition is met, the small-signal or unsaturated population difference $\Delta N_0$ between the two laser levels is then directly proportional to the pump rate $R_p$ times an effective "population integration time," which is basically the upper-level lifetime $\tau_2$ reduced by the factor $(1 - \tau_1/\tau_{21})$.
- The effective recovery time $\tau_{\text{eff}}$, which determines the signal saturation behavior in Equation 6.25, is in general a combination of the various inter-level relaxation rates or lifetimes in the system. If the lower-level lifetime becomes short enough, $\tau_1 \to 0$, so that little or no population

can accumulate in the lower level $E_1$, then $\tau_{\text{eff}}$ becomes just the upper-level lifetime, $\tau_{\text{eff}} \approx \tau_2$.

- Finally, in this system as in many real lasers, the saturation intensity of the inverted population depends only on the relaxation lifetimes between the atomic levels, *and does not depend directly on the pumping intensity $R_p$*. That is, the signal intensity $W_{\text{sig}}$ needed to reduce the population inversion or laser gain to half its initial value does not depend at all (at least in this example) on how hard the atoms are being pumped.

The final point implies in particular that turning up the pump intensity will not increase the saturation intensity of the material, or the signal level required to reduce the gain to half its small-signal value. Turning up the pump intensity does increase the unsaturated population inversion $\Delta N_0$ and hence the unsaturated gain, so that the laser must oscillate harder to bring the saturated gain down to match the losses; but the value of $W_{\text{sat}}$ or $I_{\text{sat}}$ is basically independent of how hard the system is pumped. This same behavior is characteristic of most real laser systems.

### The Factor of 2*

We showed earlier that in a simple two-level system with a constant total population, or $N_1 + N_2 = N$, the rate equation for the absorbing population difference $\Delta N \equiv N_1 - N_2$ takes the general form

$$\frac{d}{dt}\Delta N = -2W_{12}\Delta N - \frac{\Delta N - \Delta N_0}{T_1} \tag{29}$$

and so the saturation behavior takes the form

$$\Delta N = \Delta N_0 \frac{1}{1 + 2W_{12}T_1}, \tag{30}$$

where $T_1$ is the recovery time for the population difference.

Suppose instead that we have a pumped and possibly inverted two-level system like that just discussed, in which the total population is no longer necessarily constant; and let us suppose in addition that the relaxation rate $\gamma_{10}$ out of the lower level is extremely rapid, so that essentially no atoms ever collect in level 1, and that $N_1 \approx 0$ under all circumstances. According to the preceding results, the rate equation for the inverted population difference $\Delta N \equiv N_2 - N_1 \approx N_2$ the becomes

$$\frac{d}{dt}\Delta N \approx -W_{\text{sig}}\Delta N - \frac{\Delta N - \Delta N_0}{\tau_2} \tag{31}$$

and so the saturation behavior is

$$\Delta N \approx \Delta N_0 \frac{1}{1 + W_{\text{sig}}\tau_2}, \tag{32}$$

where $\tau_2$ is the upper-level lifetime; $W_{\text{sig}} \equiv W_{12}$; and $\Delta N_0 \equiv R_p\tau_2$. These two situations thus lead to exactly the same basic equations except that there is an additional factor of 2 in the stimulated-transition term in one case and not the other. This factor occurs in the simple two-level absorbing system because the transition of one atom between levels reduces the population difference $\Delta N$ by two.

Now, there are also many inverted laser systems in which an atom that is stimulated to make a downward transition from level 2 to level 1 remains for some considerable lifetime in level 1 before relaxing down to still lower levels. If the lifetime in level 1 is particularly long, we sometimes say that the atoms are more or less "bottlenecked" in level 1. The stimulated transition of *one* atom from $N_2$ to $N_1$ thus again reduces the population difference $\Delta N \equiv N_2 - N_1$ by *two*, and we must write the stimulated-transition term as $(d/dt)\Delta N \approx -2W_{\text{sig}}\Delta N$ here also. If level 1 is not bottlenecked, however, and its population empties out very rapidly, so that $N_1 \approx 0$ at all times, we can write the stimulated term as $(d/dt)\Delta N \approx -W_{\text{sig}}\Delta N$, where $\Delta N \approx N_2$ as above.

To handle both of these situations in a single notation, and also to take account of the fact that the population difference most often recovers to some small-signal or unsaturated value $\Delta N_0$ with an effective time constant $\tau_{\text{eff}}$, we can write the saturation equation for the inverted population in many different practical laser systems (and also the absorbing population difference in many absorbing systems) in the general form

$$\frac{d}{dt}\Delta N(t) = -2^*W_{\text{sig}}\Delta N(t) - \frac{\Delta N(t) - \Delta N_0}{\tau_{\text{eff}}}, \tag{33}$$

where $2^*$ is a numerical factor with a value somewhere between $2^* = 2$ (for strongly bottlenecked systems) and $2^* = 1$ (for systems with no bottlenecking). We will use this simple form several times later on to analyze the gain saturation and atomic dynamics in many laser problems, although a more complex rate-equation analysis may be needed to evaluate the actual values of $2^*$ and $\tau_{\text{eff}}$. The effective value of $2^*$ will turn out to make a significant difference in the energy and power outputs of laser devices. (In general, the absence of bottlenecking, or $2^* \approx 1$, is good; and the presence of bottlenecking, or $2^* \approx 2$, is not so good.)

---

Problems for 6.2

1. *Transient response in the simplified laser-pumping model.* Find the transient (time-varying) solutions for the laser populations $N_1(t)$ and $N_2(t)$ in the same simplified atomic model discussed in this section, assuming that both the pump rate $R_p$ and the signal intensity $W_{\text{sig}}$ are turned on to constant values at $t = 0$. Discuss: (a) the transient build-up of $\Delta N(t)$ when $R_p$ is first turned on, assuming $W_{\text{sig}} = 0$; (b) the transient decay when $R_p$ is turned off, again assuming $W_{\text{sig}} = 0$; and (c) the transient behavior of $\Delta N(t)$ if $W_{\text{sig}}$ is suddenly turned on or changed, assuming $R_p$ is constant.

2. *Laser inversion and saturation including degeneracies.* Repeat the pumping and saturation calculations of this section assuming the same pumping rate $R_p$ and relaxation rates $\gamma_{ji}$ as in the text, but taking into account the existence of degeneracies $g_1$ and $g_2$ of the lower and upper energy levels and their effects on the definitions of $W_{12}$, $W_{21}$, and $\Delta N_{21}$

3. *Optically pumped laser absorber.* In a three-level atomic system with levels $E_1$, $E_2$ and $E_3$, for which the optical-frequency approximation is valid, pumping radiation producing a stimulated-transition rate $W_{12} = W_{21} = W_p$ is applied to the 1→2 transition, and signal radiation with $W_{23} = W_{32} = W_{\text{sig}}$ is applied to the 2→3 transition. The pump thus lifts atoms from level 1 to level 2, and creates a

kind of optically pumped laser absorption rather than laser amplification for the signal on the 2→3 transition. (Who knows?—it might be good for something.)

Find the population difference $\Delta N_{23}$ as a function of the pumping intensity $W_p$, the signal intensity $W_{\text{sig}}$, and the various downward relaxation rates in this system. Discuss in particular the signal saturation behavior of $\Delta N_{23}$ as a function of signal intensity $W_{\text{sig}}$ at fixed pump intensity $W_p$.

4. *Simultaneous pumping into both the upper and the lower laser levels.* Consider a laser transition between two upper energy levels $E_1$ and $E_2$ in an atomic system similar to that discussed in this section. Assume, however, that there is pumping up from the ground level into *both* the laser levels $E_1$ and $E_2$, at pumping rates $R_1$ and $R_2$, respectively. (Note: these are pumping *rates*, not transition probabilities.) Assume there is also a laser signal $W_{\text{sig}}$ on the 2→1 transition, as in this section.

Solve the steady-state equations for the population difference $\Delta N_{21} = N_2 - N_1$ in this system, and put the answer into a form that illustrates the dependence on pumping rates and on signal saturation. Discuss briefly (a) the pumping-rate and relaxation-time conditions for obtaining an inversion at all in this system; and (b) the form of the signal saturation behavior, and how it compares to simpler systems.

5. *Signal saturation behavior in the ideal three-level laser system.* Consider a three-level laser system like the ruby laser analyzed in a preceding section, assuming for simplicity that the only relaxation processes present are $\gamma_{32}$, which is very fast, and $\gamma_{21}$, which is slow and purely radiative. In addition to a pump transition probability on the 1 → 3 transition, add a signal transition probability $W_{\text{sig}}$ on the 2 → 1 transition. Analyze the steady-state population inversion on the 2 → 1 transition as a function of $W_p$ and $W_{\text{sig}}$. In particular, is the signal-saturation behavior produced by $W_{\text{sig}}$, for a fixed value of $W_p$, homogeneous in form? Does the saturation value for $W_{\text{sig}}$ depend on how hard the system is being pumped? If so, is there a physical explanation why this is different from the gain saturation behavior analyzed in this section?

---

## 6.3 TRANSIENT LASER PUMPING

The full transient solution to a set of multilevel rate equations can become very complicated, as we noted in Chapter 4, since there will be in general $M - 1$ transient decay terms for an $M$-level atomic system. The transient solution for the build-up of inversion in a multilevel pulse-pumped laser can, therefore, also become a complicated problem. We will illustrate one or two such transient situations in this section, however, using very simplified models, in order to give some idea of the kind of behavior to be expected.

### Transient Rate-Equation Example: Upper-Level Laser

As a first example, let us consider the "upper-level" laser model shown in Figure 6.6 and described in Equations 6.20 through 6.23. To simplify this still further, assume that no signal is present on the $E_2$ to $E_1$ transition, and that the relaxation rate out of the lower $E_1$ level is sufficiently fast that $N_1 \approx 0$ under all conditions. The transient pumping equation for the upper laser level population $N_2(t)$ is then

$$\frac{dN_2(t)}{dt} = R_p(t) - \gamma_2 N_2(t) \tag{34}$$

where $R_p(t)$ is the (possibly) time-varying pump rate (in atoms lifted up per second) applied to the atomic system. A formal solution to this equation is

$$N_2(t) = \int_{-\infty}^{t} R_p(t') e^{-\gamma_2(t-t')} dt' \tag{35}$$

This equation says, of course, that of the number of atoms $R_p(t') dt'$ lifted up during a little time interval $dt'$, only a fraction $e^{-\gamma^2(t-t')}$ will remain in the upper level at a time $t - t'$ later. Suppose we put in a square pump pulse with constant amplitude $R_{p0}$ and duration $T_p$, i.e., $R_p(t) = R_{p0}, 0 \leq t \leq T_p$. The maximum upper-level population, reached just at the end of the pumping pulse, is then given by

$$N_2(T_p) = R_{p0}\tau_2 \left[1 - e^{-T_p/\tau_2}\right] \tag{36}$$

where $\tau_2 \equiv 1/\gamma_2$ is the lifetime of the upper laser level. This tells us that in a pulse-pumped laser of the type in which one first pumps up the upper-level population, and then "dumps" this population by $Q$-switching, it is of very little use to continue the pump pulse for longer than about two upper-level lifetimes or so, since beyond that point the upper-level population no longer increases much with further pumping. Alternatively, we might define a pumping efficiency $\eta_p$ for this case as the ratio of the maximum number of atoms stored in the upper level, just at the end of the pumping pulse, to the total number of pump photons sent in or atoms lifted up during the pump pulse. Since the total number of atoms lifted up during the pump pulse is $R_{p0}T_p$, this pumping efficiency is given by

$$\eta_p = \frac{N_2(t = T_p)}{R_{p0}T_p} = \frac{1 - e^{-T_p/\tau_2}}{T_p/\tau_2} \tag{37}$$

In other words, this efficiency depends only on the ratio of pump pulsewidth $T_p$ to upper-level time constant, $\tau_2$. A little work with your pocket calculator will show that if these time constants are equal, i.e., $T_p = \tau_2$, then the pumping efficiency is only about $\eta_p \approx 63\%$. For the pumping efficiency to reach 90% requires $T_p/\tau_2 \approx 0.2$, i.e., the total pump energy must be delivered in a pulse whose width $T_p$ is only about 1/5 of the upper-level time constant $\tau_2$.

### Transient Rate-Equation Example: Pulsed Ruby Laser

A transient solution for the simplified three-level ruby laser model given earlier in this chapter can also rather easily be obtained and used to demonstrate both the techniques of rate-equation analysis and the good agreement with experiment that can be provided by even a simple rate-equation description.

In ruby the $\gamma_{32}$ relaxation rate is so fast ($> 10^{10}$ sec$^{-1}$) that atoms pumped into level 3 may be assumed to relax instantaneously into level 2. Hence we may assume that $N_3 \approx 0$ at all times, even with the strongest practical pump powers that we can apply. The three-level rate equations for a ruby laser, including pumping but not signal terms, can then be reduced to the single rate equation

$$\frac{dN_1}{dt} = -\frac{dN_2}{dt} \approx -W_p(t)N_1(t) + \frac{N_2(t)}{\tau} \tag{38}$$

combined with the conservation of atoms condition that $N_1(t) + N_2(t) \approx N$. Here the lifetime $\tau$ is the total (and, in ruby, mostly radiative) decay time of approximately 4.3 ms for relaxation downward from level 2 to level 1.

These two equations can then be combined into a single rate equation for the inverted population difference $\Delta N(t) = N_2(t) - N_1(t)$ in the form

$$\frac{d}{dt}\Delta N(t) = -[W_p(t) + 1/\tau]\Delta N(t) + [W_p(t) - 1/\tau]N. \tag{39}$$

For the special case of constant pump intensity, this can be written in the even simpler form

$$\frac{d}{dt}\Delta N = -(W_p + 1/\tau) \times [\Delta N(t) - \Delta N_{ss}]. \tag{40}$$

This equation has exactly the same form as the relaxation of an elementary two-level system toward thermal equilibrium, except that the population difference $\Delta N(t)$ here relaxes toward a nonthermal steady-state equilibrium value $\Delta N_{ss}$ given by

$$\Delta N_{ss} \equiv \frac{W_p\tau - 1}{W_p\tau + 1} N, \tag{41}$$

and the relaxation rate toward this value is $(W_p + 1/\tau)$ rather than simply $1/\tau$. If the pumping rate is above threshold, or $W_p\tau > 1$, then $\Delta N(t)$ of course actually relaxes toward an *inverted* value of $\Delta N_{ss}$.

### Pulsed Inversion

Suppose a square pump pulse with constant pump intensity $W_p$ is turned on at $t = 0$ in this particular system. (Some sort of pulse-forming network rather than just a single charged capacitor will be required to produce such a square pulse with a standard flashlamp.) The population inversion as a function of time during the pump flash is then given by the transient solution to the preceding equations with the initial condition that $\Delta N(t=0) = -N$. This solution is

$$\frac{\Delta N(t)}{N} = \frac{(W_p\tau - 1) - 2W_p\tau \exp[-(W_p\tau + 1)t/\tau]}{W_p\tau + 1}. \tag{42}$$

Suppose that this pumping rate is left on for a pump pulse time $T_p$ which is short compared to the atomic decay time $\tau$; and that the pumping rate $W_p\tau$ is $\gg 1$, which says that if the pump rate were left on for a full atomic lifetime $\tau$, it would create a very strong inversion. The inversion just at the end of the pump pulse, or $t = T_p$, is then given to a good approximation by

$$\frac{\Delta N(T_p)}{N} \approx 1 - 2e^{-W_p T_p} \qquad (T_p \ll \tau \text{ and } W_p\tau \gg 1). \tag{43}$$

This then says that (a) the inversion at the end of the pump pulse depends only on the total energy $W_pT_p$ in the pulse, and not on its duration (or even shape);

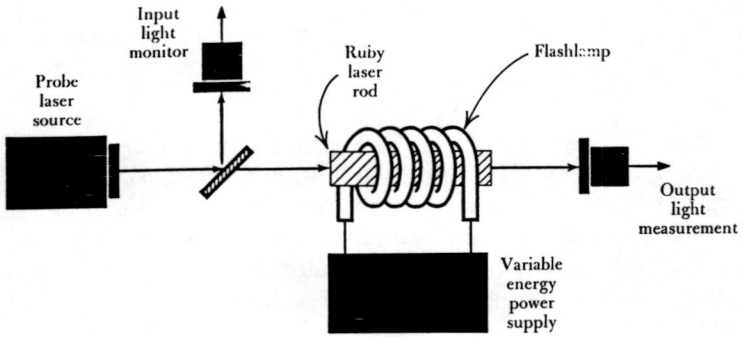

FIGURE 6.7
Pulsed laser gain measurement system.

and (b) the pulsed pump can produce complete inversion of the system, if the pumping energy is large enough ($W_p T_p \gg 1$).

Figure 6.7 illustrates a ruby-laser amplifier experiment in which a square pump pulse of length $T_p$ short compared to the atomic decay time $\tau$ was used to pump a ruby amplifier rod without end mirrors. (Practical values for the pumping circuit might be a flashlamp pulse length $T_p \approx 200$ μs, compared to a ruby fluorescent decay time of $\tau = 4.3$ ms.) A separate probe ruby laser was then used to measure the single-pass gain through the ruby rod just at the end of this pump pulse.

We will learn later that the gain or loss through a laser amplifier measured in dB is directly proportional to the laser population difference $\Delta N$. For $T_p \ll \tau$ as in these experiments, the ratio of gain $G_{\mathrm{dB}}$ just after the pump pulse to initial loss $L_{\mathrm{dB}}$ just before the pump pulse is predicted from the preceding equation to be

$$\frac{G_{\mathrm{dB}}(t=T_p)}{L_{\mathrm{dB}}(t=0)} = \frac{\Delta N(t=T_p)}{\Delta N(t=0)} \approx 1 - 2e^{-W_p T_p}. \tag{44}$$

Figure 6.8 shows experimental data for this ratio for different values of the total flashlamp pump pulse energy, which is in turn directly proportional to the normalized pump quantity $W_p T_p$. The experimental results are in excellent agreement with the simple theoretical formula. Note in particular that a sufficiently powerful and rapid pump pulse can come very close to complete inversion of the ruby transition; i.e., it can pump essentially all of the $Cr^{3+}$ atoms into the upper laser level.

### Laser Oscillation Time Delay

Figure 6.9 shows another simple experimental examination of the transient behavior of populations in a pumped laser system. The small insert in this figure shows the oscillation output from a typical flash-pumped ruby laser, including the time delay $t_d$ between the start of the pump flash and the onset of laser oscillation. The pumping pulse $W_p(t)$ requires a certain amount of time, or a certain amount of integrated pumping energy, before it can pump enough atoms up to the upper laser level to create a population inversion, especially in a three-level system such as ruby. Once a net population inversion is created, however, the laser oscillation then builds up extremely rapidly, as illustrated by the sharp

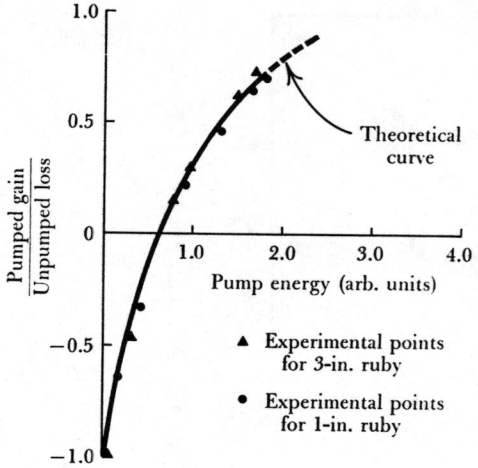

**FIGURE 6.8**
Experimental data for ruby laser gain versus pumping energy, obtained by using the experimental setup of Figure 6.7. (From J. E. Geusic and H. E. D. Scovil, *Bell. Sys. Tech. J.* **41**, 1371, July 1962.)

leading edge of the laser oscillation. (Note also the strong characteristic spiking behavior in the oscillation output.) The smoothly rising curve before the oscillation starts represents pump light leakage and preoscillation fluorescence from the laser rod.

This figure also shows how the reciprocal of the oscillation time delay varies with the total energy in a long flashlamp pulse. For a long square-topped pump pulse, the time delay $t_d$ to the onset of oscillation can be estimated from the transient solution for $\Delta N(t)$ by finding the time $t = t_d$ at which $\Delta N(t)$ passes through zero (assuming that the oscillation signal will build up very rapidly once inversion is obtained). This time delay to inversion is given for constant pumping by

$$t_d = \frac{\tau \ln[2W_p\tau/(W_p\tau - 1)]}{W_p\tau + 1}. \tag{45}$$

There is a minimum value of pumping intensity $W_p$ below which the laser will not reach oscillation threshold at all. If the pumping pulse has a pulse length $T_p$ several times longer than the upper laser level lifetime $\tau$, then this threshold pump intensity is given by $W_{p,\text{th}} \approx 1/\tau$ for $T_p \gg \tau$. For pump intensities below this value, inversion is never reached no matter how long the pump pulse continues.

The reciprocal oscillation time delay $t_d$ normalized to the upper-level lifetime $\tau$ can then be written as

$$\frac{\tau}{t_d} = \frac{r+1}{\ln[2r/(r-1)]}, \tag{46}$$

where the parameter $r$ represents the normalized pumping energy above threshold, i.e., $r \equiv W_p/W_{p,\text{th}}$. This simple expression gives a moderately good fit to the experimental data in Figure 6.9.

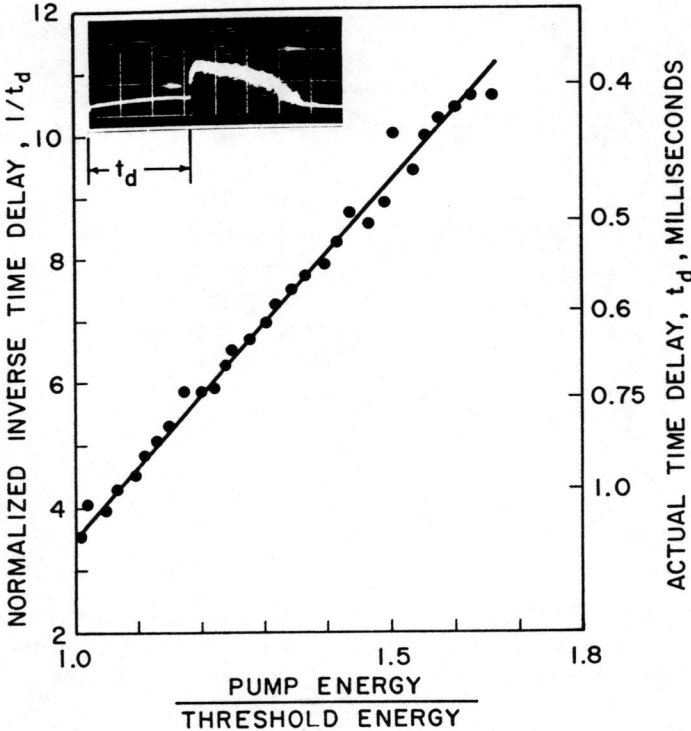

**FIGURE 6.9**
Time delay for oscillation buildup in a ruby laser versus pump pulse energy. The inset shows the power output versus time (200 $\mu$sec per division) from a flash-pumped ruby laser, and the experimental points show how the reciprocal of the oscillation time delay $t_d$ varies with energy input to the pumping flash lamp. (From A. E. Siegman and J. W. Allen, *IEEE J. Quantum Electron.* **QE–1**, 386–393, December 1965.)

---

Problems for 6.3

1. *Transient pumping with a gaussian time-varying pump pulse.* Derive an analytical formula for the upper-level population $N_2(t)$ in the simplified upper-level laser model discussed at the start of this section, assuming a gaussian rather than a square time-varying pump pulse $R_p(t)$. Hint: A very useful approximate formula for the error function $\mathrm{erf}(x)$, accurate to $\approx 0.7\%$, is $\mathrm{erf}(x) \approx [1-\exp(-4x^2/\pi)]^{1/2}$.

2. *Ditto with an exponentially varying pump pulse.* Repeat the previous problem for a single-sided exponential pump pulse, i.e., $R_p(t) = (R_{p0}/T_p)\exp(-t/T_p)$ for $t \geq 0$.

3. *Peak population inversion versus normalized pump pulsewidth.* For either of the previous problems, plot the peak inversion that is produced versus the parameter $T_p/\tau_2$. For the exponentially varying pump pulse, show that you must make the pump pulse time constant a *lot* shorter than the upper-level decay time if you want the pumping efficiency to come anywhere close to unity.

4. *Ruby laser gain analysis including degeneracy and $R_1$–$R_2$ level splitting.* In a more exact model of the ruby laser than given in the text, the ground-energy level $E_1$ is found to have a total degeneracy of $g_1 = 4$. The upper laser level $E_2$ is split into two so-called $R_1$ and $R_2$ levels, separated by $\Delta E = E_{2b} - E_{2a} = 39$ cm$^{-1}$ with separate degeneracies $g_{2a} = g_{2b} = 2$. The relaxation between these two levels is extremely rapid, and so we can always assume that the relative populations of the two levels will remain fixed in the appropriate Boltzmann ratio $\exp(-\Delta E/kT)$ corresponding to room temperature, even during laser pumping and laser oscillation. Since this Boltzmann ratio is not negligible, the $R_2$ population will be smaller than the lower-lying $R_1$ population by a significant amount, and the $R_1$ rather than the $R_2$ transition always oscillates unless special steps are taken.

Taking these splittings and degeneracy factors into account, but otherwise making the same idealized three-level assumptions as in the text, calculate the effective population difference from the $R_1$ level to the ground level as a function of pumping power. Also, develop an expression for inverted gain after a pumping pulse compared to absorption before the pumping pulse analogous to the expression given in the text, but with the degeneracy and Boltzmann effects taken into account.

# 7
# LASER AMPLIFICATION

In this chapter we begin examining the other side of the laser problem—that is, what laser atoms do to applied signals, rather than what applied signals do to atoms. This chapter and the following chapter are concerned primarily with continuous-wave or "cw" laser amplification: how inverted atomic transitions amplify optical signals; what determines the magnitude and bandwidth of this gain; how it saturates; and what phase shifts are associated with it. In Chapters 9 and 10 we will consider pulse propagation and pulsed laser amplification; and then in Chapters 11 and 12 we will add the laser mirrors to these amplifying atoms, and (finally!) be able to discuss laser oscillation and the generation of coherent laser radiation.

## 7.1 PRACTICAL ASPECTS OF LASER AMPLIFIERS

Let us begin with a few words about the practical interest in lasers as optical amplifiers, rather than as oscillators. Single-pass (and sometimes double-pass) laser amplifiers are used in many practical situations, primarily as power amplifiers, and seldom if ever as weak-signal preamplifiers. The reasons for this are generally the following.

### Laser Power Amplifiers

Large laser devices very often face severe stability problems associated with large electrical power inputs, optical damage problems, mechanical vibrations, cooling and heat-dissipation problems, acoustic noise, and other sources of what in the Soviet literature is called "technical noise." One common way to obtain high laser power output, simultaneously with good beam quality, short pulse length, excellent frequency stability, and good beam control, is to generate a stable input laser signal from a small but well-controlled laser oscillator. This signal can then be amplified through a chain of laser amplifiers, in what is commonly known as a master-oscillator-power-amplifier or MOPA system. Figure 7.1 shows, as one rather extreme example, the sequence of parallel cascaded Nd:glass laser amplifiers used in a giant laser fusion system, in which a four-story-high "space frame" supports some twenty parallel Nd:glass laser amplifier chains. Very

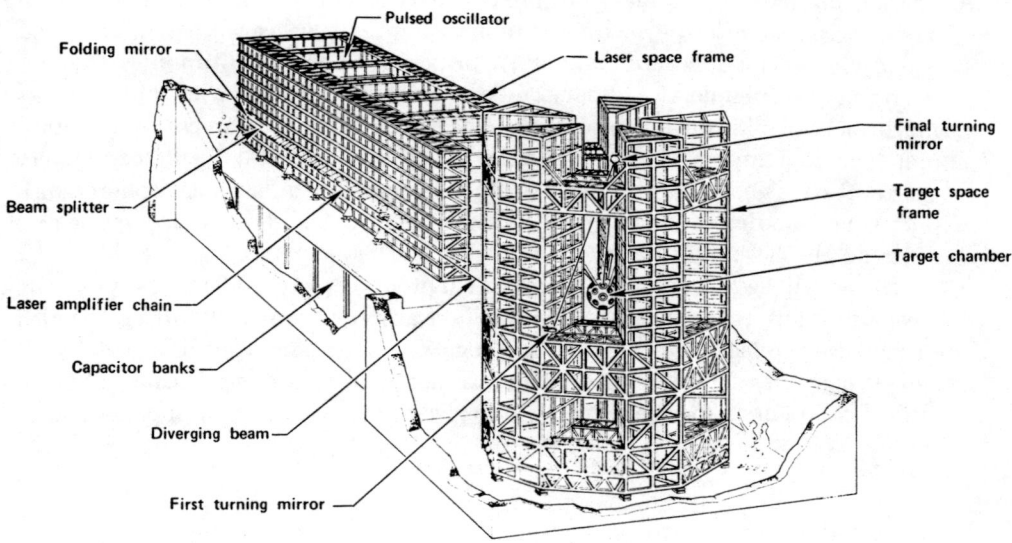

FIGURE 7.1
Laser space frame used with very large laser fusion system at Lawrence Livermore National Laboratory. The four-story-high space frame supports twenty parallel chains of cascaded Nd:glass laser amplifiers.

large high-power $CO_2$ laser amplifiers are also used in laser fusion experiments; and much smaller but very high-gain pulsed dye laser amplifiers and pulsed solid-state laser amplifiers are used to amplify tunable dye laser pulses or mode-locked solid-state laser pulses in many laboratory experiments.

The output signals in these devices will probably be much more stable than if the same laser amplifiers were converted into a single very high power laser oscillator. The primary defect in the MOPA approach, as we will see later, is that it is generally much less effective in extracting the available power in the large amplifier devices than if the large amplifiers were themselves converted into powerful but hard-to-control large oscillators.

### Lasers as Weak Signal Amplifiers

Laser amplifiers are in fact almost always used as power amplifiers, especially for pulsed input signals, almost never as preamplifiers to amplify weak signals in optical receivers, for reasons related primarily to noise figure, and secondarily to the narrow bandwidth of most laser amplifiers.

There exists, as a result of fundamental quantum considerations, an unavoidable quantum noise level in any type of coherent or linear amplifier, at any frequency, whether it be laser, maser, transistor, or vacuum tube. (By a coherent amplifier we mean one which preserves all the phase and amplitude information in the amplified signal; this includes any kind of linear coherent heterodyne detection.) This quantum noise level is roughly equivalent to one noise photon per second per unit bandwidth at the input to the amplifier.

The physical source of this quantum noise in a laser amplifier is the unavoidable spontaneous emission in the laser system, from the upper energy level of the laser transition, into the signal to be amplified. There is, however, some equivalent source of spontaneous-emission-like noise in every other linear amplification mechanism, no matter what its physical nature. (If there were not, it would be possible to use such an amplifier to make physical measurements that would violate the quantum uncertainty principle.) This "quantum noise" source is normally negligible at ordinary radio or microwave frequencies, but becomes much more significant at optical frequencies. As a result, any coherent optical amplifier, including a laser amplifier, is generally unsuitable for detecting very weak optical signals. An incoherent detection mechanism, such as a photomultiplier tube, can detect much weaker optical signals, though of course at the cost of losing all phase information contained in the signal.

As we will see in this and following chapters, laser amplifiers also generally have a quite narrow bandwidth, especially if any regenerative feedback is added to increase the laser gain. As a consequence of both noise and bandwidth considerations, therefore, laser amplifiers are not used to any significant extent in optical communications receivers or other weak-signal-detection applications.

### REFERENCES

A useful survey of the state of the art for designing and building Nd:glass laser amplifiers for pulsed laser fusion systems will be found in David C. Brown, *The Physics of High Peak Power Nd:Glass Laser Systems* (Springer-Verlag, 1980).

A good illustration of the noise properties of laser amplifiers is given by R. A. Paananen et al., "Noise measurement in an He-Ne laser amplifier," *Appl. Phys. Lett.* **4**, 149–151 (April 15, 1964). More extensive discussions of spontaneous emission and noise in laser amplifiers and electrical systems generally are given in my earlier books *Introduction to Lasers and Masers* (McGraw-Hill, 1971), Chap. 10; and *Microwave Solid-State Masers* (McGraw-Hill, 1964).

### 7.2 WAVE PROPAGATION IN AN ATOMIC MEDIUM

Our first formal step toward understanding laser amplification will be to analyze the propagation of an ideal plane electromagnetic wave through an atomic medium which may contain laser gain or loss, as well as atomic phase-shift terms and possibly ohmic losses or scattering losses.

#### The Wave Equation in a Laser Medium

We begin with Maxwell's equations for a sinusoidal electromagnetic field at frequency $\omega$. The two basic Maxwell equations are

$$\nabla \times \boldsymbol{E} = -j\omega \boldsymbol{B}, \qquad \nabla \times \boldsymbol{H} = \boldsymbol{J} + j\omega \boldsymbol{D}, \tag{1}$$

where the real vector field $\boldsymbol{\mathcal{E}}(\boldsymbol{r}, t)$ as a function of space and time is given by

$$\boldsymbol{\mathcal{E}}(\boldsymbol{r}, t) = \frac{1}{2}\left[\boldsymbol{E}(\boldsymbol{r})e^{j\omega t} + \text{c.c.}\right] \tag{2}$$

and similarly for all the other quantities. If these fields are in a linear dielectric host medium which has dielectric constant $\epsilon$, and which possibly contains laser atoms as well as ohmic losses, these quantities will also be connected by "constituitive relations" which may be written as

$$B = \mu H, \quad J = \sigma E, \quad \text{and} \quad D = \epsilon E + P_{\text{at}} = \epsilon[1 + \chi_{\text{at}}]E, \tag{3}$$

where $P_{\text{at}}$ and $\chi_{\text{at}}$ represent the contribution of the laser atoms imbedded in the host dielectric medium. The material parameters appearing in these equations include:

- The optical-frequency dielectric permeability $\epsilon$ of the host medium, not counting any atomic transitions due to laser atoms that may be present.
- The magnetic permeability $\mu$ of the host medium (which will be very close to the free-space value $\mu_0$ for essentially all common laser materials at optical frequencies).
- The conductivity $\sigma$, which is included to account for any ohmic losses in the host material.
- The resonant susceptibility $\chi_{\text{at}}(\omega)$ associated with the transitions in any laser atoms that may be present, where this $\chi_{\text{at}}(\omega)$ is defined in the slightly unconventional fashion we introduced in Section 2.4.

We will assume here that the atomic transition in these atoms is an electric-dipole transition, and thus contributes an electric polarization $P_{\text{at}}$ in the medium. A magnetic-dipole atomic transition would contribute instead an atomic magnetic polarization $M_{\text{at}}$ and thus a magnetic susceptibility $\chi_m$ in the $B = \mu H$ expression. The net result in the following expressions would be essentially the same, however, as you can verify for yourself.

Substituting Equations 7.1–7.3 into a vector identity for $\nabla \times \nabla \times E$, and then assuming that $\nabla \cdot E = 0$, gives

$$\begin{aligned}
\nabla \times \nabla \times E &\equiv \nabla(\nabla \cdot E) - \nabla^2 E \\
&= -j\omega\mu \nabla \times H \\
&= -j\omega\mu \left[\sigma + j\omega\epsilon(1 + \chi_{\text{at}})\right] E \\
&= \omega^2 \mu\epsilon \left[1 + \chi_{\text{at}} - j\sigma/\omega\epsilon\right] E.
\end{aligned} \tag{4}$$

We can assume that $\nabla \cdot E = 0$ provided only that the properties of the medium are spatially uniform (see Problems); and for simplicity we can drop the tensor or vector notation for $\chi$ and $E$. This vector equation then reduces to the scalar wave equation

$$\left[\nabla^2 + \omega^2\mu\epsilon\left(1 + \tilde{\chi}_{\text{at}} - j\sigma/\omega\epsilon\right)\right]\tilde{E}(x,y,z) = 0, \tag{5}$$

where $\tilde{E}(x,y,z)$ is the phasor amplitude of any one of the vector components of $E$.

This equation will be the fundamental starting point for the analyses in this chapter. We can immediately note as one important point that the atomic susceptibility term $\tilde{\chi}_{\text{at}} \equiv \chi' + j\chi''$ and the ohmic loss term $-j\sigma/\omega\epsilon$ appear in exactly similar fashion in this expression.

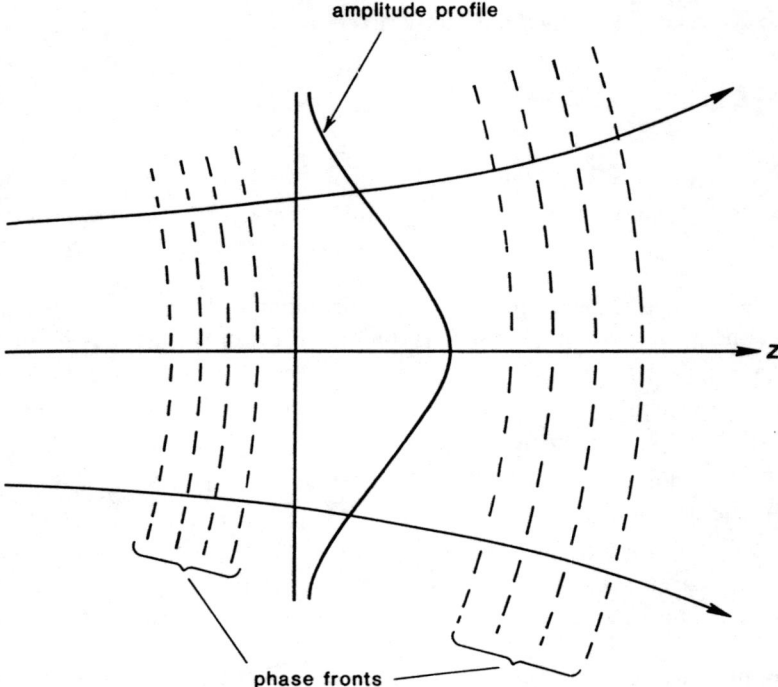

**FIGURE 7.2**
A propagating optical wave with a reasonably smooth transverse amplitude and phase profile.

### Plane-Wave Approximation

Let us consider now either an infinite plane wave propagating in the $z$ direction, so that the transverse derivatives are identically zero, i.e., $\partial/\partial x = \partial/\partial y = 0$, or a finite-width laser beam traveling in the same direction (see Figure 7.2). For a laser beam of any reasonable transverse width (more than a few tens of wavelengths) the transverse derivations will be small enough, and the transverse second derivatives even smaller, so that we can make the approximation

$$\left|\frac{\partial^2 \tilde{E}}{\partial x^2}\right|, \left|\frac{\partial^2 \tilde{E}}{\partial y^2}\right| \ll \left|\frac{\partial^2 \tilde{E}}{\partial z^2}\right| \tag{6}$$

(we will justify this in more detail in the following section).

With this approximation, Equation 7.5 will reduce to the one-dimensional scalar wave equation

$$\left[\frac{d^2}{dz^2} + \omega^2 \mu \epsilon \left(1 + \tilde{\chi}_{\text{at}} - j\sigma/\omega\epsilon\right)\right] \tilde{E}(z) = 0. \tag{7}$$

We will also for simplicity drop the subscripts on the atomic susceptibility $\tilde{\chi}_{\text{at}}$ from here on.

### Lossless Free-Space Propagation

Let us now consider the traveling-wave solutions to this equation, first of all without any ohmic losses or laser atoms. We will generally refer to this as "free-

## 7.2 WAVE PROPAGATION IN AN ATOMIC MEDIUM

space" propagation, although we are in fact including the electric and magnetic permeabilities $\epsilon$ and $\mu$ of the host dielectric medium if there is one.

For $\tilde{\chi}_{at} = \sigma = 0$ the one-dimensional wave equation reduces to

$$[d^2/dz^2 + \omega^2 \mu \epsilon]\, \tilde{E}(z) = 0. \tag{8}$$

If we assume traveling-wave solutions to this equation of the form

$$\tilde{E}(z) = \text{const} \times e^{-\Gamma z}, \tag{9}$$

where $\Gamma$ is a complex-valued propagation constant, then the wave equation reduces to

$$\left[\Gamma^2 + \omega^2 \mu \epsilon\right] \tilde{E} = 0. \tag{10}$$

The allowed values for the complex propagation factor $\Gamma$ are thus given by

$$\Gamma^2 = -\omega^2 \mu \epsilon \quad \text{or} \quad \Gamma = \pm j\omega\sqrt{\mu\epsilon} = \pm j\beta, \tag{11}$$

where the quantity $\beta \equiv \omega\sqrt{\mu\epsilon}$ is the plane-wave propagation constant in the host medium.

The complete solution for the $\mathcal{E}$ field in the medium may thus be written as

$$\begin{aligned}\mathcal{E}(z,t) =& \frac{1}{2}\left[\tilde{E}_+ e^{j(\omega t - \beta z)} + \tilde{E}_+^* e^{-j(\omega t - \beta z)}\right] \\ &+ \frac{1}{2}\left[\tilde{E}_- e^{j(\omega t + \beta z)} + \tilde{E}_-^* e^{-j(\omega t + \beta z)}\right]. \end{aligned} \tag{12}$$

In this expansion the first line on the right-hand side represents a wave traveling to the right (i.e., in the $+z$ direction) with a complex phasor amplitude $\tilde{E}_+$, and the second line represents a wave traveling to the left with phasor amplitude $\tilde{E}_-$. The student should be sure that the distinction between these two waves is clear and well-understood.

The "free-space" propagation constant $\beta$ for these waves may then be written in any of the various alternative forms:

$$\begin{aligned}\beta = \omega\sqrt{\mu\epsilon} &= \frac{\omega}{c} = \frac{n\omega}{c_0} \\ &= \frac{2\pi}{\lambda} = \frac{2\pi n}{\lambda_0}, \end{aligned} \tag{13}$$

where the refractive index $n$ of the host crystal is given by

$$n \equiv \sqrt{\mu\epsilon/\mu_0\epsilon_0} \approx \sqrt{\epsilon/\epsilon_0} \quad \text{if} \quad \mu \approx \mu_0. \tag{14}$$

Note again that in the notation used in this text, $c_0$ and $\lambda_0$ are the velocity of light and the wavelength of the radiation in vacuum, whereas $c \equiv c_0/n$ and $\lambda \equiv \lambda_0/n$ always indicate the corresponding values in the dielectric medium. When we identify particular laser transitions we normally give the value of the wavelength in air, e.g., $\lambda_0 = 1.064\ \mu$m for the Nd:YAG laser. (Note also that in very precise calculations there will even be a slight difference, typically on the order of $\sim 0.03\%$, between the exact vacuum wavelength of a transition and the commonly measured value of the wavelength in air.)

### Propagation With Laser Action and Loss

Let us now include laser action (i.e., an atomic transition) and also ohmic losses in the wave propagation calculation. The one-dimensional wave equation then becomes

$$\left[\frac{d^2}{dz^2} + \beta^2\left(1 + \tilde{\chi}_{\text{at}} - j\sigma/\omega\epsilon\right)\right]\tilde{E}(z) = 0 \quad (15)$$

If we assume a $z$-directed propagation factor $\Gamma$ in the same form as before, this propagation factor now becomes

$$\Gamma^2 = -\omega^2\mu\epsilon\left[1 + \tilde{\chi}_{\text{at}} - j\sigma/\omega\epsilon\right] = -\beta^2\left[1 + \tilde{\chi}_{\text{at}} - j\sigma/\omega\epsilon\right] \quad (16)$$

or

$$\Gamma = j\beta\sqrt{1 + \tilde{\chi}_{\text{at}} - j\sigma/\omega\epsilon} = j\beta\sqrt{1 + \chi'(\omega) + j\chi''(\omega) - j\sigma/\omega\epsilon}. \quad (17)$$

We include the specific dependence of $\chi'(\omega)$ and $\chi''(\omega)$ on frequency to emphasize that, at least for atomic transitions, this quantity will normally be complex and will have a resonant lineshape, with frequency-dependent real and imaginary parts.

Under almost all practical conditions, both the susceptibility $\tilde{\chi}_{\text{at}}(\omega)$ and the loss factor $\sigma/\omega\epsilon$ will have magnitudes that are $\ll 1$. Hence the square root in Equation 7.17 can, with negligible error, be expanded in the form $\sqrt{1+\delta} \approx 1 + \delta/2$ to give

$$\Gamma \approx j\beta \times \left[1 + \frac{1}{2}\chi'(\omega) + j\frac{1}{2}\chi''(\omega) - j\sigma/2\omega\epsilon\right]. \quad (18)$$

From here on we will separate this into the four individual terms

$$\begin{aligned}\Gamma(\omega) &= j\beta + j\beta\chi'(\omega)/2 - \beta\chi''(\omega)/2 + \sigma/2\epsilon c \\ &= j\beta + j\Delta\beta_m(\omega) - \alpha_m(\omega) + \alpha_0,\end{aligned} \quad (19)$$

where each of the factors on the first line matches up with the corresponding factor on the second line. The propagation of a $+z$ traveling wave thus takes on the form

$$\mathcal{E}(z,t) = \text{Re}\,\tilde{E}_0\exp\{j\omega t - j[\beta + \Delta\beta_m(\omega)]z + [\alpha_m(\omega) - \alpha_0]z\} \quad (20)$$

when the effects of ohmic losses and an atomic transition are included.

### Propagation Factors

The significant factors in this complex wave propagation behavior are the following.

1. *The basic plane wave propagation constant.* This is the *basic wave propagation coefficient* $\beta$ in the host medium, which is given by

$$\beta = \beta(\omega) = \omega\sqrt{\mu\epsilon} = \omega/c \quad (21)$$

This propagation constant leads to a fundamental phase variation $\phi(z,\omega) \equiv \beta z = \omega z/c = 2\pi z/\lambda$. This phase shift with distance is large (many complete

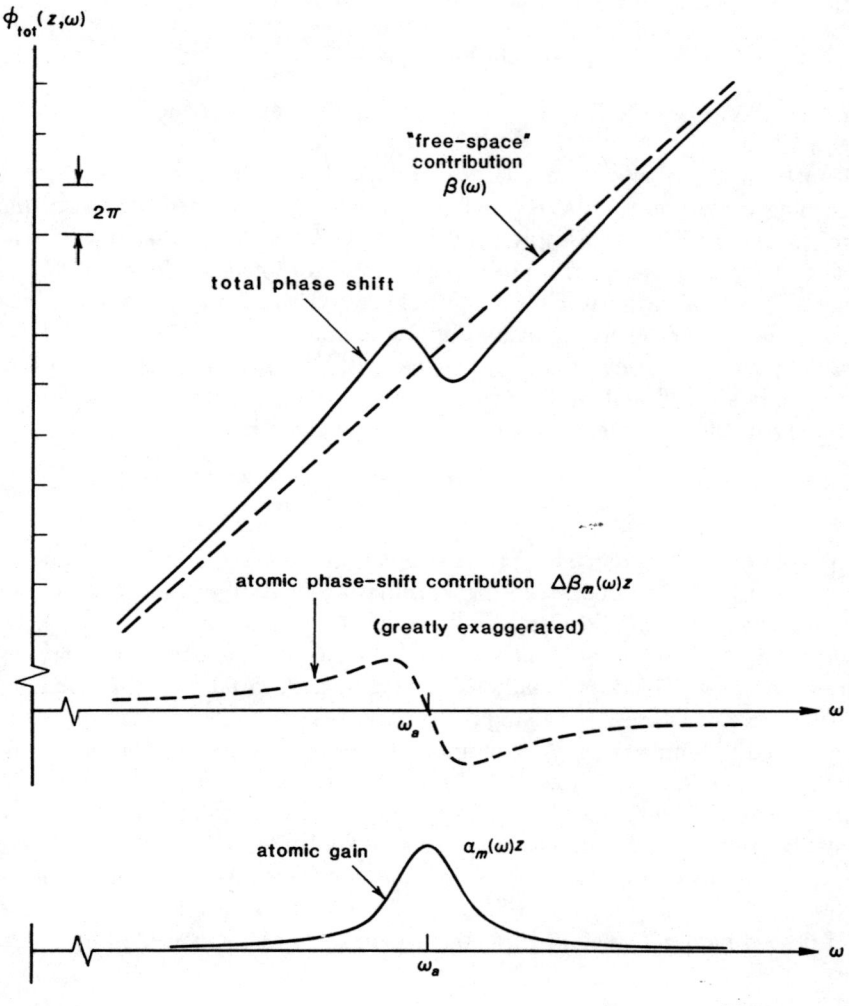

FIGURE 7.3
Phase shift and gain versus frequency in a resonant atomic medium. The atomic phase shift contribution is greatly exaggerated.

cycles) for any propagation length $z \gg \lambda$, and increases linearly (and rapidly) with frequency as shown by the dashed straight line in Figure 7.3.

2. *The additional atomic phase shift.* There is an added phase-shift factor $\Delta\phi(z,\omega) \equiv \Delta\beta_m(\omega)z$ due to the atomic transition, where $\Delta\beta_m(\omega)$ is given by

$$\Delta\beta_m = \Delta\beta_m(\omega) = (\beta/2)\chi'(\omega). \tag{22}$$

This phase shift is caused by and has essentially the same lineshape as the reactive part of the atomic susceptibility, $\chi'(\omega)$, as illustrated by the additional asymmetric contribution to the total phase shift in Figure 7.3.

Note that the sign of this term depends on the sign of the population difference $\Delta N$, just as does the atomic gain or loss coefficient $\alpha_m$. We have drawn the phase shift in Figure 7.3 assuming an inverted or amplifying population difference.

3. *The atomic gain or loss coefficient.* There is an *atomic gain (or loss) coefficient* $\alpha_m(\omega)$ *due to the atomic transition*, given by

$$\alpha_m = \alpha_m(\omega) = (\beta/2)\chi''(\omega). \tag{23}$$

This gain (or loss) has the lineshape of $\chi''(\omega)$ as illustrated in the bottom curve of Figure 7.3.

We noted in an earlier chapter that if we followed the definition $\tilde{\chi}_{at} \equiv \chi' + j\chi''$, an absorbing transition produced a negative value of $\chi''$. We see here also that an amplifying transition will imply positive values for both $\chi''$ and $\alpha_m$, but an absorbing atomic transition will imply negative values for both these quantities. Of course, we can always associate a suitable $\pm$ sign with $\alpha_m(\omega)$ to give it the proper sign for either absorbing or amplifying media.

4. *The ohmic or background loss coefficient.* Finally, there is an *ohmic or background loss coefficient* $\alpha_0$ due to the host medium itself. For pure ohmic conductivity in the host medium, this loss term is given by

$$\alpha_0 = \frac{\beta}{2}\frac{\sigma}{\omega\epsilon} = \frac{\sigma}{2\epsilon c}. \tag{24}$$

We will extend the interpretation of the coefficient $\alpha_0$ in later equations, however, to represent any kind of broadband, background absorption or loss that may be present for the signal in the laser medium, whether this loss is due to ohmic conductivity in the host crystal, or to other loss mechanisms such as scattering or diffraction losses. This loss usually has no significant variation with frequency across the range of interest for a single laser transition.

The preceding four expressions summarize the laser amplification or atomic absorption properties, as well as the phase-shift properties, of any real atomic medium. Recall that in cases of interest to us $\tilde{\chi}_{at}(\omega)$ is virtually always caused by a very narrow resonant transition, with bandwidth $\ll 1\%$. Hence the linear frequency dependence of $\beta(\omega)$ across the narrow linewidth of $\tilde{\chi}_{at}(\omega)$ can be neglected in the $(\beta/2)\chi'(\omega)$ and $(\beta/2)\chi''(\omega)$ products, and only the midband value of $\beta$ need be used. Each of these terms will show up in more detail in later sections.

### Experimental Example

A set of measurements of absorption and phase shift made by Bean and Izatt on the 694 nm laser transition in ruby, without pumping or laser inversion, will give a particularly clean and striking experimental confirmation of the results we have just derived.

Let us recall that the laser transition in the ruby energy-level system terminates on the ground level, so that this transition will have a strongly absorptive population difference in the absence of any laser pumping. We have also noted earlier that the $^4A_2$ ground state of the $Cr^{3+}$ ion in ruby is actually two energy levels which are split, even in zero magnetic field, into two closely spaced sublevels separated by $\Delta E = 0.38$ cm$^{-1}$ = 11.4 GHz, as illustrated in Figure 7.4. (Each of these two sublevels is in fact also a doublet, which can be further split into two Zeeman levels using a dc magnetic field of a few hundred to a few thousand gauss.)

At liquid-nitrogen temperature the phonon broadening in a good sample of ruby becomes small enough ($\Delta\omega_a \leq 2\pi \times 6$ GHz) that the separate absorption lines from the two ground levels can be clearly resolved in the optical absorption

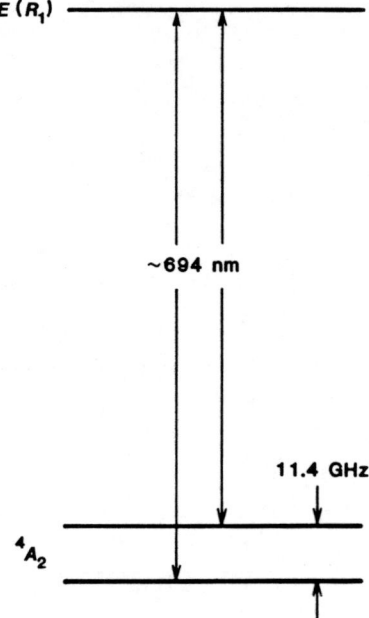

FIGURE 7.4
Absorption transitions near 694 nm from the two ground-state sublevels in ruby to the first excited level.

spectrum of ruby. Bean and Izatt have in fact made careful measurements of the transitions from this split ground state to the first excited or $R_1$ level in a ruby sample, measuring both the absorption coefficient $\alpha_m(\omega)$ versus frequency, which is directly proportional to $\chi''(\omega)$, and the change in index of refraction $\Delta n(\omega)$ relative to the background index $n_0$, which is directly proportional to $\chi'(\omega)$. Typical results of their experiments are shown in Figure 7.5.

These independent measurements of $\chi''(\omega)$ and $\chi'(\omega)$ can then be fitted very closely by simply summing two partially overlapping complex lorentzian lineshapes, as illustrated both in Figure 7.5 and in the double-lorentzian curves in Figure 7.6. These latter curves represent the sum of two elementary lorentzian lines with a relative peak amplitude of 1.29 to 1, a resonance frequency spacing $\omega_{a2} - \omega_{a1} = 2\pi \times 11.5$ GHz, and equal linewidths $\Delta\omega_a = 2\pi \times 5.88$ Ghz.

The close agreement between theory and experiment that is obtained here demonstrates both the validity of the lorentzian lineshape analysis and the close relationship between the $\chi'(\omega)$ and $\chi''(\omega)$ parts of the atomic response.

### Larger Atomic Gain or Absorption Effects

The analytical results in this section (and indeed in most of the rest of this book) are based on the approximation that $|\tilde{\chi}_{at} - j\sigma/\omega\epsilon| \ll 1$. There are in fact only a few optical situations where this approximation is not valid, and where the related Taylor approximation for the complex propagation constant $\Gamma$ will no longer be valid. These include:

1. *Absorption in metals and semiconductors.* For propagation into a semiconductor or a metal (or reflection from their surfaces) at wavelengths shorter than the band edge, or frequencies $\hbar\omega$ greater than the bandgap energy $E_g$, the effective conductivity $\sigma$ and the $-j\sigma/\omega\epsilon$ term can become very large. Exact expressions for both the propagation factor $\Gamma$ and the wave impedance must

274    CHAPTER 7: LASER AMPLIFICATION

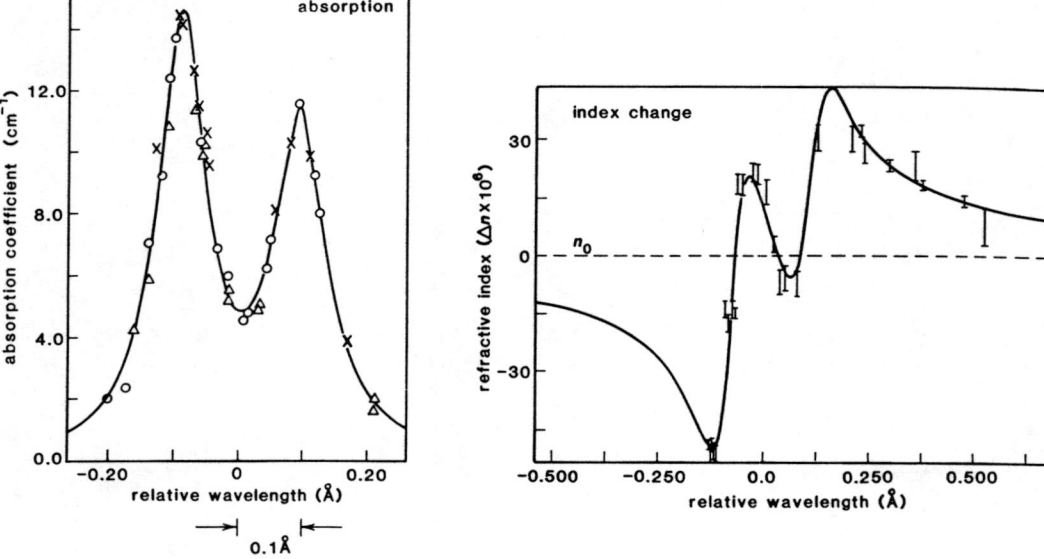

FIGURE 7.5
Measured absorption coefficient and index change for the ground-state absorption line in a ruby crystal at 95 K. (Adapted from B. L. Bean and J. R. Izatt, *J. Opt. Soc. Am.* **63**, 832–839, July 1973.)

then be employed to calculate absorption coefficients and phase shifts (as well as surface reflectivities).

2. *Absorption on strong resonance lines in metal vapors.* Another and more interesting situation where the atomic susceptibility term $\tilde{\chi}$ can become quite large compared to unity is ground-state absorption on the very strong visible or near-UV resonance lines of alkali metal vapors, such as sodium or rubidium, or other metal vapors such as Hg or Cd, at vapor pressures of a few torr or even lower. One particularly common example of this is the pair of sodium D lines at 589.0 and 589.6 nm in the green portion of the visible spectrum. The special features in these situations are that the transitions are very strongly allowed (with oscillator strengths approaching unity); they are relatively narrow, being broadened by doppler broadening only; and they are all ground-state absorption lines, so that all the atoms are in the lower level of the absorbing transition.

As a result, the absorption per unit length at line center on one of these transitions can be extremely large. For example, at the inside surface of a window in a cell containing a moderate vapor pressure of Na or Rb or Hg, the vapor will be so highly absorbing that it will appear essentially metallic and very highly reflecting. This will hold true, however, only within the very narrow range of frequencies within the atomic linewidth (typically a few GHz).

Interesting experiments on optical propagation and atomic transition phenomena can often be done in such vapors, using tunable dye lasers to tune at or very close to these transitions. The practical applications of these phenomena are somewhat limited, however, by the narrow bandwidths, and also by the voracious appetite of the alkali metal vapors for consuming and destroying almost

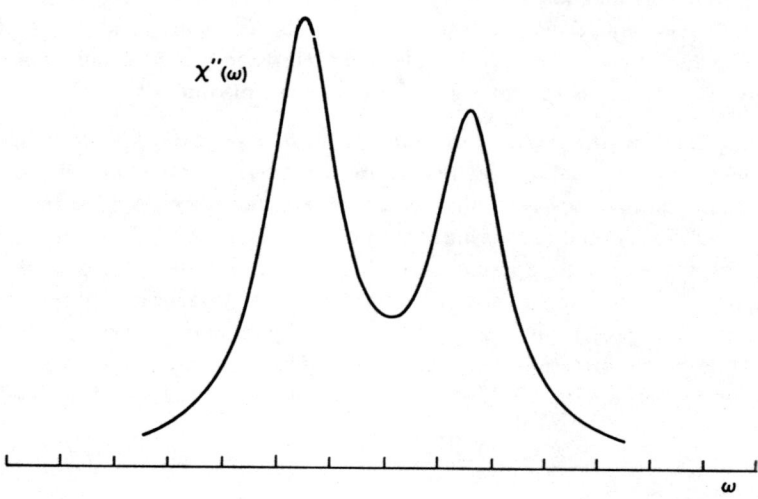

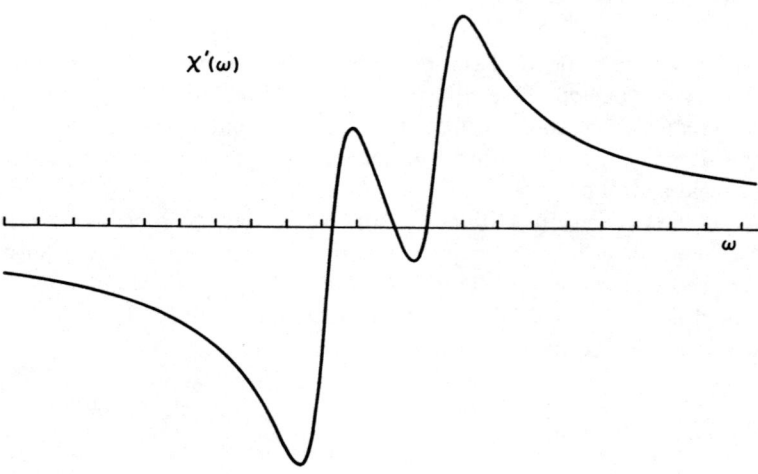

FIGURE 7.6
The summation of two complex lorentzian lines with slightly shifted center frequencies closely matches the experimental data of Figure 7.5.

any conveniently available transparent window materials (not to mention seals and even the metal walls of the vapor cells).

---

Problems for 7.2

1. *Lineshapes for absorption and phase shift in ruby.* Reproduce the theoretical double-lorentzian absorption and phase-shift curves shown for the ruby example in this section.

2. *The $\nabla \cdot \boldsymbol{E} = 0$ approximation in deriving the wave equation.* Disputatious students always question the wave-equation approximation that $\nabla \cdot \boldsymbol{E} = 0$, since the rigorous form of Gauss's law is actually $\nabla \cdot \boldsymbol{D} = \rho$, where $\rho$ is the free charge

that may be present. The free charge $\rho$ and convection current $\boldsymbol{J}$ are, however, also connected by an equation of continuity which says that $\nabla \cdot \boldsymbol{J} + \partial \rho/\partial t = 0$. Using these two equations plus the assumption of an ohmic conductivity, i.e., $\boldsymbol{J} = \sigma \boldsymbol{E}$, show that $\nabla \cdot \boldsymbol{E} = 0$ is indeed strictly valid provided only that the quantity $(\tilde{\chi} - j\sigma/\omega\epsilon)$ is spatially uniform within the medium.

3. *Extending the Taylor approximation to higher-order terms in high-loss materials.* Consider a lossy medium having a finite conductivity $\sigma$ but, for purposes of this problem, no laser susceptibility $\tilde{\chi}_{at}(\omega)$. First-order expressions for the propagation coefficient $\beta$ and the attenuation coefficient $\alpha$ for this case are derived in the text by a Taylor-series approximation in $\sigma/\omega\epsilon$. Extend this approximation to get the next higher-order corrections to both $\beta$ and $\alpha$. How large will the power attenuation have to become (in the first-order approximation) before either of these higher-order corrections amounts to 10% of the first-order expressions? Express your answer in units of dB of power attenuation per wavelength of distance traveled.

## 7.3 THE PARAXIAL WAVE EQUATION

The next step in accuracy beyond the plane-wave approximation of Section 7.2 is the *paraxial wave equation*. This equation, which we will derive in this section, in fact leads to exactly the same results for axial propagation as in Section 7.2, but also makes it possible to handle transverse variations and diffraction effects of the optical beam profile.

The paraxial wave equation is, in fact, complete enough to describe essentially all laser amplification and laser propagation problems of practical interest in lasers; so it is used in a wide variety of laser and nonlinear optical calculations. It seems worthwhile therefore to derive the paraxial equation at this point, even though we will not need to use it until later in this book.

### Paraxial Wave Derivation

The full vector form of the wave equation from Section 7.2 is

$$\left[\nabla^2 + \beta^2(1 + \tilde{\chi} - j\sigma/\omega\epsilon)\right] \boldsymbol{E}(x,y,z) = 0, \tag{25}$$

where $\beta$ is the plane-wave propagation constant in the host medium, disregarding losses and/or atomic transitions. Suppose we now write any given vector component of this complex $\boldsymbol{E}$ vector in the form

$$\tilde{E}(x,y,z) \equiv \tilde{u}(x,y,z)e^{-j\beta z}. \tag{26}$$

This says that the field $\tilde{E}(x,y,z)$ is basically a traveling wave of the form $\exp(-j\beta z)$ in the $+z$ direction. (We would, of course, write this as $e^{+j\beta z}$ if the wave were traveling instead in the $-z$ direction; so reversing the wave direction is the same thing as reversing the sign of $\beta$ in all the following equations.)

This traveling wave may, however, have a transverse amplitude and phase variation, i.e., a dependence on $x$ and $y$ as contained in $\tilde{u}(x,y,z)$; and this transverse profile $\tilde{u}(x,y,z)$ will in general change slowly with propagation distance $z$ as the wave grows, spreads, and/or changes in shape because of absorption and/or diffraction effects, as illustrated for a typical case in Figure 7.2. The very rapid phase variation $\exp(-j\beta z) = \exp(-j2\pi z/\lambda)$ due to the traveling-wave part of the propagation has, however, been factored out of $\tilde{u}(x, y, z)$.

Putting the above form into the wave equation then yields

$$\nabla^2 \tilde{E} = \left[ \frac{\partial^2 \tilde{u}}{\partial x^2} + \frac{\partial^2 \tilde{u}}{\partial y^2} + \frac{\partial^2 \tilde{u}}{\partial z^2} - 2j\beta \frac{\partial \tilde{u}}{\partial z} - \beta^2 \tilde{u} \right] e^{-j\beta z}. \tag{27}$$

Now, we know in advance (or at least we can verify shortly) that the transverse beam profile $\tilde{u}(x, y, z)$ for any reasonably well-collimated optical beam will change only rather slowly with distance $z$ along the beam. That is, the effects of both diffraction and atomic gain or loss on the beam profile $\tilde{u}(x, y, z)$ will be fairly slow, at least compared with the variation of one complete cycle in phase that occurs in one optical wavelength $\lambda$ because of the $\exp(-j2\pi z/\lambda)$ term. Hence we will make the *paraxial approximation* that the $z$ dependence of $\tilde{u}(x, y, z)$ is particularly slow, especially in its second derivative, so that

$$\left| \frac{\partial^2 \tilde{u}}{\partial z^2} \right| \ll \left| 2\beta \frac{\partial \tilde{u}}{\partial z} \right| \equiv \frac{4\pi}{\lambda} \left| \frac{\partial \tilde{u}}{\partial z} \right| \tag{28}$$

and also that

$$\left| \frac{\partial^2 \tilde{u}}{\partial z^2} \right| \ll \left| \frac{\partial^2 \tilde{u}}{\partial x^2} \right|, \quad \left| \frac{\partial^2 \tilde{u}}{\partial y^2} \right|. \tag{29}$$

We will show shortly that these approximations can in fact be very well justified for beams of interest in lasers.

Making these approximations then allows us to drop the $\partial^2 \tilde{u}/\partial z^2$ term in the preceding equation, and thus reduce the wave equation to the so-called paraxial form

$$\nabla_t^2 \tilde{u} - 2j\beta \frac{\partial \tilde{u}}{\partial z} + \beta^2 (\tilde{\chi}_{\text{at}} - j\sigma/\omega\epsilon) \tilde{u} = 0, \tag{30}$$

where the laplacian operator in the transverse plane is denoted by

$$\nabla_t^2 \equiv \frac{\partial^2}{\partial x^2} + \frac{\partial^2}{\partial y^2}. \tag{31}$$

This paraxial form is the desired and widely used *paraxial wave equation*.

### Diffraction Effects Versus Propagation Effects

The paraxial wave equation may also be turned around into the equivalent form

$$\frac{\partial \tilde{u}(x, y, z)}{\partial z} = -\frac{j}{2\beta} \nabla_t^2 \tilde{u}(x, y, z) - [\alpha_0 - \alpha_m + j\Delta\beta_m] \tilde{u}(x, y, z), \tag{32}$$

where the loss term $\alpha_0$ and the atomic susceptibility terms $\alpha_m(\omega)$ and $j\Delta\beta_m(\omega)$ are defined exactly as in Section 7.2. Equation 7.32 neatly separates the axial rate of change of the complex wave amplitude $\tilde{u}(x, y, z)$ into two terms: the $\nabla_t^2 \tilde{u}$ term, which represents *diffraction effects*; and the $\alpha_0$, $\alpha_m$ and $j\Delta\beta_m$ terms, which represent *ohmic and atomic gain, loss, and phase shift effects* caused by $\sigma$ and $\tilde{\chi}_{\text{at}}$.

Since these diffraction and gain or phase-shift effects appear in the differential equation as separate and independent terms for the $z$ variation of the beam profile, we can conclude that the atomic gain and phase-shift effects for a finite laser beam are to first order unaffected by diffraction effects, and are the same as for an infinite plane wave; and also that the diffraction effects on such a beam are to first order unaffected by *spatially uniform* atomic gain or phase-shift effects. Note also that the paraxial results for the gain and phase shift $\alpha_m$ and $\Delta\beta_m$ are exactly the same as the plane-wave results derived by expanding the square-root function for $\Gamma$ to first order in $\tilde{\chi}_{at}$ and in $\sigma/\omega\epsilon$. The paraxial approximation invokes essentially the same physical approximation concerning $z$-axis propagation as does the Taylor expansion of the square root in Equation 7.18.

### Validity of the Paraxial Approximation

A simple analytical example may give somewhat more insight into the validity of the paraxial approximation. Many real laser beams have a gaussian transverse profile of the form

$$|\tilde{u}(x)| = \tilde{u}_0 \exp\left(-\frac{x^2}{w^2(z)}\right). \tag{33}$$

where the gaussian spot size $w = w(z)$ is a slowly varying function of axial distance $z$. (The complex field will have some similar transverse phase variation or phase curvature as well, but for simplicity let us leave this out of the following discussion.) The transverse derivatives of this beam profile at any fixed plane $z$ are then given by

$$\frac{1}{\tilde{u}}\frac{\partial \tilde{u}}{\partial x} = \frac{2x}{w^2} \quad \text{and} \quad \frac{1}{\tilde{u}}\frac{\partial^2 \tilde{u}}{\partial x^2} = \left(\frac{2}{w^2} - \frac{4x^2}{w^4}\right) \approx \frac{2}{w^2}. \tag{34}$$

where the final approximation is valid both on the optic axis and over most of the main part of the gaussian beam profile.

Suppose now that the gaussian spot size $w$ equals 1 mm (which is a fairly slender beam), at a visible wavelength of $\lambda = 500$ nm. The term that represents diffraction effects in the paraxial wave equation will then have an approximate numerical magnitude

$$\left|\frac{1}{\tilde{u}}\frac{\partial \tilde{u}}{\partial z}\right| \approx -j\left|\frac{1}{\beta\tilde{u}}\frac{\partial^2 \tilde{u}}{\partial x^2}\right| \approx \frac{\lambda}{\pi w^2} \approx 10^{-1} \text{ m}^{-1}. \tag{35}$$

In other words, this small but rather smooth beam will propagate about 10 meters or so before diffraction effects cause any major change in the beam profile $\tilde{u}(x,y,z)$.

Suppose also that the amplitude gain or loss in the axial direction due to the $\alpha_m$ or $\alpha_0$ terms has an $e$-folding length somewhere between 10 cm and 1 m (which implies a rather large power gain or attenuation, of between 10 and 100 dB/meter). The gain term in the paraxial equation then has a magnitude

$$\left|\frac{1}{\tilde{u}}\frac{\partial \tilde{u}}{\partial z}\right| \approx \alpha_m \approx 1 \text{ to } 10 \text{ m}^{-1}. \tag{36}$$

In this example at least, diffraction spreading occurs somewhat more slowly than amplification.

The normalized first axial derivative $(1/\tilde{u})(\partial\tilde{u}/\partial z)$ that results from either gain or diffraction effects thus occurs at a rate somewhere between $10^{-1}$ and $10^{1}$ m$^{-1}$. The second derivative $(1/\tilde{u})(\partial^2\tilde{u}/\partial z^2)$ in the axial direction will then have a magnitude corresponding to (at most) this rate squared, say,

$$\left|\frac{1}{\tilde{u}}\frac{\partial^2\tilde{u}}{\partial z^2}\right| \approx 10^{-2} \text{ to } 10^{2} \text{ m}^{-2}. \qquad (37)$$

Therefore the axial derivative contribution produced by this second derivative term, which we dropped in deriving the paraxial equation, if expressed in the same fashion as Equation 7.37, would be about

$$\left|\frac{1}{2\beta\tilde{u}}\frac{\partial^2\tilde{u}}{\partial z^2}\right| \approx \frac{\lambda}{4\pi}\left|\frac{1}{\tilde{u}}\frac{\partial^2\tilde{u}}{\partial z^2}\right| \approx 5\times 10^{-10} \text{ to } 5\times 10^{-6} \text{ m}^{-1}. \qquad (38)$$

The normalized second derivative given in Equation 7.38 is thus many orders of magnitude smaller than the other derivative terms given in Equations 7.35 through 7.37. The basic paraxial approximation is clearly very well-justified in this example, even for a wide range of different axial growth rates.

---

### Problems for 7.3

1. *Applying the paraxial-wave approximation to gaussian beam propagation.* A more accurate form for the gaussian transverse profile in real laser beams is $\mathcal{E}(x,y,z) = A(z)\exp[-jk(x^2+y^2)/2\tilde{q}(z)]\exp(-j\beta z)$, where $A(z)$ and $\tilde{q}(z)$ are functions of $z$ only, not of $x$ or $y$. (We will see later that the parameter $\tilde{q}(z)$ is a kind of complex gaussian radius of curvature plus spot size.) Using the paraxial wave equation, including atomic susceptibility and loss terms, find differential equations for $A(z)$ and $\tilde{q}(z)$, and discuss their meaning.

    For example, will $\mathcal{E}(x,y,z)$ remain gaussian as the wave propagates? How do $A(z)$ and $\tilde{q}(z)$ change with distance, and why? Note that the factor $A(z)$ might be replaced by $A(z) \equiv \exp a(z)$, or $a(z) \equiv \ln A(z)$, and one could then get a differential equation for $a(z)$ instead.

---

## 7.4 SINGLE-PASS LASER AMPLIFICATION

Let us look next at some of the practicalities of single-pass, small-signal amplification for a wave passing through an inverted laser medium

### Laser Gain Formulas

If a quasi-plane wave propagates through a length $L$ of laser material, as in Figure 7.7, the complex amplitude gain or "voltage gain" in an inverted laser

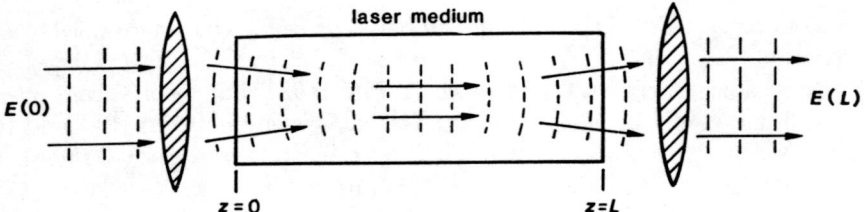

FIGURE 7.7
An elementary single-pass laser amplifier.

medium will be

$$\tilde{g}(\omega) \equiv \frac{\tilde{E}(L)}{\tilde{E}(0)} = \exp\{-j[\beta + \Delta\beta_m(\omega)]L\} \times \exp\{[\alpha_m(\omega) - \alpha_0]L\}. \qquad (39)$$

The first exponent on the right-hand side represents the total phase shift through the amplifier, and the second is the amplitude gain or loss.

Because signal power or intensity $I(z)$ is proportional to $|\tilde{E}(z)|^2$, the single-pass power or intensity gain going through the laser medium is

$$G(\omega) \equiv \frac{I(L)}{I(0)} = |\tilde{g}(\omega)|^2 = \exp[2\alpha_m(\omega)L - 2\alpha_0 L]. \qquad (40)$$

(Note again that in this text symbols like $\alpha_m$ and $\alpha_0$ will always denote gain coefficients or loss coefficients for the field amplitude or "voltage" of a wave; and hence we will always write $2\alpha$ for a power gain coefficient. In other books and papers in the literature, the symbol $\alpha$ by itself often means the power gain or loss coefficient.)

In most useful laser materials the ohmic insertion loss coefficient $\alpha_0$ will be small compared to the laser gain coefficient $\alpha_m$; so for simplicity we will leave out the $2\alpha_0 L$ loss factor in most of the following equations. Also, for many transitions the laser lineshape will be lorentzian, so that the imaginary part of the susceptibility is given by

$$\chi''(\omega) = \frac{\chi_0''}{1 + [2(\omega - \omega_a)/\Delta\omega_a]^2}, \qquad (41)$$

where $\chi_0''$ is the midband value. The power gain $G(\omega)$ then has the frequency lineshape

$$G(\omega) = \exp\left[\frac{\omega L \chi_0''}{c} \times \frac{1}{1 + [2(\omega - \omega_a)/\Delta\omega_a]^2}\right], \qquad (42)$$

where $c$ is the velocity of light in the laser medium. Note that in this gain expression, the lorentzian atomic lineshape appears in the exponent. If the atomic lineshape were inhomogeneous and gaussian, then the gaussian lineshape would similarly appear in the exponent.

The quantity $G(\omega)$ is power gain expressed as a number. To convert this to power gain in decibels, or dB, as often used in engineering discussions, we must use the definition that

$$G_{dB}(\omega) \equiv 10\ \log_{10} G(\omega) = 4.34\ \log_e G(\omega) = \frac{4.34\omega_a L}{c} \chi''(\omega). \qquad (43)$$

Therefore *the power gain measured in dB has the same lineshape as the atomic susceptibility* $\chi''(\omega)$, whether this lineshape is lorentzian, gaussian, or whatever.

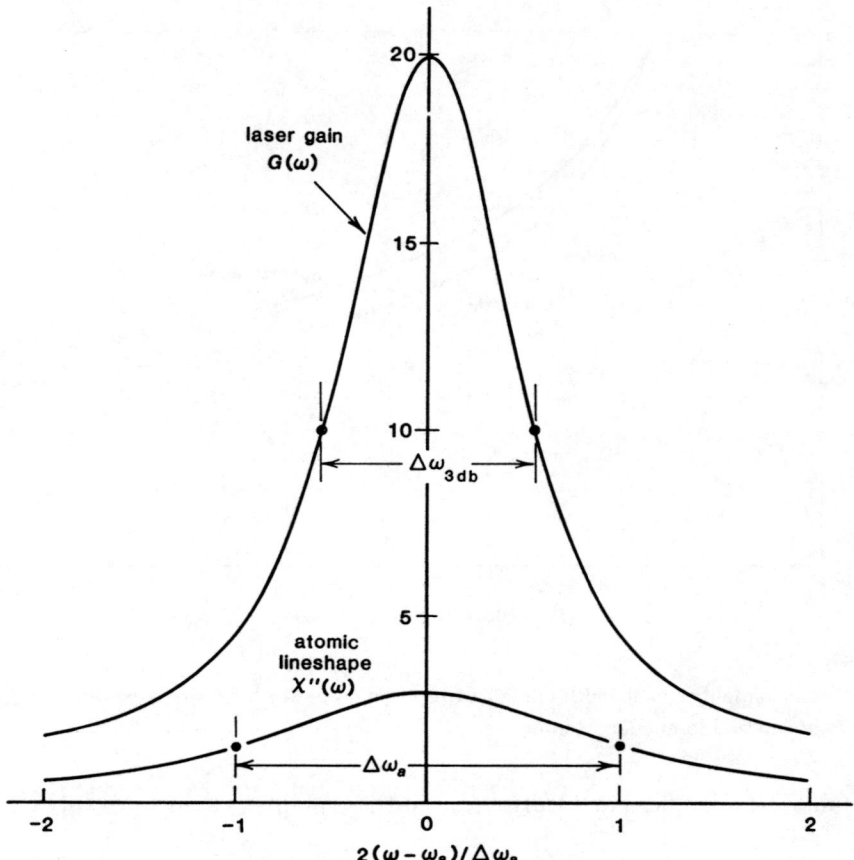

FIGURE 7.8
Gain narrowing in a single-pass laser amplifier.

### Amplification Bandwidth and Gain Narrowing

Because the frequency dependence of $\chi''(\omega)$ appears in the exponent of the gain expression, the exponential gain falls off much more rapidly with detuning than the atomic lineshape itself. The bandwidth of a single-pass laser amplifier is thus generally narrower than the atomic linewidth (see Figure 7.8); and this bandwidth narrowing increases (that is, the bandwidth decreases still further) with increasing amplifier gain.

The conventional definition for the bandwidth of any amplifier is the full distance between frequency points at which the amplifier power gain has fallen to half the peak value. This corresponds to "3 dB down" from the peak gain value in dB, if we recall that $10 \log_{10} 0.5 = -3.01$. For a lorentzian atomic line the amplifier 3 dB points are thus defined as those frequencies $\omega$ for which

$$G_{\mathrm{dB}}(\omega) = \frac{G_{\mathrm{dB}}(\omega_a)}{1 + [2(\omega - \omega_a)/\Delta\omega_a]^2} = G_{\mathrm{dB}}(\omega_a) - 3 \qquad (44)$$

or

$$(\omega - \omega_a)_{3dB} = \pm \frac{\Delta\omega_a}{2} \sqrt{\frac{3}{G_{\mathrm{dB}}(\omega_a) - 3}}. \qquad (45)$$

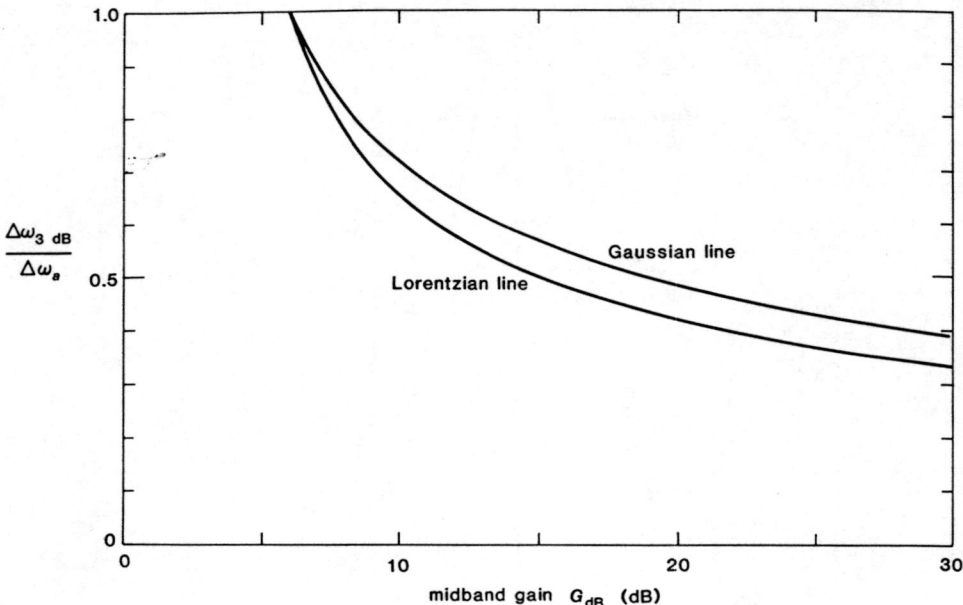

FIGURE 7.9
Reduction of amplifier bandwidth compared to atomic linewidth for single-pass amplifiers using gaussian and lorentzian atomic lines.

The full 3-dB amplifier bandwidth between these points is then twice this value or

$$\Delta\omega_{3dB} = \Delta\omega_a \sqrt{\frac{3}{G_{\text{dB}}(\omega_a) - 3}}. \tag{46}$$

Figure 7.9 plots this amplifier bandwidth, normalized to the atomic linewidth, as a function of the midband gain $G_{\text{dB}}(\omega_a)$ for both lorentzian and gaussian atomic lineshapes.

The 3-dB amplification bandwidth is substantially smaller than the atomic linewidth, dropping to only 30% to 40% of the atomic linewidth at higher gains. This so-called *gain narrowing* at higher gains is significant in reducing the useful bandwidth of a high-gain laser amplifier.

### Amplifier Phase Shift

The total phase shift for a single pass through a laser amplifier can be written as $\exp[-j(\beta + \Delta\beta_m)L] \equiv \exp[-j\phi_{\text{tot}}(\omega)]$, where the total phase shift $\phi_{\text{tot}}(\omega)$ is given by

$$\phi_{tot}(\omega) \equiv \beta(\omega)L + \Delta\beta_m(\omega)L = \frac{\omega L}{c} + \frac{\beta L}{2}\chi'(\omega). \tag{47}$$

The first term gives the basic "free-space" phase shift $\beta(\omega)L = \omega L/c = 2\pi L/\lambda$ through the laser medium. This term is large and increases linearly with increasing frequency. The second term is then the small added shift $\Delta\beta_m(\omega)L$ due to the atomic transition, as illustrated earlier in Figure 7.3.

Note that the magnitude of the added phase shift through a laser amplifier or absorber is directly proportional to the net gain or attenuation through the same atomic medium. For a lorentzian atomic transition we can in fact relate the added phase shift in radians to the amplitude gain (or loss) factor $\alpha_m L$ (the value of which is often said to be measured in units of *nepers*) by the relation

$$\Delta\beta_m(\omega)L = \left(2\frac{\omega - \omega_a}{\Delta\omega_a}\right) \times \alpha_m(\omega)L \quad (48)$$

$$= \frac{G_{\text{dB}}(\omega_a)}{20\log_{10}e} \times \frac{2(\omega - \omega_a)/\Delta\omega_a}{1 + [2(\omega - \omega_a)/\Delta\omega_a]^2}.$$

In practical terms, the peak value of the added phase shift $\Delta\beta_m L$ occurs at half a linewidth, or $\pm\Delta\omega_a/2$, off line center on each side; and the added phase shift in radians at these peaks is related to the midband gain in dB by $(\Delta\beta_m L)_{\text{max}} = G_{\text{dB}}/40\log_{10}e \approx G_{\text{dB}}/17.4$.

### Absorbing Media

The results just discussed are for an amplifying laser medium. The same formulas and physical ideas apply equally well to an absorbing (uninverted) atomic transition, however, if we simply reverse the sign of $\tilde{\chi}_{\text{at}}(\omega)$ and hence of both $\alpha_m(\omega)$ and $\Delta\beta_m(\omega)$. Figure 7.10 plots, for example, the power transmission $T(\omega) = \exp[-2\alpha_m(\omega)L]$ versus frequency through a material with a lorentzian absorbing atomic transition. [In this terminology the power transmission $T(\omega)$ is the same as the power gain $G(\omega)$, but with a magnitude less than unity, not greater than unity.] Note that for very strong absorption the transmission curve "touches bottom" and then broadens with increased absorption strength. An absorbing transition thus has "absorber broadening" rather than the "gain narrowing" discussed earlier (see Problems).

### REFERENCES

A very useful even if rather early work is E. U. Condon and G. H. Shortley, *The Theory of Atomic Spectra* (Cambridge University Press, 1935; reprinted 1963).

An early but still quite clear and detailed discussion of a laser amplifier experiment (using a ruby laser rod) is given by J. E. Geusic and E. O. Schulz-DuBois, "A unidirectional traveling-wave optical maser," *Bell Sys. Tech. J.* **41**, 1371–1397 (July 1962). An early analysis of bandwidth narrowing in a doppler-broadened laser, including saturation effects, is D. F. Hotz, "Gain narrowing in a laser amplifier," *Appl. Opt.* **34**, 527–530 (May 1965).

An interesting experimental verification of strong phase-shift effects in a high-gain, narrow-line laser amplifier can be found in C. S. Liu, B. E. Cherrington, and J. T. Verdeyen, "Dispersion effects in a high-gain 3.39 $\mu$m He-Ne laser," *J. Appl. Phys.* **40**, 3556 (August 1969).

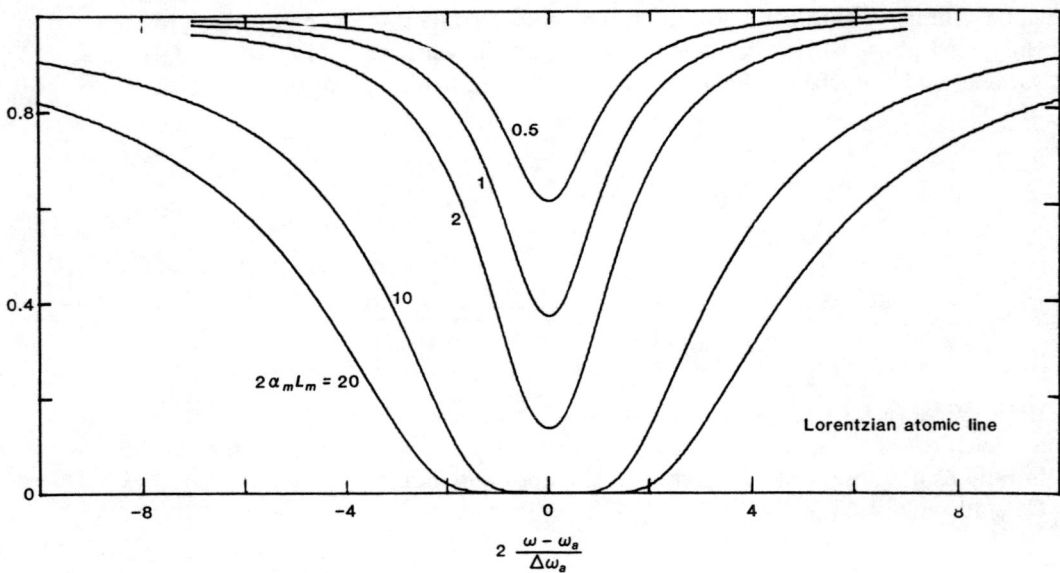

**FIGURE 7.10**
Power transmission versus frequency through an absorbing medium with a lorentzian atomic lineshape, for increasing degrees of midband absorption. (Adapted from E. U. Condon and G. H. Shortley, *The Theory of Atomic Spectra*, Cambridge Univ. Press, 1935, p. 111.)

---

Problems for 7.4

1. *Amplification bandwidth for a gaussian atomic transition.* Derive an analytic expression for the 3-dB bandwidth $\Delta\omega_{3\text{dB}}$ of a single-pass laser amplifier which has a *gaussian* rather than a lorentzian linewidth, as plotted in this section.

2. *Absorption linewidth for an absorbing atomic transition.* Consider the curves of power transmission $T(\omega) = \exp[-2\alpha_m(\omega)L]$ through an atomic medium with a lorentzian resonant transition, plotted versus normalized frequency detuning $(\omega - \omega_a)/\Delta\omega_a$ for various values of the midband absorption factor $2\alpha_m(\omega_a)L$, as shown in this section. Suppose an "absorption linewidth" $\Delta\omega_{\text{abs}}$ is defined as the full width of the power transmission profile $T(\omega)$ at a level halfway down into the dip, i.e., halfway between the midband value $T(\omega_a)$ and the far-off-resonance value $T = 1$. Derive an expression for this linewidth $\Delta\omega_{\text{abs}}$ as a function of the midband absorption $2\alpha_m(\omega_a)L$, and examine its limiting values for very small and very large absorptions.

3. *An alternative bandwidth definition for low-gain amplifiers.* The conventional definition of the 3-dB linewidth as discussed in the text ceases to have meaning for a laser amplifier whose peak gain is less than 3 dB, i.e., when $G(\omega_a) < 2$. Extend the linewidth calculation of the previous problem to the amplifying case (i.e., change the sign of $2\alpha_m L$), and compare the amplifier linewidth calculated in this way to the definition of 3 dB given in the text, assuming the atomic line has a lorentzian lineshape.

4. *Testing for a gaussian atomic lineshape.* How might you plot experimental data on the single-pass gain $G(\omega)$ of a laser amplifier versus frequency detuning $\omega - \omega_a$ to see immediately if the atomic lineshape of the amplifying medium is gaussian?

5. *Gain versus frequency for a cascaded amplifier plus absorber.* A tunable optical signal is passed through a linear single-pass laser amplifier having midband gain coefficient $\alpha_1 L$ and lorentzian atomic linewidth $\Delta\omega_{a1}$, followed by a linear single-pass laser *absorber*—that is, a collection of absorbing atoms—having midband absorption coefficient $\alpha_2 L$ and atomic linewidth $\Delta\omega_{a2}$, with both transitions centered at the same resonance frequency and with $\alpha_1 > \alpha_2$. What condition on the relative atomic linewidths $\Delta\omega_{a1}$ and $\Delta\omega_{a2}$ will just lead to a double-humped rather than single-humped curve of overall power gain versus frequency for the two atomic systems in cascade? (Hint: You can solve this problem by differentiating the power-transmission expression a couple of times, but there's an easier approach also.)

6. *Continuation of the previous problem: general evaluation of passband broadening in a laser amplifier.* Consider the approach outlined in Problem 5 in more detail, as a possible method of broadening the amplification bandwidth of a single-pass laser amplifier. Give a short analysis and summary of the passband broadening that might be obtained, what this costs in midband gain reduction and in gain variation across the passband, and what conditions on the absorber are required in some typical cases. Note: The allowable gain variation between the two peaks and the midband value in a practical amplifier depends on the application for which the amplifier is to be used; but usually cannot exceed somewhere between 1 dB and 3 dB of peak-to-peak ripple.

7. *Continuation of the previous problem: relationship between midband gain and phase shift derivatives?* In Problem 6, what relationship if any is there between having a maximally flat gain profile at line center and the slope of the phase variation $\Delta\beta_m(\omega)L$ versus $\omega$ at line center (where $\Delta\beta_m$ includes both the amplifier and absorber phase-shift contributions)?

8. *"Linewidth modulation spectroscopy": a new experimental technique.* A low-frequency pressure modulation or mechanical squeezing, when applied to certain organic host crystals, will modulate the atomic linewidth $\Delta\omega_a$ of an absorbing transition in the organic crystal by a small amount about its average value, without changing any other parameters of the transition. Suppose we modulate the linewidth $\Delta\omega_a$ in this fashion by a very small amount at some low modulation frequency, perhaps in the audio range, and then measure the resulting ac modulation of the transmitted intensity of an optical signal transmitted through the absorbing medium, while we slowly scan the optical frequency $\omega$ of the signal across the absorption line. The resulting ac signal in the detector will be proportional to the first derivative $d\chi''/d\Delta\omega_a$ at each frequency $\omega$ across the absorption profile. (This general type of technique, in which we slowly scan the optical measuring frequency $\omega$ across a line, while modulating some parameter of the line at a low modulation frequency and measuring the resulting ac output, is often called *modulation spectroscopy*.)

Derive and plot the lineshape that we will see for the magnitude and phase of this low-frequency modulation signal versus the optical frequency $\omega$, and give a brief physical explanation for its form.

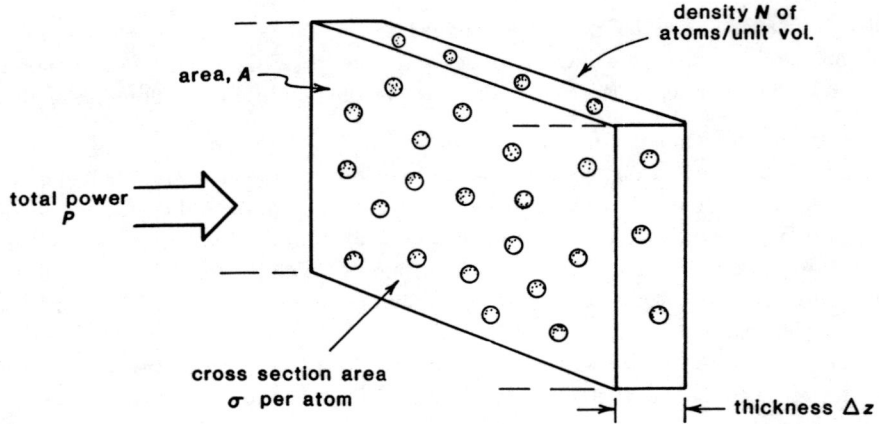

FIGURE 7.11
A collection of atoms with small but finite absorption cross sections distributed throughout a thin slab.

## 7.5 STIMULATED-TRANSITION CROSS SECTIONS

A very useful concept which we will next introduce for describing both stimulated transitions in absorbing atoms and laser amplification in laser media is the *stimulated-transition cross section* of a laser atom. If you have encountered the idea of a cross section previously in some other connection, you will find this concept straightforward; if not, you may have to pay close attention at first.

### Absorption and Emission Cross Sections

Suppose a small black (i.e., totally absorbing) particle with a capture area or "cross section" $\sigma$ is illuminated by an optical wave having intensity or power per unit area $I \equiv P/A$. The net power $\Delta P_{\text{abs}}$ absorbed by this object from the wave will then be its capture area or cross section $\sigma$ times the incident power per unit area in the wave, or

$$\Delta P_{\text{abs}} = \sigma \times (P/A) = \sigma I. \tag{49}$$

(In this text we attempt to always use $P$ to represent total power in watts, and $I$ to represent intensity in watts per unit area. When given in conversation or in texts, values for laser beam intensities are almost universally expressed in dimensions of watts/cm$^2$, although the correct mks unit for substitution into formulas is watts/m$^2$.)

Consider next a thin slab of thickness $\Delta z$ and transverse area $A$, as in Figure 7.11, containing densities $N_1$ and $N_2$ of atoms in the lower and upper energy levels of some atomic transition. Suppose we say that each lower-level atom has an effective area or cross section $\sigma_{12}$ for power absorption from the wave, and similarly each upper-level atom has an effective cross section $\sigma_{21}$ for "negative absorption" or emission back to the wave (since upper-level atoms must emit power rather than absorb it).

The total number of lower-level atoms in this slab will then be $N_1 A \Delta z$, and the total absorbing area that results from all the lower-level atoms will be the total number of atoms times the cross section per atom, or $N_1 \sigma_{12} A \Delta z$. (We assume the slab is thin enough and the atoms small enough that shadowing of any one atom by other atoms is negligible.) Similarly, the total effective "emitting area" that results from all the upper-level atoms will be $N_2\sigma_{21}A\Delta z$. The net power absorbed by the atoms in the slab from an incident wave carrying a total power $P$ distributed over the area $A$ will then be

$$\Delta P_{\text{abs}} = (N_1\sigma_{12} - N_2\sigma_{21}) \times P\,\Delta z. \tag{50}$$

Note that the area factors in the slab volume $A\,\Delta z$ and in the power density $P/A$ just cancel.

The quantities $\sigma_{12}$ and $\sigma_{21}$ that we have introduced here are the *stimulated-transition cross sections* of the atoms on the $1 \to 2$ transition, with $\sigma_{12}$ being the *stimulated-absorption cross section* and $\sigma_{21}$ the *stimulated emission cross section*. These cross sections, which have dimensions of area per atom, provide a very useful way of expressing the strength of an atomic transition, or the size of the atomic response to an applied signal.

### Cross Sections and Amplification Coefficients

The net growth or decay with distance caused by an atomic transition for a wave carrying power $P$ or intensity $I$ through an atomic medium can then be written as

$$\frac{dP}{dz} = -\lim_{\Delta z \to 0}\left(\frac{\Delta P_{\text{abs}}}{\Delta z}\right) = -(N_1\sigma_{12} - N_2\sigma_{21}) \times P. \tag{51}$$

The relationship between upward and downward cross sections on a possibly degenerate transition is in fact given by $g_1\sigma_{12} = g_2\sigma_{21}$, and so the preceding equation, if converted into units of intensity $I(z)$ and population difference $\Delta N$, can be written as

$$\frac{1}{I}\frac{dI}{dz} = -\Delta N_{12}\sigma_{21} = -[(g_2/g_1)N_1 - N_2]\sigma_{21}, \tag{52}$$

where $\Delta N_{12} = (g_2/g_1)N_1 - N_2$ as usual.

But the growth or decay rate for a wave passing through an absorbing or amplifying atomic medium may also be written as $I(z) = I(z_o)\exp[-2\alpha_m(z-z_0)]$, which corresponds to the differential relation

$$\frac{1}{I}\frac{dI}{dz} = -2\alpha_m(\omega). \tag{53}$$

Hence we obtain the simple but very useful expression for the absorption (or amplification) coefficient $\alpha_m$ on the $1 \to 2$ transition in terms of only the population difference and cross section, namely,

$$2\alpha_m(\omega) = \Delta N_{12}\sigma_{21}(\omega). \tag{54}$$

To specify the loss or gain per unit length on an atomic transition, all we need to know is the atomic density and the transition cross section. Note that the absorption coefficient $\alpha_m(\omega)$ and the cross section $\sigma_{21}(\omega)$ must necessarily have the same dependence on the atomic lineshape; i.e., there is an atomic lineshape contained in $\sigma_{21}(\omega)$, although usually only the numerical value at midband is stated as so many cm$^2$.

In practice the gain coefficient $2\alpha_m$ is commonly expressed in cm$^{-1}$, the density $N$ in atoms/cm$^3$, and the cross section $\sigma$ in cm$^2$/atom, which makes Equation 7.54 dimensionally consistent even without use of mks units. Both $\Delta N$ and $\alpha_m$ will, of course, change sign on an inverted laser transition.

### Formula for the Cross Section

One way to obtain a theoretical expression for the cross section $\sigma$ of a transition is to combine Equation 7.54 with the gain formula 7.23 to obtain

$$2\alpha_m(\omega) = \Delta N \sigma(\omega) = \frac{2\pi}{\lambda} \chi''(\omega)$$

$$= \frac{3^*}{2\pi\lambda} \frac{\Delta N \lambda^3 \gamma_{\rm rad}}{\Delta \omega_a} \times \frac{1}{1 + [2(\omega - \omega_a)/\Delta \omega_a]^2}. \tag{55}$$

The population difference $\Delta N$ can then be canceled from both sides to give the midband result

$$\sigma(\omega_a) = \frac{3^*}{2\pi} \frac{\gamma_{\rm rad}}{\Delta \omega_a} \lambda^2. \tag{56}$$

The more general form of the cross-section expression for an arbitrary $i \to j$ transition, including degeneracy, is

$$\sigma_{ji}(\omega_a) = \frac{g_i}{g_j} \sigma_{ij}(\omega_a) = \frac{3^*}{2\pi} \frac{\gamma_{{\rm rad},ji}}{\Delta \omega_a} \lambda_{ij}^2. \tag{57}$$

These expressions give the midband value of the stimulated-emission cross section for a lorentzian atomic transition. For a gaussian transition, we must replace $\Delta \omega_a$ by $\Delta \omega_d$ and put an additional numerical factor of $\sqrt{\pi \ln 2} \approx 1.48$ in front. Note again that the degeneracy factors appear in such a way that the expression $N_i \sigma_{ij} - N_j \sigma_{ji}$ converts neatly into the form $\Delta N_{ij} \sigma_{ji}$, where we use the degenerate form of the population difference $\Delta N_{ij} = (g_i/g_j) N_i - N_j$ as defined earlier, and where $\sigma_{ji}$ is the cross section in the *downward* direction.

The effective cross section $\sigma_{21}(\omega)$ then decreases off line center with precisely the same lineshape as the absorption susceptibility $\chi''(\omega)$ or the stimulated-transition probability $W_{21}(\omega)$. That is, we may have either the lorentzian expression

$$\sigma(\text{lorentzian}) = \frac{3^*}{2\pi} \frac{\gamma_{\rm rad} \lambda^2}{\Delta \omega_a} \frac{1}{1 + [2(\omega - \omega_a)/\Delta \omega_a]^2} \tag{58}$$

or the gaussian expression

$$\sigma(\text{gaussian}) = \sqrt{\pi \ln 2} \, \frac{3^*}{2\pi} \frac{\gamma_{\text{rad}} \lambda^2}{\Delta \omega_d} \exp\left[-(4\ln 2)\left(\frac{\omega - \omega_a}{\Delta \omega_d}\right)^2\right] \quad (59)$$

corresponding to the homogeneous or inhomogeneous limiting cases.

### Maximum Value of the Transition Cross Section

The stimulated-emission (or absorption) cross section provides a convenient and useful way to express the apparent "size" of an atom for interacting with an optical wave, as well as a convenient way to calculate the expected gain in laser systems. Let us look first at the maximum possible value that any such cross section can have. The cross section will be maximal for a transition that has purely radiative lifetime broadening only, and no other line-broadening effects, so that $\Delta \omega_a \equiv \gamma_{\text{rad}}$. If the atoms all have their transition axes aligned and the incident fields are optimally polarized, so that $3^* \equiv 3$, the cross section is then

$$\sigma_{\max} = \frac{3\lambda^2}{2\pi} \approx \frac{\lambda^2}{2}. \quad (60)$$

This says that the maximum cross section is roughly one wavelength square. For a visible transition this means

$$\sigma_{\max} \approx 0.5 \times (5000\text{Å})^2 \approx 10^{-9} \text{ cm}^2. \quad (61)$$

Now, the actual physical size of an atom, as measured, say, by the radius of its outermost Bohr orbit, is only a few Ångstroms; yet its effective cross section for capturing radiation can be thousands of Ångstroms in diameter. The physical explanation for this is essentially that the atom has an internal resonance which makes it act like a miniature dipole radio antenna, whose effective cross section for receiving radio or optical waves can be very much larger than the physical dimensions of the antenna, or the atom, by itself.

### Real Atomic Cross Sections

For more realistic atoms with realistic line-broadening effects and random orientations ($3^* = 1$), the effective cross section at midband is given by values more like

$$\sigma(\text{lorentzian}) \approx \frac{\lambda^2}{2\pi} \frac{\gamma_{\text{rad}}}{\Delta \omega_a} \quad (62)$$

or by

$$\sigma(\text{gaussian}) \approx \frac{\lambda^2}{4} \frac{\gamma_{\text{rad}}}{\Delta \omega_d}. \quad (63)$$

Note that the wavelength $\lambda$ in these expressions is, as always, the wavelength *in the laser medium*. Table 7.1 gives some typical cross-section values for a few of the more useful laser transitions.

TABLE 7.1
Typical Laser Transition Cross Sections

| Laser system | Transition cross section $\sigma$ |
|---|---|
| Gas lasers in the visible and near IR | $10^{-11}$ to $10^{-13}$ cm$^2$ |
| Low-pressure CO$_2$ laser (10.6 $\mu$m) | $3 \times 10^{-18}$ cm$^2$ |
| Organic dye laser (Rhodamine 6G) | 1 to $2 \times 10^{-16}$ cm$^2$ |
| Nd$^{3+}$ ion in Nd:YAG | $4.6 \times 10^{-19}$ cm$^2$ |
| Nd$^{3+}$ ion in Nd:glass | $3 \times 10^{-20}$ cm$^2$ |
| Cr$^{3+}$ ion in ruby | $2 \times 10^{-20}$ cm$^2$ |

For example, on a strong but doppler-broadened visible laser transition in a gas with oscillator strength $\mathcal{F} \approx 1$, a radiative decay rate $\gamma_{\text{rad}} \approx 10^8$ s$^{-1}$, and a doppler linewidth $\Delta\omega_d/2\pi \approx 2 \times 10^9$ Hz, the cross section will be $\sigma \approx 5 \times 10^{-12}$ cm$^2$. Experimentally this would be regarded as a very large cross section; the oscillator strengths and cross sections for transitions in real single atoms in gases are typically one to three orders of magnitude smaller.

Visible and near IR laser transitions in solid-state laser materials have much smaller cross sections, in the range of $\sigma = 10^{-18}$ to $10^{-20}$ cm$^2$. For a typical rare-earth laser transition in a solid, the wavelength might be $\lambda = \lambda_0/n \approx 0.6$ $\mu$m; the radiative decay rate might be 500 sec$^{-1}$; and the linewidth might be 4 cm$^{-1}$ or $\Delta\omega_a \approx 2\pi \times 4 \times 30$ GHz. The resulting cross section will be $\sigma \approx 4 \times 10^{-19}$ cm$^2$. Visible transitions in the organic molecules used as laser dyes have very wide linewidths, but also very strong radiative decay rates, with oscillator strengths close to unity. This leads to considerably larger cross sections, in the range of $\sigma \approx 1$ to $5 \times 10^{-16}$ cm$^2$.

### Transition Strength

The cross section $\sigma$ is a measure of the strength of an atomic transition, as modified by the line-broadening effects of whichever linewidth mechanism $\Delta\omega_a$ or $\Delta\omega_d$ is dominant in determining the linewidth of the atomic response. If the cross section as a function of frequency is integrated across the entire linewidth, however, we obtain a so-called *transition strength* given by

$$S \equiv \int \sigma(\omega)\, d\omega = \frac{3^* \gamma_{\text{rad}} \lambda^2}{4}, \qquad (64)$$

which is the direct measure of the strength of the transition and is entirely independent of the lineshape of $\sigma(\omega)$, whether it be lorentzian, gaussian, or otherwise. With degeneracy factors included this expression becomes

$$\int \sigma_{ji}(\omega)\,d\omega = \frac{g_i}{g_j}\int \sigma_{ij}(\omega)\,d\omega = \frac{3^*\gamma_{\text{rad},ji}\lambda^2}{4}, \tag{65}$$

where $j$ is the upper and $i$ the lower level. Measuring (carefully) the integrated absorption or cross section across the full linewidth of an atomic transition is thus one practical way of determining the radiative decay rate or Einstein $A$ coefficient for that transition. Calculated or measured values of the integrated line strengths for different transitions are often given in handbooks and tables of atomic properties.

---

Problems for 7.5

1. *Practical expression for atomic oscillator strength.* Develop a general formula for the cross section of a gaussian atomic transition in the form $\sigma = K\mathcal{F}/\Delta\nu$, where $K$ is a numerical constant (which you should evaluate); $\mathcal{F}$ is the oscillator strength of the atomic transition; and $\Delta\nu$ is the doppler linewidth of the transition expressed in units of wavenumbers or cm$^{-1}$.

2. *Design considerations for a high-energy-storage laser medium.* Laser designers often face the following problem. Suppose you want to build a large high-energy single-pass amplifier for laser pulses. To accomplish this you must pump the laser medium up to an inverted condition, which takes a substantial pumping time; and then "dump" this inversion into the amplified pulse in a very short time. In the inverted condition you want to have the gain coefficient fairly low, to avoid parasitic oscillations in the large inverted volume; and yet you must have a large stored energy density in the laser medium to get large energy output. If you were evaluating different laser media for this application, what specific characteristics (cross section, lifetime, etc.) of the laser atomic transition would you look for? Outline briefly the reasoning behind your choice of specifications.

3. *Measuring an inverted laser transition cross section.* Measuring the cross section $\sigma$ of an inverted laser transition by simply measuring the small-signal power gain $G = \exp(N\sigma L)$ is not straightforward, because of the difficulty of measuring in any direct way the inverted population difference $N$. One practical way of bypassing this problem is to measure the gain for a weak signal pulse of total energy $U$ passing through the amplifier, while monitoring the sidelight fluorescence intensity $I$ from the inverted laser atoms. The signal pulse is made powerful enough to cause a small but observable change $\Delta I$ in the fluorescence intensity from just before to just after the pulse passes; i.e., the pulse causes a small change in the inverted population $N$, but not enough to represent any significant gain saturation.

We then measure the pulse energy gain $G$ in the amplifier; the net energy $\Delta U$ acquired by the laser pulse; and the *fractional* change $\Delta I/I$ in the sidelight fluorescence (which does not require any absolute calibration of the fluorescent intensity). Show that the cross section for the laser transition is then given by $\sigma = (h\nu/\Delta U)(\Delta I/I)\ln G$, and discuss why this can be a practical method for a real measurement using available apparatus. (See B. S. Guba et al., "Measurement of cross section for induced transitions in neodymium glasses," *Opt. and Spectrosc.* **47**, 67–69, July 1979.)

4. *Energy storage in a Nd:YAG rod.* A Nd:YAG laser rod 6.4 mm diameter by 75 mm long is to be pumped to have a maximum one-way power gain of 20. How many joules of laser energy can this rod potentially deliver in a single short pulse (no repumping during the pulse)? [Hint: You know the transition cross section for this material].

5. *Gain through a thin atomic layer near a mirror.* A very thin layer of inverted laser atoms is located half an optical wavelength in front of a perfectly conducting metal mirror. The layer has $N$ atoms per unit cross-sectional area, and each atom has a stimulated-emission cross section $\sigma$. A signal wave is incident perpendicular to the atoms and the mirror. What will be the net gain of the wave after double-passing the thin layer?

## 7.6 SATURATION INTENSITIES IN LASER MATERIALS

The amplification coefficient for a signal wave passing through a laser amplifier is proportional to the population difference on the amplifying transition. At the same time, however, for a strong enough input signal the stimulated transition rate may become large enough to saturate the population difference, and thus reduce the gain coefficient seen by the signal. This process is commonly referred to as *saturation* of the gain (or absorption) coefficient by the applied signal.

Saturation behavior in a laser amplifier (or for that matter an atomic absorber) can be expected therefore whenever the signal strength becomes strong enough for the signal itself to reduce the signal growth or attenuation rate. Understanding this kind of saturation behavior, which is very important in determining the performance of practical laser systems, is our objective here.

### Saturation of the Population Difference

A wave traveling through an atomic medium will grow or decay in intensity with distance through the medium according to the differential formula

$$\frac{dI}{dz} = \pm 2\alpha_m I = \pm \Delta N \sigma I, \tag{66}$$

where $\sigma$ is the stimulated-transition cross section and the $\pm$ signs apply to inverted or absorbing population differences. We have also shown that the population difference $\Delta N$, whether emitting or absorbing, will often saturate with increasing signal strength in the homogeneous form

$$\Delta N = \Delta N_0 \times \frac{1}{1 + W\tau_{\text{eff}}} = \Delta N_0 \times \frac{1}{1 + I/I_{\text{sat}}}, \tag{67}$$

where $\Delta N_0$ is an unsaturated or small-signal inversion value; $\tau_{\text{eff}}$ is an effective lifetime or recovery time for the transition; and $I_{\text{sat}}$ is the *saturation intensity*, or the value of signal intensity passing through the laser medium that will saturate the gain (or loss) coefficient down to half its small-signal or unsaturated value.

The saturation intensity is thus a parameter of great importance in practical laser materials; and our first task in this section is to derive a simple formula and some typical values for this quantity. (Note also that in writing the $W\tau_{\text{eff}}$ term we have omitted the factor of 2 that appears in the $1 + 2WT_1$ denominator for the ideal two-level case, because the condition that $N_1 + N_2 =$ constant does not apply on most laser transitions as it does for a simple ideal two-level system.)

### The Stimulated-Transition Probability

Obviously, the stimulated-transition probability $W$ must be directly proportional to the signal intensity (power per unit area) $I$ inside the laser medium, with a proportionality factor that can be obtained by the following argument. The net power absorbed by the atoms in a thin slab of thickness $\Delta z$ from an incident wave carrying total power $P$ uniformly distributed over a transverse area $A$ will be

$$\Delta P_{\text{abs}} = (N_1 \sigma_{12} - N_2 \sigma_{21})\, P \Delta z. \tag{68}$$

(Note that $N_1$ and $N_2$ are, as usual, atoms per unit volume, and that the area factors in the slab volume $A \Delta z$ and the power density $P/A$ just cancel.) But from a rate-equation analysis the net power absorption by the atoms in the same slab can also be written as

$$\Delta P_{\text{abs}} = (W_{12} N_1 - W_{21} N_2)\, A \Delta z\, \hbar \omega_a, \tag{69}$$

where the energy per photon $\hbar \omega$ must be included to convert the net stimulated transition rate in atoms/second into a net power-absorption rate.

Equating these two expressions, including possible degeneracy factors, then gives the relation

$$W_{21} = \frac{g_1}{g_2} W_{12} = \frac{\sigma_{21}}{\hbar \omega} \frac{P}{A} = \frac{\sigma_{21}}{\hbar \omega} \times I \tag{70}$$

or in simple terms

$$W \equiv \frac{\sigma I}{\hbar \omega}. \tag{71}$$

This is a very useful and general relation which connects the cross section $\sigma$, intensity $I$, and stimulated transition probability $W$. For degenerate transitions the upward and downward stimulated transition cross sections must obey the same relationship $g_1 \sigma_{12} = g_2 \sigma_{21}$ as do the stimulated-transition probabilities $g_1 W_{12} = g_2 W_{21}$.

### Saturation Intensity Derivation

The gain or absorption coefficient $2\alpha_m$ for a homogeneous atomic transition will thus commonly saturate with increasing signal intensity in the form

$$2\alpha_m = \frac{2\alpha_{m0}}{1 + I/I_{\text{sat}}} = \frac{\Delta N_0 \sigma}{1 + (\sigma \tau_{\text{eff}}/\hbar \omega) I}. \tag{72}$$

The *saturation intensity* that reduces the small-signal absorption coefficient $2\alpha_{m0} \equiv \Delta N_0 \sigma$ down to half its small-signal value is thus given by

$$I = I_{\text{sat}} \equiv \frac{\hbar \omega}{\sigma \tau_{\text{eff}}}. \tag{73}$$

From this formula, the saturation intensity is inversely proportional to the transition cross section $\sigma$; that is, the larger the cross section, the easier the transition is to saturate. The saturation intensity is also inversely proportional to the recovery time $\tau_{\text{eff}}$, because the longer the recovery time (the slower the recovery rate), the easier the transition is to saturate. In fact, an intensity $I = I_{\text{sat}}$ basically means one photon incident on each atom, within its cross section $\sigma$, per recovery time $\tau_{\text{eff}}$.

Of course, if the signal being applied to either an amplifying or an absorbing atomic transition is tuned off line center, the stimulated transition rate and hence the degree of saturation produced by that signal will decrease in proportion to the atomic lineshape, since for a given intensity $I$ the applied signal will be less effective in inducing transitions and thus causing saturation. Suppose the transition has a homogeneous lorentzian lineshape, and suppose we use the normalized variable $y = 2(\omega - \omega_a)/\Delta\omega$ as a shorthand for the frequency detuning. The effective saturation of the atomic gain or loss coefficient $\alpha_m$ by a signal of intensity $I$ applied off line center will then be given by

$$2\alpha_m(\omega, I) = \frac{2\alpha_{m0}(\omega)}{1 + (I/I_{\text{sat}}) \times \dfrac{1}{1+y^2}}, \tag{74}$$

where $I_{\text{sat}}$ is the saturation intensity appropriate to a signal at midband. (Note that $\alpha_{m0}$ here indicates the unsaturated or small-signal gain, not the midband gain.)

We must then take this frequency dependence into account either by retaining the explicit frequency dependence $1/(1+y^2)$ in this formula in all further calculations, or by assuming that the saturation intensity itself becomes frequency dependent, with the effective saturation intensity for an off-resonance signal increasing by the amount

$$I_{\text{sat}}(\omega) = I_{\text{sat}}(\omega_a) \times \left[1 + \left(2\frac{\omega - \omega_a}{\Delta\omega_a}\right)^2\right]. \tag{75}$$

The effective saturation intensity goes up off line center, because the applied signal fields are less effective in inducing transitions; so a larger signal intensity is needed to produce a given amount of saturation. The most common procedure is to give a number for the midband-saturation intensity value, and then to include the frequency dependence explicitly in Equation 7.74.

### Saturation Broadening or Power Broadening

Suppose we tune a signal of fixed intensity $I$ across an absorption line or a gain profile, and measure the saturated loss or gain coefficient $\alpha_m(\omega, I)$ versus $\omega$ using this fixed-intensity signal. Then the complete frequency dependence for the gain coefficient (or the absorption coefficient) on a homogeneous atomic transition will include both the real lorentzian lineshape or frequency dependence of the atomic response itself, which will have the form $1/(1+y^2)$, and the frequency dependence of the saturation behavior, which we have given in Equation 7.74. Suppose we include both of these frequency dependences explicitly in the gain coefficient $\alpha_m(\omega, I)$. We can then rewrite Equation 7.74 in terms of the midband gain coefficient and saturation intensity, with the explicit frequency dependences

$$2\alpha_m(\omega, I) = \frac{2\alpha_{m0}(\omega_a)}{1+y^2} \times \frac{1}{1+(I/I_{\text{sat}})(1/1+y^2)}$$
$$= \frac{2\alpha_{m0}(\omega_a)}{1+I/I_{\text{sat}}+y^2}. \tag{76}$$

This can then be further rewritten in the equivalent form

$$2\alpha_m(\omega, I) = \frac{2\alpha_{m0}(\omega_a)}{1+I/I_{\text{sat}}} \times \frac{1}{1+[2(\omega-\omega_a)/\Delta\omega_b]^2}, \tag{77}$$

where $\Delta\omega_b$ is a *power-broadened* or *saturation-broadened linewidth* given by

$$\Delta\omega_b \equiv \sqrt{1+I/I_{\text{sat}}} \times \Delta\omega_a. \tag{78}$$

That is, the measured lineshape for $\alpha_m(\omega, I)$ will still be lorentzian, but it will now appear to have a broadened linewidth given by $\Delta\omega_b$ rather than $\Delta\omega_a$. Note that the homogeneous linewidth of the transition has not really been broadened in any fundamentally new way; but the absorption lineshape measured by means of a tunable signal of fixed intensity $I$ appears to be broadened because of stronger saturation and hence flattening down of the gain or loss coefficient at the middle of the line. This type of *power broadening* of the atomic response appears in other laser situations as well.

### Numerical Values for Saturation Intensities

This saturation intensity, measured in watts per unit area, is very important in determining the large-signal saturation behavior of laser amplifiers and oscillators, as well as saturable absorbers. A laser amplifier will become saturated and give little or no additional gain when the signal intensity passing through the laser material becomes of the order of the saturation intensity. Similarly, the power level in a laser oscillator, at least under steady-state conditions, is going to build up to at most a few times the saturation intensity, at which point the gain in the laser medium will be saturated down to equal the losses in the laser cavity. The saturation intensity is thus a very important measure of the amount of power per unit cross-sectional area that can be extracted from a practical laser device.

In practical terms a visible gas-laser transition might have, very approximately, $\hbar\omega \approx 10^{-19}$ J, $\sigma \approx 10^{-13}$ cm$^2$, $\tau_{\text{eff}} \approx 10^{-6}$ s, and hence $I_{\text{sat}} \approx 1$ W/cm$^2$. The oscillation power outputs from visible cw gas lasers do typically range from milliwatts to perhaps a few watts at most. A solid-state laser, on the other hand, might have $\sigma \approx 10^{-19}$ cm$^2$, $\tau_{\text{eff}} \approx 10^{-3}$ sec, and hence $I_{\text{sat}} \approx 1$ kW/cm$^2$. A good cw Nd:YAG laser oscillator with an area $A \approx 0.3$ cm$^2$ can have a cw power output of a few hundred watts. Note that a typical liquid-dye laser might have $\sigma \approx 10^{-16}$ cm$^2$, and $\tau_{\text{eff}} \approx 10^{-9}$ sec, giving $I_{\text{sat}} \approx 1$ MW/cm$^2$.

Note also that *the saturation-intensity value in general does not depend on the pumping intensity applied to the laser medium*, since neither the cross section $\sigma$ nor the effective recovery time (in most materials) depends directly on the pumping rate. Pumping a laser medium harder generally creates more small-signal gain, which has to be saturated down further; but it does not change the saturation intensity involved in the saturation expression.

## Problems for 7.6

1. *Saturation intensity for a three-level atomic absorber.* A system with three energy levels has transition frequencies $\omega_{32}$, $\omega_{21}$, and $\omega_{31}$; total decay rates $\gamma_{32}$, $\gamma_{21}$, and $\gamma_{31}$; and stimulated-transitions probabilities $\sigma_{32} = \sigma_{23}$, $\sigma_{21} = \sigma_{12}$, and $\sigma_{31} = \sigma_{13}$ between its levels. The "optical frequency approximation" is valid for all transitions. What will be the saturation intensity $I_\text{sat}$ for a signal passing through this collection of atoms with frequency $\omega$ tuned to the transition frequency $\omega_{31}$?

2. *Saturation lineshape for the reactive part of a homogeneous two-level atomic transition.* How will the atomic phase shift $\Delta\beta_m L$, as contrasted to the atomic gain or absorption coefficient $\alpha_m L$, saturate in a homogeneous atomic medium?

    To examine this, suppose that a signal wave having fixed intensity $I$ but variable frequency $\omega$ is transmitted through a thin slab of lorentzian, homogeneously saturable absorbing medium of thickness $L$; and the added phase shift $\Delta\beta_m(\omega)L$ caused by the atoms is measured as a function of $\omega$. Let the absorption in the cell be fairly small, say, $2\alpha L = 0.1$, so that the intensity $I(z)$ is essentially constant throughout the cell. Plot the variation of $\Delta\beta_m(\omega)L$ versus $\omega$ for selected values of $I/I_\text{sat}$ both $< 1$ and $> 1$. How does the spacing between the peaks of $\Delta\beta_m(\omega)L$ change with increasing intensity?

3. *Saturation behavior in a two-level absorber including excited-state absorption.* A certain molecule has two lowest energy levels $E_1$ and $E_2$ with stimulated-transition cross section $\sigma_{12} = \sigma_{21}$ between them. The energy decay time from level $E_2$ back to level $E_1$ is $T_1$.

    These same molecules also absorb at the same wavelength, with a stimulated-transition cross section $\sigma_{23} = \sigma_{32}$, from level $E_2$ up to a higher level (or group of levels) $E_3$. (This would be referred to as an *excited-state absorption transition.*) One can assume, however, that the relaxation rate from the upper levels $E_3$ back to $E_2$ is so fast as to be essentially instantaneous, so that the approximation $N_3 \approx 0$ prevails under all conditions.

    Evaluate the absorption through a thin slab of this medium as a function of the incident signal intensity $I$ and find its saturation behavior with increasing intensity.

    If you could measure the intensity transmission $T(I)$ through a thin slab of this material over a wide range of incident intensities $I$, what information could you gain (from this data alone) about the relative cross sections $\sigma_{12}$ and $\sigma_{23}$?

4. *Power balance versus intensity in a two-level saturable absorber.* This problem combines the saturation-intensity concepts of this section with the fundamental rate equation discussed in earlier chapters. Suppose an optical signal with adjustable intensity $I$ is applied to a collection of elementary two-level atoms in an optically thin slab (net attenuation through the slab is small). The slab has total volume $V$, and the atoms have cross section $\sigma$, relaxation time $T_1$, and total density of $N$ atoms per unit volume. The optical approximation does *not* apply.

    Evaluate the steady-state power balance in this slab by evaluating (a) the net power absorbed *by* the atoms, *from* the signal, (b) the net power absorbed by the atoms from their "thermal surroundings," and (c) the net power spontaneously

## 7.7 HOMOGENEOUS SATURATION IN LASER AMPLIFIERS

As an optical signal passes through a laser amplifier, the signal intensity $I(z)$ grows more or less exponentially with distance along the length of the amplifier. However, when the signal intensity begins to approach the saturation intensity for the laser medium, the population difference and hence the gain coefficient in the laser material begin to be saturated; the rate of signal growth with distance begins to decrease; and the signal intensity thus grows more slowly with distance.

In a single-pass laser amplifier such saturation effects begin first at the output end of the amplifier (see Figure 7.12), but only when the input signal is large enough that the amplified signal level at the output end has approached the saturation intensity of the laser medium. This saturation at the output end then causes the growth rate to decrease near the output end, and this in turn reduces the overall saturated gain from input to output as compared to the small-signal or unsaturated gain of the amplifier.

As we increase the input intensity to a laser amplifier, the intensity $I(z)$ will reach the saturating range at an earlier point along the amplifier: the saturation region moves toward the input end as the input power is increased. The net result of this saturation behavior is that large-signal output is not a linear function of large-signal input. In this section we will analyze this behavior in a simple lossless single-pass amplifier, assuming cw signals and homogeneous saturation of the laser gain coefficient.

### Homogeneous Saturation Analysis

Suppose the laser gain coefficient in a single-pass laser amplifier saturates homogeneously, with unsaturated gain coefficient $2\alpha_{m0}$, saturation intensity $I_{\text{sat}}$ and, for simplicity, no linear losses, so that $\alpha_0 = 0$. The basic differential equation governing the growth rate for the signal intensity along the amplifier thus becomes

$$\frac{1}{I(z)}\frac{dI(z)}{dz} = 2\alpha_m(I) = \frac{2\alpha_{m0}}{1 + I(z)/I_{\text{sat}}}, \qquad (79)$$

where $\alpha_{m0}$ is the unsaturated gain coefficient and $I_{\text{sat}}$ the saturation intensity of the laser medium. Obviously we can *not* simply integrate this equation to obtain an overall gain $G = \exp(2\alpha_m L)$, since the gain coefficient $\alpha_m$ varies with intensity $I$ and hence with distance $z$ along the amplifier length.

If, however, we assume an input intensity $I_{\text{in}}$ at the input end $z = 0$ and an output intensity $I_{\text{out}}$ at the output end $z = L$, then this equation can be rearranged into the form

$$\int_{I=I_{\text{in}}}^{I=I_{\text{out}}} \left[\frac{1}{I} + \frac{1}{I_{\text{sat}}}\right] dI = 2\alpha_{m0} \int_{z=0}^{z=L} dz. \qquad (80)$$

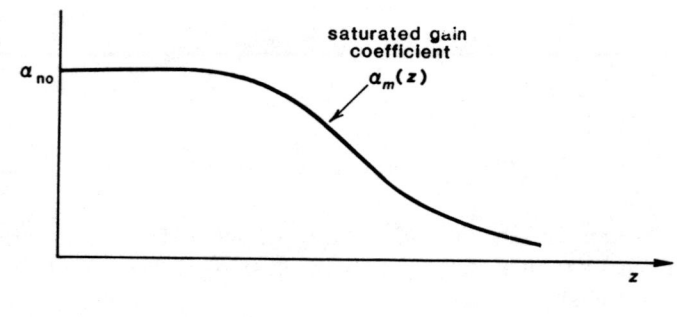
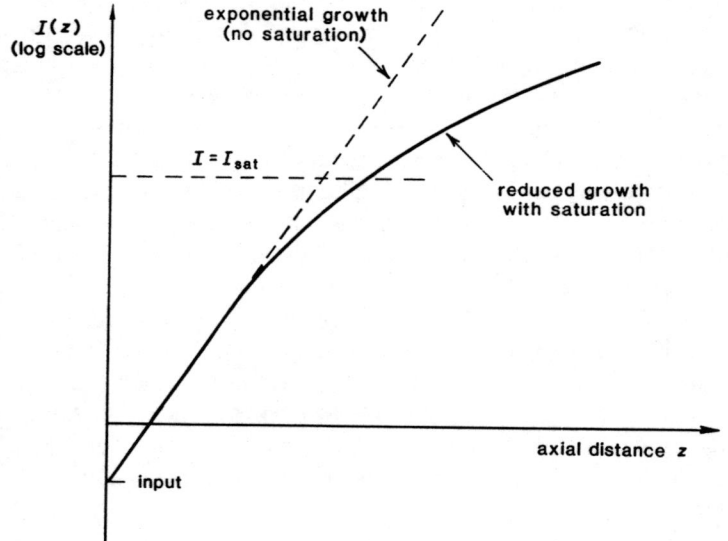

FIGURE 7.12
Gain saturation as a function of distance along a single-pass laser amplifier.

Both sides of this equation can then be integrated to obtain the expression

$$\ln\left(\frac{I_{\text{out}}}{I_{\text{in}}}\right) + \frac{I_{\text{out}} - I_{\text{in}}}{I_{\text{sat}}} = 2\alpha_{m0}L = \ln G_0, \quad (81)$$

where $G_0 \equiv \exp(2\alpha_{m0}L)$ is the small-signal or unsaturated power gain through the amplifier.

As it stands this result gives us an implicit relationship between the input and output intensities, the saturation intensity, and the unsaturated power gain $G_0$ for the amplifier. We can then obtain useful numbers from this equation in the following fashion. Suppose we define the actual power gain of the amplifier as the ratio of output over input, or $G \equiv I_{\text{out}}/I_{\text{in}}$, under arbitrary saturation conditions. This gain cannot be written as $\exp(2\alpha_m L)$, and its value in fact depends on the intensities $I_{\text{in}}$ or $I_{\text{out}}$. We can, however, use Equation 7.81 to write this overall power gain in the form

$$G \equiv \frac{I_{\text{out}}}{I_{\text{in}}} = G_0 \times \exp\left[-\frac{I_{\text{out}} - I_{\text{in}}}{I_{\text{sat}}}\right], \quad (82)$$

which says that the value of the saturated gain $G$ at a given value of $I_{in}$ (or $I_{out}$) is reduced below the unsaturated value $G_0$ by a factor that depends exponentially on the extracted intensity $I_{out} - I_{in}$ relative to the saturation intensity $I_{sat}$.

These results can then be manipulated into a variety of forms that can be used in different ways. For example, Equation 7.82 can be rewritten in the forms

$$G \equiv \frac{I_{out}}{I_{in}} = G_0 \times \exp\left[-\frac{(G-1)I_{in}}{I_{sat}}\right] = G_0 \times \exp\left[-\frac{(G-1)I_{out}}{G I_{sat}}\right]. \quad (83)$$

The first of these forms can then be turned around to give the relationship

$$\frac{I_{in}}{I_{sat}} = \frac{1}{G-1} \ln\left(\frac{G_0}{G}\right), \quad (84)$$

which gives the input intensity in terms of the unsaturated gain $G_0$ and saturated gain $G$. But using the second form (or just multiplying both sides of Equation 7.84 by $G$) also gives the result

$$\frac{I_{out}}{I_{sat}} = \frac{G}{G-1} \ln\left(\frac{G_0}{G}\right) \quad (85)$$

which gives the output intensity as a function of the same quantities. For any given value of unsaturated gain $G_0$ we can then plug different values of saturated gain in the range $1 < G < G_0$ into Equations 7.84 and 7.85 to obtain paired values of normalized input intensity $I_{in}/I_{sat}$ and output intensity $I_{out}/I_{sat}$.

Figure 7.13 illustrates the resulting amplifier input-output intensity curves for two different small-signal gain values. Note that for each value, the actual gain $G$ begins to be saturated below its small-signal value $G_0$ even at output intensities well below the saturation intensity. At high enough input intensities the gain always saturates down toward the limiting value $G = 1$, or 0 dB. For high intensities the amplifier transmission saturates down, not toward zero transmission, but toward unity transmission—that is, the amplifier (which is assumed to have zero ohmic losses) becomes essentially transparent at high enough input intensities.

### Power Extraction and Available Power

We might next ask *how much intensity, or how much power per unit cross-section area, can be extracted from such an amplifier at different input-signal levels?* By manipulating the preceding equations we can find that the power per unit area extracted from the amplifier—that is, the output power minus the input power, or the power really supplied to the wave by the amplifier—is given by

$$I_{extr} \equiv I_{out} - I_{in} = \ln\left(\frac{G_0}{G}\right) \times I_{sat}. \quad (86)$$

The values of this quantity are illustrated by the dashed lines in Figures 7.13 and 7.14.

Note that for low input intensity and high gain ($G \approx G_0$), the output power and the extracted power are essentially the same—that is, we are putting in very little power at the input end compared to what we are getting out at the output end. As the amplifier begins to saturate, however, the extracted power approaches a limiting value, which is the maximum power available to be extracted from the

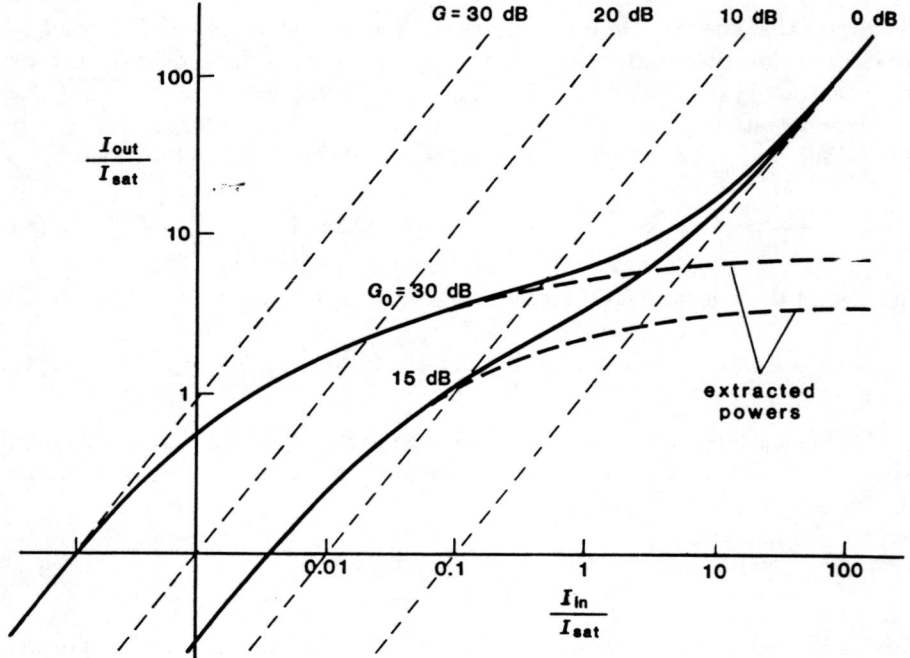

FIGURE 7.13
Amplifier output versus input curves for two different values of small-signal gain.

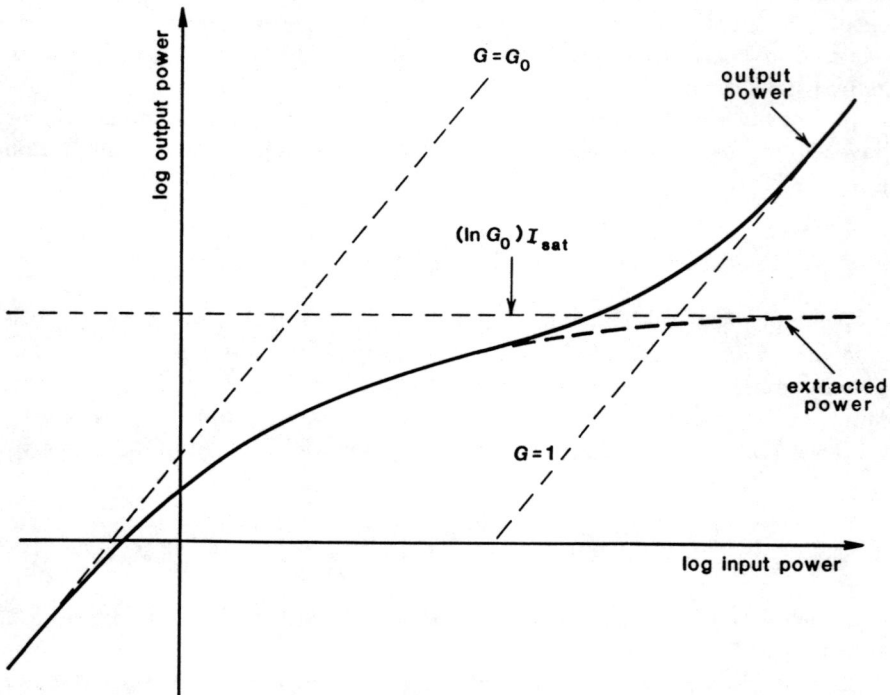

FIGURE 7.14
Power input, power output, and power extraction from a saturating laser amplifier.

amplifying medium. This *maximum available power from the amplifier* (per unit area) is given by the limiting value

$$I_{\text{avail}} = \lim_{G \to 1} \ln\left(\frac{G_0}{G}\right) \times I_{\text{sat}} = (\ln G_0)\, I_{\text{sat}}. \tag{87}$$

This is the maximum power per unit area that is available in the laser medium to be given up to the amplified signal.

This expression for the available intensity in the laser medium has a simple physical interpretation. It can be rewritten, using earlier formulas, as

$$I_{\text{avail}} = 2\alpha_{m0} L \times I_{\text{sat}} = (\Delta N_0 \sigma L) \times \left(\frac{\hbar\omega}{\sigma \tau_{\text{eff}}}\right). \tag{88}$$

Since intensity is already power per unit area, we can reduce this to available power per unit volume by writing it as

$$\frac{I_{\text{avail}}}{L} \equiv \frac{P_{\text{avail}}}{V} = \frac{\Delta N_0\, \hbar\omega}{\tau_{\text{eff}}}. \tag{89}$$

This says that *the maximum power output per unit volume that we can obtain from the laser medium is given by the initial or small-signal inversion energy stored in the medium, or $\Delta N_0 \hbar\omega$, times an effective recovery rate $1/\tau_{\text{eff}}$*. In other words, we can obtain the initial inversion energy $\Delta N_0 \hbar\omega$ once in every effective relaxation or gain recovery time $\tau_{\text{eff}}$, which makes good physical sense.

### Power-Extraction Efficiency

A major practical problem, however, is that this available power can be fully extracted only by heavily saturating the amplifier, in essence, by saturating the amplifier gain down until its saturated gain is reduced close to $G = 1$. Suppose we calculate the input and output power, and the associated gain and extracted power, for a hypothetical single-pass amplifier having an unsaturated gain $G_0 = 1{,}000 = 30$ dB and an available power $P_{\text{avail}} = (\ln G_0) A I_{\text{sat}} = 1$ kW/cm$^2$. With this amplifier we might hope to put in an input of 1 W and obtain an amplified output of 1,000 times larger, or close to 1 kW. The actual numbers for this case are, however, those shown in Table 7.2. Note in particular that by the time the device is putting out 800 W, the actual gain has already been reduced from a small-signal gain of 1,000 or 30 dB down to 9 dB or approximately 8. Hence, to obtain this output of 800 W, we must drive the amplifier with an input not of 0.8 W but of 100 W. To get an actual output power equal to the nominally available 1,000 W, we must provide 220 W of input; that is, we already need a fairly high-power preamplifier, just to extract the available power from this power amplifier.

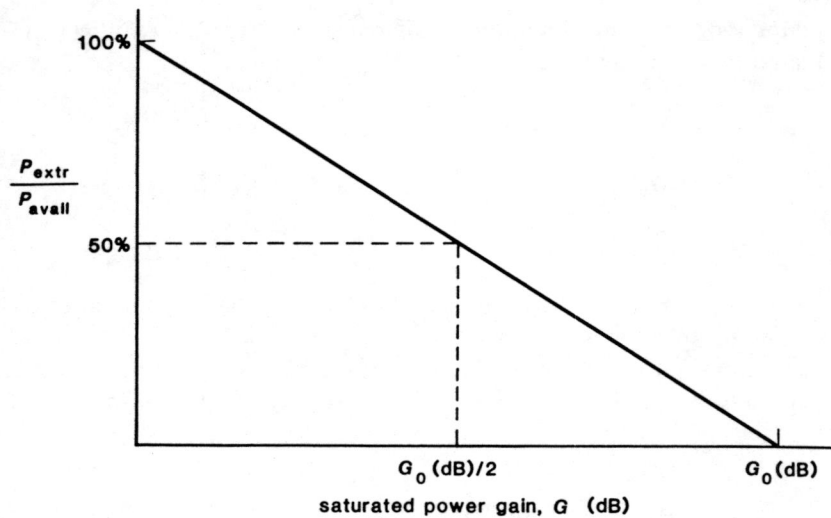

**FIGURE 7.15**
Power-extraction efficiency versus saturated power gain in a homogeneously saturating laser amplifier.

**TABLE 7.2**
Actual gain versus input power in a $CO_2$ amplifier[a]

| $P_{in}$ | $P_{out}$ | $P_{extr}$ | Saturated gain G | Gain G in dB |
|---|---|---|---|---|
| $\sim 0$ | $\sim 0$ | $\sim 0$ | 1,000 | 30 dB |
| 0.1 | 64 | 64 | 643 | 28 dB |
| 1 | 220 | 219 | 220 | 23 dB |
| 10 | 457 | 447 | 46 | 17 dB |
| 100 | 800 | 700 | 8 | 9 dB |
| 220 | 1,000 | 780 | 4.5 | 7 dB |
| $\sim \infty$ | $\sim \infty$ | 1,000 | 1 | 0 dB |

[a] $G_0 = 30$ dB, $P_{avail} \equiv (\ln G_0) P_{sat} = 1,000\,W$.

An instructive way to demonstrate this same point is to define an energy-extraction efficiency as the ratio of actually extracted power to available power, or

$$\eta_{extr} \equiv \frac{I_{extr}}{I_{avail}} = \frac{\ln G_0 - \ln G}{\ln G_0} = 1 - G_{dB}/G_{0,dB}. \tag{90}$$

This energy-extraction efficiency, if plotted versus actual or saturated gain $G$ in dB, is thus a straight line (see Figure 7.15). To extract even half the power potentially available in a cw amplifier, one must give up half the small-signal dB gain of the amplifier.

Driving a single-pass amplifier hard enough to extract most of the energy potentially available in the laser medium is thus a difficult problem, and this difficulty in obtaining full energy extraction is the principle defect in MOPA applications. One way around this difficulty is to use multipass amplification, with the same beam sent through the amplifier medium several times, possibly from slightly different directions. Another solution (with its own difficulties) is to convert the amplifier into an oscillator, since oscillators are generally more efficient at extracting the available energy from a laser medium, as we will see in a later chapter.

### Saturable Amplifier Phase Shift

If the intensity $I(z)$ at any plane in an amplifier is sufficient to cause saturation of the population difference and thus the gain coefficient $\alpha_m$, then it will also cause a similar saturation of the atomic phase-shift coefficient $\Delta\beta_m$. The total phase shift through a linear amplifier may thus also change under saturation conditions. To analyze this, we can note that the differential equation for the net phase shift $\phi(z)$ for a signal of frequency $\omega$ as a function of distance $z$ along the amplifier can be written as

$$d\phi(\omega, z) = \frac{\omega\, dz}{c} + \Delta\beta_m(\omega, z)\, dz. \tag{91}$$

The atomic phase-shift term $\Delta\beta_m$ will depend on the degree of saturation or on the intensity $I(z)$, and thus will vary with distance $z$ along the amplifier. For a homogeneous lorentzian atomic transition, by using the relationship between saturated gain and the saturated value of $\Delta\beta_m$ we can integrate this equation and show that the overall phase shift is related to the input and output intensities in the form

$$\phi_{\text{tot}}(\omega, L) = \frac{\omega L}{c} + \frac{\omega - \omega_a}{\Delta\omega_a} \times \ln\left(\frac{I_{\text{out}}}{I_{\text{in}}}\right) \tag{92}$$

where $I_{\text{in}}$ and $I_{\text{out}}$ are the actual (saturated) values of input and output intensity at frequency $\omega$. There is, of course, no atomic phase-shift contribution exactly at line center.

## REFERENCES

All the results presented in this section, plus more detailed analyses taking into account finite-bandwidth applied signals, are discussed by A. Y. Cabezas and R. P. Treat, "Effects of spectral hole-burning and cross relaxation on the gain saturation of laser amplifiers," *J. Appl. Phys.* **37**, 3556–3563 (August 1966).

More detailed analyses of saturation effects can also be found in W. W. Rigrod, "Gain saturation and output power of optical masers," *J. Appl. Phys.* **34**, 2602–2609 (September 1963), and "Saturation effects in high-gain lasers," *J. Appl. Phys.* **36**, 2487–2490 (August 1965).

A representative example of the measurement of saturation intensity in a laser material is T. S. Lomheim and L. G. DeShazer, "Determination of optical cross sections by the measurement of saturation flux using laser-pumped laser oscillators," *J. Opt. Soc. Am.* **68**, 1575–1579 (November 1978).

A good source on saturable absorption effects, including experimental results, is A. Zunger and K. Bar-Eli, "Nonlinear behavior of solutions illuminated by a ruby laser. II," *IEEE J. Quantum Electron.* **QE–10**, 29–36 (January 1974).

Suppose two independent signals are sent through an amplifier separately but simultaneously. Then the saturation effects due to one signal can change the gain seen by the other signal, and vice versa. This can lead to interesting and potentially useful cross-modulation or cross-saturation effects, some of which are discussed and analyzed by R. W. Gray and L. W. Casperson in "Optooptic modulation based on gain saturation," *IEEE J. Quantum Electron.* **QE–14**, 893–900 (November 1978).

Problems for 7.7

1. *Input-output intensity curves for a saturable amplifier off line center.* The typical curves of output intensity versus input intensity for a homogeneous single-pass laser amplifier given in this section assume a signal tuned to the atomic line center. Explain clearly (with sketches) exactly how these curves will change, for the same amplifier and the same plot, as the signal frequency is tuned off line center. How will the plot of energy extraction efficiency $\eta$ versus net power gain $G$ change with the same detuning?

2. *Input versus output intensities for a saturable atomic absorber.* All the equations in this section should apply equally well, with an appropriate change of sign, to saturable atomic absorbers. Consider a single-pass, saturable, traveling-wave *atomic attenuator* (rather than amplifier) using a homogeneously saturating atomic system. Analyze the output intensity versus input intensity for this device, and plot intensity out versus intensity in on log scales for several different unsaturated loss values, say 10 dB, 20 dB, and 30 dB.

3. *Signal penetration and saturation depth versus signal intensity in a homogeneous saturable absorber.* A weak cw laser beam injected into one end of a very long cell containing a homogeneously saturable absorbing medium will be almost totally attenuated within a few absorption lengths. A strong enough input signal, with $I_{in} > I_{sat}$, on the other hand, will saturate the absorption and "burn" its way some longer distance into the absorbing cell before being absorbed. Calculate and plot the signal intensity $I(z)$ on a log scale, and the saturated population difference $\Delta N(z)/\Delta N_0$ on a linear scale, both versus distance $z$ through the cell, for input signals that are 1, 10, 100, and 1,000 times the saturation intensity. Can you give an approximate rule of thumb for how far into the cell a strong beam will "burn" its way if $I_{in} \gg I_{sat}$?

4. *Obtaining an amplifier output intensity just equal to the available intensity of the laser medium.* A single-pass, homogeneously saturable laser amplifier has an available intensity $I_{avail} = 1$ kW/cm$^2$. Suppose you want to obtain an actual output intensity equal to the same value, i.e., $I_{out} = 1$ kW/cm$^2$, with an actual (saturated) power gain of $G = 10 = 10$ dB. What must be the unsaturated gain $G_0$ of the amplifier (either as a number, or in dB)? /itemIn general, if you want to obtain an actual output $I_{out}$ from an amplifier which is just equal to the available intensity $I_{avail}$, with a specified actual gain $G$, what is a relationship for the required unsaturated gain $G_0$?

5. *Power output stabilization factor in a partially saturated laser amplifier.* It is sometimes desirable that the output power $I_\text{out}$ from a laser amplifier remain as nearly constant as possible when the input power $I_\text{in}$ fluctuates by a small amount $\Delta I_\text{in}$. A useful measure of this output-power stabilization is the stabilization factor $S$, defined as $\Delta I_\text{in}/I_\text{in}$ divided by $\Delta I_\text{out}/I_\text{out}$ evaluated in the limit as $\Delta I_\text{in} \to 0$. Evaluate this stabilization factor as a function of operating conditions for a single-pass homogeneously saturating laser amplifier. Suppose you want to achieve a specified output level $I_\text{out}/I_\text{sat}$, a specified saturated gain $G$, and a specified stabilization factor $S$ in a practical laser system. Discuss the design procedure you would follow.

6. *Cross-saturation of a transversely double-pass laser amplifier.* A laser beam with input power $I_1$ is sent through a single-pass laser amplifier going in the $z$ direction; and the output beam $I_2$ is then transversely expanded in one direction and brought around to pass through the amplifier sideways in, say, the $x$ direction. Assume the rectangular laser medium has length $L$ in the $z$ direction and width $d$ in the $x$ direction, and the length-width $(L/d)$ ratio of the amplifier is such that it will have a high unsaturated gain $G_0$ in the $z$ direction, but only small net gain in the transverse direction. Develop an expression relating the input and output intensities $I_1$, and $I_2$ in this system, taking saturation into account. Indicate how you could calculate and plot a curve of $I_2$ versus $I_1$ for the system.

7. *Amplifier input-output curves for other forms of laser saturation.* Suppose that certain unusual laser media saturate with the intensity dependences either $1/[1+(I/I_\text{sat})^2]$ or $1/[1+(I/I_\text{sat})]^2$ instead of the usual $1/[1+I/I_\text{sat}]$. Evaluate and plot the output versus input intensities (on log scales) for single-pass amplifiers using these media, and compare to a standard homogeneous laser system for $G_0 = 30$ dB.

8. *Signal transmission through two intermingled saturable absorber transitions.* Suppose a saturable absorber system contains a mixture of two independent homogeneously saturating atomic absorbers, with absorption coefficients $\alpha_{m1}$ and $\alpha_{m2}$ and saturation intensities $I_{s1}$ and $I_{s2}$, respectively. Show that the power transmission $G$ through a length $L$ of this dual absorber (where $G \leq 1$) is related to the input intensity $I_\text{in}$ by the implicit relation $\ln(G/G_0) = C_1 \ln[(1+C_2 I_\text{in})/(1+C_2 G I_\text{in}] + C_3(1-G)I_\text{in}$, where $G_0 \equiv \exp(2\alpha_{m1}L + 2\alpha_{m2}L)$ is the unsaturated transmission through the absorber cell. (The algebra involved in this calculation is undeniably a bit messy. The problem is also treated by L. Huff and L. G. DeShazer, "Saturation of optical transitions in organic compounds by laser flux," *J. Opt. Soc. Am.* **60**, 157–165, February 1970.)

9. *Saturation effects on transverse beam profiles.* Suppose a collimated laser beam with a smooth transverse amplitude profile (for example, a gaussian transverse amplitude profile) is passed through a thin layer of a rather highly absorbing, homogeneously saturable atomic absorbing medium. Let us consider what happens not to the power level but to the *shape* and the *angular spreading* (or focusing) of the beam.

In experiments like this, we find that in fact both the shape and the focusing properties of the beam coming out of the absorbing slab depend on both the intensity of the incident beam and its frequency $\omega$ relative to the resonance frequency $\omega_a$ of the absorbing atoms.

List the significant physical effects or processes that might be responsible for these sorts of effects; and predict in general terms the sort of behavior (e.g., additional beam spreading, or beam focusing) that might be expected under various experimental conditions.

# 8

# MORE ON LASER AMPLIFICATION

We extend the discussion of laser amplification in this chapter to examine a few more advanced aspects of cw laser amplification, including the transient response of laser amplifiers; spatial hole-burning or standing-wave grating effects in laser amplifiers; and some more details on saturation in laser amplifiers.

## 8.1 TRANSIENT RESPONSE OF LASER AMPLIFIERS

Let us look first at the very interesting topic of the *linear transient response* of a laser amplifier, for example, to a step-function or a delta-function type of signal input.

To understand what a laser amplifier does on a transient basis, we have to recall what the atoms in the laser amplifier do on a transient basis. The induced polarization on an atomic transition is linear in the applied signal field, at least within the rate-equation approximation. A laser amplifier is thus a linear system in its response to an applied signal, at least at low enough signal levels that no saturation effects occur. The impulse response of a laser amplifier to a delta-function-like input pulse should therefore be the Fourier transform of the transfer function, or the complex voltage gain function $\tilde{g}(\omega)$, of this linear system; and the response to a fast-rising step-function input should be the integral of this impulse response. In this section we will illustrate what this means for both passive absorbers and laser amplifiers.

### Step Response of an Atomic Absorption Cell

We can obtain a very instructive picture of the transient response both of an atomic cell and of the atoms themselves, by examining an ingenious optical-pulse generation experiment carried out by Eli Yablonovitch at Harvard University, using a $CO_2$ laser and an atomic absorption cell filled with absorbing (that is, unpumped) hot $CO_2$ vapor.

Imagine first that a step-function optical signal with carrier frequency tuned to the resonance frequency $\omega_a$ is sent into an absorption cell having a large total attenuation $2\alpha_m L$, where $\alpha_m$ is the midband absorption coefficient for the atomic transition in the cell. Suppose that the rise time for the leading edge of

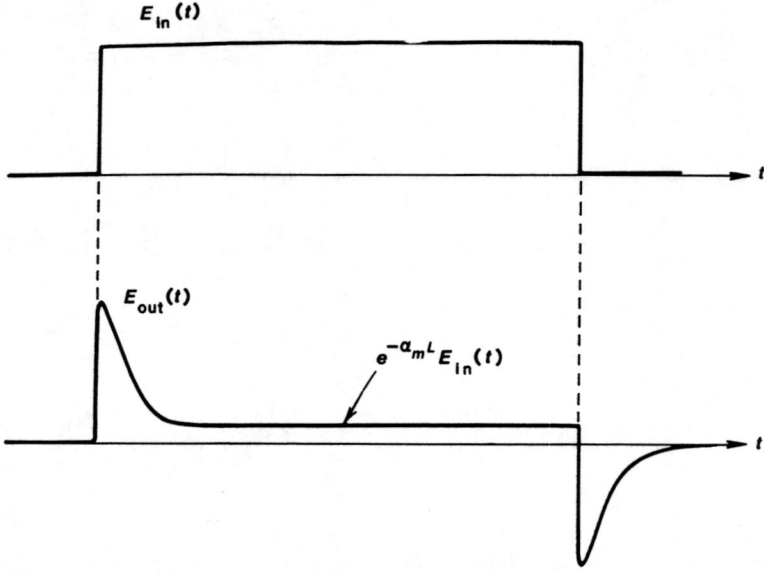

**FIGURE 8.1**
The step response of an atomic absorber to an input signal which is suddenly turned on, and just as suddenly turned off.

this optical signal is fast compared to the inverse linewidth $T_2 \equiv 1/\Delta\omega_a$ of the atomic transition (but not compared to the optical carrier frequency $\omega_a$ of the signal). As the leading edge of the signal pulse sweeps past each point in the cell, there will then be a time delay of order $\approx T_2$ before the absorbing atoms at that cross section can begin to respond to the applied signal, that is, before the induced sinusoidal polarization $\tilde{P}(\omega)$ can build up to its steady-state value.

As a result, the leading edge of the pulse will sweep through the full length of the cell with essentially no attenuation. Only after the atoms have begun to respond can the cell begin to function as an absorber, and begin to attenuate the applied signal. The output response of an absorbing cell to a step input with fast enough rise time will thus be a short pulse as shown in Figure 8.1, with essentially the same peak amplitude as the leading edge of the input signal, and with a duration that corresponds roughly to the time constant $T_2$ before the atomic response and hence the attenuation of the cell can develop.

### Transient Response to Signal Turn-Off

An even more interesting observation is that an essentially similar pulse, again of duration $\approx T_2$ but now of opposite sign, will also emerge from the cell *just after the termination of the input signal*, if we can suddenly turn off the input signal with a similarly fast fall time. The physical interpretation of this trailing-edge pulse is particularly instructive.

As one way of understanding it, consider the total signal that will be seen by an optical detector looking at the total transmitted output from the absorption cell. This detector in our example will see essentially no signal under steady-state conditions, during the main part of the input pulse, since we assume that the attenuation through the absorber cell is large.

One instructive (and correct) way of describing this situation is to say that the detector is actually seeing the superposition of the full original signal field that would be generated at the detector location *by the input signal source without the absorber cell present*, plus the additional fields that are produced at the detector *by radiation from the induced polarization $p(r,t)$ in the absorber cell itself*. This induced polarization is coherently related to the applied signal; and because we assume an absorbing medium or an absorbing transition, the induced polarization reradiates in a way that will cancel (or nearly cancel) the original applied signal at the detector. (For a strongly absorbing cell this induced polarization exists, with steadily decreasing amplitude, only in the first few absorption lengths at the input end of the cell, after which the total field, applied plus reradiated, becomes small to negligible for the rest of the cell length.)

Suppose we suddenly turn off the applied signal, with a very fast fall time. After an appropriate time delay, corresponding to the travel time at the velocity of light from the applied signal source to the detector, the component of the applied signal field due to the applied signal source itself will thus suddenly vanish. The atomic dipoles, however, will at that instant still be oscillating and radiating in coherent fashion; and, as we have described earlier, they will continue to oscillate and reradiate until the coherent polarization dies out with time constant $T_2$ because of an appropriate combination of dephasing and energy decay.

Just before the applied signal is turned off, the net signal reaching the detector is essentially zero, since the applied signal field is almost totally canceled by the fields radiated by the absorber-induced polarization (for high insertion loss). Just *after* the signal turnoff the applied signal component is gone; but the atomic polarization $p(t)$ and its dipole radiation contribution remains, at least for a time of order $T_2$. *The net signal at the detector, or at the absorber output, will thus suddenly jump up to an amplitude essentially equal to the unattenuated input signal, but with a phase 180° out of phase with the applied signal*, as shown in Figure 8.1. In other words, suddenly turning off the input will also produce a short transient pulse at the output.

### Experimental Results

Turning an optical signal on or off with a rise time short compared to an atomic response time $T_2$ requires an unusually fast electrooptic modulator and/or a very narrow atomic absorption line; so experiments of the type described here are not in general easy. Yablonovitch developed an ingeniously simple and also useful way to carry out such a demonstration, using the experimental system shown in Figure 8.2.

In this experiment the output from a pulsed TEA $CO_2$ laser, which generates a 10.6 $\mu$m laser pulse with a pulsewidth of about 100 ns and a peak power of about 100 MW, was passed first through a lens pair which focused the incident beam down to a focal spot less than two wavelengths in diameter. The beam diverging from this focal spot was then recollimated by the second lens, and transmitted through an absorption cell several meters long and containing hot $CO_2$ vapor, which automatically absorbs at the $CO_2$ laser wavelength. The absorbing cell was heated in order to thermally populate the lower level of the $CO_2$ absorption line, and the pressure could be changed in order to vary the pressure-broadened linewidth and hence the response time $T_2$.

When this type of TEA laser is fired, the input intensity to the absorbing cell rises quite slowly, following the build-up time of the TEA laser itself, which has a rise time much too slow ($\approx$ 100 ns) to produce any of the transient pulse effects we have discussed here. At a certain power level, however, the optical intensity

(a)

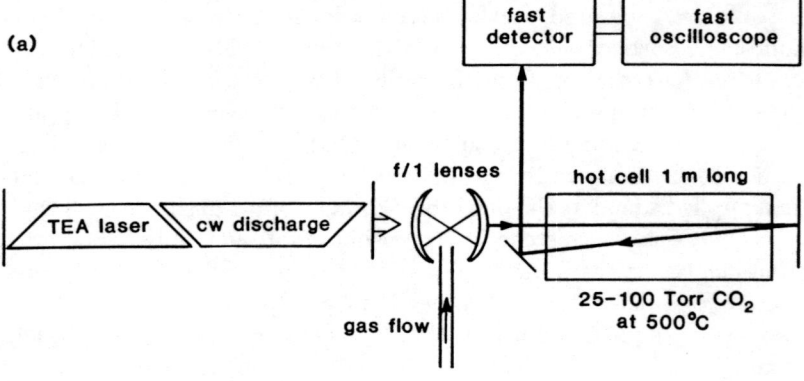

(b)

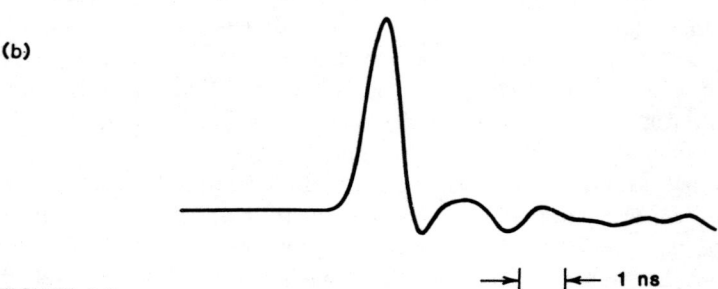

**FIGURE 8.2**
(a) Apparatus employed to demonstrate fast signal truncation with a gas breakdown cell, followed by short pulse generation in a linear absorber cell. (b) Short pulse generated in the $CO_2$ absorber cell (rise time of the pulse is limited by the oscilloscope response time). (From E. Yablonovitch and J. Goldhar, *Appl. Phys. Lett.* **25**, 580–582, November 15, 1974.)

in the focal spot can become sufficient to produce very rapid gas breakdown in air, leading to the almost instantaneous creation of a high-density plasma or "laser spark" (accompanied by a very loud sound wave). The electron density in this plasma almost immediately becomes so high that its index of refraction drops almost instantaneously to zero. This plasma spark then acts like a tiny but highly reflecting ball, which scatters essentially all the laser energy out of the beam. The gas breakdown thus provides in essence a self-actuated optical switch, which can completely shut off the transmitted laser beam with a fall time of less than 30 psec in practice.

Shutting off the incident signal using this technique produced the transient output pulse from the $CO_2$ absorber cell shown in Figure 8.2(b). This pulse not only is interesting as a demonstration of laser dynamics, but can also be useful for subsequent experiments, since it has a peak power nearly equal to the incident TEA laser signal, and a pulsewidth which can be varied simply by changing the gas pressure in the absorber cell in order to change the time constant $T_2$.

## Mathematical Analysis: The Step-Function Spectrum

To obtain a simple mathematical description of this pulse-generation process, consider a sinusoidal signal of the form $\mathcal{E}_1(t) = E_s(t)\exp(j\omega_a t)$, where $E_s(t)$ is a unit step-function, so that the carrier signal is suddenly turned on (or off) with zero rise time at $t = 0$. Such a signal has an optical spectrum, or Fourier transform, of the form

$$\tilde{E}_s(\omega) = \int_0^\infty e^{-j(\omega-\omega_a)t}\, dt = \frac{1}{j(\omega-\omega_a)}. \tag{1}$$

When this spectrum passes through an absorber cell with a lorentzian lineshape, the output spectrum is this input spectrum multiplied by the transfer function of the absorber cell or

$$\tilde{E}_2(\omega) = \tilde{E}_s(\omega) \times \exp\left[\frac{-\alpha_m L}{1 + jT_2(\omega-\omega_a)}\right], \tag{2}$$

where we use the simplified formula that $\Delta\omega \equiv 2/T_2$ to define $T_2$. The output signal is then given by the inverse Fourier transform

$$\begin{aligned}\mathcal{E}_2(t) &= \frac{1}{2\pi}\int_{-\infty}^\infty \tilde{E}_2(\omega)e^{j\omega t}\, d\omega \\ &= e^{j\omega_a t}\int_{-\infty}^\infty \frac{\exp[j2\pi st - \alpha_m L/(1+j2\pi T_2 s)]}{j2\pi s}\, ds.\end{aligned} \tag{3}$$

If we write this as $\mathcal{E}_2(t) = E_2(t)\exp(j\omega_a t)$, where $E_2(t)$ is the output envelope for a step-function input, then Yablonovitch and Goldhar have noted that this integral can be evaluated in terms of Bessel functions $J_m$ in the form

$$E_2(t) = e^{-\alpha_m L} - e^{-t/T_2}\sum_{m=0}^\infty \left(\frac{t}{T_2\alpha_m L}\right)^{m/2} J_m\left(2\sqrt{\frac{\alpha_m L t}{T_2}}\right), \quad t \geq 0, \tag{4}$$

for the boundary conditions corresponding to the fast turn-off case.

(In doing this analysis we have left out the $e^{-j\omega L}$ phase shift through the amplifier length $L$, since this would merely produce a propagation time delay $t = L/c$ in the output pulse: i.e., the output signal given in Equation 8.21 should really occur starting at $t \geq L/c$ and not $t \geq 0$.)

## Step-Function Spectrum

This Fourier analysis says in physical terms that turning a monochromatic signal on or off very rapidly will give the signal a spectrum with frequency components extending far into the wings on both the high-frequency and the low-frequency side of the carrier frequency $\omega_a$. This spectral broadening following the breakdown point was confirmed in the Yablonovitch experiments by using an infrared spectrometer to show that the sharply truncated signal following the breakdown point had a much wider power spectrum than the input TEA laser signal. Figure 8.3 illustrates the resulting long and approximately $1/\omega^2$ tails in the measured spectral density on both high-frequency and low-frequency sides of the $CO_2$ laser wavelength.

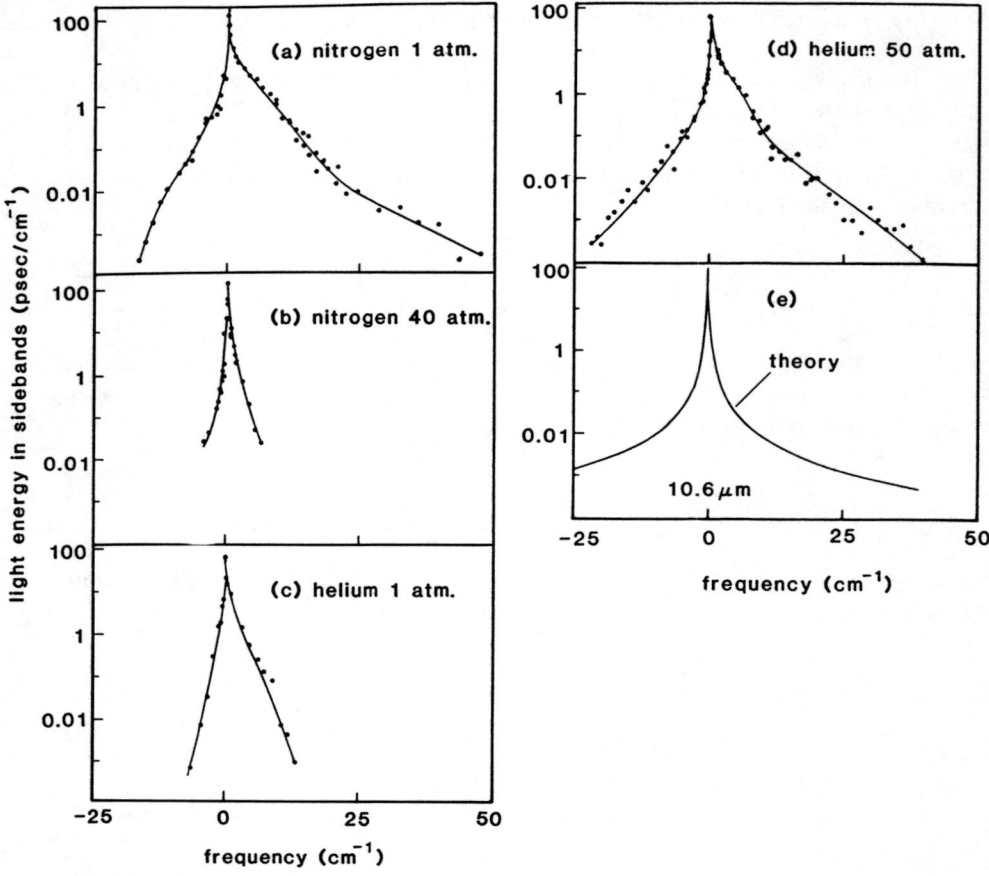

FIGURE 8.3
Spectral broadening of an initially monochromatic $CO_2$ laser signal that is suddenly truncated by plasma breakdown in various gases. (From E. Yablonovitch, *Phys. Rev. Lett.* 32, 1102–1104, May 20, 1974.)

An alternative physical approach to understanding the pulse-formation process on either the leading or trailing edges is thus the following. When the spectrally broadened signal shown in Figure 8.3 is transmitted through the absorber cell, the comparatively narrowband $CO_2$ absorption cuts out only the narrow central portion of the spectrum, within about one linewidth $\Delta\omega_a$ about the carrier frequency. In frequency terms, it is the unattenuated spectral wings that are transmitted through the absorber cell, with a hole cut out of the center, that produce the leading and/or trailing edge pulses.

It is also important to realize that this pulse behavior results entirely from the *linear or small-signal transient behavior* of the absorber cell—no nonlinear or large-signal or Rabi flopping phenomena are involved in these particular results. The trailing-edge pulse in particular is a classic example of what is generally referred to as *free induction decay*—that is, after the signal turnoff the oscillating dipoles are still freely oscillating, initially in phase but with decaying total macroscopic polarization, and in the process are inducing a matching pulse of decaying radiation out the end of the absorption cell. The Yablonovitch experiment thus gives a highly instructive illustration of the linear response behavior of an atomic transition, and of how this behavior can be variously interpreted using

## Impulse and Step Responses of Laser Amplifiers

Essentially the same transient effects can also be produced in an inverted laser amplifier using an input signal that is either a short enough pulse (quasi delta function) or a step-function having fast enough rise or fall times. For the amplifier it is perhaps more instructive, in fact, to begin with the impulse response of the amplifier to a very short input pulse.

If the input signal to a laser amplifier has the form $\mathcal{E}_1(t) = E_i(t)e^{j\omega_a t}$, where the envelope $E_i(t) \approx \delta(t)$ is a very short delta-function-like pulse, then the Fourier spectrum of this input signal will be essentially flat. The output signal, or the impulse response of the laser amplifier, will then be essentially the Fourier transform of the amplifier's complex voltage gain function, or

$$\tilde{E}_2(\omega) = \exp\left[\frac{+\alpha_m L}{1 + jT_2(\omega - \omega_a)}\right] \tag{5}$$

(assuming that the amplifier has a lorentzian lineshape). An analysis by Bridges, Haus, and Hopf then shows that the inverse transform of this spectrum is given by $\mathcal{E}_2(t) = E_2(t)e^{j\omega_a t}$, where the output envelope $E_2(t)$ is given by

$$E_2(t) = \delta(t) + \sqrt{\frac{\alpha_m L}{T_2 t}}\, I_1\left[2\sqrt{\alpha_m L t/T_2}\right] e^{-t/T_2}, \qquad t \geq 0, \tag{6}$$

with $I_1$ being the modified Bessel function of first order. (Changing the sign of the atomic absorption from $-\alpha_m$ to $+\alpha_m$ in the $\sqrt{\alpha L t/T_2}$ argument changes the Bessel function from the oscillatory $J_m$'s given in Equation 8.4 to a growing $I_1$-type modified Bessel function.)

The physical interpretation of the first term in this analytical result, as shown in Figure 8.4, is that the impulse function itself will travel through the amplifier essentially unchanged—that is, neither attenuated nor amplified—since the atoms simply do not have time to respond or to build up oscillation and begin reradiating in steady-state fashion as the impulse rushes past. The sinusoidal fields during the impulse do, however, give a finite impulse or "kick" to the atomic dipoles even during the brief passage of the pulse, so that the atoms are left with some induced oscillation or polarization $p(t)$ following the passage of the pulse. The atomic dipoles then continue radiating, and thus produce the decaying free-induction tail which follows the impulse, as shown in Figure 8.4. Because the atoms are inverted or amplifying, this induced tail is in phase with the impulse function, rather than 180° out of phase like the absorber turn-off tail. (In a certain sense, all the gain experienced by the input impulse occurs *after the impulse itself has swept past*.)

## Amplifier Step Response

The step response of a laser amplifier to a fast-rising input signal will be given by the integral of the impulse response. In other words, if $E_2(t)$ as given in Equation 8.6 is the impulse response, then the output response produced by

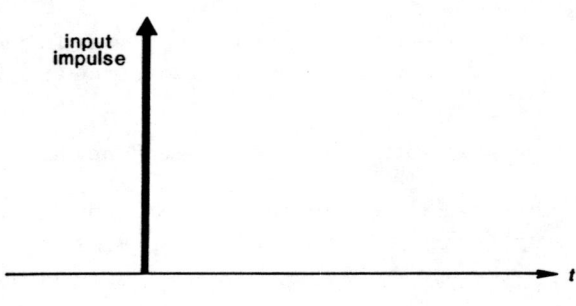

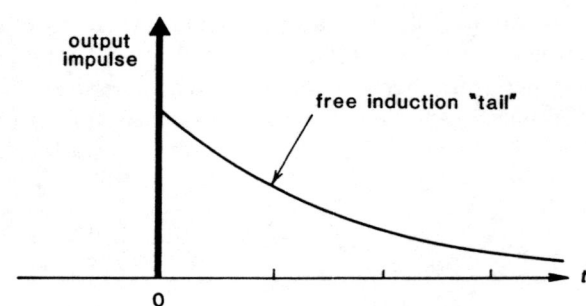

FIGURE 8.4
Linear impulse response of a laser amplifier. The free-induction "tail" on the output pulse actually contains all the energy or power gain which the amplifier gives to the input pulse.

a step input will be given by

$$E_2(t)|_{\text{step}} = \int_0^t E_2(t')|_{\text{impulse}} \, dt'. \tag{7}$$

Integrating over the delta-function part of the impulse response will cause the output step response to jump instantaneously from 0 to the input value at $t = 0$, as shown in Figure 8.5. This obviously represents the fast-rising leading edge sweeping through the amplifier with neither gain nor attenuation, just as in the absorber.

Integrating over the free-induction "tail" will then cause the output signal to continue to grow up to the steady-state amplified output value of $e^{\alpha_m L}$, also as shown in Figure 8.5. Again, obviously the net gain of the amplifier comes in integrating over the area of the tail; and in fact the net voltage gain is just the total area in the impulse plus the tail, compared to the area in the impulse only.

### Experimental Results

Careful measurements of the step response of a laser amplifier have been made and compared with a similar but more detailed analysis in work done by Bridges, Haus, and Hopf. These measurements were made using a low-pressure $CO_2$ laser amplifier having a pressure-broadened homogeneous atomic linewidth of around 120 MHz or a $T_2$ of $\approx$ 2.6 ns. A fast-rising input pulse with less than 1 nanosecond rise time was obtained by passing a low-power cw laser beam through an electrooptic light modulator, and this signal was then reflected back and forth for five passes through the amplifier in order to obtain a net gain of greater than 20 dB. Since the gain-narrowed 3 dB bandwidth of the low-pressure $CO_2$ amplifier was reduced to about 50 MHz, the 1 ns pulse rise time was short enough to approximate a step-function input, and the $\approx$ 15 ns rise time of the

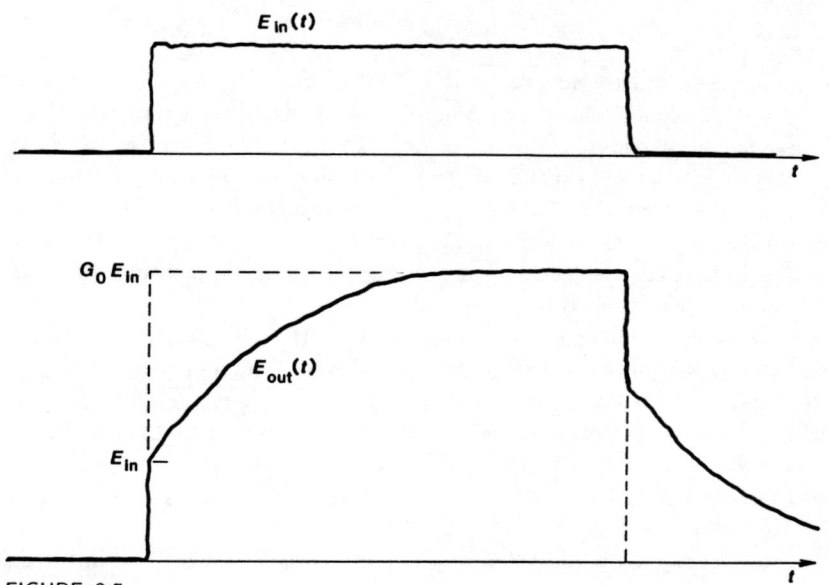

FIGURE 8.5
Linear step response of a laser amplifier. It requires a time $\approx T_2$ for the output signal to build up to the full amplified input value.

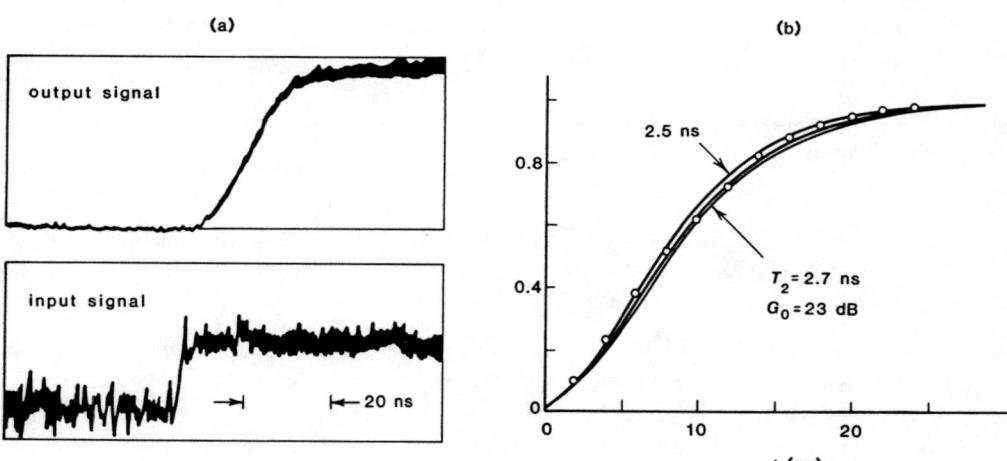

FIGURE 8.6
(a) Measured step response of a $CO_2$ laser amplifier (input signal, lower trace; output signal, upper trace). (b) Comparison of measured output with theory for two different assumed values of $T_2$. (From T. J. Bridges, H. A. Haus, and P. W. Hoff, *IEEE J. Quantum Electron.* **QE–4**, 777–782, November 1968.)

amplifier's output signal could then be observed with a fast infrared detector and oscilloscope, as illustrated in Figure 8.6.

The results in this experiment are closely fit by a more exact version of the theory outlined above, as illustrated by the theoretical plot in Figure 8.6(b). In fact, several measurements of this type using different amplifier gas pressures yielded both the pressure-broadening coefficient for the $CO_2$ laser transition, and the zero-pressure Doppler-broadened intercept of $\approx 59$ MHz.

## REFERENCES

The transient turn-off experiments for the $CO_2$ absorber cell described in this section come from E. Yablonovitch and J. Goldhar, "Short $CO_2$ laser pulse generation by optical free induction decay," *Appl. Phys. Lett.* **25**, 580–582 (November 15, 1974). The background physics of gas breakdown and the resultant spectral broadening with a $CO_2$ laser are described in E. Yablonovitch, "Self-phase modulation of light in a laser-breakdown plasma," *Phys. Rev. Lett.* **32**, 1101–1104 (May 20, 1974); and "Self-phase modulation and short-pulse generation from laser-breakdown plasma," *Phys. Rev. A* **10**, 1888–1895 (November 1974).

Further measurements on the same system are in H. S. Kwok and E. Yablonovitch, "30-psec $CO_2$ laser pulses generated by optical free induction decay," *Appl. Phys. Lett.* **30**, 158–160 (February 1, 1977), and "$CO_2$ oscillator-pulse shaper-amplifier system producing 0.1 J in a 500 psec laser pulse," *Rev. Sci. Instrum.* **46**, 814–816 (July 1975). The same scheme is applied to the 1.315 $\mu$m iodine laser by E. Fill, K. Hohla, G. T. Schappert, and R. Volk, "100-ps pulse generation and amplification in the iodine laser," *Appl. Phys. Lett.* **29**, 805–807 (December 15, 1976).

The earlier work on amplifier step-response measurements discussed in this section comes from T. J. Bridges, H. A. Haus, and P. W. Hoff, "Small-signal step response of laser amplifiers and measurement of $CO_2$ laser linewidth," *IEEE J. Quantum Electron.* **QE-4**, 777–782 (November 1968), and "$CO_2$ laser linewidth by measurement of step response," *Appl. Phys. Lett.* **13**, 316–318 (November 1, 1968).

Interesting extensions to the impulse response of a laser amplifier, including double-pulse inputs and nonlinear large-signal responses, are discussed and illustrated in S. M. Hamadani, N. A. Kurnit, and A. Javan, "Coherent optical pulse evolution in a $CO_2$ amplifier," *Opt. Commun.* **17**, 32–37 (April 1976).

---

Problems for 8.1

1. *The leading-edge pulse in a Yablonovitch-type experiment.* Find the analytical expression for the leading-edge pulse in a Yablonovitch-type experiment, assuming an infinitely fast signal turn-on. Hint: Use superposition.

2. *Asymptotic expressions for amplifier or absorber transient responses.* Using the asymptotic expressions for Bessel functions for very small or very large arguments, show that the Yablonovitch and Goldhar analytical result checks out for (a) a weakly absorbing cell, $\alpha_m L \ll 1$; and (b) a very strongly absorbing cell, $\alpha_m L \gg 1$, for the limit as $t \to \infty$.

---

## 8.2 SPATIAL HOLE BURNING, AND STANDING-WAVE GRATING EFFECTS

When two waves traveling in different directions are simultaneously present in a laser medium, interference between these two waves will produce both frequency beating effects and standing-wave patterns in the optical intensity. These interference effects in turn may produce both temporal modulation and spatial variations in the amount of saturation in the laser medium.

Interference between two waves at the same frequency but traveling in different directions in particular can produce spatial hole burning effects, which can

## 8.2 SPATIAL HOLE BURNING, AND STANDING-WAVE GRATING EFFECTS

modify the saturation behavior of each wave independently, as well as induced grating effects, which can couple the two initially independent waves to each other. Nonlinear coupling between waves can in turn significantly modify the behavior of certain laser systems. In this section we will therefore introduce the fundamentals of spatial hole burning, and analyze the first-order coupling effects that spatial hole burning can produce in elementary two-wave situations.

### Wave Interference Effects

Consider a general situation in which two propagating waves with complex amplitudes $\tilde{E}_1$ and $\tilde{E}_2$, frequencies $\omega_1$ and $\omega_2$, and propagation vectors $\beta_1$ and $\beta_2$ are simultaneously present in an atomic medium. The total $\mathcal{E}$ field intensity at any point in the atomic medium must then be written as

$$\mathcal{E}(z,t) = \mathcal{E}_1(z,t) + \mathcal{E}_2(z,t)$$
$$= \mathrm{Re}\left[\tilde{E}_1(z)\exp j(\omega_1 t - \beta_1 z) + \tilde{E}_2(z)\exp j(\omega_2 t - \beta_2 z)\right] \quad (8)$$

and so the total optical intensity $I(z,t)$, at any point $z$ and any instant of time $t$, must then in general be written in the form

$$I(z,t) = |\mathcal{E}(z,t)|^2 = \left|\tilde{E}_1(z)\right|^2 + \left|\tilde{E}_2(z)\right|^2$$
$$+ \tilde{E}_1^*(z)\tilde{E}_2(z)e^{j[(\omega_2-\omega_1)t-(\beta_2-\beta_1)z]} + \mathrm{c.c.} \quad (9)$$

We see that the local intensity will contain, in addition to the average intensities $|\tilde{E}_1|^2$ and $|\tilde{E}_2|^2$ associated with the two waves separately, an interference term proportional to the dot product $\tilde{E}_1^* \tilde{E}_2$. This interference term contains both a time-variation, at the "beat frequency" or difference frequency $\omega_{\mathrm{beat}} = \omega_2 - \omega_1$ between the two signals, and a spatial variation, with a spatial periodicity given by $\beta_2 - \beta_1$.

### Temporal Interference Terms

The interference between two signals with different frequencies $\omega_1$ and $\omega_2$ will thus produce a time-varying intensity at each point in the atomic medium, with a sinusoidal frequency equal to the beat frequency $\omega_{\mathrm{beat}}$. What this time-varying intensity does to the atomic medium, and particularly to the local population difference $\Delta N(t)$, depends on the difference frequency $\omega_{\mathrm{beat}}$ and especially on its value relative to the atomic time constants $T_1$ and $T_2$.

Often this sinusoidal modulation can be neglected, for several reasons. Suppose the difference frequency $\omega_2 - \omega_1$ between the two modes is large compared to any of the population recovery times $\tau$ or $T_1$, as it often is. (This difference frequency may, for example, represent an axial-mode beat frequency of several hundred megahertz or larger.) Then the time-varying part of this modulation will be so rapid that the atomic population difference will simply not respond to this frequency; and hence all the terms oscillating sinusoidally in time can be ignored.

In other situations the two waves $\tilde{E}_1$ and $\tilde{E}_2$ may have orthogonal polarizations, so that the vector dot product between them is zero. The interference terms that vary in time and space will then all be identically zero.

Finally, sometimes there may be not just two such ideal sine waves but in fact many such waves, with a significant spread in frequency. If this spread in frequency is sufficiently large—in other words, if the overall temporal coherence of the optical signal is not large—then the temporal interference effects between the multiple signals will tend to be washed out on the average, and only the total time-averaged intensity of the signals will be important.

### Cross-Modulation Effects

Note, however, that if any of the atomic properties, such as the gain or loss or phase shift in the atomic medium, do become significantly modulated at the difference frequency $\omega_{\text{beat}} = \omega_2 - \omega_1$, either by time-varying saturation effects or by other nonlinear mixing effects in the atomic medium, then the resulting modulation effects will produce frequency sidebands on both of the applied signals. In fact, the modulation of the $\omega_2$ optical signal by the time-variations at $\omega_{\text{beat}}$ will produce both an upper sideband at $\omega_2 + \omega_{\text{beat}} = 2\omega_2 - \omega_1$ and a lower sideband at $\omega_2 - \omega_{\text{beat}} = \omega_1$; while the $\omega_1$ signal will similarly acquire an upper sideband at $\omega_1 + \omega_{\text{beat}} = \omega_2$ and a lower sideband at $\omega_1 - \omega_{\text{beat}} = 2\omega_1 - \omega_2$.

In other words, *any type of nonlinear modulation or cross-saturation effects in the atomic response produced by the two signals will react back to couple or cross-modulate the two signals to each other* (as well as to produce new nonlinear mixing frequencies in the system). These nonlinear mixing or cross-modulation effects can become extremely complex, and also quite important in coupling together different frequency signals either in a laser medium or in other kinds of nonlinear optical materials.

### Standing-Wave Interference Effects

Even if the two optical waves are at the same frequency, they will still produce spatial (though not temporal) cross-modulation and cross-coupling effects. That is, even if $\omega_1 = \omega_2$ the intensity $I$ in Equation 8.26 will have a spatial variation of the form

$$I(z) = I_1(z) + I_2(z) + 2\sqrt{I_1 I_2}\cos[(\beta_2 - \beta_1)z + \phi], \qquad (10)$$

where $I_1$ and $I_2$ are the intensities of the two waves separately, and the sinusoidal standing-wave portion has a spatial phase angle $\phi$ related to the relative phases of the two $E$ fields.

If an intensity pattern of this form is present in a homogeneously saturable atomic medium, it will presumably produce a spatially varying saturation of the form

$$\frac{\Delta N(z)}{\Delta N_0} = \frac{1}{1 + I(z)/I_{\text{sat}}} = \frac{1}{1 + [I_1 + I_2 + 2(I_1 I_2)^{1/2}\cos(\Delta\beta z)]/I_{\text{sat}}}, \qquad (11)$$

as illustrated in Figure 8.7. This spatial variation can then considerably complicate the analysis of gain saturation, as well as introduce complex wave-coupling effects in laser problems. The spatial variation of the gain (or loss) saturation in an atomic medium, as illustrated in Figure 8.7(b), is commonly referred to as *spatial hole burning* in the medium.

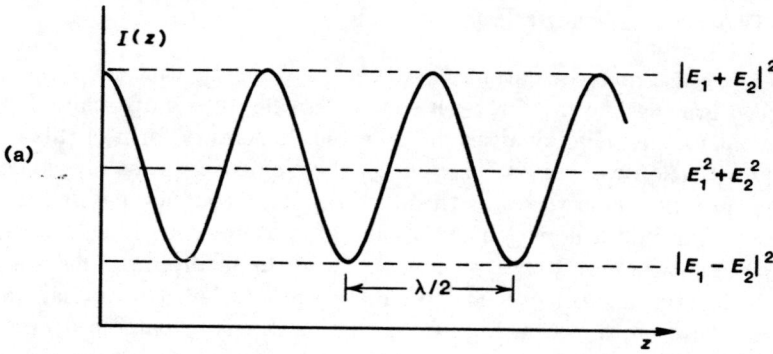

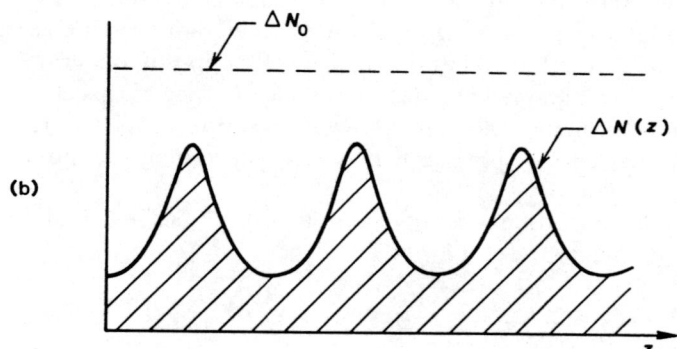

FIGURE 8.7
(a) Spatial intensity pattern produced by interference between two oppositely traveling waves having the same optical frequency. (b) Resulting spatial hole burning or saturation grating pattern of the population difference $\Delta N(z)$.

There are many ways in which the spatial interference effects between two or more waves can be washed out, however, so that we can merely add intensities—that is, merely write $I(z) = I_1 + I_2$, without the cross term—in order to calculate the total atomic saturation, as we will do in a number of later analyses.

First of all, if the waves are in fact at different frequencies, then the interference fringes or standing-wave patterns produced by the two beams will move, or sweep through the material, because of the temporal part of the interference effect. If the beat frequency is large and the material response is slow—that is, if $\omega_{\text{beat}} \gg 1/T_1$, as it often is—then the spatial saturation effects tend to be washed out. Crossed polarizations will also eliminate interference effects—in isotropic though not in anisotropic materials.

Finally, if not just two ideal plane waves are present, but instead many components with different $k$ vectors, then the standing-wave patterns between different waves can become sufficiently complex that many of the cross-coupling effects tend to be washed out on the average, and again only the average intensity in all the waves is significant.

### Two-Wave Coupling Effects

There are many situations, however, in which spatial coupling effects, or "induced grating effects," between signals can be quite important. Let us look somewhat further at an analysis of the most elementary form of this coupling.

The situation we will analyze here is the elementary case of two coherently related uniform plane waves at the same optical frequency passing in opposite directions through a homogeneously saturable atomic medium. The superposition of these two oppositely traveling waves in the medium produces a standing wave with intensity fringes whose period is equal to half the optical wavelength of either wave, as shown in Figure 8.7. If this intensity pattern occurs inside a saturable amplifying or absorbing medium, the net result is to create a greater degree of saturation at the peaks of the intensity profile and a lesser degree of saturation at the nulls.

The saturable medium thus develops a stratified character, and becomes in effect a volume interference grating or a volume hologram. (Such a grating produced by two waves traveling in more or less exactly opposite directions is sometimes referred to as a Lippman grating.) A simple analysis of this case will both demonstrate how to analyze nonlinear wave-interaction problems, and lead to some useful and not entirely obvious conclusions concerning the interaction between these two waves.

The general one-dimensional wave equation that applies in this situation can be written as

$$\frac{d^2 \tilde{E}(z)}{dz^2} + \beta^2 \tilde{E}(z) = -\omega^2 \mu \tilde{P}(z), \tag{12}$$

where $\tilde{E}(z)$ is the total electric field and $\tilde{P}(z)$ is the polarization in the atomic medium. The two waves traveling in the $+z$ and $-z$ directions are then written in the form

$$\mathcal{E}(z,t) = \text{Re}\left[\tilde{E}_1(z)e^{j(\omega t - \beta z)} + \tilde{E}_2(z)e^{j(\omega t + \beta z)}\right], \tag{13}$$

where we allow the possibility that each complex wave amplitude $\tilde{E}(z)$ may change with distance. Note that the plus sign in front of the propagation constant $\beta$ in the second term means that this wave is traveling to the left or toward $-z$.

We also assume that the atomic medium is a homogeneously saturable gain medium (or absorption medium) in which the signal is exactly on resonance. Hence the sinusoidal polarization $\tilde{P}(z)$ at any position $z$ can be written as

$$\tilde{P}(z) = \tilde{\chi}(\omega_a, z)\epsilon\tilde{E}(z) = j\chi''(\omega_a, z)\epsilon\tilde{E}(z), \tag{14}$$

where the susceptibility $\chi''(\omega_a, z)$ at any point will be the saturated value given by the expression

$$\chi''(\omega_a, z) = \frac{\chi_0''}{1 + I(z)/I_{\text{sat}}}. \tag{15}$$

Note that $I(z)$, the total intensity at position $z$, will contain a sinusoidal standing-wave variation of the type we have written above.

### Small-Saturation Approximation

To proceed further at this point, we must make the approximation that the degree of saturation produced in the atomic medium is comparatively weak, so that we can use the mathematical approximation $1/(1 + I/I_{\text{sat}}) \approx 1 - I/I_{\text{sat}}$ for $I/I_{\text{sat}} \ll 1$. We can then write the saturated susceptibility as

$$\chi''(\omega_a, z) \approx \chi_0'' \times [1 - I(z)/I_{\text{sat}}], \qquad I/I_{\text{sat}} \leq 0.2. \tag{16}$$

We make this approximation partly because it is often physically reasonable, but also because it would be much more difficult to proceed if we did not make it.

Putting the exact form for the intensity as given in Equation 8.9 into the wave equation 8.12 then expands this equation into the form

$$\left[\frac{d^2 \tilde{E}_1}{dz^2} - 2j\beta \frac{d\tilde{E}_1}{dz}\right] e^{-j\beta z} + \left[\frac{d^2 \tilde{E}_2}{dz^2} + 2j\beta \frac{d\tilde{E}_2}{dz}\right] e^{+j\beta z} \approx -\beta^2 \chi_0''$$
$$\times \left[1 - \frac{|\tilde{E}_1|^2 + |\tilde{E}_2|^2 + \tilde{E}_1^* \tilde{E}_2 e^{+2j\beta z} + \tilde{E}_1 \tilde{E}_2^* e^{-2j\beta z}}{I_{\text{sat}}}\right] \tag{17}$$
$$\times \left[\tilde{E}_1 e^{-j\beta z} + \tilde{E}_2 e^{+2\beta z}\right].$$

We can drop both of the second-derivative terms on the left-hand side of this equation, on the basis of the slowly varying envelope approximation, and then multiply out and match up the $e^{-j\beta z}$ and $e^{+j\beta z}$ traveling-wave terms on each side of this equation.

When we do this, we note that there is a product term between the $\tilde{E}_1 \tilde{E}_2^* e^{-2j\beta z}$ interference term in the saturation expression for $\chi''(\omega, z)$ and the left going wave term $\tilde{E}_2 e^{+j\beta z}$, and that this product term leads to an additional right going term $\tilde{E}_1 \tilde{E}_2 \tilde{E}_2^* e^{-j\beta z}$ on the right-hand side of the equation. Similarly, there is a product of $\tilde{E}_1^* \tilde{E}_2 e^{+2j\beta z}$ times $\tilde{E}_1 e^{-j\beta z}$, which leads to an additional $\tilde{E}_1^* \tilde{E}_1 \tilde{E}_2 e^{+j\beta z}$ term on the right-hand side. When all these cross-coupling and saturation terms are sorted out, the result is the pair of coupled equations

$$\frac{d\tilde{E}_1}{dz} \approx \pm \alpha_{m0} \left[1 - \frac{|\tilde{E}_1|^2 + 2|\tilde{E}_2|^2}{I_{\text{sat}}}\right] \times \tilde{E}_1 \tag{18}$$

and

$$\frac{d\tilde{E}_2}{dz} \approx \mp \alpha_{m0} \left[1 - \frac{2|\tilde{E}_1|^2 + |\tilde{E}_2|^2}{I_{\text{sat}}}\right] \times \tilde{E}_2, \tag{19}$$

where we have used $(1/2)\beta \chi_0'' = \alpha_{m0}$, and where the upper or lower signs apply depending on whether the atomic medium is an amplifying or absorbing medium.

These equations are sometimes referred to as a *third-order expansion* for the atomic response, since the derivative terms for, say, the wave amplitude $\tilde{E}_1$ contain not only a linear or first-order term proportional to $\tilde{E}_1$, but also third-order nonlinear terms of the form $\tilde{E}_1 \tilde{E}_1^* \tilde{E}_1$ and $\tilde{E}_2 \tilde{E}_2^* \tilde{E}_1$. If we want to keep track of intensities only, we can also use these equations to obtain

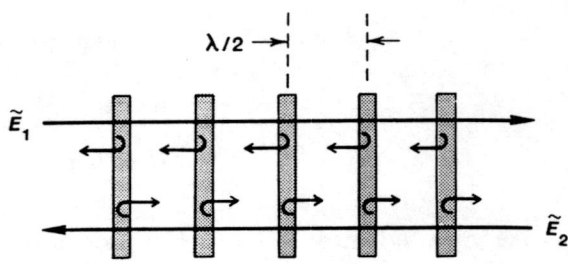

FIGURE 8.8
Two waves traveling in opposite directions through a nonlinear medium can create an induced grating (as in Figure 8.7) which can scatter each wave into the direction of the other wave, as shown in this figure.

$$\frac{dI_1}{dz} \approx \pm 2\alpha_{m0} \left[1 - \frac{I_1 + 2I_2}{I_{\text{sat}}}\right] \times I_1 \tag{20}$$

and

$$\frac{dI_2}{dz} \approx \mp 2\alpha_{m0} \left[1 - \frac{2I_1 + I_2}{I_{\text{sat}}}\right] \times I_2. \tag{21}$$

The most interesting thing to note here is the extra factor of 2 that appears in each of the cross-saturation terms, as compared to the self-saturation terms, in the preceding equations. These terms represent an important (and perhaps unexpected) result that emerges from this analysis. For some reason the cross-saturation between waves—that is, for example, the degree of gain saturation for wave #1 produced by the intensity of wave #2—is exactly twice the self-saturation of each wave by its own intensity.

### Grating Backscattering Effects

It is evident, in fact, both physically and from the way in which these terms arise in the equations, that the excess part of this cross-saturation effect is not additional "saturation" caused by the other wave, but results from a *grating backscattering effect*. That is, in physical terms the oppositely traveling waves $\tilde{E}_1$ and $\tilde{E}_2$ combine to produce a standing wave, which in turn produces a set of stratified layers or a thick grating in the partially saturated medium. This grating has exactly the correct spacing so that a portion of wave #1 gets backscattered into wave #2 and vice versa (see Figure 8.8).

This physical picture of backscattering from a standing-wave grating also makes it physically reasonable that the excess cross-saturation effect should wash out if waves $\tilde{E}_1$ and $\tilde{E}_2$ are sufficiently incoherent. If waves $\tilde{E}_1$ and $\tilde{E}_2$ contain different frequency components, or have a partially incoherent spatial pattern, the clean standing-wave grating pattern will tend to wash out or average out, and the associated backscattering or wave-coupling effects will be reduced or eliminated.

Standing-wave grating effects can play a significant role in saturable absorber mode locking of lasers, and especially in the mode competition between simultaneously oscillating modes traveling in opposite directions in a ring laser cavity, as well as in other laser experiments.

## REFERENCES

For a typical discussion of spatial hole burning and its effects, see T. Kimura, K. Otsuka, and M. Saruwatari, "Spatial hole-burning effects in a $Nd^{3+}$:YAG laser," *IEEE J. Quantum Electron.* **QE-7**, 225–230 (June 1971).

---

### Problems for 8.2

1. *Effects of drift or spatial offset in a grating or wave-coupling experiment.* Suppose for purposes of analysis that in a certain atomic medium the saturation effects are basically homogeneous, except that the saturation effect produced by the local signal intensity $I(z)$ at point $z$ tends to drift in space by a small amount $\epsilon$, so that the saturation produced by $I(z)$ actually occurs at point $z + \epsilon$. (Something like this might happen in a real system if the atomic medium is moving or flowing, so that the atoms move by a distance $\approx \epsilon$ before they decay; or if the medium is, for example, a semiconductor with strong internal fields that cause carriers to move in a certain direction.)

   For the coherent standing-wave grating discussed in this section, this would mean a spatial displacement by distance $\epsilon$ between the intensity standing wave $I(z)$ and the resulting spatial grating of $\tilde{\chi}(\omega, z)$. Analyze what this would do to the coupling between the two oppositely traveling waves, and discuss any new or interesting effects that may arise from this. (Note: certain so-called photo-refractive materials have refractive index gratings that are spatially displaced from the light intensity which produces them, and can as a result produce very interesting wave-coupling and wave-interaction effects.)

---

## 8.3 MORE ON LASER AMPLIFIER SATURATION

This section will present some additional details on laser amplifier saturation, including the effects of nonsaturable amplifier losses, saturation behavior in inhomogeneously broadened amplifiers, and transversely varying saturation effects.

### Amplifiers With Saturable Gain and Loss

Many practical laser amplifier systems will have both a homogeneously saturable gain coefficient $\alpha_m$ and a smaller but nonsaturating "ohmic" loss coefficient $\alpha_0$ due to host-crystal absorption, impurities, scattering losses, excited-state absorption, and other effects. The differential equation for signal growth along the amplifier then becomes

$$\frac{dI}{dz} = \frac{2\alpha_{m0}I}{1 + I/I_{\text{sat}}} - 2\alpha_0 I. \tag{22}$$

If this equation is integrated through an amplifier of length $L$, as we did for the lossless amplifier of Section 7.7, the more complicated relation connecting input and output signal intensities becomes

$$\ln\left[\frac{I_{\text{out}}}{I_{\text{in}}}\right] = 2\alpha_{m0}L - 2\alpha_0 L + \left(\frac{\alpha_{m0}}{\alpha_0}\right)\ln\left[\frac{\alpha_{m0} - \alpha_0(1 + I_{\text{out}}/I_{\text{sat}})}{\alpha_{m0} - \alpha_0(1 + I_{\text{in}}/I_{\text{sat}})}\right] \qquad (23)$$

or, in an alternative form,

$$\ln\left[\frac{G_0}{G}\right] = \left(\frac{\alpha_{m0}}{\alpha_0}\right)\ln\left[\frac{\alpha_{m0} - \alpha_0(1 + I_{\text{in}}/I_{\text{sat}})}{\alpha_{m0} - \alpha_0(1 + I_{\text{out}}/I_{\text{sat}})}\right], \qquad (24)$$

where $G \equiv I_{\text{out}}/I_{\text{in}}$ is the saturated gain at a given input intensity and $G_0 \equiv \exp(2\alpha_{m0}L - 2\alpha_0 L)$ is the net unsaturated gain minus loss. Either of these equations must be solved implicitly to find the input-output intensity relationship for an amplifier with given small-signal gain coefficient $2\alpha_{m0}L$ and loss coefficient $2\alpha_0 L$.

### Maximum Effective Amplifier Length

The deleterious effects of even relatively small amounts of loss in a laser amplifier can be appreciated, however, without making detailed input-output plots, by noting that if we make an arbitrarily long amplifier using such a medium, the effective gain coefficient given by

$$\frac{dI}{dz} = \left(\frac{2\alpha_{m0}}{1 + I/I_{\text{sat}}} - 2\alpha_0\right) I = 2\alpha'_m I \qquad (25)$$

will eventually saturate down to give zero further growth, i.e., $\alpha'_m \to 0$, as the intensity flux in the amplifier approaches a maximum value $I_{\text{max}}$ given by

$$I_{\text{max}} = (\alpha_{m0}/\alpha_0 - 1)\, I_{\text{sat}}. \qquad (26)$$

In other words, for a given small-signal gain $\alpha_{m0}$ and loss $\alpha_0$, there is a maximum possible signal flux which can build up in the amplifier. Alternatively, there is a maximum useful amplifier length given approximately by

$$L_{\text{max}} \approx \frac{1}{2(\alpha_{m0} - \alpha_0)} \ln\left[\frac{\alpha_{m0} - \alpha_0}{\alpha_0} \times \frac{I_{\text{sat}}}{I_{\text{in}}}\right]. \qquad (27)$$

Beyond this length the signal intensity never increases, because all the additional energy given to the signal by the inverted medium through stimulated emission is immediately absorbed and dissipated by the ohmic loss mechanisms.

Ohmic loss considerations can become important in solid state amplifiers when we wish to obtain particularly high gains and high pulsed flux densities. Losses due to scattering from materials imperfections, or to absorption by either weak ground-state or excited-state absorptions, can then limit the available gain and power output. Similar considerations apply also to dye laser amplifiers, and to high-power visible and ultraviolet gas laser devices, such as excimer lasers, in which both high power and high efficiency are very desirable.

High-power gas lasers in particular often employ moderately high-pressure mixtures of several different gases which are very highly excited by intense transverse arc discharges or electron-beam pumping. There may well be previously unknown or unexpected excited-state absorption lines in these mixtures that unfortunately overlap the desired laser transition; more than one originally promising high-power gas laser system has been eliminated by such unpleasant discoveries.

### Inhomogeneously Saturating Amplifiers

Although many high-power lasers do have the kind of homogeneous broadening that we have assumed in the discussion of saturation thus far, other useful laser materials (including most low-pressure doppler-broadened gas lasers) can be inhomogeneously broadened. We will learn in Chapter 30 that when a strongly inhomogeneous transition is subjected to a strong monochromatic signal, the gain (or loss) for that signal saturates in the inhomogeneous form

$$\frac{dI}{dz} = \frac{2\alpha_{m0}I}{(1+I/I_\text{sat})^{1/2}}. \tag{28}$$

The square root in the denominator comes about (as we will show in Chapter 30) because in an inhomogeneous line the applied signal saturates not only the spectral packet with which it is in exact resonance, but also other adjoining packets with which it has weaker interactions, thus "burning a hole" in the inhomogeneous line.

The analog to the input-output relationship for the homogeneous amplifier that we derived in the previous section then becomes

$$\int_{I=I_\text{in}}^{I=I_\text{out}} \left(\frac{1}{I^2} + \frac{1}{I I_\text{sat}}\right)^{1/2} dI(z) = \int_{z=0}^{z=L} 2\alpha_{m0}\, dz. \tag{29}$$

Performing the complicated integration in this case leads to the result

$$\frac{\sqrt{1+I_\text{out}/I_\text{sat}}-1}{\sqrt{1+I_\text{out}/I_\text{sat}}+1} = \frac{\sqrt{1+I_\text{in}/I_\text{sat}}-1}{\sqrt{1+I_\text{in}/I_\text{sat}}+1}$$

$$\times \exp\left[2\alpha_{m0}L - 2\sqrt{1+I_\text{out}/I_\text{sat}} + 2\sqrt{1+I_\text{in}/I_\text{sat}}\right] \tag{30}$$

This expression provides an implicit (and somewhat complex) way of computing $I_\text{out}$ versus $I_\text{in}$ for specified values of $G_0$ and $I_\text{sat}$ in the inhomogeneous case.

### Transversely Varying Saturation

The discussions of amplifier saturation thus far have also all been phrased in terms of the intensity (power per unit area) of the amplifying beam. If a beam has a flat transverse intensity profile, then we can simply multiply the intensity $I(z)$ by the cross-sectional area $A$ of the beam to get the total power $P(z)$ at any plane.

Real laser beams. however, typically have nonuniform transverse intensity profiles, with the gaussian transverse intensity profile being a common example. When a nonuniform beam passes through a saturable amplifier, the more intense parts of the beam saturate more rapidly than the weaker portions. The beam profile can thus be distorted, with in general the higher-intensity peaks being flattened out relative to the weaker parts.

The transverse profile of the amplifying beam may also change with distance because of diffraction effects as the beam propagates through the amplifier. Usually, however, amplifiers will be short enough and/or the beam diameters large enough that the beam will not have significant diffraction spreading; so diffraction effects will be far less important than spatially nonuniform saturation effects in the amplifier. To gain some insight into the latter, let us consider a few simple points about such saturation effects, using an elementary gaussian beam profile.

### Gaussian Beam Saturation

The transverse intensity distribution in a cylindrically symmetric gaussian beam with spot size $w$ may be written as

$$I(r) = \frac{2P}{\pi w^2} \exp\left(-\frac{2r^2}{w^2}\right). \tag{31}$$

(The factor of 2 appears in the exponent because $w$ is conventionally defined as the $1/e$ radius for the $E$ field amplitude.) The peak intensity of the gaussian is thus the same as if the total power $P$ were uniformly distributed over an area $A = \pi w^2/2$; and indeed we can show that if we consider a uniform-intensity beam having the same total power and the same intensity as the central peak intensity of the gaussian, then the effective area for this equivalent uniform beam appears to be $A_{\text{eff}} = \pi w^2/2$. This may not, however, be the best choice of effective area for a gaussian beam when we want to calculate saturation and power extraction effects.

Let us assume, as is often reasonable, that the amplifying beam is collimated and that diffraction effects are small. Then in essence each elemental cross-sectional area of the beam amplifies and saturates according to its own local intensity $I(r,z)$, independently of all other points in the cross section. The equation for local intensity in a homogeneously saturating amplifier will be

$$\frac{\partial I(r,z)}{\partial z} = \frac{2\alpha_{m0}I(r,z)}{1 + I(r,z)/I_{\text{sat}}}, \tag{32}$$

and the relation between input and output intensity profiles will be the same as in Equation 7.81, namely,

$$\ln\left[\frac{I_{\text{out}}(r)}{I_{\text{in}}(r)}\right] + \frac{I_{\text{out}}(r) - I_{\text{in}}(r)}{I_{\text{sat}}} = \ln G_0. \tag{33}$$

Given an input beam profile $I_{\text{in}}(r)$, we must in general solve this equation numerically for $I_{\text{out}}(r)$, and then integrate to find the total input and output powers $P_{\text{in}}$ and $P_{\text{out}}$.

Suppose, however, that an amplifier with a gaussian input profile is either short enough or heavily saturated enough that its overall saturated gain is small. Then the beam profile will not change greatly with distance, and we may assume that the beam remains gaussian at every plane $z$ through the amplifier. This differential gain equation may then be integrated over the amplifier cross section to give the result

$$\frac{dP(z)}{dz} = \int_0^\infty \frac{\partial I(r,z)}{\partial r} 2\pi r\, dr$$

$$= \frac{8\alpha_{m0}P(z)}{w^2} \int_0^\infty \frac{r \exp(-2r^2/w^2)\, dr}{1 + [2P(z)/\pi w^2 I_{\text{sat}}]\exp(-2r^2/w^2)} \quad (34)$$

$$= \pi w^2 \alpha_{m0} I_{\text{sat}} \ln\left[1 + \frac{2P(z)}{\pi w^2 I_{\text{sat}}}\right].$$

Consider first the short and weakly saturated case, where the gaussian beam power $P$ is small compared to the quantity $\pi w^2 I_{\text{sat}}/2$. Then by expanding the logarithm to second order, we can calculate that the power extraction in a short length $\Delta z$ will vary with incident power $P$ in the form

$$\Delta P \approx 2\alpha_{m0}P\left[1 - P/\pi w^2 I_{\text{sat}}\right]\Delta z \quad \begin{cases} \text{gaussian profile,} \\ \text{weak saturation.} \end{cases} \quad (35)$$

The analogous result for power extraction by a uniform beam having area $A$ and total power $P$, in the small-saturation limit, would be

$$\Delta P = \frac{2\alpha_{m0}P\Delta z}{1 + P/AI_{\text{sat}}} \approx 2\alpha_{m0}P\left[1 - P/AI_{\text{sat}}\right]\Delta z \quad \begin{cases} \text{uniform profile,} \\ \text{weak saturation.} \end{cases} \quad (36)$$

By comparing these two equations, we may conclude that *in the weak-saturation limit the effective area of a gaussian beam for power extraction is not $\pi w^2/2$ but $A_{\text{eff}} \approx \pi w^2$*. In physical terms, the outer wings of the gaussian beam (where much of the power is carried) are at low intensity, and thus do not saturate the laser medium. The gaussian beam therefore acts as if its area were larger than we might expect.

Consider next the heavily saturated (and hence still low-gain) gaussian amplifier when $P(z) \gg (\pi w^2/2)I_{\text{sat}}$. The power extracted from the laser medium in an incremental length $\Delta z$ may then be written as

$$\Delta P \approx \pi w^2 \alpha_{m0} I_{\text{sat}} \ln\left[1 + \frac{2P_1}{\pi w^2 I_{\text{sat}}}\right]\Delta z \quad \begin{cases} \text{gaussian profile,} \\ \text{strong saturation.} \end{cases} \quad (37)$$

The corresponding expression for a uniform beam in the high-saturation limit will be

$$\Delta P \approx A \times I_{\text{avail}} \Delta z \approx 2\alpha_{m0} I_{\text{sat}} A \Delta z \quad \begin{cases} \text{uniform profile,} \\ \text{strong saturation.} \end{cases} \quad (38)$$

By comparing these we can get a rough idea of how the effective saturation area of the gaussian beam increases at high intensities, as more and more of the gaussian beam profile rises above the saturation-intensity level.

## REFERENCES

The effects of amplifier losses and the concept of maximum useful amplifier length are discussed in A. Y. Cabezas, G. L. McAllister, and W. K. Ng, "Gain saturation in neodymium:glass laser amplifiers," *J. Appl. Phys.* **38**, 3487–3491 (August 1967).

An extensive analysis plus experimental results for the combination of saturable gain plus nonsaturating loss will be found in S. M. Curry, R. Cubeddu, and T. W. Hänsch, "Intensity stabilization of dye laser radiation by saturated amplification," *Appl. Phys.* **1**, 153–159 (1973).

Another reference on amplifier saturation, including inhomogeneous systems with hole broadening, is Kazantsev, Rautian, and Surdutovich, "Theory of a gas laser with nonlinear saturation," *Sov. Phys. JETP* **27**, 756 (1968).

---

Problems for 8.3

1. *Maximum available power in an inhomogeneously broadened laser amplifier.* Evaluate the available power that can be extracted from an inhomogeneously broadened single-pass laser amplifier in the limit of very large signals when $I_{in}$ and $I_{out} \gg I_{sat}$. Comment on differences between this case and the homogeneous case, and give a physical explanation of the difference. (You may need to refer to the discussions of hole burning and inhomogeneous saturation given in later chapters.)

2. *Power output versus power input for a gaussian beam profile in a homogeneously saturable amplifier.* Consider a collimated laser beam with a gaussian transverse profile passing through a homogeneously saturable single-pass laser amplifier. Assume that diffraction effects in passing through the amplifier length are small, so that each element of the beam cross section saturates essentially independently based on its local intensity. Set up a computer program to integrate the output intensity $I_{out}(r)$ across the output beam cross section to get the total output power $P_2$ for general values of unsaturated gain $G_0$ and normalized input power $P_1/\pi w^2 I_{sat}$. Evaluate in particular the extracted power $P_{extracted} \equiv P_2 - P_1$ versus $P_1$, and discuss the variation of the effective cross section of the gaussian beam for extracting energy as a function of the input power for some typical values of $G_0$.

3. *Behavior of a combined saturable amplifier and saturable absorber system.* Suppose an atomic medium contains both a saturable atomic gain with small-signal gain coefficient $\alpha_m$ and saturation intensity $I_m$; and a *saturable atomic loss* with small-signal loss coefficient $\alpha_0$ and saturation intensity $I_0$. Describe with the aid of appropriate sketches the behavior of the intensity $I(z)$ of a wave passing through such a medium (especially the behavior at large $z$) for all possible relative values of $\alpha_m$ and $\alpha_0$, and $I_m$ and $I_0$, and for an arbitrary initial input intensity $I(0)$. (A quantitative and graphic description, rather than any mathematical analysis, is what is wanted here.)

4. *Improving laser amplifier energy extraction by reshaping the laser medium: continuous.* Suppose a single-pass homogeneous cw laser amplifier is to be operated with a fixed input-signal power that falls in the saturating range for the laser amplifier. For such a situation the unsaturated input end of the amplifier gives full gain but inefficient energy extraction, whereas the more heavily saturated output end of the amplifier gives efficient energy extraction but not much gain.

It might seem that we could improve the overall performance of an amplifier using the same total volume of laser material by arranging to have a constant degree of gain saturation and energy extraction all through the amplifier. This might be done by tapering the cross-sectional area of the signal beam and the amplifier in such a way that, as the signal grows in power, it also grows in area, so that the intensity $I(z)$ just stays constant.

Suppose appropriate optics are provided to let us continuously expand the diameter of both the laser beam and the laser medium along the amplifier, in such a way that the laser intensity $I(z)$ (i.e., the power per unit area) stays constant with distance along the amplifier. Analyze this case to find the necessary change in amplifier cross section with distance and the resulting saturated power input-output relationship for the amplifier.

Compare the performance of this variable-cross-section amplifier to a constant-cross-section amplifier, assuming the same laser material with the same unsaturated gain coefficient $\alpha_{m0}$ and saturation intensity $I_{\text{sat}}$, the same overall amplifier length $L$, the same total volume $V$ of laser material, and the same signal power input $P_1$.

[For a related reference, see J. H. Jacob, et al., "Expanding beam concept for building very large excimer laser amplifiers," *Appl. Phys. Lett.* **48**, 318–320 (3 February 1986).]

5. *Improving laser amplifier energy extraction by reshaping the laser medium: in two steps.* Repeat the previous problem, but use only two fixed-diameter amplifier stages rather than a continuous variation.

That is, suppose a single-pass, homogeneous, cw, saturating amplifier of length $L$ is divided into two sections each of length $L/2$, with the same basic gain medium and hence the same total unsaturated gain through the two sections, but with different cross-sectional areas for the two sections. Assume the output beam from the first stage is appropriately magnified or demagnified by a suitable telescope between stages to match the change in area between the two stages. Since the cost of the power supplies and pumping hardware for a laser is likely to be more or less directly proportional to the volume of laser material that must be pumped, assume that the total volume of the two stages remains constant and equal to the original single-stage amplifier, but that the relative areas of the two stages may be changed.

Develop an analysis for the power output versus power input of this two-stage amplifier in the presence of saturation, allowing for different distributions of area between the two stages. Suppose the original single-stage amplifier is intended to operate with a certain specified power input sufficient to produce significant saturation in the single-stage device. Is it possible by going to two stages to increase the saturated gain and power output, as compared to the single-stage values, for the same specified power input?

6. *General analysis of output-power improvement by amplifier reshaping (research problem).* As an extension of the preceding problems, consider the more general question: Can we redistribute a given volume of laser material to obtain better overall performance than we get from a constant-cross-section amplifier, and how much better a performance can be obtained in this way? As design constraints for this question, assume: (a) a homogeneously saturating laser material with fixed gain coefficient $\alpha$ and saturation intensity $I_{\text{sat}}$; (b) a fixed overall amplifier length

$L$ (and hence fixed total small-signal gain); (c) fixed total amplifier volume $V$ (and hence fixed total materials cost and pumping-power requirements); and (d) fixed signal input power $P_1$. Assume, however, that optics can be provided to achieve any desired change in beam cross section along the amplifier. Consider then what sorts of tapered or axially varying amplifier cross sections might improve amplifier performance, especially for situations you can analyze either numerically or in closed form. What progress can you make in obtaining improved performance (at least in theory), either by increasing the saturated gain for a given energy extraction, or by increasing the energy extraction for a given saturated gain? (Note: this question leads to some interesting and not entirely expected answers.)

# 9
# LINEAR PULSE PROPAGATION

Extraordinarily short optical pulses can be generated in mode-locked lasers, and these pulses can then be amplified to very large energies in subsequent laser amplifiers. Such pulses can be used for laser ranging (laser radar, or lidar); for pulse-modulated optical communications, both in free space and especially along optical fibers; and as measurement probes for studying a very wide variety of ultrafast physical, chemical, and biological processes, in what has come to be known as picosecond spectroscopy.

Pulse propagation both in passive optical propagation systems and in laser amplifiers is therefore a subject of considerable practical interest. Understanding the propagation of optical pulses through both linear and nonlinear systems is also important in mode-locked lasers, in optical fibers and other propagation systems, and in picosecond spectroscopic applications.

In this chapter we first introduce some of the fundamental ideas of pulse propagation in linear systems, including the concepts of group and phase velocities, and pulse compression and broadening in linear dispersive systems and laser amplifiers. In the following chapter we will discuss some elementary concepts in the amplification and distortion of optical pulses caused by saturation in laser amplifiers, and also some of the interesting and useful effects, such as nonlinear pulse compression and soliton propagation, that occur in nonlinear dispersive fibers and other propagation systems.

## 9.1 PHASE AND GROUP VELOCITIES

In this section we will analyze some of the fundamental effects that can arise in pulse propagation through linear systems, especially systems with either group velocity or gain dispersion. Fundamental concepts we will examine include *pulse delay*, *group velocity*, and *pulse compression* or *group velocity dispersion effects*.

### Gaussian Pulses

The concepts we will introduce in this section occur for pulses of any shape. The analysis of these effects becomes particularly simple, however, with little if any of the physics being lost, if we analyze these effects using primarily gaussian pulses. Such pulses are simple, mathematically tractable, and clearly exhibit all the essential physical features. Many real systems such as actively mode-locked lasers generate pulses that are very close to complex gaussian pulses.

We consider as our basic model, therefore, an optical pulse with a carrier frequency $\omega_0$ and a complex gaussian envelope written in the form

$$\mathcal{E}(t) = \exp(-at^2)\exp j(\omega_0 t + bt^2) = \exp(-\Gamma t^2)\exp j\omega_0 t. \tag{1}$$

The complex gaussian parameter describing this pulse is thus

$$\Gamma \equiv a - jb. \tag{2}$$

(This $\Gamma$ has nothing to do with the axial wave propagation constant $\Gamma$ we used in an earlier chapter.) The instantaneous intensity $I(t)$ associated with this pulse can be written as

$$I(t) = |\mathcal{E}(t)|^2 = \exp(-2at^2) = \exp[-(4\ln 2)(t/\tau_p)^2], \tag{3}$$

so that the pulsewidth $\tau_p$, defined in the usual FWHM fashion, is related to the parameter $a$ by

$$\tau_p = \sqrt{\frac{2\ln 2}{a}}. \tag{4}$$

Note that this is the FWHM pulsewidth for the intensity $I(t)$, and not for the signal amplitude $\mathcal{E}(t)$.

### Instantaneous Frequency

The time-varying phase shift or phase rotation of the sinusoidal signal within this gaussian pulse is given by

$$\mathcal{E}(t) \propto \exp j(\omega_0 t + bt^2) = \exp[j\phi_{\text{tot}}(t)], \tag{5}$$

so that the total instantaneous phase of the signal is

$$\phi_{\text{tot}}(t) = \omega_0 t + bt^2 \tag{6}$$

What then is the "instantaneous frequency" $\omega_i(t)$ of this sinusoidal signal?

The total phase variation in this case can obviously be written in the form $\phi_{\text{tot}}(t) = \omega_0 t + bt^2 = (\omega_0 + bt)t$. We might therefore be led to say that the instantaneous frequency of the pulse at time $t$ should be written as $\omega_i(t) = \omega_0 + bt$. This is not, however, a correct interpretation.

The instantaneous radian frequency of an oscillating signal, in the way this term is usually interpreted, should instead be *the rate at which the total phase of the sinusoidal signal rotates forward*, or alternatively $2\pi$ times the number of cycles completed per unit time, as measured in any small time interval. In other words, the instantaneous frequency in radians per second is properly defined as

$$\omega_i(t) \equiv \frac{d\phi_{\text{tot}}(t)}{dt}. \tag{7}$$

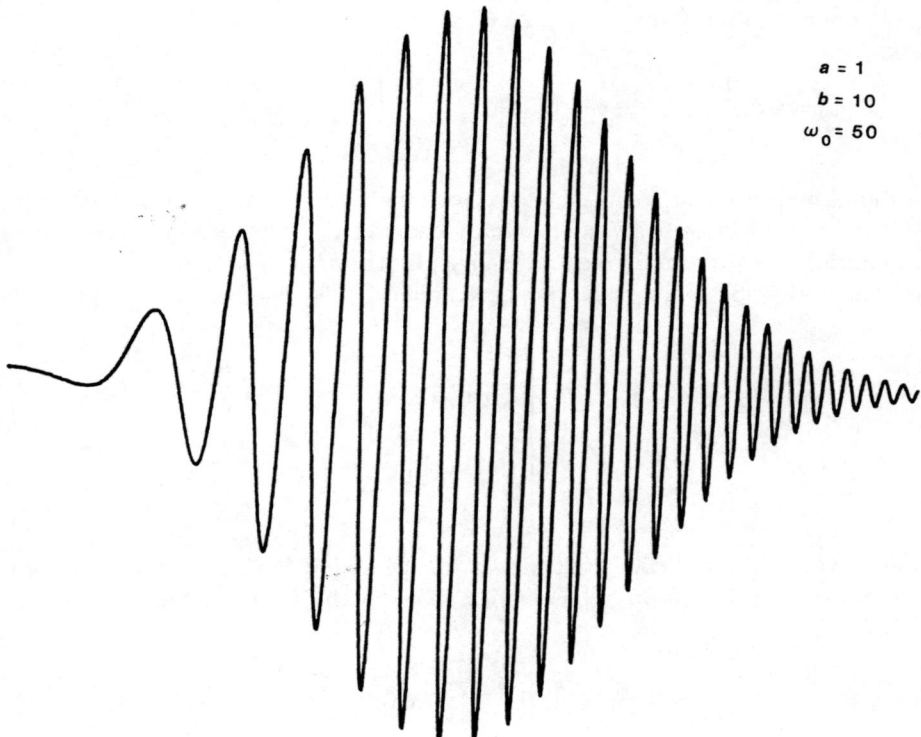

**FIGURE 9.1**
A chirped gaussian signal pulse.

In the complex gaussian case this gives

$$\omega_i(t) \equiv \frac{d}{dt}(\omega_0 t + bt^2) = \omega_0 + 2bt. \tag{8}$$

The factor of 2 in the time-varying part of this expression is important.

A gaussian pulse with a nonzero imaginary part $b$ thus has a linearly time-varying instantaneous frequency. Such a signal is often said to be *chirped*, with the parameter $b$ being a measure of this chirp. Figure 9.1 shows a rather strongly chirped gaussian pulse with a sizable variation of the instantaneous frequency within the pulse.

### Gaussian Pulse Spectrum

One of the major virtues of the gaussian pulse approach is that a gaussian pulse in time immediately Fourier-transforms into a gaussian spectrum in frequency, in the form

$$\mathcal{E}(t) = \exp(-\Gamma t^2 + j\omega_0 t) \quad \Rightarrow \quad \tilde{E}(\omega) = \exp\left[-\frac{(\omega - \omega_0)^2}{4\Gamma}\right]. \tag{9}$$

The exponent in the Fourier transform thus has a complex-quadratic dependence on frequency of the form

$$\tilde{E}(\omega) = \exp\left[-\frac{1}{4}\left(\frac{a}{a^2+b^2}\right)(\omega-\omega_0)^2 - j\frac{1}{4}\left(\frac{b}{a^2+b^2}\right)(\omega-\omega_0)^2\right]. \quad (10)$$

A signal with a linear frequency chirp (or quadratic phase chirp) in time automatically also has a quadratic imaginary component or quadratic phase shift of its Fourier spectrum in frequency, as given by the $b/(a^2+b^2)$ factor.

The power spectrum, or power spectral density, of this pulse is then given by

$$\left|\tilde{E}(\omega)\right|^2 = \exp\left[-\frac{1}{2}\left(\frac{a}{a^2+b^2}\right)(\omega-\omega_0)^2\right]$$
$$= \exp\left[-(4\ln 2)\left(\frac{\omega-\omega_0}{\Delta\omega_p}\right)^2\right], \quad (11)$$

where $\Delta\omega_p$ is the FWHM spectral width (in radians/second) of the pulse. We can convert this into a pulse bandwidth measured in Hz by writing

$$\Delta f_p \equiv \frac{\Delta\omega_p}{2\pi} = \frac{\sqrt{2\ln 2}}{\pi}\sqrt{a[1+(b/a)^2]}. \quad (12)$$

For a given pulsewidth in time as determined by the real parameter $a$, the presence of a frequency chirp as determined by the imaginary parameter $jb$ increases the spectral bandwidth $\Delta\omega_p$ by a ratio $\sqrt{1+(b/a)^2}$, as compared to an unchirped pulse with the same pulsewidth in time.

### Time-Bandwidth Products, and Transform-Limited Pulses

Combining the preceding equations shows in fact that the gaussian pulse has a *time-bandwidth product* given by

$$\Delta f_p \tau_p = \left(\frac{2\ln 2}{\pi}\right) \times \sqrt{1+(b/a)^2} \approx 0.44 \times \sqrt{1+(b/a)^2}. \quad (13)$$

The minimum or unchirped value of time-bandwidth product for a gaussian pulse is thus $\Delta f_p \tau_p \approx 0.44$. The presence of chirp increases this time-bandwidth product to a value given, in the limit of large chirp, by $\approx b/a$ times the minimum value.

This particular time-bandwidth product is the gaussian-pulse, FWHM version of a general Fourier theorem which says the time-bandwidth product of any pulsed signal is constrained by the uncertainty principle $\Delta f_{\rm rms}\Delta t_{\rm rms} \geq 1/2$, where $\Delta f_{\rm rms}$ and $\Delta t_{\rm rms}$ are the root-mean-square widths of the signal in frequency and in time. If one uses the rms definitions of $\Delta f$ and $\Delta t$, the time-bandwidth product for a chirped gaussian pulse is in fact the same as Equation 9.13, except that the 0.44 factor is replaced by exactly 0.5. More generally, the exact value of time-bandwidth product $\Delta f \Delta t$ for an arbitrary pulse shape depends on:

- the exact shape of the pulse (gaussian, square, exponential etc.);
- how $\Delta f$ and $\Delta t$ are defined (rms, FWHM, etc.); and
- especially on the amount of chirp or other amplitude or phase substructure within the pulse.

Pulses with little chirp or other internal substructure will have a time-bandwidth product close to the value of $\approx 0.5$. Such pulses are often referred to as *transform-limited pulses*. If separate measurements of pulsewidth and spectral width on a pulsed signal give a time-bandwidth product close to this limit, the pulsed signal must have little or no amplitude or phase substructure within the pulse duration. (See Problem 9.1-1 for some other examples of time-bandwidth products.)

### Dispersive Systems and "Omega-Beta Curves"

Consider now a dispersive atomic medium, or any other kind of dispersive wave-propagating system, such as a transmission line, waveguide, or optical fiber. By "dispersive" in this context we mean any linear system in which the propagation constant $\beta(\omega)$ as a function of frequency has any form other than a straight line through the origin, i.e., $\beta = \omega/c$.

We have shown plots in earlier chapters of the propagation constant $\beta(\omega)$, or the total phase shift $\phi(\omega) = \beta(\omega)L$, plotted versus frequency $\omega$ for various atomic systems. In discussing dispersive systems, however, it is convenient to plot $\omega$ versus $\beta$, rather than $\beta$ versus $\omega$, as shown in Figure 9.2. Such an "omega-beta plot" may represent the dispersive effect of an atomic transition or of the background index in a host medium, in which case it is called *material dispersion*. Alternatively, it may represent the propagation characteristics of a guided mode in some waveguiding system such as a microwave waveguide, an optical fiber, or a general filter network, in which case the dispersion is commonly referred to as *waveguide dispersion* or *modal dispersion*.

Suppose we are concerned with narrowband signals having frequency components primarily near some center frequency $\omega_0$. Then the propagation constant of a dispersive system can be conveniently expanded about its value at $\omega_0$ in the form

$$\beta(\omega) = \beta(\omega_0) + \beta' \times (\omega - \omega_0) + \frac{1}{2}\beta'' \times (\omega - \omega_0)^2, \tag{14}$$

where the derivatives $\beta' \equiv d\beta/d\omega$ and $\beta'' \equiv d^2\beta/d\omega^2$ are both evaluated at $\omega = \omega_0$.

Besides the frequency-dependent propagation constant $\beta(\omega)$, we might at the same time consider the effects of a frequency-dependent gain or loss coefficient $\alpha = \alpha(\omega) = \alpha(\omega_0) + \alpha' \times (\omega - \omega_0) + \frac{1}{2}\alpha'' \times (\omega - \omega_0)^2$ in the same system. Both of these frequency variations will distort or modify a pulse propagating through the system. We wish to focus at this point, however, on pulse propagation and pulse-compression phenomena due only to velocity dispersion. We will therefore ignore for now any frequency-dependent gain coefficient, and assume that any gains or losses are either zero or at least independent of frequency.

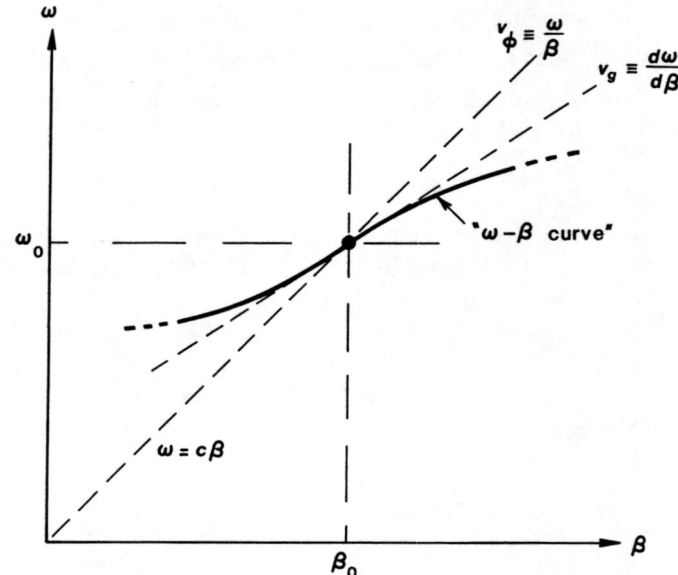

FIGURE 9.2
An "omega-beta" diagram for a dispersive wave-propagating system.

### Gaussian Pulse Propagation Through a Dispersive System

Suppose then that we put a gaussian pulse of the form

$$\mathcal{E}_0(t) = \exp(-\Gamma_0 t^2 + j\omega_0 t), \qquad \tilde{E}_0(\omega) = \exp\left[-\frac{(\omega-\omega_0)^2}{4\Gamma_0}\right], \qquad (15)$$

into such a dispersive system, where $\Gamma_0 \equiv a_0 - jb_0$ is the initial pulse parameter at the input to the system. The output pulse spectrum $\tilde{E}(z,\omega)$ after propagating a distance $z$ through such a system will be the input spectrum $\tilde{E}_0(\omega)$ of Equation 9.9, multiplied by the frequency-dependent propagation through the system, or

$$\tilde{E}(z,\omega) = \tilde{E}_0(\omega) \times \exp[-j\beta(\omega)z]$$
$$= \exp\left[-j\beta(\omega_0)z - j\beta' z \times (\omega-\omega_0) - \left(\frac{1}{4\Gamma_0} + \frac{j\beta'' z}{2}\right) \times (\omega-\omega_0)^2\right]. \quad (16)$$

The output pulse in time from this system will be the inverse Fourier transform of the output spectrum, or

$$\mathcal{E}(z,t) \equiv \int_{-\infty}^{\infty} \tilde{E}(z,\omega) e^{j\omega t}\, d\omega. \qquad (17)$$

With some minor manipulations, this integral can be put into the form

$$\mathcal{E}(z,t) = \frac{e^{j[\omega_0 t - \beta(\omega_0)z]}}{2\pi} \int_{-\infty}^{\infty} \exp\left[-\frac{(\omega-\omega_0)^2}{4\Gamma(z)} + j(\omega-\omega_0)(t-\beta' z)\right] d(\omega-\omega_0). \quad (18)$$

where $1/\Gamma(z) \equiv 1/\Gamma_0 + 2j\beta''$. In this form, the carrier-frequency time and space dependence have been moved out in front, so that the integral part of this expression gives the time and space dependence of the output pulse envelope. The output pulse is still a gaussian pulse, but with an altered gaussian pulse parameter $\Gamma(z)$ at the output of the system.

To interpret this mathematical result, we can carry out the integration explicitly using "Siegman's lemma," namely,

$$\int_{-\infty}^{\infty} e^{-Ay^2 - 2By}\, dy \equiv \sqrt{\frac{\pi}{A}} e^{B^2/A}, \qquad \mathrm{Re}[A] > 0 \tag{19}$$

or we can simply note that the integral in Equation 9.18 is obviously the Fourier transform of a gaussian pulse of the form $\exp(-\Gamma t^2)$, with a shift in time by $t - \beta' z$ included. From either approach, the output pulse after traveling any arbitrary distance $z$ through the system is given by

$$\begin{aligned}\mathcal{E}(z,t) &= \exp[j\,(\omega_0 t - \beta(\omega_0) z)] \times \exp[-\Gamma(z) \times (t - \beta' z)^2] \\ &= \exp\left[j\omega_0 \left(t - \frac{z}{v_\phi(\omega_0)}\right)\right] \times \exp\left[-\Gamma(z) \times \left(t - \frac{z}{v_g(\omega_0)}\right)^2\right],\end{aligned} \tag{20}$$

where $\Gamma(z)$ is the modified gaussian pulse parameter after traveling a distance $z$, and where $v_\phi(\omega_0) \equiv \omega_0/\beta(\omega_0)$ and $v_g(\omega_0) \equiv 1/\beta'(\omega_0)$.

### Phase Velocity

The first exponent in each line of Equation 9.20 says that in propagating through the distance $z$, the phase of the sinusoidal carrier frequency $\omega_0$ is delayed by a midband phase shift $\beta(\omega_0) z$, or by a midband phase delay $t_\phi$ (in time) given by

$$\text{phase delay,} \quad t_\phi = \frac{z}{v_\phi(\omega_0)} = \frac{\beta(\omega_0)}{\omega_0} z. \tag{21}$$

This says that the carrier-frequency cycles, or the sinusoidal waves within the pulse envelope, will appear to move forward with a *midband phase velocity* $v_\phi(\omega_0)$ given by

$$\text{phase velocity,} \quad v_\phi(\omega_0) = \frac{z}{t_\phi} = \frac{\omega_0}{\beta(\omega_0)}. \tag{22}$$

The midband phase velocity is thus determined by the propagation constant $\beta(\omega_0)$ at the carrier frequency $\omega_0$.

### Group Velocity

The second exponent in each line of Equation 9.20 says, however, that the *pulse envelope*, which remains gaussian but with a modified pulse parameter $\Gamma(z)$, is delayed by the *group delay time* $t_g$ given by

$$\text{group delay,} \quad t_g = \frac{z}{v_g(\omega_0)} = \beta' z. \tag{23}$$

That is, the pulse envelope appears to move forward with a *midband group velocity* $v_g(\omega_0)$ given by

$$\text{group velocity,} \quad v_g(\omega_0) = \frac{1}{(d\beta/d\omega)}\bigg|_{\omega=\omega_0} = \left(\frac{d\omega}{d\beta}\right)_{\omega=\omega_0}. \tag{24}$$

If we could take instantaneous "snapshots" of the pulse fields $\mathcal{E}(z,t)$ from Figure 9.1 as the pulse propagated through the system, we would see the (invisible) pulse envelope moving forward at the group velocity $v_g(\omega_0)$, while the individual cycles within the pulse envelope moved forward at the phase velocity $v_\phi(\omega_0) \equiv \omega_0/\beta(\omega_0)$. For $v_g(\omega_0) < v_\phi(\omega_0)$, for example, we would appear to see cycles of the carrier frequency walking into the pulse envelope from the back edge and disappearing out through the front edge of the pulse envelope, while the envelope itself moved forward at a slower velocity.

### Pulse Compression

Finally, the gaussian pulse parameter $\Gamma(z)$ at distance $z$, relative to the value $\Gamma_0$ at the input, is given from Equations 9.16 and 9.18 by

$$\frac{1}{\Gamma(z)} = \frac{1}{\Gamma_0} + 2j\beta''z. \tag{25}$$

The change in $\Gamma(z)$ is determined by $\beta''$, the second derivative of the propagation constant at line center. We will discuss the meaning of this formula, which includes *pulse compression* in particular, in much more detail in the following section.

### Summary

The successive coefficients in the power series expansion of $\beta(\omega)$ thus have the meanings

$$\begin{aligned}
\beta &\equiv \beta(\omega)|_{\omega=\omega_0} = \frac{\omega_0}{v_\phi(\omega_0)} \equiv \frac{\omega_0}{\text{phase velocity}}, \\
\beta' &\equiv \frac{d\beta}{d\omega}\bigg|_{\omega=\omega_0} = \frac{1}{v_g(\omega_0)} \equiv \frac{1}{\text{group velocity}}, \\
\beta'' &\equiv \frac{d^2\beta}{d\omega^2}\bigg|_{\omega=\omega_0} = \frac{d}{d\omega}\left(\frac{1}{v_g(\omega)}\right) \equiv \text{"group velocity dispersion."}
\end{aligned} \tag{26}$$

The physical interpretation of these coefficients in terms of **group and phase velocities**, although derived here using gaussian pulses, is very **general, and applies** to any sort of pulse signal. If a pulsed signal has a carrier frequency $\omega_0$ within a pulse envelope of any shape, and this pulse propagates through **any lossless linear** system that has a midband propagation constant $\beta(\omega_0)$ and **a first-order linear** variation $\beta' \times (\omega - \omega_0)$ across the pulse spectrum, then the **carrier-frequency** cycles within the **pulse will move** forward at the phase velocity $v_\phi$, while the pulse envelope itself **will move** forward at the group velocity $v_g$ evaluated at the center of the pulse spectrum. The pulse envelope itself may also **change in shape** with distance because of the $\beta''$ term, as we will discuss in the **following section.**

## REFERENCES

For an excellent survey of ultrashort optical pulses and the **various ways of gener**ating, artificially compressing, and applying them, see the **various chapters included** in *Ultrashort Light Pulses*, ed. by S. L. Shapiro (Topics in Applied Physics, Vol. 18, Springer-Verlag, 1977).

There are many discussions in the literature of dispersive wave propagation, group and phase velocities, and pulse velocity and pulse distortion. A collection of classic early papers by A. Sommerfeld and L. Brillouin is given in L. Brillouin, *Wave Propagation and Group Velocity* (Academic Press, 1960).

For a list of more recent references, work backward starting from S. C. Bloch, "Eighth velocity of light," *Am. J. Phys.* **45**, 538–549 (June 1977).

## Problems for 9.1

1. *Time-bandwidth products for various optical pulseshapes.* Evaluate the time-bandwidth products, using both rms and FWHM definitions of pulsewidth and bandwidth, for (a) a square pulse of width $T$ in time; (b) a double-sided exponential pulse $\mathcal{E}(t) = \exp(-|t/T|)$; (c) a single-sided exponential pulse $\mathcal{E}(t) = \exp(-t/T)$ for $t > 0$ and $\mathcal{E}(t) = 0$ for $t < 0$; and (d) a secant-squared pulse $\mathcal{E}(t) = \text{sech}^2(t/\tau_p)$. (Note: There are some unanticipated difficulties in this problem.)

## 9.2 THE PARABOLIC EQUATION

There is an alternative and somewhat more general way to derive the linear pulse propagation results we are presenting in this chapter, by using the so-called "parabolic wave equation." This parabolic equation is widely used in the professional literature, and it also brings out an interesting analogy between dispersive pulse spreading and diffractive optical beam spreading. We will therefore give a brief derivation of the parabolic equation in this section, although we will not make any further direct use of it here. Readers in a hurry for results may therefore want to skip over this short section.

### Derivation of the Parabolic Equation

The basic wave equation for a one-dimensional signal in a dispersive medium, or on a dispersive transmission line, may be written as

$$\frac{\partial^2 \mathcal{E}(z,t)}{\partial z^2} - \mu_0 \epsilon_0 \frac{\partial^2 \mathcal{E}(z,t)}{\partial t^2} = \mu \frac{\partial^2 p(z,t)}{\partial t^2}, \qquad (27)$$

where $p(z,t)$ is the potentially dispersive but linear polarization of the medium or transmission line. (In more sophisticated problems, a nonlinear polarization may be included here also.) Suppose we write this field $\mathcal{E}(z,t)$ in the form

$$\mathcal{E}(z,t) \equiv \text{Re}\, \tilde{E}(z,t) e^{j[\omega_0 t - \beta(\omega_0) z]}, \qquad (28)$$

where $\omega_0$ is again a carrier or midband frequency for the signal, with propagation constant $\beta(\omega_0)$ at this midband frequency, and $\tilde{E}(z,t)$ is taken to be the complex envelope of the pulsed signal.

We can write the polarization $p(z,t)$ in terms of its Fourier transform $\tilde{P}(z,\omega)$ in the form

$$p(z,t) = \frac{1}{2\pi} \int_{-\infty}^{\infty} \tilde{P}(z,\omega) e^{j\omega t} \, d\omega. \tag{29}$$

Assume this polarization arises from a linear but possibly dispersive response in the medium or transmission line. Then, we may write it in terms of the electric field in the form

$$\tilde{P}(z,\omega) = \tilde{\chi}(\omega) \epsilon_0 \tilde{E}(z,\omega), \tag{30}$$

where $\tilde{E}(z,\omega)$ is the Fourier transform of $\mathcal{E}(z,t)$ given by

$$\begin{aligned}\tilde{E}(z,\omega) &= \int_{-\infty}^{\infty} \mathcal{E}(z,t) e^{-j\omega t} \, dt \\ &= \int_{-\infty}^{\infty} \tilde{E}(z,t) e^{j(\omega_0 - \omega)t} \, dt \end{aligned} \tag{31}$$

and where $\tilde{\chi}(\omega)$ is the dispersive susceptibility of the propagation system.

By using these definitions, plus standard Fourier transform theorems, we can write the polarization term on the right-hand side of Equation 9.27 as

$$\begin{aligned}\frac{\partial^2 p(z,t)}{\partial t^2} &= -\frac{1}{2\pi} \int_{-\infty}^{\infty} \omega^2 \tilde{P}(z,\omega) e^{j\omega t} \, d\omega \\ &= -\frac{\epsilon_0}{2\pi} \int_{-\infty}^{\infty} \omega^2 \tilde{\chi}(\omega) e^{j\omega t} \, d\omega \int_{-\infty}^{\infty} \tilde{E}(z,t') e^{j[\omega_0 t' - \beta(\omega_0)z]} \, dt' \end{aligned} \tag{32}$$

The derivation of the parabolic equation then proceeds by expanding the quantity $\omega^2 \tilde{\chi}(\omega)$ in Equation 9.32 about its midband value in the form

$$\begin{aligned}\omega^2 \tilde{\chi}(\omega) &\approx \omega_0^2 \tilde{\chi}(\omega_0) + \frac{d}{d\omega}\left[\omega^2 \tilde{\chi}(\omega)\right] \times (\omega - \omega_0) \\ &\quad + \frac{1}{2}\frac{d^2}{d\omega^2}\left[\omega^2 \tilde{\chi}(\omega)\right] \times (\omega - \omega_0)^2 + \cdots. \end{aligned} \tag{33}$$

with all derivatives evaluated at $\omega = \omega_0$. This is of course exactly the same approximation as in the expansion of $\beta(\omega)$ in the previous section. It is then possible to evaluate the polarization integral of Equation 9.32 by making use of the convenient identities

$$\frac{1}{2\pi} \int_{-\infty}^{\infty} e^{j(\omega-\omega_0)(t-t')} \, d\omega \equiv \delta(t-t') = \delta(t'-t) \tag{34}$$

as well as

$$\frac{1}{2\pi} \int_{-\infty}^{\infty} (\omega - \omega_0) e^{j(\omega-\omega_0)(t-t')} \, d\omega \equiv j\delta^{(1)}(t-t') = -j\delta^{(1)}(t'-t) \tag{35}$$

and

$$\frac{1}{2\pi} \int_{-\infty}^{\infty} (\omega - \omega_0)^2 e^{j(\omega-\omega_0)(t-t')} \, d\omega \equiv -\delta^{(2)}(t-t') = -\delta^{(2)}(t'-t). \tag{36}$$

## 9.2 THE PARABOLIC EQUATION

In these relations, $\delta^{(n)}(t)$ indicates an $n$-th order derivative of the Dirac delta function, with the useful property that

$$\int_{-\infty}^{\infty} \delta^{(n)}(t-t_0) f(t)\, dt \equiv \left. \frac{d^n f(t)}{dt^n} \right|_{t=t_0} \tag{37}$$

when applied to any reasonable function $f(t)$. We will also use the various identities that

$$\beta^2(\omega) \equiv \omega^2 \mu_0 \epsilon_0 + \mu_0 \epsilon_0 \omega^2 \tilde{\chi}(\omega) \tag{38}$$

as well as

$$2\beta \frac{d\beta}{d\omega} \equiv 2\omega \mu_0 \epsilon_0 + \mu_0 \epsilon_0 \frac{d}{d\omega}\left[\omega^2 \tilde{\chi}(\omega)\right] \tag{39}$$

and

$$\left(\frac{d\beta}{d\omega}\right)^2 + \beta \frac{d^2 \beta}{d\omega^2} \equiv \mu_0 \epsilon_0 + \frac{\mu_0 \epsilon_0}{2} \frac{d^2}{d\omega^2}\left[\omega^2 \tilde{\chi}(\omega)\right], \tag{40}$$

where $[d\beta(\omega)/d\omega]_{\omega=\omega_0} \equiv v_g(\omega_0)$ is the midband group velocity in the system; and we will use the slowly varying envelope approximation to drop second derivatives of the pulse envelope $\tilde{E}(z,t)$ with respect to distance $z$.

When all these approximations are inserted and all the algebra is cleared away, the basic wave equation for $\mathcal{E}(z,t)$ given in Equation 9.27 reduces to the *parabolic equation for the pulse envelope* $\tilde{E}(z,t)$ given by

$$\frac{\partial \tilde{E}(z,t)}{\partial z} + \frac{1}{v_g} \frac{\partial \tilde{E}(z,t)}{\partial t} - j\frac{\beta''}{2} \frac{\partial^2 \tilde{E}(z,t)}{\partial t^2} = 0. \tag{41}$$

where $v_g$ and $\beta''$ are both evaluated at midband, $\omega = \omega_0$. This equation is called the parabolic equation both because of the parabolic expansion of $\omega^2 \tilde{\chi}(\omega)$ used in deriving it, and because of the second-derivative term in $t$ which appears as a consequence of this approximation.

### Group Velocity and Group Velocity Dispersion

We can note first that if the second-derivative term $\beta'' \equiv 0$, then this equation is obviously satisfied by any solution of the form $\tilde{E}(z,t) \equiv \tilde{E}(z - v_g t)$, where $v_g$ is the midband value at $\omega = \omega_0$ as defined in Equations 9.23 and 9.24. This shows that the group velocity concept for propagation of the pulse envelope $\tilde{E}(z,t)$ applies to much more than just the gaussian pulses discussed in the preceding section. For any reasonably narrowband pulsed or modulated signal, the pulse or modulation envelope moves forward at velocity $v_g$, whereas the individual optical cycles move forward at velocity $v_\phi$.

If the $d^2\beta/d\omega^2$ term is nonzero, however, the propagation system will have a "group velocity dispersion," or a variation of group velocity with frequency, as we will discuss in the following sections. The $j(\beta''/2)\, \partial^2 \tilde{E}/\partial t^2$ term in Equation 9.41 then acts like a kind of generalized complex diffusion term for the pulse envelope $\tilde{E}(z,t)$ in the time coordinate. This "complex-valued diffusion" leads to pulse broadening, pulse compression, and pulse reshaping effects that we will discuss in more detail in the following sections.

### Alternative Form

There is also an alternative form for the parabolic equation, in which we begin with a pulseshape defined as

$$\mathcal{E}(z,t) \equiv \operatorname{Re} \tilde{E}(z,\eta)e^{j(\omega_0 t - \beta(\omega_0)z)}$$
$$= \operatorname{Re} \tilde{E}(z, t - z/v_g)e^{j(\omega_0 t - \beta(\omega_0)z)}, \qquad (42)$$

so that $\eta \equiv t - z/v_g$ is a displaced time coordinate whose origin $\eta = 0$ is centered at the time of arrival of the pulse at each plane $z$. The parabolic equation (9.41) then simplifies to the form

$$\frac{\partial \tilde{E}(z,\eta)}{\partial z} - \frac{j}{2}\left(\frac{d^2\beta}{d\omega^2}\right)\frac{\partial^2 \tilde{E}(z,\eta)}{\partial \eta^2} = 0. \qquad (43)$$

Obviously if the dispersion term $d^2/d\omega^2 \equiv 0$, then the pulse shape $\tilde{E}(z,\eta)$ becomes independent of $z$, or $\tilde{E}(z,\eta) \equiv \tilde{E}_0(\eta) \equiv \tilde{E}_0(t-z/v_g)$, as we have discussed. This form offers a slightly simpler way to express the same ideas.

### Space-Time Analogy

The parabolic equation derived in this section has exactly the same mathematical form as the paraxial wave equation used in optical beam propagation analyses (Chapter 16), if we identify the delayed time coordinate $t - z/v_g$ (or $\eta$) in the parabolic equation with either of the transverse spatial coordinates $x$ or $y$ in the paraxial equation. The dispersive or second-derivative term that leads to broadening (or compression) of a pulse's time envelope with distance $z$ in the parabolic equation then plays exactly the same role as the diffractive term that leads to transverse spreading of a laser beam's transverse profile with distance $z$ in the paraxial equation. There is thus a very close analogy between signal pulse distortion with propagation distance in the dispersive equation, and changes in transverse beam profile with propagation distance due to diffraction effects in the paraxial situation. An optical wavefront with positive or negative wavefront curvature (imaginary quadratic dependence on $x$ or $y$ in the exponent) is directly analogous to an optical signal with positive or negative chirp (imaginary quadratic dependence on $t$ in the exponent); and this wavefront curvature may lead to a converging or diverging optical-beam profile, just as chirp may lead to pulse compression or expansion with distance.

As another example, an initially square signal pulse propagating on a dispersive transmission line will broaden into a $(\sin t)/t$ pulseshape after a long enough distance, exactly as a uniform plane wave coming through a rectangular slit in the near field will broaden into a $(\sin x)/x$ beam pattern in the far field. This general approach can give useful insights into the relationship between pulse-distortion effects on dispersive lines and beam-spreading effects in diffractive propagation.

### REFERENCES

The space-time analogy between dispersive pulse compression in time and optical-beam focusing in space is clearly illustrated in E. B. Treacy, "Optical pulse compression with diffraction gratings," *IEEE J. Quantum Electron.* **QE–5**, 454–458 (September 1969).

This same analogy is also developed in more detail, and applied to both linear and nonlinear examples, by S. A. Akhmanov, A. P. Sukhorukov, and A. S. Chirkin, "Stationary phenomena and space-time analogy in nonlinear optics," *Sov. Phys. JETP* **28**, 748-757 (April 1969). Another paper by the same group is "Nonstationary nonlinear optical effects and ultrashort light pulse formation," *IEEE J. Quantum Electron.* **QE-4**, 598-605 (October 1968).

---

Problems for 9.2

1. *Parabolic equation derivation.* Carry through the detailed steps leading from Equation 9.27 to 9.41 in this section.

---

## 9.3 GROUP VELOCITY DISPERSION AND PULSE COMPRESSION

If a pulse propagates through a system in which the group-velocity dispersion term $\beta'' \times (\omega - \omega_0)^2$ has a significant amplitude, then we must consider not only the phase and group velocities as discussed in the preceding sections, but also the fact that the pulseshape itself will be significantly changed in propagating through the system. Interesting and useful effects, such as pulse compression, pulse spreading, and pulse reshaping, can result from such second-order dispersion effects. Once again it is very convenient to derive and illustrate such effects using a chirped gaussian pulse model, as we will show in this section.

### Gaussian Pulse Propagation

From the analysis of the preceding section, if we put a pulse with initial pulse parameter $\Gamma_0 = a_0 - jb_0$ through a dispersive propagation system whose propagation constant has nonzero second derivative $\beta''$ at the carrier frequency of the pulse, then the change in the complex pulseshape parameter $\Gamma(z)$ with propagation distance $z$ through the system will be given by

$$\frac{1}{\Gamma(z)} = \frac{1}{\Gamma_0} + 2j\beta''z = \frac{a_0}{a_0^2 + b_0^2} + j\left(\frac{b_0}{a_0^2 + b_0^2} + 2\beta''z\right)$$

$$= \frac{1}{a(z) - jb(z)} = \frac{a(z)}{a^2(z) + b^2(z)} + j\frac{b(z)}{a^2(z) + b^2(z)}. \tag{44}$$

This result can be interpreted graphically by noting that the quantity $1/\Gamma(z)$ moves along a vertical straight-line trajectory in the complex $1/\Gamma$ plane with increasing propagation distance $2\beta''z$, starting from an initial point $1/\Gamma_0$, as indicated in Figure 9.3.

Since the real part of $1/\Gamma(z)$ determines the pulse bandwidth, it is evident from the left-hand part of Figure 9.3 that the pulse bandwidth stays constant, as it obviously should do in the absence of gain narrowing. This trajectory in the $1/\Gamma$ plane can then be converted to a trajectory in the $a - jb$ plane (or more conveniently in the $a + jb \equiv \Gamma^*$ plane) by simply inverting each complex point through the origin. An inversion of this form always converts a straight line in

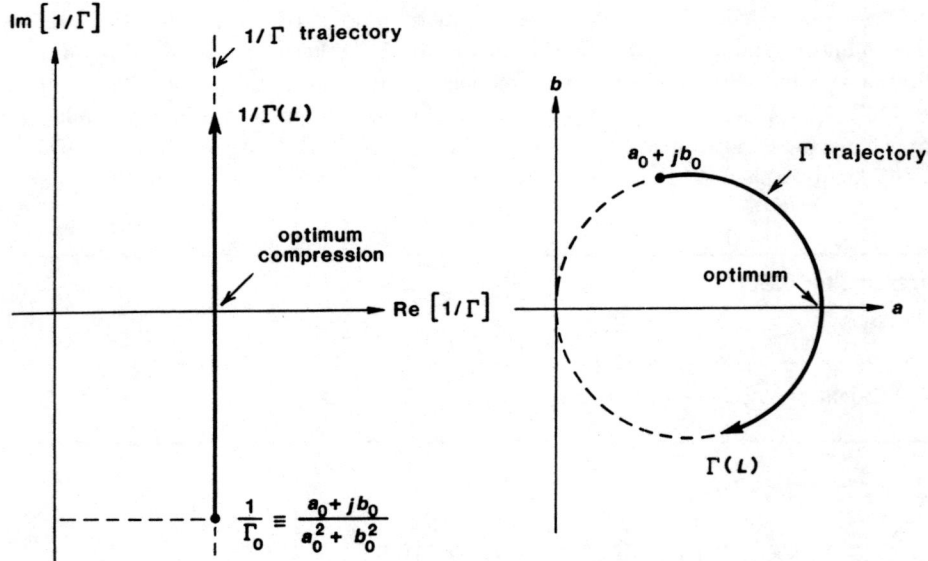

**FIGURE 9.3**
Trajectories for dispersive pulse propagation and pulse compression in the $\Gamma$ and $1/\Gamma$ planes.

the $1/\Gamma$ plane to a circle in the $\Gamma$ or $\Gamma^*$ plane. Propagation through a distance $2\beta''z$ now represents propagation about an arc of this circle, as illustrated on the right-hand side of Figure 9.3.

### Differential Approach

These same results can also be derived using a differential approach. If we differentiate the complex beam parameter $\Gamma(z)$ with respect to distance along the system, we obtain the differential equation

$$\frac{d\Gamma(z)}{dz} = -2j\beta'' \times \Gamma^2(z) \tag{45}$$

and this in turn separates into the two equations

$$\frac{da(z)}{dz} = -4\beta'' a(z)b(z), \qquad \frac{db(z)}{dz} = 2\beta'' \left[a(z)^2 - b(z)^2\right]. \tag{46}$$

The solutions to these equations are a family of circular trajectories, such as the trajectory indicated in Figure 9.3.

### Pulse Compression

The pulse parameters at the output of a length $z$ of the dispersive line are

$$a(z) = \frac{a_0}{(1 + 2\beta'' z b_0)^2 + (2\beta'' z a_0)^2} \tag{47}$$

and

$$b(z) = \frac{b_0(1 + 2\beta'' z b_0) + 2\beta'' z a_0^2}{(1 + 2\beta'' z b_0)^2 + (2\beta'' z a_0)^2}. \tag{48}$$

## 9.3 GROUP VELOCITY DISPERSION AND PULSE COMPRESSION

The point where the trajectory of $\Gamma(z)$ crosses the positive real axis in either the $\Gamma$ or $1/\Gamma$ plots obviously corresponds to the maximum value of the output parameter $a(z)$, and hence to the minimum value of the output pulsewidth $\tau_p(z)$. As the pulse propagates from the initial value $\Gamma_0$ to this point, the pulse is being *compressed in width*, or *shortened in time*, at least with the choice of parameters we have used in Figure 9.3. Beyond this point, the pulse broadens again in time. This pulse compression for a chirped pulse passing through a dispersive propagation system can be very useful in a wide variety of not only optical but also microwave and radio frequency applications, as we will see later.

### Optimum Compression Length

Suppose we can adjust the total dispersion $2\beta''z$, by changing either the group velocity dispersion $\beta''$ or the distance $z$ that is traveled. The size of this parameter is related to the distance traveled along the straight-line trajectory or, in a more complicated way, to the arc length traveled along the circle in Figure 9.3. What dispersion length $2\beta''z$ is then needed to reach the minimum pulsewidth point, starting with a given input pulse parameter $\Gamma_0$? Differentiating the quantity $a(z)$ with respect to the parameter $2\beta''z$ shows that the maximum value of $a(z)$ occurs for an optimum propagation distance related to the input pulse parameters by

$$(2\beta''z)_{\text{opt}} = -\frac{b_0}{a_0^2 + b_0^2} \approx -\frac{1}{b_0} \quad \text{if} \quad b_0 \gg a_0. \tag{49}$$

The output pulse parameters at this optimum distance are given by

$$a_{\text{opt}} = a_0\left[1 + (b_0/a_0)^2\right] \approx b_0^2/a_0 \quad \text{and} \quad b_{\text{opt}} \equiv 0. \tag{50}$$

After propagating an optimum distance through the system, the output pulse is compressed in time down to a minimum pulsewidth $\tau_{p,\text{min}}$ which is related to its input pulsewidth $\tau_{p0}$ and to its initial pulse parameters $a_0$ and $b_0$ by

$$\frac{\tau_{p,\text{min}}}{\tau_{p0}} = \sqrt{\frac{1}{1 + (b_0/a_0)^2}} \approx \left|\frac{a_0}{b_0}\right| \quad \text{if} \quad b_0 \gg a_0. \tag{51}$$

A large initial chirp compared to the pulsewidth, or $b_0 \gg a_0$—which is the same thing as a large initial time-bandwidth product—leads to the possibility of large pulsewidth compression (Figure 9.4); whereas an initial condition such that $b_0 \leq a_0$ permits only a negligible pulse compression. It is also evident that at the optimum compression point $b_{\text{opt}} = 0$, meaning that all the chirp has been removed. A gaussian pulse will be compressed all the way down to its minimum time-bandwidth product $\Delta f_p \tau_{p,\text{min}} = 0.44$ at the optimum point.

Again these results, although derived for a gaussian pulse, are in fact quite general conclusions, which by no means apply only to gaussian pulses. All signals with large initial time-bandwidth products are potentially compressible; signals with near-transform-limited initial time-bandwidth products are not.

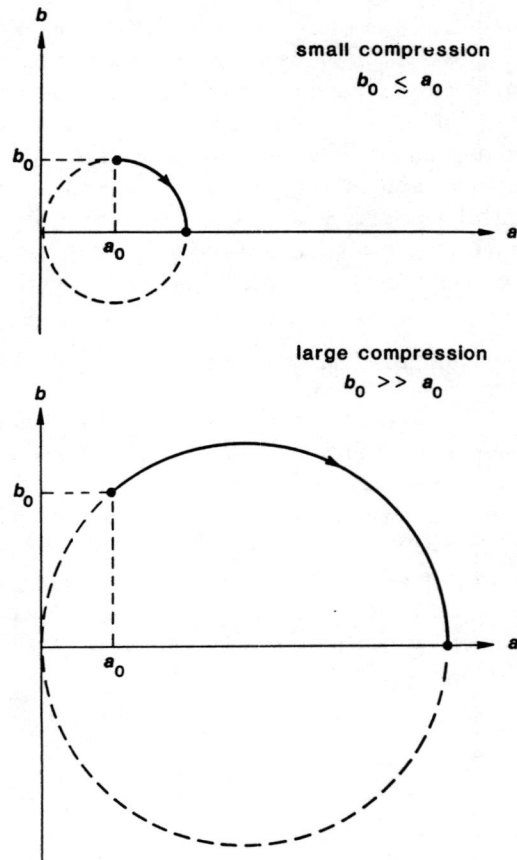

FIGURE 9.4
Optimum pulse compression with the same initial pulsewidths but smaller and larger initial time-bandwidth products.

### Physical Interpretation

We can give a physical interpretation of the preceding results as follows. A quadratic variation in $\beta(\omega)$ means that the group velocity, which is related to the linear variation of $\beta(\omega)$, must itself vary significantly across the pulse spectrum. The dispersion parameter $\beta''$ is related in fact to the so-called *group-velocity dispersion* $dv_g(\omega)/d\omega$ by

$$\beta'' = \frac{d}{d\omega}\left(\frac{1}{v_g(\omega)}\right) = -\frac{1}{v_g^2(\omega_0)}\frac{dv_g(\omega)}{d\omega}. \tag{52}$$

Hence the group velocity $v_g(\omega)$ as a function of the frequency deviation of a signal away from $\omega_0$ will be given by

$$v_g(\omega) \approx v_g(\omega_0) - \beta'' v_g^2(\omega_0) \times (\omega - \omega_0); \tag{53}$$

i.e., the group velocity itself is frequency dependent.

Consider then a strongly chirped pulse whose instantaneous frequency varies with time in the form $\omega_i(t) = \omega_0 + 2bt$ at the input plane to a linear system. Instead of thinking of this as a single pulse with carrier frequency $\omega_0$, let us mentally break this pulse up into a number of segments or subpulses, each with a slightly different carrier frequency, and hence a slightly different group velocity.

## 9.3 GROUP VELOCITY DISPERSION AND PULSE COMPRESSION

Suppose in particular that the center portion of the pulse, which has the central frequency $\omega_0$, leaves $z = 0$ at $t_0 = 0$, and travels a distance $z$ with a group delay given by $t_{d0} = z/v_g(\omega_0)$. Any other part of the pulse starting at some slightly earlier or later time $t_1$ has an instantaneous carrier frequency $\omega_1 \approx \omega_0 + 2b(t_1 - t_0)$. Hence we can, in a crude way, say that this other portion of the pulse will travel with slightly different group velocity $v_g(\omega_1)$.

Let us assume that the chirp $b_0$ is $> 0$, so that the instantaneous frequency $\omega_1$ is greater than $\omega_0$ for $t_1 > t_0$ (i.e., the part of the pulse that starts late). Then we can say that this part of the pulse travels at a velocity

$$v_{g1} \approx v_g(\omega_0) - 2\beta'' v_g^2(\omega_0) b_0 (t_1 - t_0). \tag{54}$$

Hence it travels the distance $z$ in a time

$$t_{d1} \approx \frac{z}{v_g(\omega_0) - 2\beta'' v_{g0}^2 b_0 (t_1 - t_0)} \approx \frac{z}{v_g(\omega_0)} \left[1 - 2\beta'' v_g(\omega_0) b_0 (t_1 - t_0)\right]. \tag{55}$$

In order for the reduction in travel time for this portion of the pulse to just match the amount $t_1 - t_0$ by which it started late, so that it will exactly catch up with the center of the pulse, we should have

$$t_{d0} - t_{d1} = t_1 - t_0 \tag{56}$$

Substituting the above equations into this leads to the condition

$$2\beta'' L \approx -1/b_0. \tag{57}$$

which is the same as the optimum result for large chirp given above.

We can thus view the pulse-compression process as one in which different parts of a chirped pulse, which start out down the line at slightly earlier or later times, also have slightly different frequencies. They can then travel slightly more slowly or rapidly down the line because of group-velocity dispersion, in such a way that they just exactly catch up with the central portion of the pulse.

### Pulse Compression With Other Pulseshapes: Chirp Radars

We have discussed pulse compression for the analytically tractable case of a gaussian pulse. In other cases, however, it may be necessary to work with other pulseshapes.

In certain radar systems, for example, it is easy to generate rectangular pulses which have constant output amplitude during a long time duration, thus combining low peak power with large total energy per pulse. Suppose we then give these same pulses a linear frequency chirp within the pulse as shown in Figure 9.5. A pulse like this, after propagating through a properly designed dispersive system, can also be substantially compressed in time (by roughly its initial time-bandwidth product). However it will also be distorted in shape, and generally will acquire side-lobes something like a sinc function, as illustrated in Figure 9.5.

Pulses much like this are often used in microwave chirped radar systems, since they can combine a comparatively long low-intensity pulse, easily obtainable from a microwave transmitter, with the much sharper range resolution achieved by using substantial pulse compression in the microwave receiver. Such systems are commonly referred to as *chirped radar systems*. The name dates back to an early classified memo during World War II which described such systems under

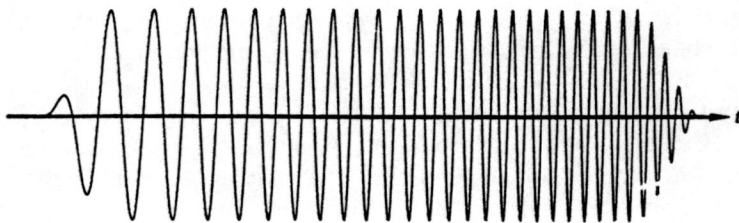

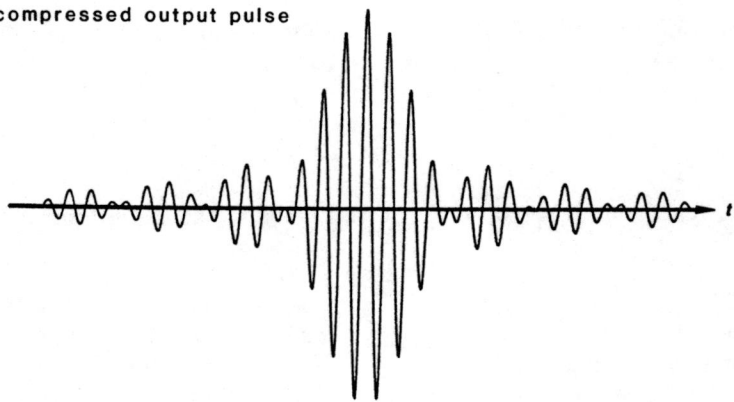

FIGURE 9.5
Pulse compression and reshaping with a square input pulse envelope.

the title "Not With a Bang But a Chirp." The echo-locating properties of bats are also related to a form of sonic chirped radar.

### Other Dispersive Optical Systems

In addition to dispersive atomic media, and to dispersive propagation effects in waveguiding systems such as optical fibers, various other dispersive optical systems have been invented and used, particularly to compress naturally or deliberately chirped laser pulses.

One such system is the Gires-Tournois interferometer (Figure 9.6). This device is simply a lossless etalon with a partially reflecting front surface and a 100% reflecting back surface, so that there is regenerative interference between the front and back surfaces. If the back surface is truly 100% reflecting, and the material between the surfaces is sufficiently lossless, then this device must have a reflectivity magnitude equal to unity at all frequencies. The interference between the front and back surfaces leads, however, to a periodic phase-versus-frequency curve for the complex reflectivity, which varies periodically with the axial mode spacing of the etalon. Portions of this phase-versus-frequency curve can then exhibit the correct dispersion to be used as a pulse-compression method. It is

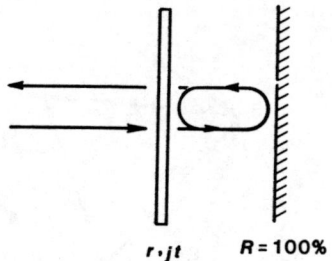

FIGURE 9.6
A Gires-Tournois interferometer has 100% amplitude reflectivity, but a frequency-dependent phase variation.

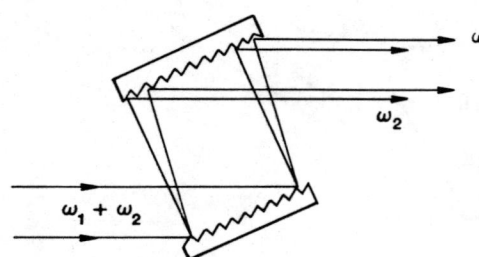

FIGURE 9.7
A pair of diffraction gratings used as a dispersive optical element.

physically obvious, however, that the interferometer itself must be physically very thin compared to the physical length of the laser pulse, or it will obviously break a single laser pulse into two or more multiply reflected pulses, rather than compressing it.

Another and much more useful dispersive system consists of a pair of gratings arranged as shown in Figure 9.7. Different frequencies or wavelengths have a different geometrical propagation distance through this system because they diffract from the gratings at slightly different angles. Such grating pairs have been used very successfuly to compress short chirped optical pulses in several different experiments, as will be discussed later.

Still another kind of dispersive system is a sequence of prisms as shown in Figure 9.8. Since the angular dispersion from a prism is generally smaller than from a diffraction grating, prism systems generally produce considerably smaller dispersion effects than grating pairs. On the other hand the insertion losses are also much smaller with prisms. Systems such as Figure 9.8 can thus be placed inside laser cavities to provide small corrections to the round-trip group velocity dispersion which are important in controlling the mode-locking behavior in very short-pulse lasers.

### Sign of the Group-Velocity Dispersion

The second derivative of the dispersion parameter $\beta$ with respect to frequency can be related to other frequency or wavelength derivatives in the forms

$$\beta'' \equiv \frac{d^2\beta(\omega)}{d\omega^2} = \frac{4\pi^2 c_0}{\omega^3} \frac{d^2 n(\lambda_0)}{d\lambda_0^2} = -\frac{1}{v_g^2} \frac{dv_g(\omega)}{d\omega}, \qquad (58)$$

where $c_0$ and $\lambda_0$ are the velocity of light and the optical wavelength in free space, and $n(\lambda_0)$ is the index versus wavelength in a dispersive medium.

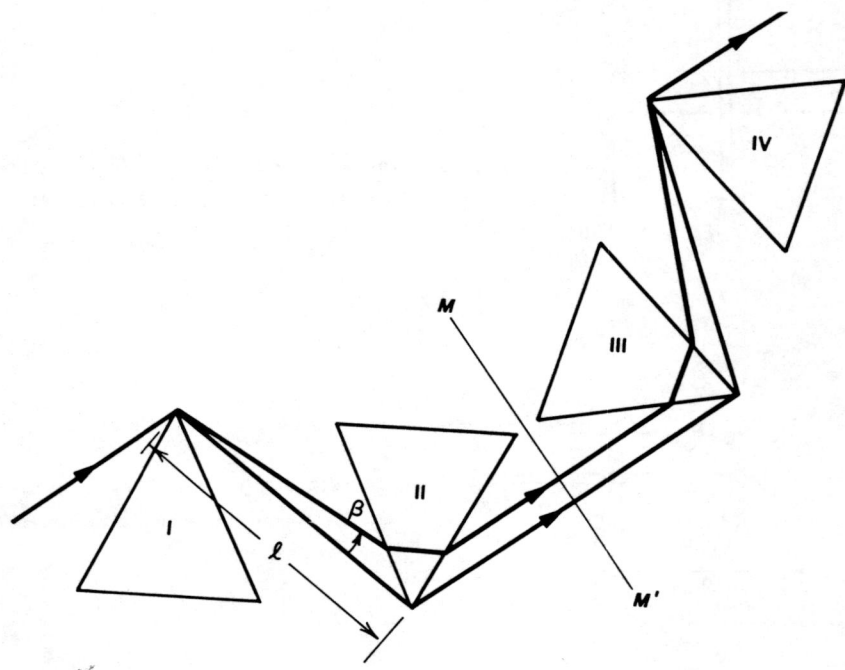

**FIGURE 9.8**
A sequence of prisms which can be adjusted to give a negative value of group velocity dispersion.

The usual practice in optics texts has been to plot index of refraction $n(\lambda_0)$ of a dispersive medium versus free-space wavelength $\lambda_0$, and then to speak of regions where this plot is concave upward (positive $d^2n/d\lambda_0^2$) as regions of *positive dispersion*, and regions of opposite sign as regions of *negative dispersion*. Positive dispersion in this sense thus means positive values for both $\beta''$ and $d^2n/d\lambda_0^2$, but a negative derivative for $dv_g/d\omega$, i.e., for group velocity versus frequency. Most common optical materials exhibit positive dispersion in the visible region, turning to negative dispersion somewhere in the near infrared.

With the recent advent of extremely short (femtosecond-time-scale) pulses, as well as growing realization of the role that self-chirping and pulse compression play in the mode-locked lasers that generate these pulses, there is growing interest in low-loss dispersive systems which can generate dispersion of either sign. The prism configuration in Figure 9.8 can be designed to give either negative or positive dispersion, with the additional advantages that the beams enter and leave the prisms at Brewster's angle, and there is neither displacement nor deviation of the input and output beam paths.

## REFERENCES

A detailed survey of microwave chirped radar concepts is given by J. R. Klauder, A. C. Price, S. Darlington, and W. J. Albersheim, "The theory and design of chirp radars," *Bell Sys. Tech. J.* **39**, 745–808 (July 1960). See also C. E. Cook, "Pulse compression—key to more efficient radar transmission," *Proc. IRE* **48**, 310–316 (1960); M. L. Skolnik, *Introduction to Radar Systems* (McGraw-Hill, 1962), pp. 493–500; and S. C. Bloch,

"Introduction to chirp concepts with a cheap chirp radar," *Am. J. Phys.* **41**, 857–864 (July 1973).

The first suggestions for the chirping of optical pulses and then their subsequent compression seem to have come from F. Gires and P. Tournois, "Interféromètre utilisable pour la compression d'impulsions lumineuses modulées en fréquence," *Compt. Rend. Acad. Sci. (Paris)* **258**, 6112–6115 (June 1964); and J. A. Giordmaine, M. A. Duguay, and J. W. Hansen, "Compression of optical pulses," *IEEE J. Quantum Electron.* **QE–4**, 252–255 (May 1968).

One of the earliest experiments was by M. A. Duguay and J. W. Hansen, "Compression of pulses from a mode-locked He-Ne laser," *Appl. Phys. Lett.* **14**, 14–16 (January 1969).

For a description of optical pulse compression using diffraction grating pairs, see E. B. Treacy, "Optical pulse compression with diffraction gratings," *IEEE J. Quantum Electron.* **QE–5**, 454–458 (September 1969); or J. Desbois, F. Gires, and P. Tournois, "A new approach to picosecond laser pulse analysis, shaping and coding," *IEEE J. Quantum Electron.* **QE–9**, 213–218 (February 1973).

The prism configuration shown in Figure 9.8. and other related designs, are outlined in R. L. Fork, O. E. Martinez, and J. P. Gordon, "Negative dispersion using pairs of prisms," *Opt. Lett.* **9**, 150–152 (May 1984), and J. P. Gordon and R. L. Fork, "Optical (ring) resonator with negative dispersion," *Opt. Lett.* **9**, 153–155 (May 1984).

Problems for 9.3

1. *Phase shift versus frequency analysis for the Gires-Tournois interferometer.* Assuming that the voltage reflection coefficient of a Gires-Tournois interferometer can be written as $\tilde{g}_{\text{refl}}(\omega) = \exp[-j\phi(\omega)]$, calculate and plot a few curves of $\phi''(\omega) \equiv d^2\phi(\omega)/d\omega^2$ versus $\omega$ over one full free spectral range for two or three different values of the front mirror power reflectivity $R \equiv r^2$. Derive approximate analytical formulas for the maximum value of $\phi''(\omega)$, and the value of $\omega T$ at which this occurs, where $T = 2L/c$ is the round-trip transit time inside the interferometer.

2. *Usefulness of the Gires-Tournois interferometer?* Extend the results of the previous problem to show that the Gires-Tournois interferometer is in fact a fairly lousy pulse-compression device in practical terms.

## 9.4 PHASE AND GROUP VELOCITIES IN RESONANT ATOMIC MEDIA

Particularly strong and interesting dispersion effects can occur when signals are tuned close to the transition frequency of a narrow atomic resonance in an absorbing or amplifying atomic medium. In this section we give a brief description of these atomic dispersive effects, showing how they confirm both the absorption and especially the phase-shift properties of a resonant atomic transition.

## Phase and Group Velocities Near Atomic Transitions

The total phase shift for a wave making a single pass through a laser amplifier (or an absorbing atomic medium) can be written as $\exp[-j\phi_{tot}(\omega)] = \exp[-j(\beta + \Delta\beta_m)L]$, where the total phase shift consists of

$$\phi_{tot}(\omega) \equiv [\beta(\omega) + \Delta\beta_m(\omega)]L = \frac{\omega L}{c} + \frac{\beta L}{2}\chi'(\omega), \tag{59}$$

The first term gives the basic "free-space" phase shift $\omega L/c$ through the laser medium, a phase shift which is large and linear in frequency; the second term is the small added shift $\Delta\beta_m(\omega)L$ due to the atomic transition. The phase velocity $v_\phi(\omega)$ of the wave in the medium is then given by

$$v_\phi(\omega) = \frac{\omega L}{\phi_{tot}(\omega)} \tag{60}$$

and the group velocity $v_g(\omega)$ by

$$v_g(\omega) = \frac{L}{d\phi_{tot}(\omega)/d\omega} = \frac{L}{(d/d\omega)[\beta(\omega) + \Delta\beta_m(\omega)]}. \tag{61}$$

The phase velocity in a medium with an atomic transition is thus given by

$$v_\phi(\omega) = \frac{\omega}{\beta(\omega) + \Delta\beta_m(\omega)} = \frac{c}{1 + \chi'(\omega)/2}, \tag{62}$$

and the group velocity is given by

$$v_g(\omega) = \frac{v_\phi(\omega)}{1 - (\omega/v_\phi)(dv_\phi/d\omega)}. \tag{63}$$

Figure 9.9 shows these quantities for a wave passing through a resonant amplifying laser medium, with the atomic or $\chi'(\omega)$ contributions very much exaggerated.

The group velocity over the central portion of the amplifying bandwidth in a laser medium is slightly *slower* than the free-wave velocity in the medium. A physical explanation for this is that as a pulse travels through the medium, the leading edge of the pulse must first build up a coherent induced polarization in the inverted atomic transition, before this polarization can begin radiating back into the input pulse to amplify it. This build-up, however, requires a short but finite build-up time, on the order of $T_2$, as described in the preceding chapter. The leading edge of the pulse thus gets slightly "under-amplified" compared to the steady-state gain of the medium, and by similar arguments the continuing reradiation of the oscillating atoms slightly "over-amplifies" the trailing edge of the pulse. The net pulse envelope in an amplifying medium thus appears to travel slightly more slowly than the free-space wave velocity.

## Group Velocities Faster Than the Velocity of Light?

The phase and group velocities in Figure 9.9 are associated with an amplifying atomic medium. Going from an amplifying to an absorbing medium will reverse the signs of both $\chi''$ and $\chi'$, and thus reverse the sign of all the atomic phase-shift contributions in this figure.

The careful reader may then note that for a strongly absorbing atomic transition the group velocity at the center of the transition can apparently become

## 9.4 PHASE AND GROUP VELOCITIES IN RESONANT ATOMIC MEDIA

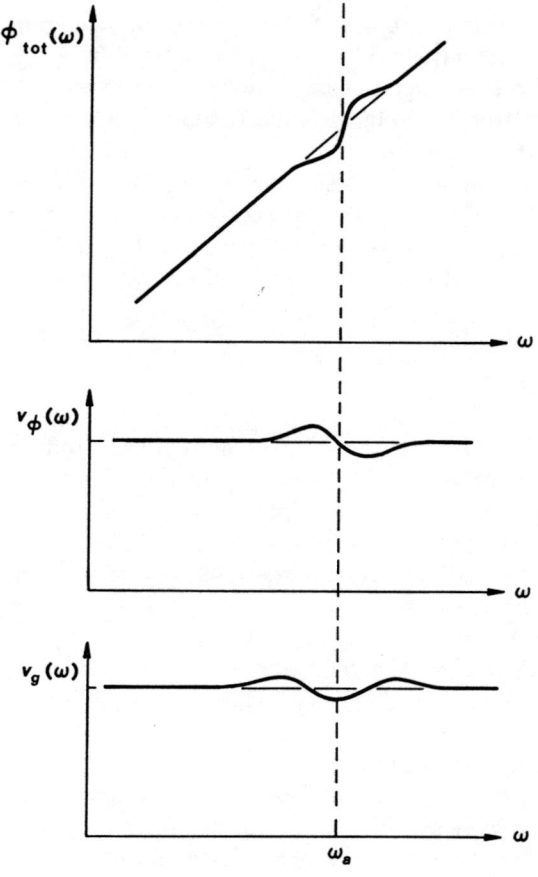

FIGURE 9.9
Phase shift, phase velocity, and group velocity versus frequency for propagation through an inverted atomic transition.

*faster* than the velocity of light $c$ in the host medium, and also the group velocity off line center can become negative for strong enough absorbing transitions. In fact, for a strong enough transition in a gas, $v_g$ might even become greater than the velocity of light $c_0$ in vacuum. But this would seem to be in serious conflict with a fundamental axiom of relativity, that no signal or information can ever be transmitted at a velocity greater than $c_0$.

The resolution to this apparent paradox lies in the fact that the group velocity $v_g$, as defined here, can significantly exceed $c_0$ only at the center of a very strongly absorbing transition. But this is also a situation in which any incident signal will be both strongly attenuated and distorted by this attenuation. Both detailed calculations and experimental measurements for a pulsed signal sent into a strongly absorbing atomic medium, with a carrier frequency anywhere within the absorption line, have been carried out in recent years (see References). These results have shown that when a smooth pulse is sent into a strongly absorbing medium, the observed signal-pulse envelope will indeed appear to travel at very close to the group velocity $v_g \equiv d\omega/d\beta$ in almost all cases, with very little distortion of the pulse shape, even when $v_g$ has values that are greater than $c_0$, or even negative (i.e., the peak of the pulse appears to come out of the absorbing sample before it goes in.

This is not a violation of special relativity, however, nor does it mean that the signal is transmitted at greater than the (vacuum) speed of light. The pulses used for all such calculations (and measurements) necessarily always have long tails that are weak but finite. In passing through the lossy medium, different portions of the pulse spectrum are attenuated and phase-shifted very differently, in such a way that the peak of pulse envelope appears to move faster than $c$. In reality, however, the pulse is being severely modified under the envelope; and no part of it is actually moving faster than light.

When such calculations are done for an input signal with a sharp discrete leading edge, it is always found that no output ever emerges at the output face at a time earlier than the transit time $L/c_0$ through the system at the vacuum velocity of light. (For more discussion of this point, see some of the references at the end of this section.)

### Group Velocities Much Slower Than Light

The dispersion in the wings of a strongly absorbing transition can produce a group velocity much less than the phase velocity $c$ in the medium (and also obviously less than $c_0$ in free space); and if the entire signal spectrum of a pulse is sufficiently far out in the wings, the net absorption for the pulse may simultaneously be made very small. With a sufficiently strong and narrow transition, one can thus obtain a very large slowing of the group velocity of a pulse, even while there is very small net absorption of the pulse energy.

As an example of this, cells filled with alkali vapor (such as rubidium or sodium) have been used to produce some rather large (though comparatively narrowband) dispersive effects at wavelengths on the sides of the strong resonance absorption lines in these atoms. In experiments done at the IBM Laboratories, Yorktown Heights, N.Y., the pulse-envelope group delay was measured when short pulses from a tunable dye laser were sent through a gas absorption cell one meter long. By using pulses whose center frequencies were located far out on the wings of a very strongly absorbing transition (i.e., by operating at $|\omega - \omega_a| \gg \Delta\omega_a$), the experimenters were able to obtain a power absorption loss of only $\approx 20\%$ per pass, and yet have a group velocity as slow as 1/10 to 1/15 the velocity of light.

## REFERENCES

A detailed analysis of pulse propagation and apparent group velocity in strongly absorbing or amplifying media is given by C. G. B. Garrett and D. E. McCumber, "Propagation of a gaussian light pulse through an anomalous dispersion medium," *Phys. Rev. A* **1**, 305–313 (February 1970).

The series of striking experiments demonstrating group velocities much less than the velocity of light, obtained by operating far from line center on very strongly absorbing atomic transitions in sodium and rubidium vapor cells, are described by D. Grischkowsky in "Adiabatic following and slow optical pulse propagation in rubidium vapor," *Phys. Rev. A* **7**, 2096–2102 (June 1973). See also *ibid.*, "Compression of low-intensity, phase modulated light pulses," *IEEE J. Quantum Electron.* **QE–10**, 723 (September 1974), and "Optical pulse compression," *Appl. Phys. Lett.* **25**, 566–568 (November 15, 1974); and J. K. Wigmore and D. Grischkowsky, "Temporal compression of light," *IEEE J. Quantum Electron.* **QE–14**, 310–315 (April 1978).

Similarly striking experiments showing greatly modified and even negative pulse velocities produced by a strong exciton absorption line in a semiconductor (nitrogen-doped GaP) are reported by S. Chu and S. Wong. "Linear pulse propagation in an absorbing medium," *Phys. Rev. Lett.* **48**, 738–741 (March 15, 1982).

The strong dispersion in the refractive index of a semiconductor near the band gap is also demonstrated in J. P. van der Ziel and R. A. Logan, "Dispersion of the group velocity refractive index in GaAs double heterostructure lasers," *IEEE J. Quantum Electron.* **QE–19**, 164–169 (February 1983).

Problems for 9.4

1. *Analysis of group-velocity slowing in the wings of a strong atomic transition.* Analyze the type of experiment discussed in the text, in which the group delay of a gaussian pulse is measured when the pulse is sent through a strongly absorbing atomic transition with the pulse carrier frequency tuned quite far out on the wings of the absorption line. What conditions are necessary (i.e., what values of midband absorption, atomic linewidth, and detuning off line center) to obtain a absorption loss of only about 20% per pass, and yet have a group velocity as slow as 1/10 or 1/15 the velocity of light? Note: You may want to refer back to the discussion in Section 3.7 which pointed out that far enough from line center, even inhomogeneous doppler-broadened transitions appear to be essentially homogeneous or lorentzian in their lineshapes.

2. *Phase and group velocity versus frequency in a mixed laser amplifier and atomic absorber medium.* In an earlier chapter we considered a linear single-pass laser amplifier with midband gain coefficient $\alpha_1 L$ and lorentzian atomic linewidth $\Delta\omega_{a1}$, followed by a linear single-pass laser *absorber*—that is, a collection of absorbing atoms—having midband absorption coefficient $\alpha_2 L$ and atomic linewidth $\Delta\omega_{a2}$, with both transitions centered at the same resonance frequency; and we asked what relative gain coefficients $\alpha_1$ and $\alpha_2$ and relative atomic linewidths $\Delta\omega_{a1}$ and $\Delta\omega_{a2}$ would just lead to a double-humped rather than single-humped curve of overall power transmission versus frequency for the two atomic systems in cascade.

   Suppose that these two atomic media are mixed together to produce a medium with both an amplifying and an absorbing transition at the same frequency. Consider the variation in phase and group velocity versus frequency across this compound transition, and especially the derivatives of phase shift and group velocity versus frequency at line center. Does there seem to be any fundamental connection between having a maximally flat gain profile at line center and having either the group velocity or phase velocity derivatives be zero at line center?

3. *Sensitivity of pulse compression to disperser length.* A chirped gaussian pulse whose initial time-bandwidth product is $N$ times the transform-limited value, where $N \gg 1$, is to be compressed by passage through an optimum length of some dispersive system. Evaluate the sensitivity of the pulse-compression process to the length $L$ of the dispersive system by finding the length tolerance about the optimum value $L_{\text{opt}}$ that will cause the output pulse length to increase by $\sqrt{2}$ from the optimum value.

## 9.5 PULSE BROADENING AND GAIN DISPERSION

Pulses can be broadened as well as compressed in a propagation system having significant group-velocity dispersion, and this broadening can be very important both in measurement applications of picosecond optical pulses and in the transmission of such pulses through optical fibers. In this section, therefore, we extend the discussion of the previous sections to cover pulse broadening in such dispersive systems.

In addition, pulses will also be broadened—and more rarely narrowed—in passing through systems with *gain dispersion*, that is, in passing through amplifiers with a finite bandwidth. In this section therefore we also consider the complementary effects of gain or absorption dispersion in a propagating system.

### Dispersive Pulse Broadening

Gaussian pulses starting out in the lower half-plane in Figure 9.3 will be initially compressed as they propagate through the dispersive system. On the other hand, pulses starting out—or moving into—the upper half-plane will be broadened (unless, of course, the sign of $\beta''$ is reversed, in which case the arrowheads on the trajectories must be reversed). Group-velocity dispersion can thus either *compress a pulse* with the right initial chirp, or *broaden a pulse* with the wrong initial chirp. In particular, a pulse with no initial chirp will begin to acquire a growing amount of chirp and then be broadened in any such system.

Consider for example an initially unchirped pulse, with $b_0 = 0$. Its gaussian pulse parameter after propagating a distance $z$ through a dispersive medium becomes

$$a(z) = \frac{a_0}{1 + (2\beta'' z)^2 a_0^2}. \tag{64}$$

Reduction in $a(z)$ means that the pulsewidth $\tau_p(z)$ has broadened, as given by the expression

$$\tau_p^2(z) \equiv \frac{2\ln 2}{a(z)} = \tau_{p0}^2 + \left(\frac{(4\ln 2)\beta'' z}{\tau_{p0}}\right)^2 = \left[1 + (z/z_D)^2\right] \times \tau_{p0}^2. \tag{65}$$

The initial unchirped pulsewidth $\tau_{p0}$ will increase by a factor of $\sqrt{2}$ after a propagation distance $z_D$ given by

$$z = z_D \equiv \frac{\tau_{p0}^2}{(4\ln 2)\beta''}. \tag{66}$$

This "dispersion length" $z_D$ is a kind of Rayleigh length for pulse broadening in time, analogous to the Rayleigh range for transverse beam spreading we will meet in a later chapter.

It is convenient to rewrite this dispersion length in the form

$$z_D = \frac{(\omega_0 \tau_{p0})^2}{8 \ln 2} \times \frac{\lambda}{D}, \tag{67}$$

where the quantity

$$D \equiv \frac{\omega_0^2 \beta''}{\beta(\omega_0)} \tag{68}$$

is a dimensionless group-velocity dispersion parameter for a propagating system or an atomic medium. The first factor in the $z_D$ expression depends only on pulse parameters—it is essentially the number of optical cycles in the pulse squared—and the second factor is the wavelength in the medium, $\lambda = \lambda_0/n$, divided by the dimensionless dispersion parameter $D$.

### Dispersive Broadening in Real Materials

We might want to evaluate this dispersion parameter, for example, for an ultrashort optical pulse propagating through a typical optical material with a frequency-dependent index of refraction, so that the propagation constant is given by

$$\beta(\omega) = \frac{n(\omega)\omega}{c_0}. \tag{69}$$

A common form for the variation of the index of the refraction across the visible region in transparent optical materials is the Sellmeier equation

$$n^2(\omega) - 1 = \frac{A\omega_e^2}{\omega_e^2 - \omega^2}. \tag{70}$$

For typical optical glasses and crystals $A$ has a value between 1 and 2, and $\omega_e$ corresponds to an effective resonant frequency or absorption band edge for the material, often located in the ultraviolet at a wavelength $\lambda_e$ somewhere between 1000 and 2000 Å. (For semiconductors this wavelength generally corresponds to the band-gap wavelength, and the Sellmeier equation then gives the index of refraction for the semiconductor in the transparent region at wavelengths longer than the band-gap wavelength.)

We can then find that for many typical transparent materials the magnitude of the dimensionless dispersion parameter is given by

$$D \equiv \frac{\omega_0^2 \beta''}{\beta(\omega_0)} \approx 0.10 \text{ to } 0.20. \tag{71}$$

If we assume a value of $D \approx 0.10$ and an index of refraction $n = 1.5$, the approximate $\sqrt{2}$ broadening lengths for initially unchirped pulses of different initial pulsewidths $\tau_{p0}$ are given by

| $\tau_{p0}$ | 100 fs | 1 ps | 10 ps | 100 ps |
|---|---|---|---|---|
| $z_D$ | $\sim 1$ cm | $\sim 100$ cm | $\sim 100$ m | $\sim 10$ km |

(72)

This type of dispersive pulse broadening becomes very important in the use of picosecond and femtosecond optical pulses from mode-locked lasers. Pulses $\leq 20$ femtoseconds long have already been generated in mode-locked dye lasers. Such a pulse obviously cannot propagate more than a few mm through a typically dispersive medium before it will be significantly broadened by the self-broadening effect.

### Dispersive Broadening in Optical Fibers

The pulse-broadening effects caused by group-velocity dispersion in optical fibers are also of great importance in determining the maximum distance that a pulse of given width can be propagated through an optical-fiber communications systems before being significantly broadened. A 100 ps visible-wavelength pulse, for example, can propagate only a few km in a typical single-mode fiber before being significantly broadened (the situation in multimode fibers is much worse, because of mode-mixing effects). Dispersive pulse broadening can become the primary factor limiting the potential data rate for long-distance communication in low-loss high-capacity optical fibers.

The dispersive behavior of an optical fiber is usually a combination of *materials dispersion*, associated with the index variation of the glass in the fiber, and *waveguide or modal dispersion* associated with the propagating normal mode patterns in the fiber. In typical fibers this net dispersion passes through zero at a wavelength around 1.3 $\mu$m, so that in principle very short pulses tuned to this wavelength could be propagated for very long distances without dispersive spreading.

It can be difficult to match the transmitted wavelength exactly to the zero dispersion point, however, and in addition the lowest-loss wavelength for optical fibers is typically closer to $\lambda = 1.5$ $\mu$m (at which wavelength the absorption and scattering losses in real fibers can have values as extraordinarily small as $\approx 0.2$ dB/km). To transmit a pulse with the minimum possible input and output pulsewidth through a given length of such a fiber, we should launch a pulse with an input pulsewidth $\tau_p \equiv \sqrt{2}\tau_{p0}$ and just the right amount of initial compressive chirp into the fiber, where the fiber length is given by $z = 2z_D$, and $\tau_{p0}$ and $z_D$ are connected by the analytical relations given in Equations 9.65 and 9.66. This pulse will then compress down to $\tau_{p0}$ at the middle of the fiber and broaden back to $\tau_p$ at the output end. For a small enough value of normalized dispersion—perhaps $D \approx 0.005$, which might be typical of a single-mode quartz fiber at $\lambda_0 = 1.5$ $\mu$m—the relationship between fiber length and minimum input-output pulsewidth will have typical values given by

| $z = 2z_D$ | 2 | 5 | 10 | 20 | 50 | 100 | km |
|---|---|---|---|---|---|---|---|
| $\tau_p = \sqrt{2}\tau_{p0}$ | 11 | 16 | 23 | 33 | 53 | 74 | ps |

(73)

Note that a data transmission rate of 10 Gbits per second, such as optical-fiber communications designers hope to achieve, requires a pulsewidth at least as short as 100 ps, and preferably somewhat less.

(One very interesting alternative approach for accomplishing the propagation of very much shorter optical pulses in fiber communications systems is the use of nonlinear solitons in optical fibers, as described briefly in the following chapter.)

### Pulse-Broadening Effects of Gain Dispersion

Velocity or phase-shift dispersion, which is produced by a frequency-dependent propagation constant $\beta(\omega)$, causes one set of pulse distortion effects. Gain dispersion, by which we mean a frequency-dependent gain coefficient $\alpha(\omega)$, produces a different set of effects. The primary effects that result when a pulse passes through a linear but frequency-dependent gain medium (now leaving out dispersion effects) are pulse-broadening in time, due to finite bandwidth of the gain medium, and possibly more complex frequency-shifting and time-shifting effects that can occur if the gain medium has a linear variation of gain with frequency across the pulse spectrum.

In this section we will consider only quadratic or pulse-broadening effects, since they are the most fundamentally important, leaving the more complex effects of linear frequency dependence to an exercise. We therefore consider again a gaussian pulse with carrier frequency $\omega_0$ and initial pulse parameter $\Gamma_0$, which now passes through a linear gain medium whose gain coefficient has the quadratic frequency dependence

$$\alpha_m(\omega) = \alpha_{m0} - \frac{1}{2}\alpha_m''(\omega_0) \times (\omega - \omega_0)^2, \qquad (74)$$

where $\alpha_m'' \equiv -d^2\alpha_m(\omega)/d\omega^2$ evaluated at midband. The first term in this expansion gives a uniform amplitude gain which applies equally to all frequency components and hence simply increases the pulse amplitude uniformly without changing its shape. The $\alpha_m''$ term, however, leads to a change in the gaussian pulse parameter given by

$$\frac{1}{\Gamma(z)} = \frac{1}{\Gamma_0} + 2\alpha_m''(\omega)z. \qquad (75)$$

The trajectory of $\Gamma(z)$ now moves horizontally in the $1/\Gamma$ plane, rather than vertically as in Figure 9.3. For $\alpha_m'' > 0$ (that is, a gain peak at line center) this increases the real part of $1/\Gamma$ and hence, as is physically obvious, decreases the spectral bandwidth of the pulse. The result is commonly, though not universally, to broaden the pulse in time as it propagates through the amplifier.

### Pulse Broadening in Amplifiers

Consider as a particular example a lorentzian atomic transition with linewidth $\Delta\omega_a$ and a spectrum centered at $\omega_0 \equiv \omega_a$. The gain variation around line center may be written to a first approximation as

$$\alpha_m(\omega) = \frac{\alpha_{m0}}{1 + [2(\omega - \omega_a)/\Delta\omega_a]^2} \approx \alpha_{m0} - \alpha_{m0} \times \left(\frac{2}{\Delta\omega_a}\right)^2 \times (\omega - \omega_0)^2. \qquad (76)$$

A length $z$ of such an amplifier then produces an output pulse parameter given by

$$\frac{1}{\Gamma(z)} = \frac{1}{\Gamma_0} + \frac{16\alpha_{m0}z}{\Delta\omega_a^2}. \qquad (77)$$

Laser amplifiers with finite bandwidths will then usually broaden pulses in time, just as occurs in any other finite bandpass system.

Consider, for example, an input pulse with no initial chirp, i.e., with $\Gamma_0 = a_0$ and $b_0 = 0$. Equations 9.75–9.77 then convert into the simple pulsewidth broadening result

$$\tau_p^2(z) = \tau_{p0}^2 + \frac{(16\ln 2)\ln G_0}{\Delta\omega_a^2}. \qquad (78)$$

where $G_0 = \exp(2\alpha_{m0}z)$. In a Nd:YAG laser amplifier with $\Delta\omega_a/2\pi \approx 120$ GHz and the very high (possibly multipass) gain $G_0 = 10^5$, this gives approximately

$$\tau_p^2 \approx \tau_{p0}^2 + (\sim 15 \text{ ps})^2. \qquad (79)$$

Such an amplifier will convert an ideal delta-function input pulse into an $\sim$15 ps output pulse, and will broaden a 50 ps input pulse to $\sim$52 ps at its output.

Also, the initially unchirped pulse develops an added chirp in passing through the amplifier. These kinds of results are important in understanding the amplification of short pulses in a laser amplifier, and, as we will see in a later chapter, in understanding mode-locking in laser oscillators.

### Pulse Narrowing in a Chirped Laser Amplifier

Laser amplifiers with finite bandwidths can also, under certain special circumstances, shorten chirped pulses in time. Suppose the input pulse also has a significant initial chirp $b_0$. The general result for the pulsewidth parameter $a(z)$ after an amplification distance $z$ is then

$$a(z) = \frac{a_0(1 + Ka_0) + Kb_0^2}{(1 + Ka_0)^2 + (Kb_0)^2}, \tag{80}$$

where $K \equiv 2\alpha_m'' z = 8 \ln G_0 / \Delta \omega_a^2$. Suppose for simplicity that the bandwidth-broadening factor $K$ is small compared to $1/\Gamma_0$ or $1/\Gamma$. Equation 9.80 expanded to first order in $K$ then becomes

$$a(z) \approx a_0 \left[1 - Ka_0 + Kb_0^2/a_0\right]. \tag{81}$$

This shows the first-order pulsewidth-broadening effect due to the $-Ka_0$ term, but also a pulsewidth-*narrowing* term in the $Kb_0^2/a_0$ term.

The physical interpretation of this effect is the following. If the pulse has a sizable chirp during its time duration (i.e., $b_0 \gg a_0$), we may think approximately of the pulse frequency sweeping across the gain profile of the amplifying transition. The center section of the pulse (in time) is at line center and hence gets maximum amplification, whereas the frequencies in both the leading and trailing edges of the pulse are somewhat off line center and get less amplification; hence the pulseshape gets somewhat narrowed in time.

This last explanation mingles time and frequency descriptions in a way that is not rigorously correct, but which still gives a reasonably correct physical picture of the result for $b_0 \gg a_0$. Note that the pulse narrowing or compression here is independent of the sign of the chirp, as is compatible with our physical reasoning.

---

Problems for 9.5

1. *Pulse broadening on passing through a Fabry-Perot etalon.* A paper by Albrecht and Mourou [*IEEE J. Quantum Electron.* **QE–17**, 1709–1712 (September 1981)] describes a laser pulse that circulates around repeatedly inside a laser cavity containing an intracavity Fabry-Perot etalon, and gives the formula

$$\tau_{\text{final}}^2 = \tau_{\text{initial}}^2 + 16(\ln 2)\left[\frac{g}{\Delta\omega_p^2} + \left(\frac{Fd}{\pi c}\right)^2\right] \times N$$

for the FWHM pulsewidth after $N$ round trips. (The $-$ rather than $+$ sign that appears inside the brackets in the original reference must be incorrect.) In this formula the etalon is characterized by its finesse $F$ and thickness $d$, and the laser gain medium by its gain coefficient per pass $g$ and bandwidth $\Delta\omega_p$.

Derive this same formula; interpret what the authors must mean by the symbols employed; and also explain the next equation in the paper, which says that the

## 9.5 PULSE BROADENING AND GAIN DISPERSION

final pulsewidth after many round trips will be given by $\tau_{\text{final}} \approx 3.5 \times 10^{-11} F d \sqrt{N}$, which depends only on the etalon and not on the laser medium.

2. *Pulse propagation through mixed group-velocity dispersion and gain dispersion.* Consider in more detail the propagation of a complex gaussian pulse with an arbitrary input pulse parameter $\Gamma_0$ through a long transmission line (or an extended laser medium) that has both finite group-velocity dispersion $\beta''$ and also finite gain dispersion $\alpha''$. Illustrate this by plotting contours of pulse propagation in the $1/\Gamma$ (or $1/\Gamma^*$) plane for various ratios of $\alpha''$ to $\beta''$.

Under which conditions can a system that produces bandwidth narrowing in the frequency domain (i.e., a system with $\alpha'' > 0$) still produce pulsewidth narrowing in the time domain?

3. *Pulse propagation and distortion tuned on the side of an amplifying atomic transition.* The carrier frequency $\omega_a$ of a gaussian pulse might be tuned off to the side of an amplifier's passband (that is, $\omega_0 \neq \omega_a$), so that the gain dispersion across the pulse spectrum would need to be written as

$$\alpha(\omega) = \alpha_0 + \alpha' \times (\omega - \omega_0) - \frac{1}{2}\alpha'' \times (\omega - \omega_0)^2$$

with $\alpha' \equiv d\alpha/d\omega$ at $\omega = \omega_0$, and $\alpha_0$ being the gain value at $\omega = \omega_0$ (not the midband value at $\omega = \omega_a$). Analyze and describe the resulting pulse-propagation effects in this situation, including trajectories in the $\Gamma$ and $1/\Gamma$ planes, and physical effects on the pulse parameters.

# 10
# NONLINEAR OPTICAL PULSE PROPAGATION

All the propagation phenomena described in Chapter 9 are *linear propagation effects*, produced by the linear response of the propagating systems. In this chapter we will give a brief survey of a few of the most important *nonlinear propagation phenomena* that occur with optical pulses. These effects include in particular: gain saturation in pulsed amplifiers (which is a relatively weak form of nonlinearity); optical pulse propagation through nonlinear dispersive systems in general; and the especially interesting topic of nonlinear pulse propagation in optical fibers, including the fascinating topic of soliton propagation in optical fibers.

## 10.1 PULSE AMPLIFICATION WITH HOMOGENEOUS GAIN SATURATION

As we mentioned earlier, laser amplifiers are much more commonly used for amplifying optical pulses than for amplifying c̄w optical signals. Common examples of pulsed laser amplification include flash-pumped Nd:YAG and Nd:glass amplifiers at 1.06 $\mu$m; electron-beam-pumped TEA $CO_2$ laser amplifers at 10.6 $\mu$m; excimer lasers in the visible; and pulsed dye laser amplifiers, which are themselves often pumped by another pulsed laser, and which can amplify across broad bandwidths in the visible and near infrared.

Short pulses passing through laser amplifiers will be broadened and distorted by the effects we discussed in Chapter 9. These effects are, however, entirely linear effects, and generally require quite short pulses and sizable dispersions to be significant. Let us now consider an entirely separate form of weakly nonlinear pulse distortion that can arise with much longer pulses, as a result of *time-varying gain saturation effects* when a higher energy pulse is amplified in a homogeneously saturable laser amplifier.

### Pulse Energy Saturation in Amplifiers and Absorbers

In order to obtain efficient energy extraction from a laser amplifier, an amplified pulse must be intense enough to cause significant saturation of the population inversion during its passage through the amplifier. But this means that the amplifier gain must necessarily be reduced from a large initial value to a small residual value during the passage of the pulse; hence this time-dependent saturation during the passage of the pulse must also lead to time-varying gain reduction and pulseshape distortion.

In the same fashion, when a strong enough pulse is sent through a saturable absorber medium, the signal energy in the pulse may partially or completely saturate the atomic absorption and thus increase the energy transmission, leading to an analogous though oppositely directed pulse distortion. Such pulse propagation through saturable absorbers is widely used to shorten mode-locked pulses in passively mode-locked lasers. Again, the pulse itself must change the transmission of the saturable absorber during the passage of the pulse. The fundamental physics is the same as it is for the saturable amplifier, except for a change of sign in going from saturable amplification to saturable absorption.

In order to explain both of these effects, this section presents an analysis of the population saturation and the pulseshape distortion that results when a sufficiently intense pulse passes through a homogeneously saturable, single-pass laser amplifier, or saturable absorber. The physical approximations made in this section are thus significantly different from the linear pulse propagation analysis of Chapter 9. The pulsewidths we are concerned with here are generally long enough, and the propagation lengths short enough, that pulse compression or expansion effects due to finite amplification bandwidths or to group velocity or gain dispersion effects are generally of minor importance. Our emphasis is thus entirely on the time-varying *saturation effects* in the atomic material.

### Homogeneous Saturation Approximations

Two physical approximations help to simplify this analysis. First, although pulse amplification often involves short pulses with fast time-variation and high intensities, usually the rate-equation approximations are still valid, and a purely rate-equation analysis can be employed. Second, in most situations of interest for laser pulse amplification, the amplified pulse durations are short enough that we can neglect both any pumping effects and any upper-level relaxation during the transit time of the amplified pulse. Hence in this section we will analyze a short pulse propagating through a prepumped and inverted laser medium, without including any pumping or relaxation effects during the pulse interval.

### Analysis of Homogeneous Pulse Amplification

We consider therefore a short pulse with signal intensity $\hat{I}(\hat{z},\hat{t})$ traveling in the $+\hat{z}$ direction through a laser medium with inverted population difference $\Delta\hat{N}(\hat{z},\hat{t})$, where $\hat{z}$ and $\hat{t}$ are the usual laboratory coordinates. (The reason for the "hats" on all these quantities will become apparent in a moment.) We neglect any transverse intensity variations to simplify the analysis.

The basic differential equations for this situation can then be developed as follows. Let us denote the electromagnetic energy density in the optical signal pulse, measured in J/m$^3$, by $\hat{\rho}_{\text{em}}(\hat{z},\hat{t})$. The instantaneous intensity $\hat{I}(\hat{z},\hat{t})$ in W/m$^2$ being carried by the pulse through any plane $\hat{z}$ at time $\hat{t}$ is then given by $\hat{I}(\hat{z},\hat{t}) = \hat{\rho}_{\text{em}}(\hat{z},\hat{t}) \times v_g$, where $v_g$ is the group velocity in the laser medium. This velocity is normally very close to the phase velocity $c$ in most laser media; so for simplicity we will write $v_g = c$ in the following analysis.

Consider then a short segment of length $\Delta\hat{z}$ in the laser medium, as shown in Figure 10.1. The rate of change of stored signal energy in the length $\Delta\hat{z}$ is given by the energy flux into one end minus the energy flux out the other end of

364     CHAPTER 10: NONLINEAR OPTICAL PULSE PROPAGATION

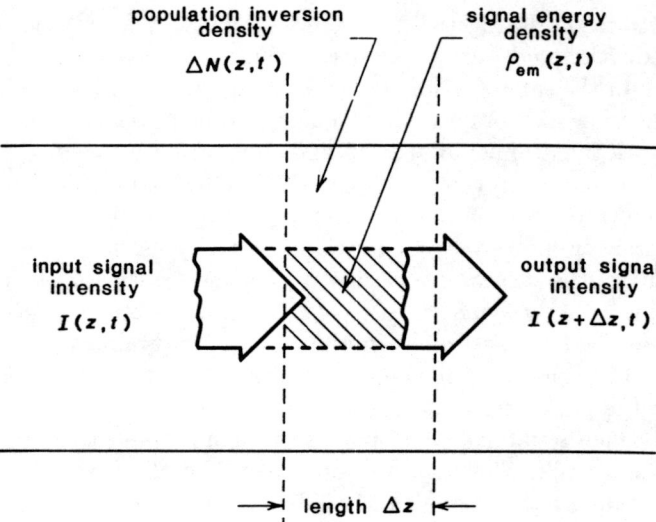

FIGURE 10.1
Optical intensity passing through a short segment of a saturable pulse amplifier.

the segment, plus the net rate of stimulated emission within the segment, or

$$\frac{\partial}{\partial \hat{t}}[\hat{\rho}_{\text{em}}(\hat{z},\hat{t})\,\Delta\hat{z}] = \hat{I}(\hat{z},\hat{t}) - \hat{I}(\hat{z}+\Delta\hat{z},\hat{t}) + \sigma\,\Delta\hat{N}(\hat{z},\hat{t})\,\hat{I}(\hat{z},\hat{t})\Delta\hat{z}, \qquad (1)$$

where $\sigma$ is the stimulated-transition cross section of the laser medium. Using $\hat{I}(\hat{z},\hat{t}) = c\hat{\rho}_{\text{em}}(\hat{z},\hat{t})$ and combining this with the rate equation for the inverted population inside the same segment then gives the two basic equations of this pulse saturation analysis, namely,

$$\frac{\partial \hat{I}(\hat{z},\hat{t})}{\partial \hat{t}} + c\frac{\partial \hat{I}(\hat{z},\hat{t})}{\partial \hat{z}} = \sigma c\,\Delta\hat{N}(\hat{z},\hat{t})\,\hat{I}(\hat{z},\hat{t}) \qquad (2)$$

and

$$\frac{\partial}{\partial \hat{t}}\Delta\hat{N}(\hat{z},\hat{t}) = -\left(\frac{2^*\sigma}{\hbar\omega}\right)\Delta\hat{N}(\hat{z},\hat{t})\,\hat{I}(\hat{z},\hat{t}). \qquad (3)$$

Note that no pumping or relaxation terms are included in the atomic rate equation. We use the convention mentioned earlier that the "saturation factor" $2^* \equiv 1$ if the lower laser level empties out rapidly compared to the pulse duration; but $2^* \equiv 2$ if the lower-level population accumulates or "bottlenecks" during the pulse.

### Transformation to Moving Coordinates

These two basic equations can then be solved with the aid of a few minor tricks, as follows. We first make a change of variables to a coordinate system *that moves with the forward-traveling pulse*, as defined by the transformation

$$z \equiv \hat{z} \qquad \text{and} \qquad t \equiv \hat{t} - \hat{z}/c. \qquad (4)$$

That is, whereas $\hat{z}$ and $\hat{t}$ are ordinary laboratory coordinates, for the remainder of this section $z$ and $t$ refer to coordinates in the moving pulse frame. Note that the delayed time coordinate $t$ is essentially centered on the pulse's arrival time at each plane $z$. For example, if the pulse starts out at an input plane $\hat{z} = 0$ centered at time $\hat{t} = 0$, and arrives at some plane $\hat{z}$ centered about time $\hat{t} = \hat{z}/c$, then the pulse written in the delayed time coordinate $t$ is centered on $t = 0$ at every plane along the amplifier.

We also rewrite the pulse intensity and the population difference in the new coordinate system in the modified forms

$$I(z,t) \equiv \hat{I}(\hat{z},\hat{t}) \quad \text{and} \quad N(z,t) \equiv \Delta\hat{N}(\hat{z},\hat{t}), \tag{5}$$

where we use $N(z,t)$ instead of $\Delta N(z,t)$ from here on merely to simplify the formulas. The basic equations for the pulse intensity and the population inversion are then transformed into the significantly simpler forms

$$\frac{\partial I(z,t)}{\partial z} = \sigma N(z,t) I(z,t) \tag{6}$$

and

$$\frac{\partial N(z,t)}{\partial t} = -\left(\frac{2^*\sigma}{\hbar\omega}\right) N(z,t) I(z,t), \tag{7}$$

where these equations are now expressed in the *transformed* or *moving* coordinate system.

### Solution of the Pulse Equations

The first of these equations can then be rearranged and integrated over the length of the amplifier in the form

$$\int_{I=I_{\rm in}(t)}^{I=I_{\rm out}(t)} \frac{dI}{I} = \sigma \int_{z=0}^{z=L} N(z,t)\,dz, \tag{8}$$

where $I_{\rm in}(t)$ is the input pulse intensity at the input plane to the amplifier, and $I_{\rm out}(t)$ is correspondingly the signal intensity at the output plane, both measured in the delayed time coordinate $t$. It will then be convenient to define the integral on the right-hand side of this equation as a kind of "total number of atoms" $N_{\rm tot}(t)$ in the amplifier, in the form

$$N_{\rm tot}(t) \equiv \int_{z=0}^{z=L} N(z,t)\,dz. \tag{9}$$

The first of the basic equations can then be expressed in the simple form

$$I_{\rm out}(t) = I_{\rm in}(t)\,e^{\sigma N_{\rm tot}(t)} = G(t)\,I_{\rm in}(t), \tag{10}$$

where $G(t) \equiv \exp[\sigma N_{\rm tot}(t)]$ is the time-varying or partially saturated gain at any instant within the pulse.

The second of the basic pulse equations can also be integrated over the amplifier length and then rewritten, using the first equation, in the form

$$\frac{\partial}{\partial t}\int_{z=0}^{z=L} N(z,t)\,dz \equiv \frac{dN_{\rm tot}(t)}{dt} = -\left(\frac{2^*}{\hbar\omega}\right)\int_{z=0}^{z=L} \frac{\partial I(z,t)}{\partial z}\,dz, \tag{11}$$

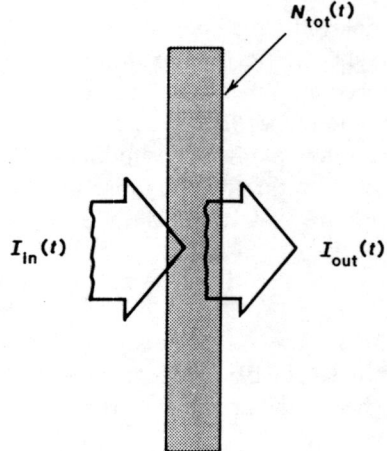

**FIGURE 10.2**
Thin-slab model for a homogeneously saturating laser pulse amplifier.

which simplifies to

$$\frac{dN_{\text{tot}}(t)}{dt} = -\frac{2^*}{\hbar\omega}\left[I_{\text{out}}(t) - I_{\text{in}}(t)\right]. \tag{12}$$

The two basic equations thus reduce to two coupled equations in the delayed time coordinate $t$ only. These will be the fundamental working equations from this point on.

### Physical Interpretation

With this change of coordinates *the equations take on the same form as if the total number of atoms $N_{\text{tot}}(t)$ were condensed into an arbitrarily thin slab,* as in Figure 10.2. Equation 10.10 is then a standard exponential gain equation, showing that the input signal $I_{\text{in}}(t)$ is exponentially amplified by $N_{\text{tot}}(t)$ to become the output signal $I_{\text{out}}(t)$; and Equation 10.12 is essentially a conservation of energy equation, showing how $N_{\text{tot}}(t)$ is burned up by the amplification of $I_{\text{in}}(t)$ to produce $I_{\text{out}}(t)$.

Note that the output pulse $I_{\text{out}}(t)$ is actually delayed in real time $\hat{t}$ with respect to the input pulse $I_{\text{in}}(t)$ by the transit time through the amplifier, since a particular point $t$ in the output pulse occurs $L/c$ later than the same point $t$ in the input pulse. Also, the "number of atoms" $N_{\text{tot}}(t)$ is actually not a physical number of atoms that exists at any instant of time $\hat{t}$, or that could be seen by some snapshot of the amplifier at any single time $\hat{t}$. It is rather a measure of the total space-integrated (or time-integrated) population difference $\int N(z,t)\,dz$ seen by any one small segment of the pulse centered at time $t$, as that segment passes through each successive plane $z \equiv \hat{z}$ of the amplifier at the local time $t$. The net result, however, still reduces to an equivalent thin slab in which everything seems to happen simultaneously in the delayed time coordinate $t$.

### Analytic Solutions

Several useful explicit solutions to these equations can be obtained as follows. We start by substituting Equation 10.10 into Equation 10.12 to obtain either of the alternative forms

$$\frac{dN_{\text{tot}}(t)}{dt} = -\frac{2^*}{\hbar\omega}\{\exp[\sigma N_{\text{tot}}(t)] - 1\} \times I_{\text{in}}(t)$$
$$= -\frac{2^*}{\hbar\omega}\{1 - \exp[-\sigma N_{\text{tot}}(t)]\} \times I_{\text{out}}(t). \tag{13}$$

Suppose the total initial inversion $N_0$ in the laser medium, at a time $\hat{t}_0$ prior to the arrival of any input pulse energy, is given by

$$N_0 \equiv \int_{\hat{z}=0}^{\hat{z}=L} \hat{N}(\hat{z}, \hat{t}_0)\, d\hat{z}. \tag{14}$$

The initial single-pass power gain of the amplifier is then $G_0 = \exp[N_0\sigma]$.

We will also write the accumulated signal energies per unit area $U_{\text{in}}(t)$ and $U_{\text{out}}(t)$ in the input and output pulses, from starting time $t_0$ up to normalized time $t$ as

$$U_{\text{in}}(t) \equiv \int_{t_0}^{t} I_{\text{in}}(t)\, dt \quad \text{and} \quad U_{\text{out}}(t) \equiv \int_{t_0}^{t} I_{\text{out}}(t)\, dt. \tag{15}$$

These pulse energies per unit area are sometimes referred to as *energy fluences*. It is also convenient to define a saturation energy per unit area $U_{\text{sat}}$ (sometimes called a *saturation fluence*) for the atomic medium by

$$U_{\text{sat}} \equiv \frac{\hbar\omega}{2^*\sigma}. \tag{16}$$

This quantity is clearly the pulsed analog to the saturation intensity $I_{\text{sat}} \equiv \hbar\omega/\sigma\tau$ we saw for continuous amplification. If a signal energy fluence $U_{\text{sat}}$ flows by an atom in a time much less than the atom's recovery time $\tau$, the atom has essentially a 50% chance of making a stimulated transition from one level to the other during the pulse. (As an example, a Nd:YAG laser, with cross section $\sigma \approx 5 \times 10^{19}$ cm$^2$, has a saturation fluence of $\approx 0.4$ Joules per cm$^2$.)

The first of the differential forms in Equation 10.13 can then be integrated in the form

$$\int_{N_0}^{N_{\text{tot}}(t)} \frac{dN_{\text{tot}}}{\exp(\sigma N_{\text{tot}}) - 1} = -\frac{2^*}{\hbar\omega}\int_{t_0}^{t} I_{\text{in}}(t)\, dt = -\frac{2^*}{\hbar\omega} U_{\text{in}}(t) \tag{17}$$

to give the useful relation

$$U_{\text{in}}(t) = U_{\text{sat}} \times \ln\left\{\frac{1 - \exp[-\sigma N_0]}{1 - \exp[-\sigma N_{\text{tot}}(t)]}\right\} = U_{\text{sat}} \times \ln\left[\frac{1 - 1/G_0}{1 - 1/G(t)}\right], \tag{18}$$

where again $G(t) = \exp[\sigma N_{\text{tot}}(t)] = I_{\text{out}}(t)/I_{\text{in}}(t)$ is the time-varying partially saturated gain within the pulse interval. This expression thus connects the cumulative input energy $U_{\text{in}}(t)$ to the net remaining inversion $N_{\text{tot}}(t)$ or the time-varying power gain $G(t)$ at any instant of (normalized) time $t$ within the pulse.

The second differential relation in Equation 10.13 can be similarly integrated to give the complementary relation

$$U_{\text{out}}(t) = U_{\text{sat}} \times \ln\left\{\frac{\exp[\sigma N_0] - 1}{\exp[\sigma N_{\text{tot}}(t)] - 1}\right\} = U_{\text{sat}} \times \ln\left[\frac{G_0 - 1}{G(t) - 1}\right], \tag{19}$$

where $U_\text{out}(t)$ is similarly the cumulative energy in the output pulse up to time $t$ (in delayed time coordinates).

### Gain Saturation

Either one of these results can then be inverted to express the instantaneous inversion and gain within the pulse in terms of either of the input or output pulseshapes, $U_\text{in}(t)$ or $U_\text{out}(t)$. For example, Equation 10.18 can be rewritten in the form

$$\sigma N_\text{tot}(t) = \ln\left[\frac{G_0}{G_0 - (G_0 - 1)\exp[-U_\text{in}(t)/U_\text{sat}]}\right], \quad (20)$$

which gives

$$G(t) = \exp[\sigma N_\text{tot}(t)] = \frac{G_0}{G_0 - (G_0 - 1)\exp[-U_\text{in}(t)/U_\text{sat}]}. \quad (21)$$

For a given input pulseshape $I_\text{in}(t)$ and a given initial gain $G_0$ we can use this to calculate the output pulseshape $I_\text{out}(t) = G(t) \times I_\text{in}(t)$.

Alternatively, if we want to specify a desired *output* pulseshape $I_\text{out}(t)$ in the presence of saturation, we can calculate the necessary gain versus time from the output pulseshape, using

$$G(t) = 1 + (G_0 - 1)\exp[-U_\text{out}(t)/U_\text{sat}], \quad (22)$$

and then find the required input pulseshape from $I_\text{in}(t) = I_\text{out}(t)/G(t)$. How to synthesize the required input pulseshape—which will be different for different output pulse energy levels or degrees of saturation—is, of course, a separate problem.

### Pulseshape Distortion

Figure 10.3 illustrates the kind of output pulse distortion that is produced by amplifier gain saturation assuming typical input pulseshapes. In Figure 10.3, where we assume a square input pulse with a perfectly sharp leading edge, the initial gain right at the leading edge of the pulse is the unsaturated value $G_0$. This gain immediately begins to saturate, however, falling rather slowly for a weak input pulse and thus producing only a certain amount of "droop" in the output pulse, as in (a), but dropping much more strongly for a strong input pulse, as in (b). The result is then a rapid decrease in the output signal, leaving a large spike on the leading edge of the amplified pulse.

This short pulse formation might seem potentially useful as a means of pulse sharpening, in order to obtain a shortened output pulse from a much longer input pulse. Its practicality is limited, however, because in order to obtain strong pulse sharpening the leading edge of the input pulse must have a rise time substantially shorter than the desired output pulse length. If a practical modulator is unable to create the desired short pulse to begin with, it may be no more capable of generating an input pulse with the required rise time on the leading edge.

A gaussian pulseshape, or any other shape with rounded leading and trailing edges, is generally a more realistic model for saturable pulse amplification. Figure 10.4 illustrates how the gaussian pulseshape (plotted on a log scale) changes as we increase the input energy level to an amplifier with an initial gain $G_0 = 10{,}000$

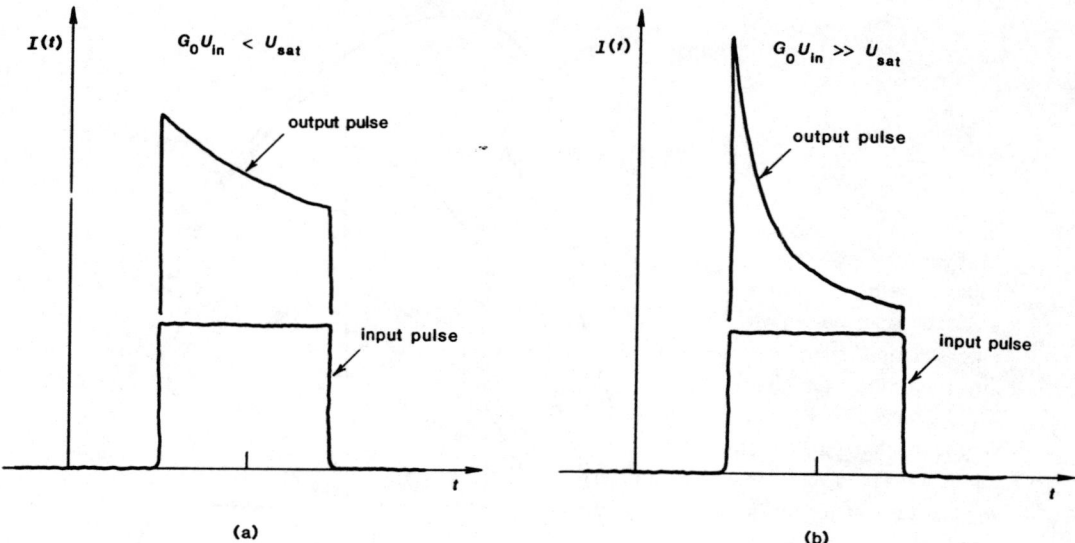

**FIGURE 10.3**
Output pulseshape distortion for square input pulses with small and large input energy.

($\equiv$ 40 dB), keeping the input gaussian pulseshape constant. Significant saturation effects begin to occur when the amplified output energy disregarding saturation, or $G_0 U_{\text{in}}$, begins to approach the saturation energy $U_{\text{sat}}$. Even for input energies well above this value, however, the output pulsewidth appears to be very little changed from the input pulsewidth, even with quite strong saturation, although the pulse does seem to move forward slightly in time.

The output pulse is substantially changed, nonetheless, since there clearly is substantial gain saturation during the pulse. This saturation shows up primarily as an apparent advance in time of the peak of the output pulse. The pulse is not really advanced in time, but merely appears so because the leading edge receives essentially full amplification, whereas the gain is substantially reduced during the peak and trailing-edge portions of the pulse.

Figure 10.5 also plots the total pulse output energy $U_{\text{out}}$ integrated over the full pulsewidth versus total pulse input energy $U_{\text{in}}$ for the particular case of a homogeneously saturable amplifier with $G_0 = 1{,}000 = 30$ dB. This plot clearly shows how the pulse energy gain saturates down as the output pulse energy increases much above $U_{\text{sat}}$. (Note that these results are independent of the actual shape of the pulses.)

### Pulse Energy Extraction and Energy Gain

The efficiency with which a signal pulse extracts the available energy from a laser pulse amplifier can be calculated in a simple fashion as follows. Subtracting the two earlier expressions for $U_{\text{in}}(t)$ and $U_{\text{out}}(t)$ gives

$$\frac{U_{\text{extr}}(t)}{U_{\text{sat}}} \equiv \frac{U_{\text{out}}(t) - U_{\text{in}}(t)}{U_{\text{sat}}} = \ln\left[\frac{G_0}{G(t)}\right] \qquad (23)$$

370   CHAPTER 10: NONLINEAR OPTICAL PULSE PROPAGATION

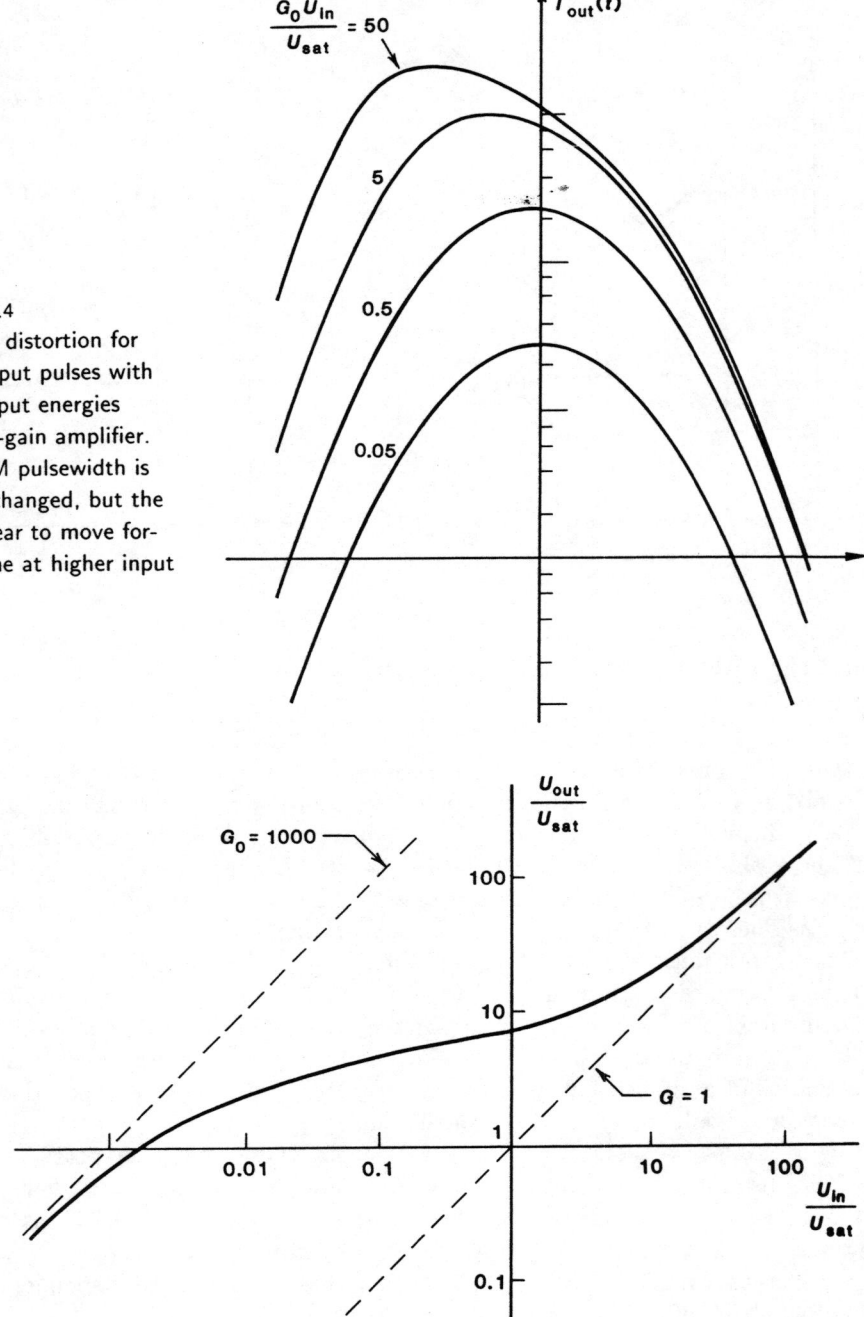

FIGURE 10.4
Pulseshape distortion for gaussian input pulses with different input energies into a high-gain amplifier. The FWHM pulsewidth is very little changed, but the pulses appear to move forward in time at higher input energies.

FIGURE 10.5
Pulse output energy versus pulse input energy for a homogeneously saturable amplifier with initial gain $G_0 = 1,000 = 30$ dB.

Let $U_{\text{in}}$ and $U_{\text{out}}$ without a specific time-dependence henceforth denote the *total energies* in the complete input and output pulses, i.e., the limits of $U_{\text{in}}(t)$ and $U_{\text{out}}(t)$ as $t \to +\infty$; and similarly let $G_f$ denote the final value of $G(t)$ after the pulse has passed, i.e., the limit of $G(t)$ as $t \to \infty$. The total energy extracted

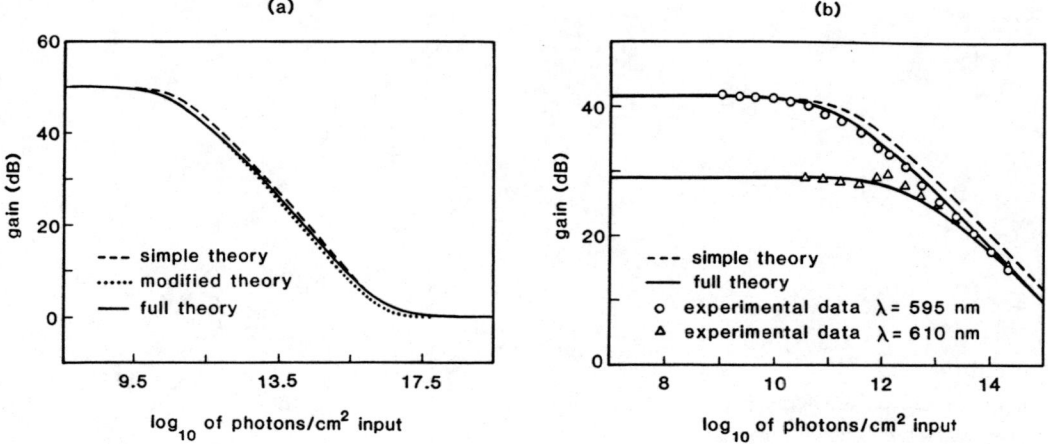

FIGURE 10.6
Pulse energy gain (integrated over the full pulse) versus input pulse energy: (a) theoretical curves for an initial gain $G_0 = 50$ dB; (b) experimental results and theoretical curves for two somewhat lower initial gains in a dye-laser pulse amplifier (from T. L. Koch, L. C. Chiu, and A. Yariv, *Opt. Commun.* 40, 364–368, February 1982).

from the gain medium by the complete pulse is then given by

$$U_{\text{extr}} \equiv U_{\text{out}} - U_{\text{in}} = U_{\text{sat}} \times \ln\left(\frac{G_0}{G_f}\right). \tag{24}$$

The maximum available energy from the amplifier is obviously obtained when the residual gain is saturated all the way down to $G_f \to 1$. The maximum available energy that can be extracted from the amplifier, assuming an input pulse strong enough to completely saturate the initial inversion, is thus given by

$$U_{\text{avail}} = U_{\text{sat}} \times \ln G_0 = \frac{N_0 \hbar \omega}{2^*}. \tag{25}$$

The right-hand expression confirms the physically obvious result that the available energy from the amplifier is either the total initial inversion energy, $N_0 \hbar \omega$, or half that value, depending on whether the lower level does or does not empty out rapidly during the pulse.

We might also define an overall or pulse-averaged "pulse energy gain" $G_{\text{pe}}$ as the ratio of the total pulse energy output $U_{\text{out}}$ to the total pulse energy input $U_{\text{in}}$, or

$$G_{\text{pe}} \equiv \frac{U_{\text{out}}}{U_{\text{in}}} = \frac{\ln\left[(G_0-1)/(G_f-1)\right]}{\ln\left[(G_0-1)/(G_f-1)\right] - \ln\left[G_0/G_f\right]}. \tag{26}$$

Figure 10.6 shows theoretical and experimental examples for the reduction in laser pulse energy gain with increasing input pulse energy for one particular experiment using a picosecond-pulse dye laser amplifier.

372   CHAPTER 10: NONLINEAR OPTICAL PULSE PROPAGATION

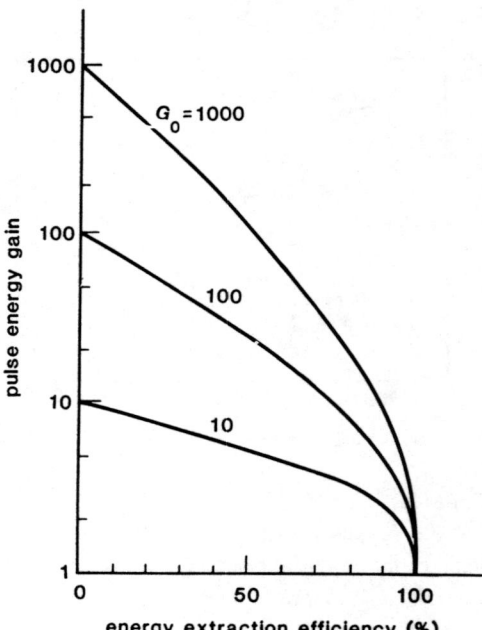

FIGURE 10.7
Integrated pulse energy gain versus energy-extraction efficiency for pulse amplifiers with different initial gains $G_0$.

### Pulse Energy Extraction Efficiency

Equations 10.23 through 10.26 can also be manipulated in various other ways that may be useful. For example, we might define an energy extraction efficiency $\eta$ as the ratio of the energy actually extracted from the laser medium to the maximum energy available in the medium, or

$$\eta \equiv \frac{U_{\text{out}} - U_{\text{in}}}{U_{\text{avail}}} = \frac{\ln G_0 - \ln G_f}{\ln G_0}. \tag{27}$$

Inverting this tells us that the final gain $G_f$ can be related to the initial gain $G_0$ and the energy extraction efficiency $\eta$ in a simple fashion by

$$G_f = G_0^{1-\eta}. \tag{28}$$

Obviously, if the energy extraction efficiency approaches anywhere near 100%, the final saturated gain $G_f$ at the end of the pulse will be much less than the unsaturated gain $G_0$ at the beginning of the pulse.

Putting this result into Equation 10.26 then gives a general relationship between the initial or unsaturated power gain $G_0$, the time-averaged pulse energy gain $G_{\text{pe}}$, and the energy extraction efficiency $\eta$, entirely independent of input or output pulseshapes. Figure 10.7 illustrates how the pulse energy gain is rapidly saturated downward as we attempt to obtain increased energy extraction from an amplifier (i.e., by putting in stronger and stronger input pulses).

### Summary

This section has presented the simplest possible rate-equation analysis of pulsed amplifier saturation (often referred to as the Frantz-Nodvik analysis; see the References). More complicated analyses can, of course, be developed to take into account transverse intensity variations, finite pumping and relaxation times, large-signal and Rabi-flopping effects in the atoms, dispersive wave propagation effects, and other complications; some of these effects are treated in the references.

All of these results for pulse amplification are obviously similar in character to the power extraction and efficiency results we obtained for continuous amplification in Chapter 7. As usual, efficient energy extraction is obtained only at the cost of substantial reduction in effective energy gain integrated over the full pulse. Except for relatively smooth pulses, like gaussian pulses, efficient energy extraction can also mean severe distortion of the output pulseshape.

All the results given in this section also apply directly to pulse propagation through a saturable absorber, if we simply invert the signs of $N_0$ and $N_{\text{tot}}(t)$ in all the expressions. We will apply these results to passive saturable absorber mode locking in a later chapter.

## REFERENCES

An extensive and detailed discussion of the behavior of a pulse sent into a strongly absorbing or amplifying medium, including "negative group delay effects," is given in a lengthy paper by E. O. Schulz-DuBois, "Pulse sharpening and gain saturation in traveling-wave masers," *Bell Sys. Tech. J.* **43**, 625–658 (1964). One of the most complete and detailed reviews of nearly all aspects of laser pulse amplification and saturation, using both rate-equation and resonant-dipole approaches, and covering both simple saturable-absorption aspects and more complex coherent-pulse and self-induced-transparency aspects, is given in P. G. Kryukov and V. S. Letokhov, "Propagation of a light pulse in a resonantly amplifying (absorbing) medium," *Sov. Phys. Uspekhi* **12**, 641–672 (March-April 1970).

There are numerous other analyses of optical pulse amplification and propagation in the published literature. Some of the more useful articles include the following.

R. Bellman, G. Birnbaum, and W. G. Wagner, "Transmission of monochromatic radiation in a two-level material," *J. Appl. Phys.* **34**, 780–783 (April 1963). A compact rate-equation analysis of pulse absorption and/or amplification in saturable two-level absorbers and/or amplifiers.

L. W. Davis and Y. S. Lin, "Propagation of optical pulses in a saturable absorber," *IEEE J. Quantum Electron.* **QE-9**, 1135–1138 (December 1973). A resonant-dipole calculation of pulse propagation and attenuation in a saturable absorber, going beyond the rate-equation limits, and showing how coherent ringing and optical-nutation effects also occur at large enough input pulse intensities.

E. Fill and W. Schmid, "Amplification of short pulses in $CO_2$ laser amplifiers," *Phys. Lett.* **45A**, 145–146 (September 10, 1973). A resonant-dipole calculation of short pulse amplification including both coherent-pulse effects and effects of rotational relaxation in $CO_2$. Shows how the pulse leading edge steepens and then breaks up into optical nutations at large enough intensity.

L. M. Frantz and J. S. Nodvik, "Theory of pulse propagation in a laser amplifier," *J. Appl. Phys.* **34**, 2346–2349 (August 1963). A detailed rate-equation analysis of energy amplification and pulse sharpening in saturable two-level amplifiers, similar to the discussions in this section.

A. Icsevgi and W. E. Lamb, Jr., "Propagation of light pulses in a laser amplifier," *Phys. Rev.* **185**, 517–545 (September 10, 1969). Extensive and detailed analysis of pulse propagation using density matrix atomic equations and including inhomogeneous (doppler) broadening effects, with extensive numerically calculated solutions.

A. E. Siegman, "Design considerations for laser pulse amplifiers," *J. Appl. Phys.* **35**, 460–461 (February 1964). A simplified rate-equation derivation of the energy extraction efficiency results for pulse amplification.

J. P. Wittke and P. J. Warter, "Pulse propagation in a laser amplifier," *J. Appl. Phys.* **35**, 1668–1672 (June 1964). Analysis of pulse propagation in homogeneously saturable laser amplifiers with a nonsaturable loss also present, using the Bloch form for the atomic equations. Shows that with loss included a steady-state pulseshape develops at large enough gain or long enough propagation distance.

One practical application of pulsed saturable absorbers is in stabilizing very high-gain pulsed amplifier systems against parasitic oscillations. A good example of the practical complexities this entails in a real application can be found, for example, in R. F. Haglund, Jr., A. V. Nowak, and S. J. Czuchlewski, "Gaseous saturable absorbers for the Helios $CO_2$ laser system," *IEEE J. Quantum Electron.* **QE–17**, 1799–1808 (September 1981). Another practical application of pulsed saturable absorbers is in nonlinear pulse compression, with a good practical example being a paper by J. E. Murray, "Temporal compression of mode-locked laser pulses for laser-fusion diagnostics," *IEEE J. Quantum Electron.* **QE–17**, 1713–1723 (September 1981).

---

### Problems for 10.1

1. *Pulse input energy to saturate the pulse energy gain down to just half the initial unsaturated gain.* A laser amplifier with large initial power gain, $G_0 \gg 1$, is to be used to amplify a pulse having just enough energy so that the final gain after the pulse has passed is half the initial (numerical) gain. Show that the required input energy is $U_{\text{in}} \approx U_{\text{sat}}/G_0$, the resulting output energy is $U_{\text{out}} \approx (\ln 2)U_{\text{sat}}$, and hence the pulse energy gain is $G_{\text{pe}} \approx (\ln 2)G_0$.

2. *Leading-edge spike width for an infinitely sharp square input pulse.* A square input pulse with an infinitely sharp leading edge when sent into a high-gain saturable pulse amplifier produces a sharp spike on the leading edge of the output pulse. Develop an expression for the width of this spike (FWHM) as a function of the unsaturated amplifier gain, amplifier initial stored energy or available energy, and the power level of the input pulse step.

3. *Calculating input-output pulse profiles for gaussian input pulses of varying pulse energy.* A pulse with total energy $U_{\text{in}}$ and gaussian pulse width $\tau_p$ (FWHM) is sent through a homogeneous saturable laser amplifier with small-signal gain $G_0 = \exp(2\alpha_0 L)$. Plot the input and output pulseshapes $I_{\text{in}}(t)$ and $I_{\text{out}}(t)$ on log scales versus $t$ for gains $G_0 = 3, 10$, and $100$, and for various values of input energy, such as $U_{\text{in}}/U_{\text{sat}} = 0.001, 0.01, 0.1$, and $1.0$. Repeat for a saturable absorber with small-signal transmission $T_0 = \exp(-2\alpha_0 L) = 0.1$ and $0.01$. Try square instead of gaussian pulse with the same parameters.

4. *Pulse energy transmission through a saturable atomic absorber.* A pulse of input energy $U_{\text{in}}$ is sent through a homogeneous saturable absorber having a small-signal transmission $T_0 = \exp(-2\alpha_0 L)$. Plot the net energy transmission $T =$

$U_{\text{out}}/U_{\text{in}}$ through the absorber versus input pulse energy $U_{\text{in}}$ for $T_0 = 0.1$, $0.01$, and $0.001$.

5. *Measuring pulse saturation energies using photoacoustic spectroscopy.* If a short pulse of laser energy is passed through a very weakly absorbing gas mixture ($2\alpha_m L \ll 1$), then a technique called photoacoustic or optoacoustic spectroscopy can be used to measure accurately, at least on a relative scale, the very small amount of energy $\Delta U$ absorbed by the gas. This technique works in essence by measuring the sound impulse that this sudden heat input produces, using an ordinary microphone inside the atomic cell.

Suppose we plot the measured $1/\Delta U$ (in arbitrary units) versus the reciprocal input pulse energy $1/U_{\text{in}}$ in the range $U_{\text{in}} \leq U_{\text{sat}}$. Show that the results of these measurements should be a straight line, which we can extrapolate to find the saturation energy $U_{\text{sat}}$ without needing to know either $\alpha_0 L$ or the absolute calibration factor on the measurement of $\Delta U$. (See E. A. Ryabov, "Method for measuring the saturation energy of weakly absorbing gases," *Sov. J. Quantum Electron.* **5**, 81–82, July 1975.)

6. *Penetration depth versus energy for pulses traveling into a saturable absorber.* A short laser pulse is injected into one end of a long cell containing a homogeneously saturable absorbing medium. Make plots of the saturation absorption coefficient versus depth into the cell just after the pulse has gone past, for different amounts of input pulse energy, from well below to well above the saturation energy density of the absorbing medium. What is an approximate "rule of thumb" for how far into the cell a strong pulse will penetrate?

7. *Pulse input-output and pulse energy extraction for a partially bottlenecked lower energy level (research problem).* Consider a saturable pulse amplifier of the type analyzed in this section, but assume that the relaxation rate $\gamma_1$ out of the lower laser level may be on the same time scale as the pulsewidth (although the pulse is still assumed very short compared to the upper-level relaxation or pumping times). Develop the necessary set of three coupled rate equations (cavity, upper laser level, lower laser level) to describe this situation, and see if you can make any progress on solving these equations in order to predict, for example, pulse output versus pulse input.

## 10.2 PULSE PROPAGATION IN NONLINEAR DISPERSIVE SYSTEMS

When an optical pulse propagates through any kind of nonlinear system, we can expect to see at least some nonlinear distortion of the pulseshape with propagation distance, with stronger effects for larger amplitude pulses. When such nonlinear distortion is combined, however, with linear dispersive pulse distortion effects such as those discussed in the Chapter 9, even more complex and interesting effects can be expected to occur—especially since the nonlinear distortion effects may in general tend either to *combine with* or to *cancel out* the linear dispersive effects. In this section we will introduce several such nonlinear and dispersive effects that are of particular importance in real laser systems.

### Nonlinear Pulse Propagation in Atomic Systems

Let us first make some general observations about the large-signal propagation of an optical pulse through an atomic medium which contains a resonant atomic transition, such as a typical laser medium.

When a pulsed signal with an electric field variation $\mathcal{E}(z,t)$ propagates through any kind of atomic medium, the electromagnetic aspects of the pulse behavior are governed by the electromagnetic wave equation. This equation is a fundamentally *linear equation* for the electric field $\mathcal{E}(z,t)$ in terms of the polarization $p(z,t)$ in the atomic medium. If the polarization term on the right-hand side of this wave equation in turn represents an induced polarization which is also *linear in the applied signal field*, then the overall response of this system will be entirely linear; and the pulse propagation and distortion behavior in the system can be completely described by the dispersion curve for the atomic medium, or the $\omega$-$\beta$ curve for the wave-propagating system.

Suppose however that the polarization $p(z,t)$ arises from a resonant atomic transition, and also that the applied signal fields are strong enough to produce significant nonlinear or "Rabi flopping" behavior (as described in Chapter 5). The polarization response $p(z,t)$ is then more complicated and no longer linear in the applied signal. The polarization response must be described instead by (at least in simple cases) a resonant dipole equation for the polarization response, together with an additional equation for the population difference $\Delta N(z,t)$ on the relevant transition as a function of space and time. Both of these equations are basically nonlinear equations, at least at larger signal levels.

To find the complete pulse propagation behavior for a large-amplitude wave passing through a resonant atomic system, all three of these equations must then be solved simultaneously and in a self-consistent fashion, taking full account of both the effects of the pulse fields on the atoms and the effects of the atomic polarization back on the pulses. For larger signals where these equations become more strongly nonlinear, the resulting solutions will generally be quite complicated, in themselves and in how they depend on the pulse intensity. As a result there are many complicated analyses of such phenomena in the scientific literature. (Only the atomic equations are nonlinear; Maxwell's equations and hence the electromagnetic wave equation are entirely linear in $\mathcal{E}$ and $p$.)

The results that come out of these analyses (and experiments) include purely linear behavior, such as free induction decay, and other types of pulse propagation and distortion behavior, such as Rabi flopping behavior, "self-induced transparency," and "$\pi$ and $2\pi$ pulse propagation." We will not attempt to review any of these resonant-atom pulse phenomena in detail here, since such discussions are unavoidably lengthy and complex, and since the resulting large-signal phenomena, though sometimes experimentally interesting, do not seem to have major practical applications. (We have given a brief description of large-signal pulse propagation in Chapter 5, where we introduced the concept of Rabi flopping behavior).

### Nonlinear Optical Polarization: The Optical Kerr Effect

There is, however, another fundamental type of nonlinear polarization response that occurs in essentially all transparent optical materials, not just on resonant atomic transitions, and that can be of considerable practical importance in many laser situations. This is a nonlinear change in the dielectric constant or index of refraction of almost any optical material with increasing optical intensity, often referred to as an *optical Kerr effect*. In the rest of this section we will examine several of the important propagation effects produced by this optical Kerr effect.

When an electric field $\mathcal{E}$ is applied to a transparent dielectric medium, the force associated with this field produces a distortion of the electron-charge clouds in that medium, and also a possible reorientation of the molecular axes of molecules in a liquid medium, since such molecules generally like to line up with an applied field. Both effects in turn lead to a macroscopic polarization $p$ in the medium that in first order will be linear in the applied $\mathcal{E}$ field. This linear response in a low-loss or transparent dielectric is, of course, just the linear dielectric constant or index of refraction of the medium.

If the applied field is strong enough, however, the polarization response of the medium may become nonlinear in the applied field. (The distortion of the electron-charge clouds, or the realignment of the molecular axes, becomes nonlinear—usually weakly nonlinear—in the applied field strength.) This is very often expressed in a somewhat simplified but still fairly general fashion by writing the polarization as a series expansion in the applied field in the form

$$p = \chi_{(1)}\epsilon_0\mathcal{E} + \chi_{(2)}\mathcal{E}^2 + \chi_{(3)}\mathcal{E}^3 + \cdots, \qquad (29)$$

where $\chi_{(1)}$ is the linear susceptibility, and $\chi_{(2)}$ and $\chi_{(3)}$ (which have quite different dimensions) represent weak higher-order nonlinearities in the dielectric response of the medium. (In a more accurate picture, all three of the $\chi$ quantities should be tensor quantities, and all three should have frequency dependences that become increasingly complex for the higher-order terms.)

### Second-Order Susceptibility: Harmonic Generation and Modulation

The $\chi_{(2)}\mathcal{E}^2$ term—sometimes written in an alternative notation as $d_2\mathcal{E}^2$—represents a second-order nonlinearity, which can be responsible for second-harmonic generation, optical rectification, optical parametric amplification, and other useful nonlinear effects. By symmetry arguments, however, this effect must be identically zero in any material that has a centrosymmetric arrangement of atoms. Effects of this type are found therefore primarily in certain special crystals having a noncentrosymmetric crystal structure—in essence, only in those materials that are also piezoelectric.

This includes, for example, barium titanate or $BaTiO_3$, crystal quartz, potassium dihydrogen phosphate (KDP), ammonium dihydrogen phosphate (ADP), cesium dihydrogen arsenate (CDA), and lithium niobate ($LiNbO_3$); these are some of the nonlinear optical crystals most widely used in optical modulators and harmonic generators.

### The Third-Order Susceptibility

The third-order susceptibility term $\chi_{(3)}\mathcal{E}^3$ can be present, however, with varying strength, in essentially *all optical materials of any crystal structure or class*, including liquids and gases. If we include this term in the polarization $p$, the total electric displacement $d$ in the medium can then be related to the applied field $\mathcal{E}$ in the form

$$d = \epsilon_0[1 + \chi_{(1)}]\mathcal{E} + \chi_{(3)}\mathcal{E}^3 = \epsilon_0[1 + \chi_{(1)} + \epsilon_0^{-1}\chi_{(3)}\mathcal{E}^2]\mathcal{E}. \qquad (30)$$

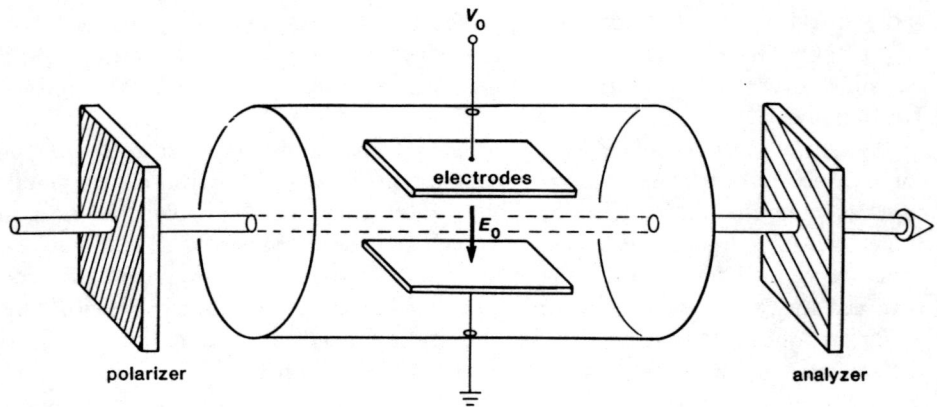

**FIGURE 10.8**
A Kerr cell light modulator.

Hence the dielectric constant $\tilde{\epsilon}$ is now a nonlinearly varying quantity given by

$$\tilde{\epsilon} = \epsilon_1 + \epsilon_2 \mathcal{E}^2, \tag{31}$$

in which $\epsilon_1 \equiv \epsilon_0(1 + \chi_{(1)})$ is the linear or first-order dielectric constant, and $\epsilon_2 \mathcal{E}^2 \equiv \chi_{(3)} \mathcal{E}^2$ is the nonlinear change in the dielectric constant, produced by the applied field.

Since the optical index of refraction $n$ is related to the optical-frequency value of $\epsilon$ by $n = \sqrt{\epsilon/\epsilon_0}$, we can also view this as a nonlinear dependence of the index of refraction on applied signal strength, as given by

$$n = n_0 + n_{2E}\mathcal{E}^2, \tag{32}$$

where $n_0 = \sqrt{\epsilon_1/\epsilon_0}$ is the linear value and $n_{2E}\mathcal{E}^2$ the nonlinear variation of the index of refraction.

### Kerr Cell Light Modulators

Suppose we construct a liquid cell containing $CS_2$ or nitrobenzene or some similar liquid, as in Figure 10.8, to which we apply both a strong low-frequency modulation field $E_0$ (by using suitable electrodes) and a much weaker optical-frequency field $\mathcal{E}$ (in the form of a traveling optical wave). A strong enough modulation field will then change the index of refraction seen by the optical wave according to the relationship

$$n = n_0 + n_{2E}E_0^2, \tag{33}$$

and this provides a way of phase modulating the light beam.

To be slightly more accurate, the modulation field $E_0$ usually causes an *increase* in index of refraction for optical $\mathcal{E}$ fields polarized *parallel* to the dc field, and a *decrease* in index of refraction for fields polarized *perpendicular* to $E_0$. This then creates an *induced birefringence* in the modulation cell, with a magnitude proportional to the modulating voltage squared. This birefringence can in turn be converted into amplitude modulation by placing the modulation cell between suitable crossed polarizers. This physical effect is known as the *Kerr effect*, and the resulting device is a *Kerr cell modulator*.

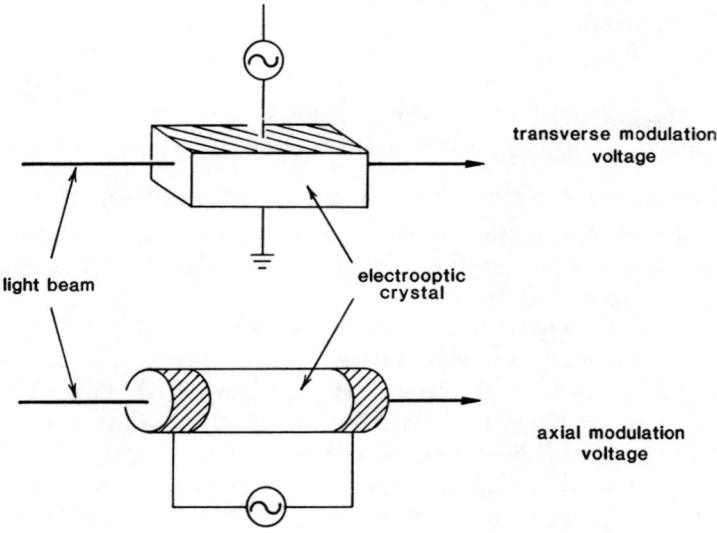

**FIGURE 10.9**
Alternative ways of constructing a Pockels cell light modulator.

### Pockels Cell Light Modulators

Practical Kerr cells, when they are used at all, usually employ one of the molecular liquids mentioned earlier, since the reorientation of the molecules in these liquids under the influence of the applied field $E_0$ produces the strongest available Kerr coefficient, or change of index with voltage. Even with these liquids, however, voltages on the order of 25,000 volts are necessary to produce sizable amplitude-modulation effects.

Most of the electrooptic modulators used with lasers, therefore, are instead *Pockels cell modulators*, which use one of the noncentrosymmetric crystals mentioned earlier, and produce the index change or birefringence through the second-order nonlinearity term $\chi_{(2)} E_0^2$. The induced birefringence is then linear rather than quadratic in the modulation field $E_0$, and adequate modulation can be obtained in practice with modulation voltages of a few thousand volts, or even less in some especially favorable cases. Figure 10.9 shows two examples of simple Pockels cell modulator designs.

### Optical Kerr Effect

Suppose, however, we consider an optical signal with a sufficiently strong optical field strength that we can have a significant $\chi_{(3)} \mathcal{E}^3$ or $n_{2E} \mathcal{E}^2$ term *produced by the optical beam itself.* If $\mathcal{E}(t)$ is in fact an optical-frequency signal, at frequency $\omega$, then this term will produce two effects. On the one hand the $\chi_{(3)} \mathcal{E}^3$ term will produce a third-harmonic polarization $p(t)$ at frequency $3\omega$; this may radiate—typically very weakly—at the third harmonic of the applied optical frequency $\omega$. We will neglect this third-harmonic generation process here, since it is usually very weak; is not of direct interest at this point; and will not be properly phase-matched in most situations.

On the other hand, this same $n_{2E} \mathcal{E}^2$ effect will also produce a zero-frequency or time-averaged signal-induced change in the refractive index for the signal field, which can be written as

$$n = n_0 + n_{2E} \langle \mathcal{E}^2 \rangle = n_0 + n_{2I} I, \qquad (34)$$

where $\langle \mathcal{E}^2 \rangle$ represents the time-averaged value of the field squared, and $I \equiv \sqrt{\epsilon/\mu} \langle \mathcal{E}^2 \rangle$ represents the optical intensity. This change in the average value of the optical refractive index produced by the optical signal itself is commonly referred to as the *optical Kerr effect*.

Such an optical Kerr effect will be present, with a positive sign ($n_{2I} > 0$), in nearly all optical materials. A representative value for the optical Kerr coefficient in a typical glass (such as might be used in an optical fiber) might be $n_{2E} \approx 10^{-22}$ m$^2$/V$^2$ or $n_{2I} \approx 10^{-16}$ cm$^2$/W. Certain strongly polarizable molecular liquids, such as $CS_2$ or certain long-chain organics, can have values 10 to 20 times larger than this. (Essentially all condensed materials have an *electronic optical Kerr effect* of roughly comparable magnitude; the strongly enhanced response in molecular liquids comes from an *orientational Kerr effect* similar to that produced by low-frequency electric fields).

We will review several nonlinear phenomena produced by this optical Kerr effect in subsequent paragraphs. We can estimate, however, that this optical Kerr effect might produce significant effects if it produces an additional path length $\Delta n L$ of half a wavelength, or an additional half cycle of phase shift, in a path length of, say, $L = 1$ cm, or $2\pi n_{2I} I L/\lambda = \pi$. If we use $L = 1$ cm and $\lambda = 0.6$ nm, then a significant nonlinear effect will occur for an optical intensity of $I \approx 1$ to 10 GW/cm$^2$, depending on the value of $n_{2I}$. It is in fact in this range of intensities that significant optical Kerr effects do occur in optical systems like high-power laser rods, focusing lenses, and other optical elements.

Note, however, that intensities in this range will occur for a total input power of less than 1 watt in a 4 μm-diameter optical fiber; moreover, in an optical fiber it may take kilometers rather than centimeters for the resulting phase-modulation effects to accumulate. This can make possible very strong and useful nonlinear optical effects in optical fibers, as we will see in more detail in the following section.

### Whole-Beam Self-Focusing Effects

As a first illustration of an important effect produced by the optical Kerr effect, we can consider so-called *self-focusing* of an optical beam. Suppose an optical beam with a moderate intensity $I$ and a smooth transverse profile passes through a medium having a finite and positive optical Kerr coefficient $n_{2I}$, as in Figure 10.10. The higher intensity in the center of the beam will then cause an increase in the index of refraction seen by the center of the beam, as compared to the wings; in other words, the optical medium will be given a focusing power, or converted into a weak positive lens.

If this self-focusing effect in propagating through a given length of the medium exceeds the diffraction spreading of the optical beam in the same length, the optical beam profile will begin to be focused inward as the beam propagates. But, inward focusing then increases the intensity in the center of the beam, and makes the sides of the beam profile steeper; and this in turn increases the strength of the self-induced lens. The beam will then continue to be focused ever more strongly inward, in an essentially runaway fashion.

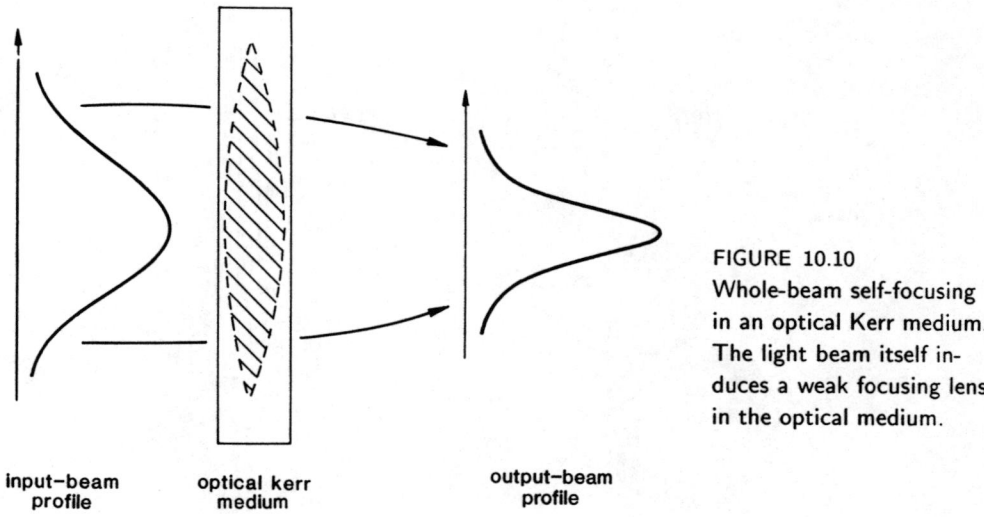

**FIGURE 10.10**
Whole-beam self-focusing in an optical Kerr medium. The light beam itself induces a weak focusing lens in the optical medium.

This type of self-focusing of the entire beam profile is known as *whole-beam self-focusing*. A more detailed analysis shows that, when the effects of self-focusing and diffraction spreading are both taken into account, runaway whole-beam self-focusing will begin to occur when the total power in a beam with a smooth transverse profile exceeds a certain critical power $P_{\text{crit}}$, independent of the diameter of the beam. Typical values of $P_{\text{crit}}$ range from a few tens of kilowatts in strong Kerr liquids to a few megawatts in materials with typical weak Kerr coefficients.

### Small-Scale Self-Focusing Effects

There is a similar phenomenon known as *small-scale self-focusing*, in which any small-amplitude variations or ripples on a transverse beam profile will begin to grow in amplitude exponentially with distance because of the optical Kerr effect. In essence the transverse spatial variation of the optical beam intensity produces a transverse spatial variation in refractive index, or a refractive index grating. This grating diffracts some of the optical beam energy into small-angle scattering, and this diffracted light interferes with the original beam in exactly such a way as to make the initial intensity ripples on the beam profile grow in amplitude with distance.

Figure 10.11 shows the dramatic results of an experiment in which initially small-amplitude ripples were put on the transverse amplitude profile of an optical beam, and the beam then sent through a strong optical Kerr medium with an intensity sufficient to cause significant growth in these ripples after a few tens of cm. The runaway growth of the periodic amplitude variations is evident.

In general, if either whole-beam or small-scale self-focusing becomes significant, this self-focusing will continue in a runaway fashion. The optical beam may then rapidly collapse with distance into one or several very small filaments or self-focused focal spots. Once this happens, not only does the beam become badly distorted in its transverse profile, but the power density also usually becomes large enough to cause optical damage, optical breakdown, or other undesirable nonlinear effects to ensue.

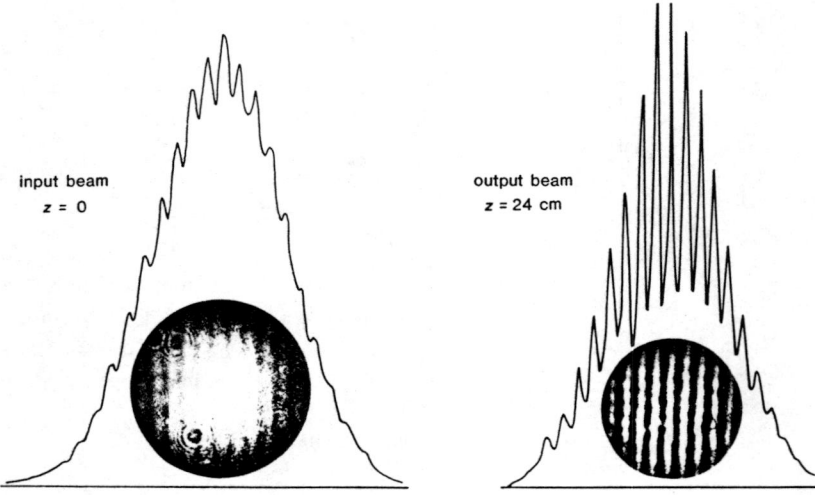

FIGURE 10.11
An experimental demonstration of small-scale self-focusing, or the exponential growth with distance of small periodic intensity ripples on the transverse profile of an optical beam.

The type of nonlinear self-focusing produced by the optical Kerr effect can thus be a major problem in many high-power lasers, especially pulsed lasers, as well as in scientific experiments using focused high-power laser beams.

### Self-Phase Modulation

The optical Kerr effect can also produce a very similar *self-phase-modulation effect*, which occurs for pulsed or modulated signals in the time rather than the spatial domain. To demonstrate this, we can next suppose that an optical pulse with some given intensity variation $I(t)$ in time propagates through a certain length $L$ of a medium with a finite optical Kerr coefficient $n_{2I}$, and that the pulse amplitude is large enough to produce a significant index change $\Delta n(t) \equiv n(t) - n_0 = n_{2I}I(t)$ and a significant change in optical path length $\Delta n(t)L$, at least near the peak of the pulse. (Assume for now that the transverse beam profile is uniform, or that we somehow otherwise avoid the self-focusing effects just discussed.)

The pulse fields will then experience a time-varying phase shift or phase modulation $\exp[j\Delta\phi(t)] = \exp[-j2\pi\Delta n(t)L/\lambda] = \exp[-j2\pi n_{2I}I(t)L/\lambda]$ produced by the intensity variation of the pulse itself. If the optical Kerr coefficient is positive ($n_{2I} > 0$), as it usually is, this self-phase modulation will represent in effect a lowering of the optical frequency of the pulse during the rising or leading edge of the pulse, since $dn/dt > 0$ and hence $\Delta\omega_i(t) = (d/dt)\Delta\phi(t) < 0$. (In physical terms, the medium is getting optically longer; so the arrival of optical cycles is delayed or slowed down.) Similarly there will be an increase in the instantaneous frequency of the pulse signal during the trailing or falling edge of the pulse. The maximum frequency shift will occur at the points of maximum slope or maximum $dI(t)/dt$.

A pulse with a smooth time envelope, as in Figure 10.12, will thus acquire a more or less linear frequency chirp across the central region of the pulse, as

## 10.2 PULSE PROPAGATION IN NONLINEAR DISPERSIVE SYSTEMS

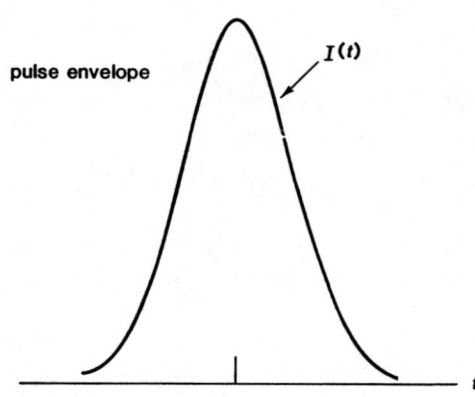

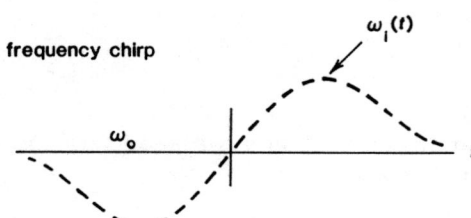

**FIGURE 10.12**
The initial effect of intensity-dependent self-phase modulation is to lower the frequency on the leading edge and raise the frequency on the trailing edge of a pulse, thus producing a chirp.

shown in the lower plot. The magnitude of this chirp will increase more or less linearly with distance through the medium, at least as long as the pulse intensity profile remains unchanged.

In most practical situations, however, the optical medium will also have a certain value of *group-velocity dispersion* $dv_g(\omega)/d\omega$. The different portions of the pulse, with their slightly different optical frequencies, will thus begin to travel at slightly different group velocities; and as a result the pulseshape will begin to change by an increasing amount with increasing distance. Depending on circumstances, this effect can lead to at least three different types of behavior: severe pulse distortion and breakup, soliton formation and propagation, or pulse broadening and enhanced frequency chirping. We can discuss each of these in turn.

### Approximate Analysis of Self-Phase Modulation

We can develop an approximate analysis to indicate the magnitude of these self-phase modulation effects as follows. Suppose an initially unchirped **gaussian** input pulse has the time-variation

$$\mathcal{E}(t) = \mathcal{E}_0 e^{-at^2} \quad \text{or} \quad I(t) = I_0 e^{-2at^2}. \tag{35}$$

The net phase shift for this pulse in passing through a length $L$ of nonlinear medium will then be

$$\phi(t) = \frac{2\pi(n_0 + n_{2I}I)L}{\lambda}, \tag{36}$$

and hence the phase-shift derivative will be

$$\frac{d\phi}{dt} = \frac{2\pi n_{2I}L}{\lambda}\frac{dI(t)}{dt} \approx \frac{4\pi a n_{2I}LI_0}{\lambda} \times te^{-2at^2}, \tag{37}$$

where we assume for simplicity that to first order the pulseshape $I(t)$ is not changed in passing through the length $L$.

But this phase modulation corresponds to giving the pulse a frequency chirp at the center of the pulse which is given (in our earlier gaussian pulse notation) by

$$\frac{d\phi}{dt} = 2bt \approx \frac{4\pi a n_{2I}LI_0}{\lambda}t. \tag{38}$$

This means that the pulse will acquire a chirp parameter $b = a$, and thus increase its time-bandwidth product by a factor of $\sqrt{2}$, after passing through a length of nonlinear medium given by

$$\frac{2\pi n_{2I}I_0 L}{\lambda} = 1. \tag{39}$$

That is, the pulse will acquire a significant amount of self-phase modulation in length $L$ if its peak intensity exceeds a value given by

$$I_0 = \frac{\lambda}{2\pi n_{2I}L}. \tag{40}$$

If we take $n_{2I} = 3 \times 10^{-16}$ cm$^2$/W, $L = 10$ cm, and $\lambda = 0.5$ $\mu$m, this gives a threshold intensity for significant self-phase modulation of $I_0 \approx 3$ GW/cm$^2$, which is close to the damage threshold in most optical materials. Suppose, however, we consider a single-mode optical fiber with a diameter of 4 $\mu$m and a length of $L = 10$ m. Its threshold intensity of 30 MW/cm$^2$ is reached with a total power in the fiber of $\approx 3$ W.

### Pulse Distortion and Breakup Effects

We can recall that the group velocity in a dispersive medium is given by $1/v_g(\omega) = d\beta(\omega)/d\omega \equiv \beta'(\omega)$. Hence the variation of group velocity with frequency is given by

$$\frac{dv_g}{d\omega} = -v_g^2 \frac{d^2\beta}{d\omega^2} = -v_g^2 \beta'', \tag{41}$$

where $\beta'' \equiv d^2\beta/d\omega^2$ is commonly referred to as the *group-velocity dispersion* of the medium. A negative value of $\beta''$, which corresponds to negative dispersion according to this conventional definition, thus means that the group velocity increases with increasing frequency.

Suppose that an optical Kerr medium in fact has such a negative dispersion, so that $v_g$ increases with increasing $\omega$. This means physically that the leading edge of the chirped pulse in Figure 10.12 will begin to travel more slowly, and to fall back against the main part of the pulse, while the trailing edge of the pulse will begin to travel faster and to catch up with the main part of the pulse. In other words, the pulse will generally become compressed as it propagates, as a consequence of the self-phase-modulation process.

As the pulse becomes more compressed, however, its peak intensity will increase, on the one hand, and its rise and fall times will become shorter, on the other hand. Both effects will then combine to greatly increase the self-chirping effects on the leading and trailing edges of the pulse, and this in turn will increase the pulse compression, in another runaway type of process.

If the dispersion in the medium is of opposite sign, an initially smooth pulse will become broadened in time rather than compressed (as we will discuss in more detail in a later section). Even in this situation, however, the pulse will also acquire a growing amount of chirp, and its spectrum will be continuously broadened by the combination of nonlinear effects plus dispersion. (Note also that the analog to dispersion for pulse distortion in time is diffraction for pulse distortion in space—i.e., in the transverse coordinates—and this diffractive dispersion always has a sign corresponding to pulse compression in space for positive $n_{2I}$. Self-focusing thus always leads to beam compression in the transverse direction.)

In either situation, if the time envelope of either a pulsed or a cw signal contains any significant amount of initial amplitude (or phase) modulation or pulse substructure—that is, if either the phasor amplitude or the phase angle of the signal field has significant time modulation within the overall pulse envelope—this will increase these nonlinear distortion effects. The phase substructure will represent additional chirp, and the amplitude structure will reinforce the self-phase-modulation effect. The result will often be that the envelope of a high-power laser beam will not retain a smooth shape, if indeed it initially has one, but will begin to break up into increasingly complex subpulses within the main pulse envelope. This increasingly strong phase and amplitude modulation will also broaden the frequency spectrum of the pulse (but not the overall time envelope) by an amount that can increase rapidly with increasing distance.

### Pulse Breakup in Practical Laser Systems: The $B$ Integral

These pulse-breakup and spectral-broadening effects, especially when acccmpanied and intensified by self-focusing effects, can be a source of considerable difficulty in many high-power lasers, and particularly in mode-locked lasers, where the peak power can be very high even though the total energy or the average power may be quite low. As we have said, the effects of nonlinear modulation and dispersion will grow exponentially as a laser signal begins to break up, because the time-variation becomes faster across the substructure within the pulse, and because the pulse energy gets compressed into shorter subpulses with higher peak intensities.

Self-phase-modulation and self-focusing effects are thus especially strong "runaway" effects in such devices as multistage Nd:glass laser amplifier chains and mode-locked Nd:glass laser oscillators. In both the pulse is continually being further amplified, and the laser medium has a broad enough atomic linewidth to continue amplifying the pulse even after its spectrum has been substantially broadened by the nonlinear effects. When a mode-locked Nd:glass laser is pumped too strongly, for example, the early pulses in the mode-locked and $Q$-switched burst may be reasonably clean and well-formed, but the pulses near and after the peak of the $Q$-switched burst often become severely distorted and spectrally broadened.

Self-phase modulation of this type is often accompanied by, and reinforced by, self-focusing effects in the same system. It is also a common characteristic of such self-phase modulation that the pulse spectrum gets greatly broadened, generally in a one-sided fashion, to the low-frequency side of the original carrier frequency; and catastrophic optical damage may occur, often in small self-focused spots, if the peak intensity is not limited.

As a generalization of the self-phase-modulation criterion we developed a few paragraphs back, it has become conventional to define the "$B$ integral" for a multipass laser system as a cumulative measure of the nonlinear interaction, where this integral is given by

$$B \equiv \frac{2\pi}{\lambda} \int_0^L n_{2I}(z)\, I(z)\, dz, \qquad (42)$$

taking into account the changes in diameter and power level of the laser beam through the complete system. A generally accepted criterion for high-power laser systems is that the cumulative $B$ integral must be kept somewhere below the value $B \leq 3$ to 5 to avoid serious nonlinear damage and distortion effects due to either self-phase modulation or self-focusing.

## REFERENCES

The literature on self-focusing and self-phase-modulation effects is very extensive. A review article covering many aspects of the subject is S. A. Ahkmanov, R. V. Khokhlov, and A. P. Sukhorukov, "Self-focusing, self-defocusing and self-modulation of laser beams," in *Laser Handbook*, edited by F. T. Arecchi and E. O. Schulz-Dubois (North-Holland, 1972), pp. 1151–1228.

An early paper on the concept of whole-beam self-focusing is P. L. Kelley, "Self-focusing of optical beams," *Phys. Rev. Lett.* **15** 1005–1008 (December 27, 1965). For a recent and rather clean experimental example of whole-beam self-focusing effects, see J. E. Bjorkholm and A. Ashkin, "CW self-focusing and self-trapping of light in sodium vapor," *Phys. Rev. Lett.* **32**, 129–132 (January 28, 1974).

The earliest discussion of small-scale self-focusing effects seems to be by V. I. Bespalov and V. I. Talanov, "Filamentary structure of light beams in nonlinear liquids," *Sov. Phys.—JETP* **3**, 307–310 (1966), and the associated analysis is often referred to as the "Bespalov-Talanov analysis." See also V. I. Talanov, "Focusing of light in cubic media," *Sov. Phys.—JETP* **11**, 199-201 (1970).

An extended version of this theory is given by B. R. Suydam, "Self-focusing of very high power laser beams: II," *IEEE J. Quantum Electron.* **QE–10**, 837–843 (November 1974); and the deleterious effects of both whole-beam and small-scale self-focusing are discussed in B. R. Suydam, "Effect of refractive-index nonlinearity on the optical quality of high-power laser beams," *IEEE J. Quantum Electron.* **QE–11**, 225–230 (June 1975).

Definitive experimental demonstrations of small-scale self-focusing are given by A. J. Campillo, S. L. Shapiro, and B. R. Suydam, "Periodic breakup of optical beams due to self-focusing," *Appl. Phys. Lett.* **23**, 628–630 (December 1, 1973); and "Relationship of self-focusing to spatial instability modes," *Appl. Phys. Lett.* **24**, 178–180 (February 15, 1974).

One of the first experiments to combine nonlinear chirping via self-phase modulation with subsequent pulse recompression is R. A. Fisher, P. L. Kelley, and T. K. Gustafson, "Subpicosecond pulse generation using the optical Kerr effect," *Appl. Phys. Lett.* **14**, 140–143 (February 15, 1969).

An extensive discussion of the $B$ integral and its application in multistage amplifier design is given in D. C. Brown, *The Physics of High Peak Power Nd:Glass Laser Systems* (Springer-Verlag, 1980).

## 10.3 THE NONLINEAR SCHRÖDINGER EQUATION

The basic equation of motion for analyzing signal propagation through a weakly nonlinear optical medium, or along a nonlinear transmission line (such as an optical fiber with an optical Kerr coefficient), is a nonlinear extension of the parabolic equation we derived in Chapter 9. We can derive this nonlinear form in a simplified manner as follows.

### Derivation of the Nonlinear Schrödinger Equation

Consider an optical signal of the form $\mathcal{E}(z,t) \equiv \tilde{E}(z,t)\exp j[\omega_0 t - \beta(\omega_0)z]$ traveling in the $+z$ direction, where $\tilde{E}(z,t)$ is the slowly varying amplitude of this signal. If we Fourier-transform this signal $\tilde{E}(z,t)e^{j\omega_0 t}$ into its frequency spectrum $\tilde{E}(z,\omega)$ at any arbitrarily chosen plane $z$, propagate each frequency component forward by a small distance $dz$ using the frequency-dependent and intensity-dependent propagation constant $\beta(\omega)$, and then Fourier-transform these components back into the time domain, we can find that the signal envelope $\tilde{E}(z,t)$ at the plane $z = z + dz$ is given by

$$\tilde{E}(z+dz,t) = \frac{1}{2\pi}\int_{-\infty}^{\infty} d\Delta\omega \int_{-\infty}^{\infty} dt'\, \tilde{E}(z,t')\, e^{j\Delta\omega(t-t')}\, e^{-j\Delta\beta\, dz}, \quad (43)$$

where $\Delta\omega \equiv \omega - \omega_0$ and $\Delta\beta \equiv \beta(\omega) - \beta(\omega_0)$.

The partial derivatives of the signal $\tilde{E}(z,t)$ with respect to time can be calculated by multiplying the integrand in Equation 10.43 by $\partial/\partial t \equiv j\Delta\omega$, and with respect to distance by either expanding in powers of $dz$ or by multiplying the integrand by $\partial/\partial z \equiv -j\Delta\beta$. We can also suppose that the propagation constant $\beta(\omega)$ has the weakly nonlinear and dispersive form

$$\beta(\omega) = \beta(\omega_0) + \beta_2 \langle \mathcal{E}^2 \rangle + \beta'(\omega_0) \times (\omega - \omega_0) + \frac{1}{2}\beta''(\omega_0) \times (\omega - \omega_0)^2, \quad (44)$$

where $\beta'$ and $\beta''$ are the first and second derivatives of $\beta$ with respect to $\omega$, and where

$$\beta_2 \langle \mathcal{E}^2 \rangle \equiv \frac{2\pi\omega_0 n_{2E} \langle \mathcal{E}^2 \rangle}{c} \quad (45)$$

gives the (small) change in midband propagation constant due to the optical Kerr effect. We can then show, after some algebra, that the integral form for $\tilde{E}(z,t)$ in Equation 10.43 is equivalent to the differential equation

$$\left[\frac{\partial}{\partial z} + \beta'\frac{\partial}{\partial t} - j\frac{\beta''}{2}\frac{\partial^2}{\partial t^2} + j\frac{\beta_2|\tilde{E}|^2}{2}\right]\tilde{E}(z,t) = 0, \quad (46)$$

where we have used $\langle \mathcal{E}^2 \rangle \equiv \frac{1}{2}|\tilde{E}|^2$ for a sinusoidal signal.

### Discussion

Equation 10.46 is a generalization of the parabolic equation 9.41 derived in Chapter 9, with a nonlinear optical Kerr effect term added. (This approach to its derivation also illustrates another way of arriving at Equation 9.41.) Equation 10.46 has the form of a *nonlinear Schrödinger equation*, with a nonlinear potential function, except that $z$ and $t$ are interchanged from the roles they usually play in the conventional Schrödinger equation.

This same equation arises in other physical situations, including deep-water wave propagation, ion-acoustic waves in plasma physics, superconductivity, and vortex motions; so many techniques for its solution have been developed. Note again that the group-velocity dispersion term $\beta''$ in the propagation constant translates into what is essentially a complex diffusion term $j\beta'' \partial^2/\partial t^2$ in the differential equation. Since this diffusion coefficient can have either sign, it can correspond to either pulse spreading or pulse compression in different situations. Its effects must, however, be balanced against the nonlinear propagation effects, as we will see further in the following section.

### REFERENCES

For references on the solutions to this equation, see the works by Akhmanov and coworkers cited at the end of Section 9.2, and the works on the analysis of solitons cited in the final sections of this chapter.

## 10.4 NONLINEAR PULSE BROADENING IN OPTICAL FIBERS

To illustrate the importance of self-phase-modulation effects in fiber optics, we can consider what happens when a low- to moderate-power optical pulse (e.g., a few hundred milliwatts to a few watts peak power) of very short time duration (picoseconds to femtoseconds) is injected into a very low-loss single-mode optical fiber, typically a few microns in diameter.

Self-focusing effects will then be effectively eliminated by the strong waveguiding properties of the optical fiber; at the same time, the low losses and small area of the fiber will permit strong self-phase-modulation and dispersion effects to accumulate over very long distances in the fiber, at energy and power levels well below those that will produce optical damage. This permits the demonstration of some very interesting and useful nonlinear propagation effects, including in particular pulse self-chirping and subsequent compression, and optical soliton propagation.

### Dispersive Effects in Optical Fibers

We need first to describe the dispersion effects versus wavelength in an optical fiber, especially a single-mode optical fiber. Figure 10.13 shows the typical variation of the index of refraction versus wavelength or frequency across the visible and near-infrared regions, and the resulting variations of $\beta'$ and $\beta''$ across the same regions, for typical transparent optical materials such as quartz or glass. The group-velocity dispersion parameter $\beta''$ changes from being positive at shorter wavelengths or higher frequencies to negative at longer wavelengths.

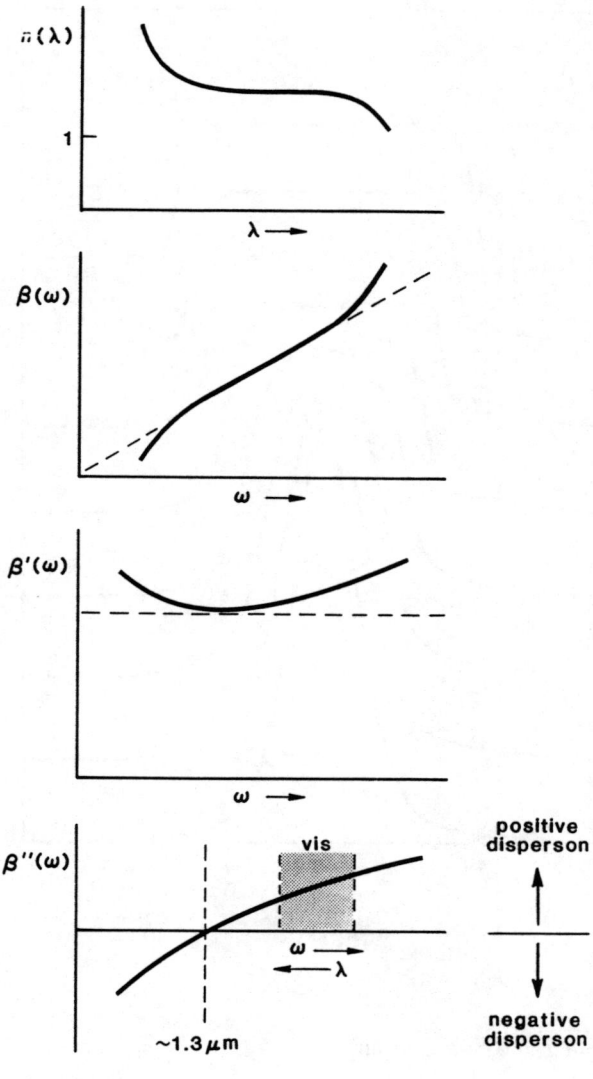

FIGURE 10.13
Optical dispersion versus frequency or wavelength in a typical optical fiber. The dispersion parameter $\beta''$ is usually positive in the visible region, becoming negative somewhere in the near infrared.

There is some confusion in the literature over how dispersive properties should be labeled; but the situation where $\beta'' > 0$, which is equivalent to $dv_g/d\omega < 0$ or $dv_g/d\lambda > 0$, is usually referred to as *positive* or *normal dispersion*, and the opposite case is referred to as *negative* or *anomalous dispersion*.

The effective dispersion values for a propagating wave in a single-mode optical fiber will differ somewhat from the purely material dispersion properties, because of *modal dispersion effects*, or waveguide propagation effects, which depend in essence on the mode shape, and on how the fields of the propagating mode are distributed between the core and the cladding of the optical fiber. The general rule for quartz optical fibers, however, is that the group-velocity dispersion is positive (according to the preceding definition) across the visible region, goes through zero in the vicinity of 1.3 $\mu$m, and becomes increasingly negative at longer wavelengths. Note that the lowest-loss region for such optical fibers occurs, however, in the vicinity of 1.5 $\mu$m, where the dispersion has become significantly negative.

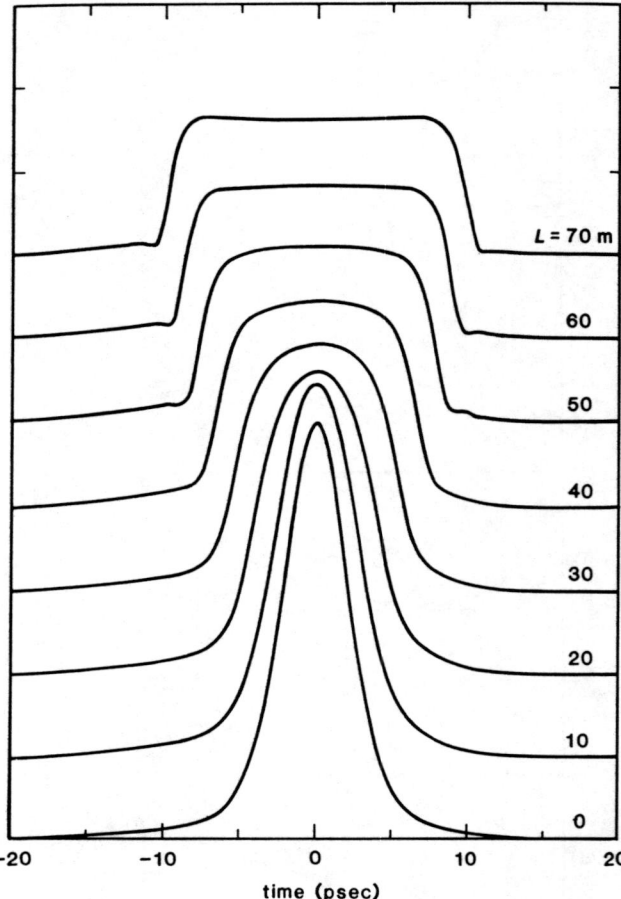

FIGURE 10.14
Pulse broadening produced by self-phase modulation plus positive dispersion for a 5.5 ps, 10 W input pulse at $\lambda_0$ = 590 nm traveling through increasing lengths of single-mode fiber.

### Nonlinear Pulse Broadening and Self-Chirping

Let us first consider the propagation properties of an optical fiber in the visible region, where the group-velocity dispersion is positive. From the arguments given two sections back, this means that the frequency chirp produced by the optical Kerr effect across the center of an optical pulse with a smooth time envelope will lead to *broadening* of the pulse envelope in time. It also turns out that such a pulse, as it propagates, will not only broaden, but acquire a growing amount of frequency chirp.

Suppose a short optical pulse propagates through an optical fiber at a wavelength in the visible region where the group-velocity dispersion in glass fibers is positive. Propagation of this pulse can then be calculated by the nonlinear Schrödinger equation derived in the preceding section, with an appropriate sign for the group-velocity-dispersion parameter. It is found that an initially smooth input pulse gradually broadens out to acquire an essentially rectangular shape, with increasing width and increasingly sharp rising and falling edges at increasing distances. Figure 10.14 shows predicted pulseshapes if we transmit a 5.5-ps pulse with 10 W peak power at 590 nm through increasing lengths ranging from 0 to 70 meters of a typical 4 $\mu$m-diameter single-mode optical fiber. The self-broadening effect on the pulse profile is evident.

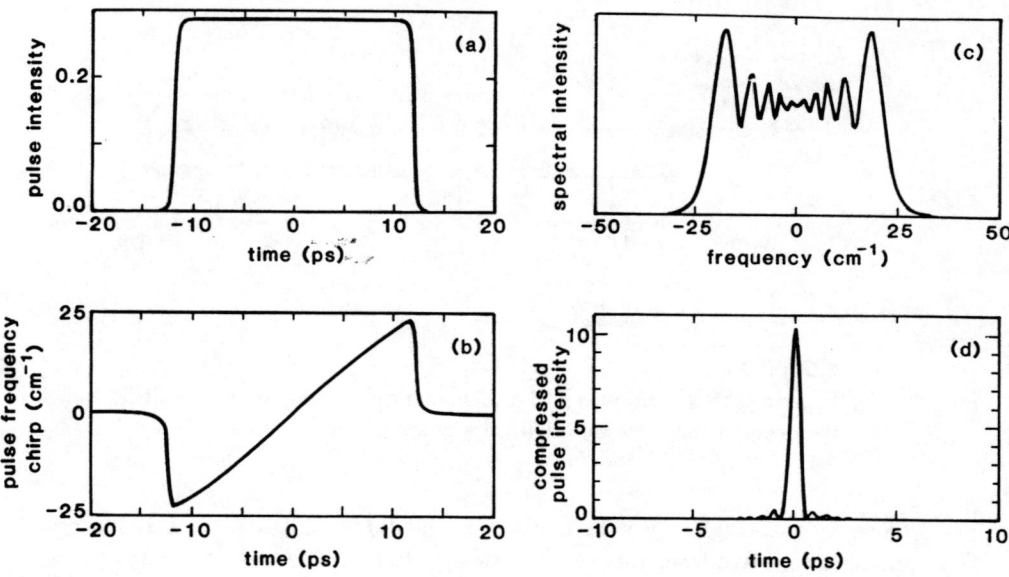

**FIGURE 10.15**
Self-broadening of an initial 6-ps 100-W pulse after propagation through 30 m of single-mode fiber. (a) Output pulse intensity versus time. (b) Output frequency chirp. (c) Output pulse spectrum. (d) Result of linear dispersive compression of this chirped pulse.

Figure 10.15 shows more details from a similarly calculated result for an initially 6-ps 100-W pulse propagated through 30 m of single-mode fiber. The pulse has broadened from a initial smooth hyperbolic secant pulse with initial pulsewidth of 6 ps to a rectangular pulse $\approx 24$ ps in duration as shown in (a). This pulse has also acquired a nearly linear frequency chirp over the full pulse duration as shown in (b). In agreement with this, the pulse spectrum has broadened from the initial transform limit of $\approx 2.5$ cm$^{-1}$ to a characteristic phase-modulation spectrum nearly 50 cm$^{-1}$ wide, as shown in (c). Plot (d) shows the greatly shortened pulse that could result from taking the self-chirped pulse in this particular theoretical example and compressing it externally using an optimum linear dispersion element.

### Chirped Pulse Recompression

It was in fact realized and demonstrated by Grischkowsky and co-workers at the IBM Research Laboratories that this kind of strongly self-chirped pulse is an essentially ideal input signal for subsequent pulse recompression using any type of auxiliary dispersive medium following the fiber, such as the diffraction grating pair shown earlier.

Figure 10.16 illustrates an experiment in which an initial pulse from a mode-locked laser is first self-chirped and broadened using a length of optical fiber, and then compressed to less than a tenth of its initial pulsewidth by using a simple grating pair of the type described in an earlier section as the auxiliary linear dispersive element. (Note that Grischkowsky and colleagues in fact used a retroreflective prism to achieve the desired dispersion with only a single grating.)

By cascading two stages of this type of self-broadening and linear recompression, this group has in fact converted initial 5.9-ps pulses into 0.09-ps (or 90-fs) pulses, as illustrated in Figure 10.17. In other experiments, Shank and co-workers

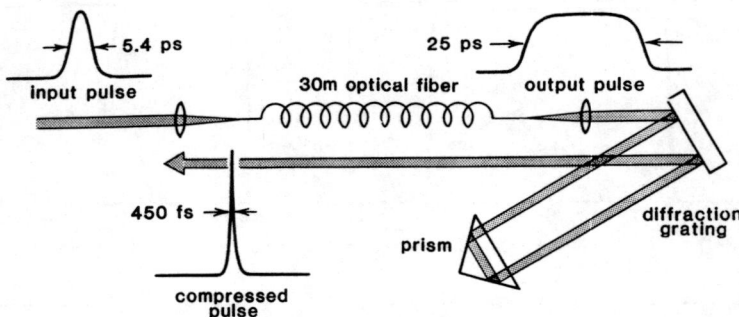

**FIGURE 10.16**
Experimental system for first broadening a pulse using an optical fiber, and then compressing it with a diffraction grating system.

have used a single stage to convert initial 90-fs pulses from a colliding-pulse mode-locked dye laser into 30-fs optical pulses. These pulses are approximately 14 optical cycles long, and are the shortest pulses known to date.

## REFERENCES

One of the earliest articles on pulse self-broadening and compression in optical fibers is H. Nakatsuka, D. Grischkowsky, and A. C. Balant, "Nonlinear picosecond-pulse propagation through optical fibers with positive group-velocity dispersion," *Phys. Rev. Lett.* **47**, 910–913 (September 28, 1981).

Other recent articles from this same group include D. Grischkowsky and A. C. Balant, "Optical pulse compression based on enhanced frequency chirping," *Appl. Phys. Lett.* **41**, 1–3 (July 1, 1982); B. Nikolaus and D. Grischkowsky, "12× pulse compression using optical fibers," *Appl. Phys. Lett.* **42**, 1–3 (January 1, 1983); and B. Nikolaus and D. Grischkowsky, "90 fsec tunable optical pulses obtained by two-stage pulse compression," *Appl. Phys. Lett.* **43**, 228–230 (August 1, 1983).

Application of the same technique to femtosecond dye laser pulses is also described by C. V. Shank et al., "Compression of femtosecond optical pulses," *Appl. Phys. Lett.* **40**, 761–763 (May 1, 1982).

## 10.5 SOLITONS IN OPTICAL FIBERS

In 1834 John Scott Russell, then a young Scottish university scientist and later to become a famous Victorian engineer and shipbuilder, recorded the following observations from the banks of the Glasgow-Edinburgh canal, where he first developed many of the fundamental principles of hydrodynamics and of ship's hull design: "I was observing the motion of a boat which was rapidly drawn along a narrow channel by a pair of horses, when the boat suddenly stopped—not so the mass of water in the channel which it had put in motion; it accumulated round the prow of the vessel in a state of violent agitation, then suddenly leaving it behind, rolled forward with great velocity, assuming the form of a large solitary elevation, a rounded, smooth and well-defined heap of water, which continued its course along the channel apparently without change of form or diminution of speed. I followed it on horseback, and overtook it still rolling on at a rate of some

## 10.5 SOLITONS IN OPTICAL FIBERS

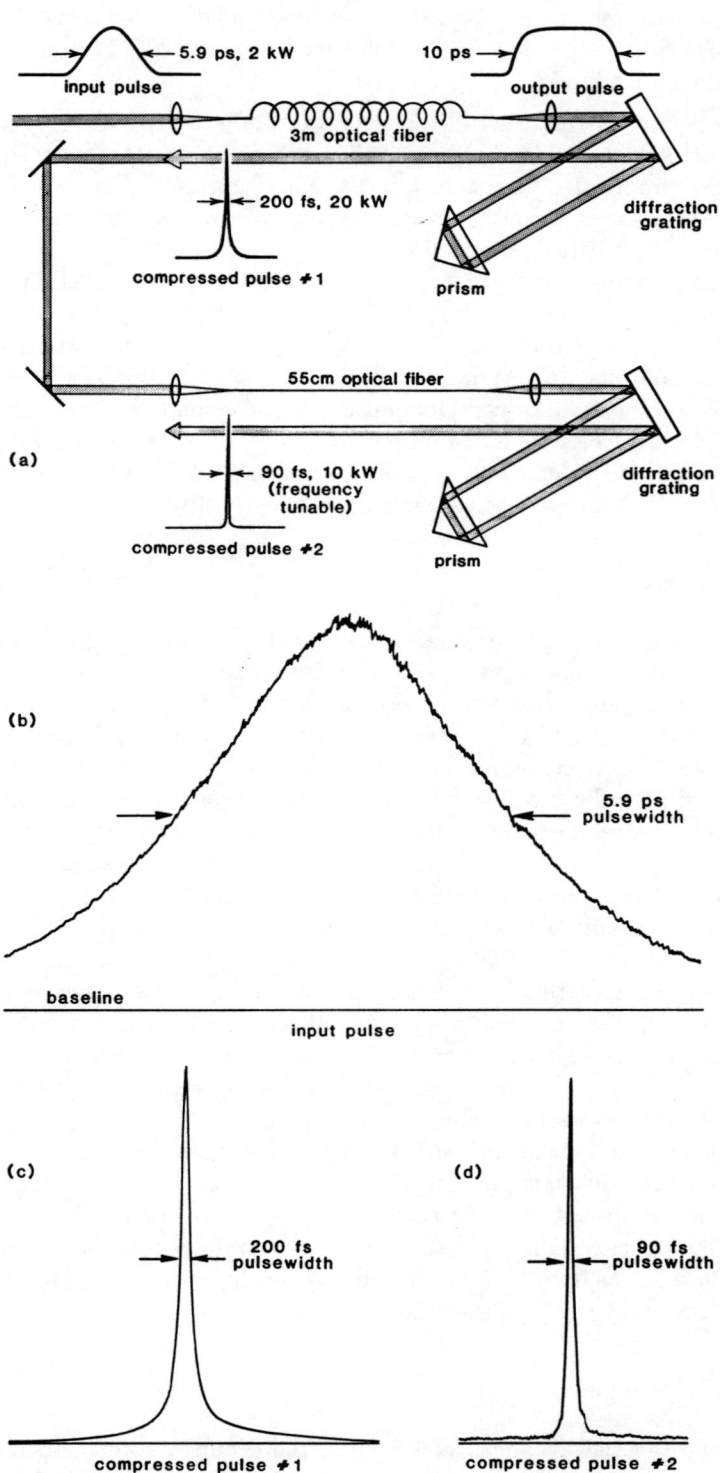

**FIGURE 10.17**
(a) Two-stage pulse compression system. (b) Autocorrelation trace of 5.9-ps initial input pulse. (c) Output of first compression stage (450 fs). (d) Output of second stage (90 ps).

eight or nine miles an hour, preserving its original figure some thirty feet long and a foot to a foot and a half in height. Its height gradually diminished, and after a chase of one or two miles I lost it in the windings of the channel. Such, in the month of August 1834, was my first chance interview with that singular and beautiful phenomenon ..."

This represented one of the first detailed observations of a *solitary wave*, or *soliton*, a special large-signal solution to some nonlinear dispersive propagation equation, which either propagates with a certain fixed and unchanging steady-state pulseshape over very long distances, or else displays a slow periodic oscillation with distance between a certain set of similar characteristic pulseshapes.

Solitons having these properties represent a very interesting and useful physical phenomenon, with applications in many fundamental areas of physics. It has very recently been realized that very short optical pulses in optical fibers can also propagate as solitons, at very modest power levels; and that these soliton pulses may be very important for carrying pulsed optical communications signals through such fibers over very long distances and at very high data rates. In this section therefore we will briefly review some of the basic concepts of solitons, and particularly of their properties in single-mode optical fibers.

### Solitons in General

A soliton is in general any member of a class of solutions to some nonlinear equation or nonlinear propagation problem, in which each such solution is characterized by a certain amplitude or power level and a certain pulseshape, with these two usually being interrelated; and in which these solutions can either propagate with an unchanging pulseshape over an indefinite distance, or else display a slow periodic oscillation with distance through a set of recurring characteristic pulseshapes. Depending on the particular nonlinear equation, the soliton pulses may have different shapes; and the velocities of propagation and the distances for periodic recurrence generally depend on both the nonlinear equation and the pulse amplitude.

(According to a more precise classification, any solution to a nonlinear equation which will propagate with unchanged shape, or will repeat its shape periodically in distance, is known as a *solitary wave*; but only those classes of solitary waves which can collide or pass through each other and then resume their solitary propagation without change of shape after such a collision are called *solitons*. Hence not all solitary-wave solutions are solitons.)

Three nonlinear equations which are known to have soliton solutions are the *Korteweg-deVries equation*, the *Sine-Gordon equation*, and the *nonlinear Schrödinger equation* already introduced in this chapter. Equations like these arise in many physical situations, including shallow- and deep-water wave propagation, waves in plasmas, lattice waves in solids, superconductivity, vortex motions in liquids, and propagation in optical fibers.

### Solitons in Optical Fibers

At wavelengths longer than $\lambda_0 \approx 1.35$ $\mu$m, the group-velocity dispersion in quartz single-mode optical fibers has the appropriate sign (namely, $\beta'' < 0$) such that the chirp produced by self-phase modulation through the optical Kerr effect will lead, at least at first, to time-compression of the central part of the optical pulse, rather than to pulse broadening as described in a preceding section.

A smooth pulse of sufficient amplitude will thus be steadily compressed, and also progressively distorted in shape, at least until it becomes sufficiently short that higher-order nonlinear effects begin to compete with the dispersive pulse compression.

It is found, in fact, that such a pulse can approach a limiting pulseshape which does not change further with distance, and which represents in fact the *lowest-order soliton solution* to the nonlinear Schrödinger equation that governs the nonlinear wave propagation in the fiber. This solution has a sech dependence of the pulse amplitude on time, in the form

$$\mathcal{E}(z,t) = \mathcal{E}_0 \operatorname{sech}\left(\frac{t - t_0 - z/v_g'}{\tau_0}\right) \exp[j(\Omega t - \kappa z)], \tag{47}$$

where the peak amplitude $\mathcal{E}_0$, the pulsewidth $\tau_0$, the modified group velocity $v_g'$, the (small) frequency shift $\Omega$, and the wavenumber shift $\kappa$ are all interrelated. Different values for these parameters thus describe a continuous family of solutions that have the same basic shape but different amplitudes and pulsewidths, that carry different amounts of energy per pulse, and that travel at very slightly different group velocities.

For example, such a pulse in a single-mode fiber might have a width of 3 ps and a peak power of 100 mW, and thus carry a total energy of $\approx 0.3$ pJ. Lower peak powers mean longer pulsewidths, and the converse.

### Higher-Order Soliton Solutions

There also turn out to be higher-order soliton solutions, characterized by an index $N \geq 1$, which do not propagate with constant shape, but which instead, if launched with a proper initial shape and amplitude will return to that same initial shape at periodic distances along the fiber. Analytical solutions for these periodically recurring solitons are difficult to obtain, and they are often studied by means of large-scale numerical simulations.

These higher-order solutions generally require higher amplitudes and energies than the lowest-order soliton, and the soliton period generally decreases with increasing pulse amplitude. They have also now been seen experimentally (see References).

### Fermi-Pasta-Ulam Recurrence

One of the counterintuitive properties of these higher-order periodic soliton solutions in certain systems—including optical fibers—is that the frequency spectrum of the nonlinear pulse signal, starting from a narrow and even transform-limited initial pulse, can first broaden out substantially with distance (or with time), but then can later condense back again to the same narrow initial spectrum. This seems quite counter to what might be an initial expectation, that nonlinear and intermodulation effects should generally always act to continually broaden a signal spectrum.

This spectral broadening and subsequent recondensation is sometimes referred to as "Fermi-Pasta-Ulam recurrence"; these three men carried out an early set of calculations on the first large-scale computers at Los Alamos in order to trace the long-term dynamics of a computer model of nonlinear springs and discrete masses having many nonlinearly coupled resonant modes. Instead of displaying a long-term trend from initial order toward eventual quasi random thermalization, as had been expected, these computer simulations frequently revealed a mysterious periodic recurrence behavior, which was not understood, and which is now sometimes explained as the excitation of periodically recurring solitons in the nonlinear system.

### The Soliton Laser

The circulating pulses in some of the narrowest-pulse mode-locked lasers are in fact probably solitons, in the sense that nonlinear chirping and dispersion in the various laser elements begins to play a significant role in the reshaping of the pulses. This reshaping can be beneficial, if it leads to narrower pulses, or deleterious, as in the pulse break-up effects discussed earlier. There is an even more interesting way of using solitons in a mode-locked laser, which is accomplished as follows.

Suppose a length of fiber is connected to one end of a laser cavity in such a fashion that a pulse can come out the end of the laser, propagate down the fiber and reflect at the far end, and come back and enter the laser cavity again; and suppose the fiber length corresponds to the periodic recurrence or reshaping distance for a certain optical soliton. If the laser operates at a wavelength where the fiber supports soliton propagation, and if the laser energy is properly adjusted, it is possible for the pulse to travel through the laser medium as a comparatively wide pulse in time, with a correspondingly narrow bandwidth that remains within the amplification bandwidth of the laser medium; but then it enters the fiber to propagate as a higher-order soliton ($N > 1$). As the pulse travels down the fiber and back, it can thus narrow in time and broaden in bandwidth, and then reverse this process as it returns back to the laser. This makes it possible to generate mode-locked pulses which are much narrower than can normally be supported by the finite amplification bandwidth of the mode-locked laser medium. This important (and very recent) development is referred to as the "soliton laser."

## REFERENCES

The quotation from John Scott Russell which opens this section comes from his "Report on Waves," *Proceedings of the Royal Society of Edinburgh*, 319 (1844). It is reprinted in a recent and extensive review paper with many references on "The soliton: A new concept in applied science." by A. C. Scott, F. Y. F. Chu, and D. W. McLaughlin, *Proc. IEEE* **61**, 1443–1483 (October 1973).

The initial suggestion for soliton propagation in optical fibers was made by A. Hasegawa and F. Tappert, "Transmission of stationary nonlinear optical pulses in dispersive optical fibers: I, Anomalous dispersion; II, Normal dispersion," *Appl. Phys. Lett.* **23**, 142–144 and 171–172 (August 1 and 15, 1973). More detailed theoretical discussions are given in A. Hasegawa and Y. Kodama, "Signal transmission by optical solitons in monomode fiber." *Proc. IEEE* **69**, 1145–1150 (September 1981).

The first experimental confirmation of their prediction was L. F. Mollenauer, R. H. Stolen, and J. P. Gordon, "Experimental observation of picosecond pulse narrowing and solitons in optical fibers," *Phys. Rev. Lett.* **45**, 1095–1098 (September 29, 1980). More recent experimental observations include R. H. Stolen, L. F. Mollenauer, and W. J. Tomlinson, "Observation of pulse restoration at the soliton period in optical fibers," *Opt. Lett.* **8**, 186–188 (March 1983); and L. F. Mollenauer, R. H. Stolen, J. P. Gordon, and W. J. Tomlinson, "Extreme picosecond pulse narrowing by means of soliton effect in single-mode optical fibers," *Opt. Lett.* **8**, 289–291 (May 1983). See also L. F. Mollenauer and R. H. Stolen, "The soliton laser," *Opt. Lett.* **9**, 13–15 (January 1984).

The literature on solitons in other physical systems unfortunately seems to be far more mathematical than physical in approach. For further general discussions of nonlinear waves and solitons, see, for example, G. B. Whitham, *Linear and Nonlinear Waves* (Wiley, 1974), *Solitons in Action*, edited by K. Lonngren and A. Scott (Academic Press, 1978), or G. L. Lamb, Jr., *Elements of Soliton Theory* (Wiley, 1980).

The original Fermi-Pasta-Ulam recurrence phenomena can be found in *The Collected Papers of Enrico Fermi* (University of Chicago Press, 1962), II, 978–988.

# 11
## LASER MIRRORS AND REGENERATIVE FEEDBACK

In previous chapters we have described what optical signals do to laser atoms, including resonant atomic response, laser pumping, and population inversion; and what laser atoms do back to optical signals, including the amplification of signals in a single pass through a laser medium. In this and the following two chapters we will bring in the laser mirrors and resonant cavities needed to provide regenerative feedback and eventually oscillation in laser devices.

Because the mirrors are the critical new elements, and regenerative feedback is the critical new physical process, we will first examine in this chapter some basic physical properties of mirrors and optical beam splitters, as well as the resonant properties of optical cavities, or etalons, or interferometers. We then explore the regenerative feedback and amplification effects that occur in a resonant laser cavity *below* threshold. Besides introducing some useful new concepts and devices, this will give us an understanding of cavity resonant frequencies and axial modes, and show us how lasers behave just below the oscillation threshold that we will (finally!) reach in the following chapter.

## 11.1 LASER MIRRORS AND BEAM SPLITTERS

Laser mirrors and beam splitters have certain fundamental properties that are important to understand. Before discussing the use of mirrors in laser cavities, let us therefore review the more important of these properties.

### Single Dielectric Interface

The simplest example of a partial mirror or beam splitter is the interface between two dielectric media, as shown in Figure 11.1. Suppose we write the normalized fields for the incident and reflected waves on the two sides of this interface (labeled by subscripts $i = 1, 2$) in the form

$$\mathcal{E}_i(z,t) = \mathrm{Re}\left\{\tilde{a}_i \exp[j(\omega t \mp \beta_i z)] + \tilde{b}_i \exp[j(\omega t \pm \beta_i z)]\right\}, \quad i = 1, 2, \quad (1)$$

where $\beta_1$ is the propagation constant in the dielectric medium on the left side of the interface, $\tilde{a}_1$ is the normalized wave amplitude of the *incident* wave and

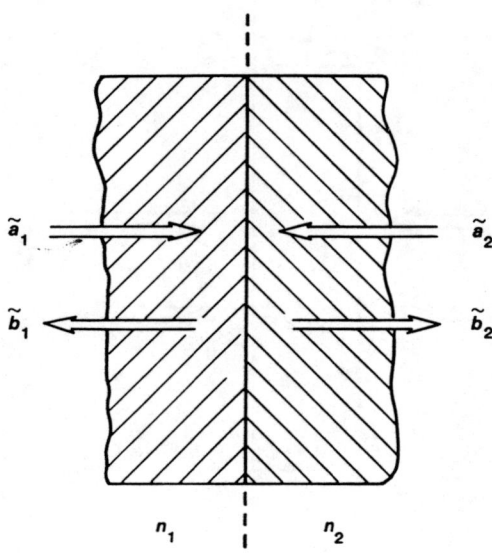

FIGURE 11.1
Reflection and transmission of optical waves at a dielectric interface.

$b_1$ is the normalized wave amplitude of the *reflected* wave on that side of the interface, and the same expressions with subscript $i = 2$ apply on the right-hand side of the interface. The field amplitudes $\mathcal{E}_i$ are normalized so that $|\mathcal{E}_i|^2$ gives the intensity or power flow in the medium on either side of the interface.

Note that the upper signs in front of the $\beta_i$ terms in the exponents apply on the left-hand side of the interface, so that $\tilde{a}_1$ represents the complex amplitude of the incident wave traveling to the right (toward $+z$), and $\tilde{b}_1$ represents the reflected or left-traveling wave. The lower signs apply on the right-hand side of the interface, so that $\tilde{a}_2$ is again the incident but now left-traveling wave and $\tilde{b}_2$ the right-traveling wave.

The amplitude reflection and transmission properties for a simple dielectric interface at normal incidence can then be written as

$$\tilde{b}_1 = r\,\tilde{a}_1 + t\,\tilde{a}_2,$$
$$\tilde{b}_2 = t\,\tilde{a}_1 - r\,\tilde{a}_2, \quad (2)$$

or in matrix notation

$$\begin{bmatrix} \tilde{b}_1 \\ \tilde{b}_2 \end{bmatrix} = \begin{bmatrix} r & t \\ t & -r \end{bmatrix} \times \begin{bmatrix} \tilde{a}_1 \\ \tilde{a}_2 \end{bmatrix}, \quad (3)$$

where the reflection and transmission coefficients $r$ and $t$ are given for this particular interface by

$$r = \frac{n_1 - n_2}{n_1 + n_2} \quad \text{and} \quad t = \frac{2\sqrt{n_1 n_2}}{n_1 + n_2}, \quad (4)$$

and the lossless nature of the interface is expressed by $r^2 + t^2 = 1$. Note that in writing these relations we are implicitly locating the $z = 0$ plane, or the reference plane for the fields given in Equation 11.1, exactly at the dielectric interface. The $\tilde{a}_i$ and $\tilde{b}_i$ coefficients thus express the complex wave amplitudes exactly at the interface between the two dielectrics.

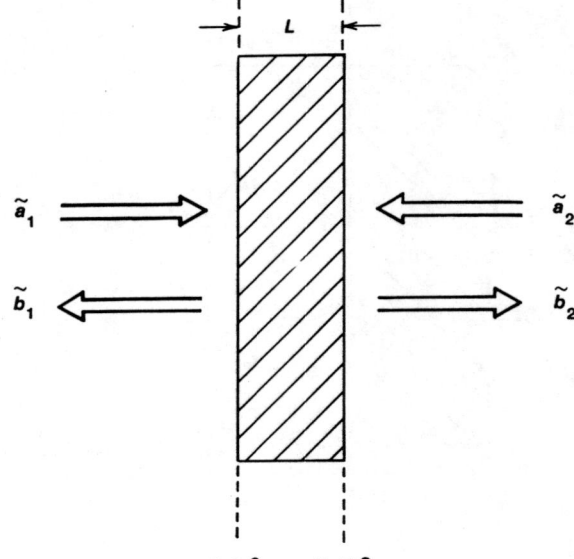

**FIGURE 11.2**
Reflection and transmission of optical waves from a thin dielectric slab.

For this particular interface, and this particular choice of reference plane, the coefficients $r$ and $t$ are purely real numbers. The reflection coefficients have opposite signs, however, depending on the direction from which the wave approaches the interface. This change of sign can be understood in physical terms because a medium with a very high index of refraction acts essentially like a metallic surface or like a very large shunt capacitance across a transmission line. Going from a low-index to a high-index medium ($n_2 > n_1$) is thus like the reflection from the end of a short-circuited transmission line, with a 180° phase shift on reflection or $r < 0$, whereas reflection from a low index medium acts like an open-circuited transmission line, with a reflection coefficient $r > 0$.

### Thin Dielectric Slab

As a more realistic model for a real laser mirror, we might consider next the reflection and transmission properties of a thin lossless dielectric slab of thickness $L$ and index $n$ as shown in Figure 11.2, assuming, for simplicity, air or vacuum on both sides of this slab.

We will again write the fields on both sides of the slab as in Equation 11.1, using $\tilde{a}_i$ and $\tilde{b}_i$ coefficients for the incident and reflected waves, except that we will now measure the incident waves and the reflected waves using separate $z = 0$ planes that are located at the outer surfaces of the slab on each side, so that the reference planes on opposite sides of the slab are offset by the thickness of the slab.

With a little calculation we can find that the general relationship between incident and reflected waves in this case can be written in the general complex matrix form

$$\begin{bmatrix} \tilde{b}_1 \\ \tilde{b}_2 \end{bmatrix} = \begin{bmatrix} \tilde{r}_{11} & \tilde{t}_{12} \\ \tilde{t}_{21} & \tilde{r}_{22} \end{bmatrix} \times \begin{bmatrix} \tilde{a}_1 \\ \tilde{a}_2 \end{bmatrix}, \tag{5}$$

where the complex reflection coefficients are now given by

$$\tilde{r}_{11} = \tilde{r}_{22} = r_0 \frac{1 - e^{-j\theta}}{1 - (r_0 e^{-j\theta})^2} \tag{6}$$

and the complex transmission coefficients by

$$\tilde{t}_{12} = \tilde{t}_{21} = e^{-j\theta} \frac{1 - r_0^2}{1 - (r_0 e^{-j\theta})^2}. \tag{7}$$

In these expressions $\theta = n\omega L/c_0$ is the optical thickness of the slab, and $r_0 = (1-n)/(1+n)$ is the single-surface reflection coefficient at either surface of the slab. The scattering matrix is now *complex but symmetric*, and the coefficients now obey the complex condition $|\tilde{r}|^2 + |\tilde{t}|^2 = 1$ representing zero losses in the slab.

### Purely Real Reflectivity

As one particularly simple example of this type of mirror, we might adjust the optical thickness $nL$ of the slab to be an odd number of quarter wavelengths, so that $\theta = n\omega L/c_0$ is an odd integer multiple of $\pi/2$. The reflection and transmission coefficients for the mirror then take on the particularly simple form

$$\begin{bmatrix} \tilde{b}_1 \\ \tilde{b}_2 \end{bmatrix} = \begin{bmatrix} r & jt \\ jt & r \end{bmatrix} \times \begin{bmatrix} \tilde{a}_1 \\ \tilde{a}_2 \end{bmatrix}, \tag{8}$$

where $r$ and $t$ are again purely real and subject to $r^2 + t^2 = 1$.

The reflection coefficients from the two sides of this slab are now symmetric and purely real. In writing down the reflection equations we no longer have to worry about whether we are approaching this mirror or beamsplitter from its "high index side" or its "low index side." The transmission factors in this example, however, now have an additional factor of $j$, or a phase shift of 90°, associated with them. This phase shift arises essentially from the fact that we measure the waves at two different reference planes, on opposite sides of the slab and separated by the (small) thickness of the slab.

Note that we can, at least in principle, always adjust the index $n$ and the thickness $L$ of such a slab separately to obtain any desired values of $\theta$ and $r_0$, and hence any desired values for the magnitudes of $r$ or $t$. This thin-slab model might therefore be a particularly simple and symmetric model for any real lossless dielectric mirror.

### Scattering Matrix Formalism

The formalism we have been using here is obviously a *scattering matrix formalism* of the kind used in circuit theory or in describing waveguide or transmission-line junctions. From this viewpoint a partially transmitting mirror is simply a two-port network connected between two waveguides or transmission lines; and the matrices we have written represent simple examples of the $2 \times 2$ scattering matrix $S$ that might describe such a two-port network in transmission-line theory.

This scattering matrix approach is of course not limited to two-port cases. Figure 11.3 shows, for example, a partially transmitting mirror or optical beam

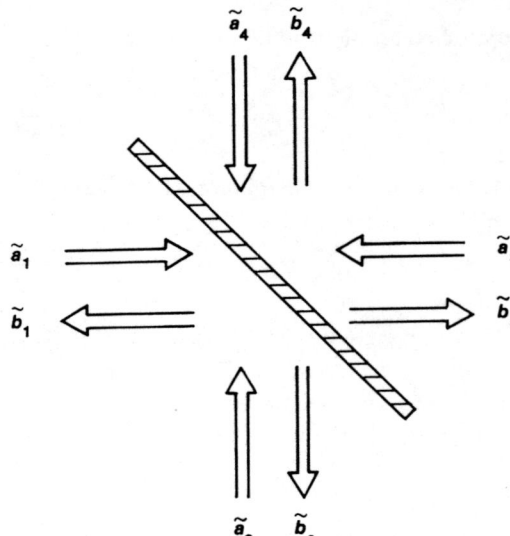

FIGURE 11.3
A partial mirror or optical beamsplitter at off-normal incidence.

splitter used at off-normal incidence, so that it now has four incident and outgoing waves. This is equivalent to a general four-port network, and would require a 4 × 4 scattering matrix, which we would have to write in the form

$$\begin{bmatrix} \tilde{b}_1 \\ \tilde{b}_2 \\ \tilde{b}_3 \\ \tilde{b}_4 \end{bmatrix} = \begin{bmatrix} \tilde{r}_{11} & \tilde{t}_{12} & \tilde{t}_{13} & \tilde{t}_{14} \\ \tilde{t}_{21} & \tilde{r}_{22} & \tilde{t}_{23} & \tilde{t}_{24} \\ \tilde{t}_{31} & \tilde{t}_{32} & \tilde{r}_{33} & \tilde{t}_{34} \\ \tilde{t}_{41} & \tilde{t}_{42} & \tilde{t}_{43} & \tilde{r}_{44} \end{bmatrix} \times \begin{bmatrix} \tilde{a}_1 \\ \tilde{a}_2 \\ \tilde{a}_3 \\ \tilde{a}_4 \end{bmatrix}. \qquad (9)$$

In more compact notation we can write Equations 11.2, 11.5 or 11.9 as

$$\bm{b} = \bm{S} \times \bm{a}. \qquad (10)$$

The column vectors $\bm{a}$ and $\bm{b}$ then contain the incident and reflected wave amplitudes. The diagonal elements of the matrix $\bm{S}$ give the (generally) complex reflection coefficients $\tilde{r}_{ii}$ looking into each port of the system, and the off-diagonal elements $\tilde{t}_{ij}$ give the amplitude transmission coefficients from, say, the wave going into port $j$ to the wave coming out of port $i$.

### Multilayer Dielectric Mirrors

We now need to discuss some subtleties concerning the effective reference planes of real laser mirrors, and how we should choose these reference planes in writing scattering matrices and carrying out laser analyses.

So far we have discussed two particularly simple examples of reflecting systems and their resulting scattering matrices. Figure 11.4, however, illustrates a more typical multilayer dielectric mirror of the type often used in lasers. Such a mirror may have as many as twenty or more quarter-wavelength-thick dielectric layers of alternating high and low index of refraction, evaporated onto a transparent substrate. (The opposite surface of this substrate usually has a high-quality antireflection coating; and the two surfaces of the substrate are often wedged by a few degrees to avoid etalon effects from any residual back-surface reflection.)

## 11.1 LASER MIRRORS AND BEAM SPLITTERS

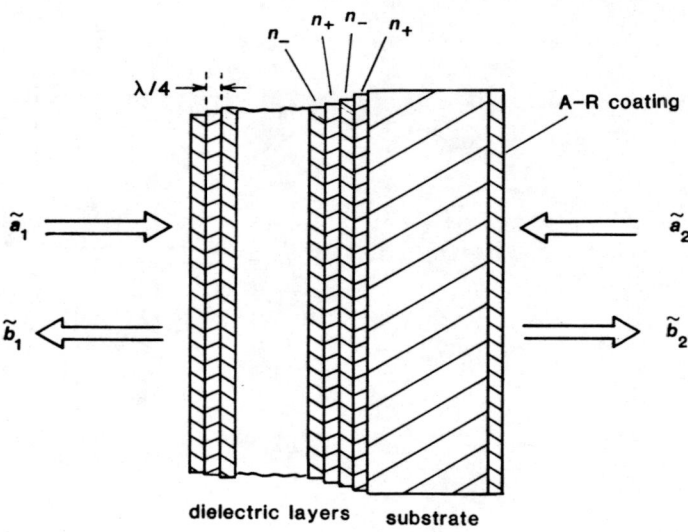

**FIGURE 11.4**
Reflection and transmission of optical waves from a multilayer dielectric mirror.

In the simple examples we have discussed so far, it may have seemed physically obvious that the reflecting surfaces are located at the physical surfaces of the dielectrics. But where should the effective mirror surface, or the effective reflecting plane $z = 0$, be located in a thick multilayer mirror like Figure 11.4 that is several optical wavelengths thick?

### Mirror Reference Planes

In fact, there really is *no* unique plane which we can (or need to) identify as the exact reflecting surface, or the unique reference plane, for such a multilayer mirror. The total reflection of the mirror, as seen from outside the coating layers on either side, builds up gradually through the series of layers, which can be several wavelengths thick overall. The choice of where to locate the reference plane in this (or any other) mirror is entirely arbitrary. We can pick *any* reasonable reference plane within (or even outside) the multilayer mirror as the reference plane for defining the scattering matrix coefficients. In physical terms, picking such a reference plane simply means choosing the $z = 0$ origin for measuring the electric fields $\mathcal{E}(z,t)$ well outside the mirror, assuming the fields are expanded as in the opening equation of this section.

Even with the single dielectric interface, it is not essential that the mirror reference surface be chosen right at the physical surface between the dielectrics—especially since we very seldom if ever can position any mirror or optical element with an absolute position accuracy of better than a few optical wavelengths. So long as we are concerned only with the amplitudes and phases of the waves $\mathcal{E}(z,t)$ at larger distances — say, more than a few wavelengths—away from the reflecting surface, we can choose the reference surface anywhere near the physically reflecting structure. Shifting the choice of reference plane from one axial position to another location a distance $\Delta z$ away then merely rotates the phase angles of the complex wave amplitudes $\tilde{a}_i$ and $\tilde{b}_i$ by phase shifts $\exp(\pm j\beta_i \Delta z)$, without changing their amplitudes. This rotation of the phase angles of $\tilde{a}_i$ and $\tilde{b}_i$ in turn merely attaches different phase angles to each of the scattering coefficients $\tilde{S}_{ij}$.

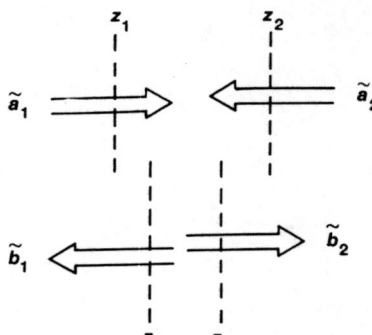

FIGURE 11.5
The reference planes for incident and reflected waves need not be at the same positions.

We can in fact even choose *different reference planes at which to measure the complex incoming and outgoing waves in each of the arms*, as illustrated in Figure 11.5 (although why this would be a useful choice may be open to question). Regardless of the choice of reference plane(s) and the associated phase shifts, however, the scattering matrix elements still have certain fundamental properties, which we must next consider.

### Hermitian Matrix Notation

To understand a bit more about the basic properties of mirrors and beam splitters and their scattering matrices, it will be useful to introduce some *hermitian matrix notation*.

We are using the notation $b$, for example, to denote a column vector with complex elements $\tilde{b}_1, \tilde{b}_2, \tilde{b}_3, \ldots$ running from the top down. We can then define the *hermitian adjoint* or *hermitian conjugate* to this vector, denoted by $b^\dagger$, as the row vector with complex conjugate elements $\tilde{b}_1^*, \tilde{b}_2^*, \tilde{b}_3^*, \ldots$ running horizontally. Hermitian conjugation converts a column vector into a row vector, or vice versa, and takes the complex conjugate of each individual element.

More generally, suppose we have an $m \times n$ matrix $S$ with complex elements $\tilde{S}_{ij}$ as given above. Then, the hermitian adjoint or hermitian conjugate $S^\dagger$ of this matrix will be an $n \times m$ matrix with complex elements given by $(S^\dagger)_{ij} = (S)_{ji}^*$. The hermitian adjoint or hermitian conjugate is a kind of matrix generalization of the complex conjugate of an ordinary quantity. To calculate the hermitian adjoint of a complex matrix or vector, interchange subscripts and take the complex conjugate. Just as with ordinary complex conjugation, applying this operation twice restores the original matrix or vector, i.e., $S^{\dagger\dagger} \equiv S$.

### Power Flow Into and Out of a Scattering Matrix System

Matrix notation can then be used to write the total power flowing into or out of a multiport scattering system in a particularly compressed form. We have assumed that the wave amplitudes in our examples are normalized, so that the time-averaged power flowing into or out of any single port can be written as $|\tilde{a}_i|^2$ or as $|\tilde{b}_i|^2$, respectively. Assuming we pick the right units for $\tilde{a}_i$ and $\tilde{b}_i$, the total power flowing out of an optical element, or out of all the ports in an $N$-port network, can then be written in the simplified form

$$P_{\text{out}} = \sum_{j=1}^{N} \tilde{b}_j^* \tilde{b}_j = \left[\tilde{b}_1^*, \tilde{b}_2^*, \tilde{b}_3^*, \ldots\right] \times \begin{bmatrix} \tilde{b}_1 \\ \tilde{b}_2 \\ \tilde{b}_3 \\ \ldots \end{bmatrix} = b^\dagger \times b, \tag{11}$$

where the final term implies matrix multiplication of $b^\dagger$ times $b$, carried out, as in all hermitian adjoint formulas, using the usual rules for matrix multiplication.

By using this notation, plus the usual rules for matrix multiplication, we can then relate the total power flowing out of any scattering system—for example, any optical mirror or beam splitter—to the input waves and the scattering matrix in the form

$$\begin{aligned} P_{\text{out}} &= b^\dagger b = (Sa)^\dagger (Sa) \\ &= (a^\dagger S^\dagger)(Sa) = a^\dagger (S^\dagger S) a. \end{aligned} \tag{12}$$

In going from the third to fourth and fourth to fifth terms, we have made use of the basic rules that (i) matrix multiplication is associative, i.e., $A(BC) = (AB)C$, and (ii) the hermitian adjoint of any product is the product of the individual adjoints taken in reverse order, i.e., $(ABC)^\dagger \equiv C^\dagger B^\dagger A^\dagger$.

### Scattering Matrices for Lossless Systems

But, if a mirror or other scattering element is to be lossless, then the output power in Equation 11.12 must equal the input power, which is just $P_{\text{in}} = a^\dagger a$. The only way this can be true in general, for arbitrary input signals $a$, is for the product $S^\dagger S$ in Equation 11.12 to equal the identity matrix; i.e.,

$$S^\dagger S = I \quad \text{or} \quad S^\dagger \equiv S^{-1} \tag{13}$$

In this equation $I$ is the identity matrix (unity elements on the diagonal and all other elements zero), and $S^{-1}$ denotes the matrix inverse of the $S$ matrix. In matrix terms, this says that the scattering matrix $S$ for a lossless network must be a *unitary matrix*, since Equation 11.13 is the definition of unitarity.

### Matrix Forms for Lossless and Reciprocal Twoport Networks

It is also a general theorem that the scattering coefficients of a *reciprocal system* must obey $|\tilde{S}_{ij}| \equiv |\tilde{S}_{ji}|$. This result comes from the symmetrical behavior of Maxwell's equations if we reverse either the $E$ or the $H$ fields and also reverse the sign of the time $t$. Most common optical elements are in fact reciprocal; only elements such as optical isolators containing Faraday rotators or similar elements containing a dc magnetic field can be nonreciprocal.

If all these constraints are applied to a lossless reciprocal two-port network, the result is the set of conditions

$$\begin{aligned} |\tilde{t}_{12}| &= |\tilde{t}_{21}|, \qquad |\tilde{r}_{11}| = |\tilde{r}_{22}|, \\ |\tilde{r}_{11}|^2 &+ |\tilde{t}_{21}|^2 = |\tilde{r}_{22}|^2 + |\tilde{t}_{12}|^2 = 1, \\ \tilde{r}_{11}\tilde{t}_{12}^* &+ \tilde{t}_{12}\tilde{r}_{22}^* = 0 \end{aligned} \tag{14}$$

These are general conditions that must be obeyed by the complex reflection and transmission coefficients of any lossless two-port mirror or beam splitter.

The purely real and the complex symmetric examples we derived earlier in this section, namely,

$$S = \begin{bmatrix} r & t \\ t & -r \end{bmatrix} \quad \text{and} \quad S = \begin{bmatrix} r & jt \\ jt & r \end{bmatrix} \tag{15}$$

with $r$ and $t$ real, are two of the possible ways in which the four conditions of Equation 11.14 can be satisfied for a two-port system. One the other hand, the real *and* symmetric matrix

$$S = \begin{bmatrix} r & t \\ t & r \end{bmatrix}, \tag{16}$$

is *not an allowable scattering matrix* for a lossless optical mirror or beam splitter.

The conditions of unitarity plus reciprocity will always lead to a set of relationships like Equations 11.14 between the coefficients $\tilde{S}_{ij}$ of any lossless $N \times N$ scattering matrix. We can always rotate the complex phase angles of the different matrix elements for a given physical system by choosing different reference planes in the various input and output arms. The magnitudes of the scattering coefficients of course will not change in this process, since the power transfer from any one arm to any other arm is not changed by a different choice of reference planes. No matter how the reference planes are chosen, however, certain phase relationships between the different coefficients must be maintained, at least for lossless systems.

The exact form of the scattering matrix $S$ for a real mirror or beamsplitter thus depends on where we pick the reference planes; and there is in general no unique or preferred place to pick the reference plane in a real mirror. *For all future analyses of laser cavities and interferometers in this book, however, we will arbitrarily choose the complex symmetric form $S = [r, jt, jt, r]$, with $r$ and $t$ purely real, as the scattering matrix form to describe all mirrors and beam splitters.* This arbitrary choice will make no difference in any of the physical conclusions we reach about laser devices. It seems easier, however, to remember that transmission coefficients always have a factor of $j$ associated with them than to remember which side of each mirror in a laser system is the $+r$ and which is the $-r$ side.

### Polarization Effects and Transverse Mode Effects

In all the examples discussed so far, we have implicitly assumed a single sense of polarization in each of the input and output directions. Optical waves can, however, have two orthogonal senses of polarization for the wave in each direction. These may be, for example, two orthogonal linear polarizations, or positive and negative circular polarization, or whatever. If two orthogonal polarizations are present, each is in essence a separately measurable wave, with a separate wave amplitude. If both polarizations are considered separately, therefore, the total number of ports in the scattering matrix must be doubled; i.e., a system with $N$ input and output directions will require a $2N \times 2N$ scattering matrix.

In addition, if we go to more realistic optical beams (or fibers), in which there may be both a lowest-order transverse mode and various higher-order transverse modes, then in essence each such transverse mode is a separate port or beam;

and the dimensionality of the scattering matrix must be expanded to include a separate port for each different transverse mode (with possible coupling between transverse modes inside the scattering system).

### Further Discussion

All the analysis in this section may seem an overly complicated approach to the scattering properties of a simple mirror or beam splitter. If we failed to include the unitary properties of beam splitters when we analyze more complex configurations, such as Michelson interferometers or ring-laser cavities, however, it would be easy to invent cavities or optical devices that do not conserve energy, or have other useful properties. (It is by no means unknown for such inventions to be suggested, and even to appear in research proposals.)

The dielectric mirrors and beam splitters used in most laser applications are in fact almost perfectly lossless (though the partially transmitting metal films which were often used as output mirrors for early solid-state lasers were by contrast quite lossy). Higher-power lasers require nearly lossless mirrors if the mirrors are not to be destroyed by the power they absorb; and low-gain lasers need high reflectivity and low mirror losses for good efficiency. Even a lossy mirror can usually be described as a lossless mirror which obeys the preceding restrictions, sandwiched between two thin absorbing layers.

Finally, we might also emphasize that once we choose a specific reference plane in a multilayer mirror, the phase shifts associated with the scattering matrix for that mirror are fixed at any one frequency, *but may have different values at different frequencies*. The different phase shifts in reflecting from a mirror at two different optical wavelengths can be quite significant when we intercompare two optical frequency standards using interferometric methods, since the exact optical length of an interferometer cavity (between the reference planes of the two end mirrors) need *not* be the same for two different wavelengths.

Mirror phase shifts can also be significant in nonlinear optics experiments, such as double-pass harmonic-generation experiments. Suppose a fundamental wave passes through a phase-matched nonlinear crystal, generating second-harmonic radiation; and both the fundamental and the harmonic then reflect off a mirror and back through the crystal again. This is *not* necessarily equivalent to a nonlinear crystal twice as long, if the relative phases of the fundamental and the harmonic are shifted in bouncing off the mirror.

### REFERENCES

Multilayer dielectric mirrors have been produced with accurately measured power reflectivities as high as $R = 99.975\%$ in the visible. Measuring a mirror reflectivity to this accuracy is far from an easy task, as discussed, for example, by J. M. Herbelin and J. A. McKay, "Development of laser mirrors of very high reflectivity using the cavity-attenuated phase-shift method," *Appl. Optics* **20**, 3341-3344 (October 1, 1981).

---

### Problems for 11.1

1. *Scattering matrix for a general dielectric slab.* Derive the general scattering matrix results for a thin dielectric slab, as given in the text.

2. *Changes in the scattering matrix for different reference planes.* Show in detail how the scattering matrix for a planar interface between two different dielectric media can be converted to complex symmetric form by choosing a different reference plane or planes. Indicate specifically where the new reference plane(s) should be located.

3. *Derivation of the necessary matrix element relationships for a lossless reciprocal two-port.* Write out the hermitian adjoint and inverse matrices $S^\dagger$ and $S^{-1}$ in full for the general two-port $S$ matrix given in the text, using the $\tilde{r}_{ij}$ and $\tilde{t}_{ij}$ notations, and show that reciprocity and unitarity lead to the conditions that are stated there.

4. *Scattering matrix for a transmission line junction.* Work out the scattering matrix for a transmission line of characteristic impedance $Z_{01}$ connected to a second transmission line of characteristic impedance $Z_{02}$, using the connection point as the reference plane; and show that it takes the purely real form given in the text.

5. *Transmission-line junction with a lumped shunt capacitance.* Repeat this calculation assuming a lumped capacitance of value $C$ is connected across the two lines in shunt right at the connection point. Is the scattering matrix still unitary?

6. *Three-port and five-port optical scattering systems?* Invent and sketch some real, physical 3-port and 5-port optical beam splitters.

7. *Impossibility of a completely matched three-port network.* When a wave is sent into any one of the ports of the four-port beam splitter shown in this section, there is no reflected wave directly back out the same port, so that $\tilde{r}_{ii} = 0$ for all $i$. In transmission line jargon this system is said to be *matched* looking into each of the four ports. Show that it is impossible in principle to devise any kind of lossless three-port network in which all three ports are similarly matched.

8. *Conditions for an $N$-port equal-amplitude beam splitter.* An $N$-port optical element is to function as a lossless equal-amplitude beam splitter, dividing the beam coming into any port into two equal beams coming out the next two adjacent ports moving clockwise around the network. All inputs are to be matched, i.e., there is no reflection of the input beam looking into any of the ports. Can you develop an all-real and symmetric form of the scattering matrix for such a system with $N = 4$ ports?

9. *Synthesizing an arbitrary complex optical two-port (research problem).* Suppose you are given the values of all four elements of an arbitrary complex 2×2 scattering matrix, which may be in general neither lossless nor reciprocal. Synthesizing a lumped electrical circuit which will produce this scattering matrix is a classic electric-circuit design problem. How about the optical analog? If you wanted to synthesize an arbitrary 2×2 optical scattering matrix, which elementary "building blocks" might you employ, and how could you synthesize a given arbitrary scattering matrix $S$ from them?

## 11.2 INTERFEROMETERS AND RESONANT OPTICAL CAVITIES

In this section, we will introduce some of the key ideas concerning the resonance behavior of passive optical cavities, without laser gain. We will use a plane-wave

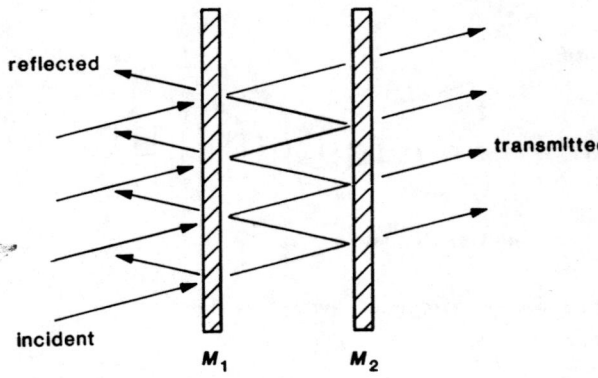

FIGURE 11.6
A Fabry-Perot interferometer (old style) with a slightly off-axis incident wave.

or transmission-line model for the resonators, and discuss both standing-wave and ring optical resonators on an equal footing. In later sections we will then introduce laser gain to produce regenerative amplification and, eventually, laser oscillation in such structures.

### Fabry-Perot Interferometers and Etalons

A common optical element, widely used since long before the advent of lasers is the *Fabry-Perot interferometer* or *Fabry-Perot etalon* sketched in Figure 11.6. In its original form, a Fabry-Perot interferometer consisted of two closely spaced and highly reflecting mirrors, with mirror surfaces adjusted to be as flat and parallel to each other as possible. An alternative but conceptually equivalent element is a solid etalon made from some very low-loss material such as fused quartz or sapphire, with its two faces polished flat and parallel and perhaps coated with a metal or dielectric mirror coating. As we will see, such Fabry-Perot interferometers or etalons can have sharp resonances or transmission passbands at discrete optical frequencies. Fabry-Perot interferometers or etalons have thus long been used as narrowband optical filters for measuring the frequency spectrum of particularly narrow optical lines, especially lines whose width was below the resolving power of prism or grating spectrometers.

In their original form, such interferometers used only flat or planar reflecting surfaces, and the spacings between the mirrors were usually smaller than, or at most on the same scale as, the transverse diameters of the mirrors. Moreover, it was usual to illuminate such an interferometer with a converging or diverging beam having a spread of angular directions, and then look at the "Fabry-Perot rings" transmitted through the interferometer in certain discrete angular directions.

Interferometers used in this manner were generally analyzed using an infinite plane-wave model, with the plane waves assumed to be arriving either at normal incidence or at some specified angle to the normal, as in Figure 11.6. The standard formulas in optics texts, as a result, consider the resonant frequencies and the transmission properties of Fabry-Perot etalons as a function of the mirror spacing, the optical wavelength or frequency, and the angle of incidence. The transverse width or shape of the two mirrors is generally not taken into account, and transverse field variations are neglected.

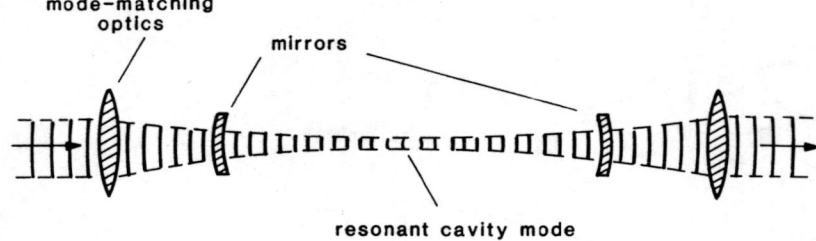

**FIGURE 11.7**
A typical optical resonator or passive interferometer cavity (new style) with an on-axis resonant cavity mode.

### Optical Resonators

As the fundamental ideas of laser devices began to emerge, however, researchers began to consider the properties of interferometers formed by setting up rather small mirrors, spaced by distances large compared to the mirror sizes, as in Figure 11.7. The waves in such long, narrow optical cavities must travel at very small angles to the optical axis of the cavity, or else the waves will very rapidly "walk off" past the edges of the mirrors; hence the off-axis angular properties of such structures are of little interest. It was soon realized, in fact, that such structures are better thought of as *optical resonators* or *optical cavities*, with properties related as much to microwave waveguide resonators as to optical interferometers. The ideas of transverse as well as longitudinal modes in such structures, and of using curved as well as planar mirror surfaces, then began to be developed.

### Modes in Planar and Curved-Mirror Cavities

Figure 11.8(a) shows, for example, a typical optical cavity formed by two partially transmitting mirrors set up facing each other, such as might be used in a regenerative laser amplifier or oscillator, or in a modern optical interferometer, along with two lenses being used to focus a collimated external optical beam into and out of this cavity. The one slightly unrealistic aspect of this drawing is that real laser cavities are often even longer and relatively more slender than shown here.

Most modern optical resonators and laser cavities are also designed using mirrors which are slightly curved rather than planar, as illustrated in Figures 11.7 and 11.8. The use of such curved mirrors generally leads (as we will study in much more detail in later chapters) to the existence of very well-defined and well-behaved, low-loss *transverse-mode patterns* in such cavities. (By loss we mean here the *leakage or diffraction losses* caused by the loss of energy out the open sides of the cavity or past the edges of the finite-diameter mirrors.) The transverse-mode patterns in many, though not all, curved-mirror optical resonators take the form of quite smooth and regular transverse patterns that resemble Hermite-gaussian or Laguerre-gaussian cross sections, and that depend to first order only on the curvature and spacing, and not on the transverse size, of the end mirrors.

Optical resonators formed by finite flat or planar mirrors have definite transverse modes also. These modes are generally more irregular than in curved-mirror cavities, and not gaussian in profile; they also typically have somewhat larger

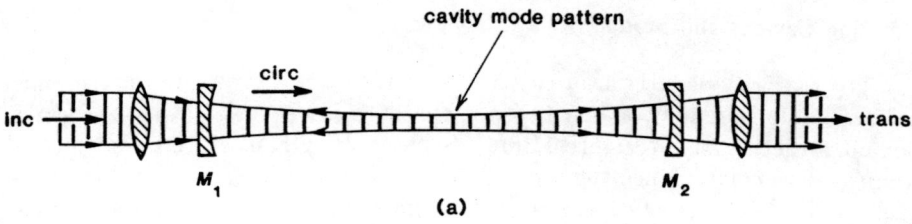

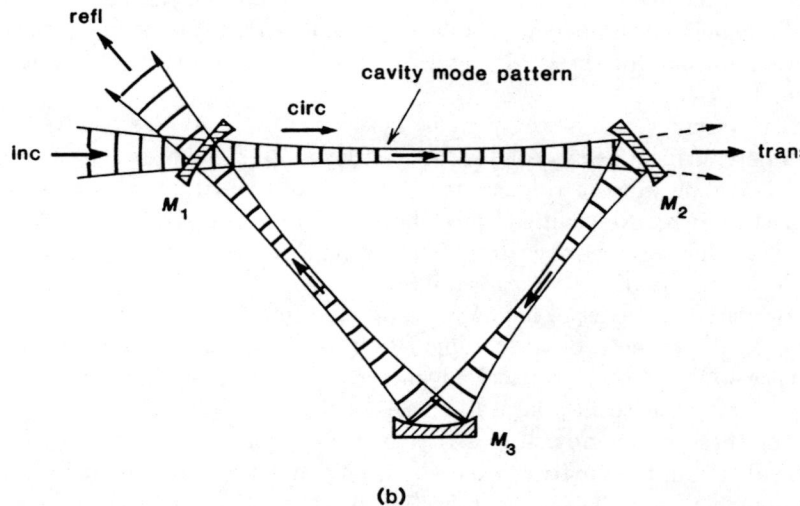

FIGURE 11.8
Simple examples of (a) a linear or standing-wave optical cavity, and (b) a ring or traveling-wave optical cavity.

diffraction losses or power leakage. The details of the transverse mode profiles in planar-mirror cavities also depend rather more critically on the exact size and transverse shape of the mirrors (e.g., circular, square, or whatever).

There also exist so-called *unstable optical resonators*, whose mirrors have a negative or divergent curvature so that the optical waves tend to be spread outward as they bounce back and forth between the end mirrors. These unstable resonators still have definite transverse mode patterns, and in fact are very useful for high-gain lasers, though their mode properties are more complicated and their diffraction losses much higher than in planar or convergent-mirror resonators.

The basic idea that it is possible set up two aligned mirrors to create a resonant optical cavity with clearcut resonant modes may seem straightforward and obvious now, but was in fact one of the key ideas in the development of the laser. The antecedents to this idea were the passive resonant etalons and interferometers used in classical optics; and there are many useful devices in optics today which involve passive optical cavities, or resonant mirror structures without optical gain.

### Ring Cavities and Standing-Wave Cavities

The majority of early laser cavities, as well as passive resonant interferometers and etalons, employed just two mirrors set up facing each other to form a resonant structure, as in Figure 11.8(a). Such a cavity is often referred to as a *standing-wave cavity*, since the two waves traveling in the forward and reverse directions in such a cavity form an optical standing wave. In such a standing-wave system, the signal $E$ and $H$ fields will have periodic spatial variations along the axis, with a period equal to one-half the optical wavelength.

(This is strictly true, of course, only if the signal inside the cavity is at a single frequency. If multiple frequencies are present, the standing waves associated with different frequency components will have different periods and spatial locations, and the summation of these will tend to wash out some of the standing-wave character.)

In more recent years, however, many laser cavities, as well as passive optical interferometers, have been designed as *ring resonators*, such as that in Figure 11.8(b). Traveling-wave or ring cavities are not really different in principle from linear or standing-wave cavities, since the round-trip optical path in going once down a standing-wave cavity and back is essentially equivalent to going once around a ring cavity of the same overall path length. Ring resonators do have the special property, however, of having separate and independent resonances in the two opposite directions around the ring. Despite their slightly greater complexity, ring cavities offer several practical advantages in different applications, and are being increasingly used in practical devices (see Section 13.5).

One of these major advantages is that when a ring resonator is driven by an external signal, the cavity is excited with signals going in only one direction around the ring, and there is no reflection directly back into the external signal source. This can be important because many laser devices do not function well when looking directly into a back reflection. Ring laser oscillators can also, with proper design, be made to oscillate in one direction only, so that there is no standing-wave character to the fields inside the cavity; and this can give advantages in power output and in mode stability.

### Mode-Matching Optics

To excite any such optical cavity in just one of its transverse modes, it is necessary to shape and focus the input beam using lenses and other so-called *mode-matching optics* in order to couple properly into the desired transverse mode of the cavity. The desired mode is usually the lowest-order transverse mode of the cavity, since this is (by definition) the transverse mode with the most highly confined transverse field pattern and the lowest leakage or diffraction losses.

If the input beam is not properly aligned and mode-matched to the transverse pattern of the lowest-order mode, the input wave will excite some mixture of lowest-order and higher-order transverse modes in the cavity. Since these higher-order transverse modes usually have slightly different resonance frequencies, tuning the input signal may excite a number of separate and frequency-shifted resonances in different transverse modes as the frequency is varied; but since the higher-order modes often have larger diffraction losses and thus lower $Q$ values, the cavity response in the higher-order modes is often weaker than in the lowest-order transverse mode.

### Uniform Plane-Wave (Transmission-Line) Approximation

The transverse field patterns inside most practical laser resonators and interferometer cavities, even when excited in a single transverse mode, are still very close to ideal plane waves. The fields propagate along the axial direction of the resonator essentially like uniform plane waves, with only minor or second-order effects due to the finite transverse width and transverse mode profile of the fields.

We will, therefore, in this and several following chapters, disregard all these transverse-mode complications and analyze the resonant properties of the signals inside and outside such cavities or interferometers using only a simple on-axis plane-wave approach. That is, we will consider the variations of the fields only in the axial or $z$ direction, and ignore any variations in the transverse or $x$ and $y$ directions. This is equivalent to using essentially a *transmission-line model* to describe all the cavity resonance effects.

## REFERENCES

For much more detailed information on the transverse modes and mode properties of optical resonators and interferometers, see Chapters 14–23 of this text, and the references in those chapters.

## 11.3 RESONANCE PROPERTIES OF PASSIVE OPTICAL CAVITIES

Let us develop therefore an elementary analysis for the resonance properties of either a linear (standing-wave) or a ring (traveling-wave) optical cavity, using the plane-wave or transmission-line analytical models shown in Figure 11.9.

### Basic Cavity Analysis: The Circulating Intensity

To do this, we will suppose that a steady-state sinusoidal optical signal is incident on one of the cavity mirrors, call it mirror $M_1$, using the notations $\tilde{E}_{\text{inc}}$ and $\tilde{E}_{\text{refl}}$ to denote the *incident* and *reflected complex signal amplitudes*, respectively, as measured just *outside* this mirror. We will also use $\tilde{E}_{\text{circ}}$ to denote the *circulating signal amplitude* inside the cavity, as measured just *inside* the same mirror.

The circulating signal just inside the input mirror then consists of the vector sum of that portion of the incident signal which is transmitted through the input mirror, and thus has the value $jt_1\tilde{E}_{\text{inc}}$; plus a contribution representing the circulating signal $\tilde{E}_{\text{circ}}$ which left this same point one round-trip time earlier, traveled once around the cavity, and has returned to the same point after passing through all the elements (twice, in the standing-wave model) and bouncing off mirror $M_1$ as well as all the other mirrors in the cavity. The total circulating signal just inside mirror $M_1$ can thus be written in the form

$$\tilde{E}_{\text{circ}} = jt_1\tilde{E}_{\text{inc}} + \tilde{g}_{\text{rt}}(\omega)\,\tilde{E}_{\text{circ}}, \tag{17}$$

**FIGURE 11.9**
Elementary models for the incident, reflected, and circulating waves in resonant optical cavities or interferometers.

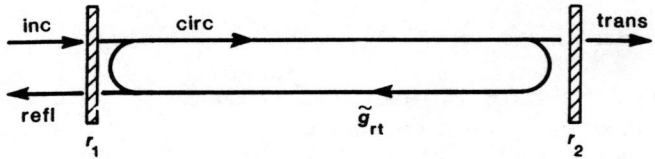
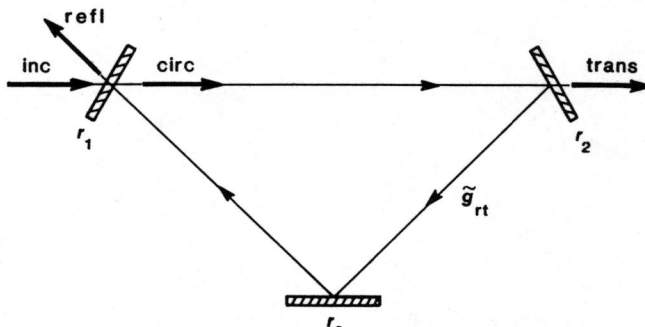

where $\tilde{g}_{rt}(\omega)$ is the net complex round-trip gain for a wave making one complete transit around the interior of the resonant cavity, whether it be a standing-wave or a ring-type cavity. *Equation 11.17 is the key equation for calculating the resonance properties of any resonant optical cavity, optical interferometer, or oscillating laser system.*

### Passive Lossy Optical Cavities

In analyzing optical cavities we will consistently use $L$ for the one-way length of a standing-wave cavity, and $p$ ($\equiv 2L$) for the *perimeter* or the round-trip *path length* in either the ring or the standing-wave cavities. By using this notation, and by always considering *round-trip* gains, losses, and phase shifts, we can develop a unified analysis that treats standing-wave or ring cavities on an equal footing.

Suppose then that the round-trip optical path in either type of cavity contains material with voltage absorption coefficient $\alpha_0$ (or possibly other kinds of internal losses), so that the attenuation of the signal amplitude or signal voltage in one round trip is $\exp(-\alpha_0 p) = \exp(-2\alpha_0 L)$, and the round-trip power reduction is $\exp(-2\alpha_0 p) = \exp(-4\alpha_0 L)$.

We are, of course, considering here sinusoidal optical signals with frequency $\omega$ and propagation constant $\beta = \beta(\omega) = \omega/c$, where $c$ is the velocity of light in the material inside the cavity. Hence there is also a phase shift or propagation factor $\exp(-j\omega p/c)$ associated with the round trip. (Let's leave out any atomic phase shifts $\Delta \beta_m$ for the moment.)

The circulating signal after one complete round trip in either type of cavity will then return to the reference plane just inside mirror $M_1$ with a net round-trip transmission factor, or complex round-trip gain, which is given for a passive lossy cavity by

$$\tilde{g}_{\rm rt}(\omega) \equiv r_1 r_2 (r_3 \ldots) \times \exp[-\alpha_0 p - j\omega p/c]. \tag{18}$$

In writing this expression we put the $(r_3 \ldots)$ factor inside brackets because there may or may not be a third or fourth mirror in the cavity, depending on whether we are considering a simple two-mirror standing-wave resonator or some kind of multimirror ring (or folded linear) cavity. We refer to $\tilde{g}_{\rm rt}$ as the "complex round-trip gain" inside the cavity, even though of course in any passive optical cavity (or even in any laser cavity below oscillation threshold) the magnitude of this round-trip gain will be less than unity, i.e., $|\tilde{g}_{\rm rt}| < 1$.

We can then write Equation 11.17 as

$$\tilde{E}_{\rm circ} = jt_1 \tilde{E}_{\rm inc} + r_1 r_2 (r_3 \ldots) \exp[-\alpha_0 p - j\omega p/c] \tilde{E}_{\rm circ}. \tag{19}$$

This expression then applies equally well to either a ring or a standing-wave cavity, if we simply replace $p$ by $2L$ for the standing wave.

### Cavity Resonances

This derivation says we can relate the circulating signal inside the cavity to the incident signal outside the cavity by

$$\frac{\tilde{E}_{\rm circ}}{\tilde{E}_{\rm inc}} = \frac{jt_1}{1 - \tilde{g}_{\rm rt}(\omega)} = \frac{jt_1}{1 - r_1 r_2 (r_3 \ldots) \exp[-\alpha_0 p - j\omega p/c]}. \tag{20}$$

What does this equation tell us? To help answer this question, Figure 11.10(a) shows several examples of how the circulating intensity $I_{\rm circ} \equiv |\tilde{E}_{\rm circ}|^2$ inside such an optical resonator varies with frequency $\omega$ or round-trip phase shift $\omega p/c$, assuming unit incident intensity, a round-trip internal power loss of $2\alpha_0 p = 2\%$, and symmetric mirror reflectivities $R_1 = R_2 = R$ which vary from $R = 70\%$ to $R = 98\%$.

It is obvious from these plots, as well as from Equation 11.20, that the signal inside the optical resonator exhibits a strong resonance behavior each time the round-trip phase shift $\omega p/c$ equals an integer multiple of $2\pi$, i.e., each time $\omega = \omega_q \equiv q \times 2\pi \times (c/p)$, with $q$ being an integer. In fact, the circulating intensity inside the cavity at these resonances becomes many times larger than the intensity incident on the cavity from outside. As we will discuss in more detail in the following sections, these resonant frequencies are known as *cavity axial modes*, and the frequency interval between resonances is known as the *axial mode spacing* or the *free spectral range* of the cavity.

### Rotating Vector Interpretation

The resonance behavior that is evident in these plots can perhaps be most easily understood from the following graphical analysis. The denominator in the ratio of circulating to incident signal amplitudes in Equation 11.20 is given by the complex factor $1 - \tilde{g}_{\rm rt}(\omega) \equiv 1 - r_1 r_2 (r_3 \ldots) \exp[-\alpha_0 p - j\omega p/c]$. The quantity $\tilde{g}_{\rm rt}(\omega)$ is a complex vector with a magnitude that is less—but perhaps not much less—than unity. This complex gain has a phase angle $\omega p/c$ such that $\tilde{g}_{\rm rt}(\omega)$ rotates through one complete revolution in the complex plane every time $\omega p/c$ increases by $2\pi$. Since the cavity perimeter $p$ is many optical wavelengths in length, the

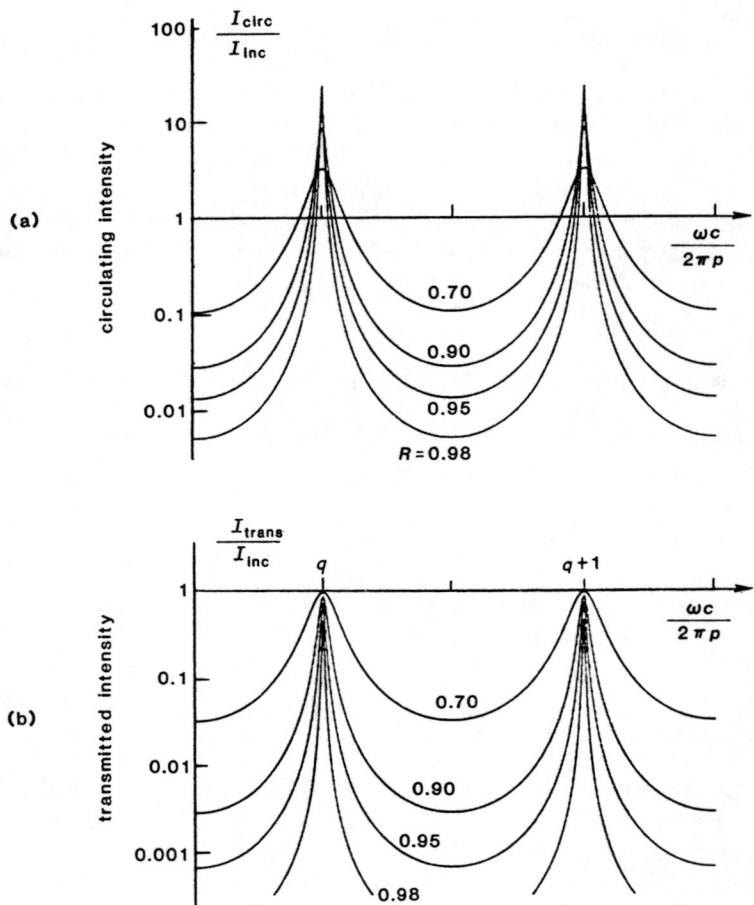

**FIGURE 11.10**
Circulating (a) and transmitted (b) power in an optical resonator plotted versus frequency or round-trip phase shift $\omega p/c$ and mirror reflectivity $R_1 = R_2 = R$, assuming a fixed internal power loss of 2% per round trip.

rotation of $\omega p/c$ through many complete cycles with increasing frequency is quite rapid.

Suppose we plot this denominator in a complex plane. The complex vector representing $\tilde{g}_{\rm rt}(\omega)$ then rotates about the point $1 + j0$ as shown in **Figure 11.11**. Every time the tip of the rotating $1 - \tilde{g}_{\rm rt}(\omega)$ vector sweeps close to the origin in this sketch, the denominator $1 - \tilde{g}_{\rm rt}(\omega)$ of the $\tilde{E}_{\rm circ}/\tilde{E}_{\rm inc}$ ratio becomes very small, and the value of the circulating field becomes correspondingly large. This occurs, of course, every time $\omega p/c$ passes through another integer multiple of $2\pi$.

### Circulating Intensity Magnification

Let us examine how large the circulating signal inside the cavity can become at one of these peaks. Consider as a simple example a symmetric linear cavity with equal end-mirror reflectivities $r_1 = r_2 = r$, and assume negligible internal losses, or $\alpha_0 p \approx 0$. The peak value of the circulating field at resonance is then

## 11.3 RESONANCE PROPERTIES OF PASSIVE OPTICAL CAVITIES

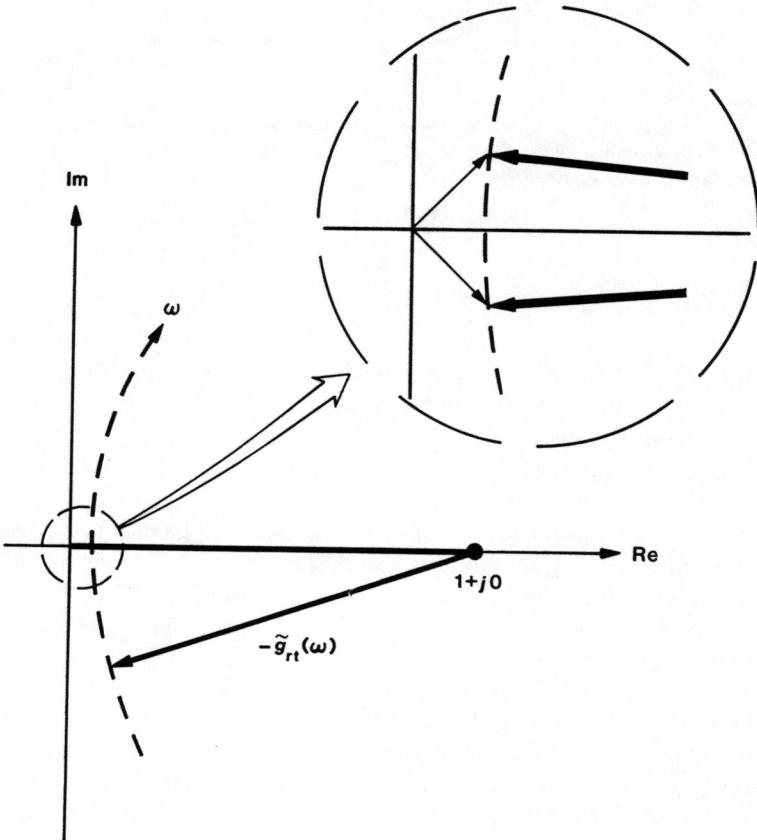

FIGURE 11.11
Graphical diagram to help explain the behavior of an interferometer cavity near resonance.

given by

$$\left.\frac{\tilde{E}_{\text{circ}}}{\tilde{E}_{\text{inc}}}\right|_{\omega=\omega_q} = \frac{jt}{1 - r_1 r_2 e^{-\alpha_0 p}} \approx \frac{jt}{1 - r^2} = \frac{j}{t}, \qquad (21)$$

where we have used $t_1 = t_2 = \sqrt{1-r^2}$ for lossless mirrors. (Note that the 90° phase shift between the incident and circulating fields is inherent in our choice of matrix representation for the partially transmitting mirror.)

The ratio of circulating intensity to incident intensity for a symmetric cavity with negligible internal losses is thus given by

$$\left.\frac{I_{\text{circ}}}{I_{\text{inc}}}\right|_{\omega=\omega_q} \approx \left|\frac{1}{t}\right|^2 = \frac{1}{T}, \qquad (22)$$

where $T \equiv t^2$ is the power transmission of either end mirror. If we assume, for example, end mirrors which are 99% reflecting and 1% transmitting, so that $T = 1\% = 0.01$ (note that mirrors are usually characterized by their *power* reflection and transmission values), then this gives

$$I_{\text{circ}} \approx 100 \times I_{\text{inc}} \qquad \text{for} \qquad \begin{cases} R_1 = R_2 = 0.99, \\ \alpha_0 p \ll 0.01. \end{cases} \qquad (23)$$

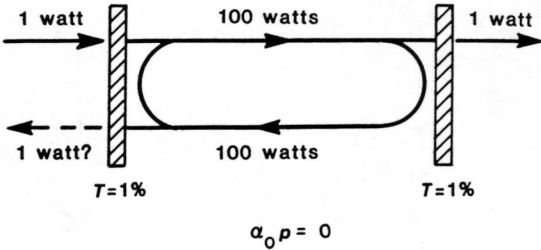

FIGURE 11.12
Magnification of the circulating signal level in a lossless optical cavity at resonance.

In other words, 1 watt of laser power incident on this cavity from outside will build up a circulating power of ≈100 watts traveling in each direction inside the laser cavity, as illustrated in Figure 11.12.

The circulating power inside a passive cavity resonator can thus be much larger than the power incident on the cavity end mirror from outside. There is, of course, no way that this magnified circulating power can be usefully extracted (at least not continuously), since energy conservation still must be obeyed! This form of power magnification can be used, however, in testing the damage thresholds of low-loss optical elements placed inside such a cavity. This circulating stored energy can also be switched out of the cavity on a transient basis, using a fast switch, to give a short pulse of energy at the magnified intensity level. The latter technique is sometimes referred to as "cavity dumping".

The intensity magnification in this symmetric and lossless example has a maximum value of $1/t^2 = 1/(1-R)$, where $R$ is the mirror reflectivity at each end. Finite internal losses or increased mirror transmission will in general reduce this resonance enhancement, the circulating intensity at resonance being given more generally by

$$\left.\frac{I_{\text{circ}}}{I_{\text{inc}}}\right|_{\omega=\omega_q} = \frac{t_1 t_2}{[1 - r_1 r_2 (r_3 \ldots) e^{-\alpha_0 p}]^2}. \tag{24}$$

If, for example, $r_1 r_2 = 0.99$ (and $r_3 \equiv 1$), so that the round-trip power loss through the end mirrors is 1%, then giving the internal losses the same value of 1% by making $\alpha_0 p = 0.01$ will cut the circulating field amplitude in half, and decrease the circulating intensity by four times, or a reduction of approximately 6 dB.

### Transmitted Cavity Fields

When the 100 watts of power circulating inside the cavity in Figure 11.12 impinge on the 1% transmitting output mirror, this means that a net transmitted power of 1 watt—equal to the incident signal—must be transmitted out through the output mirror at the opposite end of the cavity, as shown in Figure 11.12. In other words, this particular cavity, at resonance, has a resonant transmission from input to output of essentially unity. As the frequency $\omega$ is tuned off resonance, however, both the circulating and the transmitted signal intensities will drop rapidly, as illustrated in Figure 11.10.

To develop a more general formula for the transmitted field $\tilde{E}_{\text{trans}}$ coming out the other end or through the other mirror $M_2$, in either the ring or the linear example, we can suppose that a portion $p_1$ of the cavity path is in the leg between mirrors $M_1$ and $M_2$ ($p_1 \equiv L$ in the linear example). The transmitted signal intensity coming out through mirror $M_2$ will then be given by

$$\tilde{E}_{\text{trans}} = jt_2 \exp[-\alpha_0 p_1 - j\omega p_1/c] \times \tilde{E}_{\text{circ}}. \quad (25)$$

Hence the net transmission through the cavity or interferometer, from input to output, is given by

$$\frac{\tilde{E}_{\text{trans}}}{\tilde{E}_{\text{inc}}} = \frac{-t_1 t_2 \exp[-\alpha_0 p_1 - j\omega p_1/c]}{1 - r_1 r_2 (r_3 \ldots) \exp[-\alpha_0 p - j\omega p/c]} = \frac{-t_1 t_2 \exp[-\alpha_0 p_1 - j\omega p_1/c]}{1 - \tilde{g}_{\text{rt}}(\omega)} \quad (26)$$

for the ring example, or by the essentially equivalent formula

$$\frac{\tilde{E}_{\text{trans}}}{\tilde{E}_{\text{inc}}} = \frac{-t_1 t_2 \exp[-\alpha_0 L - j\omega L/c]}{1 - r_1 r_2 \exp[-2\alpha_0 L - 2j\omega L/c]} = -\frac{t_1 t_2}{\sqrt{r_1 r_2}} \frac{\sqrt{\tilde{g}_{\text{rt}}(\omega)}}{1 - \tilde{g}_{\text{rt}}(\omega)} \quad (27)$$

for the standing-wave cavity. The minus sign in front of either expression is a basically irrelevant additional phase shift of $\pi$ that arises because we insist on making a certain choice of reference planes at each mirror, as discussed in the preceding section.

Both Figure 11.10(b) and Figure 11.13(a) plot the transmitted intensity versus frequency for various choices of mirror reflectivities and losses. It is evident that the resonant cavity acts as a narrowband transmission filter, with a periodically spaced set of transmission passbands whose bandwidth and peak transmission depend on the cavity losses and the balance between input and output coupling to the cavity. Figure 11.13(b) also shows the transmission phase angle versus frequency for a typical case.

### Dielectric Etalons

The resonant transmission properties of optical interferometers or Fabry-Perot etalons have long been used as passive optical filters for incoherent light sources. In the laser field, thin dielectric etalons, with or without additional reflective coatings, are also often used as filters *inside* laser cavities, in order to tune the laser, to obtain wavelength or frequency selection, or to reduce the gain bandwidth and thus limit the number of oscillating axial modes inside a laser cavity. The intracavity application of such an etalon is illustrated in Figure 11.14.

In this application the etalon is usually tilted to an angle large enough that the external reflected beams from the etalon are deflected away from the cavity axis, so that they do not set up any unwanted resonances with the other mirrors in the resonator. At the same time the angle is made small enough that the waves bouncing back and forth inside the etalon are shifted transversely by only a very small amount compared to the beam diameter on each bounce, thus keeping the "walk-off losses" of the etalon interferometer small.

The resonant transmission peaks of the etalon can then be tuned by small changes in the etalon angle. (Warning: the peaks tune "the wrong way" with change in angle; see Problem 11.3–2.) By using several etalons of different thickness in cascade, it is possible to combine the narrow linewidth but small free spectral range obtained from a longer etalon, with the wider linewidth but also wider free spectral range of a thinner etalon, as shown in Figure 11.14(b).

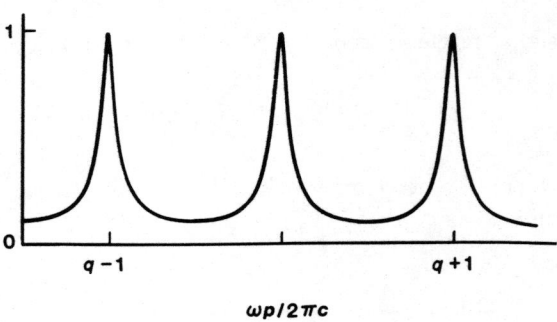

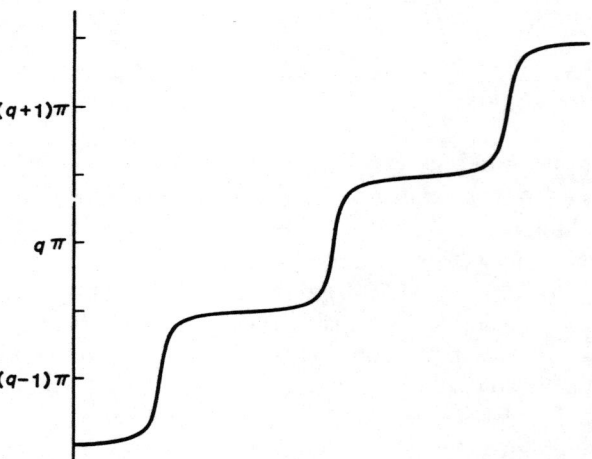

FIGURE 11.13
Transmitted intensity (top curve) and phase shift (bottom curve) versus frequency through a typical interferometer or etalon.

### Reflected Cavity Fields

Let us examine finally the *reflected* wave that comes back from a resonant cavity or etalon at the cavity input mirror, as illustrated in Figure 11.15. Note that in in a standing-wave cavity (Figure 11.15(a)) the reflected wave goes straight back along the same direction as the incident wave, and hence presumably straight back into whatever source generated the incident wave; whereas in the ring cavity (Figure 11.15(b)) this reflected wave goes off in a different direction, like a specular reflection from mirror $M_1$. This can be a major advantage of a ring as compared to a standing-wave cavity, since many laser devices do not function well when looking into even a relatively weak back-reflection.

Suppose we look first at the reflected wave from the symmetric cavity example with the 100 watts of circulating power in Figure 11.12. Since the input mirror $M_1$ in this example also has a 1% power transmission, it might appear that at resonance another 1 watt of power must be transmitted back out of the cavity in the reverse direction, because of transmission from the 100 watts of circulating power back through the 1% mirror at the input end. This seems to say that with 1 watt of incident power, 1 watt of power can appear in the reflected wave as well as in the transmitted wave coming out of the cavity. Obtaining 2 watts of total output power in the transmitted plus reflected waves from the cavity, with only 1 watt of input power, does pose some conceptual difficulties, however; and

## 11.3 RESONANCE PROPERTIES OF PASSIVE OPTICAL CAVITIES

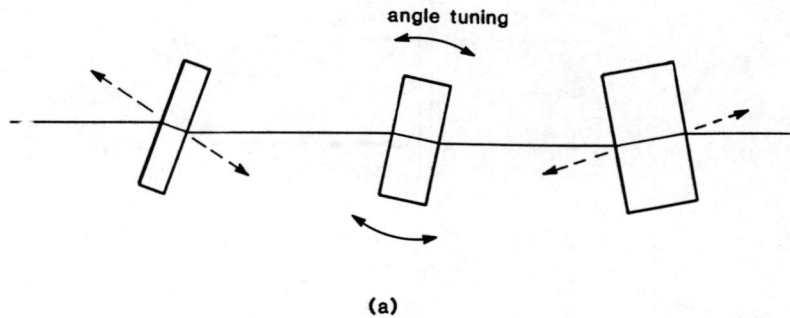

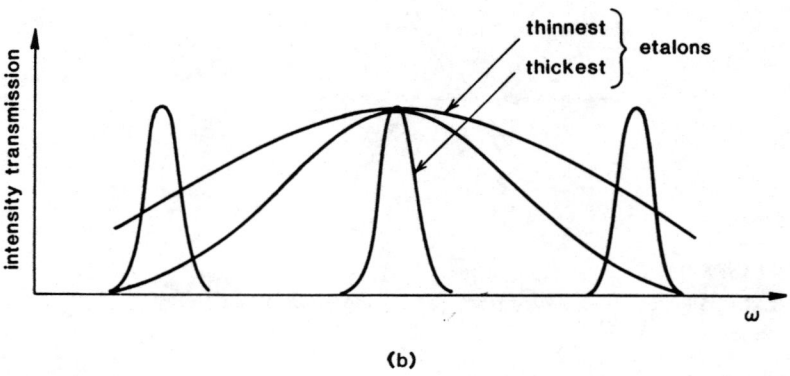

**FIGURE 11.14**
(a) Tilted Fabry-Perot etalons as intracavity filters or tuning elements, and (b) the sequential transmission curves for etalons of different thicknesses, adjusted so that their transmission peaks coincide at one frequency.

it would seem that $\approx 0$ watts in the reflected wave would be a more reasonable result for this example.

### Reflected Signal Formulas

The significant point here is, of course, that for both the traveling-wave and standing-wave cases the total "reflected" wave $\tilde{E}_{\text{refl}}$ coming from mirror $M_1$ must consist of a component $r_1\tilde{E}_{\text{inc}}$ that is due to straightforward reflection from the outer surface of mirror $M_1$, plus a second component that represents the circulating signal $\tilde{E}_{\text{circ}}$ inside the cavity that is transmitted out through the mirror $M_1$ into the same direction, as illustrated graphically in Figures 11.15(a) and (b). The latter component comes from the circulating signal $\tilde{E}_{\text{circ}}$ that left the reference plane one round-trip time earlier; traveled once around the cavity *except* for bouncing off mirror $M_1$; and then is transmitted out through the input mirror. The value of this component is thus given by $jt_1(\tilde{g}_{\text{rt}}/r_1) \times \tilde{E}_{\text{circ}}$. (The round-trip gain must be divided by $r_1$ because the wave does not bounce off mirror $M_1$, it goes through it.)

The total reflected wave thus consists of

$$\tilde{E}_{\text{refl}} = r_1\tilde{E}_{\text{inc}} + jt_1(\tilde{g}_{\text{rt}}/r_1)\tilde{E}_{\text{circ}}. \tag{28}$$

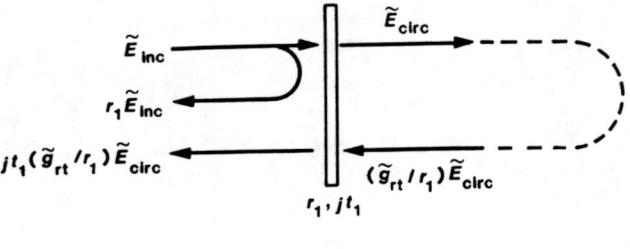

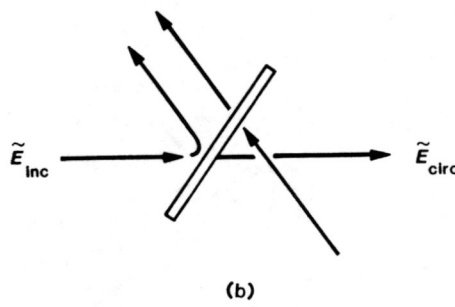

**FIGURE 11.15**
Externally reflected waves from (a) standing-wave and (b) ring-laser cavities.

Using our earlier expressions for $\tilde{E}_{\text{circ}}$, we can then write the total reflection coefficient from mirror $M_1$ as

$$\frac{\tilde{E}_{\text{refl}}}{\tilde{E}_{\text{inc}}} = r_1 - \left[\frac{t_1^2 r_2 e^{-\alpha_0 p - j\omega p/c}}{1 - r_1 r_2 (r_3 \ldots) e^{-\alpha_0 p - j\omega p/c}}\right] = r_1 - \frac{t_1^2}{r_1} \frac{\tilde{g}_{\text{rt}}(\omega)}{1 - \tilde{g}_{\text{rt}}(\omega)}. \quad (29)$$

These expressions, with their two separate terms, are useful in emphasizing that the reflected signal does consist of the directly reflected component, plus a transmitted component coming from the circulating signal inside the cavity, as shown in Figure 11.15. By using the lossless mirror expression that $r_1^2 + t_1^2 = 1$, however, we can also convert these expressions into the slightly simpler forms

$$\frac{\tilde{E}_{\text{refl}}}{\tilde{E}_{\text{inc}}} = \frac{r_1 - r_2 e^{-\alpha_0 p - j\omega p/c}}{1 - r_1 r_2 e^{-\alpha_0 p - j\omega p/c}} = \frac{1}{r_1} \times \frac{r_1^2 - \tilde{g}_{\text{rt}}(\omega)}{1 - \tilde{g}_{\text{rt}}(\omega)}. \quad (30)$$

The second form of this expression makes it evident that the reflectivity from mirror $M_1$ of the cavity depends only on the amplitude reflectivity $r_1$ of that mirror and the round-trip gain $\tilde{g}_{\text{rt}}(\omega)$, and that the reflectivity can go to zero if these become exactly equal at resonance.

Figures 11.16 and 11.17 show several curves of the power reflectivity $I_{\text{refl}}/I_{\text{inc}}$ from a resonant cavity or interferometer versus frequency, assuming a fixed reflectivity $R_1$ for the front mirror and varying values of the additional losses due to $R_2 e^{-2\alpha_0 p}$. Note that only the product $R_2 e^{-2\alpha_0 p}$ counts; it makes no difference to the total reflectivity at mirror $M_1$ how the additional losses in the rest of the cavity are divided between the second mirror reflectivity $R_2$ and the internal losses $e^{-2\alpha_0 p}$.

In plotting Figure 11.17, we have plotted the intensity-reflection curves for $R_1 \geq R_2 e^{-2\alpha_0 p}$ above the axis and the curves for $R_1 \leq R_2 e^{-2\alpha_0 p}$ below the

## 11.3 RESONANCE PROPERTIES OF PASSIVE OPTICAL CAVITIES

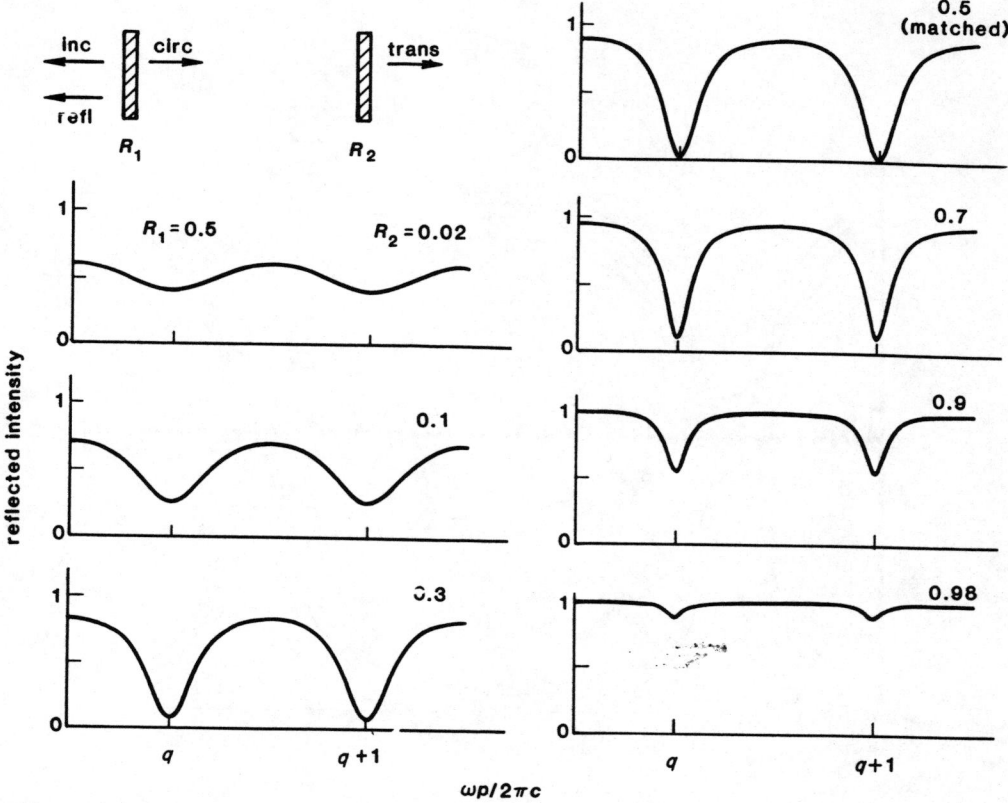

**FIGURE 11.16**
Reflected intensity versus frequency from one mirror of an interferometer cavity, for different values of the internal cavity loss or the reflectivity of the other mirror.

axis, to indicate that there is indeed a 180° phase shift in the total reflectivity at resonance as the interferometer goes from one situation to the other. We have also shown some examples of the rather complex phase-angle variations with frequency exhibited by the reflected wave.

### Matched Input Conditions

The special situation when the input mirror reflectivity $R_1$ exactly equals the additional loss terms given by $R_2 e^{-2\alpha_0 p}$ causes the two terms in the reflection expression to exactly cancel, and the net reflection coefficient to become exactly zero at resonance. This is often called the *impedance-matched* situation, since it corresponds to looking into an impedance-matched resonant circuit or cavity (i.e., load impedance = characteristic impedance) on an ordinary transmission line. The numerical example that we considered in Figure 11.12, with the 100 watts of circulating power, was an input-matched situation, as is any symmetric interferometer (i.e., $r_1 = r_2$) with very small internal losses.

### Etalon Mirrors

Even when the reflectivities from each individual mirror surface are comparatively low, i.e., $R_1$ and $R_2 \ll 1$, the combined reflectivity in the backward

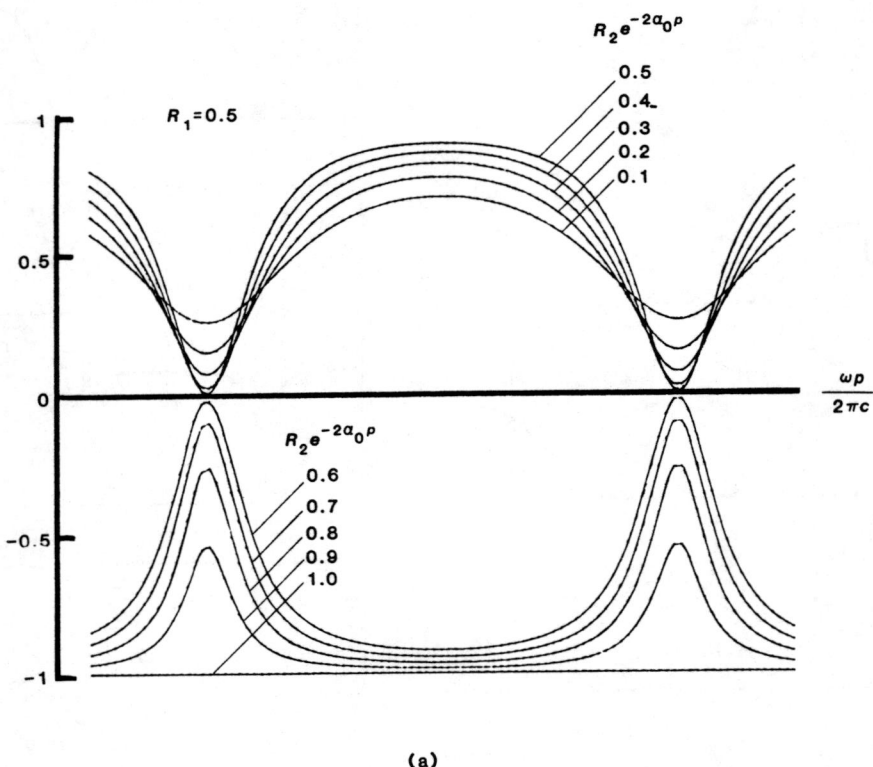

**FIGURE 11.17(a)**
Intensity of the reflected signal plotted versus frequency for a cavity with fixed front-mirror reflectivity and various values of the additional cavity losses plus back-mirror reflectivity.

direction from an interferometer cavity over the off-resonance part of the reflection curve can be considerably larger than the individual reflections from either surface alone, as illustrated in Figure 11.18.

As one application of this, polished etalons made from dielectric materials such as quartz ($n \approx 1.46$) or clear sapphire ($n \approx 1.76$) with highly parallel faces are often used as the output mirrors for pulsed high-power solid-state lasers—that is, the lasers are operated with a 100% mirror on one end and a polished etalon a few mm or a cm thick, generally with no additional reflective coatings, as the output mirror on the other end. Since these lasers typically have large round-trip gains, they operate best with low-reflectivity output mirrors, and the uncoated dielectric etalon provides a simple way of achieving the necessary output mirror reflectivity. These uncoated etalons are also simple to fabricate, can have very high optical-damage thresholds, and the reflectivity peaks can serve a useful purpose in narrowing the oscillation bandwidth of a wide-line laser medium, such as a Nd:glass or dye laser. Note that the reflectivity peaks in these mirrors occur not at resonance, but rather half-way between the axial-mode resonances of the etalon itself.

## 11.3 RESONANCE PROPERTIES OF PASSIVE OPTICAL CAVITIES

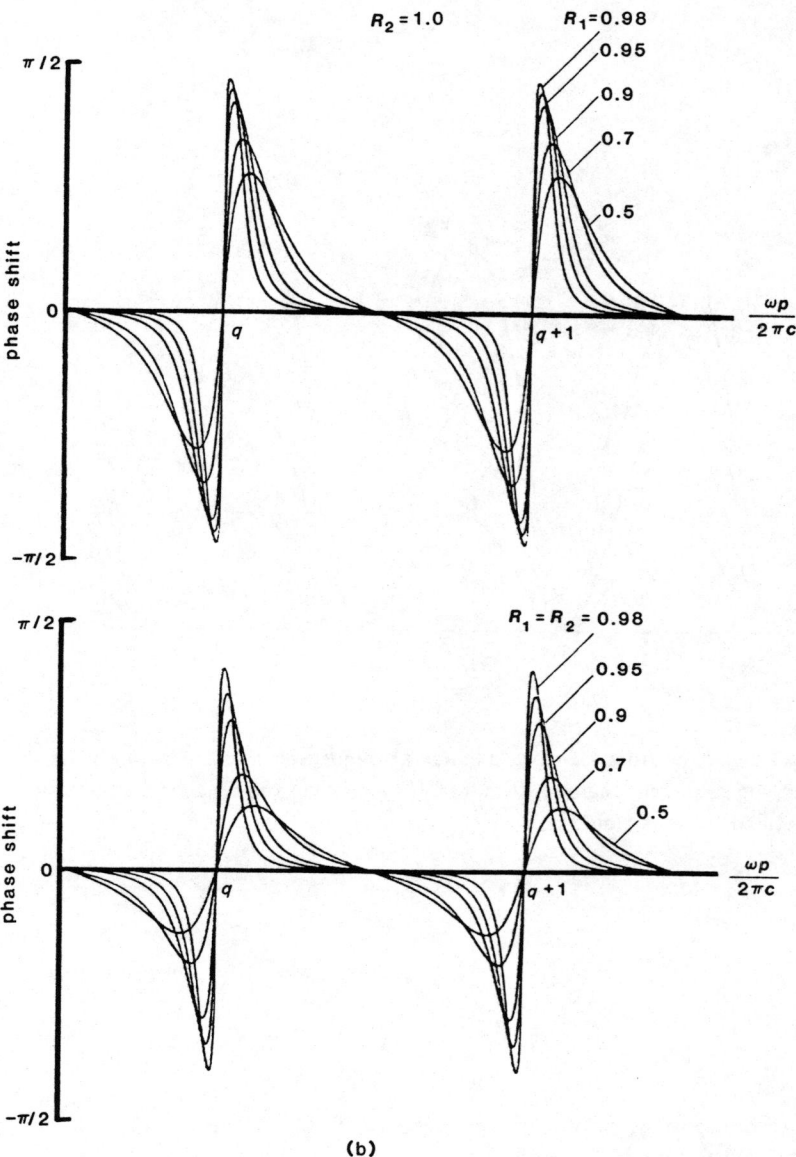

(b)

**FIGURE 11.17(b)**
Phase angle of the reflected signal plotted versus frequency for a cavity with fixed front-mirror reflectivity and various values of the additional cavity losses plus back-mirror reflectivity.

### Transient Cavity Reflections

The fact that the total reflected signal $\tilde{E}_{\text{refl}}$ from a resonant cavity is formed from the vector combination of the directly reflected input signal $r_1 \tilde{E}_{\text{inc}}$, plus a transmitted portion of the circulating signal, or $jt_1(\tilde{g}_{\text{rt}}/r_1)\tilde{E}_{\text{circ}}$, means that the transient response of the cavity reflection, if we suddenly change either one of these signals, can be rather complex. If an input signal is very suddenly turned on, for example, the directly reflected component appears immediately; but the

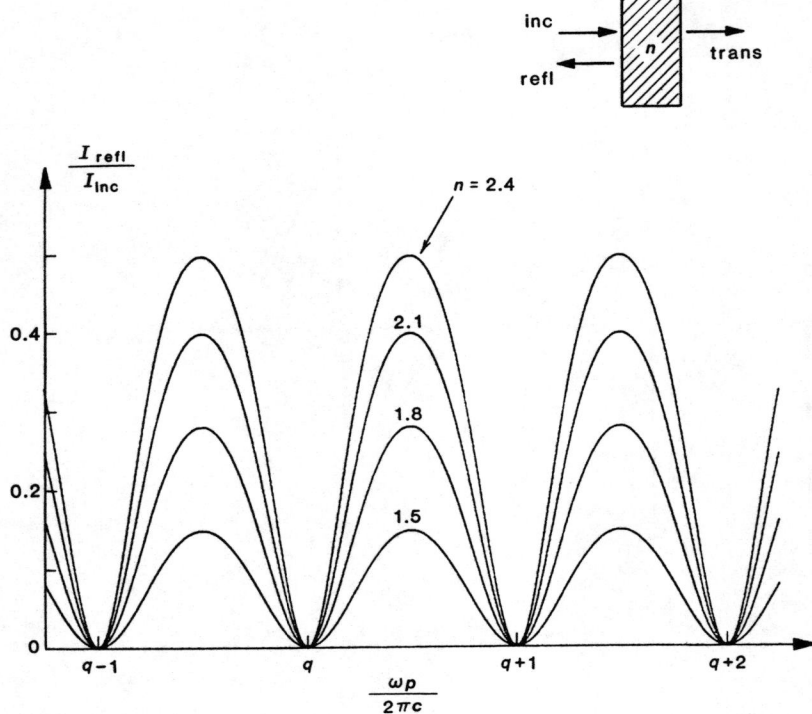

**FIGURE 11.18**
The back-reflection from a dielectric etalon at frequencies midway between the transmission resonances can be substantially larger than the reflection from either of the etalon surfaces alone.

circulating component only appears more gradually, after the circulating field inside the cavity has time to build up.

This makes it possible to devise various clever schemes for modulating the reflected signal on a transient or pulsed basis. Suppose, for example, that a matched steady-state on-resonance situation with these two terms essentially canceling each other has been established, and that the incident signal is then suddenly turned off by means of some kind of fast electro-optic modulator. The net reflected signal will then suddenly jump from near zero to a value equal to $-r_1$ times the originally incident signal, with a step-function leading edge; and will then gradually die away, with the cavity decay rate $\omega/Q_c$ as the stored energy drains out of the resonant cavity.

If we can instead suddenly shift the phase of the incident signal by 180°, the total reflected signal will suddenly jump up to a value $\tilde{E}_{\text{refl}} \approx -2r_1\tilde{E}_{\text{inc}}$, or a reflected power equal to four times the incident power, at least in the leading edge of the resulting transient response.

## Problems for 11.3

1. *Design specifications for a transmission etalon.* A Fabry-Perot etalon is to be used as a transmission filter inside an oscillating laser cavity, as mentioned in this section. A very high peak transmission ($\geq 99.0\%$) is required in order to avoid excessive losses inside the laser, and a finesse $\geq 30$ is also needed. What specifications must be given for the etalon mirror reflectivities $R_1$ and $R_2$, and for the internal round-trip power loss $\delta_0 \equiv 2\alpha_0 p$ in the etalon?

2. *Angle tuning of a transmission etalon.* Analyze (or look up in an optics text) the transmission function of a resonant etalon as a function of angle; and discuss how the transmission peak of an intracavity etalon will tune with angle.

   At larger angles the apparent thickness or path length through the etalon increases; therefore the etalon should tune to lower frequencies or longer wavelengths with increasing angle—right?

3. *Calculating cavity parameters from measured transmission-reflection curves.* Suppose that by using a tunable laser you can measure fairly accurately the magnitudes of both the transmission coefficient and the reflection coefficient versus frequency for a passive interferometer cavity (i.e., internal losses but no gain medium) across one full axial-mode resonance. Develop formulas by which you can work backward to calculate the reflectivities $r_1$ and $r_2$ of the two end mirrors, and also the internal cavity loss $\alpha_0 p$ of the cavity, in terms of the measured midband transmission and reflection factors and their measured bandwidths.

4. *Reflection properties of uncoated dielectric etalon mirrors.* Carry out an analysis of the dielectric-etalon mirrors mentioned in this section, assuming a simple dielectric slab a few mm to a cm thick with polished and parallel front and back surfaces, negligible internal losses, and no additional mirror coatings. Analyze the reflectivity properties versus frequency of these dielectric etalons, and find typical values of peak reflectivity for such etalons both on resonance and midway between resonances. (Note: typical values of index of refraction range from $n \approx 1.45$ for simple typical glasses to $n \approx 1.54$ for quartz, $n \approx 1.76$ for sapphire or $n \approx 3.3$ for GaAs.) Does the maximum reflectivity occur at resonance, or midway between resonances, and how large is this reflectivity compared to the single-surface reflectivities of the same etalons? Discuss in physical terms how the reflection versus frequency behavior can be explained.

5. *Linewidth of a power reflectivity dip.* Suppose we define the "50% linewidth" $\Delta\omega_{50}$ for the power reflection curve from a Fabry-Perot interferometer or lossy regenerative cavity as the full frequency width between the points halfway down into the resonance dip. Find an expression for this linewidth versus the midband loss of the interferometer in dB.

6. *Reflection phase angle versus frequency.* Compute and plot the phase angle versus frequency for the complex interferometer reflectivity $\tilde{E}_{\text{refl}}/\tilde{E}_{\text{inc}}$ for some typical choices of $R_1$, $R_2$ and $\delta_0$. Consider in particular the case of the *Gires-Tournois interferometer*, which has $R_1$ finite, $R_2 = 100\%$, and $\delta_0 = 0$, so that the magnitude of the overall reflectivity is constant and equal to 1 at all frequencies, but the phase angle of the reflection changes sharply with frequency. (Hint: Consider the phase angles versus frequency for the numerator and denominator separately, and then combine.)

7. *Field magnification inside a resonant cavity.* How much larger is the electric field strength inside the resonant cavity of Figure 11.11, compared to the field strength at any point in the incident laser beam outside the cavity?

---

## 11.4 "DELTA NOTATION" FOR CAVITY GAINS AND LOSSES

Before we continue our general discussion of cavity resonance properties, let us introduce in this section a unified notation for describing cavity gain and loss factors that will be useful throughout the rest of this book. We can also then use this notation to simplify some of the formulas from the preceding section.

### Mirror Reflectivities: The Delta Notation

The usual practice in optics is to describe mirrors by their *power reflection and transmission* values; "95% reflectivity," for example, means a mirror with $R_1 = 0.95$ and hence, for a lossless mirror, $T_1 = 1 - R_1 = 0.05$.

In the early days of lasers, the available gains in most laser systems were very small, and oscillation could be obtained only with very high-reflectivity mirrors. It then became conventional to describe the small difference between the mirror power reflectivity and unity by the symbol $\delta$, so that a mirror with 95% reflectivity would be described by $R_1 = 1 - \delta_1 = 0.95$, or $\delta_1 = T_1 = 0.05$.

As a more convenient and general definition, however, one useful for both high- and low-reflectivity mirrors, and hence for either small or large output coupling, we will in the remainder of this text relate the reflectivity $R_1$ of any mirror to a "mirror coupling coefficient" $\delta_1$ by means of the definition

$$\begin{aligned} R_1 &\equiv e^{-\delta_1} \quad &\text{(exact definition, arbitrary } \delta_1\text{),} \\ &\approx 1 - \delta_1 \quad &\text{(approximate definition, } \delta_1 \ll 1\text{).} \end{aligned} \tag{31}$$

Thus, if we have a laser cavity with two end mirrors having reflectivities $R_1$ and $R_2$, we will write these as $R_1 \equiv r_1^2 \equiv e^{-\delta_1}$ and $R_2 \equiv r_2^2 \equiv e^{-\delta_2}$. The general definition of $\delta_i$ is thus

$$\delta_i \equiv \ln\left(\frac{1}{R_i}\right) = 2\ln\left(\frac{1}{r_i}\right) \tag{32}$$

for mirrors with arbitrarily low reflectivity and thus arbitrarily large output coupling.

In the high-reflectivity, low-coupling limit we can still write the mirror transmission as $T = 1 - R$, with the approximation that $T \approx \delta$ and hence $t \approx \sqrt{\delta}$ for $\delta \ll 1$. For a mirror having, say, $R = 80\%$ reflectivity and $T = 20\%$ power transmission, the exact value of $\delta$ is given by $\delta = \ln(1.25) = 0.223$, not far distant from the approximate value of $1 - R = 0.20$. Hence the approximation that $\delta \approx 1 - R = T$ remains reasonably accurate even for mirror reflectivities as low as $R \approx 80\%$ and $T \approx 20\%$.

### Internal Cavity Gain and Loss Factors

If we use this notation for the end-mirror coupling factors $\delta_1$ and $\delta_2$, the round-trip power gain inside any cavity or interferometer with a perimeter $p$, internal loss coefficient $\alpha_0$, and internal gain coefficient $\alpha_m$, can be conveniently condensed into the form

$$|\tilde{g}_{\rm rt}|^2 = R_1 R_2 e^{2\alpha_m p_m - 2\alpha_0 p} = e^{\delta_m - \delta_0 - \delta_1 - \delta_2}. \qquad (33)$$

That is, we can make a natural extension of the "$\delta$ notation" by also writing the total round-trip power gain and loss in the cavity due to the internal loss coefficient $\alpha_0$ and the laser gain medium in the forms

$$\delta_0 \equiv 2\alpha_0 p \quad \text{and} \quad \delta_m \equiv 2\alpha_m p_m. \qquad (34)$$

As a further extension we can include within the internal loss coefficient $\delta_0$ not only the round-trip power reduction arising from any distributed attenuation $\alpha_0 p$, but also any additional discrete losses that may occur inside the cavity because of lossy interfaces, imperfect Brewster windows, internal scattering elements, or whatever. The significant quantity is thus not $\alpha_0$ or $p$ separately, but the total internal power loss in one complete round trip, as expressed by $e^{-\delta_0}$.

From here on we will thus generally express any kind of round-trip power gain or power loss in an optical cavity by the notation

$$\delta_x \equiv \ln[\text{power gain, or power loss, ratio per round trip}]. \qquad (35)$$

In the small gain or loss limit, $\delta_x$ is essentially the fractional power gain or loss per round trip due to mechanism $x$, and we will often speak loosely of $\delta_x$ in those terms, e.g., $\delta_x = 0.20$ means $\approx 20\%$ power gain or loss per round trip.

### Total Cavity Gains and Losses

As one final bit of notation, it will often be convenient to combine all the round-trip losses contained in the factor $\delta_0$—which we will call *internal cavity losses*—with all the loss factors $\delta_1$ and $\delta_2$ due to the cavity mirrors—which we will call *external coupling losses*—to give a *total cavity-loss factor* $\delta_c$ defined by

$$\delta_c \equiv \delta_0 + \delta_1 + \delta_2 = 2\alpha_0 p + \ln\left(\frac{1}{R_1 R_2}\right) \qquad (36)$$

(plus additional mirror reflectivities $R_3$ if needed). With this notation the round-trip power gain inside any cavity also containing a laser gain medium can then be written in the simple form

$$|\tilde{g}_{\rm rt}|^2 = e^{\delta_m - \delta_c} \approx 1 + \delta_m - \delta_c \quad \text{if} \quad |\delta_m - \delta_c| \ll 1. \qquad (37)$$

The net growth (or decay) rate for a signal circulating around inside the laser cavity (with no injected signal) is thus simply the difference between the total (saturated) laser gain factor $\delta_m$ and the total cavity loss factor $\delta_c$.

With all of these gain and loss factors $\delta_x$ defined in terms of a *round trip*, most of our formulas will apply equally well to either standing-wave or ring-type laser cavities. Note that similar notation is used in much of the laser literature, but published papers are not always consistent about whether $\delta_x$ means power gain or loss *per one-way pass* or *per round trip*. In consulting the literature, watch out for possible factors of two, depending on which definition is employed.

### Cavity Q Values

It can also be useful in some situations to relate cavity gain or loss factors to cavity $Q$ factors, defined in the following manner.

Suppose some initially injected energy is circulating around inside a laser cavity, with no further injected signal being applied. If we consider only a "cold" laser cavity (no gain present), this circulating energy will decrease after a number $N$ of round trips in the exponential fashion

$$I_{\text{circ}}(t) = I_{\text{circ}}(t_0) \times \exp[-N\delta_c] = I_{\text{circ}}(t_0) \times \exp\left[-\frac{\delta_c}{T_{\text{rt}}}(t-t_0)\right], \qquad (38)$$

where $T_{\text{rt}} \equiv p/c$ is the round-trip transit time in the laser cavity, and $N = (t-t_0)/T_{\text{rt}}$ is the number of round trips in time $t-t_0$. Another way of expressing this exponential decay that is commonly used in many engineering fields is to write it in the form

$$I_{\text{circ}}(t) = I_{\text{circ}}(t_0) \times \exp\left[-\frac{\omega_a}{Q_c}(t-t_0)\right], \qquad (39)$$

where $Q_c$ is sometimes called the "cold cavity $Q$" of the laser cavity due to internal losses plus external coupling. This $Q$ value plays the same role in an optical cavity as does, for example, the familiar $Q = \omega L/R$ value characteristic of a series $RLC$ electrical circuit.

The $Q$ factor (sometimes called the "quality factor") of an optical cavity due to its internal losses plus external coupling through the mirrors can thus be calculated from

$$Q_c = \frac{\omega_a T_{\text{rt}}}{\delta_c} = \frac{2\pi p}{\lambda}\frac{1}{\delta_c}. \qquad (40)$$

(We could also define a negative $Q_m$ value representing the laser gain, by replacing the loss factor $\delta_c$ by $\delta_m$ in this expression.) Real laser cavities typically have very large $Q_c$ values, even in very lossy optical cavities.

Suppose for example that an optical cavity is very lossy, with 90% power loss per round trip, corresponding to $\delta_c = \ln(1/0.1) \approx 2.3$. The $Q_c$ value will then still be very large, even though 90% of the circulating energy is lost out of the cavity on every round trip, because the cavity perimeter $p$ will (except in very special cases) be larger than the optical wavelength $\lambda$ by a factor typically somewhere between $10^4$ and $10^6$. In physical terms, the power loss *per round trip* is large, but the fractional power loss *per cycle* (which determines the $Q_c$ value) is very small.

### Field Values in Low-Loss Cavities

By using the delta notation, plus the low-loss approximations, we can write some useful simplified forms of the analytical results given earlier in this chapter. The on-resonance value of the denominator $1 - \tilde{g}_{\text{rt}}$ that appears in all the resonant-cavity expressions can first be simplified to the form

$$1 - \tilde{g}_{rt} \equiv 1 - r_1 r_2 e^{\alpha_m p_m - \alpha_0 p} \approx \frac{\delta_c - \delta_m}{2}. \tag{41}$$

The peak value for the circulating intensity in a purely passive cavity ($\delta_m = 0$) at resonance can then be written as

$$\left.\frac{I_{\text{circ}}}{I_{\text{inc}}}\right|_{\omega=\omega_q} \approx \frac{4\delta_1}{(\delta_1 + \delta_2 + \delta_0)^2} \approx \begin{cases} 4/\delta_1 & \text{if } \delta_2 + \delta_0 \ll \delta_1 \text{ and } \delta_1 \ll 1, \\ 1/\delta_1 & \text{if } \delta_2 + \delta_0 = \delta_1. \end{cases} \tag{42}$$

The power increase of the signals inside the cavity at resonance is thus of order $4\delta_1/\delta_c^2 \approx 1/\delta_c$, where again $\delta_c \equiv \delta_1 + \delta_2 + \delta_0$. For maximum enhancement, the only loss mechanism in the cavity should be the external mirror transmission or coupling $\delta_1$ through which the external signal is injected.

Similarly the peak signal transmission through a passive low-loss interferometer or cavity at resonance can be written in the form

$$\left.\frac{I_{\text{trans}}}{I_{\text{inc}}}\right|_{\omega=\omega_q} \approx \frac{4\delta_1\delta_2}{(\delta_1 + \delta_2 + \delta_0)^2} = \frac{4\delta_1\delta_2}{\delta_c^2}. \tag{43}$$

A little examination shows that this gives $I_{\text{trans}}/I_{\text{inc}} \approx 1$ if $\delta_1 \approx \delta_2$ and $\delta_0 \ll \delta_1, \delta_2$. The peak transmission through a Fabry-Perot etalon can thus approach 100%, provided that (a) the end-mirror reflectivities are closely enough matched, and (b) the internal loss $\delta_0$ is small compared to the end-mirror couplings. The actual mirror-transmission values $\delta_1$ and $\delta_2$ are not important for high peak transmission (though of course they have a critical effect on the *bandwidth* of the transmission peak).

### Reflected Waves For Low-Loss Cavities

The somewhat more complex behavior of the reflected waves for a passive cavity or interferometer can be emphasized for the low-loss situation, where all the $\delta$'s are $\ll 1$, by writing the on-resonance voltage reflectivity in the form

$$\left.\frac{\tilde{E}_{\text{refl}}}{\tilde{E}_{\text{inc}}}\right|_{\omega=\omega_q} \approx \frac{\delta_2 + \delta_0 - \delta_1}{\delta_2 + \delta_0 + \delta_1} \qquad \text{if all } \delta\text{'s} \ll 1. \tag{44}$$

This gives the limiting values

$$\left.\frac{\tilde{E}_{\text{refl}}}{\tilde{E}_{\text{inc}}}\right|_{\omega=\omega_q} \approx \begin{cases} +1 & \text{if } \delta_2 + \delta_0 \gg \delta_1, \\ 0 & \text{if } \delta_2 + \delta_0 = \delta_1, \\ -1 & \text{if } \delta_2 + \delta_0 \ll \delta_1. \end{cases} \tag{45}$$

These three limiting cases may be described as follows.

1. If the internal cavity losses plus output mirror coupling are significantly larger than the input mirror coupling, i.e., $\delta_2 + \delta_0 \gg \delta_1$, then this represents an *undercoupled cavity*. The circulating intensity inside the cavity does not build up to a large value, and the net reflectivity for the external signal is essentially just the normal reflectivity $r_1 \approx +1$ due to the input mirror alone.

2. If the input coupling is exactly equal to all the other cavity losses, i.e., if $\delta_1 = \delta_2 + \delta_0$, then this is the *impedance-matched situation*, in which the

normal reflection component of $+r_1$ from the mirror itself is just matched by a net component of $-r_1$ from the circulating energy inside the cavity. The input reflectivity is zero, and all the power delivered by the external source onto mirror $M_1$ goes into either the internal cavity losses or the transmitted output through mirror $M_2$.

3. Finally, in the *overcoupled situation*, all the other losses and coupling are small compared to the coupling at mirror $M_1$, i.e., $\delta_2 + \delta_0 \ll \delta_1$. The on-resonance circulating intensity builds up to its largest possible value $\left(\tilde{E}_{\text{circ}}/\tilde{E}_{\text{inc}} \approx 4j/\delta_1\right)$; and the out-coupled portion of this circulating intensity completely reverses the $r_1 \approx +1$ term to give a total reflectivity of $\approx -1$ instead.

Obviously, all these reflected and transmitted cavity field expressions can be generalized to active laser cavities if we use $\delta_c - \delta_m$ rather than simply $\delta_c$ in the denominator; but we leave the examination of these formulas as an exercise for the reader.

## 11.5 OPTICAL-CAVITY MODE FREQUENCIES

Since the resonance frequencies of optical cavities and interferometers are of particular interest, let us also examine in somewhat more detail the frequency properties of passive optical resonators, including the axial-mode spacing, the resonance bandwidths, and the frequency tuning or scanning possibilities in optical resonators.

### Axial-Mode Spacing

The axial modes in a passive optical resonator (without laser gain) occur, as we have already seen, at those frequencies $\omega$ which satisfy the round-trip phase condition $\phi(\omega) \equiv \omega p/c = q \times 2\pi$ in a ring cavity of perimeter $p$, or $\phi(\omega) \equiv 2\omega L/c = q \times 2\pi$ in a standing-wave cavity of length $L$, with $q$ being a (large) integer. (We have left out any atomic pulling effects at this point, but will include them in the next section.)

The resonant frequencies of the optical cavity are thus given by

$$\omega = \omega_q = q \times 2\pi \times \frac{c}{p} = q \times 2\pi \times \frac{c}{2L}, \qquad q = \text{integer.} \tag{46}$$

These axial modes form an equally spaced comb of resonant frequencies labeled by index $q$, as in Figure 11.19, with each mode separated by the *axial-mode spacing* or *axial-mode interval*

$$\Delta\omega_{ax} \equiv \omega_{q+1} - \omega_q = 2\pi \times \frac{c}{p} = 2\pi \times \frac{c}{2L}. \tag{47}$$

The frequency spacing between the axial modes of a standing-wave cavity, expressed in Hz or cycles/second, is thus $\Delta f_{\text{ax}} \equiv \Delta\omega_{\text{ax}}/2\pi = c/p$ or $c/2L$. (Since most cavities in the earlier days of lasers were standing-wave cavities, many laser workers routinely speak of the "$c/2L$ mode spacing" in a laser cavity.) These quantities must be written as $c_0/np$ or $c_0/2nL$ if it is necessary to take explicitly into account the index of refraction of any dielectric material inside the cavity.

## 11.5 OPTICAL-CAVITY MODE FREQUENCIES

(a)

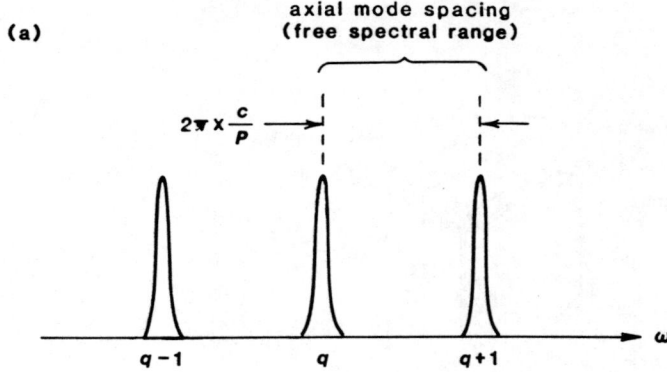

(b)

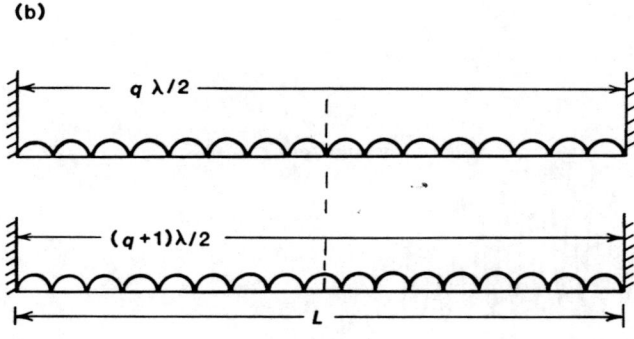

FIGURE 11.19
(a) Axial-mode resonances in an interferometer or laser cavity, and (b) the corresponding electric field distributions along the axis of the cavity.

For typical cavities in laser oscillators, this axial mode spacing will have values in the range

$$\Delta f_{\text{ax}} = \frac{c}{2L} \approx \begin{cases} 150 \text{ MHz} & \text{if } L = 1 \text{ m}, \\ 500 \text{ MHz} & \text{if } L = 30 \text{ cm}, \\ 2{,}000 \text{ MHz} & \text{if } L = 5 \text{ cm and } n = 1.5. \end{cases} \qquad (48)$$

The axial-mode intervals for laser cavities are thus typically a hundred Mhz or less for a one- or two-meter-long argon-ion or $CO_2$ laser cavity; rising to 500 MHz for a shorter, 30-cm-long He-Ne or Nd:YAG laser; and increasing to several GHz for a very short laser a few cm long. An example of the latter category might be a very simple solid-state laser with the mirror coatings evaporated directly on the ends of the rod.

Note also that semiconductor injection lasers (and also certain very short dye-laser cavities) can have cavity lengths $L$ of only a few tens to a few hundreds of microns. These axial modes become so widely spaced that it may make more sense to specify their axial-mode spacing as a wavelength spacing in Å than as a frequency spacing in MHz. A typical GaAs semiconductor diode laser with length $L = 100$ $\mu$m and index of refraction $n = 3.6$ has an axial-mode spacing of $\Delta f_{\text{ax}} \approx 4.2 \times 10^{11}$ Hz, corresponding to a wavelength interval $\Delta \lambda \approx 10$ Å at a center wavelength of $\lambda \approx 8600$ Å. Figure 11.20 shows, as a similar example, the amplified spontaneous emission spectrum from a thin film of optically pumped organic dye

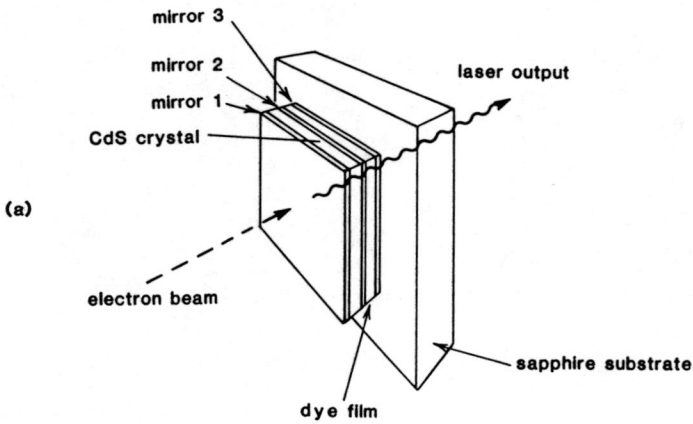

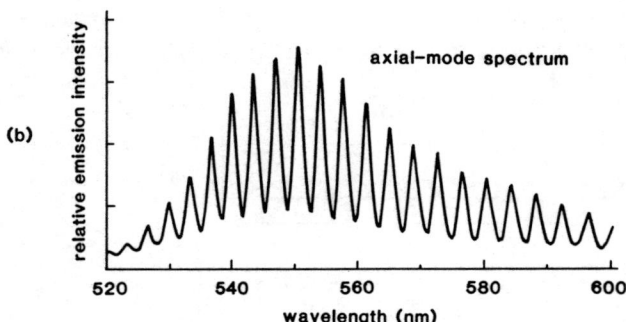

FIGURE 11.20
(a) A "Dagwood sandwich" laser, in which an electron beam pumps a thin CdS semiconductor laser, which in turn pumps a very thin film of dye laser material. (b) Amplified spontaneous emission spectrum from the dye laser segment below oscillation threshhold, showing the very widely spaced axial mode resonances. (Adapted from J. R. Onstatt, *Appl. Phys. Lett.* **31**, 818–820, December 15, 1977.)

medium filling the space between two very closely spaced high-reflectivity mirrors (see Problems).

### Interferometers and Free Spectral Range

In the jargon of optical interferometry, the $c/p$ or $c/2L$ frequency spacing is commonly called the *free spectral range*, since it represents the frequency interval between transmission peaks of a resonant interferometer. Fabry-Perot etalons, which are often used in laser experiments as resonant mirrors, filters, or bandwidth narrowing elements, typically have lengths ranging from $L = 1$ cm down to 100 $\mu$m, and are often made of materials like fused quartz, with an index of refraction $n \approx 1.46$, or sapphire, with $n \approx 1.76$. Their axial-mode spacings or

## 11.5 OPTICAL-CAVITY MODE FREQUENCIES

free spectral ranges therefore typically have values more like

$$\Delta f_{\text{ax}} = \frac{c_0}{2nL} \approx \begin{cases} 0.33 \text{ cm}^{-1} \approx 10 \text{ GHz} & \text{if } L = 1 \text{ cm,} \\ 3.3 \text{ cm}^{-1} \approx 100 \text{ GHz} & \text{if } L = 1 \text{ mm,} \\ 33 \text{ cm}^{-1} \approx 10^{12} \text{ Hz} & \text{if } L = 100 \text{ } \mu\text{m.} \end{cases} \quad (49)$$

assuming $n = 1.5$ in each case. When mode spacings become this large, it is more convenient to express them in wavenumber units, or

$$\Delta \nu_{\text{ax}} \equiv \Delta \left( \frac{1}{\lambda_0} \right)_{\text{ax}} = \frac{c_0^2}{np} \quad \text{or} \quad \frac{c_0^2}{2nL}. \quad (50)$$

A convenient rule of thumb to remember is that 1 wavenumber or 1 cm$^{-1}$ equals 30 GHz.

### Axial-Mode Number

The axial mode index $q$ in an optical cavity or interferometer is given by

$$q = \frac{\omega_q}{\Delta \omega_{\text{ax}}} = \frac{p}{\lambda_q} = \frac{L}{\lambda_q/2}. \quad (51)$$

Since the perimeter $p$ of an optical cavity is typically much longer than the optical wavelength $\lambda$, the value of $q$ is typically a very large integer, on the order of $10^6$ to $10^7$ for typical laser cavity lengths. In a ring cavity the index $q$ represents simply the number of optical wavelengths or optical cycles around the cavity perimeter, and in a standing-wave cavity it represents the number of half-optical-wavelengths along the cavity axis as illustrated in Figure 11.19(b). The values of the mode integer $q$ in very short cavities or thin etalons become more like $10^3$ to $10^5$.

Going from mode $q$ to mode $q + 1$ thus corresponds to increasing the optical frequency or decreasing the optical wavelength $\lambda$ just enough to squeeze one more half-wavelength into the standing-wave cavity length, as shown in Figure 11.19(b). Note that the standing-wave patterns of two adjacent axial modes in a linear cavity are spatially in phase at the ends of the cavity, but exactly out of phase at the center of the cavity. The spatial offset of the fields can have important implications for axial-mode competition in various kinds of lasers.

### Bandwidth, Resolving Power, and Finesse

Let us next look at the resonance bandwidths of optical cavities or interferometers. The dominant frequency dependence for both the circulating and the transmitted signals in a resonant cavity is obviously contained in the resonance denominator $1 - \tilde{g}_{\text{rt}}(\omega) \equiv 1 - r_1 r_2 (r_3 \ldots) e^{-\alpha_0 p - j\omega p/c}$. The magnitudes of $\tilde{E}_{\text{circ}}/\tilde{E}_{\text{inc}}$ and $\tilde{E}_{\text{trans}}/\tilde{E}_{\text{inc}}$ will be decreased by $\sqrt{2}$, or the corresponding intensities reduced to half their maximum values, at those frequencies for which the quantity $|1 - \tilde{g}_{\text{rt}}(\omega)|^2$ doubles compared to its value at a resonance peak. With a little algebra, this gives a FWHM bandwidth for the resonance peaks of

$$\Delta\omega_{\rm cav} = \frac{4c}{p} \sin^{-1}\left[\frac{1-g_{\rm rt}}{2\sqrt{g_{\rm rt}}}\right] \tag{52}$$

$$\approx \frac{2\pi c}{p} \times \left[\frac{1-g_{\rm rt}}{\pi\sqrt{g_{\rm rt}}}\right] = \left[\frac{1-g_{\rm rt}}{\pi\sqrt{g_{\rm rt}}}\right] \times \Delta\omega_{\rm ax},$$

where the assumption in the second line is that the magnitude of the round-trip gain, $g_{\rm rt} \equiv |\tilde{g}_{\rm rt}(\omega)|$, is not too much less than unity. The resonance bandwidth in general is obviously only a fraction of the axial-mode spacing or free spectral range, and becomes narrower the closer the round-trip gain $\tilde{g}_{\rm rt}$ comes to unity.

In classical optics the power transmission through an etalon or interferometer is often written in the form

$$\left|\frac{\tilde{E}_{\rm trans}(\omega)}{\tilde{E}_{\rm inc}(\omega)}\right|^2 = \frac{T_{\max}}{1 + (2\mathcal{F}/\pi)^2 \sin^2(\pi\omega/\Delta\omega_{\rm ax})}, \tag{53}$$

where $T_{\max}$ is the peak transmission through the etalon; $\Delta\omega_{\rm ax} \equiv 2\pi c/p$ is the free spectral range or axial-mode interval between resonances; and $\mathcal{F}$ is the so-called *finesse* of the interferometer. By comparing this with Equation 11.27 for $\tilde{E}_{\rm trans}/\tilde{E}_{\rm inc}$, we can see that the finesse is in fact just the ratio of the free spectral range to the cavity bandwidth, as given by

$$\text{finesse}, \ \mathcal{F} \equiv \frac{\pi\sqrt{g_{\rm rt}}}{1-g_{\rm rt}} \approx \frac{\Delta\omega_{\rm ax}}{\Delta\omega_{\rm cav}}. \tag{54}$$

The finesse thus gives the *resolving power* of the etalon used as a transmission filter. This resolving power obviously becomes largest in the limit of mirror reflectivities approaching unity ($r_1, r_2 \to 1$) and very small internal losses ($\alpha_0 p \to 0$). As a practical matter, a finesse of $\mathcal{F} \approx 100$ for a passive interferometer or optical cavity in the visible is considered extremely good; and a finesse this large clearly requires $1 - g_{\rm rt} \leq 0.03$, or less than 3% round-trip voltage loss.

If we use the delta factors defined in the preceding section, and include gain as well as cavity losses, the finesse can be written as

$$\mathcal{F} \equiv \frac{\pi\sqrt{g_{\rm rt}}}{1-g_{\rm rt}} \approx \frac{2\pi}{\delta_c - \delta_m}, \tag{55}$$

where $\delta_c \equiv \delta_1 + \delta_2 + \delta_0$ is the total fractional power loss per one round trip in the cavity due to *all* the cavity-loss mechanisms—mirror reflectivities plus internal losses. The laser gain, if any is present, then appears as a kind of "negative loss" term. The resonance bandwidth for the circulating signals then becomes

$$\Delta\omega_{\rm cav} \approx \frac{\Delta\omega_{\rm ax}}{\mathcal{F}} = \frac{\delta_c - \delta_m}{2\pi} \times \Delta\omega_{\rm ax}. \tag{56}$$

Obviously by adding laser gain $\delta_m$ to a passive cavity with total losses $\delta_c$ we can make the finesse $\mathcal{F}$ approach infinity, and the resonance bandwidth approach zero.

### Axial Modes in Dispersive Optical Cavities

The resonance frequency formulas given in Equation 11.46 above become slightly more complicated for *dispersive optical cavities*—cavities in which the velocity of light $c$ or the index of refraction $n$ are themselves functions of frequency. The round-trip phase-shift condition for the $q$-th axial mode in this case (again with atomic pulling or $\Delta\beta_n$ effects neglected) becomes

$$\frac{n(\omega)\omega p}{c_0} \equiv \frac{2n(\omega)\omega L}{c_0} = q \times 2\pi, \tag{57}$$

where $c_0$ is the velocity of light in free space and $n(\omega)$ the frequency-dependent refractive index.

Since the fractional spacing between axial modes is normally very small, and the index variation with frequency is also small, we can almost always expand the index of refraction about its value at some central mode $\omega_q$ in the form $n(\omega_{q+1}) \approx n(\omega_q) + n'(\omega_q) \times \Delta\omega_{\text{ax}}$, where $n' \equiv dn(\omega)/d\omega$. The axial-mode spacing is then given, to first order of approximation in $n(\omega)$, by

$$\Delta\omega_{\text{ax}} \approx \frac{2\pi c_0}{(n+n'\omega)p} = 2\pi \times \frac{1}{1+(\omega/n)(dn/d\omega)} \times \frac{c}{p}, \tag{58}$$

where $n$ and $n' \equiv dn/d\omega$ are midband values.

The correction term $(\omega/n)(dn/d\omega)$ for transparent dielectrics is usually positive, so that the effective axial-mode spacing is slightly reduced by this term. The resulting correction factor can become as large as a 10-percent reduction in axial-mode spacing over the $2\pi c/p$ value for the special case of GaAs injection lasers, in which the GaAs crystal fills the entire cavity and has an unusually large dispersion at the lasing wavelength. For most other lasers, even solid dielectric etalons, this correction is very small and is usually neglected.

### Optical Cavity Tuning

Very small changes in the length of an optical cavity can be used to tune the resonant frequencies of the cavity by sizable amounts. From the resonant-frequency expressions given earlier, we can see that changing the cavity perimeter by a small amount $\delta p$ at fixed $q$ tunes each of the axial-mode resonant frequencies by an amount

$$\frac{\delta\omega_q}{\omega_q} \approx -\frac{\delta p}{p} \approx -\frac{\delta L}{L}, \tag{59}$$

which can be rewritten as

$$\delta\omega_q \approx -\frac{\delta p}{\lambda} \times \Delta\omega_{\text{ax}} \approx -\frac{\delta L}{\lambda/2} \times \Delta\omega_{\text{ax}}. \tag{60}$$

In other words, changing the ring-cavity perimeter by one optical wavelength, or the standing-wave cavity length by one half-wavelength, shifts each of the axial modes over by an amount just equal to the spacing between axial modes. A round-trip length change of $\lambda$ causes mode $q$ to be tuned over to the frequency previously occupied by mode $q \pm 1$, depending on whether the cavity is shortened or lengthened.

### Cavity Tuning Methods

Interferometer cavities and laser oscillators are commonly tuned (or stabilized) in absolute frequency by a combination of temperature tuning (to be described below), plus the use of a piezoelectric mounting on one cavity mirror to move the mirror back and forth by a few optical wavelengths, thus scanning the absolute frequency of each axial mode by a few axial-mode intervals.

With typical piezoelectric "stacks," a few hundred volts applied to the piezoelectric element is usually sufficient to tune each resonance through one axial-mode interval. (Note that moving one of the mirrors in a ring cavity by a distance $\Delta z$ actually increases the ring perimeter by an amount $\approx 2\Delta z$, depending on how the ring is laid out; so a mirror motion of $\Delta z$ accomplishes approximately the same frequency tuning in either the ring or the standing-wave cavity.) The absolute amount of frequency tuning for an increase of one wavelength in the cavity perimeter, namely, $\Delta\omega_{ax}$, is itself inversely proportional to the cavity length; so the absolute amount of frequency tuning can become very large for very short interferometer cavities.

Magnetic drivers—in the simplest case, converted loudspeaker coils—can also be used to obtain larger mirror motions, for example, for long-wavelength infrared lasers. To first order the spacing $\Delta\omega_{ax}$ between adjacent axial modes is hardly changed by adding a few half-wavelengths $\lambda/2$ to the cavity length $L$ or perimeter $p$. Hence, to first order, changing the cavity length by a few wavelengths simply tunes the entire comb of axial modes back and forth underneath the atomic line, without noticeably changing the axial-mode spacing.

### Temperature Tuning and Thermal Drifts

Note also that in a typical optical cavity or laser structure a temperature change $\delta T$ of a few degrees or less will produce enough thermal expansion of the cavity to give a $\delta p$ of one half-wavelength or more. Optical cavities thus generally have a large thermal tuning or thermal drift rate, unless carefully stabilized in temperature.

Highly stable laser cavities are often made with the mirror spacing controlled by a rod of Invar, a steel alloy having small or even zero expansion coefficient at room temperature. Rods of quartz, carbon fiber, or zero-expansion ceramics can also be used for the same purpose. Unwanted tuning and frequency jitter of lasers caused by mechanical vibrations and acoustic noise is another very serious issue in any laser where high-frequency stability is required, and careful shock mounting and acoustic isolation may be required for highest stability.

### Scanning Optical Interferometers

Tunable interferometer cavities are often used as passive tunable filters, or as so-called scanning interferometers, for measuring laser output spectra or other optical signals.

To measure a laser signal in this fashion, we can send the signal through a passive optical cavity or scanning interferometer as illustrated in Figure 11.21, and then scan the axial modes of the interferometer back and forth in frequency across the laser spectrum by changing the passive cavity length (typically at an audio frequency rate or slower). A strong optical signal is transmitted through the passive cavity each time one of its axial-mode resonances coincides with an

## 11.5 OPTICAL-CAVITY MODE FREQUENCIES 439

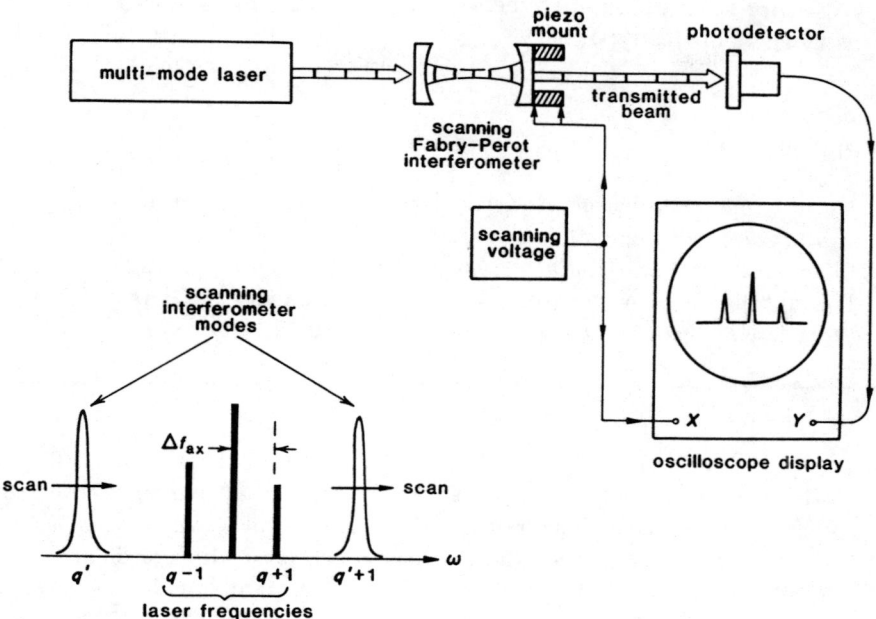

FIGURE 11.21
A scanning Fabry-Perot interferometer used as a tunable filter for observing the frequency output of a multifrequency laser. The laser oscillator has axial modes $q-1$, $q$, and $q+1$, and the interferometer cavity has axial modes $q'$ and $q'+1$.

input laser signal. This transmitted signal is then detected and displayed on an oscilloscope as illustrated in Figure 11.21.

Note that if the scanning interferometer is scanned by more than one of its free spectral ranges, another transmission resonance will be observed each time another axial mode of the scanning interferometer cavity scans across any one of the incident laser frequencies. Hence to prevent confusion or ambiguity in the results, the axial-mode spacing or free spectral range of the interferometer should be wider than the full oscillation range of the laser signal being measured, as illustrated in Figure 11.21. This requires that the interferometer cavity be shorter than the cavity of the laser generating the signals—and often considerably shorter. The shorter the scanning cavity, however, the wider its resonance linewidth and the poorer its frequency resolution for a given amount of loss. Scanning interferometer design is thus a tradeoff between free spectral range and resolving power, with a high premium given to minimizing the losses in the scanning cavity.

### Confocal Fabry-Perot Interferometers

With a conventional Fabry-Perot interferometer using planar or only slightly curved mirrors, the incident laser beam must be very precisely aligned with the axis of the interferometer in order not to excite many higher-order transverse modes of the interferometer cavity, and thus obtain very confused resonance signals. For interferometer cavities which are exactly *confocal* (meaning that the center of curvature of each mirror lies exactly on the other mirror) this difficulty does not occur, for reasons we will explain in a later chapter. Hence confocal optical cavities, because of their relative freedom from alignment restrictions, are widely used for scanning interferometers, including several commercially available instruments of this type.

## REFERENCES

Very strong second-order dispersion effects on the mode spacing in semiconductor injection lasers are demonstrated, for example, by L. F. Johnson, "Mode locking a diode laser," *J. Appl. Phys.* **51**, 6413–6414 (December 1980); or by J. P. van der Ziel and R. A. Logan, "Dispersion of the group velocity refractive index in GaAs double heterostructure lasers," *IEEE J. Quantum Electron.* **QE–19**, 164–169 (February 1983).

### Problems for 11.5

1. *Axial-mode spectrum for an optical cavity with an internal dielectric section.* Evaluate the axial-mode spectrum of a standing-wave cavity of total length $L = L_1 + L_2$ assuming that length $L_1$ of the cavity is filled with a medium having index of refraction $n_1$ and length $L_2$ is filled with medium of index $n_2$. Neglect any dispersion effects in the two dielectics, and also any reflections at the interface between the two dielectrics.

2. *Axial-mode spectrum including dispersion.* Repeat the previous problem assuming that dispersion effects in both dielectrics are in fact significant.

3. *Resonance properties of an equilateral triangular dielectric prism.* Suppose a prism in the shape of an equilateral triangle is used as a three-mirror reflective ring-type solid etalon, in which an optical beam enters at the midpoint of one face, bounces around inside the prism, reflecting at the midpoint of each of the three faces with an internal angle of incidence 30° of the normal to surface, and emerges at the same point it entered. What will be the finesse of this interferometer (a) if it is made of quartz ($n = 1.46$) and depends only on the air-dielectric reflection at each of the three faces; and (b) if the two faces other than the input-output face are silvered to give $\approx 100\%$ reflectivity?

4. *Mirror spacing in an optically pumped thin dye laser.* Figure 11.20 shows the fluorescence emission spectrum from a thin "sandwich" of optically pumped organic dye molecules filling the space between two high-reflectivity mirrors, as observed normal to the mirror surfaces. When optically pumped by another laser beam, the dye molecules become inverted and regeneratively amplify their own spontaneous emission. What is the spacing $L$ between the mirrors in this experiment? Assume the dye solution has index of refraction $n \approx 1.5$.

## 11.6 REGENERATIVE LASER AMPLIFICATION

In this section we will finally add laser gain as well as mirrors to a laser cavity, and thus finally achieve true regenerative feedback and regenerative amplification in a laser amplifier. Doing this will bring us closer to the threshold of laser oscillation—a threshold we will finally cross in the following chapter.

## Regenerative Gain Formula

Suppose that we add a laser gain medium with gain coefficient $\alpha_m(\omega)p_m$ and added phase shift $-j\Delta\beta_m(\omega)p_m$ to the interferometer model we have already analyzed. Then the formulas we have already developed will all remain valid, except that the round-trip gain $\tilde{g}_{\rm rt}(\omega)$ inside the laser cavity will be modified to

$$\tilde{g}_{\rm rt}(\omega) = r_1 r_2 (r_3 \ldots) \times \exp[\alpha_m p_m - \alpha_0 p - j\omega p/c - j\Delta\beta_m(\omega)p_m]. \qquad (61)$$

The length $p_m$ here is the total length of the active laser medium in a ring laser cavity, or twice the length of the laser medium (i.e., $p_m = 2L_m$) in a standing-wave laser cavity.

The circulating power in the laser cavity will then still be given by Equation 11.20, except that $\tilde{g}_{\rm rt}(\omega)$ will now be given by Equation 11.61. The overall regenerative gain through the cavity, or the transmission from input to output, will thus be given by

$$\begin{aligned}\frac{\tilde{E}_{\rm trans}}{\tilde{E}_{\rm inc}} &= -\frac{t_1 t_2 \exp[(\alpha_m p_m - \alpha_0 p - j\omega p/c - j\Delta\beta_m p_m)/2]}{1 - r_1 r_2 \exp[\alpha_m p_m - \alpha_0 p - j\omega p/c - j\Delta\beta_m p_m]} \\ &= -\frac{t_1 t_2}{\sqrt{r_1 r_2}} \times \frac{\sqrt{\tilde{g}_{\rm rt}(\omega)}}{1 - \tilde{g}_{\rm rt}(\omega)}.\end{aligned} \qquad (62)$$

This is the formula for *transmission gain* through the regenerative amplifier. We could write another, slightly more complex formula for the *reflection gain* $\tilde{E}_{\rm refl}/\tilde{E}_{\rm inc}$ coming back out the input end of the amplifier (and this reflection gain would in fact be a more useful way to employ a regenerative ring-laser amplifier); but we leave this task as an exercise for the reader (see Problems).

## Gain Properties of Regenerative Amplifiers

We are now going to demonstrate that as we turn up the magnitude of the round-trip gain inside the interferometer or laser cavity toward a limiting value of unity, the peak value of the transmission (and also the reflection) gain through the laser cavity will shoot upward toward infinity, as a result of regeneration in the laser cavity.

Figure 11.22 plots the log of the transmitted power gain $|\tilde{E}_{\rm trans}/\tilde{E}_{\rm inc}|^2$ versus frequency, as given by Equation 11.62, assuming a fixed gain medium with the rather largish midband gain value $\exp[\alpha_m p_m - \alpha_o p_o] = 2$, and with increasing end-mirror reflectivities ranging from $R_1 = R_2 = 0$ (i.e., no mirrors) to $R_1 = R_2 \approx 35\%$.

Two aspects of the amplification behavior in this system are immediately apparent. First, when mirrors with even rather small reflectivity are added, the overall power gain $|\tilde{g}(\omega)|^2$ at certain frequencies within the atomic-gain curve can become larger, and eventually very much larger, than the single-pass gain of the laser medium itself; and second, these high-gain frequencies occur only in very narrow bands located at the regularly spaced *axial-mode resonances* of the cavity.

The round-trip cavity length $p$ for these particular calculations has been chosen so that one of the axial-mode resonances lies exactly at the line-center frequency $\omega_a$, and several adjoining axial modes are nearby within the atomic linewidth. The regenerative amplification of these off-line-center axial modes

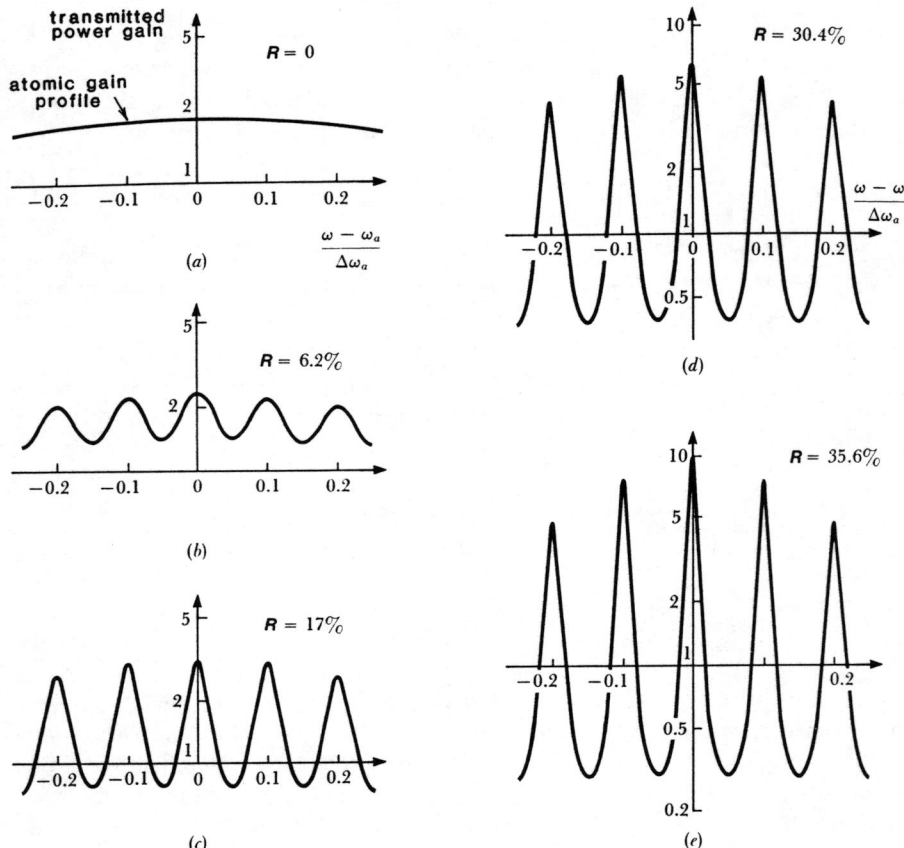

**FIGURE 11.22**
Regenerative power gain (on log scales) through a regenerative laser cavity versus frequency detuning from atomic line center. Plot (a) shows the single-pass gain through the laser medium alone, without mirrors. Plots (b) through (d) show the overall transmission gain for increasing values of end-mirror reflectivity.

is clearly reduced relative to the centermost mode, especially at higher mirror reflectivities, because of the decreasing atomic gain away from line center. We have used a fairly high atomic-gain value in order to make all the regenerative gain peaks broader and thus easier to plot; and we have used log scales in order to exhibit the large increases that the overall gain can acquire.

Figure 11.23 plots a similar set of transmission gain curves versus frequency, but this time on a linear scale and assuming fixed mirror reflectivities of $R_1 = R_2 = 40\%$, with increasing amounts of internal laser gain. Figure 11.23(a) shows the overall power gain or intensity transmission through the cavity with no internal laser gain and 4% internal round-trip power losses, demonstrating the typical resonance behavior of a passive interferometer cavity or etalon. The peak overall transmission is slightly less than unity, and the transmission peaks are spaced in frequency by the usual free spectral range of the interferometer. Adding small amounts of internal gain then rapidly converts these resonance peaks into overall regenerative gain peaks, with peak gain substantially larger than unity, as shown in Figure 11.23(b).

## 11.6 REGENERATIVE LASER AMPLIFICATION

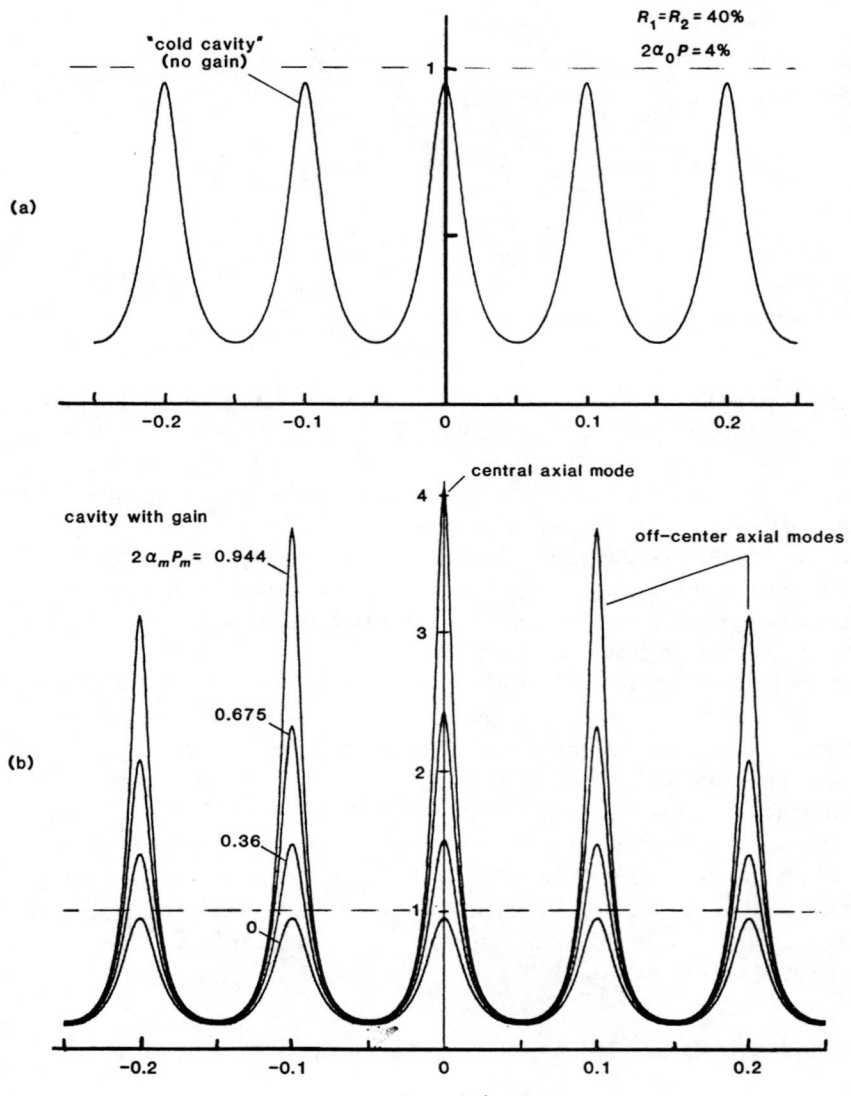

**FIGURE 11.23**
Plots of regenerative power gain similar to Figure 11.22, but on a linear scale and with fixed mirror reflectivities of $R_1 = R_2 = 40\%$. (a) Transmission through the "cold cavity" or passive interferometer without laser gain. (b) Transmission through the amplifier cavity with increasing amounts of intracavity laser gain.

### Regenerative Feedback Model

To readers familiar with regenerative feedback systems, the reasons for the behavior shown in Figures 11.22 and 11.23 will seem obvious. The laser cavity can be modeled by a typical control-system or feedback-system block diagram, as in

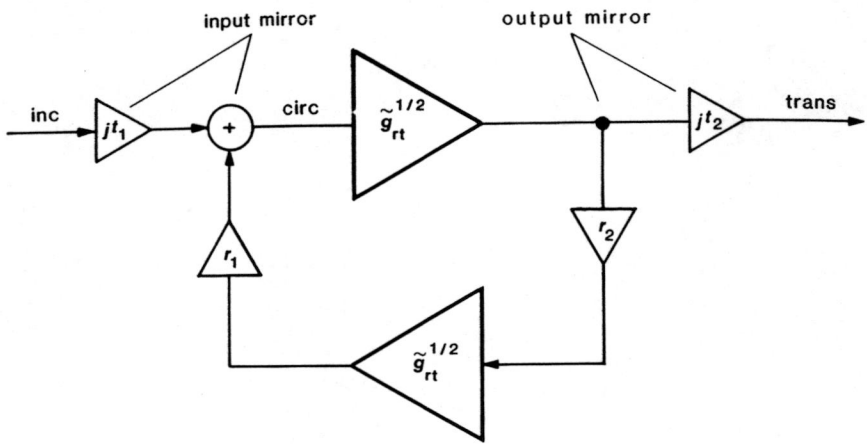

**FIGURE 11.24**
Feedback diagram describing a regenerative laser cavity.

Figure 11.24. Near the axial-mode resonances the system has a round-trip phase shift which passes through an integer multiple of $2\pi$, thus producing *positive* or *regenerative* feedback. This condition reoccurs at each axial mode; and because the round-trip path length in a laser is very long in units of wavelengths, these axial modes are very closely spaced in frequency.

As the magnitude of the round-trip gain in this feedback loop approaches unity, the overall gain from input to output of the feedback system approaches infinity; and the system in fact becomes unstable and breaks into self-oscillation when the round-trip gain just becomes unity.

This same kind of regenerative feedback can always be used to obtain large overall gain in any kind of amplifying system, using a single-pass amplifier with comparatively small gain, but applying positive feedback to obtain a very large overall gain. This feedback mechanism is effective, however, only over a very limited bandwidth, where the feedback signal has the correct phase. For example, at frequencies halfway between the axial modes, the round-trip phase shift changes so that the feedback produces instead *negative feedback* or *degeneration*. This in turn actually reduces the overall gain below what would otherwise be the net transmission through the two mirrors and the gain medium. (Note the demonstration of this in Figure 11.22.)

### Physical Interpretation: The Approach to Threshold

Another physical interpretation of the large regenerative gain observed at resonance can be given as follows. Suppose as an extreme example that the input mirror to a regenerative amplifier has a large reflectivity, say, $R = 98\%$. It may then appear that 98% of an input signal is immediately reflected back from the laser input and wasted, with only 2% entering the amplifier to be amplified.

The high-reflectivity mirrors plus the internal gain inside the cavity, however, permit any energy inside the cavity to recirculate or reverberate inside the cavity many times, extracting energy from the laser medium on each bounce, so that the recirculating wave also builds up to very large amplitudes inside the cavity, relative to the incident wave amplitude outside the cavity. This build-up plus coherent reinforcement leads to a very large increase of the internal circulating energy relative to the incident energy striking the mirror from outside. On each

## 11.6 REGENERATIVE LASER AMPLIFICATION

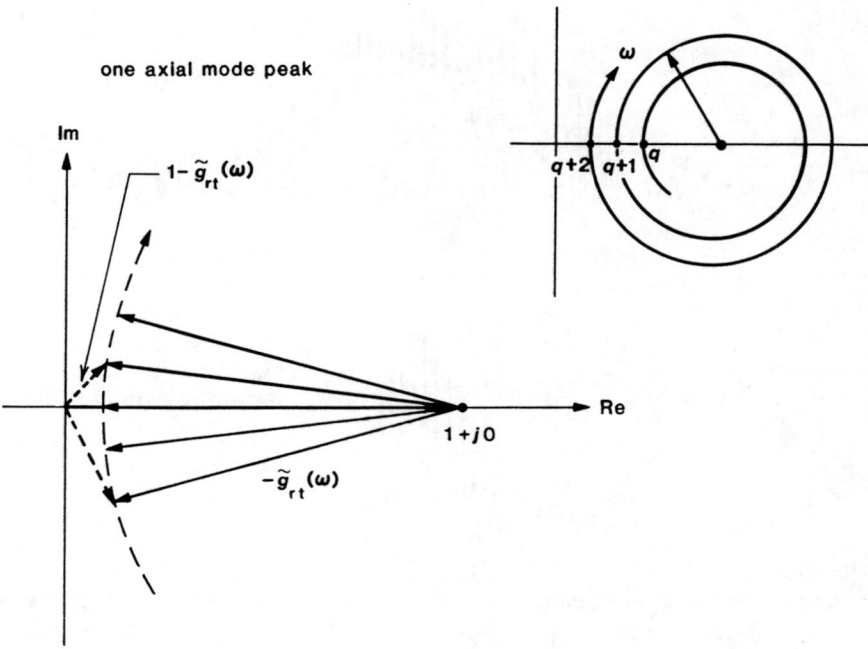

FIGURE 11.25
Geometric interpretation of axial-mode gain peaks.

end-mirror reflection, a portion of this very large circulating energy is also transmitted back out through both the input and the output mirrors, leading to the large overall transmission and reflections gains that occur at resonance.

### Geometric Interpretation

Looking again at the vector or geometric interpretation given in Figure 11.11 may also be helpful in explaining the high overall gain obtained at resonance. The crucial aspect of the overall gain expression is again the feedback denominator $1 - \tilde{g}_{rt}(\omega)$. The round-trip gain $\tilde{g}_{rt}(\omega)$ has a magnitude just less than unity (for cavities below threshold), and a phase angle which rotates rapidly with frequency in the complex plane. The length of this vector may also change slowly as it rotates because of the change in laser gain as the frequency is tuned off line center.

Figure 11.25 shows again how this vector is pivoted at the point $1 + j0$ in the complex plane, and rotates rapidly about that point. The cavity transmission gain is inversely proportional to the distance from the origin to the tip of this vector. Hence each time the tip of $1 - \tilde{g}_{rt}(\omega)$ sweeps close to the origin the gain becomes very high—but over only a brief section of the rotation cycle—and another axial-mode resonance is generated.

### Experimental Illustration

The existence of axial cavity modes, and especially the regenerative amplification which occurs at axial-mode peaks in a laser cavity below threshold, shows up most dramatically perhaps in semiconductor diode lasers. The widely

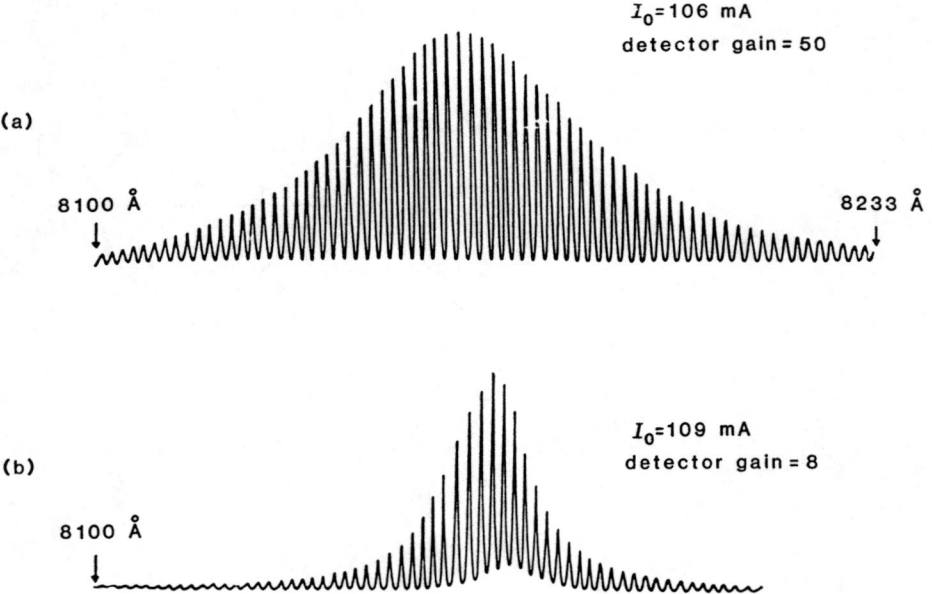

**FIGURE 11.26**
Regeneratively amplified spontaneous emission spectra from a GaAs semiconductor laser at two levels of regenerative gain just below oscillation threshold. In curve (b) the cavity is closer to oscillation threshold, and the central modes have become much stronger relative to the outer axial modes.

spaced axial modes in these lasers ($\Delta\omega_{ax} \approx 10$Å) are located within an even wider atomic transition (atomic linewidth greater than 100Å), and it becomes possible to observe these modes with a relatively low-resolution optical spectrometer.

Moreover, rather than making a measurement of regenerative laser gain versus frequency (which requires a tunable signal source and other complexities), we can simply measure the *amplified internal spontaneous emission* coming from within the laser diode itself, as this radiation is regeneratively amplified inside the laser cavity and transmitted through the end mirrors. Measurements of this type are also aided by the particularly strong spontaneous emission in semiconductor lasers.

A typical example of this kind of measurement is shown in Figure 11.26. Note that the curve in part (b) of this figure represents a slightly higher driving current through the laser, which produces slightly more internal gain, thus bringing the laser closer to oscillation threshold. (Note also the difference between the vertical scales of the two parts.) The regenerative increase of the centermost modes relative to the modes further out on the atomic gain curve thus becomes quite marked in part (b).

Note also that this kind of axial-mode structure will develop only in the emission traveling along the axial direction and coming out through the ends of the laser cavity. The spontaneous emission from the atoms coming out through the sides of the laser cavity will not exhibit this structure (except insofar as internal defects or spurious reflections may scatter some of the axial radiation out through the sides of the cavity).

**Problems for 11.6**

1. *Reflection gain of a regenerative laser amplifier.* The "reflection gain" looking into the input end of a regenerative laser amplifier may be defined as $\tilde{g}_{\text{refl}}(\omega) = \tilde{E}_{\text{refl}}/\tilde{E}_{\text{inc}}$, using the same notation as in the text. Derive a general expression for this reflection gain $\tilde{g}_{\text{refl}}(\omega)$ as a function of the mirror parameters and the round-trip gain of the laser medium, and discuss its behavior as compared to the transmission gain derived in this section.

2. *Phase angle versus frequency for a regenerative laser cavity amplifier.* The variation of the *magnitude* of the overall transmission gain $|\tilde{E}_{\text{trans}}/\tilde{E}_{\text{inc}}|$ versus frequency over several axial modes has been plotted and discussed in this section. Give a similar analysis and description of how the *phase angle* of this transmission gain versus frequency varies across a similar range spanning two or three axial modes. You may ignore any small frequency pulling effects due to the $\Delta beta_m$ term in doing this. (Hints: To understand the phase variation with frequency, consider the rotating-vector picture of regenerative feedback, and evaluate the total phase shift by evaluating the phase shifts of numerator and denominator separately and then subtracting them.)

3. *Enhanced feedback diagram.* Extend and complete the feedback diagram of Figure 11.24 to include reflection gain from the interferometer cavity, and possible input signals that might be sent into the "output" as well as "input" ends of the interferometer cavity.

## 11.7 APPROACHING THRESHOLD: THE HIGHLY REGENERATIVE LIMIT

As the laser gain is turned up (or the cavity losses are turned down) in a regenerative laser cavity, and the laser cavity approaches oscillation threshold, we can observe that:

- the regenerative gain peaks become *very high* (especially the centermost one);
- these regenerative gain peaks also become *very narrow*; and
- each regenerative gain peak approaches (as we will now show) a *fixed gain-bandwidth product*.

This section analyzes this limiting situation when an optical cavity is highly regenerative and just below threshold.

### The Approach to Threshold

Let us note once again that the quantity $1 - \tilde{g}_{\text{rt}}(\omega)$ appearing in the denominator of Equation 11.62 is one minus the round-trip voltage gain for a wave circulating around inside the laser cavity, including bouncing off the mirrors at each end. If this round-trip gain has magnitude less than unity, then the laser cavity is below threshold. This means that unless new energy is continually injected into the cavity by an external injected signal, the recirculating signal energy inside the cavity will decay in amplitude on successive round trips, so that any signals in the cavity will gradually die out. The cavity can thus not oscillate so long as $\tilde{g}_{rt} < 1$. It can, however, function as a regenerative amplifier, with potentially very high transmission or reflection gain.

If the magnitude of the internal round-trip gain approaches and then exceeds unity, however, then any circulating signals inside the cavity will grow in amplitude on each successive round trip, eventually building up to unlimited amplitudes. Of course when the signal amplitude inside the cavity grows to a large enough value, the signal fields will begin to saturate the population inversion and reduce the atomic gain. The round-trip gain will then be driven back down toward the value of exactly unity, at which point the circulating signals inside the cavity neither grow nor decay on successive round trips. The laser can then maintain a steady-state self-sustained oscillation, without any externally injected signal. The condition for the build-up of such a steady-state self-sustained oscillation in a laser cavity (starting from an injected signal, or just from spontaneous emission noise) is thus $|\tilde{g}_{rt}| > 1$.

The line where the round-trip gain magnitude $|\tilde{g}_{rt}|$ becomes just equal to unity, as shown in Figure 11.27, thus marks a boundary line between the stable, below-threshold, finite regenerative-gain region, and the unstable, above-threshold region where no steady-state operation is possible. This boundary line thus represents both *oscillation threshold* (where oscillation can just start) *and the steady-state oscillation condition for an oscillating laser*.

### The High-Gain Near-Threshold Limit

Some interesting calculations can then be made as the round-trip gain in the cavity approaches the threshold limit from below. To show this, suppose we write the round-trip gain inside a regenerative cavity in the phase-amplitude form

$$\tilde{g}_{rt}(\omega) \equiv g_{rt}(\omega) e^{-j\phi(\omega)}. \tag{63}$$

Then we can generally assume that the *round-trip gain magnitude* $g_{rt}(\omega)$ will be essentially constant across any one axial-mode peak at $\omega = \omega_q$, although the value of $g_{rt,q} \equiv g_{rt}(\omega_q)$ will change from one axial mode peak to the next, depending on where each individual peak is located within the atomic linewidth.

This is equivalent to saying that the laser gain coefficient $\alpha_m(\omega)$ within any one axial-mode peak may be approximated by its value at the center of that peak; i.e., $\alpha_m(\omega) \approx \alpha_m(\omega_q) \equiv \alpha_{mq}$ for $\omega \approx \omega_q$ for the $q$-th axial mode. We must, however, keep track of the slightly different gain values $\alpha_{mq}$ at different axial modes $q$, because very small differences in $\exp(2\alpha_{mq}p_m)$ between different axial modes can lead to large differences in the height of the overall gain peaks, especially as the centermost mode approaches threshold.

Near any single high-gain axial-mode peak located at $\omega = \omega_q$, we can also approximate the *round-trip phase shift* inside the cavity by

$$\phi(\omega) \approx \frac{\omega p}{c} = \frac{\omega_q p}{c} + \frac{(\omega - \omega_q)p}{c} = q \times 2\pi + \delta\phi(\omega), \tag{64}$$

## 11.7 APPROACHING THRESHOLD: THE HIGHLY REGENERATIVE LIMIT

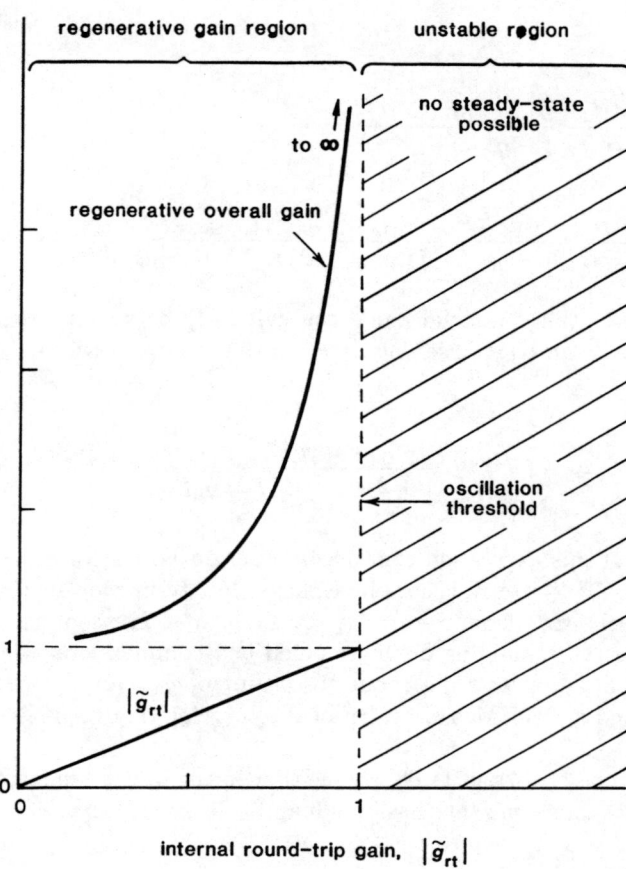

FIGURE 11.27
Regenerative gain versus internal round-trip gain, showing the approach to oscillation threshold.

where $q \times 2\pi$ is the phase shift exactly at the axial-mode peak, and the additional phase shift $\delta\phi(\omega)$ given by

$$\delta\phi(\omega) \equiv \frac{\omega - \omega_q}{c} p \approx 2\pi \times \frac{\omega - \omega_q}{\Delta\omega_{\text{ax}}} \tag{65}$$

is the small phase deviation as we tune away from resonance. In writing these expressions, we have supposed that any small atomic pulling effects due to the $\Delta\beta_m p_m$ term are incorporated into a slightly pulled value for the axial-mode frequency $\omega_q$.

### Gain and Bandwidth Near Any One Axial-Mode Peak

Suppose we consider only those frequencies within a narrow bandwidth about one such axial mode, so that $\omega \approx \omega_q$ and $|\omega - \omega_q| \ll \Delta\omega_{\text{ax}}$. We may then make the approximation that

$$e^{-j\phi(\omega)} = e^{-j\delta\phi(\omega)} \approx 1 - j\delta\phi(\omega) = 1 - j2\pi\frac{\omega - \omega_q}{\Delta\omega_{\text{ax}}}. \tag{66}$$

The transmission gain given by Equation 11.62 near this one axial-mode peak (assuming for simplicity that the round-trip path length is evenly divided between the forward and reverse paths, as in a standing-wave cavity) can then be put into the form

$$\left.\frac{\tilde{E}_{\text{trans}}}{\tilde{E}_{\text{inc}}}\right|_{\omega \approx \omega_q} = -\frac{t_1 t_2}{\sqrt{r_1 r_2}} \frac{g_{\text{rt}}^{1/2}(\omega) e^{-j\phi(\omega)/2}}{1 - g_{\text{rt}}(\omega) e^{-j\phi(\omega)}} \qquad (67)$$

$$\approx -\frac{t_1 t_2}{\sqrt{r_1 r_2}} \frac{g_{\text{rt},q}^{1/2} e^{-j\phi(\omega)/2}}{1 - g_{\text{rt},q} + j\left(2\pi g_{\text{rt},q}/\Delta\omega_{\text{ax}}\right) \times (\omega - \omega_q)}.$$

The overall gain profile for this one axial mode can evidently be well approximated by a *complex lorentzian resonance lineshape*, so that we can rewrite this expression in the form

$$\left.\frac{\tilde{E}_{\text{trans}}}{\tilde{E}_{\text{inc}}}\right|_{\omega \approx \omega_q} = -e^{-j\phi(\omega)/2} \frac{g_{0,q}}{1 + 2j(\omega - \omega_q)/\Delta\omega_{\text{3dB},q}}. \qquad (68)$$

The minus sign in front of this expression comes from our convention for mirror transmissions, and the $e^{-j\phi(\omega)/2}$ term is simply a phase shift term representing the net optical path length $\omega L/c$ from mirror $M_1$ to mirror $M_2$. The significant part of this expression is the remaining portion, which is a complex lorentzian lineshape with a peak voltage gain from input to output of $g_{0,q}$ for the $q$-th axial-mode gain peak, and a FWHM bandwidth of $\Delta\omega_{\text{3dB},q}$ for that same gain peak.

By comparing Equations 11.67 and 11.68, we see that in the highly regenerative limit any single axial-mode peak thus has a midband voltage gain magnitude given by

$$g_{0,q} \equiv \frac{t_1 t_2}{\sqrt{r_1 r_2}} \frac{g_{\text{rt},q}^{1/2}}{1 - g_{\text{rt},q}} \qquad (69)$$

and a 3 dB amplification bandwidth given by

$$\Delta\omega_{\text{3dB},q} \approx \frac{1 - g_{\text{rt},q}}{g_{\text{rt},q}} \times \frac{\Delta\omega_{\text{ax}}}{\pi}. \qquad (70)$$

As the round-trip gain magnitude $g_{\text{rt},q}$ inside the cavity approaches unity, the corresponding regenerative transmission gain through the cavity obviously becomes very high, so that $g_{0,q} \to \infty$, and the bandwidth of that gain peak becomes very narrow, so that $\Delta\omega_{\text{3dB},q} \to 0$.

### Gain-Bandwidth Product

But more than this, as the peak gain becomes very large and the bandwidth very small, their product approaches a *fixed gain-bandwidth product* given by

$$[g_0 \Delta\omega_{\text{3dB}}]_q \approx g_{\text{rt},q}^{-1/2} \times \frac{t_1 t_2}{\sqrt{r_1 r_2}} \times \frac{\Delta\omega_{\text{ax}}}{\pi}. \qquad (71)$$

But since in the high-gain limit $g_{\text{rt},q} \to 1$, we can further simplify this to

$$g_0 \Delta\omega_{\text{3dB}} \approx \frac{t_1 t_2}{\sqrt{r_1 r_2}} \times \frac{\Delta\omega_{\text{ax}}}{\pi}. \qquad (72)$$

## 11.7 APPROACHING THRESHOLD: THE HIGHLY REGENERATIVE LIMIT

In this final result, therefore, the dependence on the atomic gain and cavity losses, and even on the axial-mode coefficient $q$, drops out entirely, leaving only the cavity external-coupling parameters $r_1, r_2$ and $t_1, t_2$ in the formula.

We conclude that there is a fixed *gain-bandwidth product* in the high-gain limit for each axial-mode peak. Moreover, this gain-bandwidth product is the same for all axial modes, and depends only on the *cavity coupling* parameters, i.e., on $r_1$, $r_2$, $t_1$, and $t_2$, and not on either the laser gain or the internal cavity losses.

Note that the peak transmission gain values $g_{0,q}$ for the different axial modes across an atomic-gain curve will have significantly different values, because of the slightly different values of $\alpha_{mq}$ or $g_{\mathrm{rt},q}$ in the resonance denominators. The centermost mode will rapidly outstrip the off-center modes as its value of roundstrip gain $g_{\mathrm{rt},q}$ comes closest to unity. Earlier figures have illustrated how the peak gain of the most favored mode races up to infinity, and the bandwidth heads toward zero, as the mode approaches threshold. The off-center axial modes will not come quite as close to the threshold limit, but all will have the same gain-bandwidth product.

### Numerical Example

Does this kind of regenerative gain enhancement have practical applications in laser devices? The answer is generally no, first because the practical gain-bandwidth products are too small to be useful, and second because the necessary adjustments of the round-trip gain to achieve high overall gain are too delicate to be controlled in practical applications. Unless a laser medium has enough *single-pass* gain to be used without regeneration, it is probably not useful as a laser amplifier. On the other hand, the theory of regenerative laser amplification is very useful in understanding laser physics and particularly in understanding the manner in which laser oscillators approach oscillation threshold.

As a representative numerical example for gain-bandwidth product, we might consider a typical low-loss laser cavity with the parameters

$$\left.\begin{array}{l} r_1 r_2 = R = 0.97 \\ t_1 t_2 = T = 0.03 \\ L = 30 \text{ cm} \\ \Delta\omega_{\mathrm{ax}} = 2\pi \times 500 \text{ MHz} \end{array}\right\} \quad g_0 \Delta f_{3dB} \approx 5 \text{ MHz}. \qquad (73)$$

Suppose we want to place a 30-cm-long He-Ne laser tube, which might be able to produce somewhere between 5% and 10% power gain per one-way pass, inside this cavity, and then magnify this up by regeneration to obtain a peak-transmission gain for the centermost axial mode of $g_0 = 10$ or $g_0^2 = 100 = 20$ dB. Since this cavity has about 6% power loss through the end mirrors per round trip, and perhaps a few percent more of internal losses, the He-Ne laser tube will easily be able to bring the cavity arbitrarily close to threshold, and produce the desired 20 dB of overall gain from input to output on the centermost axial mode.

The amplification bandwidth of this axial mode will then, however, turn out to be only $\Delta f_{3dB} \approx 500$ kHz! The usefulness of a bandwidth of a few hundred kHz, even though it may be at an optical carrier frequency, seems dubious. In fact, even to measure this bandpass will require a frequency stability of $\Delta f/f_0 \approx 5 \times 10^5 / 5 \times 10^{14} \approx 1 \times 10^{-9}$ between the signal source and the laser amplifier.

Regenerative optical amplifiers are thus useful as a source of insight into the physics of laser oscillation, but seem not to have practical applications.

The concept of a fixed gain-bandwidth product which we have derived here applies, of course, not only to laser amplifiers, but also to any type of regenerative amplifier in any frequency range. Given any kind of electronic or acoustic or mechanical amplification process, no matter how weak its gain, we can always employ positive feedback to increase the overall gain by any desired amount. If the feedback loop has a long time delay, however, as is inherent in a laser merely from the propagation time around the cavity, the total phase shift in the feedback loop will be large and will change rapidly with frequency. This in turn will inherently limit the bandwidth or, more precisely, the gain-bandwidth product of the regeneratively magnified amplification.

### Regenerative Noise Amplification in a Laser

The fixed gain-bandwidth product for a regenerative laser amplifier operating just *below threshold* can be used to derive, at least in a heuristic fashion, one of the most famous formulas in laser theory, the so-called Schawlow-Townes formula for the spectral linewidth caused by quantum noise of a laser oscillator operating *above threshold*.

To derive this, we must first note that a regenerative laser amplifier—or indeed any other kind of coherent optical amplifier—will have a certain finite amount of noise because of spontaneous emission from the upper-laser-level atoms inside the amplifier. For a laser amplifier of any kind, regenerative or single pass, this finite amount can be represented by an equivalent input noise power, which we view as coming into the input of the amplifier, with an input noise power spectral density given by

$$\frac{dP_n}{d\omega} = \frac{N_2}{N_2 - N_1} \times \hbar\omega. \tag{74}$$

In other words, the equivalent input noise power to the amplifier is equivalent to one input photon per second per cycle of bandwidth, multiplied by an *excess noise factor* $N_2/(N_2 - N_1)$. This excess noise factor is unity if the lower laser level is empty, so that $N_2 - N_1 = N_2$. It becomes larger than unity if $N_1$ is finite, because then more upper-level atoms and hence more spontaneous emission will be present for the same net inversion and gain.

Consider now a regenerative laser amplifier pumped right up to the oscillation threshold point, with no coherent input signal applied. Even with no input signal present, this laser will still amplify the equivalent input noise within its 3 dB amplification bandwidth $\Delta\omega_{3dB}$. (Really, of course, it is regeneratively amplifying its own spontaneous emission generated *within* the laser cavity.) The effective rectangular bandwidth of a lorentzian amplifier with FWHM bandwidth $\Delta\omega_{3dB}$ is in fact $(\pi/2) \times \Delta\omega_{3dB}$. Hence the total amplified noise output power from the amplifier very close to threshold will be given by

$$\begin{aligned} P_{\text{out}} &= G_0 \times \frac{\pi \Delta\omega_{3dB}}{2} \times \frac{dP_n}{d\omega} \\ &= \frac{N_2}{N_2 - N_1} \times \frac{\pi G_0 \Delta\omega_{3dB} \hbar\omega}{2}, \end{aligned} \tag{75}$$

where $G_0 \equiv g_0^2$ is the midband overall transmission gain through the laser cavity.

## 11.7 APPROACHING THRESHOLD: THE HIGHLY REGENERATIVE LIMIT

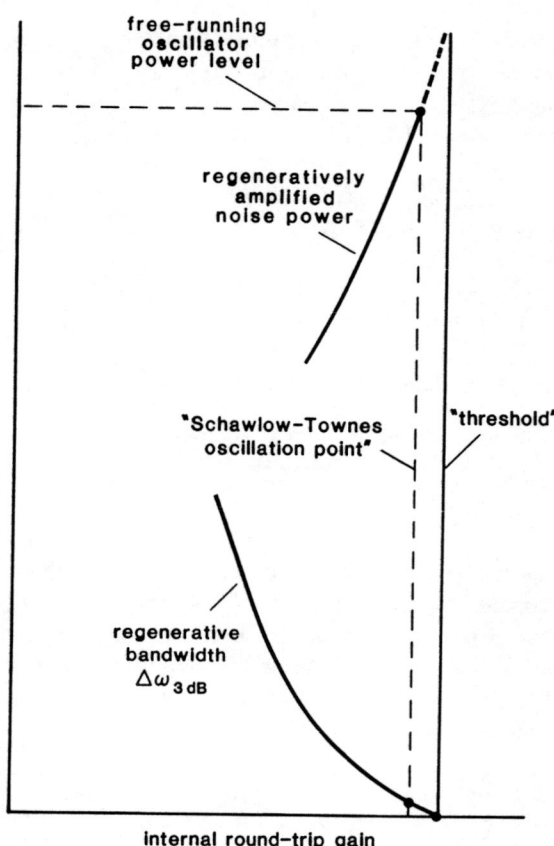

FIGURE 11.28
Schawlow-Townes model of a laser oscillator as a very high-gain, very narrowband regenerative noise amplifier operating just below the threshold point.

As we bring the cavity closer and closer to threshold, the overall gain $G_0$ will become very large; the bandwidth $\Delta\omega_{3dB}$ will become very narrow; and the noise output from the laser will become an increasingly powerful but increasingly narrowband amplified noise signal. Suppose we bring the cavity extremely close to oscillation threshold, as shown in Figure 11.28. One partially correct, but incomplete, way of describing what happens as the laser comes very close to the threshold point is the following: When the amplified noise output power comes close to the available power that can be extracted from the laser gain medium, then the laser gain medium begins to saturate. As a result of this gain saturation, the overall power gain $G_0$ will no longer increase beyond the point where the (very) narrowband amplified noise output equals the potential output oscillation power from the laser.

### The Schawlow-Townes Formula

From this (partially correct) viewpoint, the laser oscillator is simply a very high-gain, very narrow bandwidth, amplified spontaneous emission noise source operating just the slightest bit below the exact threshold point, with an overall power output given by Equation 11.75, with $P_{\text{out}} = P_{\text{osc}}$, where $P_{\text{osc}}$ is the free-running power output that the laser oscillator can deliver. But we also have a constant gain-bandwidth expression connecting $g_0 \equiv G_0^{1/2}$ and $\Delta\omega_{3dB}$. For simplicity let us assume a laser cavity with very small internal losses and reasonably small external coupling, so that we can write $r_1 r_2 \approx 1$ and $t_1 t_2 \approx \delta_c$, where $\delta_c$ is the cavity loss factor derived earlier. By using the cavity $Q_c$ definition from Equation 11.40, we can then rewrite the gain-bandwidth product in the form

$$G_0^{1/2} \Delta\omega_{3\text{dB}} \approx \frac{2\omega_a}{Q_c} = 2\Delta\omega_c, \tag{76}$$

where $\Delta\omega_c \equiv \omega/Q_c$ is the "cold cavity" bandwidth of any one axial mode in the laser cavity due to its external coupling.

Combining Equations 11.75 and 11.76 then gives the interesting result that

$$\Delta\omega_{\text{osc}} = \Delta\omega_{3\text{dB}} \approx (2) \times \frac{N_2}{N_2 - N_1} \times \frac{\pi\hbar\omega\Delta\omega_c^2}{P_{\text{osc}}}, \tag{77}$$

where $\Delta\omega_{\text{osc}}$ is to be interpreted as the spectral width of the highly amplified noise coming out of the laser operating at or above threshold. This formula for the "noise bandwidth" of the laser output is commonly referred to as the "Schawlow-Townes formula" for a laser oscillator. Note that this linewidth depends only on the cold-cavity bandwidth $\Delta\omega_c$ of the laser cavity, and on the power level $P_{\text{osc}}$ at which the laser oscillates above threshold.

### More Correct Description

We have put the factor of 2 in Equation 11.77 in brackets to emphasize the way in which this formula is partially correct and partially incorrect. When a laser reaches its oscillation threshold, there is, as we noted in Chapter 1, a qualitative change in the character of the laser output spectrum. The laser changes over just at threshold from being a very narrowband but still essentially *incoherent gaussian noise source*, with large (but slow) fluctuations in both amplitude and phase, to being a *coherent sinusoidal oscillator*, with a highly stabilized phasor amplitude, but still with random noise-like but very slow fluctuations or drifts in the oscillation phase.

If we simply delete the bracketed factor of 2 in Equation 11.77, *this equation still correctly predicts the spectral bandwidth of the coherent laser oscillation above threshold*, as caused by random phase fluctuations in the laser output. In other words, the Schawlow-Townes result, reduced by a factor of two, is still correct in the nonlinear region above threshold, even though it is derived by using a linear below-threshold model.

The phase fluctuations and the consequent oscillation spectral broadening caused by spontaneous emission in a laser oscillator above threshold (along with some small but still observable residual amplitude fluctuations) are commonly referred to as *quantum noise fluctuations* in the laser. These quantum amplitude and frequency fluctuations in ordinary lasers are usually completely masked by much larger fluctuations due to mechanical vibrations, acoustic noise, thermal drift, and other "technical noise sources." Quantum amplitude and frequency fluctuations have been measured, however, in excellent agreement with the Schawlow-Townes formula, by careful measurements both on highly stabilized gas lasers, and on semiconductor injection lasers, where these fluctuations can be substantially more noticeable.

## REFERENCES

Despite the technical difficulties in dealing with such high gains and narrow linewidths, painstaking measurements demonstrating in detail the fixed gain-bandwidth product in a He-Ne laser with a gain-bandwidth product of ≈1 MHz were once carried out by G. Herziger, G. Makosch, and J. Weber, "Verstärkung, Bandbreite und Photonendichte beim He-He-Laserverstärker für die Wellenlaenge $\lambda = 6328$Å," *Z. Physik* **228**, 89–98 (1969). Peak gain in the experiment was varied from $G = 1$ up to $G = 40{,}000$, and the corresponding bandwidth varied from 1 MHz down to 5 kHz.

## Problems for 11.7

1. *Power transmission through a laser cavity halfway between axial modes.* Suppose one sets up two high-reflectivity mirrors in series, but with the mirrors misaligned in angle by enough to destroy any regenerative or resonance effects. The total power transmission through these two mirrors in cascade is then simply $T_1 T_2$. Examination of the results given in this chapter will show that the power transmission through a highly regenerative traveling-wave amplifier at a frequency halfway between two axial-mode resonances turns out to be approximately 6 dB lower than this value for the end mirrors by themselves with no regenerative effects—in other words, regeneration with the wrong phase actually reduces the power transmission from input to output. Verify this, and explain why the reduction factor is roughly 6 dB.

2. *Gain sensitivity of a regenerative laser amplifier.* The gain sensitivity of a laser amplifier to small changes in the active laser medium may be defined as the fractional change $\delta G/G$ in the overall midband power gain divided by the fractional change $\delta\alpha_m/\alpha_m$ in the laser-medium gain coefficient, for small changes $\delta\alpha_m$ and $\delta G$. (The change in $\alpha_m$ might be caused, for example, by a small change in the laser's pumping rate.) Calculate and compare the overall gain sensitivities for a round-trip laser amplifier and for a highly regenerative laser amplifier, as a function of the overall midband power gain in each case. (Disregard saturation effects in both cases.)

3. *Skirt selectivity of a regenerative laser amplifier.* In some applications it can be important to know not only the 3 dB bandwidth of an amplifier, but also how fast the amplifier gain falls off outside this bandwidth (for example, in order to find out how much a strong interfering signal outside the amplifier passband will be suppressed). Analyze and discuss this so-called "skirt selectivity" performance for a highly regenerative laser amplifier, considering the maximum rejection halfway between axial modes, and the linewidths for the gain to drop 10, 15, 20 dB, etc., below the peak gain.

4. *Output versus input for a regenerative laser amplifier with saturable internal gain.* The following problem is slightly tricky, but also instructive.

    Suppose a highly regenerative ring interferometer cavity has an input mirror reflectivity $R_1 = 95\%$, an internal round-trip voltage attenuation $\alpha_m p = -4\% = -0.04$ due to an *absorbing* atomic transition, and no other internal losses (i.e., $\alpha_0 = 0$). The output mirror reflectivity is $R_2 = 100\%$. The internal atomic absorption $\alpha_m$ saturates in homogeneous fashion with a saturation intensity $I_{\text{sat}}$.

An input signal with intensity $I_{inc}$ tuned to the resonant frequency of the resonant cavity is sent against the input mirror (mirror $M_1$) of the cavity from outside. The problem is to plot the reflected intensity $I_{refl}$ that is reflected back from the input mirror as a function of the input intensity $I_{inc}$ of the input signal.

Hints: (1) You clearly need to establish a relationship between the input intensity $I_{inc}$ and the intensity $I_{circ}$ inside the cavity, since it is the latter intensity that saturates the atomic absorption. (2) Since the round-trip gain inside the cavity is always close to unity, you can assume that the intensity inside the cavity has essentially the same value all around the cavity. (3) The system may exhibit bistable or bivalued behavior in its output-versus-input relationship.

5. *Regenerative gain peaks for off-line-center axial modes.* In a certain laser the axial-mode spacing $2\pi c/2L$ is exactly one-fifth the lorentzian atomic linewidth $\Delta\omega_a$, and the centermost axial mode (the $q$-th axial mode, say) is located exactly at line center. The mirror reflectivity at each end of the laser is $R_1 = R_2 = 0.95$, and there are no internal cavity losses. Find and plot the overall power gains at the peaks of the three closest off-line-center axial modes (i.e., the $q+1$, $q+2$, and $q+3$ modes) versus the peak power gain of the centermost axial mode, as the round-trip gain inside the laser cavity is slowly turned up to unity.

You should discover that the peak gains of the off-center modes go to finite values as the gain of the centermost mode goes to infinity (i.e., to oscillation threshold). What gain values in dB do the off-center modes approach as the gain of the central mode approaches infinity?

6. *Transient reflection from a resonant cavity.* Suppose a sinusoidal signal with a step-function turnon (that is, $\mathcal{E}_{inc}(t) = 0$ for $t < 0$ and $\mathcal{E}_{inc}(t) = \sin(\omega_0 t)$ for $t \geq 0$) is incident on a highly regenerative optical cavity or interferometer, with the signal carrier frequency $\omega_0$ tuned to one of the axial mode resonances of the cavity. Use the highly regenerative approximation to find the complex voltage reflection coefficient of the cavity as a function of frequency for $\omega$ near $\omega_0$; and then use Fourier transform or Laplace transform methods to find the reflected signal $\mathcal{E}_{refl}(t)$ from the cavity as a function of time for $t \geq 0$. Discuss the physical significance of your result, and compare the descriptions of the reflected signal in the time and frequency domains.

(Hint: A signal with a sharp step-function leading edge will have a frequency spectrum which actually spreads out over several axial modes. Using the highly regenerative approximation and considering only the one axial mode with which the signal is in resonance, as in this problem, will in essence filter out the discrete step-wise variation of the reflected signal as the circulating signal travels around and builds up on successive round trips.)

7. *Approach to threshold in the Schawlow-Townes model.* For a typical set of parameters, how close is the round-trip gain magnitude $g_{Rt}$ to unity at the operating point implied by the Schawlow-Townes laser model illustrated in Figure 11.28. What is the fractional deviation of $g_{Rt}$ from unity?

# 12

# FUNDAMENTALS OF LASER OSCILLATION

This chapter brings us finally to the complete laser oscillator: atoms, plus pumping and population inversion, plus signals and amplification, plus mirrors to provide feedback and oscillation.

In this chapter we will develop formulas for some of the simpler aspects of laser operation, including the population inversion required to reach oscillation threshold; the pumping power density required to produce this inversion; the laser power output, and its dependence on output coupling and pumping power in simple cases; the difference between homogeneously and inhomogeneously broadened lasers; and the atomic-frequency pulling effects in a laser oscillator.

In Chapter 13 we will then develop a set of coupled rate equations which link cavity photons to laser atoms, and laser atoms to cavity photons. Using these equations we will explore laser oscillation buildup and the remarkable threshold properties characteristic of the laser oscillator.

## 12.1 OSCILLATION THRESHOLD CONDITIONS

The basic requirement either for just reaching laser oscillation threshold, or for maintaining steady-state laser oscillation, is that *the round-trip gain inside the laser cavity, including mirror reflections, must be exactly unity*, modulo an integer number of multiples of $e^{-j2\pi}$. Only if the round-trip gain is exactly unity can the system maintain steady-state oscillation, in which the circulating signal inside the cavity neither grows nor decays on successive round trips. (This assertion does leave out some extremely minute effects due to spontaneous emission or noise in the laser cavity, which are totally negligible in any of the following discussions.)

In the notation developed in Chapter 11, unity round-trip gain requires that

$$\tilde{g}_{rt}(\omega) \equiv r_1 r_2 (r_3 \ldots) \times \exp\left[\alpha_m(\omega) p_m - \alpha_0 p - j\frac{\omega p}{c} - j\Delta\beta_m(\omega) p_m\right]$$
$$= \exp[-jq2\pi], \qquad (1)$$

where $q$ is an integer. This can in turn be separated into an *amplitude or magnitude condition*, which says that at steady-state the round-trip gain must have magnitude unity, or

$$r_1 r_2 (r_3 \ldots) \times \exp[\alpha_m(\omega) p_m - \alpha_0 p] = 1, \quad (2)$$

and a *phase or frequency condition*, which says that at steady state the round-trip phase shift must be an integer multiple of $2\pi$, or

$$\frac{\omega p}{c} + \Delta\beta_m(\omega) p_m = q \times 2\pi. \quad (3)$$

The first of these conditions determines the population inversion density, and hence the pumping rate, needed to reach oscillation threshold. The second condition determines primarily the frequency $\omega$ at which the laser must oscillate.

### Threshold Inversion Density

Suppose we rewrite the amplitude condition in terms of the round-trip power gains and losses, since we usually speak of power gains and mirror power reflectivities $R_i$ rather than voltage reflectivities $r_i$ in practical discussions. The gain coefficient required to just reach threshold in the laser cavity is then given by

$$2\alpha_m(\omega) p_m = 2\alpha_0 p + \ln\left[\frac{1}{R_1 R_2 (R_3 \ldots)}\right], \quad (4)$$

or, in terms of the "delta notation" we introduced in the previous chapter,

$$2\alpha_m(\omega) p_m] \equiv \delta_m(\omega) = \delta_0 + \delta_1 + \delta_2 + \cdots] \equiv \delta_c. \quad (5)$$

Now, we can recall from earlier chapters that the laser gain coefficient for a lorentzian atomic transition is given by

$$\alpha_m(\omega) = \frac{3^*}{4\pi} \frac{\gamma_{\text{rad}} \lambda^2}{\Delta\omega_a} \frac{\Delta N}{1 + [2(\omega - \omega_a)/\Delta\omega_a]^2} \quad (6)$$

or by a very similar expression for a gaussian atomic transition. The inversion density required either to reach threshold, or to maintain steady-state oscillation, in a cavity mode located at midband ($\omega = \omega_a$) on a lorentzian atomic transition, is thus given by

$$\Delta N = \Delta N_{\text{th}} \equiv \frac{2\pi}{3^*} \times \frac{\Delta\omega_a}{\gamma_{\text{rad}}} \times \frac{1}{\lambda^2} \times \frac{\delta_c}{p_m}. \quad (7)$$

In order to have achieve oscillation with the lowest possible inversion density we want to have a laser system with the following characteristics:

- A narrow atomic linewidth $\Delta\omega_a$.
- A strong radiative decay rate $\gamma_{\text{rad}}$.
- A long wavelength $\lambda$.
- Low cavity losses and output coupling, $\delta_c$.
- A long gain medium $p_m$.

The dependence on wavelength in particular agrees with the general observation that infrared lasers are usually fairly easy to obtain, whereas visible and UV lasers become progressively more difficult.

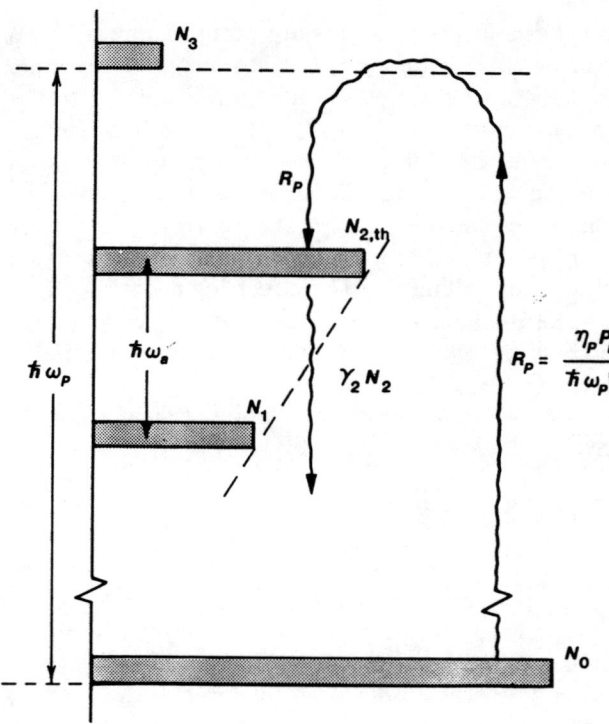

FIGURE 12.1
A general model for calculating pumping power density required in a typical four-level laser oscillator.

Not all of these criteria are essential; some are not even always desirable. For example, many useful laser materials have wide linewidths and small radiative decay rates, and many lasers work best with very large output couplings. They do, however, at least indicate which properties will make achieving laser action more or less difficult.

This same threshold condition can also be expressed much more simply in terms of the transition cross section $\sigma$ given by $2\alpha_m = \Delta N \sigma$, which leads to the particularly simple result

$$\Delta N_{\text{th}} = \frac{\delta_c}{\sigma p_m}. \tag{8}$$

A large transition cross section and a small threshold inversion obviously go together.

### Threshold Pump Power Density

Although the threshold inversion density is important, of more practical importance is the *pump power density* required to achieve this threshold inversion in a practical laser. We can express this threshold pump power density in a fairly general form, more or less independent of the particular pumping mechanism that is employed, using the general laser pumping model shown in Figure 12.1.

First of all, obtaining the threshold inversion density given by Equation 12.7 will require achieving an upper-laser-level population density $N_{2,\text{th}}$ that is greater than this threshold inversion $\Delta N_{\text{th}}$ by some ratio that depends, as shown in Figure 12.1, on how much population accumulates in the lower laser level, which depends in turn on how rapidly the lower laser level empties out.

Suppose that in order to achieve inversion atoms are pumped upward by a pumping power density (power per unit volume) $P_p/V$ into one or more upper pump levels $E_3$, from which some fraction of these atoms fall down into the upper laser level $E_2$, with a net pumping rate of $R_p$ atoms per unit volume per second into this upper laser level. Let the effective energy gap across which atoms must be lifted by the pumping mechanism be given by $\hbar\omega_p$, where $\omega_p$ is the "pumping frequency" (whether or not the pumping is actually done optically using photons of energy $\hbar\omega_p$ or by some other mechanism); and let the fraction of excited atoms that actually end up falling into the upper laser level be given by a *pumping efficiency* $\eta_p$. (The remainder of the pumping power is wasted, either in lifting up atoms which drop back down through other paths, or simply in added heat dissipation in the medium.)

The effective pumping rate $R_p$ and the upper level population density $N_2$ which is created by this pumping rate are then given by

$$N_2 = \frac{R_p}{\gamma_2} = \frac{\eta_p P_p}{\gamma_2 \hbar \omega_p V}, \tag{9}$$

where $P_p/V$ is the total pumping power density (power per unit volume) going into the laser medium, and $\gamma_2$ is interpreted as the total downward decay rate out of level 2 due to all decay mechanisms.

By combining this expression with the threshold inversion expression 12.7, we can find that a quite general expression for threshold pump power density is

$$\frac{P_{p,\text{th}}}{V} = \frac{1}{\eta_p} \times \frac{N_{2,\text{th}}}{\Delta N_{\text{th}}} \times \frac{\omega_p}{\omega_a} \times \frac{\gamma_2}{\gamma_{\text{rad}}} \times \frac{4\pi^2}{3^*} \times \frac{\hbar \Delta\omega_a}{\lambda^3} \times \frac{c\delta_c}{p_m}. \tag{10}$$

This is a very important expression for calculating—or at least estimating—the performance characteristics and pumping requirements of a given laser system.

### Practical Laser Pumping Requirements

The first four factors in this expression are all dimensionless ratios with values which may vary greatly from laser system to laser system, but which are never smaller than unity. Each factor can thus provide a criterion for searching for good laser systems. Each ratio does in fact come close to unity in certain particularly favored laser systems, though each of them is also more commonly much worse than unity in other systems.

If we look at each of these factors in turn, we can see that the criteria for a good laser medium—or at least one with low pump-power requirements—include, first, a good pumping efficiency $\eta_p$ in terms of atoms lifted up per unit pump power applied to the medium. Systems which show up well on this criterion include optically pumped dye lasers, solid state lasers, semiconductor lasers, and some gas lasers such as chemical lasers and the $CO_2$ laser. Systems not favored by this criterion include many common gas-discharge lasers, such as the He-Ne or ion lasers, where only a small part of the pumping energy goes into lifting atoms into the desired upper laser levels.

In a He-Ne or Argon-ion gas laser, for example, the number of atoms actually pumped into the upper laser level is very small compared to photon units of electrical energy dissipated in the gas discharge medium, so that $\eta_p \ll 1$. In these lasers most of the electrical energy input into the laser gas goes either into exciting unwanted atomic levels or just into heating up the gas atoms and the electrons. In a laser-pumped dye laser, by contrast, the ratio of dye molecules pumped up to the upper laser level to laser pumping photons absorbed can be almost unity. Of course, these laser-generated pumping photons are themselves rather expensive photons to obtain.

A good laser system should also have a lower laser level that empties out rapidly, so that $N_1 \approx 0$ and $\Delta N \approx N_2$ (which, of course, works against three-level lasers like ruby). A good laser system should also have its upper pumping level not far above the upper laser level, and its lower laser level close to (but not right at) the ground level, so that the pump photons can be as small as possible. Again this favors dye lasers, most solid-state lasers, and some gas lasers, and works strongly against other gas lasers like He-Ne and argon lasers, where we must lift the atoms up to levels that are $\approx 20$ eV or more above ground level, in order to get out $\approx 2$ eV photons across the laser transition.

Finally, we want a laser transition which is as close to purely radiative on the laser transition as possible, with no other radiative or nonradiative decay rates to leak off atoms, so that $\gamma_2 \approx \gamma_{rad}$. This favors the ruby laser, organic dye lasers, and to a lesser extent other solid-state and gas lasers.

Note in particular that if the condition $\gamma_2 \approx \gamma_{rad}$ is satisfied, then the actual value of the transition strength $\gamma_{rad}$ drops out of the pump power density expression. A strong transition with a large $\gamma_{rad}$ needs a smaller population inversion, according to our above results; but at the same time the faster decay makes this inversion harder to maintain continuously. Hence ruby and Rhodamine 6G are both very good visible laser systems, even though their values of $\gamma_{rad}$ differ by some 6 orders of magnitude.

After all these factors are taken into account, the remaining factors in obtaining laser inversion are the final two terms in Equation 12.10. Laser action is always harder to obtain the wider the atomic linewidth (though at the same time wide linewidth is essential if we want to have tunable laser action). We also see once again that the difficulty in obtaining laser action goes up very rapidly as the laser wavelength gets shorter, with the pump power density rising, other factors being equal, at least proportional to $1/\lambda^3$. In fact, in doppler-broadened gas lasers the doppler linewidth itself tends to increase as $1/\lambda$, and hence the wavelength dependence of $P_{p,th}$ will be more like $1/\lambda^4$. A genuine X-ray laser will be very difficult to obtain, both or this reason and because of other factors as well, not the least of these being the lack of good X-ray mirrors.

Finally, the last term in the threshold pumping condition contains the cavity factors, i.e., to reach oscillation in a weak laser system we want the lowest possible cavity losses, and the longest possible gain medium.

## Problems for 12.1

1. *Off-resonance regenerative amplification through an oscillating laser?* A standing-wave laser cavity with mirror reflectivities $R_1 = R_2 = R$ is oscillating in steady state at its centermost axial mode. A separate signal tuned off by one-half of an axial-mode spacing is sent into one end of this cavity. What is the overall transmission gain for this signal out the opposite end of the cavity? (The laser medium is homogeneous, and its atomic linewidth is wide compared to the axial mode spacing.) What happens to this transmission gain if the mirror reflectivity $R$ is small? Explain physically.

## 12.2 OSCILLATION FREQUENCY AND FREQUENCY PULLING

Let us next look at the frequency characteristics of laser oscillators: whether a laser will oscillate only in a single mode and at a single frequency, or in many modes at once, and also what the exact frequencies of these oscillations will be if atomic pulling effects are included.

The axial cavity mode whose frequency is located closest to the center of an atomic transition will of course normally see the highest gain, and will thus normally reach oscillation threshold first, before other modes located further from the atomic line center. Suppose, however, that we turn up the gain or the pump power still further, beyond the point where the first cavity mode reaches threshold. Will additional axial (or transverse) modes then also begin oscillating? The answer to this question is that there are in general two idealized or limiting types of laser oscillation behavior for the remaining cavity modes, depending on whether the laser transition is *homogeneously* or *inhomogeneously broadened*. We will first consider these two general classes of laser oscillation, before discussing the atomic pulling effects that occur for either class.

### Ideal Homogeneous Lasers: Single Frequency Oscillation

In an ideally *homogeneous laser transition*, the atomic lineshape is fixed and identical for all the atoms in the laser medium. The magnitude of the gain and phase shift measured at any given frequency will move up and down as the population inversion $\Delta N$ varies; but the lineshapes of $\chi''(\omega)$ and $\chi'(\omega)$ versus frequency will remain unchanged—in essence the whole lineshape moves up and down together.

Suppose the midband gain in such a homogeneous laser medium is increased until the axial mode closest to line center just reaches threshold (i.e., gain just equals losses), as illustrated in Figure 12.2(a). This mode $q$ can then begin to oscillate, whereas all the other modes ($q - 1$, $q + 1$, and $q + 2$) are still below threshold and cannot oscillate. (Note that the gain actually exceeds the losses in the center portion of the atomic line, but there is no cavity mode located there to build up to oscillation.)

Even if we pump this laser harder, we cannot push the gain profile further up so as to cause the $q + 1$ mode to oscillate, as illustrated by the dashed gain profile in Figure 12.2(a). Such oscillation is not possible, at any rate, on a cw or steady-state basis, because then gain would exceed loss for the $q$-th mode, and the amplitude of this mode would grow continuously on successive round trips.

It may of course be possible to push the gain for several modes above the steady-state or threshold value on a transient basis, during initial turn-on or pulsed operation of the laser. Note also that the centermost axial mode may not be the first or preferred mode to oscillate, if special mode-control methods are used to increase the losses of this mode relative to another axial mode further out on the atomic gain profile.

An ideally homogeneous laser, therefore, should oscillate under steady-state conditions in only *one preferred mode*, the first mode to reach threshold; and the gain in the laser medium will be clamped at the level that just causes that mode to reach threshold. Pumping harder will make that preferred mode oscillate more strongly, as we will see very shortly, but will not increase the gain or start new modes oscillating.

## 12.2 OSCILLATION FREQUENCY AND FREQUENCY PULLING

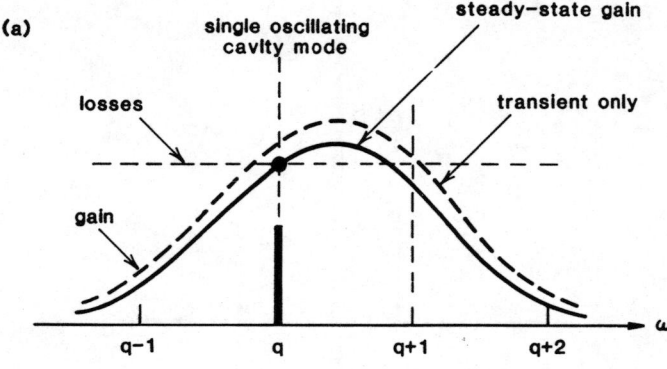

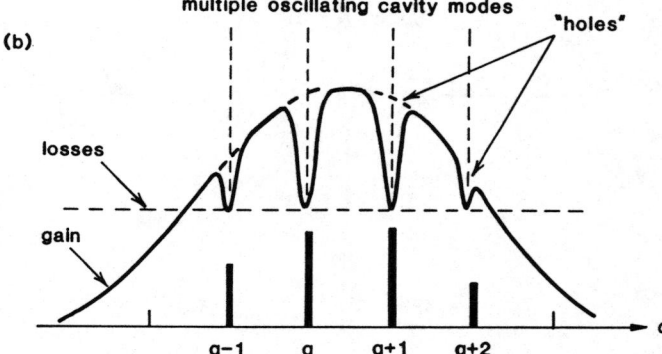

FIGURE 12.2
(a) In an ideal homogeneously broadened laser, the gain profile cannot be pushed above the threshold point for the first oscillating mode—at least not on a steady-state basis—because the first mode would then have a permanently positive growth rate. (b) Multiple axial modes can, however, oscillate with each mode, burning a separate and independent "hole" in the gain profile to make gain equal loss for each mode.

Several practical factors, such as spatial hole burning (Section 8.2), tend to weaken this conclusion in real lasers. In an ideal system, however, and to a sizable extent in many real lasers, *a homogeneously broadened laser will tend to oscillate at only a single frequency, on its centermost (or most preferred) axial and transverse mode.*

The experimental spectra in Figure 12.3, taken on a semiconductor injection laser, give an excellent illustration of how a single oscillating mode can emerge just above threshold from a cluster of regeneratively amplified axial-mode noise peaks just below threshold. Note the sharp change in the character and intensity of the output spectrum as the diode injection current is increased from $I_o = 155$ mA, just below threshold, to $I_o = 162.5$ mA, just above threshold. The special characteristics of semiconductor diode lasers, including their very small cavity volume, high gain, strong spontaneous emission, broad linewidth, and wide axial-mode spacing, make it comparatively easy to obtain this type of experimental result. In most other types of lasers the below-threshold output is relatively much weaker, and the threshold transition very much sharper, so that similar experiments become very much more difficult.

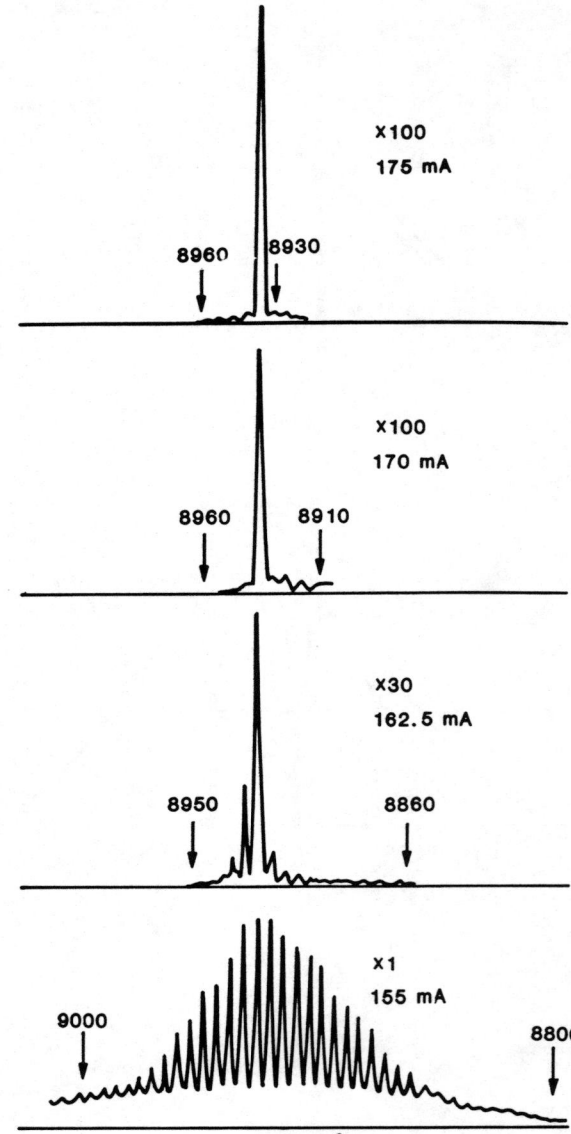

FIGURE 12.3
Single-mode oscillation rises up out of multimode amplified noise as the excitation current is increased in a homogeneous semiconductor diode injection laser. Note changes in vertical scale in the successive curves.

### Inhomogeneous Lasers: Multi-Axial-Mode Oscillation

Doppler-broadened gas lasers, and other lasers with *strongly inhomogeneous transitions*, by contrast, can easily oscillate simultaneously on multiple frequencies or multiple axial modes within the atomic linewidth.

As we will show in a later chapter, when the atomic gain in an inhomogeneous transition exceeds the loss, each axial mode for which this occurs saturates only that subgroup of atoms, or that particular spectral packet, whose atomic frequencies are in resonance with that particular oscillation frequency. As a result, the laser "burns a hole" in the gain curve, and saturates the gain down to equal the loss, at each oscillating axial mode separately, as illustrated in Figure

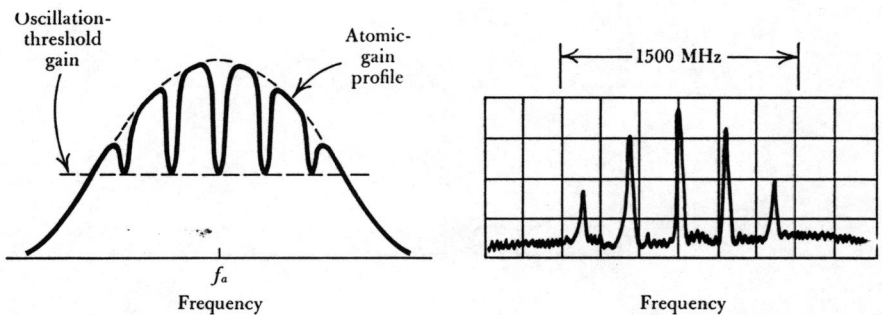

**FIGURE 12.4**
Multiple simultaneously oscillating axial modes in a He-Ne laser oscillator.

12.2(b). *Inhomogeneous lasers can thus oscillate simultaneously in many axial modes, with each mode oscillating almost independent of all the other modes.*

For many common laser systems, there can be a substantial number of axial cavity modes within the atomic gain profile of the laser. Helium-neon lasers, for example, with a doppler linewidth $\Delta f_d \approx 1{,}500$ MHz and axial-mode spacings $\Delta f_{\text{ax}} \approx 150$ to 500 MHz, will typically have three to ten axial modes within the atomic linewidth; and because the line is strongly inhomogeneous, the laser can oscillate in all of these modes at once. Figure 12.4 shows, for example, five simultaneous axial-mode oscillation frequencies from a typical cw He-Ne laser.

A low-pressure $CO_2$ laser, on the other hand, with only 60 to 100 MHz of combined doppler and pressure broadening, may have only one axial mode within its atomic linewidth. Far-infrared and submillimeter molecular gas lasers also generally have very narrow atomic lines (because they operate at low gas pressures and because the doppler broadening decreases as the transition frequency decreases); and so they usually have one (or even fewer) axial modes within their atomic linewidth.

A neodymium-YAG laser with an atomic linewidth of $\Delta f_a \approx 4$ cm$^{-1} \approx 120$ GHz will typically have hundreds of axial modes within the atomic gain curve. As a result this type of laser will often exhibit highly multimode oscillation under the transient conditions associated with short-pulse operation. At the same time this laser can oscillate in only one mode or more often a few simultaneous axial modes under continuous-wave or cw conditions, because of the strongly homogeneous character of the atomic transition.

### Spatial Hole Burning

The most significant effect leading to multimode operation even in spectrally homogeneous lasers is *spatial inhomogeneity*, and especially *spatial hole burning*, as described previously in Section 8.2, and illustrated in Figure 12.5.

Suppose a linear or standing-wave laser is initially oscillating in the $q$-th axial mode. This leads to a standing-wave pattern for the field amplitude or optical intensity along the $z$ axis, with peaks and nulls spaced by one-half optical wavelength (between each null). The inverted population in this laser will then be saturated in a similar spatially periodic fashion, as illustrated in Figure 12.5.

One of the effects of this saturation will be to produce a spatial inverted-population grating or gain grating, which will introduce cross-coupling between the forward and backward-traveling wave components of the $q$-th axial mode. Of more importance at this point, however, is the fact that, at least near the center

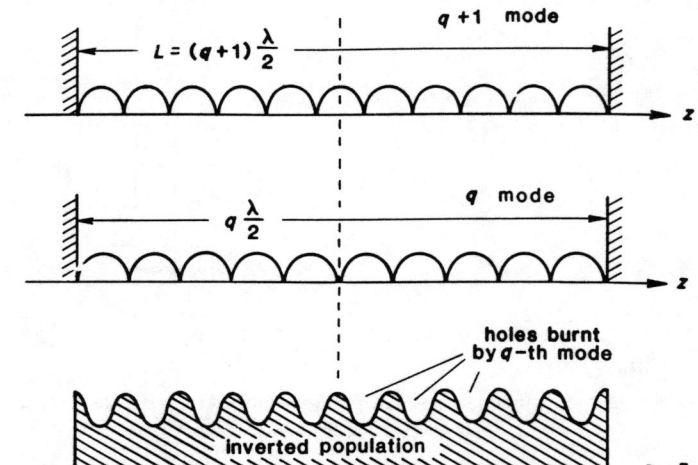

FIGURE 12.5
Spatial hole burning.

of the cavity, the standing-wave pattern of the $(q+1)$-th mode—which squeezes one more half optical wavelength into the cavity length—will have its maximum intensity located just at the points that are left unsaturated by the $q$-th mode. [The same point is, of course, equally true for the $(q-1)$-th axial mode.]

As a result of this, the gain competition between the two adjacent axial modes is much reduced; and both axial modes may well be able to oscillate simultaneously, even with a strongly homogeneous laser medium, by using in essence different groups of atoms. Oscillation with any two adjacent axial modes at equal amplitudes will then saturate the population uniformly, at least in the center of the cavity, possibly discouraging the oscillation of any further axial modes. This behavior is sometimes seen, for example, in solid-state lasers, such as the Nd:YAG laser, which often seem to prefer to oscillate at steady state in just two axial modes.

Unidirectional oscillation in a ring-laser cavity is one way of eliminating this kind of hole burning, thus giving a better chance of obtaining single-frequency operation. Placing a comparatively short section of active laser medium close to one of the end mirrors is another way to reduce the effectiveness of the spatial hole-burning process.

### Exact Oscillation Frequencies, and Frequency Pulling Effects

Laser oscillation normally occurs in only a few preferred longitudinal and transverse modes of a laser cavity. The exact oscillation frequency of a laser will, however, be shifted away by a small amount from the resonance frequency of the corresponding "cold cavity" mode—that is, the resonance frequency of the cavity mode without laser material—because of small frequency pulling effects associated with the $\chi'$ part of the atomic susceptibility. Let us next look at how these pulling effects can be calculated.

The round-trip phase shift $\phi(\omega)$ in a laser cavity, with the gain medium present, must satisfy the phase shift condition

$$\phi(\omega) \equiv \frac{\omega p}{c} + \Delta\beta_m(\omega) p_m = q2\pi, \tag{11}$$

## 12.2 OSCILLATION FREQUENCY AND FREQUENCY PULLING

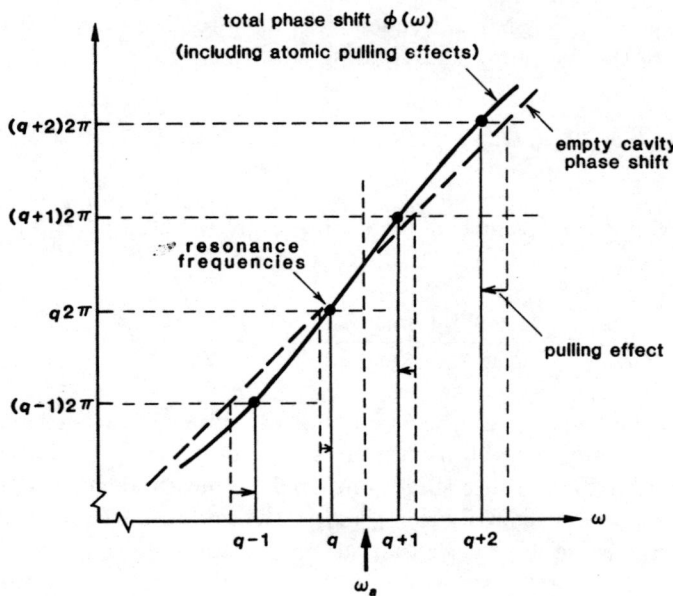

**FIGURE 12.6**
Atomic frequency pulling effects. (An inverted transition always "pulls" the cavity resonance frequencies toward the atomic line center.)

where the atomic phase shift term is normally given by

$$\Delta\beta_m(\omega) = \frac{\beta\chi'(\omega)}{2} = \frac{\omega\,\chi'(\omega)}{2c}. \tag{12}$$

We thus obtain a general condition on the laser frequency given by

$$\frac{\omega p}{c} \times \left[1 + \frac{p_m}{2p}\chi'(\omega)\right] = q2\pi. \tag{13}$$

Figure 12.6 shows, in greatly exaggerated form, how the $\Delta\beta_m(\omega)$ or $\chi'(\omega)$ contribution appropriate to an amplifying transition causes a shift in the exact frequency at which the $\phi(\omega)$ curve intersects the $q2\pi$ resonance values.

### Frequency-Pulling Expression

The magnitude of the $\chi'(\omega)$ term in Equation 12.13 is usually small compared to unity; and the size of the pulling effect will be still further reduced if the length of the atomic medium $p_m$ is small compared to the overall cavity perimeter $p$. Given this assumption, we can invert Equation 12.13 and solve for the pulled laser oscillation frequency—let's call it $\omega'_q$—in the form

$$\omega = \omega'_q = \frac{q2\pi c/p}{1 + (p_m/2p)\,\chi'(\omega'_q)}$$

$$\approx \frac{q2\pi c}{p} \times \left[1 - \frac{p_m}{p}\frac{\chi'(\omega'_q)}{2}\right] \tag{14}$$

$$= \omega_q + \delta\omega_q,$$

where $\omega_q \equiv q2\pi c/p$ is the unpulled or "cold cavity" resonance frequency, and $\delta\omega_q$ is the pulling of the resonance frequency by the atomic phase-shift effects. We can then write this (usually) small pulling effect as

$$\delta\omega_q \approx -\frac{p_m}{2p}\omega_q \chi'(\omega_q') \approx -\frac{\Delta\beta_m(\omega_q')p_m}{p/c}. \tag{15}$$

But since the axial-mode spacing in the cavity is given by $\Delta\omega_{ax} = 2\pi c/p$, we may rewrite this as

$$\frac{\delta\omega_q}{\Delta\omega_{ax}} \equiv \frac{\text{pulling amount}}{\text{axial mode spacing}} \approx -\frac{\Delta\beta_m p_m}{2\pi}. \tag{16}$$

Since the atomic phase-shift term $\Delta\beta_m p_m$ will usually be small compared to $2\pi$, the pulling of each mode will usually be small compared to an axial-mode interval. If this pulling term is also small compared to the atomic linewidth, then we can evaluate the pulling contribution $\chi'(\omega)$ at the *unpulled* frequency $\omega = \omega_q$ (which is much simpler to do) rather than at the pulled frequency $\omega = \omega_q'$.

The magnitude of the reactive susceptibility $\chi'(\omega)$ in an oscillating laser will, of course, depend on the degree of saturation of the atomic transition, and thus on how strongly the laser is oscillating, as well as on where the oscillating mode (or modes) are located within the atomic linewidth. The phase-shift expressions 12.13 to 12.16 serve primarily, however, to determine the exact frequency at which the laser must oscillate, rather than its power level or other characteristics.

### Linear Dispersion Region

Figure 12.6 shows how the $\chi'(\omega)$ term for an amplifying transition tends to shift each axial mode in toward the atomic transition frequency $\omega_a$ by an amount $\delta\omega_q$ that depends on distance from the line center. Amplifying transitions always tend to "pull" cavity frequencies toward line center; absorbing transitions have $\chi'$ values of opposite sign, and thus tend to "push" the cavity resonances away from line center.

For axial modes within the central part of an atomic line, where the value of $\chi'(\omega)$ increases essentially linearly with the mode offset, modes that are further from line center tend to be pulled proportionately more strongly, in such a way that the axial-mode spacing between successive modes remains nearly constant, though decreased by the mode-pulling effect. This is sometimes referred to as the *linear dispersion region* of the atomic transition (Figure 12.7), in contrast to the *nonlinear dispersion region* further out on the atomic gain profile, where the value of $\chi'(\omega)$ begins to bend over and no longer increases linearly with frequency.

### Frequency Pulling for Lorentzian Atomic Transitions

The frequency-pulling term for a lorentzian atomic transition can be rewritten in a particularly simple form by noting that for a lorentzian transition the atomic gain coefficient $\alpha_m(\omega)$ and the atomic phase shift $\Delta\beta_m(\omega)$ are related by

$$\Delta\beta_m(\omega) = 2\frac{\omega - \omega_a}{\Delta\omega_a} \times \alpha_m(\omega). \tag{17}$$

## 12.2 OSCILLATION FREQUENCY AND FREQUENCY PULLING

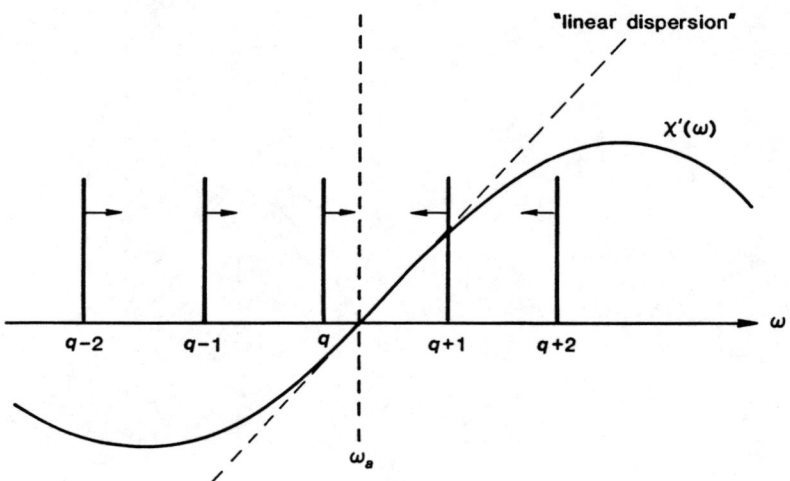

**FIGURE 12.7**
The pulling effects increase linearly with distance away from line center in the central linear-dispersion region of the atomic line.

Hence the frequency-pulling correction can be written as

$$\frac{\delta\omega_q}{\Delta\omega_{ax}} \approx -\frac{2\alpha_m(\omega_q)p_m}{2\pi} \times \frac{\omega_q - \omega_a}{\Delta\omega_a}. \tag{18}$$

In most lasers the round-trip power gain coefficient $2\alpha_m p_m$ will be considerably smaller than $2\pi$; and for those axial modes near line center the quantity $\omega_q - \omega_a$ will be only a few axial-mode intervals, and hence usually small compared to the atomic linewidth $\Delta\omega_a$. Hence the frequency pulling of each mode will be only a small fraction of the axial-mode spacing.

As a numerical example, we might consider a He-Ne laser with 10% power gain per round trip ($2\alpha_m p_m = 0.1$) and ten axial modes within the atomic linewidth ($\Delta\omega_{ax}/\Delta\omega_a = 0.1$). The fractional pulling for the first axial mode on either side of line center will then be

$$\frac{\delta\omega_q}{\Delta\omega_{ax}} \approx -\frac{0.1}{2\pi} \times \frac{1}{10} \approx -1.6 \times 10^{-3}. \tag{19}$$

This corresponds to $\approx$100 kHz pulling out of a 150 MHz axial-mode spacing.

It is also possible, however, for pulling effects to become much larger in lasers that have both very high gain and very narrow atomic linewidth. One example of this is the very high gain 3.39 $\mu$m transition in narrow-bore He-Ne laser tubes.

### Still Another Frequency-Pulling Formulation

Still another version of the frequency-pulling formula is worth presenting briefly, because of the additional insight it gives into frequency-pulling effects. For the lorentzian case we have discussed, we can also rewrite the round-trip phase-shift condition into the form

$$\frac{\omega p}{c} + 2\alpha_m p_m \frac{\omega - \omega_a}{\Delta\omega_a} = q\,2\pi \equiv \frac{\omega_q p}{c}, \tag{20}$$

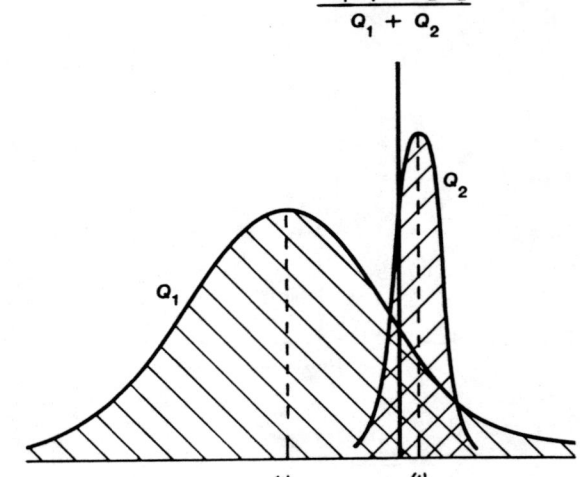

**FIGURE 12.8**
The oscillation frequency for detuned cavity and atomic resonances will lie between the two resonance frequencies and closer to whichever one has the higher $Q$ value.

where $\omega_q \equiv q2\pi c/p$ is the unpulled or cold cavity frequency. We can then argue that at steady-state oscillation in a homogeneous laser the total cavity gain given by $2\alpha_m p_m$ should just equal the cavity losses, given by $\delta_c$, and so the frequency-pulling expression can be rewritten in the form

$$\frac{\omega_a p}{c\delta_c} \times (\omega - \omega_q) + \frac{\omega_a}{\Delta\omega_a} \times (\omega - \omega_a) = 0. \tag{21}$$

But the quantity $\omega_p/c\delta_c$ that multiplies the first frequency difference in this expression is just the "cold cavity" $Q_c$ value that we have defined earlier (in Section 11.4); and we can for the sake of symmetry define the ratio multiplying the second term as a kind of "linewidth $Q_a$" given by $Q_a \equiv \omega_a/\Delta\omega_a$.

The frequency condition then takes on the particularly simple form

$$Q_c(\omega - \omega_c) + Q_a(\omega - \omega_a) = 0, \tag{22}$$

where, to make the notation symmetric, we have relabeled the axial-mode frequency $\omega_q$ as the "cold cavity" frequency $\omega_c$. This result says that the pulled cavity or oscillation frequency is given by the symmetric expression

$$\omega'_c = \frac{Q_c\omega_c + Q_a\omega_a}{Q_c + Q_a}. \tag{23}$$

This says that, as Figure 12.8 shows, if we couple a cavity resonance with $Q = Q_c$ to an atomic resonance with $Q = Q_a$, the resulting oscillation frequency will lie somewhere between the two resonance frequencies $\omega_a$ and $\omega_c$, closer to whichever one has the higher $Q$. The usual situation in most laser oscillators is that the cavity resonance has much the higher $Q$ value; and hence the oscillation occurs essentially at the cavity frequency $\omega_c$, but pulled slightly toward the atomic frequency $\omega_a$.

Exactly the opposite situation can also occur, for example, in certain microwave masers or atomic clocks that have an extraordinarily narrow atomic line and a much wider cavity linewidth. In these we want the oscillation frequency $\omega$ to occur as accurately as possible at the atomic frequency $\omega_a$, with as little perturbation as possible by the cavity frequency $\omega_c$. Equation 12.23 then tells

## 12.2 OSCILLATION FREQUENCY AND FREQUENCY PULLING

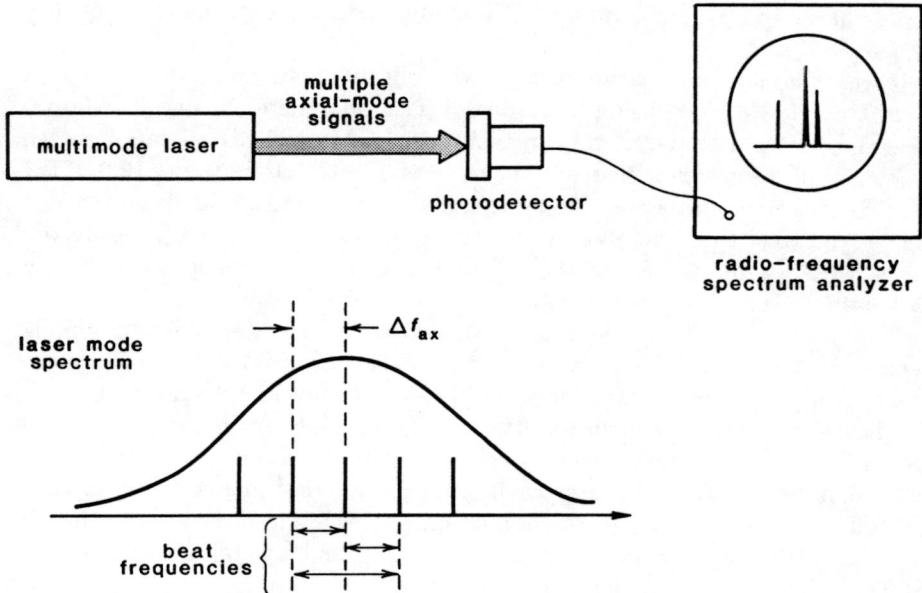

FIGURE 12.9
Measurement system for observing heterodyne beats between axial modes in a laser.

how much error may result if the cavity frequency $\omega_c$ is unavoidably detuned by some amount from $\omega_a$.

### Frequency Beating Measurements

Laser frequency-pulling effects, although typically very small, can be observed in the laboratory in a number of ways. In practice it can be quite difficult to measure the absolute frequencies of laser oscillators to very high precision (although the frequencies of highly engineered laser frequency standards can at present be measured to an absolute accuracy exceeding 1 part in $10^{10}$, and can be stabilized with relative accuracies several orders of magnitude higher). Laser frequency pulling is therefore seldom if ever measured on an absolute basis.

It is much easier, however, to measure relative laser frequencies by observing the difference or beat frequency between two different laser signals, assuming these frequencies are close enough together and stable enough with respect to each other to give a clean beat note, as is true in particular of different axial-mode resonances in the same laser cavity. Difference frequency measurements between two laser oscillations can be accomplished most easily by simply allowing the two laser beams, carefully aligned to be parallel to each other, to fall on any conventional optical detector, such as a photodiode or photomultiplier, and then looking at the photodetector output for signals at the heterodyne or "beat" frequencies between the optical frequencies, as illustrated in Figure 12.9.

Since the optical detector is a square-law device—that is, its signal current is proportional to optical intensity, or to optical $E$ field squared—the detector responds to two optical signals at, say, $f_q$ and $f_{q+1}$ in proportion not only to the dc intensities $I_q$ and $I_{q+1}$ at these two frequencies, but also to a sinusoidal heterodyne beat note of intensity $\sqrt{I_q I_{q+1}}$ at the difference frequency $f_{q+1} - f_q$. (Do not allow the heterodyne jargon here to obscure the elementary fact that the instantaneous amplitude of any signal consisting of the sum of two sinusoidal carriers at $f_q$ and $f_{q+1}$ is automatically modulated at the difference frequency $|f_{q+1} - f_q|$.)

If we make such a measurement on the output beam from, say, a typical 30-cm long He-Ne laser, using a reasonably fast photodetector and a radio receiver or spectrum analyzer, we can easily detect the 500 MHz beats between several simultaneous axial modes in the laser, as illustrated in Figure 12.9. If the experiment is done using a suitable radio-frequency spectrum analyzer, we can usually see two or three closely spaced beat notes between different pairs of axial modes, with the frequencies of the different axial-mode beats spread out by a few hundred kHz about the expected $\Delta\omega_{ax}$ value of the laser.

These multiple axial-mode beats result from the slightly different pulling effects that occur for different axial modes in the laser cavity, plus more complicated inhomogeneous cross-pulling effects we have not discussed yet. The observed beat spectral components will in fact jump about in frequency by small amounts as the axial modes shift together across the atomic gain profile because of thermal drift of the laser cavity, and as different modes suddenly turn on or off at the outer edges of the oscillation range. Mode-beating experiments are thus an effective way of observing mode-pulling effects and any other small frequency-shifting effects in laser oscillators.

### Frequency Beating Between Two Independent Lasers

Heterodyne beats can be observed between beams from two separate lasers as well, but with somewhat more difficulty. The spatial overlap and especially the angular alignment of the two laser beams must first be adjusted to a very high degree of precision to observe any beats. (In a single laser the angular alignment of different frequency modes is automatic.) The two lasers must then be tuned close enough together in frequency, and the frequency jitter of each laser must be kept small enough, so the beat frequency is within the range of the photodetector and receiver; and we must then scan either the receiver or the lasers until we find where this initially unknown beat frequency is located. Optical heterodyne measurements using sufficiently stable lasers are nonetheless commonly carried out, often with the assistance of automatic frequency control (AFC) loops to stabilize the difference frequency between the two lasers.

## REFERENCES

An early but good discussion of spatial hole-burning effects can be found in C. L. Tang, H. Statz, and G. deMars, "Spectral output and spiking behavior of solid-state lasers," *J. Appl. Phys.* **34**, 2289–2295 (August 1963).

For an illustration of mode-pulling and mode-beating effects, see U. P. Oppenheim and M. Naftaly, "Observation of mode pulling in a $CO_2$ laser," *Appl. Opt* **23**, 661–664 (March 1 1984).

---

Problems for 12.2

1. *Number of modes in a doppler-broadened laser.* A doppler-broadened He-Ne lasers ($\Delta\omega_d = 2\pi \times 1{,}500$ MHz) has a midband unsaturated gain coefficient $2\alpha_{m0}$

of 3% per meter. Assuming the intracavity power losses are 0.5% per one-way pass because of imperfect Brewster windows and the output mirror is to have $R = 99\%$, with $R = 100\%$ for the other mirror, make a plot of the number of modes oscillating versus the length $L$ of the laser. Assume the laser material fills the laser cavity except for 10 cm for the Brewster angle sections at each end, and that the centermost axial mode is located exactly at the atomic line center.

2. *Laser cavity design for at least one and not more than three simultaneous axial modes.* A certain laser system has a midband unsaturated gain coefficient $2\alpha_{m0}$, output coupling $\delta_e$, internal cavity losses $\delta_0$, and an inhomogeneous gaussian atomic lineshape with linewidth $\Delta\omega_d$. You want to be sure that this laser will always oscillate in at least one axial mode, no matter how the centermost axial mode drifts back and forth with respect to the atomic line center. How long must you make the cavity, at a minimum? What is the maximum length if you want the laser never to reach threshold for three or more modes at once?

3. *Length considerations for He-Ne laser design.* If a He-Ne laser is made too short, not only is the round-trip gain reduced, but there may be situations in which no axial mode is present within the net positive-gain region of the laser. Suppose a typical He-Ne laser has a doppler-broadened gain profile with linewidths $\Delta f_d = 1{,}500$ MHz and a midband gain coefficient $2\alpha_{m0} = 2.5 \times 10^{-4}$ cm$^{-1}$. Suppose the laser is to use mirrors with 100% and 98% reflectivity at the two ends, and that internal cavity power losses are 0.5% per one-way pass. For mechanical reasons the laser mirrors must be 3 cm beyond the end of the discharge at each end.

While the laser is running, its axial-mode frequencies will slowly drift across the atomic-gain profile because of thermal expansion. Find the allowable range of cavity lengths which will ensure that during such drifts the laser will always oscillate in one or two axial modes, but never in three or more axial modes.

4. *Mode pulling of the axial-mode spacing in different types of lasers.* Calculate by how much the beat frequency between two adjacent axial modes will differ from the "cold cavity" $c/2L$ value as a result of atomic frequency pulling in the linear dispersion regime for the cases: (a) He-Ne 6328 Å laser, $L = 10$ cm, $T = 3\%$ for the output mirror, $R = 100\%$ for the other mirror; (b) Ruby laser at 77K, $L = 5$ cm, one end fully silvered ($R = 100\%$), output end completely unsilvered (air-dielectric reflection only). Hints: The index of refraction for sapphire is $n \approx 1.76$, and data on the atomic linewidth versus temperature for the ruby laser transition is given in Figure 3.5.

## 12.3 LASER OUTPUT POWER

We will next calculate the *power output* that can be obtained from an oscillating laser, as a function of the output coupling and the pumping power, using a simple laser model. In this section we will limit the derivation to a lightly coupled laser oscillator—that is, a laser in which the reflectivities of the laser end mirrors are not too much less than unity.

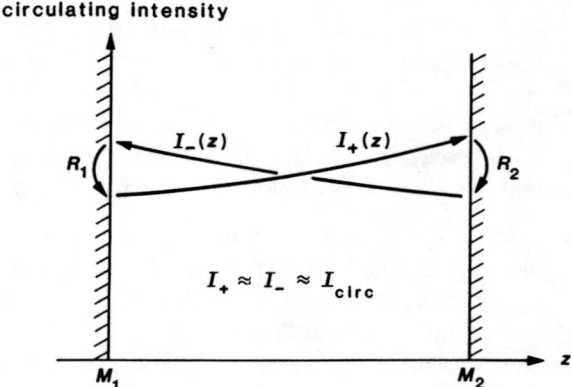

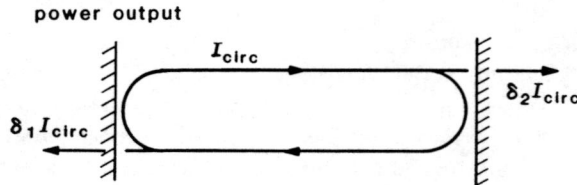

FIGURE 12.10
Left and right-traveling intensities in a laser oscillator with small output coupling.

### Steady-State Homogeneous Saturation Equations

We emphasized earlier that the round-trip power gain for the signal intensity inside a laser cavity must be exactly equal to unity under cw steady-state conditions. If we assume, for example, a standing-wave laser cavity with a simple homogeneously saturable gain medium, the growth of the two oppositely traveling waves $I_+(z)$ and $I_-(z)$ inside this cavity will be given by the two equations

$$\frac{dI_+(z)}{dz} = [2\alpha_m(z) - 2\alpha_0]\, I_+(z) \tag{24}$$

for the forward or $+z$ wave, and

$$\frac{dI_-(z)}{dz} = -[2\alpha_m(z) - 2\alpha_0]\, I_-(z) \tag{25}$$

for the reverse wave as illustrated schematically in Figure 12.10. (We will carry out all the calculations in this section for a *linear or standing-wave cavity*, leaving it to the reader to carry out the essentially similar calculations of power output and optimum output coupling for a ring-laser cavity.)

For a homogeneously saturable gain medium, which saturates at each transverse plane $z$ according to the sum of the intensities $I_+(z)$ and $I_-(z)$ at that plane, the saturated gain coefficient as a function of position along the axis will be

$$2\alpha_m(z) = \frac{2\alpha_{m0}}{1 + [I_+(z) + I_-(z)]/I_{\text{sat}}}, \tag{26}$$

where $\alpha_{m0}$ is the unsaturated gain coefficient. This assumption neglects interference or standing-wave effects between the right- and left-traveling waves, as discussed earlier in Section 8.2. Nonetheless, it serves as an excellent first approximation, and gives results that agree very well with experiment.

### Small Output Coupling Approximation

Suppose now that the end-mirror reflectivities $R_1$ and $R_2$ are both close to unity, so that the intensities $I_+$ and $I_-$ remain nearly constant along the length of the cavity, as in Figure 12.10. (The *unsaturated* gain per pass through the laser medium need not be small; but the net *saturated* gain per pass must be not much greater than unity.) We can then make the approximation that

$$I_+(z) \approx I_-(z) \approx I_{\text{circ}}, \tag{27}$$

where $I_{\text{circ}}$, the one-way circulating intensity inside the laser cavity, is to first order independent of position inside the cavity. The saturated gain coefficient is then similarly independent of position along the cavity, and can be written as

$$2\alpha_m \approx \frac{2\alpha_{m0}}{1 + 2I_{\text{circ}}/I_{\text{sat}}}. \tag{28}$$

The factor of 2 in the denominator arises, of course, because the laser medium sees equal intensities $I_{\text{circ}}$ traveling in both directions along the cavity.

### Steady-State Oscillation Condition

The threshold and/or the steady-state gain condition for the laser oscillator is then given by

$$2\alpha_m p_m \approx \frac{2\alpha_{m0} p_m}{1 + 2I_{\text{circ}}/I_{\text{sat}}} = 2\alpha_0 p + \ln\left(\frac{1}{R_1 R_2}\right) \equiv \delta_0 + \delta_1 + \delta_2. \tag{29}$$

The circulating intensity inside the cavity that must build up in order to saturate the gain factor $2\alpha_{m0} p_m$ down to where it just equals the total cavity losses $\delta_0 + \delta_1 + \delta_2$ is thus given by

$$I_{\text{circ}} = \left[\frac{2\alpha_{m0} p_m}{\delta_0 + \delta_1 + \delta_2} - 1\right] \times \frac{I_{\text{sat}}}{2}. \tag{30}$$

It is often convenient to define a threshold ratio $r$ given by

$$r \equiv \frac{2\alpha_{m0} p_m}{\delta_0 + \delta_1 + \delta_2} = \frac{\text{unsaturated round-trip laser gain}}{\text{total round-trip cavity losses}}. \tag{31}$$

The condition $r = 1$ then corresponds to threshold; and the value of $r \geq 1$ tells by how much the laser gain is pumped above threshold. The circulating intensity inside a standing-wave cavity can then be written as

$$I_{\text{circ}} = (r - 1) \times \frac{I_{\text{sat}}}{2}. \tag{32}$$

This expression, of course, has meaning only so long as the laser is above threshold, so that $r > 1$ or $2\alpha_{m0} p_m > \delta_0 + \delta_1 + \delta_2$.

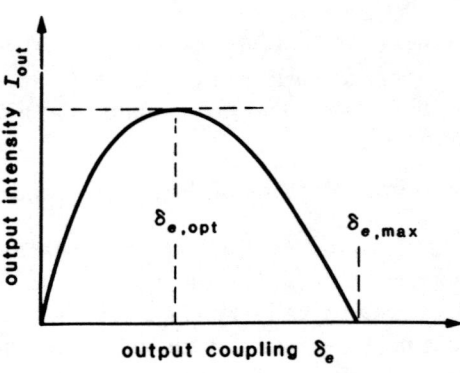

FIGURE 12.11
Useful output intensity and internal circulating intensity versus output coupling for a weakly coupled laser.

### Laser Power Output

Now, for a lightly coupled laser cavity, the end-mirror transmissions are given by $T_1 = 1 - R_1 \approx \delta_1$ and $T_2 = 1 - R_2 \approx \delta_2$, so long as both $\delta_1$ and $\delta_2$ are reasonably small compared to unity. Normally in a laser we take the power output from one end of the cavity only. The total potentially useful output intensity (power per unit area) is really the power from both ends of the laser cavity, however, or

$$I_{\text{out}} = (\delta_1 + \delta_2) \times I_{\text{circ}} = \delta_e \times I_{\text{circ}}, \tag{33}$$

where we have defined one additional delta factor $\delta_e$ (with "e" standing for "external") by the definition

$$\delta_e \equiv \delta_1 + \delta_2 = \text{external cavity coupling.} \tag{34}$$

The value of $\delta_e$ thus represents the total *external coupling*, or *output coupling*, through both ends of the laser cavity. At least for small coupling, $\delta_e$ represents the total fractional power coupled out per round trip through the external mirrors (or whatever other output coupling mechanism might be employed).

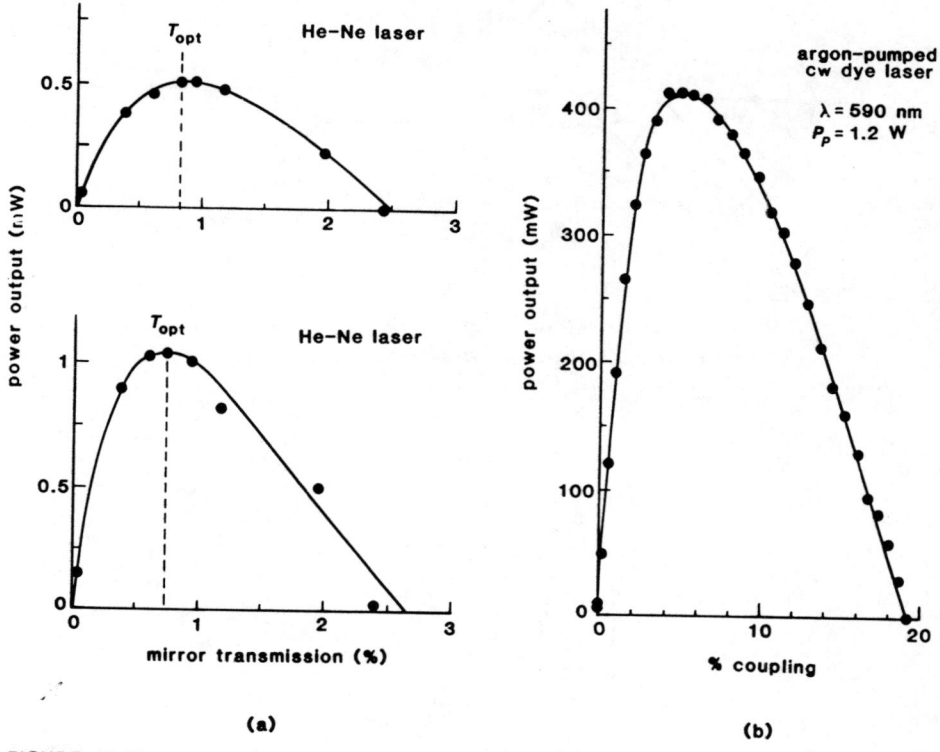

**FIGURE 12.12**
(a) Typical experimental results for laser power output versus coupling in two low-gain He-Ne lasers. (b) Similar results for a higher-gain cw dye laser. (Adapted from P. Laures, *Phys. Lett.* **10**, 61, May 15, 1964; and from C. V. Shank et al., *Opt. Commun.* **7**, 176–177, March 1973.)

The total output intensity, as a function of the unsaturated gain $2\alpha_{m0}$, the internal cavity losses $\delta_0$, and this external coupling factor $\delta_e$, then becomes

$$I_{\text{out}} = \delta_e \left[ \frac{2\alpha_{m0}p_m}{\delta_0 + \delta_e} - 1 \right] \frac{I_{\text{sat}}}{2}. \tag{35}$$

Figure 12.11 shows a typical example of how both the *circulating power* and the *output power* vary with external coupling.

### Experimental Verification

Some representative experimental results for power output versus output mirror transmission are shown in Figure 12.12, both for two very low-gain He-Ne lasers, and for a considerably higher-gain argon-laser-pumped Rhodamine 6G dye laser. Note that the results for the dye laser agree very well with the predicted form derived earlier, even though the maximum output coupling of $\approx 20\%$ is becoming significant compared to unity.

Mirrors or output couplers with continuously variable transmission or output coupling are not readily available, especially for small values of output coupling; and performing experiments like those shown in Figure 12.12 with a series of different transmission mirrors on a low-gain, low-coupling laser can be difficult,

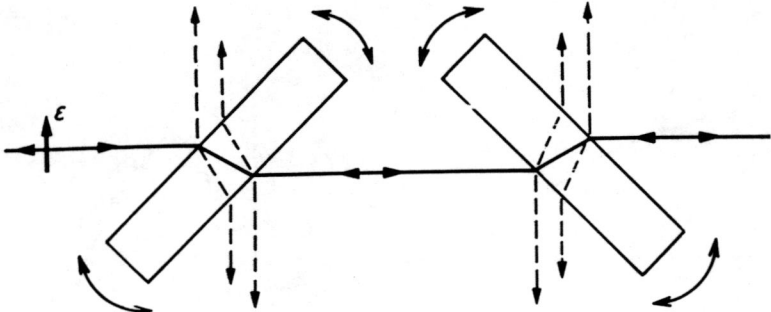

FIGURE 12.13
Device for small variable insertion loss or output coupling. The plates are operated near Brewster's angle, where all the reflections vanish.

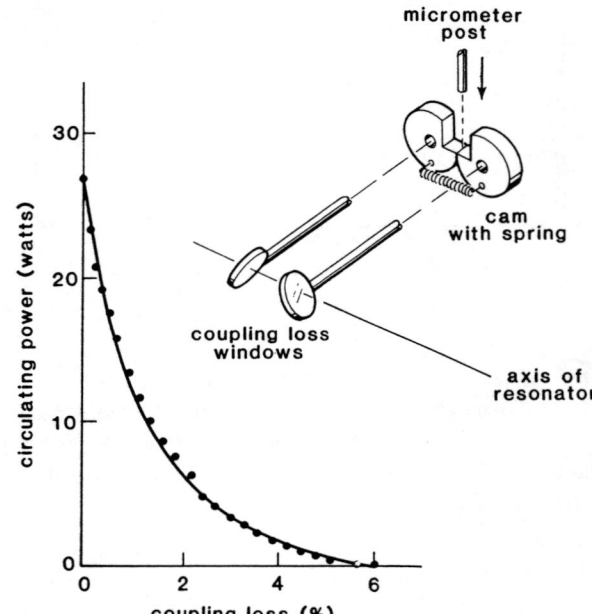

FIGURE 12.14
Circulating power versus coupling loss in a weakly coupled Nd:YAG laser.

because the laser must be readjusted and realigned with each change of mirrors, and because you can never be certain that all the mirrors are equally lossless and defect-free.

Figure 12.13 shows a device that uses two contra-rotating dielectric plates tilted near Brewster's angle, where the output reflectivity from each surface of the plates passes through zero, in order to produce a variable output coupling or insertion loss in a laser cavity. (The same device is also very useful as a separate external optical attenuator; the use of two plates means that the optical axis of the laser beam suffers no net transverse displacement as the plates are rotated in opposite directions.) Figure 12.14 then shows a careful measurement, made using one of these devices, of the circulating power inside a cw Nd:YAG laser, with results in excellent agreement with theory. Note that though the output power from this laser is only a few hundred milliwatts, the circulating power inside the cavity is several tens of Watts.

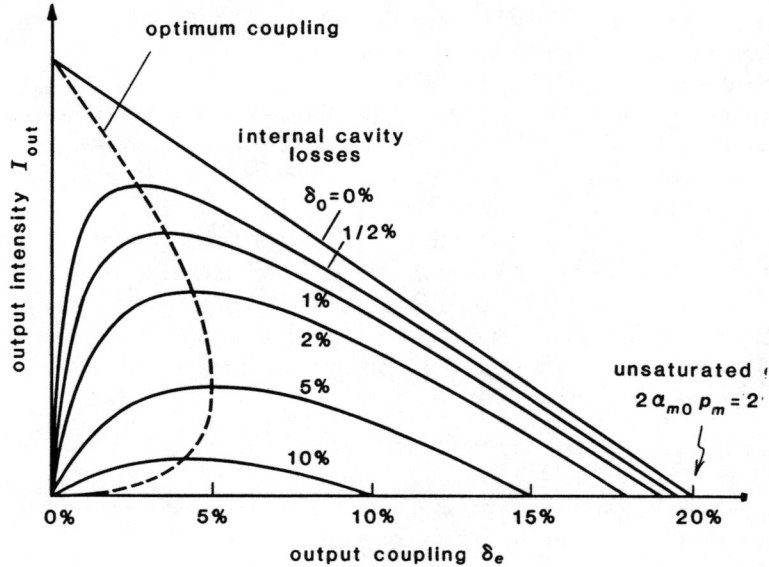

**FIGURE 12.15**
Laser output intensity versus output coupling $\delta_e$ assuming an unsaturated gain coefficient $2\alpha_{m0}p_m = 20\%$ and different values of the internal cavity loss factor $2\alpha_0 p$.

### Optimum Output Coupling Factor

For any of the lasers shown in Figures 12.11 or 12.12 there is obviously a maximum allowable output coupling, given by $\delta_{e,\max} \equiv 2\alpha_{m0}p_m - 2\alpha_0 p = \delta_{m0} - \delta_0$, beyond which the cavity is overloaded, so that total cavity losses exceed the available gain, and no oscillation is possible. As the cavity coupling or end-mirror transmission is reduced below this value, both the circulating intensity and the output intensity increase with decreasing coupling. Below a certain optimum coupling factor $\delta_{e,\text{opt}}$, however, the mirror transmission decreases faster than $I_{\text{circ}}$ increases, and the power output decreases, eventually becoming zero at zero transmission through the end mirrors. The laser at this point is, of course, still oscillating—in fact, oscillating the strongest of all—but with all its available power being uselessly dissipated in the internal cavity losses.

Figure 12.15 illustrates in more detail how the laser output intensity for a typical laser depends on the cavity output coupling, assuming a fixed value of 20% power gain per round trip, and varying amounts of internal cavity loss. It is evident that for each different value of internal cavity loss there is a different optimum output coupling which maximizes the output power. It is also apparent that the optimum output coupling is always considerably smaller than the available gain, and that even very small internal losses have a very serious effect on the maximum useful output power available from the laser.

It is a straightforward calculation to evaluate the optimum output coupling for given values of unsaturated gain and internal cavity losses. Differentiation of the expression for output intensity given in Equation 12.35 with respect to the output coupling $\delta_e$ gives for this optimum coupling

$$\delta_{e,\text{opt}} = \sqrt{2\alpha_{m0}p_m \delta_0} - \delta_0 = \left[\sqrt{\delta_{m0}/\delta_0} - 1\right]\delta_0. \tag{36}$$

where $\delta_{m0} \equiv 2\alpha_{m0}p_m$. The dashed line in Figure 12.15 indicates the locus of these optimum coupling values.

One slightly unusual aspect evident from Figure 12.15 is that the optimum coupling value apparently goes to zero as the internal cavity losses go to zero, i.e., $\delta_{e,\text{opt}} \to 0$ as $\delta_0 \to 0$. In the limiting case of zero internal losses, we would apparently get maximum output by using end mirrors with 100% reflectivity and zero transmission!

The explanation of this minor paradox is, of course, that as both $\delta_0$ and $\delta_e$ go to zero, the internal circulating power $I_{\text{circ}}$ goes to $\infty$; and the product of zero coupling times infinite circulating power leads to a finite power output. Real laser cavities will, of course, always have some small but finite losses, and so will always require an equally small but finite output coupling.

### Optimum Output Power and Power Extraction Efficiency

If one adjusts a standing-wave laser oscillator for optimum output coupling, the output intensity with this optimum coupling is then given by

$$I_{\text{out,opt}} = \left[\sqrt{2\alpha_{m0}p_m} - \sqrt{\delta_0}\right]^2 \frac{I_{\text{sat}}}{2}$$
$$= \left[1 - \sqrt{\delta_0/\delta_{m0}}\right]^2 \times [2\alpha_{m0}L_m I_{\text{sat}}], \tag{37}$$

where we have used $p_m \equiv 2L_m$. But the second term in the second line of this expression can be recognized as the same maximum available intensity from the laser medium that we obtained in our earlier discussion of laser amplification (Section 7.7), that is, $I_{\text{avail}} \equiv 2\alpha_{m0}L_m I_{\text{sat}}$.

We can then identify the remaining factors in the output intensity formula as defining the *power extraction efficiency* $\eta$ with which the laser oscillator extracts energy from the laser medium and converts it into useful power output. In particular, for a standing-wave laser cavity with arbitrary gain, loss, and output coupling, the extraction efficiency will be given in general by

$$\eta(\delta_0, \delta_e) \equiv \frac{I_{\text{out}}(\delta_0, \delta_e)}{I_{\text{avail}}} = \left[\frac{\delta_e}{\delta_0 + \delta_e} - \frac{\delta_e}{2\alpha_{m0}p_m}\right]. \tag{38}$$

The maximum value of this extraction efficiency with optimum output coupling, or $\delta_e = \delta_{e,\text{opt}}$, then becomes

$$\eta_{\text{opt}} = \left[1 - \sqrt{\frac{\delta_0}{2\alpha_{m0}p_m}}\right]^2, \tag{39}$$

which depends only on the ratio of internal losses to unsaturated gain.

The most significant aspect of these results is the extremely serious effect that even very small internal losses ( $\delta_0 \ll 2\alpha_{m0}p_m$ ) will have on the useful power output. Figure 12.16 shows how this optimum extraction efficiency rapidly decreases as the ratio of internal losses to unsaturated gain increases.

Note, for example, that internal losses only one-tenth as large as the unsaturated gain will reduce the optimized output intensity to less than 50% of its maximum value; and internal cavity losses equal to half the unsaturated gain will reduce the energy extraction efficiency to $\eta_{\text{opt}} \approx 9\%$. To put this another way, in

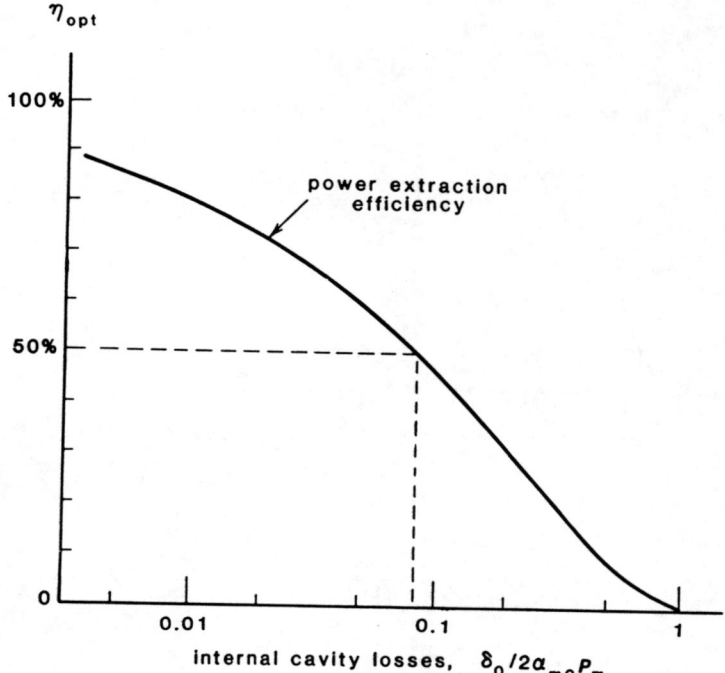

**FIGURE 12.16**
Even very small internal losses, relative to the laser gain, will cause a large reduction in power extraction efficiency in a low-gain laser oscillator.

a laser with 5% power gain per pass, to extract even 50% of the potentially available power output we must cut the internal cavity losses to $\leq 0.3\%$—something which can be very difficult to do in a real laser cavity.

If the internal losses can be made sufficiently small, however, then the extraction efficiency of a laser oscillator—in contrast to our earlier results for single-pass laser amplifiers—can approach 100% with optimum coupling. A properly coupled low-loss oscillator can extract nearly all the power that is available in a laser material, something that is much more difficult to do if the same medium is used as a single-pass laser amplifier.

### Power Output Versus Pumping

We showed in earlier chapters that in many real laser systems the unsaturated gain coefficient $2\alpha_{m0}$ increases linearly with the pumping power applied to the laser, whereas the magnitude of the saturation intensity $I_{\text{sat}}$ is most often independent of the pumping power. If this is so, then we can view the dimensionless gain factor $r$ that we defined earlier as also representing a *dimensionless pumping ratio*, which gives the amount that the laser is pumped above its oscillation threshold, i.e.,

$$r \equiv \frac{2\alpha_{m0}p_m}{\delta_0 + \delta_e} = \frac{R_p}{R_{p,\text{th}}} = \frac{\text{pumping power}}{\text{threshold pump power}}. \quad (40)$$

482        CHAPTER 12: FUNDAMENTALS OF LASER OSCILLATION

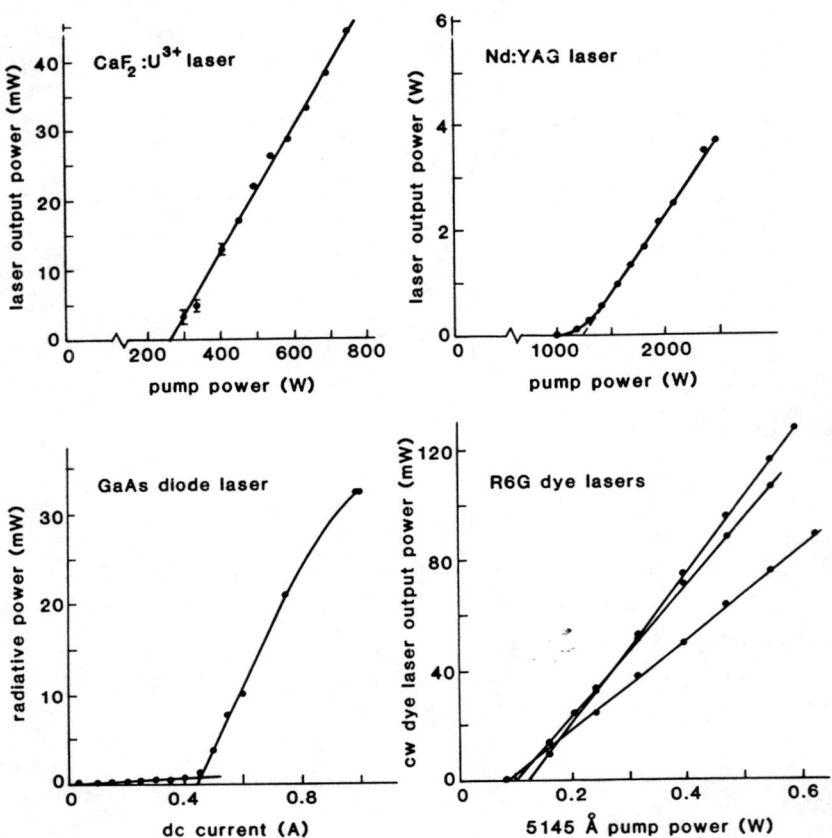

FIGURE 12.17
Laser power outputs versus pumping input for two cw arc-lamp-pumped solid-state lasers, a direct-current-pumped semiconductor diode laser, and a laser-pumped cw dye laser oscillator, all illustrating a similar linear output variation above threshold.

Equation 12.35 laser output intensity then becomes

$$I_{\text{out}} = \frac{(r-1)\delta_e I_{\text{sat}}}{2} = \left[\frac{R_p}{R_{p,\text{th}}} - 1\right] \times \frac{\delta_e I_{\text{sat}}}{2}. \qquad (41)$$

The power output versus pumping power or pumping rate, at fixed coupling, for a great many lasers will therefore be zero up to a certain threshold pumping level corresponding to $r = 1$, and will then rise more or less *linearly* with pumping rate above this threshold.

To illustrate this point, Figure 12.17 shows the oscillation power outputs versus pump power input for some very different lasers, including two cw lamp-pumped solid-state lasers; a dc-current-pumped semiconductor diode laser; and a laser-pumped cw dye laser, operating with three different end-mirror transmissions on the laser. These results are typical of many different experimental results for many different types of laser devices, all showing more or less linear variation of laser output with pump input for substantial distances above their pump thresholds.

A particularly pretty illustration of several fundamental aspects of laser physics is also shown by the experimental results for two similar narrow-strip

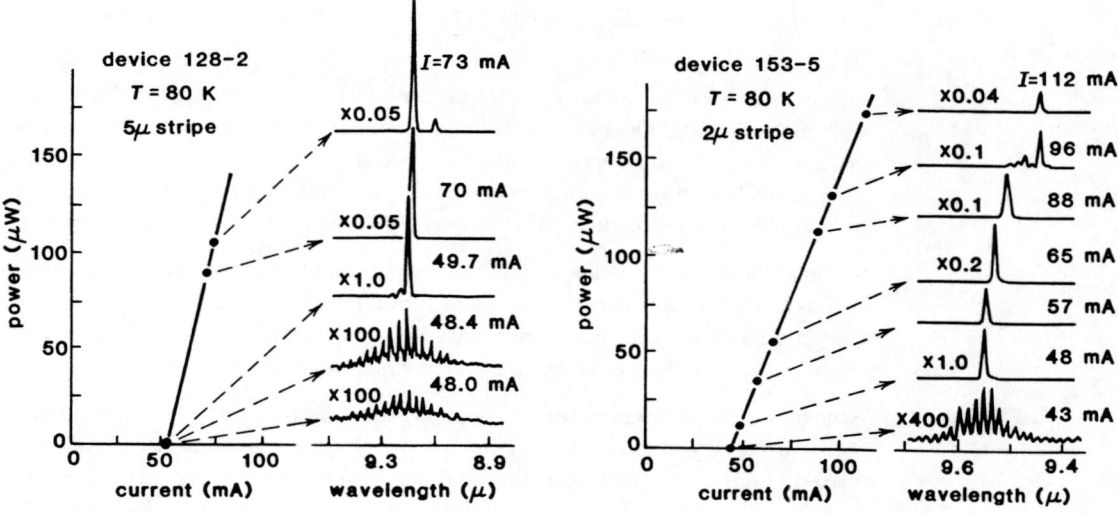

**FIGURE 12.18**
Power outputs and output spectra versus pumping current for two similar narrow-stripe PbSnTe buried heterostructure laser diodes oscillating near 9.5 μm. (Adapted from D. Kasemset and C. C. Fonstad, *Appl. Phys. Lett.*, **39**, 8720–874, December 1, 1981.)

buried heterostructure injection diode lasers shown in Figure 12.18. These lasers are both PbSnTe diodes, fabricated using liquid-phase epitaxy, with active regions 1 to 1.5 μm thick, 2 to 5 μm wide, and 250 to 450 μm long, oscillating in a single transverse mode at wavelengths near 9.5 μm. The experimental spectra clearly show (i) the regeneratively amplified spontaneous emission at the axial-mode peaks just at or below threshold: (ii) the sudden changeover to essentially a single oscillating axial mode produced by a very small change in current just at threshold (note the changes in vertical magnification between adjacent spectra); and (iii) the extreme linearity of the power output versus pumping current above threshold. The laser transition is clearly very homogeneous in its spectral behavior.

## REFERENCES

Two early laser analyses which include and extend the results of this section, and of the preceding section, are P. A. Miles and I. Goldstein, "Effects of output coupling on optical masers," *IEEE Trans. on Electron Devices* **ED-10**, 314–318 (September 1963); and A. Yariv, "Energy and power considerations in injection and optically pumped lasers," *Proc. IEEE* **51**, 1723–1731 (December 1963).

The YAG circulating power measurements in this section are from R. R. Rice, J. R. Teague, and J. E. Jackson, "Dynamic coupling characteristics of TEM Nd:YAG lasers," *J. Appl. Phys.* **46**, 2716–2720 (June 1975).

## Problems for 12.3

1. *Optimum coupling analysis for a unidirectional ring-cavity laser.* Repeat the small-coupling analysis of this section for a unidirectional ring-cavity laser (i.e., one that oscillates in only one direction around the ring). Find the power out-

put versus coupling, the optimum output coupling, and the power-extraction efficiency.

Suppose that exactly the same laser medium, with a fixed length $L_m$, fixed gain coefficient $\alpha_{m0}$, and fixed loss coefficient $\alpha_0$ (assume all the losses are in the laser medium itself), is to be used in either a standing-wave or a unidirectional ring cavity. Plot the power output versus coupling for each, assuming a 10% one-way unsaturated power gain and a 2% one-way power loss through the laser medium.

2. *Internal losses and optimum coupling in real lasers.* For the various experimental curves of power output versus coupling illustrated in Figure 12.12 can you deduce what the internal losses must have been in each situation, and hence what the power extraction efficiencies at optimum coupling must have been?

3. *Laser power output versus tuning.* Suppose that the frequency of a single axial mode can be tuned across the full gain profile of a homogeneous laser transition at fixed pump power, with all other axial modes suppressed or spaced far outside the gain profile. Show that the power output versus frequency curve for this mode can be written as

$$I_{\text{out}}(\omega) = \left[ r - 1 - \left( 2\frac{\omega - \omega_a}{\Delta\omega_a} \right)^2 \right] \times \frac{\delta_e I_{\text{sat}}}{2},$$

and that the full tuning range over which this mode will oscillate is given by $\Delta\omega_{\text{osc}} = \Delta\omega_a \times \sqrt{r-1}$, where $r$ measures how far above threshold the laser is pumped at line center.

4. *Laser oscillator with both saturable gain and saturable loss.* A laser cavity contains both a *saturable gain* medium with low-level gain coefficient $\alpha_{m0}$ and saturation intensity $I_{\text{sat},m}$; and a *saturable absorbing* medium with low-level absorption coefficient $\alpha_{a0}$ and saturation intensity $I_{\text{sat},a}$; and also some *nonsaturating* losses (such as scattering losses or output coupling losses) represented by a nonsaturating absorption coefficient $\alpha_0$. Find the steady-state internal intensity $I_{\text{circ}}$ at which this laser will oscillate (or possibly will *not* oscillate) for different relative ratios of $\alpha_{m0}$, $\alpha_{a0}$, and $\alpha_0$, and also of $I_{\text{sat},m}$ and $I_{\text{sat},a}$. Hint: there are several physically different situations that have to be considered here.

5. *Cross coupling between oscillation power and a separately injected signal.* A low-gain cavity laser with a homogeneously saturable laser medium is oscillating on an axial mode located exactly at line center. A separate laser signal of intensity $I_1$ tuned exactly to line center is also sent through the same laser medium at a very slight angle so that this external signal misses the laser mirrors but illuminates exactly the same volume of atoms as the oscillation signal inside the laser cavity. Develop an expression for the *oscillation power output* through the end mirrors of the laser cavity as a function of the externally injected signal level and the usual laser parameters.

6. *Second-harmonic output coupling.* When a laser beam passes through certain nonlinear optical crystals, such as lithium niobate (LiNbO$_3$) or potassium dihydrogen phosphate (KH$_2$PO$_4$), a portion of the incident intensity at the laser frequency $\omega$ can be converted into second-harmonic radiation at the doubled frequency $2\omega$. For small conversion efficiency, the second-harmonic power generated is given by $I(2\omega) = K_2 I^2(\omega)$, where $K_2$ is the harmonic-generation coefficient. The power converted into second harmonic is of course taken away from the fundamental intensity.

The conversion efficiency from $I(\omega)$ to $I(2\omega)$ with low-power cw laser beams is usually very small (a few percent or less), because the nonlinearity coefficient $K_2$ in real crystals is typically small. The amount of second-harmonic power obtained from a low-power laser can be increased by placing the nonlinear crystal *inside* the laser cavity, where the circulating intensity is much larger than outside the cavity. (Special mirrors are used to let the harmonic radiation escape while reflecting the fundamental-frequency laser radiation.) The second harmonic radiation then becomes the useful output coupling from the laser.

Analyze this type of second-harmonic output coupling by considering a ring-laser cavity which contains a homogeneously saturable gain medium; some small internal cavity losses at the fundamental frequency; and a square-law second-harmonic generation crystal whose conversion coefficient $K_2$ can be adjusted, for example, by changing the nonlinear crystal length or the degree of fundamental beam focusing inside the crystal. Assume the fundamental frequency output coupling is zero, i.e., 100% reflecting mirrors at $\omega$ are employed. Find the value of the nonlinearity coefficient $K_2$ that will maximize the second-harmonic power output from this laser, and discuss in general how the harmonic power output can be optimized, and how the power output at $2\omega$ can compare with the optimum fundamental power that could be obtained from the same laser medium. Will the optimum nonlinearity $K_2$ change if the laser pumping rate is changed?

For literature references on this subject, see R. G. Smith, "Theory of intracavity optical second-harmonic generation," *IEEE J. Quantum Electron.* **QE-6**, 215–223 (April 1970); D. Frölich, L. Stein, H. W. Schröder, and H. Welling, "Efficient frequency doubling of cw dye laser radiation," *Appl. Phys.* **11**, 97–101 (1976); and A. I. Ferguson and M. H. Dunn, "Intracavity second-harmonic generation in continuous-wave dye lasers," *IEEE J. Quantum Electron.* **QE-13**, 751–756 (September 1977).

---

## 12.4 LARGE OUTPUT COUPLING ANALYSIS

The power-output analysis of the previous section was based on a weak-coupling approximation. A more accurate analysis of the power output from a homogeneous laser with arbitrarily large round-trip gain and output coupling was originally developed by W. W. Rigrod at Bell Telephone Laboratories, and is often referred to as the "Rigrod analysis."

As we will demonstrate in this section, for a laser with large unsaturated atomic gain the power output remains fairly constant over a very wide range of output coupling, so that critical adjustment of the output coupling is not as essential for reasonably good energy extraction as it is for a low-gain laser.

### Analytical Formulation: The Rigrod Analysis

The analysis we will repeat here assumes a homogeneously saturable gain medium as in Section 12.3, but no distributed losses, so that $2\alpha_0 p \equiv 0$. Following Rigrod's original notation, as shown in Figure 12.19, we use $I_+(z)$ and $I_-(z)$ to indicate the intensities traveling toward $+z$ and $-z$, respectively, in the cavity.

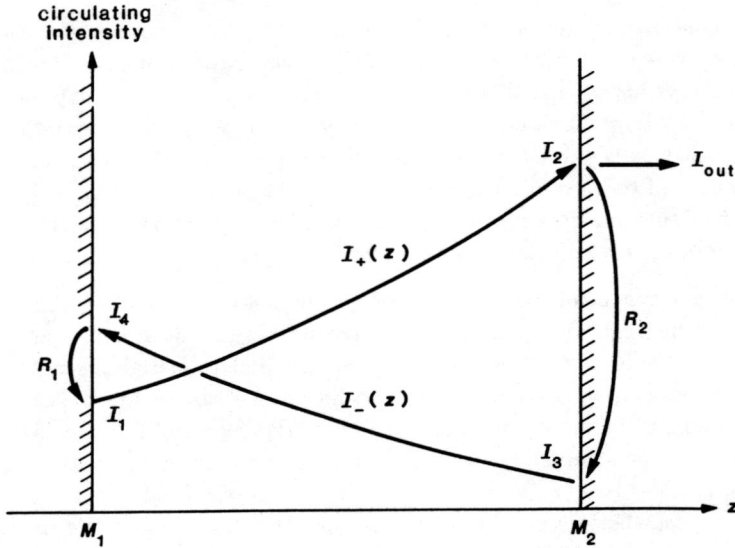

**FIGURE 12.19**
Left and right-traveling intensities in a laser oscillator with large output coupling.

These intensities then grow with distance according to the equations

$$\frac{dI_+(z)}{dz} = +2\alpha_m(z)I_+(z),$$
$$\frac{dI_-(z)}{dz} = -2\alpha_m(z)I_-(z). \tag{42}$$

(The sign changes in the second equation because the waves is traveling in the $-z$ direction.) For simplicity, let us assume that $I_+$ and $I_-$ are normalized to the saturation intensity $I_{\text{sat}}$ of the medium. The gain coefficient $\alpha_m(z)$ at any plane $z$ then saturates according to the total intensity at that plane in the form

$$\alpha_m(z) = \frac{\alpha_{m0}}{1 + I_+(z) + I_-(z)}. \tag{43}$$

Writing the gain in this form takes into account the spatial variation of both the forward- and the backward-traveling waves in a high-gain cavity, but neglects any spatial hole burning or induced-grating coupling effects caused by standing waves or by interference between the forward- and the backward-traveling waves inside the laser cavity.

By combining the two derivatives in Equation 12.42, we can see that the product of the intensities in the two directions at any plane is constant, i.e.,

$$\frac{d}{dz}[I_+(z)I_-(z)] = -2\alpha_m I_+ I_- + 2\alpha_m I_+ I_- = 0, \tag{44}$$

so that we can write at any plane

$$I_+(z)I_-(z) = \text{constant} = C. \tag{45}$$

The differential equation for, say, the $I_+(z)$ wave can then be written as

$$\frac{dI_+(z)}{dz} = \frac{2\alpha_{m0}I_+(z)}{1 + I_+(z) + C/I_+(z)}, \tag{46}$$

and this can be integrated over the length of the laser medium in the form

$$\int_{I_1}^{I_2} \left(1 + \frac{1}{I_+} + \frac{C}{I_+^2}\right) dI_+ = 2\alpha_{m0} \int_0^L dz. \tag{47}$$

Carrying out the same procedure for the $I_-(z)$ wave, and using the boundary conditions shown in Figure 12.19, then leads to the pair of expressions

$$2\alpha_{m0}L = \ln\left(\frac{I_2}{I_1}\right) + I_2 - I_1 - C\left(\frac{1}{I_2} - \frac{1}{I_1}\right),$$
$$2\alpha_{m0}L = \ln\left(\frac{I_4}{I_3}\right) + I_4 - I_3 - C\left(\frac{1}{I_4} - \frac{1}{I_3}\right). \tag{48}$$

In addition we have the mirror power reflection coefficients $I_1 = R_1 I_4$ and $I_3 = R_2 I_2$, and the two product relations at the end surfaces, namely, $I_1 I_4 = I_2 I_3 = C$.

By combining all these relations, together with some minor manipulation, we can eliminate the constant $C$ and obtain the result that, for example, the normalized intensity striking the right-hand mirror is

$$I_2 = \frac{1}{(1 + r_2/r_1)(1 - r_1 r_2)} \left[2\alpha_{m0}L - \ln\left(\frac{1}{r_1 r_2}\right)\right], \tag{49}$$

where $r_1 \equiv R_1^{1/2}$ and $r_2 \equiv R_2^{1/2}$ are the voltage reflection coefficients of the mirror.

### Power Output and Power-Extraction Efficiency

Let us now assume that mirror $M_2$ is the output mirror of the laser, with output coupling $T_2$ and reflection coefficient $R_2$, and that any finite reflectivity $R_1$ of the other mirror $M_1$ represents unwanted or unavoidable losses in that mirror. Then the useful output intensity from the laser (with all intensities measured now in real intensity units) will be

$$I_{\text{out}} = T_2 I_2 = \frac{T_2 I_{\text{sat}}}{(1 + r_2/r_1)(1 - r_1 r_2)} \left[\ln G_0 - \ln\left(\frac{1}{r_1 r_2}\right)\right]. \tag{50}$$

We know from previous sections that the maximum intensity that can be extracted from such a laser medium is

$$I_{\text{avail}} = 2\alpha_{m0} L_m I_{\text{sat}} \equiv (\ln G_0) I_{\text{sat}}, \tag{51}$$

and hence the power-extraction efficiency, or the normalized output intensity, of the laser can be written as

$$\eta \equiv \frac{I_{\text{out}}}{I_{\text{avail}}} = \frac{T_2}{(1 + r_2/r_1)(1 - r_1 r_2)} \left[1 + \frac{\ln r_1 r_2}{\ln G_0}\right]. \tag{52}$$

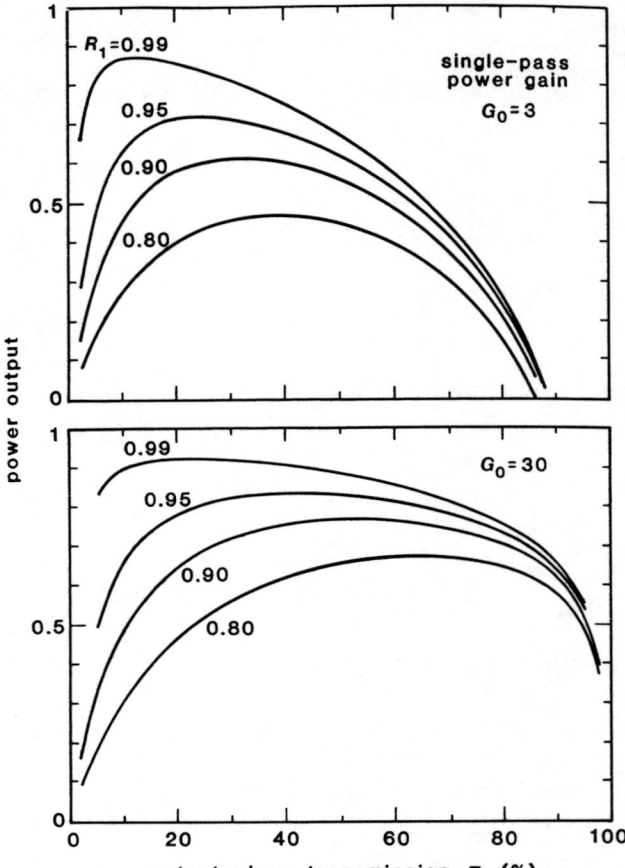

FIGURE 12.20
Normalized power output versus output mirror transmission $T_2$ for homogeneous high-gain laser oscillators with unsaturated single-pass gains of $G_0 = 3$ and $G_0 = 30$, for varying values of the mirror reflectivity $R_1$ at the opposite (nonoutput) end, according to the Rigrod analysis.

There are no small-amplitude restrictions on the unsaturated gain $G_0$ or the output coupling level in this formula.

### Typical Results

Examples of power output versus coupling as given by this formula for two comparatively large values of one-way unsaturated power gain $G_0$ are shown in Figure 12.20. For large values of $G_0$ and for reflectivity $R_1$ on the left-hand mirror not too much less than unity, the power output is roughly constant over a very wide range of output couplings. The exact output coupling level applied to a high-gain laser is thus not nearly as critical a factor as it is for a low-gain laser.

At the same time it is evident that even fairly small losses caused by the finite reflectivity $R_1 < 1$ at the left-hand mirror do have a significant effect on the useful power output from the other end. Obtaining the maximum available power output clearly requires minimum internal losses ($R_1 \to 100\%$), and power output is generally maximized by smaller rather than larger coupling ($T_2 \leq 50\%$).

We could obviously differentiate Equation 12.50 or 12.52 to find the optimum output coupling $T_{2,opt}$ and the associated optimum output power. This yields, however, a transcendental equation for $T_{2,opt}$ does not seem useful to discuss in more detail here. Internal cavity losses an internal absorption coefficient $2\alpha_0$

## 12.4 LARGE OUTPUT COUPLING ANALYSIS

FIGURE 12.21
Measured output energy versus coupling in an atomic iodine photodissociation laser at different buffer gas pressures.

could also be added to the differential equations for the laser, but these equations then become much more difficult to integrate. The general effects of small distributed losses in the laser cavity are best assessed by assuming them to be incorporated as part of the left-hand mirror reflectivity $R_1$.

Many experimental examples of measured power output versus coupling in confirmation of the Rigrod analysis can be found in the literature. Figure 12.21 shows, for example, the power output versus output mirror reflectivity for a large atomic iodine photodissociation laser, intended as an amplifier (with a one-way power gain of 200 to 300 times), but used in these tests as an oscillator.

### REFERENCES

The analysis in this section is from W. W. Rigrod, "Saturation effects in high-gain lasers," *J. Appl. Phys.* **36**, 2487–2490 (August 1965). A complementary discussion, including inhomogeneous transitions, is W.W. Rigrod, "Gain saturation and output power of optical masers," *J. Appl. Phys.* **34**, 2602–2609 (September 1963); and a more general discussion including distributed internal loss is W. W. Rigrod, "Homogeneously broadened CW lasers with uniform distributed loss," *IEEE J. Quantum Electron.* **QE–14**, 377–381 (May 1978). See also G. M. Schindler, "Optimum output efficiency of homogeneously broadened lasers with constant loss," *IEEE J. Quantum Electron.* **QE–16**, 546–549 (May 1980). A rather opaque Rigrod-type analysis for ring-laser oscillators is A. C. Eckbreth, "Coupling considerations for ring lasers," *IEEE J. Quantum Electron.* **QE–11**, 796–798 (September 1965).

The various simplifying approximations usually made in laser power analyses are examined and estimates of their validity given by L. W. Casperson, "Laser power calculations: sources of error," *Appl. Opt*, **19**, 422–434 (February 1 1980).

### Problems for 12.4

1. *Optimum output coupling for a large-gain Rigrod-type laser.* Consider a homogeneous high-gain laser which obeys the Rigrod analysis. Suppose there are no internal losses; the round-trip unsaturated power gain is 5; the non-output mirror has a power reflectivity of 95%; and the output mirror has adjustable coupling with $R+T=1$ (i.e., no loss in this mirror). What is the optimum output coupling

from this laser, and the power-extraction efficiency at that coupling? (Only the power coming through the output mirror is counted as useful power.)

2. *Total power output from a high-gain laser oscillator.* Develop an analysis for the total power output from *both* ends of a high-gain Rigrod-type laser, and plot some examples of the total power output versus mirror transmission $T_2$ for various choices of the opposite reflectivity $R_1$. How much difference in efficiency does it make to include the power output from both ends?

3. *Dual output coupling values for a high-gain laser oscillator.* The Rigrod analysis shows that a standing-wave oscillator can have the same power output through mirror $M_2$ for two very different values of output coupling $T_2$. For example, with $G_0 = 10$ and $R_1 = 0.95$ the power output will have the same value for either $R_2 \approx 0.11$ (i.e., very heavy output coupling) or for $R_2 \approx 0.949$ (i.e., very light output coupling). Examine the difference in internal behavior of the laser for these two couplings, and explain in physical terms how these two very different couplings can lead to the same useful power output.

4. *Rigrod analysis of a one-way ring-laser oscillator.* Develop a large-output-coupling analysis of laser output power versus output coupling for a ring-cavity laser oscillator. Assume a three-mirror (triangular) ring cavity with the output mirror having power reflection $R_2$. The laser medium is located between the other two mirrors, both of which have power reflection $R_1$. Only the power transmitted through the output mirror is useful. Plot power out versus coupling for $G_0 = 10$ and $R_1 = 0.99, 0.95$ and $0.90$, and compare with results for a standing-wave cavity using the same gain medium.

5. *Two-segment ring-laser oscillator.* Extend the preceding problem to a ring laser in which the gain medium is broken into two segments of equal length with a finite reflectivity mirror having reflection coefficient $R_1$ between the two segments. Assume the output mirror has reflection and transmission coefficients $R_2$ and $T_2$. Compare your results to those for a standing-wave laser, assuming the same total gain medium and same value of $R_1$ in both.

6. *Gain saturation in a high-gain, double-pass laser amplifier.* A single-pass laser amplifier can be converted to double-pass operation (with twice the dB gain) by sending the input signal through the amplifier once in one direction and then reflecting it back through the same volume of the amplifier in the opposite direction. Two polarizers and a quarter-wave plate can be used to separate the incident and return beams, or the reflected beam can simply be tilted slightly so that the incident and reflected beams overlap within the medium but can be separated externally.

Using the Rigrod formalism, develop a relationship between input and output intensities from a homogeneous cw amplifier operated in this fashion. Evaluate the maximum available power and the power-extraction efficiency, and compare to the same laser medium operated as a single-pass amplifier.

# 13

# OSCILLATION DYNAMICS AND OSCILLATION THRESHOLD

In this chapter we discuss a number of additional topics related to the elementary properties of laser oscillation. We consider first the oscillation build-up time with which coherent oscillation develops from noise in a laser cavity. From this we develop a set of simple coupled rate equations which link cavity photons to laser atoms, and laser atoms to cavity photons. Using these equations we explore further the laser oscillation buildup and the remarkable threshold properties characteristic of the laser oscillator.

We then examine briefly some of the more complex laser cavities that are useful in practice, including multimirror laser cavities, ring-cavity lasers, bistable laser systems, and "lasers" with no cavity at all.

## 13.1 LASER OSCILLATION BUILDUP

The primary question to be addressed in this section is: How fast does the coherent oscillation in a laser cavity build up from noise, when the laser is first turned on?

### Oscillation Buildup Analysis

To answer this question, consider a laser cavity in which the laser gain exceeds the cavity losses, at least temporarily; and follow any small packet of signal energy through one complete round trip within the cavity, with a round-trip time $T = p/c$, as shown in Figure 13.1. The growth in circulating intensity in one round trip, starting with intensity $I_0$ at time $t = 0$, is then

$$I(T) = I_0 \times R_1 R_2 (R_3 \cdots) \exp[2\alpha_m p_m - 2\alpha_0 p] = I_0 \exp[\delta_m - \delta_c], \quad (1)$$

where all the notation has been defined in the preceding chapters. The net growth after $N$ round trips will be given by

$$I(NT) = I_0 \times \left[R_1 R_2 (R_3 \cdots) e^{2\alpha_m p_m - 2\alpha_0 p}\right]^N = I_0 \exp[N(\delta_m - \delta_c)] \quad (2)$$

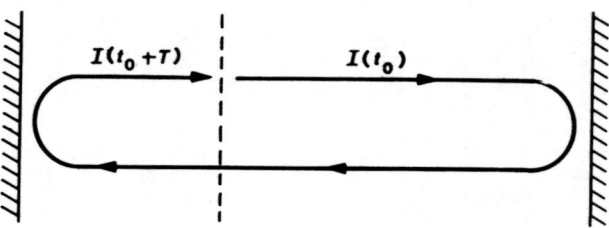

FIGURE 13.1
Round-trip intensity gain and travel time in a laser cavity.

which we can rewrite more generally as

$$I(t) = I_0 \exp\left[\frac{\delta_m - \delta_c}{T} t\right], \tag{3}$$

since the circulating intensity will make $N$ round trips in a time $t = NT$. It can also be convenient to write this as

$$I(t) = I_0 \exp\left[(\gamma_m - \gamma_c) t\right], \tag{4}$$

where we define the cavity growth and decay rates by

$$\gamma_m \equiv \frac{\delta_m}{T} = \frac{2\alpha_m p_m}{T} \quad \text{and} \quad \gamma_c \equiv \frac{\delta_c}{T} = \frac{2\alpha_0 p + \ln(1/R_{\text{tot}})}{T}, \tag{5}$$

where $R_{\text{tot}} \equiv R_1 R_2 (R_3 \cdots)$. The cavity lifetime or exponential decay time $\tau_c$ for optical signals in the cavity in the absence of laser gain is then given by

$$\tau_c \equiv \gamma_c^{-1} = \frac{T}{\delta_c}. \tag{6}$$

Since the round-trip time $T$ for a typical laser cavity is 1 to 10 ns, and the round-trip cavity losses may range from 1% ($\delta_c = 0.01$) to, say, 70% ($\delta_c \approx 1$), typical "cold cavity" decay times (with no laser gain) will range from the order of 1 ns to the order of 1 $\mu$s.

### Oscillation Buildup

The signal in a laser cavity following a sudden initial turn-on of the laser gain will thus build up exponentially with time much as shown in Figure 13.2, starting from an initial noise level $I_0$ which is usually very small, typically corresponding to only a few spontaneous-emission noise photons in the cavity. This buildup will continue until the circulating intensity reaches a steady-state level $I_{\text{ss}}$ with a very large number of photons in the cavity. This steady-state level corresponds to the oscillation level at which the laser gain is saturated down enough to just equal the total cavity losses (internal losses plus output coupling).

Figure 13.2 does assume either that the laser gain is very suddenly turned on to its full value at the start of the buildup interval, or else that some added cavity losses are suddenly turned off at this point, with a switching time short compared to the growth rate of the laser intensity. This may not be true in many real lasers. In an He-Ne laser, for example, the turn-on time for the plasma discharge and thus the laser gain will be much slower than the oscillation buildup time. In many other lasers, however, including E-beam-pumped excimer lasers, certain optically pumped lasers, and $Q$-switched lasers of all types, the gain can be switched on

## 13.1 LASER OSCILLATION BUILDUP

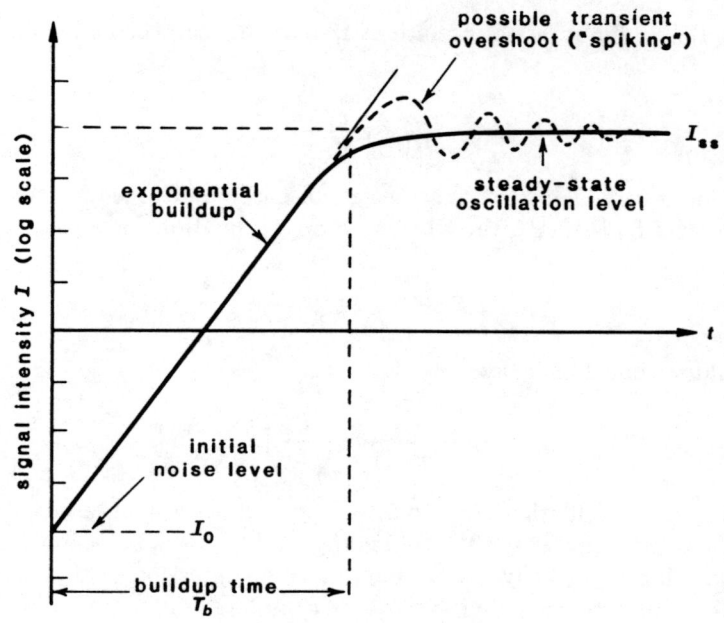

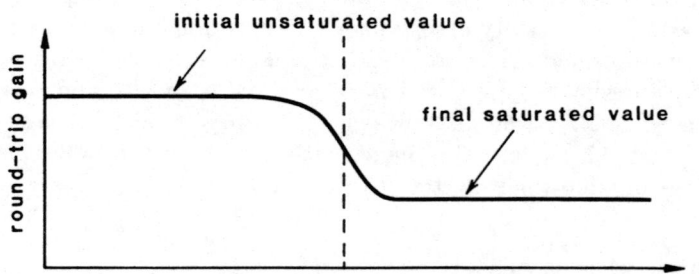

**FIGURE 13.2**
Exponential buildup of the signal intensity inside a laser cavity following sudden turn-on of the laser gain. The signal intensity starts from an initial noise level which is typically a few initial spontaneous-emission-noise photons, and grows to a very much larger steady-state oscillation level, with an approximate buildup time $T_b$. Certain lasers may approach the steady-state oscillation level with a transient overshoot or "spiking."

(or added losses switched off) in times short compared to the oscillation buildup time.

### Typical Oscillation Buildup Times

It is very convenient in discussions such as these to define a *normalized inversion ratio* $r$ as the ratio of the initial unsaturated gain coefficient $\delta_{m0}$ to the cold-cavity loss coefficient $\delta_c$, or

$$r \equiv \frac{\delta_{m0}}{\delta_c} = \frac{\gamma_{m0}}{\gamma_c} = \frac{2\alpha_{m0}p_m}{2\alpha_0 p + \ln(1/R_{\text{tot}})}, \tag{7}$$

The oscillation buildup rate of Equations 13.3 or 13.4 can then be written in the form

$$I(t) = I_0 \exp\left[\frac{r-1}{\tau_c} t\right]. \tag{8}$$

The total intensity buildup from the initial noise level $I_0$ to the final steady-state oscillation level $I_{ss}$ is then given, to a good approximation, by

$$I_{ss} \approx I_0 \exp\left[\frac{r-1}{\tau_c} T_b\right] \tag{9}$$

or the buildup time $T_b$ is given by

$$T_b \approx \frac{\tau_c}{r-1} \ln\left(\frac{I_{ss}}{I_0}\right). \tag{10}$$

The ratio of final oscillation level to initial noise level in real lasers may range from $I_{ss}/I_0 \approx 10^8$ to $I_{ss}/I_0 \approx 10^{12}$, depending on the type of laser. Since this ratio appears only logarithmically in the buildup-time expression of Equation 13.10, however, and since its logarithm varies only from $\ln(I_{ss}/I_0) \approx 18$ to $\ln(I_{ss}/I_0) \approx 28$, an exact knowledge of this ratio is not essential.

The general conclusion, in fact, is that the oscillation buildup time $T_b$ may range from $\approx 10$ to $\approx 30$ cavity decay times $\tau_c$, depending on how far the laser is pumped above threshold. Thus, for a rather long, low-loss He-Ne laser operated not far above threshold, with $L = 1$ m, $T = 6$ ns, $\delta_c = 3\%$, and $r = 1.1$, the buildup time will be $T_b \approx 50$ μs. For a short, high-gain Nd:YAG laser pumped well above threshold, on the other hand, with $L = 30$ cm, $T = 2$ ns, $\delta_c = 0.5$, and $r = 3$, the buildup time shortens to $T_b \approx 50$ ns.

### More Exact Buildup Analysis

In some but by no means all lasers the laser gain will saturate more or less immediately with increasing light intensity $I(t)$. Let us assume this, and also assume that the unsaturated growth rate $\gamma_{m0}$ is not much greater than the cold-cavity decay rate $\gamma_c$, or that the initial inversion ratio $r$ is not much greater than 1, so that the degree of saturation at steady state will remain small.

We can then write the instantaneous growth rate in a standing-wave laser oscillator cavity in the approximate form

$$\gamma_m(t) \approx \frac{\gamma_{m0}}{1 + 2I(t)/I_{\text{sat}}} \approx \gamma_{m0}\left[1 - 2I(t)/I_{\text{sat}}\right], \tag{11}$$

where $\gamma_{m0} \equiv r\gamma_c$ is the unsaturated laser growth rate. A more exact equation for the oscillation buildup, including gain saturation, can then be written in the form

$$\frac{dI}{dt} = \gamma I - \beta I^2, \tag{12}$$

where we have followed the notation commonly used in the literature, with $\gamma \equiv \gamma_{m0} - \gamma_c$ being the unsaturated growth rate and $\beta \equiv 2\gamma_{m0}/I_{\text{sat}}$ the saturation coefficient. The solution to this more exact equation is

$$I(t) = \frac{I_0 I_{ss} e^{\gamma t}}{I_{ss} + I_0(e^{\gamma t} - 1)}, \tag{13}$$

where $I_0$ is the initial intensity at turn-on, corresponding usually to a few noise photons inside the laser cavity; and $I_{ss} \equiv \gamma/\beta = (1 - \gamma_c/\gamma_{m0}) \times I_{sat}/2$ is the steady-state oscillation level at the end of the buildup period. The time delay following gain turn-on needed to reach, say, half the final intensity is then given from this expression

$$T_b = \frac{1}{\gamma} \ln\left(\frac{I_{ss} - I_0}{I_0}\right) \approx \frac{\tau_c}{r-1} \ln\left(\frac{I_{ss}}{I_0}\right), \tag{14}$$

which is essentially the same result we obtained earlier.

### Experimental Results

Figure 13.3 shows two examples of experimental results for oscillation buildup times in gas lasers. In Figure 13.3(a) a helium-neon laser which is initially oscillating at steady state is suddenly quenched by illuminating the He-Ne laser tube with a short but intense pulse of ultraviolet radiation from a xenon flashlamp. This UV radiation efficiently pumps neon atoms from a lower-lying $1s^5$ metastable level up into the lower level of the laser transition, thus destroying the laser gain and suddenly quenching the laser action, without significantly disturbing either the laser cavity or the laser discharge. The gain then recovers rapidly as these atoms relax back out of the lower laser level, and the laser oscillation builds back up again, with varying time delays for different steady-state intensities, as shown. Note that the numerical time delays agree generally with the numerical estimate of $\approx 50\mu s$ given earlier.

Figure 13.3(b) shows the buildup of oscillation in an optically pumped far-infrared laser that employs formic acid vapor as the laser medium for oscillation at a wavelength of 743 $\mu$m. The pumping power coming from the 9 $\mu$m $CO_2$ laser that excites this laser is turned on very rapidly in step-function fashion, using an acousto-optic modulator that has a 70-ns rise time, much faster than the buildup time for the far-infrared oscillation. The experimental results shown can then be fitted very accurately into Equation 13.13 using only a single intensity-scaling parameter and a laser gain coefficient $\alpha_m$ that is directly proportional to the optical pumping power.

### Spiking Behavior

In many solid-state lasers, as well as other types of lasers, the excess gain that is present during the oscillation buildup period does not saturate immediately with increasing laser intensity, but decreases only after a certain time delay required for the circulating laser intensity $I(t)$ to "burn up" the excess population inversion. Analysis of this situation requires a more exact set of equations, to describe the dynamics of the atomic populations as well as the cavity fields.

The oscillation buildup in this situation may not converge smoothly to the final steady-state value, but may instead exhibit a strong transient overshoot, followed by quasi periodic "spiking" or relation oscillation behavior, as illustrated in Figure 13.2. We will carry out a more detailed examination of this interesting but rather useless spiking behavior in a later chapter.

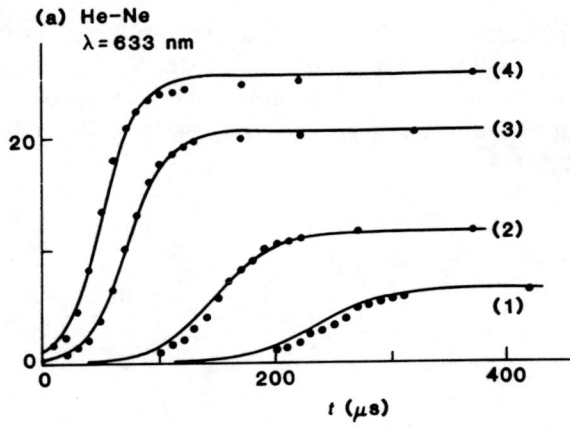
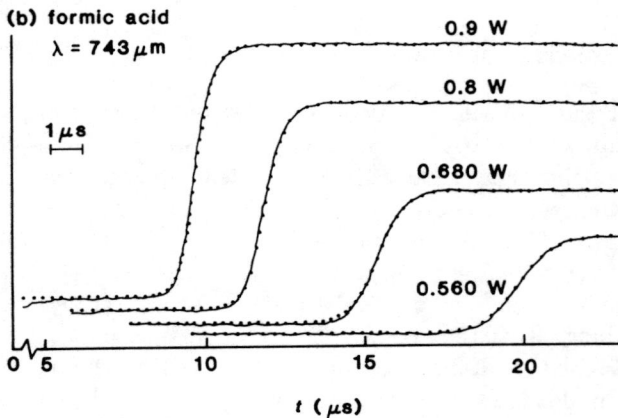

FIGURE 13.3
Laser oscillation buildup times: (a) in a helium-neon laser; (b) in a far-infrared laser.

## REFERENCES

The experimental results for laser oscillation buildup shown in this section come from B. Pariser and T. C. Marshall, "Time development of a laser signal," *J. Appl. Phys.* **6**, 232–234 (June 15 1965); and from J. Wascat, D. Dangoisse, P. Glorieux, and M. Lefebvre, "Growth of emission in a far infrared laser," *IEEE J. Quantum Electron.* **QE–19**, 92–95 (January 1983).

For another careful and detailed set of experiments, see F. T. Arecchi and V. De Giorgio, "Statistical properties of laser radiation during a transient buildup," *Phys. Rev.* **A3**, 1108–1124 (March 1971).

For still another typical illustration of laser oscillation buildup, but with a time-varying inversion and gain, see B. K. Garside, E. A. Ballik, and J. Reid, "Pulse delays in TEA $CO_2$ lasers," *J. Appl. Phys.* **43**, 2387–2390 (May 1972).

---

Problems for 13.1

1. *Discrete step behavior of laser oscillation buildup.* Suppose a very short optical pulse, like a little "bullet" of light, circulates around inside a laser cavity. The laser cavity is 120 cm long and has end mirrors with power reflectivities $R_1 = 0.35$ and $R_2 = 0.8$. In the exact center of the cavity is a very short laser rod with one-way

power gain of 3 times ($G = 3$). Plot the instantaneous energy in the circulating pulse as a function of time for 3 full round trips around the cavity, starting with unity initial energy. Also evaluate and plot the net exponential growth rate in the cavity, on the same plot. Compare this exponential growth line with the exact energy versus time plot for the circulating pulse.

2. *More complicated discrete buildup calculation.* Repeat the previous problem assuming the end mirrors have reflectivities $R_1 = R_2 = 0.5$, the one-way power gain is 4 times, and there is a uniformly distributed loss inside the laser cavity such that $2\alpha_0 p = 0.4$.

3. *Cavity lifetime in a semiconductor diode laser.* A typical GaAs injection diode laser cavity may be 200 $\mu$m long, with an index of refraction $n \approx 3.8$ and an end-mirror reflectivity $R \approx 0.36$ because of the air-dielectric interface. If the internal absorption losses in the cavity are assumed to be small (which may not in fact be true in real injection lasers), what is the cavity lifetime $\tau_c$ in such a cavity?

4. *Exact buildup solution for a typical laser.* Calculate and plot the exact oscillation buildup behavior using the fast-saturation formula (Equation 13.13) given in this section, assuming a steady-state intensity $I_{ss}$ that is $10^8$ times the initial noise intensity $I_0$ and an unsaturated growth rate $\gamma_{m0}$ that is 1.2 times the cavity decay rate $\gamma_c$. Plot both the normalized intensity $I(t)/I_{ss}$ on a log scale, and the instantaneous growth rate $\gamma_m(t)/\gamma_c - 1$ on a linear scale, versus the normalized time $t/\tau_c$.

## 13.2 DERIVATION OF THE CAVITY RATE EQUATION

In this section we will extend the cavity growth-rate calculation developed in the preceding section to derive a "photon rate equation" for the signal intensity or the number of photons in each laser cavity mode, including—for the first time—the effects of spontaneous emission. In the following section we will then combine these cavity rate equations with the atomic rate equations we have developed earlier to obtain a set of coupled cavity plus atomic rate equations which are simple, and yet extremely useful in analyzing many fundamental aspects of laser theory.

### Derivation of the Cavity Rate Equation

The exponential growth rate for the signal intensity inside a laser cavity derived in the previous section had the general form

$$I(t) = I_0 \exp\left[(\gamma_m - \gamma_c)t\right], \tag{15}$$

If either of the coefficients $\gamma_m$ or $\gamma_c$ is time-varying, however—as they well may be in real cases—then we must convert this equation to the more general differential form

$$\frac{dI(t)}{dt} = [\gamma_m(t) - \gamma_c(t)] \times I(t) \tag{16}$$

The parameters $\gamma_m$ or $\gamma_c$ might become time-varying, for example, because the gain coefficient saturates, or because we deliberately modulate the cavity losses or cavity output coupling with time. Equation 13.15 is then a correct solution to the more general Equation 13.16 only when the two gain and loss rates are constant. In particular, the gain coefficient $\gamma_m$ will be directly proportional to the inverted population difference $\Delta N(t) \equiv N_2(t) - N_1(t)$ on the laser transition; and this population difference will very likely change with time in a real laser. We can take this dependence of $\gamma_m$ on $\Delta N$ into account by writing

$$\gamma_m(t) \equiv \frac{2\alpha_m p_m}{T} \equiv K \, \Delta N(t), \tag{17}$$

where all the other geometrical and atomic parameters of the system are absorbed into the constant $K$.

At the same time that we do this, we can also conveniently express the total signal energy inside the laser cavity in dimensionless units by defining a "number of photons" $n(t)$ in the cavity by

$$n(t) = \text{"number of photons in the cavity"}$$

$$\equiv \left[ \frac{\text{total signal energy in the cavity}}{\text{quantum of energy, } \hbar\omega} \right] \tag{18}$$

$$= \text{const} \times I_{\text{circ}}(t).$$

It should be emphasized that we are not focusing any special attention on the photon nature of light by writing this equation—the emphasis in laser analyses should almost always be on the wave rather than the particle nature of light. Rather, we are simply expressing the total signal energy in the laser cavity in the convenient units of $\hbar\omega$. Also, we will not really need any explicit formula for the constant appearing in the last line of Equation 13.18, although if such a formula is wanted, the photon number $n(t)$ in a low-gain standing-wave cavity of length $L$, cross-sectional area $A$ and circulating intensity $I_{\text{circ}}$ can be calculated to a good first approximation from

$$n(t) \approx \frac{2AI_{\text{circ}}(t)L}{\hbar\omega c} = \frac{2V_c}{\hbar\omega_a c} I_{\text{circ}}(t). \tag{19}$$

where $V_c = AL$ is the volume of the cavity mode.

Equations 13.16, 13.17, and 13.18 can then be combined to give the *cavity rate equation*

$$\frac{dn(t)}{dt} = [K\Delta N(t) - \gamma_c] \times n(t) \tag{20}$$

or

$$\frac{dn(t)}{dt} = K\left[N_2(t) - N_1(t)\right] n(t) - \gamma_c n(t), \tag{21}$$

where $N_1(t)$ and $N_2(t)$ are the total number of atoms in the lower and upper levels of the laser transition. The first two terms on the right-hand side of Equation 13.21 then represent stimulated emission and absorption between the cavity mode and the atoms, while the third term represents the cavity losses plus output coupling.

Note that in earlier chapters we have consistently used $N_1(t)$ and $N_2(t)$ to indicate *atomic densities*, or numbers of atoms per unit volume. In writing the

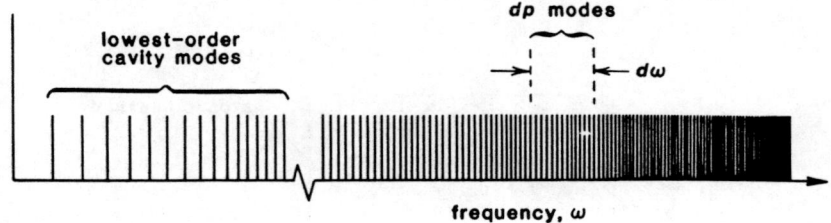

**FIGURE 13.4**
The lowest and higher-order resonant mode frequencies in a typical closed and lossless resonant cvity.

cavity and atomic rate equations in this and following chapters, however, it will be more convenient to let the symbols $N(t)$ and $\Delta N(t)$ represent the *total numbers of atoms* inside the laser cavity. The student will have to be a little cautious in interpreting these symbols, therefore, in any formulas from now on in this book.

### Value of the Coupling Constant $K$

By using the formulas derived in earlier chapters for the gain coefficient $\alpha_m$, we can rewrite the constant $K$ appearing in Equations 13.17, 13.20, and 13.21 in the form (for a lorentzian transition)

$$K \equiv \frac{2\alpha_m p_m}{T \Delta N} = \frac{3^*}{4\pi^2} \frac{\omega_a \gamma_{\text{rad}} \lambda^3}{\Delta\omega_a V_c}, \qquad (22)$$

where $V_c$ is the volume of the cavity or, more precisely, of the cavity mode with which the atoms are interacting. Note again that we are now using $N_1$ and $N_2$ to indicate the *total numbers of atoms* in the laser levels, so that $\Delta N(t)/V_c$ is the volume-averaged inversion density with which the cavity mode interacts. Equation 13.22 can be reduced to the particularly simple and useful form

$$K = \frac{3^* \gamma_{\text{rad}}}{p} \qquad (23)$$

if we define a parameter $p$, called the *cavity mode number* (no relation to the laser cavity perimeter $p$), given by

$$p \equiv \frac{4\pi^2 V_c}{\lambda^3} \frac{\Delta\omega_a}{\omega_a}. \qquad (24)$$

This parameter has a very important physical significance, as we will now show.

### Frequency Distribution of Resonant Cavity Modes

Suppose we consider some arbitrarily shaped enclosure or cavity having closed and completely reflecting walls. Let us then calculate all the theoretically possible lowest and higher-order electromagnetic modes in this cavity; and plot the resonant frequencies of these modes as tic marks on a frequency scale, as shown in Figure 13.4. Then, at some low frequency, corresponding to a wavelength on the order of the cavity dimensions, we will see the lowest-order resonant mode of the cavity, followed by a succession of higher-order resonant modes with successively higher resonant frequencies, as shown in Figure 13.4.

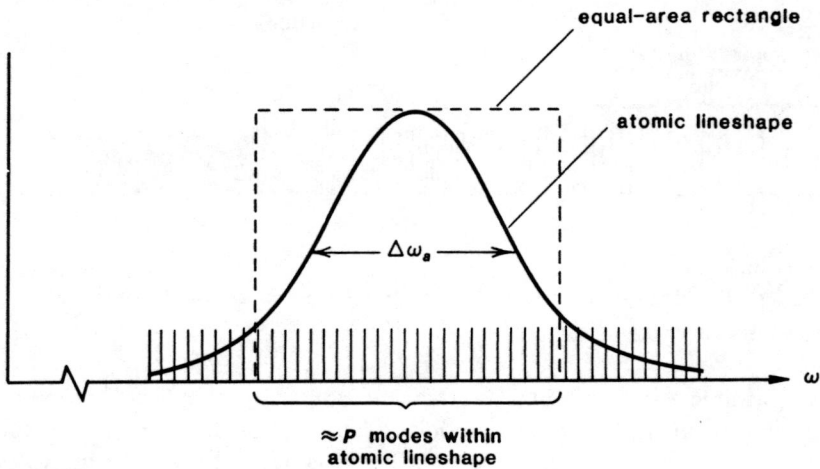

**FIGURE 13.5**
Physical interpretation of the cavity mode number $p$. Note that the areas under the lorentzian lineshape with linewidth $\Delta\omega_a$ and the rectangular box of width $(\pi/2) \times \Delta\omega_a$ are the same.

As we go to much higher frequencies, where the cavity dimensions become large compared to the resonance wavelengths, these resonant modes will become more and more closely spaced along the frequency axis, so that the mode distribution in frequency space will become very dense. In fact, it can be shown that in any such enclosure or cavity, regardless of its exact shape, the number of resonant modes falling within a unit (radian) frequency interval, or the *resonant mode density* $\rho(\omega)$ along the frequency axis, will be given by

$$\rho(\omega) = \frac{dp(\omega)}{d\omega} \equiv \left[\frac{\text{number of cavity modes}, dp}{\text{frequency range}, d\omega}\right] = \frac{8\pi V_c}{\lambda^3}\frac{d\omega}{\omega}. \quad (25)$$

This formula will hold for any cavity shape, whenever the frequencies are high enough that the cavity dimensions become large compared to the resonant wavelengths.

This distribution function can be given another interpretation that does not even require the concept of resonant modes. Suppose we wish to describe an arbitrary electromagnetic field distribution in some large rectangular volume. We can always expand such a field distribution using a Fourier-series expansion in all three spatial coordinates or, to put this in another way, we can expand the fields in a set of waves traveling in all possible directions through the volume. (This process is sometimes referred to as "box normalization" in electromagnetic theory or in quantum mechanics.) The number of independent Fourier components, or of traveling-wave terms, call this number $dp(\omega)$, needed to give a complete description of an arbitrary electromagnetic field within a large rectangular region of volume $V_c$, assuming this field distribution is made up of frequency components lying within a frequency range $d\omega$, is then given by exactly the same formula $dp(\omega) = \rho(\omega)\,d\omega$ given in Equation 13.25.

### The Cavity Mode Number p

The mode number $p$ appearing in the stimulated transition constant formula (Equations 13.23 or 13.24) can then be understood as *the effective number of laser cavity modes lying within the atomic transition linewidth* $\Delta\omega_a$, as shown in Figure 13.5. That is, this number is given by the formula

$$p \equiv \rho(\omega) \times \frac{\pi \Delta \omega_a}{2} = \frac{4\pi^2 V_c}{\lambda^3} \frac{\Delta \omega_a}{\omega_a}. \qquad (26)$$

The effective frequency bandwidth multiplying the mode-density function $\rho(\omega)$ in this equation is $(\pi/2)$ times the atomic linewidth $\Delta\omega_a$ rather than just $\Delta\omega_a$, because this is the width of an equivalent rectangular distribution having the same peak height and the same area as a lorentzian lineshape, as shown in Figure 13.5.

The linewidth $\Delta\omega_a$ of any laser transition is always small compared to the transition frequency $\omega_a$, but the cavity volume $V_c$ of a normal laser cavity is always very much larger than a single cubic wavelength. The mode number $p$ is thus normally an extremely large number for ordinary laser cavities, with values typically on the order of $p \approx 10^7$ to $p \approx 10^{10}$. We will learn more about the significance of this parameter very shortly.

### Frequency Dependence of the Coupling Coefficient $K$

Before going on to introduce the concept of spontaneous emission into a cavity mode, we should note that the value of the rate-equation coupling constant $K$ given in Equations 13.22 and 13.23 is obviously the *midband value*, appropriate to a cavity mode tuned to the center of the atomic transition. If we consider instead a cavity mode whose resonant frequency $\omega_i$ is tuned off the atomic line center, then the response of the atoms to the cavity fields will be reduced by the atomic lineshape, and hence the coupling coefficient $K(\omega_i)$ for that off-resonance mode will be reduced with a frequency dependence $K(\omega_i) = 2\alpha_m(\omega_i)p_m/T$ that is given for a lorentzian transition by

$$K(\omega_i) = K_0 \times \frac{1}{1 + [2(\omega_i - \omega_a)/\Delta\omega_a]^2}, \qquad (27)$$

or for a gaussian transition by

$$K(\omega_i) = \sqrt{\pi \ln 2} \times K_0 \times \exp\left[-4\ln 2 \left(\frac{\omega_i - \omega_a}{\Delta\omega_a}\right)^2\right]. \qquad (28)$$

where $K_0 \equiv K(\omega_a) \equiv 3^*\gamma_{\rm rad}/p$.

Suppose we sum the coupling coefficients $K(\omega_i)$ over all the cavity modes $\omega_i$ underneath an atomic transition, weighted by their frequency dependences, while also averaging the polarization factor $3^*$ over all field or atomic polarizations. This sum over all modes is, in essence, an integration over the mode density shown in Figure 13.5, weighted by the atomic transition lineshape shown in that figure, as given in Equations 13.27 or 13.28. Using either of these lineshapes, we can obtain the very fundamental result that

$$\sum_{\substack{\text{all}\\\text{modes}}} K(\omega_i) \to \int_0^\infty K(\omega) \times \rho(\omega)\, d\omega \equiv \gamma_{\rm rad}. \qquad (29)$$

This fundamental result, which says that the sum of $K(\omega_i)$ over all cavity modes under an atomic linewidth is just equal to the $\gamma_{\rm rad}$ value for that transition, is in fact a very general result, completely independent of the atomic lineshape, the

### Introduction of Spontaneous Emission

We have thus far written the cavity rate equation for a single cavity mode in the form

$$\frac{dn}{dt} = K[N_2 - N_1]n - \gamma_c n, \qquad (30)$$

where this form includes *stimulated transitions* (that is, stimulated-emission and stimulated-absorption terms), and also *cavity loss terms*, but not yet *spontaneous-emission terms*.

We must now take into account the process of *spontaneous emission* from the upper-level atoms into this cavity mode (as well as into every other cavity mode within the laser cavity volume). That is, the atoms are spontaneously emitting in a noise-like fashion, at a rate directly proportional to the number of upper-level (but not lower-level) atoms, and independent of the number of photons $n$ already in each cavity mode; and a small fraction of this spontaneous emission will have the right direction, polarization and frequency to feed directly into the cavity mode we are considering. The rate equation for each individual cavity mode must thus be extended to the more complete form

$$\frac{dn}{dt} = K[N_2 - N_1]n + K_{sp}N_2 - \gamma_c n, \qquad (31)$$

where $K_{sp}$ is a *spontaneous-emission constant* governing the rate of emission from the upper-level atoms into that particular cavity mode.

But, there is a fundamental result of quantum theory—one of the most fundamental principles of quantum electronics, in fact—which says that *the spontaneous-emission rate from any given set of atoms into any one individual cavity mode is exactly equal to the stimulated-emission rate that would be produced from those same atoms by one photon of coherent signal energy present in the same mode.* For each cavity mode with resonance frequency $\omega_i$, therefore, the *stimulated* and *spontaneous* transition constants involved in the interaction with a given set of atoms must necessarily be related by

$$K_{sp}(\omega_i) \equiv K(\omega_i) \qquad (32)$$

for each and every cavity mode. Therefore, the cavity rate equation for each separate cavity mode, including spontaneous emission, can also be rearranged into the form

$$\frac{dn}{dt} = KN_2[n+1] - KN_1 n - \gamma_c n. \qquad (33)$$

Written in this form the equation seems to say that, whereas the net atomic *absorption* rate is proportional to the instantaneous number of cavity photons $n(t)$, the net *emission* rate is proportional to $n(t) + 1$, i.e., the number of cavity photons *plus one*.

### Spontaneous Emission: The "Extra Photon"

This "plus one" factor caused by the spontaneous emission sometimes leads laser workers to speak of an "extra photon" in the cavity mode—a photon that somehow causes only downward transitions.

It is important to understand, however, that this additional spontaneous-emission term in the cavity rate equation is much more accurately viewed as an incoherent or noise-like driving term which excites the cavity mode in a random or noise-like fashion, completely uncorrelated with the coherent stimulated-emission terms or with any cavity signal that may already be present. This spontaneous-emission term thus acts as a fundamental *quantum noise source* in the cavity equations. It is this quantum noise source which is responsible for the ultimate noise figure of laser amplifiers, for example, and also for the quantum noise fluctuations in phase and amplitude that are present in even the most ideally stabilized laser oscillators or frequency standards.

One of the practical conclusions stemming from this is that it is impossible to make a laser amplifier—or, in fact, any other kind of amplifier—with an equivalent input noise power less than one noise photon per Hz of bandwidth.

### Derivation of the Spontaneous-Emission Coefficient

Let us now verify that this spontaneous emission rate for noise photons into each cavity mode corresponds exactly to the spontaneous atomic emission process that we have already discussed in earlier chapters. We can recall first that the total relaxation rate out of any upper atomic level will normally include two different relaxation processes. First of all, there will normally always be a *purely radiative relaxation rate*, or a *spontaneous emission rate*, $\gamma_{\rm rad} N_2$ on the $2 \to 1$ transition we are considering. In addition, there may be (and usually will be) both *nonradiative relaxation rates* from level 2 down to various lower levels and possibly *other purely radiative decay rates* from level 2 down to lower levels *other* than level 1.

The purely radiative or spontaneous emission part of this relaxation on the $2 \to 1$ transition can then be described in two different but physically equivalent fashions. From a "free-space" viewpoint, each upper-level atom in the cavity volume has a certain probability per unit time $\gamma_{\rm rad}$ of radiating spontaneously at some frequency within the atomic lineshape, and into some random emission direction. If we look into the open sides of the cavity from any external point, we will see this spontaneous emission coming out from all sides of the cavity, with a lineshape corresponding to the atomic transition lineshape, and with a total emission rate (in photons/second) into all directions given by $\gamma_{\rm rad} N_2$. Essentially all this spontaneous emission is emitted out through the open sides of a normal laser cavity, although a very minute portion of it is radiated into the low-loss direction exactly along the cavity axis.

From an alternative "cavity mode" viewpoint, however, the atoms can be thought of as spontaneously radiating this same energy, not out into free space, but rather *directly into each of the very large number of resonant cavity modes (mostly very lossy modes) whose frequencies lie within the atomic linewidth*. The total spontaneous-emission fields coming out of the cavity in all directions can then be viewed as the result of the very rapid leakage or diffraction loss from all of these cavity modes out the sides of the cavity.

In considering the total number of modes within a resonant cavity, we must keep in mind that all of the reasonably *low-loss* cavity modes—that is, those lowest and slightly higher-order axial-transverse modes that we have described elsewhere in this text—really represent only a very minute fraction of the total number of cavity modes associated with the cavity volume. There will typically be within an atomic linewidth only a few, or at most a few hundred, low-order axial-plus-transverse modes which describe the radiation traveling in the low-loss directions very close to the cavity axis. There are, however, some $p = 10^7$ to $10^{10}$ other potential cavity modes, most of them having enormously high losses out the cavity sides, which are needed in principle to describe all possible field configurations traveling in all directions within the cavity volume and within the atomic linewidth.

From the second viewpoint, therefore, each of these laser cavity modes within the atomic linewidth should receive spontaneous emission at a rate given by Equation 13.33, with a $K$ value $K(\omega_i)$ appropriate to that particular cavity mode. But, nearly all of these modes have extremely fast decay rates out the side of the cavity, and hence this energy radiated from the atoms into all these modes is immediately radiated on out of the cavity in all directions.

From this viewpoint we must equate the total spontaneous-emission power coming from the atoms to the total spontaneous-emission power emitted into all these cavity modes. That is, suppose we label each cavity mode by its resonance frequency $\omega_i$. Then we can write for the total spontaneous rate on the $2 \to 1$ transition

$$\sum_{\omega_i} K_{\rm sp}(\omega_i) N_2 = \sum_{\omega_i} K(\omega_i) N_2 = \gamma_{\rm rad} N_2 \qquad (34)$$

But, we have already shown in Equation 13.29 that the summation in the middle term of this equation just adds up to the radiative decay rate $\gamma_{\rm rad}$, so that this "conservation of total spontaneous emission" is indeed verified.

We can now understand better, as well, why the midband interaction constant $K$, or $K_{\rm sp}$, must have the value $3^* \gamma_{\rm rad}/p$ given in Equation 13.23. If we assume, as is reasonable, that the geometrical and atomic factors determining the spontaneous-emission rate into each mode within the atomic lineshape are likely to be essentially the same, except for polarization factors and for the atomic lineshape itself, then we can approximate the total spontaneous rate into all of the $\approx p$ modes within the main part of the atomic linewidth by

$$\gamma_{\rm rad} = \sum_{\omega_i} K_{\rm sp}(\omega_i) \approx p \times K_0, \qquad (35)$$

where $K_0 \equiv K(\omega_a) = K_{\rm sp}(\omega_a)$ refers to the on-resonance value for some preferred lowest-loss cavity mode located near the center of the atomic line. If we put in a polarization factor $3^*$, the $K_0$ value for this preferred mode located close to the line center becomes

$$K_0 = K_{\rm sp}(\omega_a) = K(\omega_a) = \frac{3^* \gamma_{\rm rad}}{p}, \qquad (36)$$

where $3^*$ has a value appropriate to that particular mode. But this is just what we started with in Equation 13.23. The $p$ in the denominator simply represents the fact that $1/p$ of the total spontaneous emission from the upper level atoms goes into that one particular cavity mode.

## REFERENCES

The spontaneous-emission intensity per mode, or the fraction $\approx 1/p$ of the total spontaneous emission coupled into each individual laser cavity mode, is a particularly important parameter in semiconductor lasers, where the mode number $p$ is comparatively small and the spontaneous emission comparatively strong. For an example of this, see W. Streifer, D.R. Scifres, and R.D. Burnham, "Analysis of diode laser properties," *IEEE J. Quantum Electron.* **QE–18**, 1918–1929 (November 1982).

The mode-density arguments, and especially the "extra photon" explanations of spontaneous emission in this section, depend in a fundamental way on the modes in question being a set of power-orthogonal electromagnetic modes. The low-order axial plus transverse cavity modes that are commonly used to describe open-sided laser cavities are, however, not power-orthogonal; and as a result the effective spontaneous-emission rate into these modes can appear to be greater than the amount corresponding to one added photon, by a so-called "excess spontaneous-emission factor," which increases the increasing diffraction loss in the cavity.

The existence of such an excess spontaneous-emission factor was first predicted for gain-guided semiconductor diode lasers by K. Petermann, "Calculated spontaneous emission factor for double-heterostructure injection lasers with gain-induced waveguiding," *IEEE J. Quantum Electron.* **QE-15**, 566–570 (July 1979). A good explanation of how this excess emission factor depends on the non-power-orthogonality of the laser modes, and how it can be reconciled with the fundamental arguments given in this section, can be found in H.A. Haus and S. Kawakami, "On the "excess spontaneous emission factor" in gain-guided laser amplifers," *IEEE J. Quantum Electron.* **QE–21**, 63–69 (January 1985).

---

## Problems for 13.2

1. *Cavity photon number in a real laser.* A certain Nd:YAG laser is 1 meter long, has internal power losses of 5% per one-way pass, and end mirrors with reflectivities $R_1 = 100\%$ and $R_2 = 95\%$. The cw power output through mirror $M_2$ is 1 watt. What is the total number of photons $n_{ss}$ in the laser cavity when it is oscillating?

2. *Effective width of a lorentzian transition.* Verify that the effective width of a lorentzian lineshape—that is, the width of a rectangular lineshape having the same peak height and the same total area—is in fact given by $(\pi/2) \times \Delta\omega_a$.

3. *Examples of the cavity-mode density formula.* Derive the cavity-mode density expression $\rho(\omega)$ given in this section for one or more specific simple cavity shapes, such as rectangular or cylindrical cavities, by starting with the standard resonant-mode formulas for microwave cavities and then taking the limit as the wavelength becomes very small compared to the cavity dimensions.

---

## 13.3 COUPLED CAVITY AND ATOMIC RATE EQUATIONS

We must now proceed to join the *cavity rate equations* developed in the preceding section to the *atomic rate equations* developed in earlier chapters. The result will be a set of coupled cavity plus atomic rate equations that are very useful

## CHAPTER 13: OSCILLATION DYNAMICS AND OSCILLATION THRESHOLD

in describing laser threshold behavior, laser amplitude modulation, laser spiking and $Q$-switching, and a wide range of other laser phenomena.

### Atomic Rate Equations

The signal and noise photons in the cavity mode discussed in Section 13.2 were assumed to be interacting with a two-level atomic system having total populations $N_1(t)$ and $N_2(t)$ in the lower and upper levels, respectively. Drawing on our results from earlier chapters, we can then write a pair of atomic rate equations for these level populations in the same form as in earlier chapters, that is

$$\frac{dN_1}{dt} = -W_{12}N_1 + W_{21}N_2 + \begin{bmatrix} \text{pumping} \\ \text{terms} \end{bmatrix} + \begin{bmatrix} \text{relaxation} \\ \text{terms} \end{bmatrix},$$
$$\frac{dN_2}{dt} = W_{12}N_1 - W_{21}N_2 + \begin{bmatrix} \text{pumping} \\ \text{terms} \end{bmatrix} + \begin{bmatrix} \text{relaxation} \\ \text{terms} \end{bmatrix}. \tag{37}$$

The $W_{12}N_1(t)$ and $W_{21}N_2(t)$ terms are the stimulated-transition terms caused by the cavity fields. The exact form of the pumping and relaxation terms in each equation will depend on the details of the particular atomic system and how it is being pumped or excited.

But, we know that the stimulated-transition probabilities $W_{12}$ and $W_{21}$ that appear in these atomic rate equations are themselves directly proportional to the signal energy, or to the cavity photon number $n(t)$, in the resonant cavity mode. We can thus write these stimulated-transition probabilities (leaving out degeneracy effects for simplicity) as being directly proportional to $n(t)$ in the form

$$W_{12} = W_{21} = K'n(t), \tag{38}$$

where $K'$ is again a proportionality constant which contains all the other geometrical and atomic parameters. The stimulated transition terms in the atomic rate equations can thus be written as

$$W_{21}N_2 - W_{12}N_1 = K'\left[N_2(t) - N_1(t)\right]n(t). \tag{39}$$

But every time an atom makes a signal-stimulated transition downward in the atomic rate equations, giving up an energy of $\hbar\omega$, this energy must be delivered into one of the cavity modes, so that the cavity photon number must simultaneously go up by one unit of $\hbar\omega$ in the cavity rate equation for that mode. The reverse argument must of course apply equally well to stimulated absorption transitions going in the opposite direction. The stimulated-transition rates $K(N_1 - N_2)n$ and $K'(N_1 - N_2)n$ in the cavity and in the atomic rate equations must therefore be numerically identical; and hence the constants $K$ and $K'$ in front of these terms must be the same, so that in fact $K' \equiv K$ (see the Problems at the end of this section for another way of deriving this same result.)

The form of the pumping and relaxation terms in Equations 13.37 will depend on the exact atomic system being considered. Suppose we now consider, as a simple but specific example, two upper atomic levels $E_1$ and $E_2$ such as we have considered in earlier chapters, with a pumping rate $R_p$ into the upper level, and with the usual relaxation rates (in the optical-frequency approximation) from the upper and lower levels downward. The complete atomic rate equations for this system will then take on the form

$$\frac{dN_2}{dt} = R_p - Kn[N_2 - N_1] - \gamma_2 N_2,$$
$$\frac{dN_1}{dt} = Kn[N_2 - N_1] + \gamma_{12} N_2 - \gamma_{10} N_1, \tag{40}$$

where the coupling coefficient $K = K(\omega_i)$ is exactly the same as in the corresponding cavity rate equation.

### Complete Coupled Cavity and Atomic Equations

We have shown in Section 13.32 how to write the cavity rate equation for any one individual cavity mode, including spontaneous emission, and have also pointed out that a real laser system will have $\approx p$ cavity modes, each one of which is, at least in principle, able to interact with the atoms contained within the cavity volume. The final result of this discussion is then that to describe properly, even within the rate-equation approximation, a laser cavity having a large number of resonant modes, each labeled by index $i$, plus a set of atoms with populations $N_1$ and $N_2$, we must, at least in principle, write down a separate rate equation for each cavity mode individually, in the form

$$\frac{dn_i(t)}{dt} = K_i N_2(t)[n_i(t) + 1] - K_i N_1(t) n_i(t) - \gamma_{ci} n_i(t), \tag{41}$$

where $n_i$ is the cavity photon number, $K_i$ the coupling constant, and $\gamma_{ci}$ the cavity decay rate for the $i$-th cavity mode. We must then also write a pair of rate equations for the atomic populations in the general form developed in this section, namely,

$$\frac{dN_2(t)}{dt} = \sum_i K_i n_i(t)[N_1(t) - N_2(t)] + + \begin{bmatrix} \text{pumping} \\ \text{terms} \end{bmatrix} + \begin{bmatrix} \text{relaxation} \\ \text{terms} \end{bmatrix},$$
$$\frac{dN_1(t)}{dt} = -\sum_i K_i n_i(t)[N_1(t) - N_2(t)] + + \begin{bmatrix} \text{pumping} \\ \text{terms} \end{bmatrix} + \begin{bmatrix} \text{relaxation} \\ \text{terms} \end{bmatrix}. \tag{42}$$

The two atomic levels are, in other words, potentially coupled to the total set of $p$ near-resonant cavity modes, as illustrated in Figure 13.6, as well as to whatever pumping and relaxation processes may be present.

Note that in these equations the stimulated-transition terms for the atoms must be summed over the total stimulated-transition effects produced by the signal fields in *all the cavity modes* acting on the atoms (or at least all those cavity modes that contain any significant number of photons). Note also that no additional spontaneous-emission terms need be added to the atomic rate equations, because the transition rate due to spontaneous emission into all the cavity modes is already included in the purely radiative part of the relaxation terms.

### Idealized Single-Mode, Single-Level Rate Equations

Writing out the complete set of cavity rate equations for $p \approx 10^8$ cavity modes would be a daunting task, with or without the assistance of a computer.

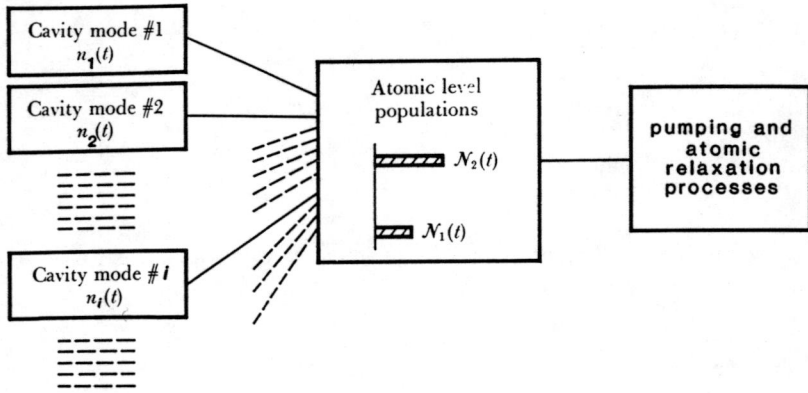

**FIGURE 13.6**
One set of atoms coupled to many cavity modes within the atomic linewidth.

Fortunately, in most real lasers we only need to write out explicitly the cavity rate equations for one or a few of the most favored or lowest-loss cavity modes, and not for the whole set of $p$ such modes. In fact, one of the most remarkable features of laser action is that a typical laser cavity having perhaps $p \approx 10^8$ individual and distinct cavity resonance modes can still oscillate in just one or a few of these cavity modes. So long as only one or a few cavity modes are excited with any significant number of photons $n_i$, we need write down the rate equations for only those few modes.

In fact, the simplest possible laser model—but one that still contains all the essential physics—is to assume that there is just *one* lowest-loss (or highest-gain) preferred cavity mode that builds up any significant photon number $n(t)$, so that we need write only *one* cavity rate equation. The atomic rate equations can also be put into their simplest form by assuming that the relaxation rate downward out of level 1 is sufficiently fast that $N_1 \approx 0$ under all circumstances, and that the pumping into level $N_2$ can be described by a simple pumping rate $R_p$.

The coupled cavity and atomic rate equations 13.41 and 13.42 will then reduce to their simplest possible combined form, namely,

$$\frac{dn}{dt} = KN_2(n+1) - \gamma_c n,$$

$$\frac{dN_2}{dt} = R_p - KN_2 n - \gamma_2 N_2. \tag{43}$$

This simple pair of equations is still surprisingly general, and we will use these two equations extensively to analyze several fundamental aspects of laser behavior in succeeding chapters.

We might also just mention some of the limitations of the coupled cavity plus atomic rate equations for analysing laser behavior, even in the case where we might write a larger number of cavity mode equations. In particular, this approach is necessarily limited to the small-signal or rate-equation atomic regime, as described in earlier chapters. No coherent-pulse effects can be included. More important, this rate-equation approach completely ignores, or hides, all the *phase information* associated with the signal fields in each resonant cavity mode. It also completely leaves out any spatial interference effects between modes, and thus any spatially inhomogeneous saturation effects or "spatial hole burning" that this may produce in the atomic level populations $N_1$ and $N_2$. Nonetheless, this rate-equation approach can be very useful in laser theory, as we will see.

---

**Problems for 13.3**

1. *Alternative derivation of the K coefficient.* Give an alternative derivation of the coefficient $K$ appearing in the coupled cavity and atomic rate equations by starting from the atomic rate equation; writing the stimulated-emission term in the form $K[N_2(t) - N_1(t)]n(t) \equiv W_{12}[N_2(t) - N_1(t)]$; and using the formulas we have developed earlier for the stimulated-transition probability $W_{12}$. Demonstrate that the final result is the same as given in this section.

2. *Thermodynamic implications of the "extra photon."* Suppose a laser cavity has no other losses or output coupling, so that $\gamma_c \equiv 0$, but suppose it does contain a set of absorbing atoms whose upper-level population $N_2$ and lower-level population $N_1$ are held in thermal equilibrium at a positive temperature $T$. These atoms then provide in effect a cavity loss, and the electromagnetic fields in the cavity should come to thermal equilibrium with the atoms at this same temperature $T$. It is also a basic result of quantum thermodynamics that the average number of photons in each resonant mode in a system at thermal equlibrium should be given by $\langle n \rangle = [\exp(\hbar\omega/kT) - 1]^{-1}$.

   Write down the cavity rate equation for any one cavity mode in such a cavity, and show that it predicts exactly this result, but only if (i) the "extra photon" is included, and (ii) the spontaneous and stimulated transition coefficients are in fact equal, or $K_{\rm sp} = K$, for each mode individually.

3. *More cavity-mode thermodynamics.* Suppose a cavity like that in the previous problem contains two different sets of atoms, both with the same transition frequency, but with their level populations held at two different temperatures. How would we calculate the effective temperature of the blackbody radiation within the cavity to which these groups of atoms are both coupled?

4. *Initial noise value for laser oscillation buildup (research problem).* We have asserted in an earlier section that oscillation in a laser cavity builds up from a small number of "initial noise photons" present in the cavity when the gain is first turned on. In fact, however, the amount of noise energy initially present in an optical cavity at ordinary temperatures, with no inverted population present, is very much less than one photon; and a more accurate picture is to say that the oscillation builds up from the effects of the spontaneous emission that always accompanies an inverted population, as described in this section. The spontaneous emission that occurs during a short time just before and just after the laser gain is first turned on, before the cavity signal builds up to more than a few stimuated photons, is particularly important in setting the effective initial noise level for the laser oscillator.

   To demonstrate this, consider a laser cavity in which the laser gain increases linearly with time in the form $\gamma_m(t) - \gamma_c = KN(t) - \gamma_c = \gamma_c \times t/t_d$. This means that the net growth rate passes upward through zero just at $t = 0$, with $t_d$ being the time it takes the laser gain to rise from zero to just equal the cavity decay rate $\gamma_c$. Using this assumption, solve separately (a) the cavity rate equation $dn(t)/dt = [KN(t) - \gamma_c]n(t)$ without spontaneous emission, but assuming an initial noise value $n_0$ at $t = 0$; and also (b) the more accurate equation $dn(t)/dt = KN(t)[n(t) + 1] - \gamma_c n(t)$ including spontaneous emission, but assuming an initial condition of zero initial photons in the cavity for $t \ll 0$.

By comparing these solutions, show that they give the same form for $n(t)$ for large positive time, $t \gg 0$, providing an initial photon number $n_0 \equiv (2\pi t_d/\tau_c)^{1/2}$ is assumed in case (a). Discuss the physical interpretation of this formula.

5. *Coupled rate-equation analysis of a transverse flow laser.* Consider a transverse-flow type of gas transport laser in which there is no laser pumping mechanism operating inside the laser cavity itself. Rather, pre-excited upper-level laser atoms flow continuously into one side of the laser cavity, and upper- and lower-level atoms flow out the other side. (Assume there is a bottleneck so that lower-level atoms cannot relax to any still lower levels.)

As a simple model for this laser, assume that:

(a) the upper- and lower-level population densities inside the laser cavity may be described by volume-averaged densities $N_2$ and $N_1$;

(b) the rate at which pre-excited upper-level atoms flow into the cavity is given by an initial density $N_0$ times the transverse flow velocity $v$;

(c) the rate at which upper- and lower-level atoms flow out of the cavity is given by the average densities inside the cavity times the same flow velocity;

(d) atoms transfer from level 2 to level 1 inside the cavity as a result of both downward relaxation with a relaxation rate $\gamma_2$ and stimulated transitions caused by a cavity photon number $n$;

(e) the cavity may be characterized by a cavity decay rate $\gamma_c$ (including both internal cavity losses and external output coupling), and a stimulated transition coefficient $K$.

Taken all together, these assumptions provide at least a rough model for certain kinds of gasdynamic and transverse-flow chemical lasers. Carry through a coupled rate-equation analysis of this system, and find the steady-state level populations $N_1$ and $N_2$ and the photon number $n$ inside the cavity as a function of the flow velocity $v$. What is the minimum or threshold flow velocity $v_{\text{th}}$ necessary to reach oscillation threshold, and the laser power output as a function of velocity above threshold? (Note: Spontaneous emission effects may be entirely neglected in this calculation.)

## 13.4 THE LASER THRESHOLD REGION

The almost discontinuous change in power output that occurs at threshold, when a laser suddenly breaks into oscillation, is one of the most remarkable feature of laser behavior. Several of the most significant aspects of this laser threshold behavior can be explained using a remarkably simple rate-equation model, as we will demonstrate in this section.

### Idealized Rate-Equation Analysis

To analyze laser threshold behavior we can use the highly idealized, and yet very realistic, laser model developed in Section 13.3, consisting of a single preferred or lowest-loss cavity mode with photon number $n(t)$, plus an ideal two-level laser transition with upper-level population $N_2(t)$. This upper level is assumed to be pumped at a steady (but adjustable) pumping rate of $R_p$ atoms/second, and to have a population decay rate $\gamma_2$. Downward relaxation out of the lower laser level is assumed to be arbitrarily fast, so that $N_1 \approx 0$ under all circumstances.

The coupled rate equations for this system, as developed in Section 13.3, are then

$$\frac{dn}{dt} = K(n+1)N_2 - \gamma_c n,$$

$$\frac{dN_2}{dt} = R_p - KnN_2 - \gamma_2 N_2, \tag{44}$$

where $\gamma_c$ is the cavity decay rate, and the coupling constant $K$ is given, as in the preceding sections, by $K \equiv 3^*\gamma_{\rm rad}/p$. As usual, $\gamma_{\rm rad}$ is the radiative decay rate on the laser transition, and the important quantity $p$ is the (very large) number of resonant cavity modes within the cavity volume and transition linewidth. To simplify the results slightly, we will set $3^* = 1$ from here on.

### Steady-State Solutions Below Threshold

The steady-state solutions to Equations 13.44, when $d/dt = 0$ in both equations, can then be manipulated in several different ways. For example, the form that is most useful for understanding below-threshold behavior is to write the steady-state solution to the first of these equations in the form

$$n_{\rm ss} = \frac{N_{\rm ss}}{\gamma_c/K - N_{\rm ss}} = \frac{N_{\rm ss}}{N_{\rm th} - N_{\rm ss}}, \tag{45}$$

and the solution to the second in the form

$$N_{\rm ss} = \frac{R_p}{\gamma_2 + Kn_{\rm ss}} = R_p\tau_2 \times \frac{1}{1 + (\gamma_{\rm rad}/\gamma_2) \times (n_{\rm ss}/p)}. \tag{46}$$

Equation 13.45 then says that the number of steady-state photons $n_{\rm ss}$ in the cavity mode will remain small, somewhere between zero and perhaps a few hundred, until the upper-level population $N_{\rm ss}$ is raised to within a fraction of a percent of a *threshold inversion value* $N_{\rm th}$, where this threshold inversion value is given by

$$N_{\rm th} \equiv \frac{\gamma_c}{K} = \frac{\gamma_c}{\gamma_{\rm rad}} p. \tag{47}$$

This value is the threshold inversion we calculated earlier, at which (or very near which) laser oscillation begins.

Equation 13.46 then says that in this same region, so long as the photon number $n_{\rm ss}$ remains very much less than $p$, the upper-level population increases essentially in direct proportion to the pumping rate; i.e., $N_2 \approx R_p\tau_2$. The threshold pumping rate, at which the population inversion $N_{\rm ss}$ will just reach the threshold

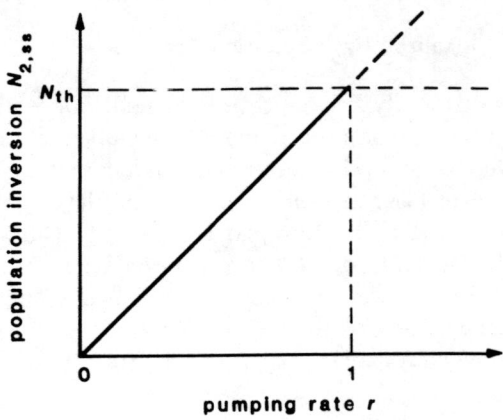

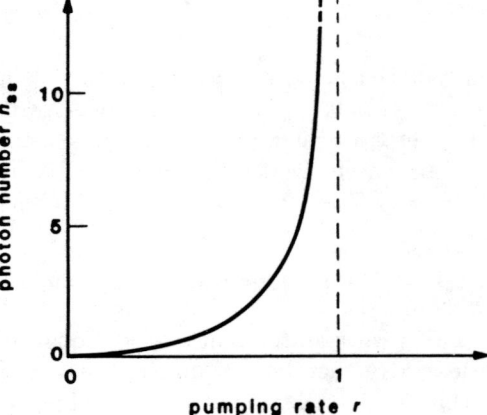

FIGURE 13.7
Laser behavior below threshold.

inversion $N_{\text{th}}$ if this continues, is given by

$$R_{p,\text{th}} = \gamma_2 N_{\text{th}} = \frac{\gamma_2 \gamma_c}{\gamma_{\text{rad}}} p \qquad (48)$$

It is convenient to define a normalized pumping rate relative to this threshold value by

$$r \equiv \frac{R_p}{R_{p,\text{th}}} = \frac{\gamma_{\text{rad}} R_p}{\gamma_2 \gamma_c p}. \qquad (49)$$

The below-threshold region ($r < 1$) is then described by the approximate results

$$\left. \begin{array}{l} n_{\text{ss}} \approx \dfrac{r}{1-r} \\[2mm] N_{\text{ss}} \approx r \times N_{\text{th}} \end{array} \right\} \quad \text{below threshold}, \quad r < 1. \qquad (50)$$

as plotted versus $r$ in Figure 13.7. It is evident that until the pumping rate $r$ becomes very close to the threshold value $r = 1$, the photon number in the cavity will be of order unity or a few orders of magnitude larger. Because the photon number $n_{\text{ss}}$ will remain $\ll p$ for $r < 1$, the saturation term $1/(1 + n_{\text{ss}}/p)$ in the denominator of the pumping Equation 13.47 will be negligible, and so $N_{ss}$ will increase linearly with pumping rate $R_p$ below threshold, as shown in Figure 13.7.

### Steady-State Behavior Above Threshold

We can also, however, rearrange the steady-state solutions to the same two rate equations 13.44 in the reversed forms

$$N_{ss} = \frac{\gamma_c}{K} \times \frac{n_{ss}}{n_{ss}+1} = \frac{n_{ss}}{n_{ss}+1} \times N_{th}, \qquad (51)$$

and

$$n_{ss} = \frac{R_p - \gamma_2 N_{ss}}{K N_{ss}} = \frac{\gamma_{rad} p}{\gamma_2} \left[ \frac{N_{th}}{N_{ss}} r - 1 \right]. \qquad (52)$$

From Equation 13.51 we can see that above threshold, or as soon as the photon number becomes very much greater than unity, the population inversion $N_{ss}$ "clamps" at the threshold value $N_{ss} \approx N_{th}$ (or, more precisely, at just a miniscule amount below $N_{th}$). At the same time, if $N_{ss}$ is clamped at $N_{th}$ then Equation 13.52 says that for $r > 1$ the cavity photon number is given by

$$n_{ss} \approx (r - 1) \times \frac{\gamma_{rad}}{\gamma_2} \times p. \qquad (53)$$

For any reasonable ratio of $\gamma_{rad}/\gamma_2$ and any pumping rate $r$ above threshold, this says that (i) the photon number $n_{ss}$ will increase *linearly* with pumping power above threshold, and (ii) the photon number will be of the same order of magnitude as the mode number $p$, which we have noted is a very large number (order of $10^8$ to $10^{10}$) in most laser cavities.

The approximate formulas for the laser behavior above threshold are thus

$$\left. \begin{array}{l} N_{ss} \approx N_{th} \\ n_{ss} \approx (r-1)\gamma_{rad} p/\gamma_2 \end{array} \right\} \quad \text{above threshold,} \quad r > 1, \qquad (54)$$

as illustrated in Figure 13.8. Note that the cavity photon number $n_{ss}$ in the below-threshold region in Figure 13.8 is orders of magnitude smaller than the value above threshold, and does not even show up on the scale of the above-threshold photon number.

### Energy Transfer Rates Below and Above Threshold

Below threshold, all of the pumping power used in lifting atoms into the upper laser level is reemitted by the atoms as *incoherent* energy, in the form of radiative relaxation processes (spontaneous emission, or fluorescence), plus non-radiative relaxation processes (lattice phonons, wall collisions, and the like), with a combined relaxation rate of $\gamma_2 N_2 \equiv \gamma_{rad} N_2 + \gamma_{nr} N_2$. The radiative part of this relaxation in particular can be pictured as a process in which the atoms spontaneously emit into all of the $\approx p$ resonant modes within the cavity linewidth, and then the energy spontaneously emitted into these cavity modes immediately leaks out of the cavity into all directions as incoherent spontaneous emission.

As soon as the laser goes above threshold, however, the upper-level population $N_{ss}$ clamps at the threshold value, and hence the incoherent relaxation out of this level (radiative plus nonradiative) also clamps just at the value it had at

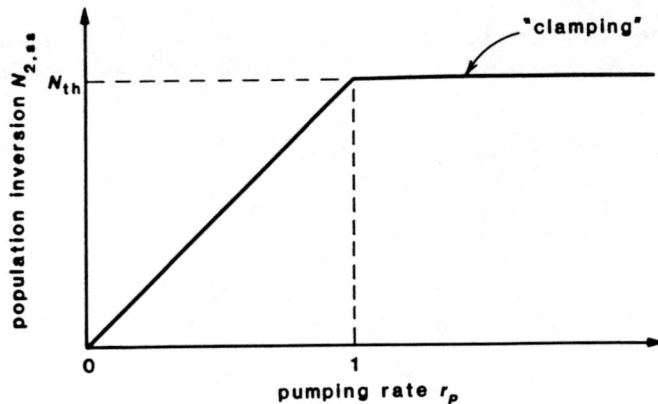

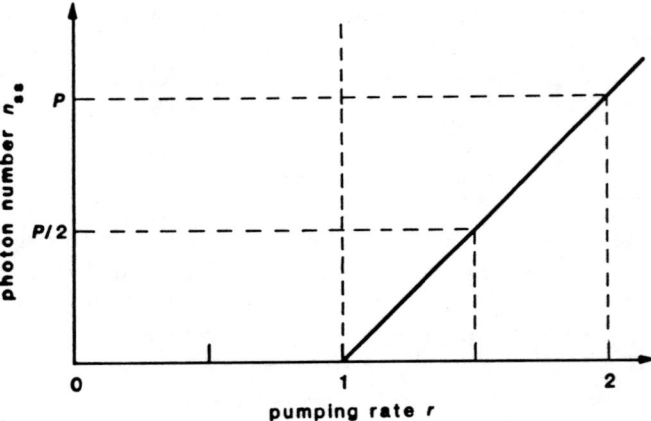

FIGURE 13.8
Laser behavior above threshold.

threshold. *All of the additional pumping power fed into the upper laser level above threshold then goes into, or is stolen by, the coherently oscillating cavity mode.* The laser thus provides a kind of optical illustration of the maxim that "the rich get richer" (or perhaps "the coherent get more coherent").

To illustrate this point, let us assume for simplicity that the upper-level relaxation is purely radiative, so that $\gamma_2 = \gamma_{\rm rad}$, and let us define $P_{\rm th} \equiv R_{p,\rm th}\hbar\omega_a$ to be the total pumping power that is fed into the upper laser level just at threshold. The total incoherent or spontaneous emission power $P_{\rm fluor}$ coming out of the atoms as they fluorescence into all directions, and the total coherent oscillation power $P_{\rm osc}$ coming out of the cavity in the one coherently oscillating cavity mode, will then be given, both below and above threshold, by the simple expressions

$$P_{\rm fluor} \equiv \gamma_2 N_{\rm ss}\hbar\omega \approx \begin{cases} rP_{\rm th} & r \leq 1, \\ P_{\rm th} & r \geq 1, \end{cases} \tag{55}$$

and

$$P_{\rm osc} \equiv \gamma_c n_{\rm ss}\hbar\omega \approx \begin{cases} 0 & r \leq 1, \\ (r-1)P_{\rm th} & r \geq 1. \end{cases} \tag{56}$$

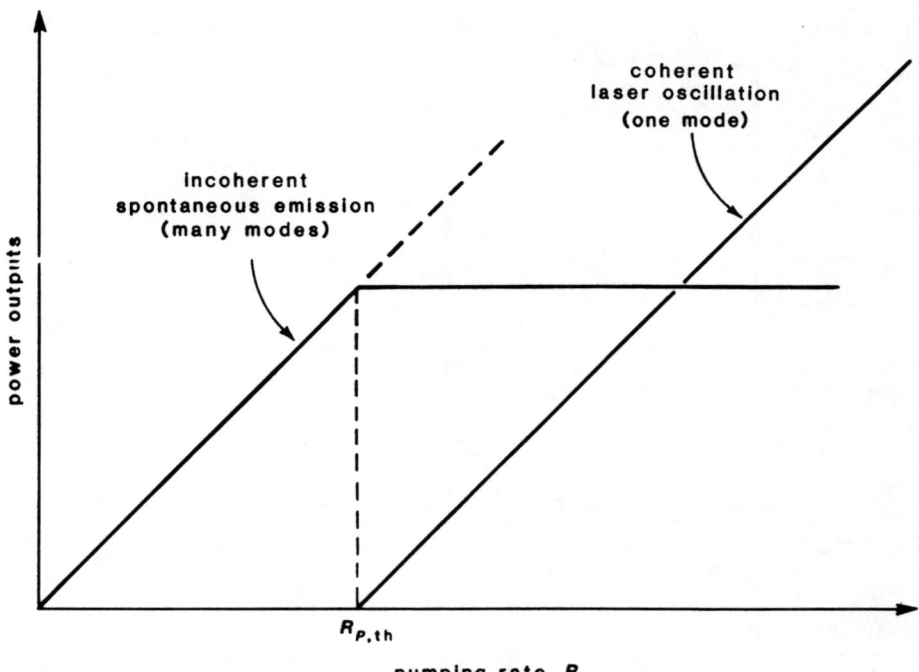

FIGURE 13.9
Coherent and incoherent power outputs.

As illustrated in Figure 13.9, below threshold all the input power goes into incoherent emission; above threshold all the additional pumping power goes into the coherent oscillation output.

Of course, in any real laser system most of the pump power input is not used directly for exciting atoms to the upper laser level, but rather is wasted in pumping atoms up into unwanted levels or in heating up the laser medium. Nonetheless, all of what does go into the upper laser level is then converted into laser oscillation above threshold.

Exact Results for the Threshold Region

The approximate results derived above are very useful for insight into the behavior of laser oscillation below and above threshold. We can, however, also obtain an exact expression for the cavity photon number $n_{ss}$ versus pumping rate $r$ that is valid for all values of $r$ (within the very mild approximations of the rate-equation approach) by eliminating $N_2$ between the two basic rate equations 13.44 and solving for $n_{ss}$ versus $r$.

Suppose for simplicity we assume again that $3^* = 1$ and in addition that $\gamma_2 = \gamma_{\text{rad}}$, i.e., that the upper level relaxes entirely by radiative relaxation into level 1. Then the *exact* steady-state solution to the two rate equations 13.44 at the start of this section is the rather innocuous-looking expression

$$n_{ss} = \left[(r-1) + \sqrt{(r-1)^2 + 4r/p}\right] \times \frac{p}{2}. \qquad (57)$$

Figure 13.10 is a plot of this expression over a range of pumping power centered about $r = 1$, showing how the cavity photon number jumps almost discontinuously from its below-threshold value of order unity, or slightly larger, to numbers

516    CHAPTER 13: OSCILLATION DYNAMICS AND OSCILLATION THRESHOLD

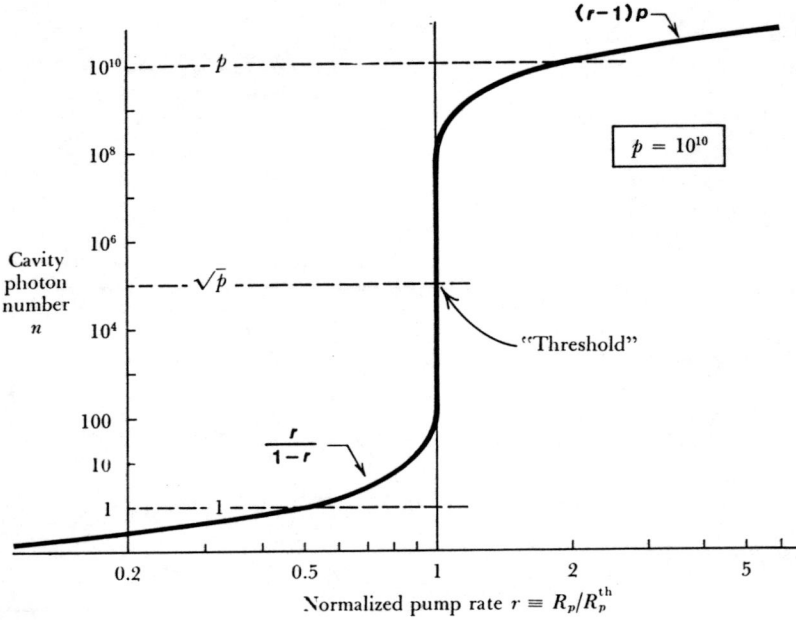

FIGURE 13.10
Cavity photon number $n_{ss}$ versus normalized pumping rate $r$ as given by the exact solution to the single-cavity-mode rate equations.

of order $p$, as the pumping rate increases by a very small amount, with a change of order $1/p^{1/2}$, at threshold. Note the widely different logarithmic scales on the two axes of this figure.

It is virtually impossible to control the pumping rate $R_p$ in a real laser to a precision of order $1/p^{1/2}$; and it is equally difficult, for that matter, to measure the cavity photon number accurately over a dynamic range covering 8 or 10 orders of magnitude. Hence it is not surprising that when we gradually turn up the pump-power knob in a real laser, the onset of oscillation at the oscillation threshold point, where $r$ passes through one, usually appears as an essentially discontinuous event.

### Oscillation Mode Discrimination

The sharpness of the photon number curve versus $r$, combined with the sudden clamping of the population inversion $N_2$ at the threshold value (really just *below* the threshold value), also helps to explain how a laser cavity having some $p \approx 10^8$ or $10^{10}$ potentially oscillating modes can actually oscillate and extract all the additional pumping input in just *one* preferred oscillating mode.

Suppose, for example, a laser cavity has a resonant mode #1 which is the "most preferred" mode, because it has the lowest losses and/or the best coupling to the laser atoms; plus a second cavity mode #2 which is slightly less preferred because it has higher losses or weaker coupling to the atoms or both. Then, as Figure 13.11 shows, when the population inversion clamps at the threshold inversion for mode #1, the less preferred mode #2 will still be slightly below threshold (at least, in an ideal picture). Hence this second mode will never be able to develop a sizable number of photons. The extraordinary sharpness of the threshold behavior and the large value of $p$ help to explain how the photon number in mode #2 can always remain $\ll p$, no matter how hard the laser is

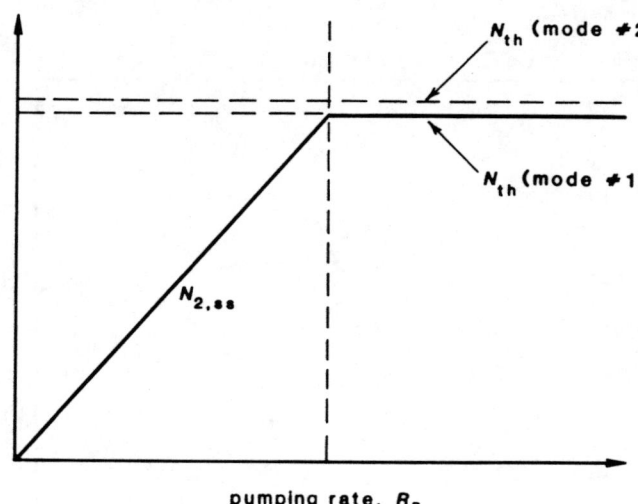

FIGURE 13.11
"Clamping" of the population inversion due to a slightly preferred mode #1, leading to suppression of oscillation for mode #2.

pumped, unless the difference in losses between the two modes is of order $1/p^{1/2}$ or smaller.

The preceding description is, of course, highly idealized. In particular it neglects the spatial and spectral inhomogeneity effects that we discuss elsewhere in this text. Only the population inversion in those atoms that are fully "seen" by mode #1 will be clamped at threshold. If there are other atoms not fully seen and saturated by mode #1, but seen by mode #2, the population inversion on these other atoms can increase with increased pumping, and can pull mode #2 above threshold.

Most practical lasers will in fact oscillate in several, or even many, cavity modes at pumping levels well above threshold; and controlling or eliminating multimode oscillation is a continuing design problem in lasers. Nonetheless, there are also real lasers which are sufficiently well controlled that they can generate large laser output powers in exactly one single laser cavity mode, in full agreement with our idealized model.

### Experimental Threshold Measurements on Injection Diode Lasers

Experimental measurements on lasers just at or below threshold are very difficult, both because of the sharpness of the threshold, and hence the extraordinary stability required in such experiments, and also because of the very weak signals emitted from a cavity containing only a few noise photons below threshold. Semiconductor diode lasers, however, because of their very small cavity volume, can have a smaller than average mode density ($p \approx 10^5 - 10^6$), giving them a comparatively "soft" threshold. Their very efficient direct-current pumping mechanism can also make threshold experiments somewhat simpler.

We have already seen in earlier chapters how a single preferred axial mode can spring into oscillation, rising out of a cluster of amplified axial-mode noise peaks in an injection laser. Figure 13.12 shows a more detailed measurement of how the output power in the dominant axial mode from an injection laser suddenly rises by a large amount as the laser current is increased by a very small amount just at threshold. The light output below threshold in this situation may not represent a fully accurate measure of the photon number in this single

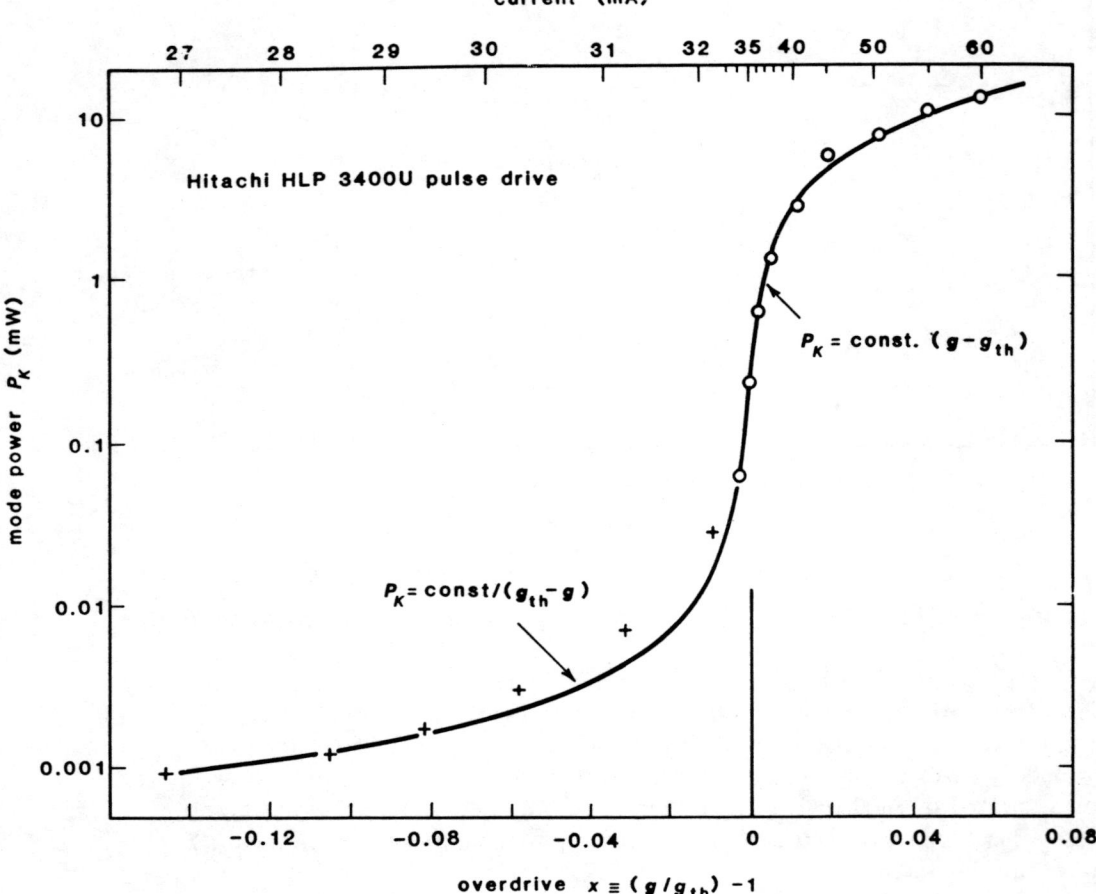

**FIGURE 13.12**
Output power in the dominant axial mode from a GaAs injection laser as the pumping current is increased through the threshold value (from Sommers).

preferred cavity mode, since the measurement apparatus may detect some of the below-threshold noise emission from other axial or near-axial cavity modes. Nonetheless, the general trend is clear.

Figure 13.13 also shows for two other injection laser diodes the sharp clamping of the upper-level population at threshold, as observed by measuring the spontaneous emission out the side or top of the diode as the diode current passes through threshold. Figure 13.13(b) shows that if the face of the diode is scratched or damaged to prevent laser oscillation, the sharp "knee" at the threshold point disappears and the sidelight fluorescence continues to increase with increasing current.

The energy-level system in a semiconductor injection laser is both very broad and much more complicated than just a simple two-level system. Hence, for example, the fluorescent emission at longer wavelengths than the laser wavelength does not clamp as sharply. This emission presumably comes from electronic levels slightly below the laser levels, whose population is not depleted or controlled as sharply by the laser action.

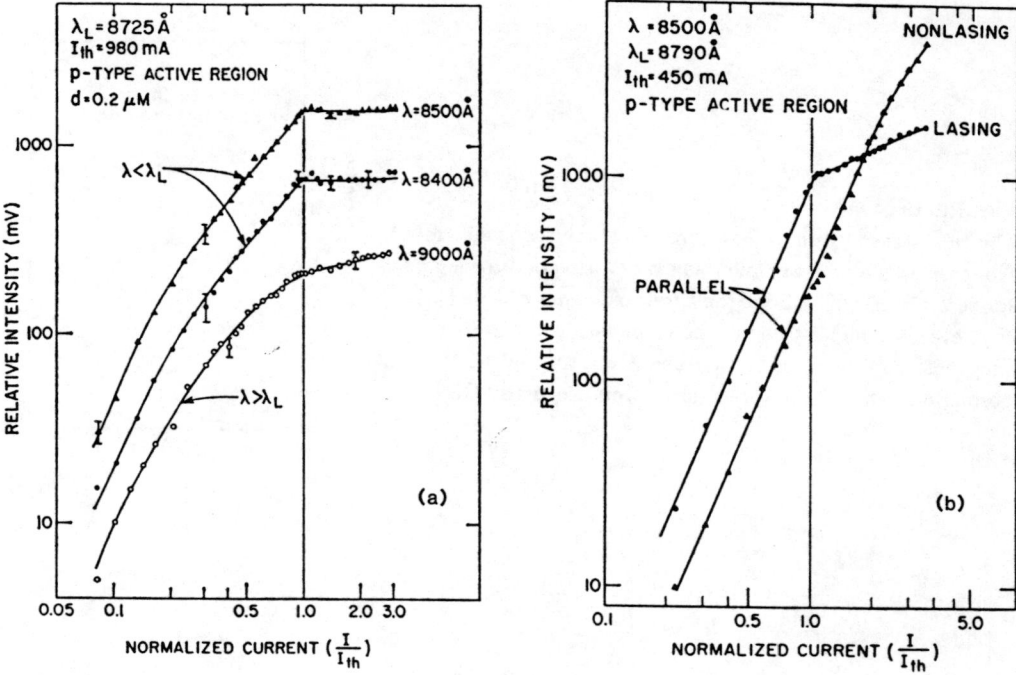

**FIGURE 13.13**
(a) Sidelight fluorescent emission from an injection diode laser as seen through the top and sides of the lasing region, versus pumping current below and above threshold. The different curves represent spontaneous emission at different wavelengths within the very broad transition characteristic of GaAs injection lasers. (b) Sidelight fluorescent emission from the same laser under lasing and nonlasing conditions (achieved by damaging the cleaved end mirror surfaces).

### Other Laser Threshold Measurements

It is also possible, with care, to make threshold measurements in other lasers, for example, in highly stabilized He-Ne lasers. One preferred technique is to stabilize the laser pumping rate at a value well above threshold at the middle of the atomic gain profile; and then tune the cavity-mode frequency out to the point on the side of the atomic-gain curve where gain just equals loss. By tuning the cavity resonance through a very small frequency range centered on this point, using piezoelectric-length tuning, we can pass smoothly and repeatedly from just below to just above threshold.

Figure 13.14 shows the normalized power output from a highly stabilized He-Ne laser (measured by photon counting techniques) as the normalized unsaturated gain or effective pumping rate $r$ is varied about the threshold by $\pm 2$ parts in $10^3$. It is possible to deduce from this data that the mode number for this cavity is $p \approx 5 \times 10^7$ (see Problems).

Finally, as another demonstration of the "clamping" phenomenon, Figure 13.15 shows the fluorescent emission below threshold and the laser emission above threshold (with greatly reduced detector sensitivity) for a group of closely adjacent transitions with a common upper level in an HgCl excimer laser.

The molecules in this laser are created in a $v' = 0$ excited state by a high-voltage electron beam passing through a high-pressure cell containing rare-gas mixtures with small traces of Hg and $CCl_4$. The left-hand diagram shows a

**520**    CHAPTER 13: OSCILLATION DYNAMICS AND OSCILLATION THRESHOLD

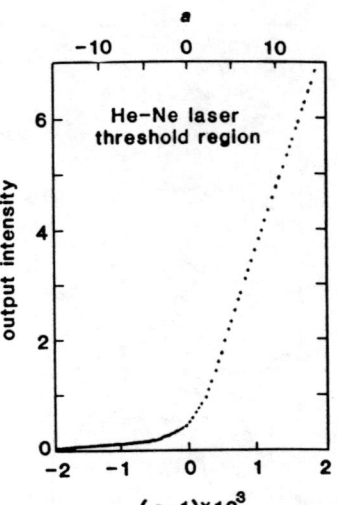

**FIGURE 13.14**
Output intensity from a very carefully stabilized small He-Ne laser as the effective pumping rate $r$ is varied through a range $\delta r = \pm 0.002$ about the threshold value $r = 1$. Note that even the highest points shown on this curve are still very close to threshold, and far below what would be the normal operating level in this laser (from Corti and Digeorgio.)

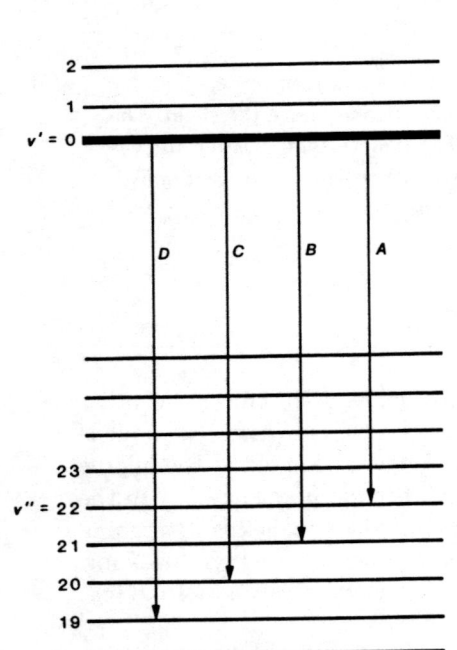
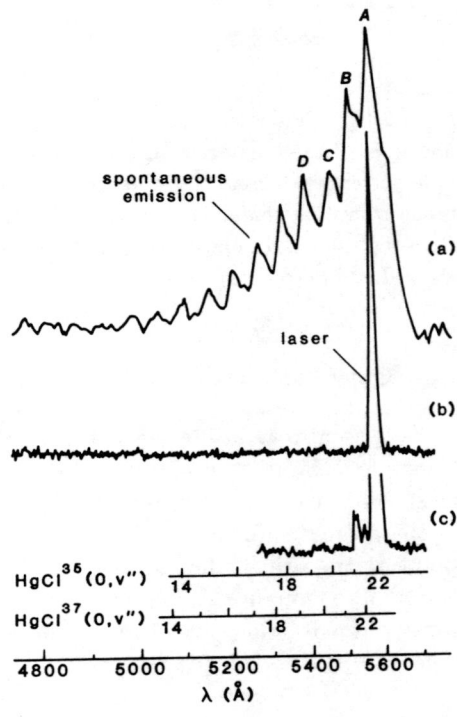

**FIGURE 13.15**
Spontaneous emission and laser oscillation from HgCl molecules excited by electron-beam pumping in an excimer laser. Curve (a): Spontaneous emission, measured with a sensitive detector below threshold, on several different vibrational-rotational transitions indicated in the adjoining partial energy diagram. Curves (b) and (c): Laser action at increased pumping rates, well above threshold.

small part of the energy-level structure of the HgCl molecule, with several of the spontaneous emission lines identified (note that these are different rotational quantum transitions or lines, not just different axial modes). The $v' = 0$ to

$v'' = 22$ transition is the strongest of these lines in fluorescence (strongest value of $\gamma_{rad}$), and it also reaches laser threshold first.

Once this transition oscillates, the population of the upper $v' = 0$ level is then clamped. (The populations of the lower $v'' < 22$ levels may also be increased once oscillations begin by cascading from the $v'' = 22$ level.) The significant point is that, even with very much higher pumping, none of the other lines can be brought to threshold, except possibly for very weak transient oscillation of the next adjoining line at the highest pump level.

### Threshold Characteristics

A summary of the changes occurring in the cavity fields and in the output beam as a laser passes through threshold for a single mode thus includes the following.

- A sudden very large rise in power output in the oscillating mode.
- Clamping, more or less completely, of the upper-level population and hence of the sidelight fluorescence.
- A sudden sharp spectral narrowing, in which the frequency width of the signal radiation suddenly changes from broadband spontaneous emission (with essentially the bandwidth of the atomic transition) to emission of all the additional energy in one or a few essentially monochromatic axial modes.
- A sudden sharp spatial or output beam narrowing, in which instead of spontaneous fluorescent emission coming out randomly in all directions, the additional energy emerges as a more or less collimated, spatially coherent beam which is describable by (depending on the cavity) only one or a few transverse cavity modes.
- A hidden but important change in the statistical character of the laser radiation, from essentially gaussian random noise to a coherent amplitude-stabilized oscillation.

None of these last three items emerges directly from the rate-equation model used in this section. However, the fact is that the signal energy in the laser cavity below threshold is essentially random noise, with random phase and with random amplitude that varies about its mean value from instant to instant. We refer to this as gaussian noise, because the instantaneous amplitudes of the cosine and sine frequency components of this noise are random variables with gaussian probability-density distributions and with no correlation between cosine and sine components. (This means that the phase of the instantaneous phasor amplitude has a uniform distribution between 0 and $2\pi$, whereas the magnitude of the phasor amplitude has a Rayleigh distribution.)

Above threshold, on the other hand, the laser oscillates (ideally) in a single mode with a coherent purely sinusoidal oscillation of the instantaneous $E$ field, just like any coherent electronic oscillator at any frequency. The *amplitude* of the oscillating $E$ field is highly stabilized (by the gain saturation feedback mechanism that stabilizes any laser oscillator), with only very small residual amplitude fluctuations about the mean value. The *phase* of the optical oscillation is random, in the sense that there is no absolute phase or absolute clock to which a free-running oscillator will be stabilized. However, the phase of a good laser oscillator will stay essentially fixed for long periods of time (an enormous number of optical cycles), changing only through a slow random walk in absolute phase caused by small residual noise effects in the laser.

## REFERENCES

The experimental results for semiconductor injection lasers in this section are from T. Paoli, "Saturation behavior of the spontaneous emission from double-heterostructure junction lasers operating high above threshold," *IEEE J. Quantum Electron.* **QE-9**, 267–272 (February 1973); and H. S. Sommers, Jr., "Spontaneous power and the coherent state of injection lasers," *J. Appl. Phys.* **45**, 1787–1793 (January 1974).

A slightly different theoretical expression for semiconductor laser power output versus pumping, which is experimentally indistinguishable from Equation 13.57 in this section, is given and tested in H. S. Sommers, Jr., "Spectral characteristics of single-mode injection lasers: The power-gain curve from weak stimulation to full output," *J. Appl. Phys.* **53**, 156–160 (January 1982). Sommers has also written a review of this subject in "Threshold and oscillation of injection lasers: A critical review of laser theory," *Solid-State Electron.* **25**, 25–44 (1982).

The He-Ne laser results of Figure 13.14 come from M. Corti and V. Degiorgio, "Analogy between the laser and second-order phase transitions," *Phys. Rev. Lett.* **36**, 1173–1176 (May 17 1976).

The excimer laser results of Figure 13.15 are from J. H. Parks, "Laser action on the $B\,^2\Sigma^+_{1/2} \to X\,^2\Sigma^+_{1/2}$ band of HgCl at 5576 Å," *Appl Phys. Lett.* **31**, 192–194 (August 1977).

The next step beyond the idealized single-mode analysis in this section would obviously be to include more than one cavity mode (perhaps many more than one) in the analysis. An expanded rate-equation analysis which includes a large number of modes has been developed by L. W. Casperson, "Threshold characteristics of multimode laser oscillators," *J. Appl. Phys.* **46**, 5194–5201 (December 1975). Casperson shows that in a cavity with a large number of very nearly equal-loss modes the laser threshold becomes "softer" and less abrupt (as might be expected).

The very next case beyond one cavity mode is, of course, two cavity modes; and one convenient way to experiment with two cavity modes is to use the two oppositely directed modes in a ring-laser cavity. Experimental results illustrating the mode competition in such a system are given by M. M. Tehrani and L. Mandel, "Mode competition in a ring laser at line center," *Opt. Lett.* **1**, 196–198 (December 1977).

The transition or change of state that a laser undergoes at threshold has many physical and theoretical similarities to the phase transitions that occur in ferromagnets, superfluids, and superconductors. A brief but readable discussion of this identification is given in R. Salomaa and S. Stenholm, "Observable manifestations of phase transitions in lasers," *Appl. Phys.* **14**, 355–360 (1977).

Other references on this analogy include M. O. Scully, "The laser-phase transition analogy — recent developments," in *Coherence and Quantum Optics*, ed. by L. Mandel and E. Wolf (Plenum Publishing Corp.); and R. Graham, "The phase transition concept and coherence in atomic emission," in *Progress in Optics, Vol. XII*, ed. by E. Wolf (North-Holland, 1974), pp. 235–288.

See also G. Marowsky and W. Heudorfer, "Second- and first-order phase transition analogy in the operation of an organic dye laser," *Opt. Commun.* **26**, 381–383 (September 1978); and A. R. Bulsara and W. C. Schieve, "First-passage time analysis of metastable states in laser phase transitions," *Opt. Commun.* **26**, 384–388 (September 1978).

## Problems for 13.4

1. *Exact solution for the inverted population.* Complete the discussion in the text by analyzing the exact variation of the upper-level population $N_{ss}$ versus $R_p$ or $r$. Use the simplifying conditions that $3^* = 1$ and $\gamma_2 = \gamma_{rad}$. Develop in addition approximate expressions for $N_{ss}/N_{th}$ to the first order in $1/p$ for $r$ both below and above threshold. How close to threshold can $r$ come (from either side) before these approximate expressions fail?

2. *Cavity mode number in a He-Ne laser.* Using the data in Figure 13.14, deduce that the mode number $p$ in the He-Ne laser cavity used in these experiments must be approximately $p \approx 5 \times 10^7$ modes within the doppler-broadened atomic linewidth. The laser employed was a Spectra-Physics Model 119 frequency-stable laser with an unusually short cavity length of 9.5 cm, bore diameter of around 1 mm, and doppler linewidth of 1,500 MHz. Does the theoretical value for $p$ in this laser check with the value estimated in the preceding (even though the laser is inhomogeneous whereas our analytical model is homogeneous)?

3. *Alternative expression for output-power variation through threshold.* A paper by Huang and Mandel, *Opt. Commun.* **32**, 345–349 (February 1980), gives a more sophisticated formula for the variation of photon number with pumping power both below and above theshold

$$n_{ss} = n_0 \left[ a + \frac{2\exp(-a^2/4)}{\sqrt{\pi}[1+\mathrm{erf}(a/2)]} \right]$$

where $a$ is some constant times $r - 1$. Relate this formula to the below and above-threshold results derived in this section, and compare its behavior in the threshold region near $r = 1$ with the "exact" formula (Equation 13.57) for the threshold region derived in this section.

4. *Threshold analysis with a partially bottlenecked lower level.* Repeat the derivations in the text for cavity photon number $n$, populations $N_1$ and $N_2$, and population inversion $N_2 - N_1$ versus pumping rate for pumping levels both below and above threshold, but use a more complicated rate-equation model in which the upper level is pumped at rate $R_p$ and both upper and lower levels have finite relaxation rates $\gamma_2$ and $\gamma_1$, with level 2 relaxing partly into level 1 (rate $\gamma_{21}$) and partly down to still lower levels (rate $\gamma$).

5. *Threshold behavior in a two-mode laser (research problem).* Extend the threshold discussion in the text by considering an idealized laser system in which just two preferred laser cavity modes, with very nearly the same losses, share the same single-level atomic system. Write the necessary rate equations, using the simplifying single-level assumption, and attempt to find exact or approximate solutions for the two cavity photon numbers versus pumping rate. Consider particularly the population in the second (more lossy) mode as the first mode goes through and above threshold. Note also whether there are any significant changes in the general behavior when the difference in mode losses becomes very small. If so, how small, in terms of $p$? Compare your results to those given by Tehrani and Mandel cited in the preceding.

6. *Threshold behavior in a partially inhomogeneous two-mode laser (research problem).* As another approach to gain some insight into multimode oscillations in a

laser, consider a cavity having two preferred cavity modes, whose losses differ by some moderate amount, say, 10 percent. As a rough way of modeling for either spatial inhomogeneity or partially overlapping hole burning in the laser transition, assume that the laser atoms are divided into three groups: those seen only by one cavity mode, those seen only by the other mode, and those seen equally by both modes. All the atoms are pumped equally.

Devise a set of laser rate equations for this system, and attempt to solve these to find the oscillation levels of the two modes as the pump level is raised so that first one and then the other mode comes above threshold (assume that each mode is either well above or well below threshold at any given pumping level, and ignore the photon density in either mode when it is below threshold). Explore the resulting behavior for various values of the cavity loss ratio and of the population distribution between the modes (i.e., for different degrees of sharing of atoms between the two modes). *Note*: It may be possible to solve this problem analytically by assuming at various pumping levels that both, one, or neither of the modes oscillates; or numerical solutions with the aid of a computer may be needed.

## 13.5 MULTIPLE-MIRROR CAVITIES AND ETALON EFFECTS

In this and the following section we consider a number of more complicated multiple-mirror cavity designs which can be used in practical lasers to help obtain various desirable laser properties such as bandwidth narrowing, axial-mode selection, or single-frequency laser operation.

### Intracavity Etalons for Frequency Tuning and Mode Control

We have already noted that in many kinds of lasers, including doppler-broadened gas lasers, most solid-state lasers, and organic dye lasers, the atomic gain profile can be much wider than the axial-mode spacing of the laser cavity; and the laser can then oscillate simultaneously over a broad spectrum of multiple axial modes, especially if the gain medium is at all inhomogeneously broadened. It is then common practice, provided the laser gain is not too small, to insert a short, tilted intracavity etalon, or even several such etalons, inside the laser cavity, as shown in Figure 13.16, so that the narrowband frequency transmission of these etalons near resonance can provide frequency tuning and axial mode selection in the laser.

We have analyzed the transmission properties of such simple passive etalons in an earlier section. In this particular situation, the tilt of the intracavity etalon must be kept small enough that it does not seriously reduce the transmission finesse of the etalon through transverse walk-off, yet large enough that the reflected waves from each side of the etalon pass out of the cavity and do not set up additional resonances with the other mirrors of the laser cavity. The center frequency of such an etalon for transmission along the axial direction of the laser can then be varied by angle tuning, temperature tuning, piezoelectric tuning, or sometimes gas-pressure tuning.

## 13.5 MULTIPLE-MIRROR CAVITIES AND ETALON EFFECTS

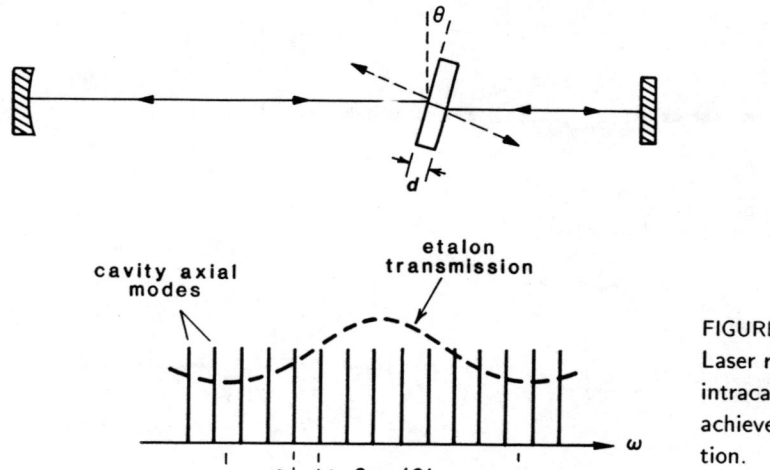

FIGURE 13.16
Laser resonator with intracavity etalon to help achieve axial-mode selection.

### Multiple-Mirror Laser Cavities and Interferometers

Sometimes additional mirrors added to laser cavities may be deliberately aligned in resonance with the existing cavity mirrors to obtain *multimirror laser cavities*. The resonance frequency properties of such cavities then become more complicated, in ways that may be useful for various laser purposes. Figure 13.17 illustrates a number of different multimirror cavity and interferometer designs that have been used or studied in connection with laser oscillators.

### Basic Analysis of the Three-Mirror Cavity

The simplest form of multimirror cavity is obviously the three-mirror cavity shown in Figure 13.17(a). The properties of such a resonant cavity with three or more mirrors are in general complex, and it may be useful to introduce briefly some of these complexities using a simple analytical model.

Suppose we consider first a general three-mirror cavity as shown in Figure 13.18. (Such a cavity, if two of the mirrors are closely enough spaced, can also be viewed as a two-mirror resonator with an etalon mirror on one end or the other.) As shown in Figure 13.18, let us assume this resonator has mirror reflectivities $R_1$, $R_2$, and $R$, and use $\tilde{g}_1$ and $\tilde{g}_2$ to describe the round-trip gains inside the two cavity segments, *leaving out the mirror reflectivities*, so that

$$\tilde{g}_1 \equiv \exp(-\alpha_1 p_1 - j\omega p_1/c) \qquad \tilde{g}_2 \equiv \exp(-\alpha_2 p_2 - j\omega p_2/c), \qquad (58)$$

with $\alpha_1 p_1$ and $\alpha_2 p_2$ representing the round-trip laser gains or losses, if any, inside each segment of the cavity. (Note that this is different from our earlier notation, where $g_{rt}$ represented the complete round-trip gain inside an interferometer, including the end-mirror reflectivities.)

One simple way to analyze such a cavity is to use our earlier results to write down the complex amplitude reflectivity, call it $r'_2$, looking into the $R$, $R_2$ section of this interferometer (of length $L_2$) from the left. We can then use this result

(a) general three-mirror cavity

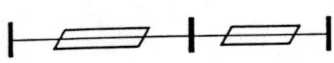

(b) etalon mirror cavity

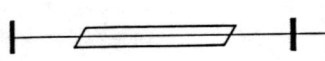

(c) Michelson-mirror cavity

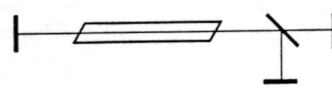

(d) Fox-Smith interferometer

(e) alternative Fox-Smith cavity

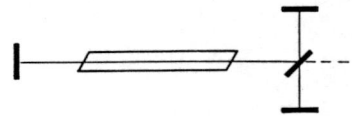

(f) Sagnac interferometer (antiresonant ring)

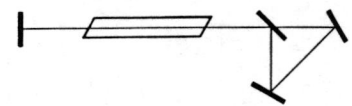

(g) Vernier Michelson cavity

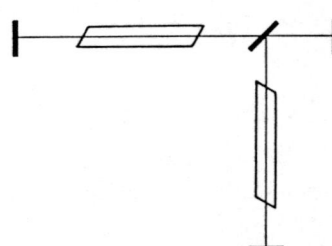

FIGURE 13.17
Examples of multiple-mirror laser cavities and interferometers used for resonance-frequency tuning and axial-mode control in lasers.

FIGURE 13.18
Analytical model for general three-mirror laser cavity.

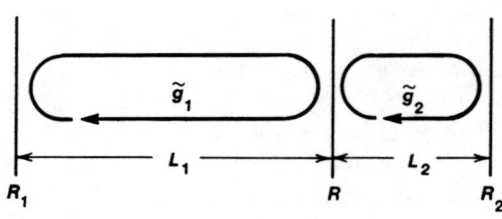

for $r_2'$ as the effective end reflectivity to write down the total reflectivity, call it $r_1'$, looking into the $R_1, R$ cavity segment (of length $L_1$) again from the left.

The end result of this is that the total reflectivity looking into the three-section cavity from outside the $R_1$ mirror can be written as

$$r_1' = \frac{r_1(1 - rr_2\tilde{g}_2) - \tilde{g}_1(r - r_2\tilde{g}_2)}{1 - rr_1\tilde{g}_1 - rr_2\tilde{g}_2 + r_1r_2\tilde{g}_1\tilde{g}_2}. \tag{59}$$

If we then consider the denominator $D(\omega)$ of this expression as a function of frequency, the complex values of $\omega$ that give the roots of this denominator, i.e., that make

$$D(\omega) \equiv r_1r_2\tilde{g}_1(\omega)\tilde{g}_2(\omega) - rr_1\tilde{g}_1(\omega) - rr_2\tilde{g}_2(\omega) + 1 = 0, \tag{60}$$

## 13.5 MULTIPLE-MIRROR CAVITIES AND ETALON EFFECTS

will define the resonance frequencies and the decay rates for the resonant modes of this cavity.

### Basic Properties of Three-Mirror Laser Cavities

Depending on the relative reflectivities and spacings of the cavity mirrors, one might view a three-mirror cavity of this type either as two semi-independent resonant cavities of lengths $L_1$ and $L_2$ coupled together by transmission through the central mirror $R$; or alternatively one might consider this as a single long cavity of total length $L_1 + L_2$ with an internal perturbation produced by the mirror $R$; or as a single cavity of length $L_1$ with an etalon mirror of length $L_2$ on one end. Various different analytical approximations can then be used to calculate the cavity resonant frequencies and losses from Equation 13.60.

As a general rule, however, the resonance properties of the three-mirror cavity are sufficiently complex that the use of numerical solutions and computer display techniques can be very helpful, if not essential, in finding and understanding the resulting cavity modes. We will show here only one or two examples of such solutions, to illustrate the type of behavior that can result.

Figure 13.19, for example, shows how the resonant frequencies of a three-mirror cavity shift and how the mode losses change in a typical situation if we vary the reflectivity $R$ of the central mirror, with fixed reflectivities $R_1$ and $R_2$ for the end mirrors. The cavity segments are assumed to be lossless in these particular plots, except for the finite mirror reflectivities, and the height of each spectral component is proportional to the energy decay rate for that resonance component.

The longer cavity segment $L_1$ in this particular situation is assumed to be three times as long as the shorter cavity segment $L_2$ on a macroscopic length scale, so that the overall axial mode spectrum will repeat periodically with a period corresponding to the axial mode spacing $2\pi \times c/2L_2$ of the shorter cavity. The spectral behavior will also depend strongly, however, on how the axial modes of the two individual cavities are "micro-tuned" with respect to each other. Parts (a) and (b) of Figure 13.19 illustrate the variation in cavity spectrum with central mirror reflectivity $R$ when the two individual cavities are adjusted so that an axial mode characteristic of the short cavity by itself falls either exactly on top of, or halfway in between, the axial modes of the longer cavity. The dashed horizontal lines indicate the mode losses that would occur for the overall cavity $L_1 + L_2$ with no central mirror.

Some examination of Figure 13.19 is worthwhile. There are, of course, four axial modes per repetition period, since the overall cavity length is four times $L_2$. It is then evident that the coupling between the two cavity segments produces both large variations in loss, and also strong frequency pulling effects on the modes, and both of these effects depend strongly on how the two cavity segments are tuned relative to each other.

Figure 13.20 similarly shows the variation in mode losses and resonant frequencies for a cavity with fixed mirror reflectivities and macroscopic lengths $L_1 \approx 4L_2$ as the shorter cavity is tuned through one of its separate axial mode intervals. This example corresponds in essence to an etalon-mirror cavity in which the etalon mirror (length $L_2$) is continuously scanned in frequency through one of its axial mode intervals.

Note that the location of the low-loss region in this spectrum tunes more or less continuously across one full mode interval of the etalon mirror as the etalon

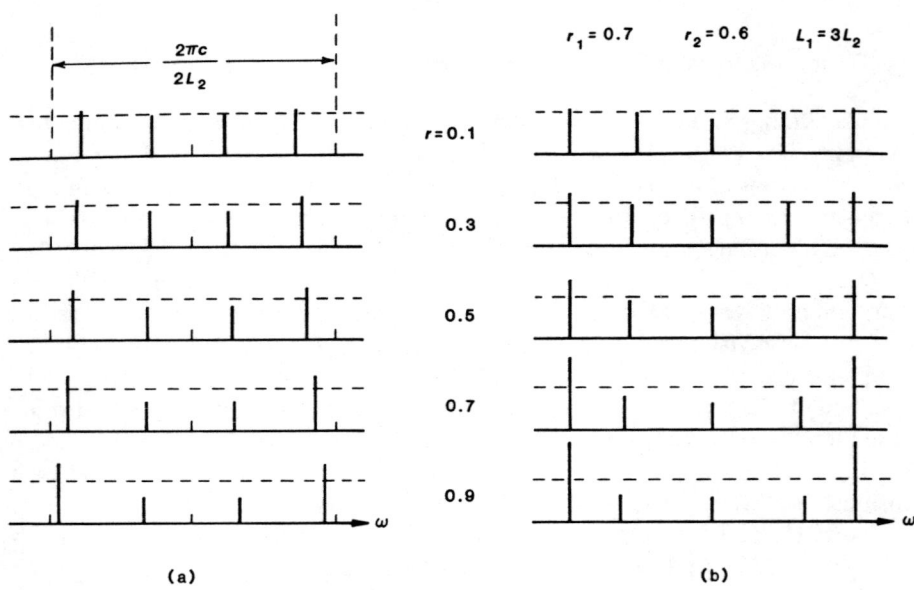

**FIGURE 13.19**
An illustration of how the mode frequencies and mode losses for a typical three-mirror cavity change as the reflectivity of the central mirror is varied. The two segments of the cavity in this situation have a 3:1 ratio of lengths ($L_1 = 3L_2$), with the two cavities being microtuned so that an axial mode of the shorter cavity is located either (a) exactly coincident with or (b) halfway in between the axial modes of the longer cavity.

is tuned. No single axial mode of the combined cavity tunes in this fashion, however, and there is clearly a discontinuous "mode jump" in the lowest-loss mode in the middle of the tuning range. If we wish to select and tune a single axial mode across the full tuning range in an etalon cavity, it is obvious that simply tuning the etalon mirror is not enough. We must instead somehow tune the lengths of both cavity segments simultaneously, so that the lowest loss mode of the overall cavity tracks the high-reflectivity region of the etalon mirror.

### Applications of Three-Mirror Laser Cavities

Simple three-mirror cavities as discussed in the preceding have found some direct applications in lasers. If, for example, we cleave a short semiconductor diode laser into two segments somewhere near the center, carefully maintaining the alignment of the two sections, and then attach separate current leads to the two sections, the result has been called the "cleaved coupled cavity" or $C^3$ type of injection laser. Both the gain and the optical length of each segment can be individually controlled in this situation; and this design has been found to have potential advantages in maintaining a single axial mode, without "mode hopping" effects, over a wide range of injection currents.

Three-mirror or etalon-mirror cavities of this type do not usually provide the optimum design for achieving single-axial-mode operation in low-gain inhomogeneously broadened gas lasers, however, although etalon mirrors are often used in high-gain pulsed solid-state lasers. One reason for this is that, as Figure 13.22

## 13.5 MULTIPLE-MIRROR CAVITIES AND ETALON EFFECTS

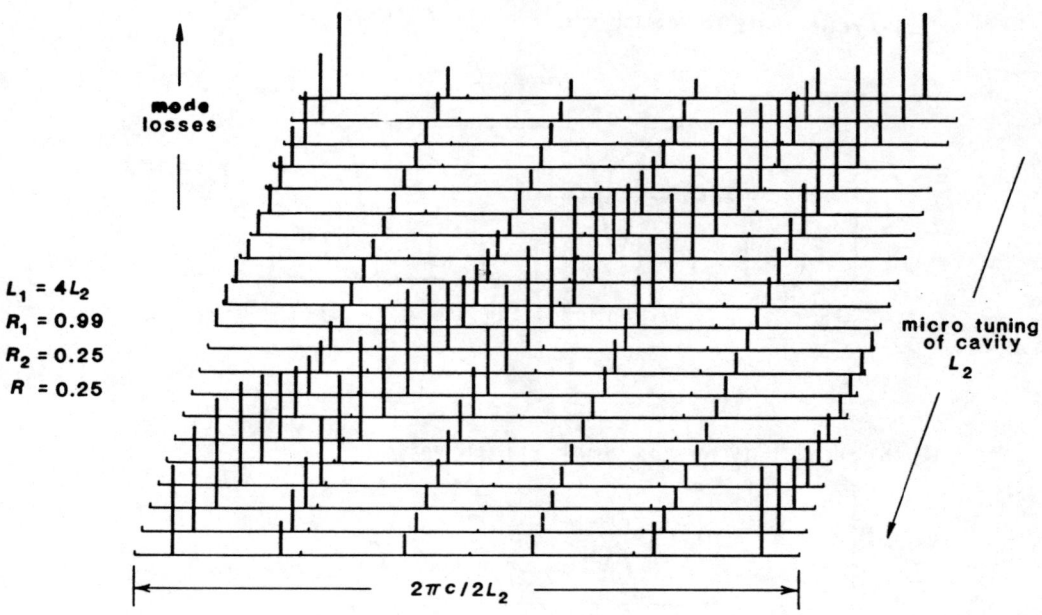

**FIGURE 13.20**
Another example showing the variations in mode losses and frequencies for a three-mirror cavity with power reflectivities $R_1 = 0.99$, $R_2 = 0.25$, $R = 0.25$, and $L_1 = 4L_2$ as the shorter cavity is tuned through one of its axial mode intervals.

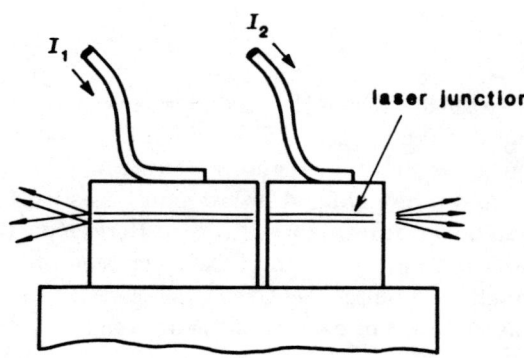

**FIGURE 13.21**
$C^3$ (cleaved-coupled-cavity) injection diode laser.

shows, a high-reflectivity or low-loss etalon normally provides a narrow transmission peak and thus a broad reflection band, whereas what is wanted for axial mode control is a narrow reflection peak. For this same reason, the Michelson-mirror cavity design shown in Figure 13.17(c), which has a sinusoidally varying reflectivity versus frequency, is also usually a less than optimum design.

### Fox-Smith Interferometers, and Other Multimirror Designs

More complex but preferred cavity designs for axial mode selection are then provided by one or another of the alternative forms of the Fox-Smith interferometer shown in Figures 13.17 (d) and (e). Note that in both forms most of the signal energy circulating in the primary cavity will be reflected out of this cavity

**(a) reflectivity for etalon mirror**

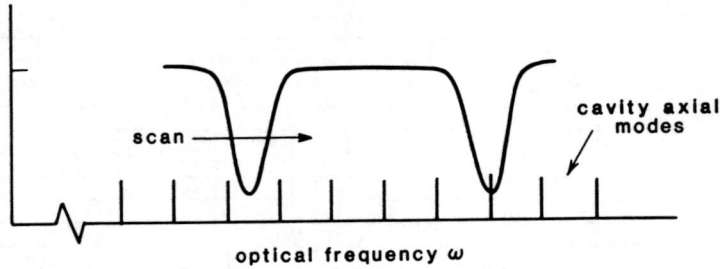

**(b) reflectivity for Fox-Smith interferometer**

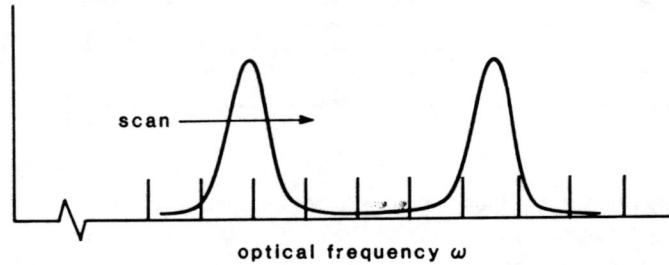

FIGURE 13.22
Reflection coefficients versus frequency for (a) etalon mirror; (b) Fox-Smith interferometer. The arrows indicate how the reflection profile scans as the etalon or interferometer is tuned.

at most frequencies, except at those frequencies where the secondary cavity becomes resonant and builds up a large internal amplitude. This design thus does provide the desired narrow reflection peak as shown in Figure 13.22(b).

Reasonably good mode selection can also be obtained in low-gain lasers using the "vernier Michelson" cavity design shown in Figure 13.17(g). Here, high selectivity is obtained by placing a laser tube in each arm of the Michelson interferometer; and the vernier action results from interference effects between the two long arms, which are made very nearly but not exactly the same length.

### Cavity Back-Reflection Effects

We might also note once again that laser cavities are often very sensitive (in power output, frequency tuning, and oscillation stability) to any back-reflection of the laser signal from external optical components directly back into the laser cavity. These back-reflection effects can of course be understood as multimirror cavity effects of the type described in this section, with the external cavity usually being both long and mechanically unstable. If the external back-reflectivity is small, this means that the external cavity segment is very lossy, or has low effective reflectivity; but the very high $Q$ of the oscillating laser cavity segment can still mean that the effects of even weak backscattered signals can be very significant.

# REFERENCES

An excellent comprehensive review of multimirror laser cavities and other resonator mode control methods is given in P. W. Smith, "Mode selection in lasers," *Proc. IEEE* **60**, 422–440 (April 1972).

The $C^3$ semiconductor laser concept and some of its advantages are discussed in W. T. Tsang, N. A. Olsson, and R. A. Logan, "High-speed direct single-frequency modulation with large tuning rate and frequency excursion in cleaved-coupled-cavity semiconductor lasers," *Appl. Phys. Lett.* **42**, 650–652 (April 15 1983).

For a recent example of three-mirror cavity-mode calculations, see M. J. Adams and J. Buus, "Two segment cavity theory for mode selection in semiconductor lasers," *IEEE J. Quantum Electron.* **QE–20**, 99–103 (February 1984).

The Sagnac interferometer cavity design is described in A. E. Siegman, "An antiresonant ring interferometer for coupled laser cavities, laser output coupling, mode locking, and cavity dumping," *IEEE J. Quantum Electron.* **QE–9**, 247–250 (February 1973).

As one example from the many papers on the effects of back-reflected signals on lasers, see J. H. Osmundsen and N. Gade, "Influence of optical feedback on laser frequency spectrum and threshold conditions," *IEEE J. Quantum Electron.* **QE–19**, 465–469 (March 1983).

## Problems for 13.5

1. *Three-mirror cavity frequency expression.* If we take the limit as the reflectivity $R$ of the internal mirror goes to zero and its transmission goes to 100%, then the expression (Equation 13.60) given in this section for the resonant denominator of a three-mirror cavity goes to $D(\omega) = 1 + r_1 r_2 g_1 g_2$ rather than to $D(\omega) = 1 - r_1 r_2 \tilde{g}_1 \tilde{g}_2$ as we might expect. How come?

2. *Energy distribution in a multimirror cavity.* The different loss rates for different resonant modes in a multimirror cavity, as illustrated in this section, must mean that the relative distribution of stored energy between the individual cavity segments is different from one resonance to another. Carry out a more detailed analysis of the three-mirror (or two-segment) cavity discussed in this section, in order to calculate the relative wave amplitudes of the circulating waves in the two sections of the cavity when the cavity is excited at one or another of its resonant frequencies.

3. *Analysis of the Fox-Smith interferometer.* Carry out a suitable analysis of the Fox-Smith interferometer in either of its forms and discuss the resulting mode selection properties. Consider in particular the requirements on the beam splitter reflection and transmission, and on the losses in the secondary cavity, if the peak reflectivity back into the primary cavity is to be very close to unity. How might we tune either of these cavity designs to obtain tunable single-frequency operation across the full axial mode spacing of the shorter secondary cavity?

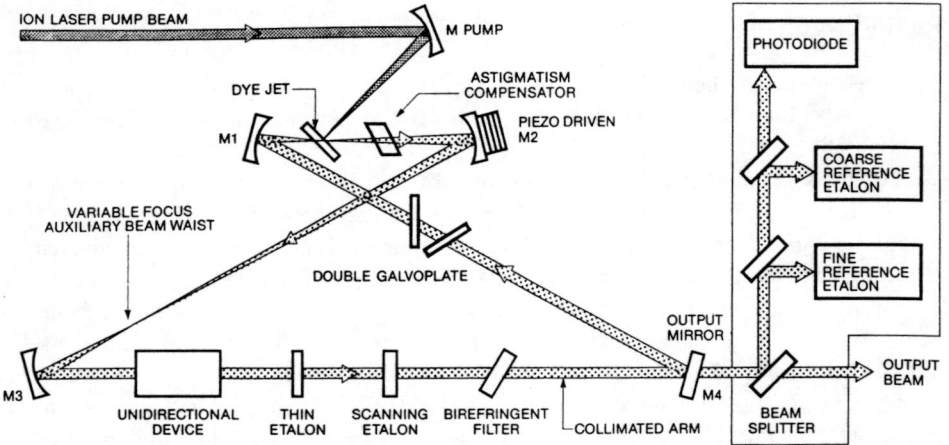

**FIGURE 13.23**
Example of a ring-laser cavity design with intracavity etalons, filters, and unidirectional devices, as used to provide single-frequency, continuously tunable, high-efficiency operation in a commercially available cw dye laser.

## 13.6 UNIDIRECTIONAL RING-LASER OSCILLATORS

Ring-laser cavities were understood and demonstrated very early on, and have since been extensively developed for application in ring-laser gyroscopes. Ring-laser cavities possess one unique capability as compared to standing-wave cavities, namely they have the ability to oscillate, simultaneously or independently, in either of two distinct counter-propagating directions.

Ring resonators also possess a number of other attributes which can be very useful in several different laser and passive interferometer applications. Full appreciation of the advantages of ring resonators in optical applications has only emerged more recently, and it seems useful therefore to give a brief summary in this section of the special properties of ring-laser resonators.

### Example of a Unidirectional Ring-laser Cavity

Figure 13.23 shows, by way of example, a typical folded ring resonator design as used in a commercially produced cw dye laser. The ring cavity in this example contains not only the dye-jet gain medium, several frequency control etalons and filters, and an astigmatism compensating element, but also a unidirectional device ("optical diode") which allows oscillation to occur in only one direction around the ring. This figure also illustrates the array of diagnostic elements which are used, in conjunction with electronic feedback loops, to control the etalon elements, the piezo mirror control, a double galvoplate, and the birefringent optical filter, all needed to give single-frequency laser operation tunable over a wide tuning range.

The primary advantage to unidirectional oscillation in a ring laser such as this is that the purely traveling-wave rather than standing-wave operation eliminates spatial hole-burning effects, making the laser medium in effect much more homogeneous. This in turn substantially increases the mode competition between adjacent axial modes, making it possible to pump the laser considerably further above threshold while maintaining single-frequency operation. In addition, because the traveling-wave mode saturates the gain medium uniformly, with no

## 13.6 UNIDIRECTIONAL RING-LASER OSCILLATORS

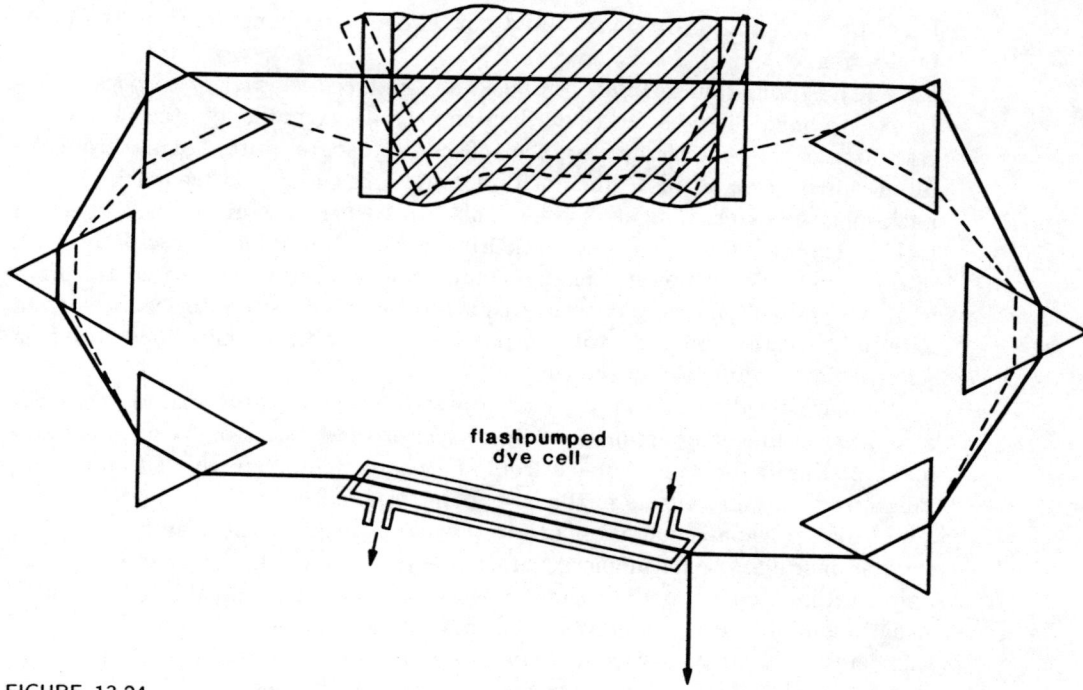

FIGURE 13.24
Prism-type ring resonator. (Adapted from Schäfer and Müller.)

spatial nodes along the axial direction, this mode can extract more power than would otherwise be obtained.

The combined effect can be an increase in single-frequency power output by more than an order of magnitude compared to what can be obtained using a standing-wave cavity in a typical dye laser example. Similar advantages can be obtained in other lasers, for example, pulsed solid-state lasers, as well.

### Other Attributes of Ring Resonators

Other potentially useful attributes of ring optical resonators include:

(1) *Increased cavity design flexibility and alignment insensitivity.* We will point out later on, in the resonator chapters of this text, that a ring optical cavity provides increased flexibility in resonator design as compared to a standing-wave cavity, especially for unstable resonator designs. In particular, a ring cavity can easily employ a comparatively short beam expansion section, using readily available short-focal-length optical elements, and then a long collimated beam section at large beam diameter, for obtaining full power extraction from large diameter laser gain media. (Figure 13.23 shows, by contrast, the way in which very small focal spots, for use with intracavity dye jets or modulation elements, can equally easily be obtained inside a ring-laser resonator.)

Ring resonators also offer the possibility of using prisms of various sorts in place of mirrors in forming the ring; and this aspect has been used in ring-laser designs such as Figure 13.24. A planar ring resonator also has the interesting attribute of being first order insensitive to misalignments in the plane of the ring. That is, when any element is misaligned by a small amount in the plane of the ring, the resonator mode can always respond by making small changes in beam position and direction to find a new closed and aligned path in the plane of the ring.

(2) *Elimination of input feedback, and reduced sensitivity to back reflection.* We have noted earlier that when an external signal is injected into a ring resonator or ring interferometer, the reflected-plus-transmitted signal from the input mirror goes off in a different direction, with no optical feedback directly back into the external signal source. This can be very useful for laser injection locking experiments, where feedback from a high-power locked oscillator back into the much lower-power injection source can be a major experimental problem. A unidirectional ring oscillator can also be less sensitive to feedback from external reflections placed in its output beam, since these reflections go into a non-oscillating direction in the ring.

Similar considerations apply when a passive ring resonator is used as a scanning interferometer or frequency filter for laser diagnostics or frequency stabilization. Elimination of feedback from the passive cavity in this situation can minimize instability effects in the laser being studied.

(3) *Single-pass operation of intracavity elements.* Intracavity elements, such as modulators, harmonic generation crystals, and the like, are excited in only one direction in a unidirectional ring-laser oscillator. While this may in many situations reduce the modulation efficiency or harmonic generation efficiency of the element, it can also simplify certain experiments and permit more accurate measurements on intracavity experimental cells or samples.

The order in which optical elements are encountered is also inherently different going in the two directions around a ring. Going in one direction, for example, a wave may see first the laser-gain medium, then a saturable absorber, and then the output coupler, whereas the order is obviously reversed in the opposite direction. This can be used to control power levels and saturation intensities in different elements, and can be a source of directional nonreciprocity in a ring-laser oscillator.

### Ring-laser Disadvantages

The primary disadvantage of the ring resonator for laser applications, leaving aside the additional complexity and structural requirements, is probably that the gain medium is traversed only once. A low-gain laser will thus operate considerably closer to threshold, and have more stringent requirements on reducing the output coupling, and especially the internal losses, if good efficiency is to be maintained. (Of course, other lossy intracavity elements other than mirrors are also encountered only once rather than twice per round trip.)

The astigmatism produced by off-axis reflection from cavity mirrors must also be taken into account in the resonator design. This astigmatism can even be an advantage in some situations, however, and can usually be compensated for when it is not.

### Techniques for Obtaining Unidirectional Oscillation

The mode competition between two potentially oscillating modes in a ring laser (or in any other multimode laser situation) is in general a complex problem. As we will show in later sections, depending on mode losses, mode cross-coupling, and mode self-saturation and cross-saturation properties, competition between two modes may lead to stable single-mode operation; to simultaneous

### 13.6 UNIDIRECTIONAL RING-LASER OSCILLATORS

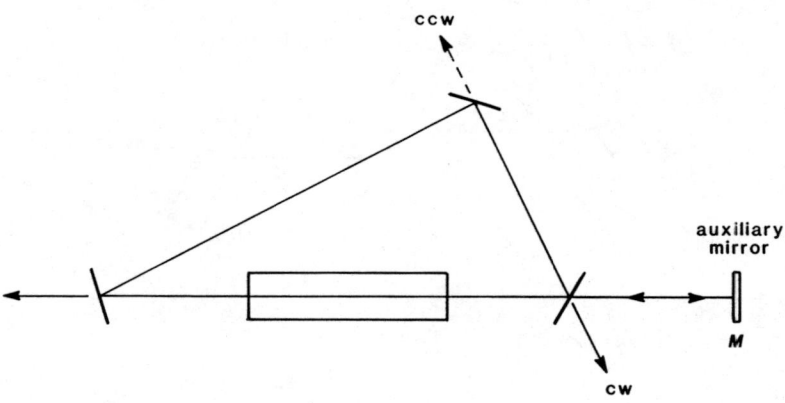

**FIGURE 13.25**
Use of an external mirror to augment one-way oscillation in a ring laser.

dual-mode operation; or to random jumping back and forth between the two potential modes. This applies especially to the oppositely traveling waves in a ring resonator, and several different techniques for obtaining or improving unidirectional operation in ring resonators have been demonstrated.

One of the simplest of these is to employ an auxiliary external mirror, having partial or complete reflectivity, to reflect part of, say the CCW circulating output back into the CW direction, as illustrated in Figure 13.25. If this cavity attempts to oscillate in the CW direction, the resulting back-reflected signal will serve as an injected signal for the CCW direction, leading to a much stronger oscillation in the CCW direction.

This technique does work as intended, at least crudely and in some situations. If the laser medium is otherwise homogeneous, the preferred CW oscillation does grow at the expense of the CCW oscillation. The CCW oscillation is not fully extinguished, however, but remains as a low-amplitude injection signal to drive the CW oscillation. Intensity ratios between the two directions in the range of 10:1 to 50:1 have been reported in typical situations.

For inhomogeneously broadened materials the technique may not work at all, especially when the centermost axial mode of the ring cavity is not at line center. In this situation the ring may oscillate with equal intensity in both directions, and may also oscillate in several axial modes at once. This scheme is also sensitive to internal backscattering inside the ring, which can interact interferometrically with the external mirror. These interactions make the basic technique fundamentally unsound when finite backscattering is taken into account.

#### Nonreciprocal Optical Diodes

A much preferable solution is to place a nonreciprocal optical isolator, sometimes referred to as an "optical diode," inside the ring cavity to introduce nonreciprocal losses in the two directions. Figure 13.26 shows the basic elements of such an optical diode.

The primary component is a Faraday rotation device using a transparent material with a finite Verdet constant placed in a dc magnetic field. When a linearly polarized optical wave passes through such an element, its plane of polarization is rotated about the optical axis, with a direction of rotation which depends on the dc magnetic field direction but not on the direction of travel of the wave. To

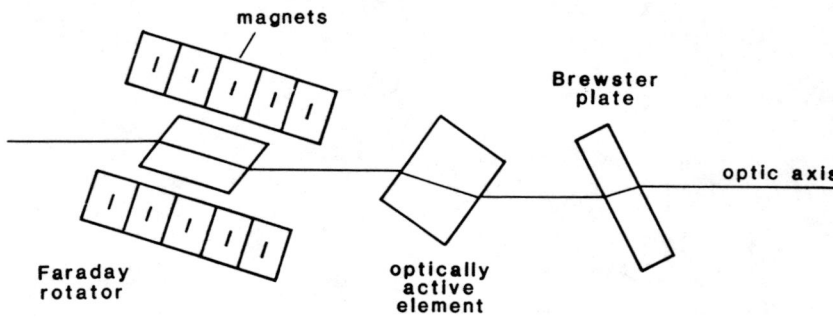

FIGURE 13.26
Nonreciprocal "optical diode" using a Faraday rotator.

make an optical isolator, a second purely reciprocal rotation element is added to cancel out the Faraday rotation going in one direction through the system. This reciprocal element may be an optically active crystal or liquid (e.g., quartz, or a sugar solution), or a birefringent crystal used as a partial wave plate.

The reciprocity properties of this system then mean that the total polarization effects due to the two elements going in the forward direction through the system can cancel each other, giving no net polarization change, whereas the effects of the two elements going in the opposite direction will add to give a finite modification of the wave polarization. A linearly polarized wave going in the forward direction through the system can then pass unattenuated through a Brewster angle plate (or some other type of polarization-sensitive element) on successive round trips, whereas a wave going in the opposite direction, because of the net polarization rotation, will experience added loss on each round trip.

(It should be noted that this additional loss is in general not given simply by calculating first the polarization rotation of a linearly polarized wave in the optically active elements and then the transmission of the resulting wave through the Brewster plate. We must instead use some form of polarization calculus to calculate the total propagation of two orthogonal polarization components around the ring, as well as the cross coupling between these polarization components in each optical element; and the use these results to find the two polarization eigenmodes and associated eigenvalues for the cavity. A cavity which contains birefringent or polarizing elements will in general have two such mixed polarization eigenmodes, neither of which will generally be as lossy as predicted by the simple approach given in the preceding.)

Useful Faraday rotators for optical wavelengths are difficult to obtain in practice, primarily because the physical basis of Faraday rotation is the anisotropic tensor response $\chi'(\omega)$ on the side of some very strong, and Zeeman split, atomic transition. Materials with large Verdet constants (i.e., large polarization rotation per unit of dc magnetic field) are thus most often also highly absorbing, whereas highly transparent materials typically have very small Verdet constants. In practice the optical diodes used in ring lasers typically have Faraday rotations of a few degrees, and differential losses between the two directions of a percent or so. This additional insertion loss in the reverse direction is, however, enough to strongly suppress oscillation in the reverse direction, and produce highly selective oscillation in the forward direction only.

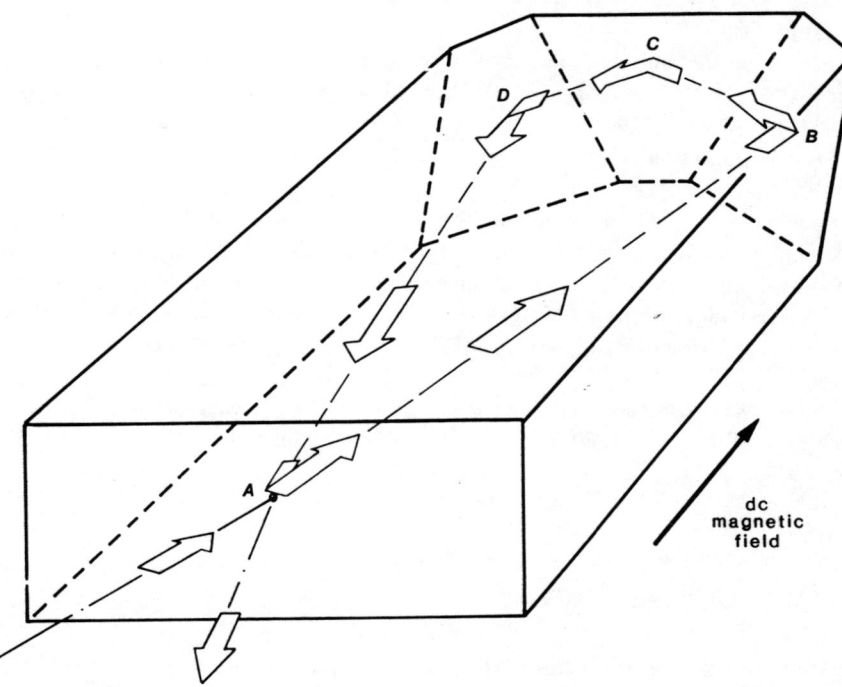

**FIGURE 13.27**
Unidirectional nonplanar solid-state ring laser. (From Byer et al.)

### Nonplanar Ring Resonators

We will also point out later in this text that *nonplanar ring resonators* can provide unique image rotation and also polarization rotation properties. These properties have been employed recently to develop a unique monolithic solid-state laser with inherent unidirectional properties, as illustrated in Figure 13.27.

The material used here is Nd:YAG, which has both laser gain and a finite Verdet constant. A small crystal cut as shown in this figure then provides a nonplanar ring resonator which employs total internal reflection at all but one of its surfaces. The polarization rotation inherent in the nonplanar ring path is then compensated in one direction, but not in the other, by the Faraday rotation produced by a dc magnetic field. The crystal thus oscillates inherently in only one direction around the ring, and as a consequence also achieves high-quality single-frequency operation.

## REFERENCES

The use of a traveling-wave ring resonator with a Faraday rotator to eliminate standing waves and thereby obtain single-frequency oscillation was demonstrated at an early date in a ruby laser by C. L. Tang, H. Statz, and G. deMars, Jr., "Spectral output and spiking behavior or solid-state lasers," *J. Appl. Phys.* **34**, 2289–2295 (August 1963).

The use of back reflection from an auxiliary mirror to improve unidirectional operation was also demonstrated not long after by M. Hercher, M. Young, and C. B. Smoyer, "Traveling-wave ruby laser with a passive optical isolator," *J. Appl. Phys.* **36**, 3351 (October 1965). The fundamental properties of this auxiliary mirror concept are also discussed in F. R. Faxvog, "Modes of a unidirectional ring laser," *Opt. Lett.* **5**, 285–287 (July 1980).

Application of the unidirectional ring resonator concept to dye lasers was first demonstrated by F. P. Schäfer and H. Müller, "Tunable dye ring-laser," *Opt. Commun.* **2**, 407–409 (January 1971); and by J. M. Green, J. P. Hohimer, and F. K. Tittel, "Traveling-wave operation of a tunable cw dye laser," *Opt. Commun.* **7**, 349–350 (April 1973).

For more recent developments in unidirectional ring lasers and Faraday rotators, see, for example, S. M. Jarrett and J. F. Young, "High-efficiency single-frequency cw ring dye laser," *Opt. Lett.* **4**, 176–178 (June 1979); or T. F. Johnston, Jr., and W. Proffitt, "Design and performance of a broad-band optical diode to enforce one-direction traveling-wave operation of a ring laser," *IEEE J. Quantum Electron.* **QE–16**, 483–488 (April 1980).

The monolithic solid-state ring laser is described by T. J. Kane and R. L. Byer, "Monolithic, unidirectional single-mode Nd:YAG ring laser," *Opt. Lett.* **10**, 65–67 (February 1985).

## 13.7 BISTABLE OPTICAL SYSTEMS

The equations of motion for a laser oscillator, or more generally for any system of coupled fields and atoms, are intrinsically nonlinear (although we often make linear approximations to these equations). It has been increasingly realized in recent years that one can often obtain in such nonlinear systems interesting and fundamental types of *bistable, multistable, self-pulsing*, and even *chaotic* behavior.

In this section, therefore, we briefly introduce some of the interesting bistability properties of lasers and also of passive optical cavities. These bistability properties may someday find practical applications in optical computers, "optical transistors," or other optical signal-processing devices, although the real practicality of any such all-optical computer systems remains at present still in considerable doubt.

### Bistable Laser Oscillation

As the simplest example of a bistable laser oscillator, we can consider a laser cavity containing both a homogeneously saturable gain medium and a homogeneously saturable atomic absorber. Suppose that at small signal levels the combined saturable and nonsaturable losses in this cavity exceed the unsaturated gain, so that the cavity cannot begin oscillating spontaneously starting from noise. The laser thus has one stable operating point in the totally quiescent condition, with no signal present.

Suppose, however, that the cavity losses saturate much more easily with increasing signal intensity than does the laser gain—that is, the absorbing atoms have a lower saturation intensity than do the amplifying atoms. At high-enough signal intensities the saturated round-trip losses can then drop below the saturated round-trip gain, as shown in Figure 13.28. If this laser can ever start oscillating, therefore—perhaps with assistance from some externally injected signal—it will build up to a large circulating intensity and remain oscillating until it is turned off.

## 13.7 BISTABLE OPTICAL SYSTEMS

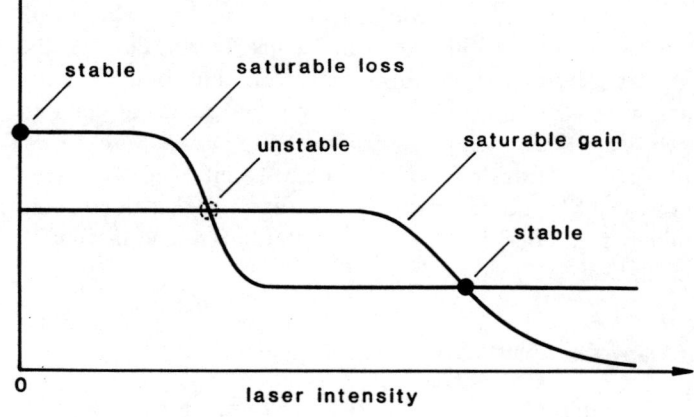

FIGURE 13.28
Gain and loss saturation versus intensity in a bistable laser oscillator.

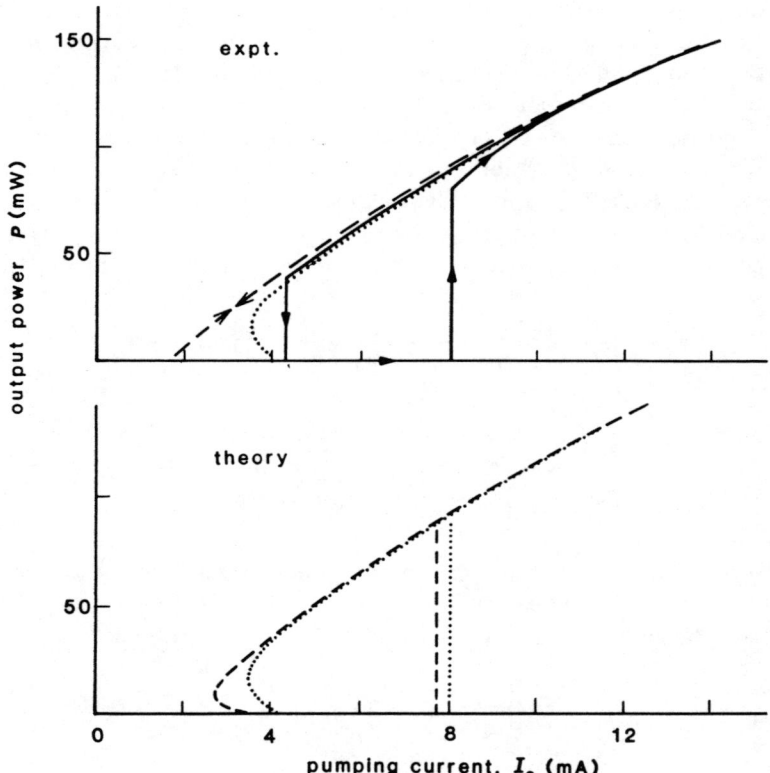

FIGURE 13.29
Hysteresis and bistable operation in the output of a $CO_2$ laser oscillator plotted versus excitation current. (Adapted from E. Aromondo and B. M. Dinelli, *Opt. Commun.*, 44, 277–282, January 15, 1983.)

This simple system thus exhibits *bistable* behavior, with two stable steady-state operating points, as shown in Figure 13.29. There is also a third potential steady-state operating point where gain also equals loss, at the first crossing of the saturable loss and saturable gain curves. However, it can readily be shown (see Problems) that this is not a stable operating point for the laser.

A laser of this type can also exhibit a strong *hysteresis* in the variation of output power with pumping power, as illustrated in Figure 13.29. The results shown there are for a cw $CO_2$ laser with an intracavity cell containing 25 mm of gaseous $SF_6$ as a saturable absorbing medium. When the pumping current $I_0$ is turned up from zero, laser action cannot start until the laser gain exceeds the laser cavity losses plus the unsaturated losses of the $SF_6$ cell. Once the laser starts oscillating, however, the $SF_6$ absorber cell is saturated, and we can then reduce the pumping current to a considerably lower value before the laser will suddenly drop out of oscillation.

### Analysis of a Nonlinearly Absorbing Cavity

As an even simpler example of a bistable optical system, we can consider a passive Fabry-Perot interferometer, of either the standing-wave or the ring-cavity type, containing only a simple passive saturable absorber, and driven by an externally applied optical signal.

Suppose such a passive Fabry-Perot cavity has input and output mirrors with reflectivities $R_1 = r_1^2 = \exp(-\delta_1)$ and $R_2 = r_2^2 = \exp(-\delta_2)$, and a round-trip power attenuation coefficient due to a saturable atomic absorber of $\delta_m \equiv 2\alpha_m p_m$. If this cavity is driven by an externally incident signal $I_{\text{inc}}$ which is tuned to the cavity resonant frequency, then the internal circulating signal $I_{\text{circ}}$ and the transmitted signal field $I_{\text{trans}}$ from the cavity will be given by the elementary interferometer relations derived in earlier sections. In particular, if we assume that all of the loss factors are small compared to unity, then the incident, circulating, and transmitted intensities from this cavity will be related by the expressions

$$\frac{I_{\text{trans}}}{I_{\text{inc}}} \approx \frac{4\delta_1 \delta_2}{[\delta_1 + \delta_2 + \delta_m(I)]^2} \equiv T(I), \tag{61}$$

and

$$\frac{I_{\text{circ}}}{I_{\text{inc}}} \approx \frac{4\delta_1}{[\delta_1 + \delta_2 + \delta_m(I)]^2} \equiv \frac{T(I)}{\delta_1}, \tag{62}$$

where $T(I) \equiv I_{\text{trans}}/I_{\text{inc}}$ is the intensity-dependent power transmission through the cavity. Let us assume that the internal atomic absorption saturates in the homogeneous fashion

$$\delta_m = \delta_m(I) = \frac{\delta_{m0}}{1 + 2^* I_{\text{circ}}/I_{\text{sat}}} = \frac{\delta_{m0}}{1 + I} \tag{63}$$

with $2^* \equiv 1$ for a ring cavity and $2^* \equiv 2$ for a standing-wave cavity, and where $I \equiv 2^* I_{\text{circ}}/I_{\text{sat}}$. Then by picking successively increasing values of the circulating intensity $I_{\text{circ}}$, we can calculate first the intensity-dependent transmission gain $T(I)$, and then calculate and plot the transmitted intensity $I_{\text{trans}}$ versus the incident intensity $I_{\text{inc}}$ for the interferometer.

If we assume for simplicity that the input and output couplings are the same, $\delta_1 = \delta_2$, then the power transmission $T(I)$ through the interferometer can be written as

$$T(I) = \left[\frac{1}{1 + R/(1+I)}\right]^2 = \left[\frac{1+I}{1+R+I}\right]^2, \tag{64}$$

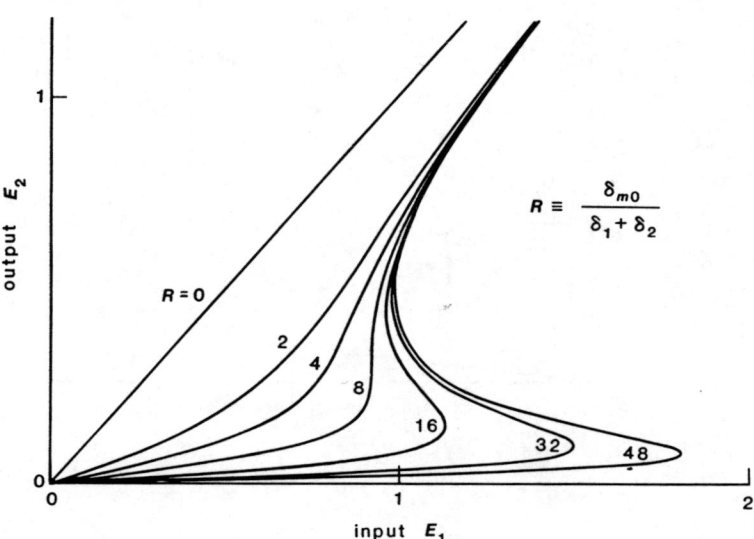

**FIGURE 13.30**
Nonlinear amplitude transmission through a Fabry-Perot interferometer containing a homogeneous saturable absorber.

where $R$ is the ratio of unsaturated gain to total coupling, as defined by

$$R \equiv \frac{\delta_{m0}}{\delta_1 + \delta_2}. \tag{65}$$

It is further convenient to define normalized input and output signal intensities for the cavity by

$$I_1 \equiv E_1^2 \equiv \frac{2^* I_{\text{inc}}}{\delta_1 I_{\text{sat}}} \quad \text{and} \quad I_2 \equiv E_2^2 \equiv \frac{2^* I_{\text{trans}}}{\delta_1 I_{\text{sat}}}, \tag{66}$$

and then to eliminate the internal circulating intensity $I \equiv 2^* I_{\text{circ}}/I_{\text{sat}}$ between the preceding equations. The input-output field relationship then takes the simple form

$$E_1 = E_2 \left[ 1 + \frac{R}{1 + E_2^2} \right]. \tag{67}$$

Figure 13.30 shows the nonlinear input-output relationship that is produced by this type of saturable interferometer transmission. A multivalued input-output relation occurs in this simple situation only if the ratio of unsaturated losses to total cavity coupling has a value $R \geq 8$.

### Absorptive Bistability

A saturable-absorber cavity of this type will, as a consequence, exhibit the general type of bistable input-output hysteresis behavior shown in Figure 13.31. That is, if we turn up the input intensity to this cavity, starting from low values, the circulating intensity inside the cavity will at first not be greatly enhanced, because of the sizable absorption losses and hence low finesse of the interferometer. When the incident signal level reaches a certain value marked

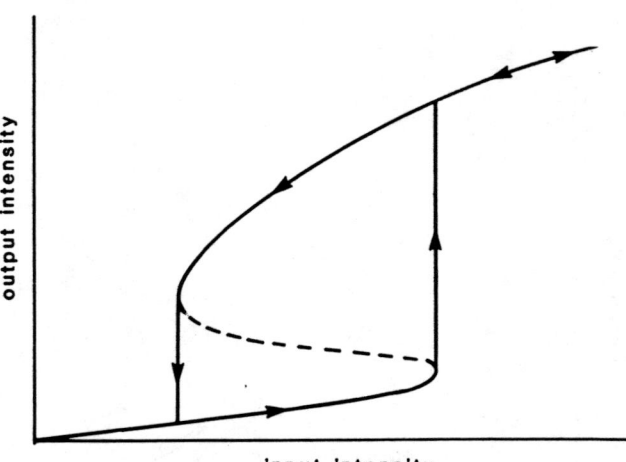

FIGURE 13.31
Hysteresis behavior in a nonlinear system such as Figure 13.30.

by the first turning point in Figure 13.31, however, the circulating intensity will become large enough to begin to saturate the absorption.

The cavity finesse will then begin to increase, which means the circulating intensity inside the cavity will also begin to increase for the same incident power level, thus causing a further increase in cavity finesse and in circulating intensity. The cavity operating point will then suddenly jump upward in a discontinuous fashion to the upper branch, where the cavity losses are essentially saturated and the cavity finesse, circulating intensity, and transmitted intensity are all much larger than on the lower branch.

If the incident intensity is then reduced, the much higher finesse of the cavity on the upper branch makes it possible for the internal circulating intensity to remain above the saturation level even with much smaller input intensity. The cavity will thus move back down along the upper branch, until at a certain point it drops discontinuously back to the lower branch. The portion of the input-output curve between these two discontinuities, marked by a dashed line, is unstable and cannot be a steady-state solution.

### Dispersive Optical Bistability

An analogous but physically different (and generally more useful) type of bistability can also occur in a passive interferometer cavity containing a nonlinearly *dispersive* rather than *absorptive* medium.

Consider, for example, a simple ring or standing-wave interferometer cavity containing an optical Kerr type of material in which the optical refractive index $n$ changes as the optical intensity is increased, in the form for example $n(I) = n_0 + n_2 I$, where $I$ is the circulating intensity inside the cavity. As the circulating intensity changes, therefore, the resonant frequency of the cavity will change, and this will in turn change the relationship between the input, circulating, and output intensities in a manner which can lead to a variety of complex bistable and multistable behavior.

The incident signal in this situation need not be tuned to the small-signal resonant frequency of the cavity. The nonlinear behavior in this system can instead be examined using the following simple graphical analysis. If we again

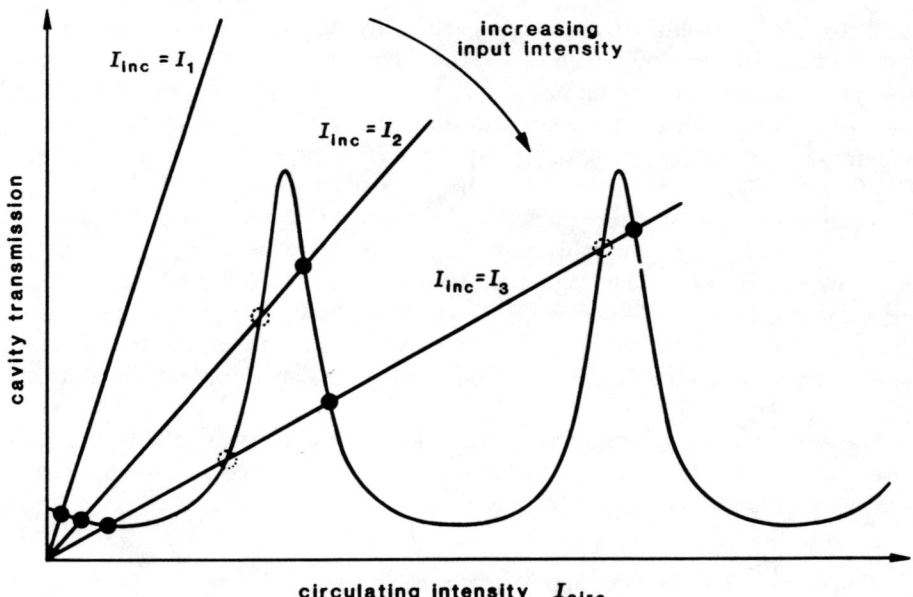

**FIGURE 13.32**
Graphical interpretation of dispersive cavity bistability.

make use of results from earlier chapters, the power transmission $T(I)$ through the cavity can be written as

$$T(I) \equiv \frac{I_{\text{trans}}}{I_{\text{inc}}} = \frac{1}{1 + F\sin^2 \phi(I)/2}, \tag{68}$$

where $F$ is the finesse of the cavity and $\phi(I)$ is the round-trip phase shift. For a cavity filled with an optical Kerr material and excited at a free-space wavelength $\lambda_0$ this phase shift will be given by

$$\phi(I) = \frac{2\pi p}{\lambda_0}[n_0 + n_2 I_{\text{circ}}], \tag{69}$$

where $p$ is the round-trip length in the cavity.

Figure 13.32 plots this transmission gain $T(I)$ through an optical cavity containing a lossless optical Kerr material, assuming that the cavity resonance is tuned well away from the applied signal frequency at small signal levels, and that the intensity is varied over a wide enough range to shift several axial modes of the cavity through the applied signal frequency. This curve will be shifted transversely, depending upon how the incident signal is tuned.

But, in addition the input, circulating, and output intensities are also related by the expressions $I_{\text{trans}} = \delta_2 I_{\text{inc}} = T I_{\text{inc}}$, or $T(I) = \delta_2 I_{\text{circ}}/I_{\text{inc}}$, which corresponds to a set of straight lines in the $T(I)$ versus $I_{\text{circ}}$ plot, with slopes which decrease with increasing input intensity $I_{\text{inc}}$.

Figure 13.32 then shows, for example, how for a low incident intensity $I_{\text{inc}} = I_1$ there is only one operating point at which these two formulas intersect. The cavity frequency in this situation is essentially unshifted from its low-intensity value, and the cavity is operating at low transmission, well off resonance.

At a larger incident intensity $I_{\text{inc}} = I_2$, however, there are three potential operating points, one at low transmission well off resonance, and two others within the high-transmission resonance peak. Only two of these points, however, indicated by the solid circles, represent stable operating points.

At the outermost operating point, for example, the cavity is operating close to one of its axial mode resonances, but with the intensity-shifted cavity resonance frequency slightly below the input signal frequency. If the incident power level increases slightly, then the circulating intensity increases, and this in turn drives the cavity resonance frequency slightly lower (by increasing the value of the intracavity index $n$). But, this moves the cavity resonance frequency slightly further away from the applied signal frequency, causing the circulating power to decrease and thus partially canceling the effect of the increased incident power. Operating points on the upper sides of the cavity resonances are thus stable, whereas operating points on the lower sides, by a similar argument, are perturbation unstable.

This same plot also shows that if the incident intensity is still further increased to the value $I_{\text{inc}} = I_3$, multistable behavior with three or more potential operating points also becomes possible.

### Fluctuations and Self-Pulsing Phenomena

These two examples illustrate the elementary properties of purely absorptive and purely dispersive optical bistability or optical multistability. More generally, if one considers an optical cavity containing a general two-level resonant atomic system, then we will have an even more complex situation, with a mixture of absorptive and dispersive nonlinear properties, depending upon how the atoms and the cavity are tuned. The atomic system may also exhibit either homogeneous or inhomogeneous saturation behavior, depending upon its type; and at larger signal levels we may have to take into account Rabi flopping effects in addition to the nonlinear saturation behavior of the atoms.

The net result of all this is a very rich and complex variety of nonlinear behavior in passively excited optical cavities containing saturable atoms. In addition to the simple bistability and hysteresis behavior illustrated in the preceding, at larger input intensities many of these nonlinear optical systems will exhibit *spontaneous periodic fluctuations*, in which the cavity intensity jumps back and forth at a regular rate between two branches of the input-output curve, thus converting a steady-state input beam into a pulsed output beam.

These jumps have a close conceptual relationship to phase transitions in atomic systems, and to limit cycle behavior in other nonlinear systems. In fact, it has been realized in recent years that many different nonlinear systems, ranging from optical cavities to mechanical systems and fluid-flow problems, can all exhibit broadly similar nonlinear properties. As we turn up the excitation intensity, or some other kind of gain parameter in a nonlinear system, we often see at first some kind of bistable or hysteresis behavior, as illustrated in the preceding. This may be followed by a periodic pulsing behavior, and this periodic behavior may at higher intensities show a discontinuous jump in the pulsation frequency, often to half the previous frequency (referred to as *period doubling*).

At each of these discontinuities if we suddenly switch the incident intensity to a value well above or well below the discontinuity point, then the transition from one form of behavior to another occurs very rapidly. If the incident intensity is only moved a very small amount beyond the transition point, however, then the transition from one branch to the other occurs only very slowly, a phenomenon referred to as *critical slowing down*.

### The Transition to Chaos

Finally, such nonlinear systems may often, as the excitation parameter is varied, suddenly jump to a very distinctive new form of behavior referred to, with good reason, as *chaos*. In the chaotic region, even though the equations of motion for the system are entirely deterministic and may contain only a few parameters, and the system input is constant, still the resulting system behavior (e.g., the cavity output intensity) fluctuates wildly with time, in what seems to be a totally random fashion. The power spectrum for the cavity amplitude fluctuations, for example, will apparently have a continuous distribution in frequency, with no observable discrete frequency components.

A passive optical cavity containing an ideal optical Kerr material, for example, with an externally applied cw signal, can pass through discrete regions of bistable behavior, then periodic fluctuations, then various sorts of period doublings, and then various sharply defined regions of chaotic behavior, as the incident signal intensity is slowly increased. The turbulence which invariably develops in a fluid flow above a sharply defined Reynolds number is another elementary illustration of chaos. These chaotic phenomena do not seem to depend on any fundamental noise sources in the system, and exhibit striking similarities across widely different physical systems.

### Experimental Results

All of the preceding-mentioned nonlinear phenomena, including bistability, multistability, periodic fluctuations, period doubling, and chaos, have in recent years been both predicted and experimentally observed, although often only with some difficulty, in optical cavities. In general, nonlinear dispersive and mixed-dispersive effects are more easily obtained (as well as more interesting) than purely absorptive effects.

For example, by combining tunable dye lasers with the very strong but narrow resonance transitions in metal vapors, such as sodium or rubidium cells, one can demonstrate many atomic nonlinear phenomena using reasonable input powers and detection speeds. It is difficult to envision practical optical signal-processing devices using this approach, however.

Optical Kerr effects are also often observed using molecular liquids such as $CS_2$, although only with comparatively high optical powers and fast pulses. Certain semiconductors, such as GaAs and InSb, and also semiconductor quantum well structures, can also exhibit strong optical nonlinearities at wavelengths near or just beyond their optical absorption edges; and there is much interest in these materials for possibly practical bistable optical systems.

## REFERENCES

For some more recent and complex results in laser bistability, see, for example, L. W. Hillman, J. Krasinski, R. W. Boyd, and C. R. Stroud, Jr., "Observations of higher order dynamical states of a homogeneously broadened laser," *Phys. Rev. Lett.* **52**, 1605–1608 (April 30 1984).

For an excellent introductory review of passive optical bistability, see E. Abraham and S. D. Smith, "Optical bistability and related devices," *Rep. Prog. Phys.* **45**, 815–885 (August 1982).

More recent but more theoretical surveys of bistability in driven optical cavities, with numerous references, are given by L. A. Lugiato, "Theory of Optical Bistability," and by J. C. Englund, R. P. Snapp, and W. C. Shieve, "Fluctuations, Instabilities and Chaos in the Laser-Driven Nonlinear Ring Cavity," in *Progress in Optics, Vol. XXI*, ed. by E. Wolf (Elsevier, 1984); pp. 69–216 and 355–428.

For an excellent introduction to the fundamental ideas of nonlinear instabilities and chaos, the student should see "Metamagical Themas," by D. R. Hofstadter in *Scientific American* **245(5)**, 16–29 (November 1981). A more formal discussion of the same ideas is given by E. Ott, "Strange attractors and chaotic motions of dynamical systems," *Rev. Mod. Phys.* **53**, 655–671 (October 1981).

A few recent research results on optical bistability and chaos include E. Abraham, W. J. Firth, and J. Carr, "Self-oscillation and chaos in nonlinear Fabry-Perot resonators with finite response time," *Phys. Lett.* **91A**, 47–51 (August 23 1982); M. Maeda and N. B. Abrahamson, "Measurements of mode-splitting self-pulsing in a single-mode Fabry-Perot laser," *Phys. Rev. A* **26**, 3395–3403 (December 1982); and H. Nakatsuka et. al., "Observation of bifurcation to chaos in an all-optical bistable system," *Phys. Rev. Lett.* **50**, 109–111 (10 January 1983).

## Problems for 13.7

1. *Bistable laser oscillator.* Using a homogeneous rate-equation model, find the potential steady-state operating points for the bistable laser oscillator discussed at the beginning of this section, and show that the first intersection between the loss and gain curves in this model is indeed unstable.

2. *Critical condition for absorptive bistability.* Show that for the simple absorptive bistable example analyzed in this section, the critical point for the onset of bistability occurs at the operating conditions $R = 8$, $I_{\text{circ}} = 3I_{\text{sat}}$, $I_{\text{inc}} = 27\delta_1 I_{\text{sat}}$, and $I_{\text{trans}} = 3\delta_1 I_{\text{sat}}$.

3. *Absorptive bistability at large nonlinearity.* Show further that for strong absorptive nonlinearity in an interferometer cavity—that is, for $R \gg 8$—the turning points in the bistability curve of $I_{\text{trans}}$ versus $I_{\text{inc}}$ are given by $I_{\text{circ}} = I_{\text{sat}}$, $I_{\text{inc}} = R^2 \delta_1 I_{\text{sat}}/4$ and $I_{\text{trans}} = \delta_1 I_{\text{sat}}$ for the first root, and by $I_{\text{circ}} = rI_{\text{sat}}$, $I_{\text{inc}} = 4R\delta_1 I_{\text{sat}}/4$ and $I_{\text{trans}} = R\delta_1 I_{\text{sat}}$ for the second root.

4. *Bistable ring absorber cavity.* Consider a ring-laser cavity which has only a single input-output mirror with reflectivity $R_1 = \exp(-\delta_1)$, all other mirrors having 100% reflectivity. Evaluate the reflected intensity $I_{\text{refl}}$ versus incident intensity $I_{\text{inc}}$ from this mirror, assuming the ring cavity contains a homogeneously saturable absorber, and no other losses.

5. *Inhomogeneous absorptive bistability* Calculate and plot the input-output intensity relationship for a purely absorptive interferometer tuned to resonance, assuming that the absorber saturates inhomogeneously—that is, as $1/(1 + I/I_{\text{sat}})^{1/2}$—rather than homogeneously.

## 13.8 AMPLIFIED SPONTANEOUS EMISSION AND MIRRORLESS LASERS

Some laser systems have such extremely high gain that they need no mirrors—they can emit very bright and more or less quasi coherent beams out each end of the laser medium simply as a result of very high-gain amplification of their own internal spontaneous emission traveling along the length of the laser-gain medium. Interstellar masers and x-ray lasers must also of necessity operate without mirrors, since no mirrors are available.

Interesting concepts that have been developed in connection with this kind of behavior include such terms as *superradiance, superfluorescence, coherence brightening,* and *amplified spontaneous emission* (ASE). In this section we will attempt to give a brief classification and explanation of each of these terms, together with a brief summary of the useful properties of mirrorless laser systems.

### Coherently Oscillating Dipoles and Free Induction Decay

In developing this topic it may be easiest to begin wih the more exotic and strongly coherent forms of behavior, and work down toward the simplest and most common kinds of mirrorless or ASE lasers. The first part of the following discussion will thus be closely related to the coherent-pulse and coherent-dipole types of behavior that we also discuss in other sections of this text.

We have pointed out elsewhere, for example, that if a collection of two-level atoms is prepared such that the individual atomic dipoles are oscillating or precessing at least partly in phase with each other, then the associated macroscopic polarization $p(t)$ in the collection of atoms will emit electromagnetic radiation in a coherent fashion—that is, the emitted radiation will be coherent or sinusoidal in time, with a time-phase determined by the initial preparation of the atoms. This radiation will also have directional or spatially coherent properties determined by the relative phases with which the radiating atoms at different points are initially set oscillating.

Such a coherently prepared atomic system may occupy a volume that is large in terms of optical wavelengths. If the initial atomic oscillations are then prepared using, for example, a traveling optical pulse of sufficiently large intensity. the resulting coherent emission will emerge in the same direction of travel as the preparing pulse.

If the degree of initial coherence imposed on the individual oscillators is comparatively small, and if the atomic populations are initially either not inverted, or at most have small gain, then this coherent radiation, although brighter and more directional than the usual spontaneous emission, will be relatively weak; and the coherently radiated signal will decay in time with the appropriate dephasing time $T_2$ in homogeneous systems, or $T_2^*$ in inhomogeneous systems, until it disappears into the incoherent spontaneous emission background from the same atoms.

This particular kind of coherent atomic radiation is often referred to as simple *free-induction decay*. Free-induction decay can be demonstrated experimentally both in low-frequency magnetic resonance systems and in optical-frequency atomic systems using pulsed laser excitation, as we describe elsewhere in this text.

### Dicke Superradiance

At a time well before the invention of the laser, R. H. Dicke also considered analytically the situation in which a sizable number of atoms contained in a small volume $V$ may all be oscillating with a very *high* degree of coherence between the individual dipole oscillations. If such a volume contains $N$ coherently oscillating atomic depoles, the macroscopic dipole moment within the volume will have magnitude $N\mu_1$, where $\mu_1$ is the oscillating moment of a single atom. The rate of coherent radiative power emission from this volume will then be proportional to $(N\mu_1)^2$, in contrast to the usual incoherent form of spontaneous emission, where the emission rate is proportional only to the number of atoms $N$. The coherent emission from this small but coherently excited volume will emerge as a short burst or pulse of radiation with a duration proportional to $1/N$, rather than as an exponential decay with a lifetime $\tau$ independent of the number of atoms.

This specific type of small-volume, coherently prepared emission, with strong initial coherence, has come to be known as *Dicke superradiance*. The atoms here are locked together, not only by their initial preparation all in phase with each other, but also by the strong coupling of all the atoms to each other through their common radiation fields. Dicke superradiance of this type has been observed in specially prepared low-frequency magnetic resonance systems, but not, at least in its simplest form, in optical-frequency systems.

### Incoherently Prepared Dicke Superradiance

Suppose next that a two-level system having $N$ atoms is initially prepared with a completely inverted population, i.e., $N_1 = 0$ and $N_2 = N$, so that each of the atoms is completely in its upper energy level to start with. The atomic system will then initially possess *no* coherent macroscopic polarization $p(t)$, since the quantum expectation value for the dipole oscillation of each individual atom is zero if the atom is entirely in its upper (or for that matter, its lower) quantum state. (As an alternative, the atoms may be prepared in an only partially inverted state, but with an *incoherent preparation method*, such that all the atomic dipoles are randomly phased and no coherent macroscopic polarization is initially present.)

Each upper-level in this situation will then begin to radiate spontaneously and incoherently through the purely quantum spontaneous emission processes that are represented by the radiative decay rate $\gamma_{\text{rad}}$. Note that this spontaneous emission process, although it can be modeled by a gaussian quantum noise source, can only be derived from a completely quantum analysis, in which the atoms and the electromagnetic field are both quantized.

Dicke then pointed out in his original paper that if the $N$ atoms prepared in this inverted but incoherent fashion were all contained in a volume small compared to the emission wavelength cubed, the atoms would all be coupled together through their overlapping radiation fields. As a consequence the individual atoms will not in fact continue to radiate independently and incoherently. Rather the initial spontaneous emission from any one atom (or, if you like, a small initial fraction of the spontaneous emission from all of the atoms) will tend to "capture" or entrain the oscillations in all the other atoms, in such a way that the inverted system can develop a very large and almost totally coherent macroscopic polarization.

As a result, this system, although initially incoherent, can still evolve into a coherent superposition, and can emit, after a certain time delay, almost exactly the same sort of $(N\mu_1)^2$ superradiant burst described in the preceding. Because the coherent emission builds up initially from spontaneous emission noise, the phase angle of the radiation will be entirely random from shot to shot, and there will also be small random fluctuations in the delay time between initial preparation of the atoms and emergence of the superradiant burst.

(In terms of the Bloch vector picture developed in a later chapter in this text, the "super Bloch vector" describing the sum of all the dipoles in the small volume is initially oriented essentially antiparallel to the effective dc magnetic field, in the highest-energy, but metastable, orientation. The effect of spontaneous emission is then to give a small initial disturbance, or effectively a small initial tilt angle to this vector. If suitable conditions are met, this Bloch vector will then precess outward, maintaining constant length, so that its tip stays on the surface of a sphere, and will eventually radiate all of its energy into a superradiant burst as the precessing vector passes from the inverted orientation through the equaorial plane and on to the lowest-energy orientation parallel to the effective dc magnetic field.)

### Optical Extensions of Dicke Superradiance

In the simplest situation, the emergence of this type of superradiant emission from an initially inverted but *incoherently prepared* atomic system depends on the atoms being contained within a volume $V \leq \lambda^3$, so that there will be very strong coupling between all the atoms through their common radiation field. (Note that this coupling occurs only through the radiation fields; the quantum wavefunctions of the individual atoms need not be overlapping.) This particular type of small-volume incoherently prepared superradiant emission has not yet been demonstrated in any optical system, largely because there seems to be no available atomic medium in which a sufficient number of suitable atoms can be assembled within a volume of the order of $\lambda^3$.

Considerable attention has been given, however, to the more general situation in which a large number of inverted atoms, with strong initial population inversion but with no initial coherent polarization, are prepared in an *extended region of space*, most often in the form of a long cylindrical region, or pencil, having a Fresnel number on the order of unity. This larger volume may then, depending on the size of the inversion, emit either simple *amplified spontaneous emission* (ASE), as we will describe below, or a kind of extended Dicke superradiance which has come to be referred to as *superfluorescence*.

### Pure Superfluorescence Behavior

Ideal or pure superfluorescence behavior, as described by a number of theories and experiments (see References), will occur in such systems only under rather specialized conditions in which the atomic transition is strong and narrow enough, and the inversion density large enough, so that the radiative coupling between atoms becomes very strong, in the same sense as in the superradiant experiments described in the preceding, even though the atoms are spread over a volume large compared to the radiation wavelength. The necessary conditions for pure superfluorescent behavior are quite complex, but a key condition seems to be that the atomic gain coefficient must be large and the sample length small compared to the distance that radiation can travel in one inverse atomic linewidth, or one atomic dephasing time $T_2$. If all the atoms are to emit cooperatively, they must be able to communicate with each other strongly in a time short compared to their dephasing time. To accomplish this, radiation coming from any one atom must be strongly amplified and transmitted to another atom, and the reverse, before either of these atoms has either radiated spontaneously or been dephased.

The principal experimental features of pure superfluorescence will then be an intense simultaneous burst of quasi coherent radiation coming out in a narrow cone from each end of the inverted pencil of atoms. This pulse will have an intensity proportional to the initial number of atoms *squared*, and an angular spread which is roughly the aspect ratio of the inverted pencil of atoms. The pulse duration will be inversely proportional to the number of atoms, and the pulse will have a definite time delay following the initial preparation of the inverted atoms. Since this emission will be initiated by random spontaneous emission in the atoms, there will again be small random variations in the time delay of the pulse from one experimental shot to another.

The essential features of pure superfluorescence are thus a delayed emission pulse, with intensity proportional to $N^2$, emerging from a large-volume atomic collection that is prepared with *no initial coherent polarization or oscillating dipole moment*.

### Superfluorescence Experiments

The conditions needed to demonstrate ideal superfluorescence are fairly hard to obtain, and only a few experiments have displayed this effect in a clear and definite fashion thus far. Perhaps the clearest experiment demonstrating pure superfluorescence was carried out on an upper-level atomic transition having a wavelength of 2.9 $\mu$m in low-pressure cesium vapor. The upper level of this transition could be selectively populated at a high density by optical pumping from the cesium ground state using a tunable pulsed dye laser at 455 nm. By using a low-pressure cesium cell it was possible to obtain a vapor with minimal collision broadening and long lifetime; and by using Zeeman splitting in a dc magnetic field together with selective pumping it was possible to populate the upper level of only a single strong transition corresponding to a near-ideal two-level atomic system. The results obtained in these experiments were then in excellent agreement with the theoretical concepts outlined preceding.

### Amplified Spontaneous Emission (ASE) Lasers

We come finally to the most common form of mirrorless laser behavior, namely, "ordinary" amplified spontaneous emission or ASE, as illustrated in Figure 13.33.

Amplified spontaneous emission as used here refers to any situation in which the spontaneous emission coming from a distribution of inverted laser atoms is linearly amplified by the same group of atoms, with a gain which is sizable in at least one direction through the atoms, but the more complex features of superfluorescence are not present. If the amplification along a long thin cylinder of inverted atoms is sufficiently large, for example, this can produce an output beam from each end of the laser medium which can be quite bright, powerful, and moderately directional, with a fair amount of spatial (but usually not temporal) coherence. This radiation may become strong enough to produce significant

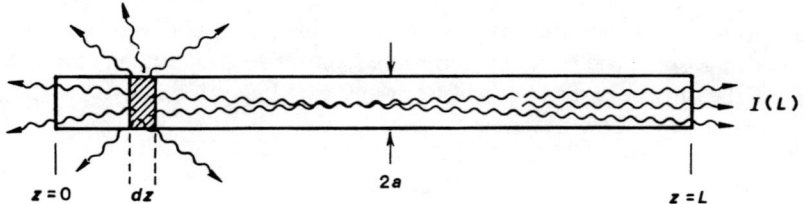

**FIGURE 13.33**
An amplified-spontaneous-emission or ASE laser.

saturation along the gain medium, and to extract the major portion of the inversion energy into the directional beams. The inverted medium thus acts as a "mirrorless laser," with output characteristics that are intermediate between a truly coherent laser oscillator and a completely incoherent thermal source.

Examples of such mirrorless lasers can include many pulsed excimer lasers and visible and ultraviolet molecular lasers, such as the $N_2$ laser at 337 nm or the $H_2$ laser at 120 nm, especially when pumped by fast transverse discharges or by electron beam pumping. Mirrorless laser action also occurs in certain very high-gain infrared gas laser lines, such as the 3.39 $\mu$m line in He-Ne or the 3.51 $\mu$m line in He-Xe; in very high-gain dye laser amplifiers; and in high-gain semiconductor diode lasers in which the mirror reflection at the end of the laser is deliberately spoiled. The enormously large and powerful natural masers and lasers which occur in interstellar space are also primary examples of mirrorless or ASE laser systems.

### Questions of Terminology

There has in the past been considerable inconsistency in the laser literature in the use of the various terms superradiance, superfluorescence, and amplified spontaneous emission; and many articles still refer to the type of mirrorless laser we are discussing here as "superradiant emission" or as a "superradiant laser system."

In most of the mirrorless lasers of practical interest, however, the laser medium can still be assumed to remain entirely in the rate-equation regime (with population saturation taken into account); and the emerging radiation can be accurately described merely as narrowband amplified gaussian noise. The kinds of pulse delays and large coherent polarizations that are characteristic of superradiance or superfluorescence, and the dependence of peak pulse intensity on $N^2$ as described in the preceding, do not appear in simple ASE lasers. It seems preferable, therefore, to refer to these simpler systems in general either as *amplified spontaneous emission* systems or as *mirrorless lasers*, and to reserve the terms *superradiance* and *superfluorescence* for the more specialized phenomena described in the preceding.

### Practical Characteristics of Mirrorless ASE Lasers

Consider a long slender rod of inverted laser medium with length $L$ and diameter $2a$, as illustrated in Figure 13.33, and assume for simplicity that the laser transition is completely inverted with inversion population density $N$. The spontaneous emission power going out into all directions from any small unit volume of this medium will then be given by $N\gamma_{rad}\hbar\omega$.

If we neglect for the minute any gain saturation effects and assume a long slender rod with $L \gg a$, the contribution to the amplified spontaneous intensity $dI$ arriving at the output end of the rod from any small length $dz$ near the input end of the rod can then be written as

$$dI = \frac{\pi a^2 N \gamma_{\text{rad}} \hbar \omega}{4 \pi L^2} e^{2\alpha_m (L-z)} \, dz, \tag{70}$$

where $2\alpha_m \equiv N\sigma$ is the power amplification coefficient in the rod. If we integrate the total spontaneous emission contribution coming from the entire length of the rod, this yields for the total ASE intensity at the output end

$$\begin{aligned} I &\approx \frac{N \gamma_{\text{rad}} \hbar \omega a^2}{4 L^2} e^{2\alpha_m z} \int_0^z e^{-2\alpha_m z} \, dz \\ &\approx \frac{\gamma_{\text{rad}} \hbar \omega a^2}{4\sigma L^2} e^{2\alpha_m L}. \end{aligned} \tag{71}$$

In writing this we have assumed that the total gain $e^{2\alpha_m L}$ along the rod is large, so that we can replace the upper limit of the integral in Equation 13.71 by infinity. Most of the ASE intensity at each end of the rod then comes from just the first gain length $(2\alpha_m)^{-1}$ at the other end of the rod, and the solid angle of this emitting volume as seen from the other end of the rod is essentially constant at $\pi a^2 / L^2$.

We have shown earlier that the stimulated transition rate $W_{12}$ produced by a wave of intensity $I$ is given by $W_{12} = \sigma I / \hbar \omega$. Thus, the ratio of the stimulated transition rate caused by the ASE to the spontaneous emission rate in the same atoms at the output end of the rod (in other words, at either end) can be written in the simple form

$$\frac{W_{12}}{\gamma_{\text{rad}}} \approx \left(\frac{a}{2L}\right)^2 e^{2\alpha_m L}, \tag{72}$$

which depends only on the aspect ratio $a/L$ of the rod, and the overall gain coefficient $2\alpha_m L$.

Although the radius to length ratio $a/L$ of a typical laser medium is normally small, the exponential gain factor in Equation 13.72 means that as soon as the gain coefficient $2\alpha_m L$ becomes larger than a few times unity, the stimulated emission rate from the atoms at each end of the laser medium due to amplified spontaneous emission from the other end will begin to exceed the purely spontaneous emission rate by a large ratio. In other words, as soon as $2\alpha_m L \gg 2\ln(2L/a)$, the presence of ASE will begin to speed up the net emission rate, and thus to shorten the effective inversion lifetime of the laser medium, by a significant amount. Obviously, this *lifetime shortening* due to ASE will become even more serious if the amplification length is large in more than one direction, for example, across the width of a flat gain slab, or across all three dimensions of a spherical or rectangular gain volume.

Since the saturation intensity in a simple homogeneous gain medium can be written at $I_{\text{sat}} = \hbar\omega / \sigma \tau_2$, where $\tau_2$ is the effective lifetime or repumping time for the upper laser level, then we can also write the ratio of the amplified spontaneous emission intensity to the saturation intensity at either end of the rod in the form

$$\frac{I}{I_{\text{sat}}} \approx \left(\frac{a}{2L}\right)^2 \left(\frac{\tau_2}{\tau_{\text{rad}}}\right) e^{2\alpha_m L}. \tag{73}$$

## 13.8 AMPLIFIED SPONTANEOUS EMISSION AND MIRRORLESS LASERS

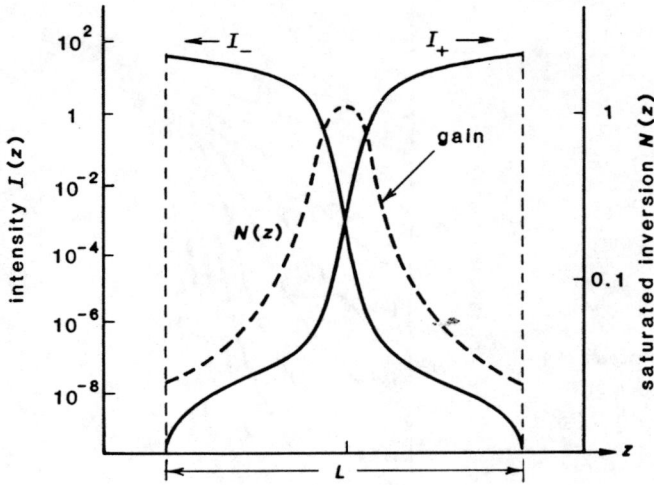

FIGURE 13.34
Gain saturation and traveling-wave intensities in a typical ASE laser.

Again, as soon as the net gain coefficient $2\alpha_m L$ becomes more than a few times unity, the ASE will surely become large enough to produce significant gain saturation and significant power extraction from the inverted gain medium.

### Saturation and Power Extraction

Calculating the total power extraction and the exact end-fire emission pattern from such a mirrorless laser system becomes a more complicated problem when gain saturation is taken into account, since we must take into account both the effects of the ASE on the atomic inversion and the effects of gain saturation back on the growth rate for the ASE. Several numerical calculations for problems of this type are listed in the References, and Figure 13.34 illustrates the general type of behavior that occurs in a long narrow high-gain ASE laser when saturation is taken into account.

The basic result illustrated here is that under high-gain conditions, ASE coming from each end of the gain medium tends to heavily saturate the inversion over a sizable region at the opposite end of the medium, leaving a relatively narrow region or band of unsaturated gain only in the central region of the rod. As we increase the pumping level or the initial unsaturated gain value in a system of this type, this central unsaturated gain region becomes narrower and narrower. The growth of the intensities in opposite directions along the rod is also, of course, no longer a simple exponential with distance, although there is still large gain from one end of the system to the other.

Casperson has further calculated the manner in which the total spontaneous emission flux from the ends of an ASE laser increases as we turn up the pumping power, or alternatively the gain length in the laser medium. Typical results for a homogeneously broadened laser medium are shown in Figure 13.35, and generally similar results are obtained for inhomogeneously broadened lasers. The parameter labeling the different curves in this figure is a dimensionless measure of the spontaneous emission rate, related to the value of $1/p$ from our earlier rate-equation analysis. This parameter thus has a value much less than unity for most typical laser systems.

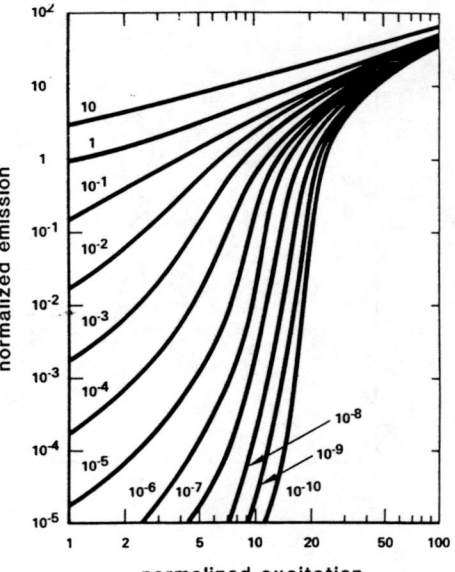

FIGURE 13.35
Total amplified spontaneous emission power output from a long cylindrical homogeneous gain medium. (From L. W. Casperson, *J. Appl. Phys.* 48, 256–262, January 1977.)

The primary interpretation to be made here is that, whereas an ASE laser cannot have the kind of extraordinarily sharp threshold behavior produced by the feedback from the mirrors in an ordinary laser cavity, mirrorless lasers can exhibit a kind of "soft threshold behavior" which may not appear too greatly different from true laser action.

### Temporal and Spatial Output

The output spectrum from a mirrorless laser, at least at low intensity, will consist of the incoherent spontaneous emission from the laser medium, which has a spectral lineshape corresponding to the atomic lineshape, as amplified by the atomic gain process. The amplification process is also characterized by the same atomic linewidth or bandwidth, however, and we have pointed out earlier that the finite linewidth of the gain medium means that the spectrum will be significantly narrowed, typically down to values 2 to 5 times narrower than the atomic linewidth in a homogeneous system at high gain values.

When saturation effects are taken into account in inhomogeneously broadened media, this narrowing can be significantly reduced, especially at higher gains, because the inhomogeneous gain profile saturates first in the center, and only more gradually in the wings of the line. Typical results have also been calculated by Casperson, as shown in Figure 13.36.

The temporal output from a mirrorless ASE laser, regardless of its spectral width, will always consist of narrowband, highly amplified but still essentially random gaussian noise, rather than any kind of coherent or amplitude-limited sinusoidal oscillation. Mirrorless lasers thus generally lack most of the important temporal coherence features associated with the sinusoidal amplitude-stabilized oscillation in a true laser oscillator.

The spatial pattern from the ends of an ASE laser will be a narrow cone with a cone angle defined by the aspect ratio of the laser rod, that is with a half-angle $\Delta\theta \approx a/L$. If the rod is very slender, so that it has a Fresnel number

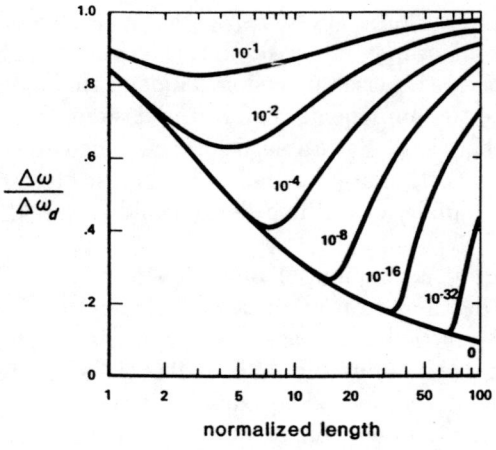

FIGURE 13.36
Normalized amplified spontaneous emission spectral width for an inhomogeneous transmission. (From Casperson, *loc. cit.*)

$N_f \equiv a^2/L\lambda \approx 1$, then all of the end-fire emission will emerge in essentially a single transverse mode. The output beam from a sufficiently slender mirrorless laser can thus have a large degree of spatial coherence (although perhaps not so much total power because of the small rod diameter).

A larger diameter rod, with $N_f > 1$, will emit its radiation into a random superposition of essentially $\pi N_f^2$ transverse modes, as we will discuss in later chapters, and will thus have considerably less ideal spatial coherence (although such a system may still compare not unfavorably with a cavity-type laser having poor mode selection and thus also a large number of transverse modes).

### Coherence Brightening and Swept-Gain Operation

This combination of significant spatial coherence with some spectral narrowing is sometimes referred to as *coherence brightening* in the mirrorless laser output. Coherence brightening is not a very precisely defined term, however, and much the same kind of coherence brightening is equally well observed in long slender superradiant or superfluorescent systems.

We can also note that in some laser systems with large spontaneous emission, high gain, and short upper-state lifetimes, the pumping process in the laser medium is carried out using some form of *traveling-wave excitation* which travels down the laser medium in one direction, at a velocity close to the velocity of light. If the total gain is large enough, the traveling excitation pulse soon becomes accompanied by a pulse of amplified spontaneous emission, which travels just behind the excitation pulse, and extracts or "dumps" the inversion energy as fast as it is created.

This type of *swept-gain laser action*, which produces a beam coming out of only one end of the laser, is likely to be characteristic of many X-ray lasers, as well as many fast pulse UV lasers, since the inversion lifetime for the laser medium in these situations may be comparable to or shorter than the transit time for a light pulse down the length of the laser medium.

### Parasitic Laser Oscillation and ASE

Amplified spontaneous emission can also play a very much unwanted role in many large high-gain laser systems. In cascaded multisection laser amplifiers, for example, such as are often used in laser fusion systems, spontaneous emission from the input end of the amplifier chain may, after amplification through the chain, become large enough to deplete the laser inversion, damage the target pellet, or even cause optical damage to components, before the desired optical signal can be sent through the amplifier chain. Saturable absorbers, which absorb the weak spontaneous emission but pass the larger signal pulses, must often be placed between the sections in such amplifier chains in order to avoid severe ASE problems.

Parasitic oscillations which arise from a combination of amplified spontaneous emission and weak unintentional reflections from various internal surfaces can also be a serious problem in any large high-gain laser system. In large glass disk amplifiers, for example, parasitic oscillations and ASE in directions running across the face of the disk, or around the rim of the disk, can be a serious problem; and in general the total energy storage volume of any large high-power laser device is often limited by a combination of parasitic oscillations and ASE.

We might note finally that the emission *improvement* and lifetime *shortening* that occurs in an inverted atomic system is the exact opposite of the emission *reduction* and lifetime *extension* that has long been known due to *radiation trapping* in strongly absorbing atomic systems. The spectral narrowing in an ASE system is thus the reverse of the *line reversal* that is well known in such radiation trapped systems.

## REFERENCES

The basic concept of superradiance was first developed in the article by R. H. Dicke, "Coherence in spontaneous radiation processes," *Phys. Rev.* **93**, 99—110 (January 1 1954).

A more detailed analysis for the optical situation is given by R. Bonifacio, P. Schwendimann, and F. Haake, "Quantum statistical theory of superradiance. I and II," *Phys. Rev. A* **4**, 302–313 and 854–864 (July and September 1971). Other extensions of this concept to optical systems are also reviewed in J. H. Eberly, "Superradiance revisited," *Am. J. Phys.* **40**, 1374–1383 (October 1972).

Some comments on the distinction between Dicke superradiance and mirrorless lasers are also given by L. Allen and G. I. Peters, "Superradiance, coherence brightening and amplified spontaneous emission," *Phys. Lett.* **31A**, 95–100 (9 February 1970).

The theory of optical superfluorescence is developed by R. Bonifacio and L. A. Lugato, "Cooperative radiation processes in two-level systems: superfluorescence. I and II," *Phys. Rev.* **A11**, 1507–1521 (May 1975) and **12**, 587–598 (August 1975).

Representative experiments demonstrating superfluorescence are described by N. Skribanowitz, I. P. Herman, J. C. MacGillivray, and M. S. Field, "Observation of Dicke superradiance in optically pumped HF gas," *Phys. Rev. Lett.* **30**, 309–311 (February 19 1973); and by H. M. Gibbs, Q. H. F. Vrehen, and H. M. J. Hikspoors, "Single-pulse superfluorescence in cesium," *Phys. Rev. Lett.* **39**, 547–550 (August 29 1977). For an extensive review of this subject, see also Q. H. F. Vrehen and H. M. Gibbs, "Superfluorescence experiments," in *Dissipative Systems in Quantum Optics*, ed. by R. Bonifacio (Springer-Verlag, 1982).

For a general discussion of lasers based on amplified spontaneous emission, see L. W. Casperson, "Threshold characteristics of mirrorless lasers," *J. Appl. Phys.* **48**, 256–262 (January 1977), and references therein.

## 13.8 AMPLIFIED SPONTANEOUS EMISSION AND MIRRORLESS LASERS

Other typical calculations on the properties of ASE systems are given by L. W. Casperson and A. Yariv, "Spectral narrowing in high-gain lasers," *IEEE J. Quantum Electron.* **QE-8**, 80–85 (February 1972); by H. Maeda and A. Yariv, "Narrowing and rebroadening of amplified spontaneous emission in high-gain laser media," *Phys. Lett.* **43A**, 383–385 (March 26 1973); and by U. Ganiel, A. Hardy, G. Neumann, and D. Treves, "Amplified spontaneous emission and signal amplification in dye-laser systems," *IEEE J. Quantum Electron.* **QE-11**, 881–892 (November 1975).

For one recent experimental example of a mirrorless ASE semiconductor laser, see C. S. Wang, *et. al.*, "High-power low-divergence superradiance diode," *Appl. Phys. Lett.* **41**, 587–589 (October 1 1982).

# 14
# OPTICAL BEAMS AND RESONATORS: AN INTRODUCTION

This chapter and the following several chapters describe the transverse mode properties of laser resonators, and the propagation properties of the optical beams generated by lasers. These are very extensive subjects, and the reader will need to pick and choose with some care, passing over those sections which treat more detailed topics than are of immediate interest. It seems a good idea therefore to give an outline at this point of the contents of these chapters.

*Chapter 14: Optical Beams and Resonators: An Introduction.* In this chapter we first give a brief overview of what we mean by the transverse modes in an optical resonator, and how these modes should be analyzed. We also summarize briefly some of the most general properties of these modes, to set the stage for the more detailed discussions to follow.

*Chapter 15: Ray Optics and Ray Matrices.* The basic concepts of ray optics, especially paraxial ray optics and the so-called ray matrices or "*ABCD* matrices," prove to be very useful in understanding both the stability properties of optical resonators and the propagation properties of optical beams. In fact the ray matrix approach—which at first seems to be limited to geometrical optics only—turns out later to provide the foundation for a sophisticated and powerful treatment of paraxial wave optics and paraxial diffraction theory in general. Chapter 15, therefore, presents a detailed review of ray optics and ray matrices.

*Chapter 16: Free-Space Wave Optics.* We then follow the ray analysis with another review of the fundamentals of wave propagation and diffraction in free space, including the paraxial wave equation and Huygens integral. We note in particular that Hermite-gaussian (or Laguerre-gaussian) beams are the "eigenmodes of free-space propagation."

*Chapter 17: Gaussian Beams in Free Space.* Because of the widespread importance of gaussian beams in lasers, Chapter 17 reviews the practical properties of these free-space gaussian optical beams in some detail.

*Chapter 18: Beam Perturbation and Diffraction Effects.* The propagation of gaussian beams or of any other beam profiles in free space will be perturbed by the diffraction effects associated either with any kind of hard-edged apertures or with any kind of scattering or grating elements through which the beam may pass. Since these effects are very important in determining the properties of real optical beams and resonators, we give a brief review of their basic properties here.

*Chapter 19: Stable Two-Mirror Gaussian Resonators.* Many practical laser resonators consist essentially of two end mirrors with only free space in between. When such a resonator is also "stable" in a certain sense, the resulting resonator modes are very close to free-space gaussian modes. A substantial body of analysis and terminology for such gaussian resonator modes has become part of the basic lore of the laser field. Chapter 19 therefore reviews the basic properties of simple two-mirror stable gaussian resonators.

*Chapter 20: Generalized Paraxial Wave Optics.* In recent years, on the other hand, a much more general and powerful approach to paraxial wave optics and to optical resonators has been developed, which includes "soft" gaussian apertures and quadratic transverse amplitude variations as part of the formalism. This generalized approach to paraxial wave optics can be expressed in a very powerful fashion using a generalized *ABCD* matrix approach, which includes complex ray matrices and complex Hermite-gaussian modes. Chapter 20 therefore develops the full complex *ABCD* matrix formalism, of which the free-space gaussian beam results are a simple limiting case.

*Chapter 21: Generalized Paraxial Resonator Analysis.* Applying this complex paraxial formalism to optical resonators then enables us to develop a very general analysis of such resonators, and in particular to see how optical resonators can be classified into "real" and "complex" resonators, into "geometrically stable" and "unstable" resonators, and into "perturbation-stable" and "unstable" resonators (which is not the same thing). Chapter 21 therefore develops this more general resonator analysis, and shows how it can be applied to important practical resonators, including multielement stable resonators and the important new class of complex-stable resonators with variable reflectivity mirrors.

*Chapter 22: Unstable Optical Resonators*, and *Chapter 23: More on Unstable Resonators*. Finally, the so-called "unstable optical resonators" that we just mentioned are, in fact, a quite different class of resonators which have emerged in recent years to provide very useful resonator designs for a wide variety of high-power and high-gain lasers. Chapters 22 and 23 therefore present an extensive review of the properties of this very useful class of resonators.

## 14.1 TRANSVERSE MODES IN OPTICAL RESONATORS

Laser cavities differ in several significant ways from the closed microwave cavities that are commonly treated in electromagnetic theory textbooks. Optical resonators first of all usually have open sides, and hence always have diffraction losses because of energy leaking out the sides of the resonator to infinity. Optical resonators are also usually described in scalar or quasi plane-wave terms, with emphasis on the diffraction effects at apertures and mirror edges, rather than in vector terms with emphasis on matching boundary conditions. The distinction between "longitudinal" and "transverse" modes in the resonator is also much sharper in optical than in microwave resonators.

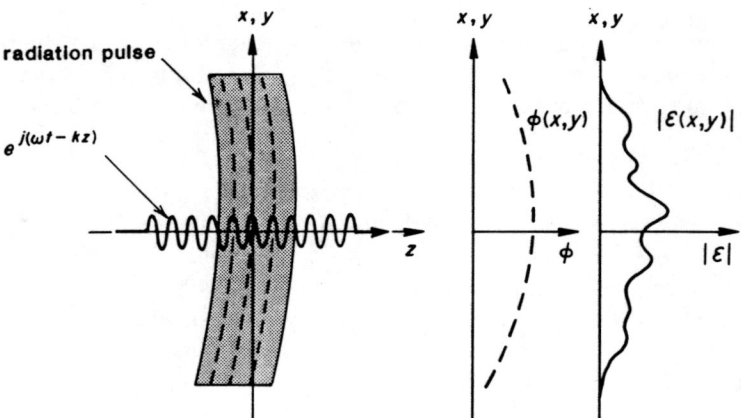

**FIGURE 14.1**
A traveling pulse or "slab" of optical radiation, propagating in the $z$ direction.

Before beginning any detailed analysis of optical resonators, therefore, it may be useful to introduce some of the fundamental concepts we will use to describe optical resonator modes in rather broad and general terms.

### The "Recirculating Pulse" Approach

In earlier chapters we have often emphasized how the optical radiation inside an optical cavity circulates repeatedly around the cavity, bouncing back and forth between the end mirrors (or circulating around the ring in a ring-laser cavity). In these earlier discussions we used only a plane-wave approximation, ignoring the transverse spatial variation of the waves.

To bring transverse variations into the discussion, let us next consider only that portion of the optical energy traveling in the $+z$ direction and contained within some short axial segment of length $\Delta z$ within the cavity. We can think of the radiation in this segment as forming a short pulse or a thin "slab" of radiation (see Figure 14.1), whose axial thickness $\Delta z$ is small compared to the length $L$ of a typical laser cavity but still very large compared to an optical wavelength $\lambda$.

The time and space variation of the $\mathcal{E}$ fields within such a circulating pulse or slab as it travels through the resonator, including transverse variations, can then be written in the form

$$\mathcal{E}(x,y,z) = \mathrm{Re}\, \tilde{E}(x,y,z)\, e^{j(\omega t - kz)}$$
$$= \mathrm{Re}\, |\tilde{E}(x,y,z)|\, e^{j(\omega t - kz) + j\phi(x,y,z)}. \quad (1)$$

By writing the fields in this fashion, we separate out the plane-wave aspects of the wave propagation as given by the $e^{j\omega t - jkz}$ factor, where $\omega$ is the optical carrier frequency and $k = \omega/c = 2\pi/\lambda$ the associated plane-wave propagation constant, from the complex phasor amplitude $\tilde{E}(x,y,z)$ which describes the *transverse* amplitude and phase variation of the beam. The transverse intensity profile of the beam within this particular pulse or slab is then given by $I(x,y,z) = |\tilde{E}(x,y,z)|^2$, whereas the transverse phase profile, or the shape of the optical wavefront is given by the transverse phase variation $\phi(x,y,z)$.

**stable resonator**

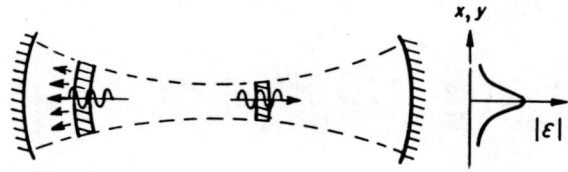

**unstable resonator**

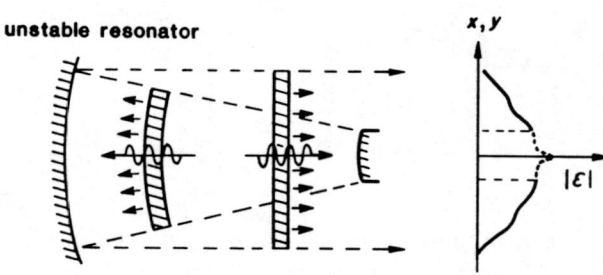

FIGURE 14.2
Circulating pulses ("slabs") in stable and unstable optical resonators.

Although we write the phasor amplitude function $\tilde{E}(x,y,z)$ as a function of $x$, $y$ and also $z$, we will see later that the variation of this transverse beam profile with the axial or $z$ coordinate is generally very slow compared to the $e^{-jkz}$ variation that we have separated out. The latter function goes through a complete $e^{\pm j2\pi}$ variation in just one optical wavelength. By contrast, the complex amplitude profile $\tilde{E}(x,y,z)$ will not change much if at all through the thickness of one "slab"; and it will also change only very slowly with distance as a particular slab propagates down the resonator, or through free space outside a resonator.

## Pulse Propagation in Stable and Unstable Resonators

If, however, we follow the transverse profile $\tilde{E}(x,y,z)$ of any one such slab as it travels (at the velocity of light) through one complete round trip around a laser cavity, we will definitely see the transverse field pattern in the slab change with distance as the slab propagates, diffracts, bounces off mirrors, and passes through rods, lenses and finite apertures. These changes in the transverse pattern $\tilde{E}(x,y,z)$ of the slab caused by propagation and diffraction are the primary effects that determine the transverse mode properties of optical beams and resonators.

We will see later that optical resonators can usually be divided into either "geometrically stable" or "geometrically unstable" categories (where these terms refer to ray stability within the resonators, and have nothing to do with whether or not the laser is or is not stable against laser oscillation). In such resonators, the recirculating slabs themselves may also acquire a certain macroscopic curvature caused by the focusing effects of the laser mirrors, as shown for either a typical "stable" resonator or an "unstable" resonator in Figure 14.2.

Each such pulse or slab of radiation as it travels around may thus be rather inelegantly described as a "recirculating pancake" of radiation within the resonator. An important point is that the propagation of each such slab is essentially

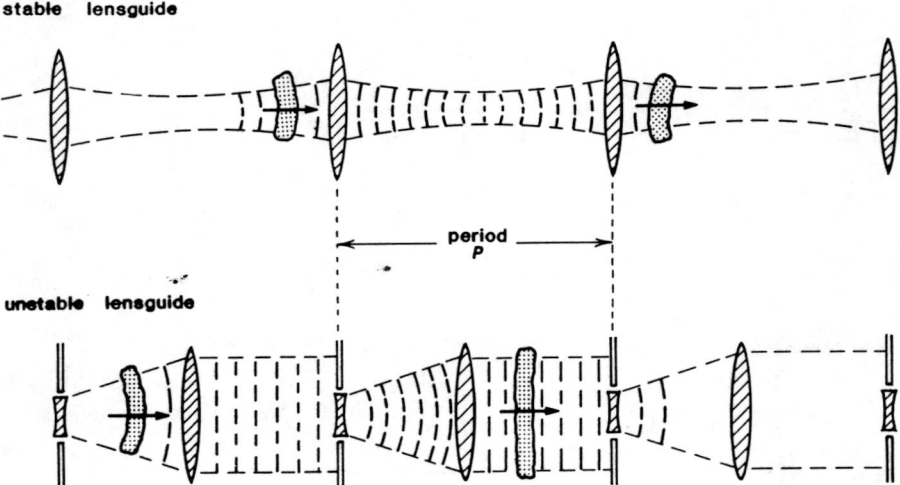

**FIGURE 14.3**
Propagation through repeated round trips in an optical resonator is physically equivalent to propagation through repeated sections of an iterated periodic lensguide. This lensguide may be, as in Figure 14.2, geometrically stable or unstable.

unaffected by the radiation in the slabs immediately in front of or behind it—the optical radiation in each axial segment or "pancake" is more or less independent of the other pancakes ahead of or behind it in the resonator.

### Optical Resonators and Equivalent Periodic Lensguides

Rather than thinking of repeated round trips within a resonator, it can be helpful to think of the pulse or "pancake" as propagating instead through repeated sections of an iterated periodic optical system as in Figure 14.3. In setting up such an iterated periodic lensguide, curved mirrors in the original resonator are replaced by thin lenses of equal focusing power, and all other elements encountered in the lensguide are made the same as those encountered in the original resonator (except that in a standing-wave cavity each element must be included twice to model a complete round trip in both directions).

The diffraction and aperturing effects that the pulse sees in a series of repeated round trips around the original laser cavity will then be the same as in propagating through an equivalent number of segments in the periodic lensguide; and the lensguide itself may be either "stable" or "unstable" in the sense discussed in the preceding. This lensguide approach obviously adds no new physics to the problem, but it does convert the resonator problem into an equivalent waveguide problem, and it can sometimes be helpful in visualizing the behavior in an optical resonator, as we shall see.

### Optical Resonator Eigenmodes and Eigenvalues

Let us now look at how this recirculating or traveling pulse approach leads to the concept of *transverse cavity modes* or *eigenmodes* in an optical resonator.

Suppose such a pulse or slab of radiation makes one complete round trip around an optical cavity, or travels through one complete period of the equivalent lensguide. After one complete round trip, the transverse field pattern $\tilde{E}^{(1)}(x, y)$ within a given slab as it arrives back at its starting plane will in general be different from its starting pattern $\tilde{E}^{(0)}(x, y)$ before the round trip, because of diffraction, reflection and aperturing effects; and after a second round trip the pattern $\tilde{E}^{(2)}(x, y)$ may again be still different. (Note that we have dropped the $z$ dependence in writing these patterns, because we are only considering the transverse variation as observed at one arbitrarily chosen reference plane somewhere within the resonator, or at a set of such planes spaced one period apart in the equivalent lensguide.)

We can then ask if, to put the question in physical terms, *there exist any transverse patterns, call them $\tilde{E}_{nm}(x, y)$, such that if a pulse or pancake starts off with one of these transverse patterns, it will return one round trip later with exactly the same pattern?* More precisely, we require that the pulse of radiation must return with exactly the same transverse form, but possibly with a reduced amplitude because of diffraction and other losses during the round trip. The wave may in general also return with an arbitrary absolute phase shift, because of the propagation distance $p$ around the resonator at the optical frequency $\omega$ of the pancake.

If we can find any such self-reproducing transverse patterns, it certainly seems reasonable to call them *transverse modes* of the resonator, or of the equivalent lensguide. That is, a pulse which is launched with an initial transverse profile matching one of these transverse modes can then propagate repeatedly around the resonator, or propagate indefinitely down the lensguide, always getting weaker in amplitude, but always maintaining the same transverse profile at the same reference plane in the resonator or the lensguide.

In fact, if we add enough laser gain within the resonator to just cancel the diffraction losses, it would seem that the resonator can oscillate indefinitely in any one of these transverse modes. (We will see shortly that this is indeed true, though with some slight complications.)

### Examples of Optical Resonator Eigenmodes

Do such lossy but self-reproducing transverse eigenmodes then really exist for open-sided and finite-diameter optical cavities—especially the very long slender cavities often used in practical lasers? The answer is that they do indeed exist, and that moreover the lowest-order transverse modes in properly designed (and aligned) laser cavities can have remarkably low diffraction losses, as well as remarkably good propagation properties.

The simplest transverse mode patterns, and the ones that are easiest to analyze, occur in the so-called *geometrically stable* optical resonators or lensguides using properly curved mirrors, such as those we have illustrated earlier. The lowest-order and higher-order modes in stable optical resonators, if expanded in rectangular transverse coordinates, are given almost (but not quite) exactly by Hermite-gaussian functions, such as those exhibited in the top part of Figure 14.4. We will discuss these gaussian modes in much greater detail in subsequent sections.

These modes, like the modes in most other optical resonators, are essentially plane waves, or slightly curved spherical waves, multiplied by the transverse amplitude and phase profiles given by the transverse mode functions $\tilde{E}_{nm}(x, y)$. Although the exact vector expressions for the associated optical beams must then necessarily have some small axial $E$ and $H$ field components, the primary

**(a) stable cavity transverse modes**

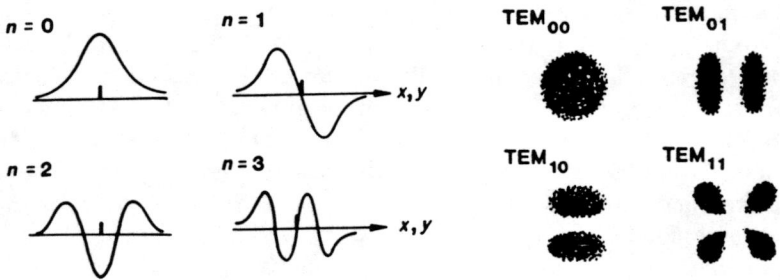

**(b) planar cavity modes**

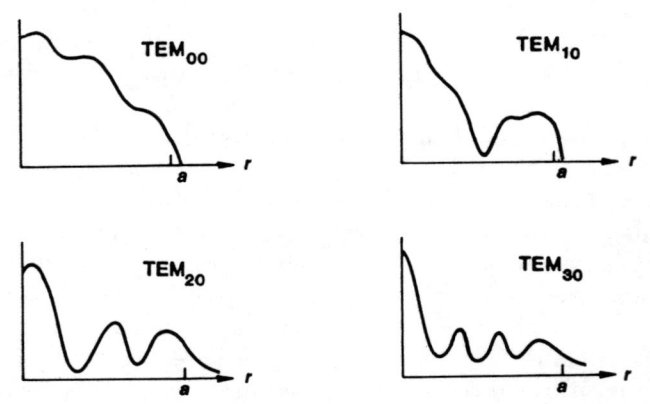

**(c) unstable cavity modes**

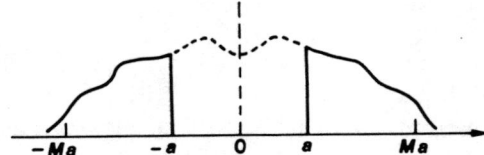

FIGURE 14.4
Examples of the lowest-order and higher-order transverse-mode intensity profiles in some typical (a) geometrically stable, (b) planar (flat mirror), and (c) geometrically unstable optical resonators.

field components in these beams are polarized transverse to the direction of propagation, just as in ideal uniform plane or spherical waves. These waves are very often referred to, therefore, as $TEM_{nm}$ optical waves, and we have used this notation in Figure 14.4. (Note, however, that a truly pure TEM electromagnetic wave can only exist in a transmission line having at least two conductors; and that these $TEM_{nm}$ optical waves must therefore always have some small axial $E$ and $H$ field components.)

## Planar and Unstable Resonator Modes

If we set up an optical resonator with two perfectly aligned flat mirrors—for example, two flat circular mirrors—the transverse mode patterns become more difficult to express analytically, but they nonetheless still exist. The first four azimuthally symmetric, or $l = 0$, modes of a typical circular plane-mirror resonator will have radial variations something like those shown in Figure 14.4(b). There also exist a large number of azimuthally varying or $l > 0$ transverse modes which we have not shown here.

Again these are essentially TEM modes, with radial amplitude patterns that in this situation look approximately like lowest and higher-order Bessel functions, with a small amount of irregular diffraction ripple added. Note, however, that these transverse mode patterns, which are viewed in this figure at the end mirror surface, do not quite go to zero at the mirror edges. The amount of energy that is lost past the mirror edges represents the diffraction loss or diffraction spillover from the ends of the resonator.

Finally, there are the even more complicated geometrically unstable resonators, which we will discuss in more detail in Chapter 22. These resonators have modes in which a large amount of energy is lost on each round trip past the edges of the smaller output mirror, as illustrated for a typical situation in the bottom part of Figure 14.4. This energy in fact forms the useful output beam in unstable-resonator lasers, which must typically have large laser gain in order to operate with such large output coupling.

Unstable resonator lasers can on the other hand have important advantages for higher-power lasers, including large mode volume, good discrimination against higher-order transverse modes, all-reflective optics (which can be, for example, water-cooled in very high-power lasers), and good output beam quality. Unstable resonators do have higher-order transverse modes, as well as the lowest-order type of mode pattern illustrated in Figure 14.4(c), but all of these modes are very difficult to express analytically and show large variations in shape with changes in the resonator length and diameter. We have therefore shown only one representative lowest-order example in Figure 14.4.

## 14.2 THE MATHEMATICS OF OPTICAL RESONATOR MODES

Let us now restate the basic problem outlined in the previous section in mathematical terms, and ask how we can calculate the propagation effects for an optical pulse through one round trip in a resonator, or one period of the periodic lensguide, and how we can find these transverse mode patterns that are self-reproducing after each such round trip or periodic step.

### The Round-Trip Propagation Integral

For essentially all the optical cavities of interest to us, the total propagation through one round trip in an optical resonator, or through one period in the equivalent lensguide, can be described mathematically by a propagation integral which will have the general form

$$\tilde{E}^{(1)}(x,y) = e^{-jkp} \int\int_{\substack{\text{Input} \\ \text{plane}}} \tilde{K}(x,y,x_0,y_0)\, \tilde{E}^{(0)}(x_0,y_0)\, dx_0\, dy_0, \qquad (2)$$

where $k$ is the propagation constant at the carrier frequency of the optical signal; $p$ is the length of one period or round trip; and the integral is over the transverse coordinates at the reference or input plane. The function $\tilde{K}$ appearing in this integral is commonly called the propagation kernel, since the field $\tilde{E}^{(1)}(x,y)$ after one propagation step can be obtained from the initial field $\tilde{E}^{(0)}(x_0,y_0)$ through the operation of the linear kernel or "propagator" $\tilde{K}(x,y,x_0,y_0)$.

Any arbitrary reference plane within the resonator, or within one period of the equivalent lensguide, may be chosen as the starting plane or reference plane for writing the preceding integral. The exact form of the kernel $\tilde{K}$ will depend on the reference plane that is chosen. If for example, the reference plane is chosen at an aperture, and the only intervening element before the next aperture is simply free space, this propagator will be simply Huygens' integral for free space, with the integral being evaluated over the aperture at the input end of each period.

More generally the propagation kernel will contain additional factors caused by intervening lenses, apertures, and other optical elements. Evaluating the form of the kernel in Equation 14.2 will be one of our major interests in the following chapters.

In doing resonator analyses, we will usually separate out from the propagation kernel the on-axis phase shift term $e^{-jkp}$, as has been done in Equation 14.2, since all the necessary information for evaluating transverse field patterns is contained in the remaining kernel $\tilde{K}(x,y,x_0,y_0)$, with the exponential term only furnishing a constant phase shift in front. We will look at propagation kernels of various types in much more detail for specific situations later on. For the present all we need understand is that there is (almost always) a linear relationship like Equation 14.2 between the input field $\tilde{E}^{(0)}(x_0,y_0)$ and the output field $\tilde{E}^{(1)}(x,y)$ after one step.

### The Eigenequation for Optical Resonator Modes

In mathematical terms the propagation integral in Equation 14.2 is a *linear operator equation*: that is, the linear propagation operator $\tilde{K}$ acts on the optical field $\tilde{E}^{(0)}(x,y)$ at a reference plane on one round trip to produce a new optical field $\tilde{E}^{(1)}(x,y)$ one round trip or one period later. Given an operator equation such as this, we may then ask whether this equation has a set of *eigensolutions*.

That is, for a given resonator or kernel, does there exist a set of mathematical eigenmodes $\tilde{E}_{nm}(x,y)$ and a corresponding set of eigenvalues $\tilde{\gamma}_{nm}$ such that each one of these eigenmodes after one round trip satisfies the round-trip propagation expression

$$\tilde{E}_{nm}^{(1)}(x,y) \equiv \iint \tilde{K}(x,y,x_0,y_0)\,\tilde{E}_{nm}^{(0)}(x_0,y_0)\,dx_0\,dy_0 = \tilde{\gamma}_{nm}\tilde{E}_{nm}^{(0)}(x,y), \qquad (3)$$

or simply

$$\tilde{\gamma}_{nm}\tilde{E}_{nm}(x,y) \equiv \iint \tilde{K}(x,y,x_0,y_0)\,\tilde{E}_{nm}(x_0,y_0)\,dx_0\,dy_0, \qquad (4)$$

where we can drop the superscript indices.

If eigensolutions that satisfy Equation 14.4 do exist, then these eigensolutions will provide exactly the self-reproducing transverse eigenmodes we seek, for either the optical resonator or the corresponding periodic lensguide. That is, if we launch a "recirculating pancake" in the form of any single one of these eigenmodes $\tilde{E}_{nm}(x,y)$ in the proper direction at the selected reference plane, then after one round trip the field at that same plane will be

$$\tilde{E}_{nm}^{(1)}(x,y) = \tilde{\gamma}_{nm} e^{-jkp} \tilde{E}_{nm}^{(0)}(x,y). \tag{5}$$

The field after one period will have exactly the same transverse form, both in its phase variation $\phi_{nm}(x,y)$ and in its amplitude variation $|\tilde{E}_{nm}(x,y)|$, although if we do not include any laser gain, the transverse mode pattern will be reduced in amplitude and shifted in absolute phase by the complex eigenvalue $\tilde{\gamma}_{nm}$. This self-reproducing behavior is the mathematical definition of a "transverse mode" of the optical resonator or the periodic lensguide.

Note that these transverse mode patterns $\tilde{E}_{nm}(x,y)$ (if any exist) will have in general a different field pattern $\tilde{E}_{nm}(x,y,z)$ at each transverse $z$ plane within the resonator, i.e., the shape of each transverse mode will change (slowly) with distance as it propagates along the resonator (or returns going in the opposite direction at the same plane in a standing-wave cavity). To put this in another way, the exact form of the kernel $\tilde{K}(x,y,x_0,y_0)$ and hence of the eigenmodes $\tilde{E}_{nm}(x,y)$ will be different if the kernel and the eigenmodes are evaluated at different reference planes, although the round-trip eigenvalues $\tilde{\gamma}_{nm}$ will be the same.

### Resonator Eigenvalues and Diffraction Losses

A transverse wave pattern that is bounded within a finite width will always spread out due to diffraction as it propagates. In an open-sided resonator with finite-diameter mirrors, therefore, some of the radiation will spread out past the mirror edges after each round trip, and the magnitudes of the transverse eigenvalues (again neglecting gain) will therefore always be less than unity, i.e., $|\tilde{\gamma}_{nm}| < 1$.

Hence even with perfectly lossless mirrors the $nm$-th eigenmode of an optical resonator will always have a power loss per round trip given by

$$\text{fractional power loss per round trip} = 1 - |\tilde{\gamma}_{nm}|^2. \tag{6}$$

These losses result from diffraction losses at the mirror edges or at apertures within the cavity, and will continue to occur on all subsequent round trips.

If no laser gain is present the amplitude of a given transverse mode will decay exponentially with successive round trips in the form

$$\frac{\tilde{E}_{nm}^{(k)}(x,y)}{\tilde{E}_{nm}^{(0)}(x,y)} = \tilde{\gamma}_{nm}^k. \tag{7}$$

If we add a laser medium with transversely uniform round-trip voltage gain $e^{\alpha_m p_m}$ inside the optical cavity, the total round-trip amplitude gain and phase shift become

$$\tilde{E}_{nm}^{(1)}(x,y) = \tilde{\gamma}_{nm} e^{\alpha_m p_m - jkp} \tilde{E}_{nm}^{(0)}(x,y). \tag{8}$$

(If the gain itself has a transverse $x,y$ variation, this must become part of the propagation kernel determining the eigenmodes.) The amplitude condition for laser threshold or for steady-state laser oscillation, as in Chapter 5, then becomes

$$\left| \frac{\tilde{E}_{nm}^{(1)}(x,y)}{\tilde{E}_{nm}^{(0)}(x,y)} \right| = |\tilde{\gamma}_{nm} e^{\alpha_m p_m - jkp}| = 1. \tag{9}$$

## CHAPTER 14: OPTICAL BEAMS AND RESONATORS: AN INTRODUCTION

The lowest-loss eigenmode, i.e., the one with the largest value of $|\tilde{\gamma}_{nm}|$ and smallest value of $\delta_{nm}$, will have the lowest threshold for oscillation and hence will (normally) be the dominant mode in the cavity.

### Existence of Resonator Eigenmodes

Many readers of this text may be familiar with the electromagnetic theory of microwave cavities or microwave waveguides, where resonant eigenmodes always do exist. That is, for closed cavities with lossless walls, such as are usually treated in electromagnetic theory texts, the wave equation describing the cavity fields is a hermitian mathematical operator; and the existence of a complete set of normal modes can therefore be rigorously proven. The completeness property then means that any arbitrary field pattern inside the cavity can always be expanded using this set of eigenmodes as the basis set.

There is a serious mathematical difficulty for open-sided optical resonators, however, in that the round-trip propagation kernel $\tilde{K}(x, y, x_0, y_0)$ for such resonators is generally found not to be a hermitian operator. This in turn means that the existence of a complete and orthogonal set of eigensolutions to Equation 14.4 is not automatically guaranteed, whereas it would be for a hermitian kernel. Such eigenmodes may exist, but we cannot guarantee in advance either their existence or, if they do exist, their completeness.

In the early days of lasers, the physical reality as well as the mathematical existence of transverse modes in open resonators was a matter of considerable debate. Even now, in fact, except for a few special situations, rigorous mathematical existence and completeness proofs for optical resonator modes do not exist. Real lasers have never had any difficulty in finding such modes in which to oscillate, however; and from a combination of empirical and experimental evidence, it is now entirely accepted that transverse eigenmodes as we have defined them in the preceding paragraphs do exist, and do provide a physically realistic and meaningful basis for describing laser oscillation in real laser resonators.

### Transverse Mode Orthogonality

A related mathematical peculiarity of optical resonator eigenmodes is that they are generally not "normal modes" in the usual sense of this term. That is, because of the nonhermitian kernel the eigenmodes $\tilde{E}_{nm}(x,y)$ of an optical resonator calculated at any plane $z$ are generally not power orthogonal in the usual fashion, i.e., for any two modes we may in general *not* write

$$\int\int \tilde{E}_{nm}(x,y)\,\tilde{E}^*_{pq}(x,y)\,dx\,dy = \delta_{np}\delta_{mq} \quad \text{(wrong)}, \tag{10}$$

where $\delta_{np}$ is the Kronecker delta function. Rather the set of modes $\tilde{E}_{nm}(x,y)$ are generally *biorthogonal* (without complex conjugation) to an *adjoint* set of modes, let's call them $\tilde{E}^\dagger_{pq}(x,y)$, in the form

$$\int\int \tilde{E}_{nm}(x,y)\,\tilde{E}^\dagger_{pq}(x,y)\,dx\,dy = \delta_{np}\delta_{mq} \quad \text{(right)}, \tag{11}$$

These adjoint functions $\tilde{E}^\dagger_{nm}(x,y)$ usually represent the transverse modes traveling in the opposite direction in the same cavity. The biorthogonality properties of general optical resonator modes are summarized at the end of Chapter 20.

It is also not in general possible to prove that the transverse eigenmodes of an optical resonator form a complete set. That is, it cannot be rigorously proven in advance that any field pattern within a given resonator can be written in the form

$$\tilde{E}(x,y) \stackrel{?}{=} \sum_{nm} c_{nm} \tilde{E}_{nm}(x,y) \quad \text{(not guaranteed)}. \tag{12}$$

However, the Hermite-gaussian or Laguerre-gaussian functions that approximate the eigenmodes in ideal stable resonators certainly do form a complete basis set; and in most practical situations people simply proceed as if the resonator eigenmodes always do form a complete set.

### Axial Versus Transverse Resonator Modes

It is important to understand that, once the axial phase shift term $e^{-jkp}$ has been factored out, the propagation kernel $\tilde{K}(x, y, x_0, y_0)$ in a typical optical resonator or lens waveguide depends only very slightly on the exact frequency $\omega$ or the exact wavelength $\lambda$ of the radiation in the recirculating pancake. In physical terms, the diffraction effects experienced by a transverse mode function $\tilde{E}_{nm}(x, y)$ in a round trip will be essentially the same for any carrier frequency (or any axial mode frequency) within the linewidth of a single atomic transition or the oscillation bandwidth of a single laser oscillator. Hence, the transverse mode properties and the axial frequency properties of a given optical resonator can be treated almost completely separately from each other.

The transverse eigenmodes for any given laser can then be calculated based only on the mean laser wavelength; and all of the axial modes within a given laser line will then have the same set of transverse eigenmodes and eigenvalues. The transverse eigensolutions, in fact, might rather be viewed as the transverse propagation modes of the equivalent lensguide, for which axial resonance frequencies have no meaning. If we shift to a different laser line which is, say, 20% different in frequency, then the diffraction effects in one round trip may change somewhat, and we can expect the form of the transverse eigenmodes to change by a noticeable amount.

By launching a continuous stream of "pancakes" one after another, nose to tail so to speak, we can fill an entire laser cavity with radiation all in one given transverse eigenmode, and all at one carrier frequency. To satisfy the round-trip phase-shift condition, or to make the axial variation of the fields continuous completely around the resonator, the carrier frequency of these pancakes would have to be one of the axial mode frequencies of the resonator; and having done this we would have filled the cavity with radiation in a single axial and single transverse mode.

## 14.3 BUILD-UP AND OSCILLATION OF OPTICAL RESONATOR MODES

Without going into details of the exact modes for any specific resonator, we can now say some additional things about how resonator transverse modes can be calculated numerically; about their exact resonance frequencies; and about how these modes build up, compete, and decay in real lasers.

### Calculating The Lowest-Loss Eigenmode

Suppose one of our pulses or "pancakes" with an arbitrary initial field distribution $\tilde{E}^{(0)}(x,y)$ is launched inside a resonator with no laser gain. We will assume, without worrying about rigorous justification, that this initial distribution can be written as a sum of the transverse eigenmodes for that particular resonator, i.e.,

$$\tilde{E}^{(0)}(x,y) = \sum_{nm} c_{nm} \tilde{E}_{nm}(x,y), \qquad (13)$$

(and we will not worry about the axial variation of the pulse, since we do not need it to calculate the round-trip propagation.)

Then, on each round trip inside the resonator each transverse mode component will be multiplied by its eigenvalue $\tilde{\gamma}_{nm}$; and hence the field at the same reference plane $k$ round trips later will be given by

$$\tilde{E}^{(k)}(x,y) = \sum_{nm} c_{nm} \tilde{\gamma}_{nm}^{k} \tilde{E}_{nm}(x,y). \qquad (14)$$

The relative amplitude of each transverse mode will thus have attenuated after $k$ successive round trips as $|\tilde{\gamma}_{nm}|^k$.

Suppose we index the transverse eigenmodes so that $nm = 00$ labels the transverse mode with the largest eigenvalue or the smallest loss per round trip. All other $nm$ combinations will then have smaller eigenvalues and hence larger mode losses. Suppose we run the field distribution $\tilde{E}(x,y)$ through many repeated round trips, letting $k$ in Equation 14.14 become large.

Then the amplitudes of the various eigenmodes will attenuate or die out with different rates on repeated round trips (see Figure 14.5). It is apparent that, whatever may be the initial mode distribution, after a sufficient number of round trips the lowest-loss or 00 mode will become dominant compared to all the other transverse modes. There is a chance that two modes will have exactly the same magnitude, and hence both will persist, but we can handle this as an unusual special situation. We can also dismiss as extremely unlikely the chance of any real initial distribution containing no initial component of the 00 mode whatsoever. Starting with any arbitrary initial transverse field pattern and following it through enough round trips in the resonator is thus a prescription for finding the lowest-order transverse mode of an optical resonator or lensguide.

### The Fox and Li Approach

This conceptual approach to finding the lowest-order resonator transverse modes is often called the "Fox and Li" approach, since it describes not only the real physical situation in an optical cavity, but also the numerical mode-calculation procedure pioneered by A. G. Fox and T. Li at Bell Telephone Laboratories around 1960, in the earliest days of the laser.

Fox and Li simulated the iterative round trips of a wavefront $\tilde{E}(x,y)$ in a resonator by using numerical computation on a digital computer. In these computations they repeatedly integrated the propagation equation (14.2) using the Huygens' integral kernels for plane-mirror resonators and other simple situations. Figure 14.6 shows some typical results from this kind of calculation.

Fox and Li's first calculations were made assuming, for simplicity, a "strip resonator," that is, a resonator with end mirrors in the form of two parallel flat

## 14.3 BUILD-UP AND OSCILLATION OF OPTICAL RESONATOR MODES

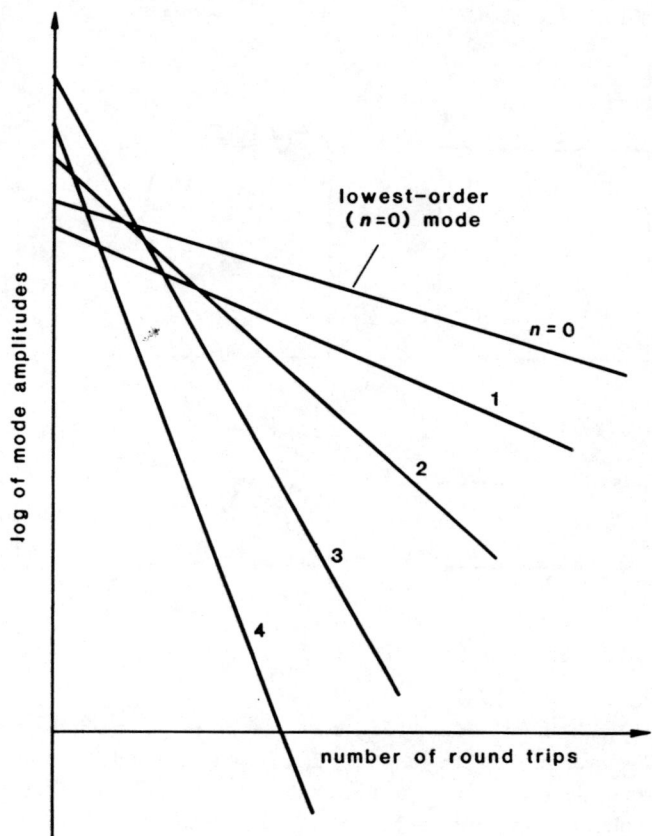

**FIGURE 14.5**
Attenuation of lowest-order and higher-order transverse modes on successive round trips.

strips having transverse variations in the $x$ direction only (strip width $= 2a$), spaced by a distance $L$ in the $z$ direction, and with no variations in the $y$ direction along the strips. The starting field on one end mirror was simply a uniform field pattern $\tilde{E}^{(0)}(x,y) = 1$ across the mirror, as in Figure 14.6(a).

The two curves in Figure 14.6(b) then show the resulting field pattern or diffraction pattern $\tilde{E}^{(1)}(x,y)$ after the first propagation step from one end of the laser cavity to the other. (Fox and Li's initial calculations involved axially symmetrical laser cavities, and were phrased in terms of propagation steps from one end to the other, rather than complete round trips; but the essential ideas remain unchanged.) A beam propagating away from an aperture with sharp edges can be expected to exhibit Fresnel diffraction ripples in its near-field pattern, and the conventional Fresnel diffraction ripples in the field pattern after this first step are very evident.

### Convergence to the Lowest-Order Mode

Initially we do not know the eigenmodes $\tilde{E}_{nm}(x)$ of the resonator and we thus have no way of separating an arbitrary starting function $\tilde{E}^{(0)}(x,y)$ into eigenmodes. After a sufficient number of bounces, however, the wavefunction $\tilde{E}^{(k)}(x,y)$ in the computer should converge in form to the lowest-order eigenmode $\tilde{E}_{00}(x,y)$, for the reasons given in the preceding; and the eigenvalue for this mode

572    CHAPTER 14: OPTICAL BEAMS AND RESONATORS: AN INTRODUCTION

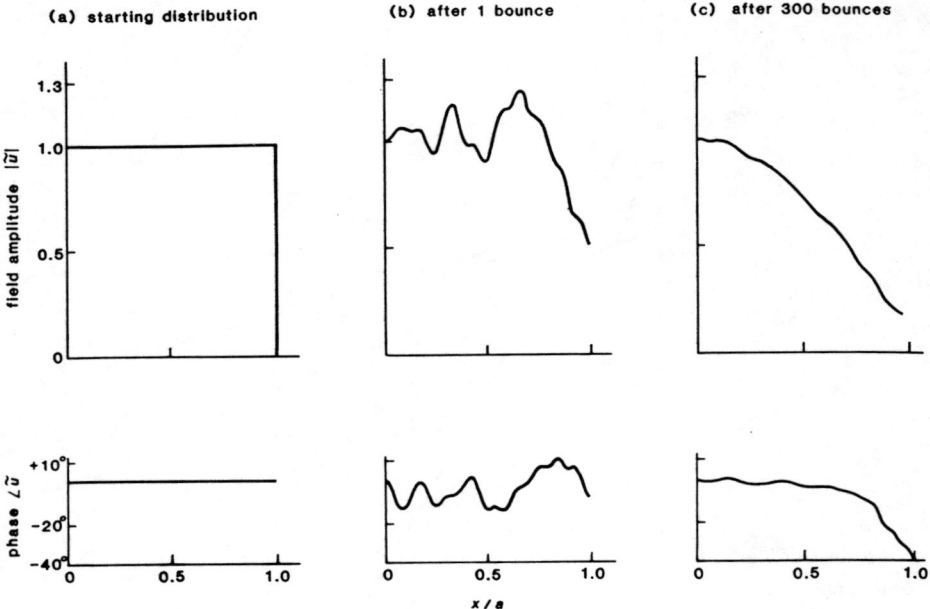

FIGURE 14.6
Typical results from Fox and Li's early numerical mode calculations, showing amplitude and phase variation of the wavefront $\tilde{E}(x)$ across one end mirror of the optical cavity. (a) Uniform initial distribution. (b) Field pattern after one bounce, showing Fresnel diffraction ripples. (c) Steady-state field pattern ($\equiv$ lowest-order mode) after 300 bounces.

should be given from the computer iterations by

$$\tilde{\gamma}_{00} = \lim_{k \to \infty} \frac{\tilde{E}^{(k+1)}(x,y)}{\tilde{E}^{(k)}(x,y)}. \qquad (15)$$

The field distribution $\tilde{E}^{(k)}(x,y)$ in the computer of course decreases steadily in overall amplitude with each successive bounce because of diffraction losses, but this is handled in the calculations simply by rescaling the overall signal level back upward by a constant amount after each iteration, or each few iterations.

Figure 14.6(c) then shows the steady-state, unchanging amplitude and phase pattern that the resonator mode in this particular example settles into after $k \approx 250$ to 300 round trips. (This is a comparatively low-loss resonator, and the higher-order modes only die out quite slowly.) The finite value of the steady-state mode pattern just at the mirror edge indicates that the mode does still have some diffraction losses past the edges of the end mirror. The smoothed shape and tapered profile of the steady-state pattern also indicate, however, that higher spatial frequency components are rapidly lost past the edges of the resonator, and that this lowest-loss transverse mode pattern has a very typical ability to "pull in its edges" and minimize its diffraction losses due to diffraction spreading. The exact shape of this mode pattern changes, and the mode losses decrease or increase, as the width of the planar end mirrors is changed, or the cavity length $L$ is changed.

The primary conclusion from this numerical simulation or "computer experiment" is that even a simple optical resonator consisting only of two flat end mirrors, with completely open sides, still has a lowest-order transverse mode

## 14.3 BUILD-UP AND OSCILLATION OF OPTICAL RESONATOR MODES

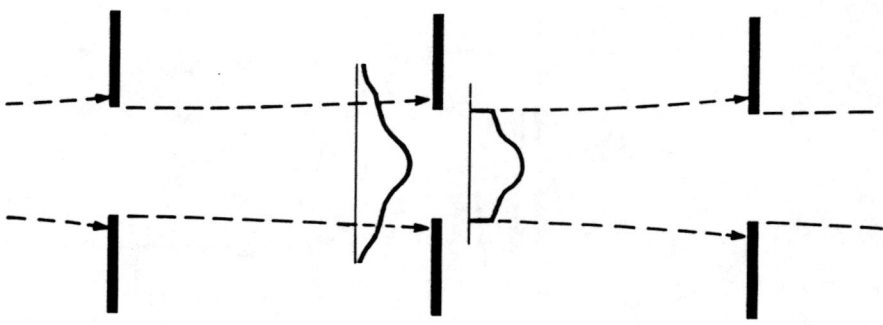

FIGURE 14.7
The field pattern in Figure 14.6 can also be interpreted as the lowest-order transverse mode in a "lensless lensguide" defined only by periodically spaced apertures.

which will reproduce itself on repeated round trips. This mode is in fact almost like a half-cosine in appearance, rather similar to more familiar waveguide cavity modes. The effects of the finite mirror edges and diffraction losses do show up, however, in the small diffraction ripples on the mode wavefront and on the mode amplitude pattern, and in the finite diffraction losses characteristic of the mode.

Note that this same analysis also means that a lens waveguide consisting simply of a series of slit apertures, without any lenses (see Figure 14.7), will also propagate exactly the same transverse mode pattern as a traveling mode pattern in the lensguide. The diffraction effects of the aperture edges in this lensguide are exactly equivalent to the cutting off of the transverse mode pattern on each bounce by the finite mirror width in the Fox and Li resonator calculation.

Fox and Li, and many others since them, have done many more such calculations for resonators with curved mirrors, mirrors of more complex shape, mirrors with central holes, planar but tilted mirrors, and so forth. In every situation a lowest-order mode with some sort of self-reproducing mode pattern and associated eigenvalue has resulted from this sort of calculation.

### Finding the Higher-Order Transverse Modes

More sophisticated numerical procedures then allow us to obtain higher-order eigenmodes from the Fox and Li iterative procedure as well. For example, even in the simple Fox and Li procedure if we reach a stage in the iterative calculation where only two dominant modes are left, then there will be only two terms left in Equation 14.14. The field amplitude at any fixed point on the mirror surface will then display a periodic beating between the two modes (see Figure 14.8).

This periodic interference occurs because the fields of the two modes combine with different phases on successive round trips, since the different eigenmodes have eigenvalues $\tilde{\gamma}_{nm}$ with different phase angles $\psi_{nm}$. The eigenvalue of the next-highest eigenmode can then be deduced from the rate and period with which this "mode beating" between the two modes dies out.

A more sophisticated procedure known as the *Prony method* is one among several numerical techniques that allow us to start with an initial distribution containing a mixture of many eigenmodes, and after $N$ iterations to deduce the $N$ lowest-loss eigenvalues $\tilde{\gamma}_{nm}$ and eigenmodes $\tilde{E}_{nm}(x,y)$.

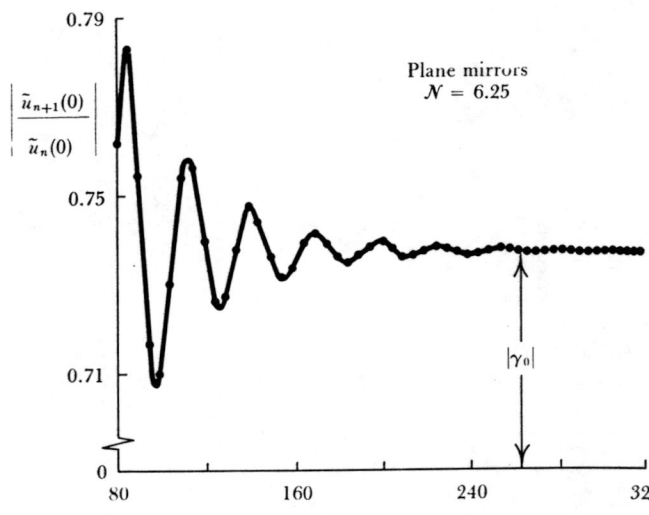

FIGURE 14.8
"Mode beating" in a Fox-and-Li mode calculation.

### Resonator Eigenfrequencies

Having found the transverse eigenmodes $\tilde{E}_{nm}(x,y)$ and eigenvalues $\tilde{\gamma}_{nm}$ of a given cavity or lensguide, we can also find the exact resonant frequencies, or axial-plus-transverse mode resonances, of the cavity in the following manner.

The exact resonance frequency of a given axial-plus-transverse mode in a laser cavity is determined by the resonance condition that the round-trip phase shift in the cavity must be an integer multiple of $2\pi$. Suppose a real regenerative laser cavity has round-trip phase shift due to the laser medium given by $\exp[-j\Delta\beta_m p_m]$, and suppose we consider a particular transverse mode $\tilde{E}_{nm}(x,y)$ with a complex eigenvalue $\tilde{\gamma}_{nm} \equiv |\tilde{\gamma}_{nm}| \exp[j\psi_{nm}]$. Regenerative feedback or laser oscillation for this particular transverse mode can occur only at frequencies for which the total round-trip phase shift is given by

$$\exp[-jkp - j\Delta\beta_m p_m + j\psi_{nm}] = \exp[-jq2\pi], \tag{16}$$

where we use $\psi_{nm}$ for the phase angle of $\tilde{\gamma}_{nm}$. The axial phase shift term $e^{-jkp}$ has been brought back into this expression, with $k = \omega/c$, and $q$ being an axial-mode integer. Equating the phase angles on opposite sides of Equation 14.16 then gives

$$\frac{\omega p}{c} + \Delta\beta_m p_m - \psi_{nm} = q \times 2\pi. \tag{17}$$

The resonance frequencies $\omega_{qnm}$ of the axial-plus-transverse modes in this cavity are thus given by

$$\omega = \omega_{qnm} \equiv \frac{2\pi c}{p} \left[ q + \frac{\psi_{nm}}{2\pi} - \frac{\Delta\beta_m p_m}{2\pi} \right]. \tag{18}$$

Since $q$ is normally a very large integer ($\approx p/\lambda$), the transverse mode factor $\psi_{nm}/2\pi$ represents only a small correction to the plane-wave resonance frequency $\omega_q \equiv q \times 2\pi(c/p)$. This correction will be in general slightly different for each specific $nm$-th transverse mode. As we already know, the $\Delta\beta_m p_m/2\pi$ factor is

## 14.3 BUILD-UP AND OSCILLATION OF OPTICAL RESONATOR MODES

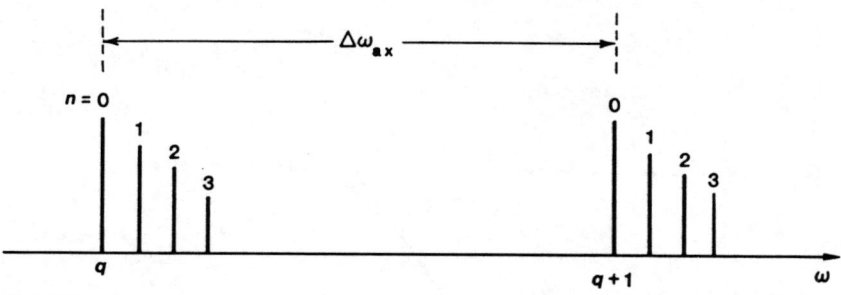

FIGURE 14.9
Transverse-mode frequencies in a typical laser resonator.

an additional (and usually still smaller) atomic frequency pulling effect caused by the reactive or $\chi'$ part of the laser susceptibility.

### Transverse Mode "Beats"

Different transverse modes $\tilde{E}_{nm}$ thus lead to slightly different resonance frequencies $\omega_{qnm}$, with small relative frequency shifts which are determined by the phase angles of the transverse mode eigenvalues $\tilde{\gamma}_{nm}$ in real laser cavities. Figure 14.9 illustrates how each axial mode frequency $\omega_q$ in the plane-wave approximation splits into a set of different axial-plus-transverse mode resonances $\omega_{qnm}$ in a typical resonator. (The actual magnitude of this splitting can be quite different for different types of real resonators.)

Heterodyne interference effects or "beats" at the difference frequencies between these transverse modes can often be detected by examining the output signal of a laser oscillator with any kind of standard square-law photodetector (i.e., any detector whose response is proportional to the optical intensity, or to the optical $E$ field squared, such as a photomultiplier tube or solid-state photodiode). These "transverse mode beats" can provide a sensitive test for the presence of multiple transverse modes. In addition, since the inter-mode beat frequency can be easily measured, and since this frequency depends in a sensitive fashion on the phase angles of the mode eigenvalues, the agreement between measured and theoretical frequencies can provide a test for the validity of the transverse mode calculations.

### The Buildup of Laser Oscillation

The Fox and Li numerical approach simulates mathematically what actually happens physically in a real optical resonator with an initially injected field distribution and no gain. Each transverse mode component circulates around and dies out at a rate determined by its eigenvalue. With a slight change in viewpoint, this same picture also describes what happens in a real laser oscillator at turn-on.

When a laser oscillator is turned on from a cold start, an initial mode distribution $\tilde{E}^{(0)}(x,y)$ (determined in most real situations by noise or spontaneous emission in the laser cavity) begins to circulate repeatedly around the cavity, and to grow in amplitude if the cavity is above threshold. If the gain medium is spatially uniform so that all modes $\tilde{E}_{nm}(x,y)$ see the same gain, then the lowest-loss or 00 mode grows the fastest, since it has the highest value of net gain minus loss.

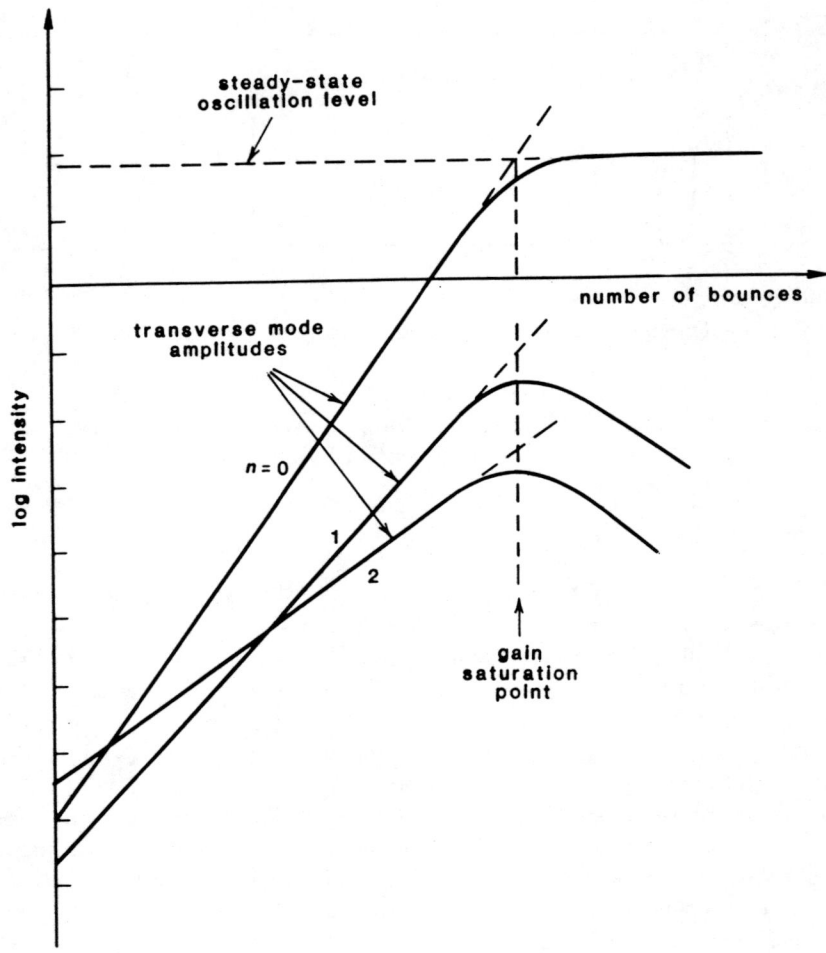

**FIGURE 14.10**
Buildup of laser transverse modes at laser turn-on.

In simple situations the dominant or 00 mode will eventually grow to a level where it saturates the gain down until the gain for this particular mode just equals the loss. This mode will then stay at a steady-state level, whereas all the higher-loss transverse modes die out, in the same way as sketched earlier. This initial growth and eventual stabilization process is illustrated in Figure 14.10.

### Transverse Mode Competition Effects

There are many factors that complicate this picture in real lasers. In a more realistic picture, for example, the $n = 0, m = 0$ mode may still build up most rapidly; but this mode will then saturate the gain medium strongly only in those regions of the transverse plane where the field amplitude $|\tilde{E}_{00}(x,y)|^2$ is large. This may leave unsaturated gain at other transverse positions $x, y$, and this may allow other higher-order transverse modes to oscillate simultaneously. In very high-gain but short-pulse lasers the entire laser pulse may be over in so few round trips that the 00 mode has insufficient time to grow to where it dominates over higher-order modes. The transverse mode selection may thus be less effective in a $Q$-switched laser than in a cw steady-state laser.

Even in a cw laser, the differences in loss and in growth rate for different eigenmodes $\tilde{E}_{nm}$ may be very small, so that the competition between modes is very weak. The gain may not be uniform across the laser, so that different eigenmodes $\tilde{E}_{nm}(x,y)$ actually see different gains. If the atomic linewidth is particularly narrow, and the laser is tuned so that the lowest-order $q00$ resonant frequency is off line center, whereas some other higher-order $qnm$ transverse mode is tuned closer to line center, the laser may then also see higher net gain in the higher-order mode even though it has higher diffraction losses. Interference effects between the fields of different transverse modes may also modulate the gain differently at different transverse positions and thus cross-couple the different transverse modes. In all these different situations, several transverse modes may then oscillate simultaneously, or the laser may jump back and forth between transverse modes.

### Single Transverse Mode Operation

All in all, it is often a considerable struggle to force a large or high-gain laser oscillator to oscillate only in a single lowest-loss transverse mode. One of the main considerations in the design of a practical laser resonator is to have simultaneously both minimum unwanted loss for the lowest-order transverse eigenmode, and also high mode discrimination—that is, a large increase in diffraction losses—for all the higher-order transverse modes. This is often accomplished by putting an adjustable aperture inside the laser cavity and reducing its size until it attenuates and if possible kills the higher-order modes, but still has negligible effect on the desired lowest-order modes. The unstable resonator provides another and somewhat different method for accomplishing the same goal.

Despite these complexities, which we will discuss in more detail in later sections, there are many lasers which do operate in a single lowest-order transverse mode according to the description presented above. Even in more complex situations the transverse eigenmode picture generally provides a solid and useful basis for describing the more complex multimode and coupled-mode phenomena that may occur in real lasers.

### Conclusions

Evaluating the round-trip wave propagation in an optical resonator, using the appropriate round-trip kernel or mathematical transformation, is obviously the primary step in evaluating and understanding the transverse modes, their losses, and their resonant frequencies, in any real laser resonator. In the following two chapters we introduce two primary tools for accomplishing this: ray matrix methods for treating ray propagation without diffraction, and paraxial wave optics for treating wave propagation including diffraction in most real laser beams and cavities.

### REFERENCES

The original (and still very instructive) reference on the Fox and Li approach to optical resonator modes is A. G. Fox and T. Li, "Resonant modes in a maser interferometer," *Bell Sys. Tech. J.* **40**, 453–458 (March 1961). Later and more extensive results are in

"Modes in a maser interferometer with curved and tilted mirrors," *Proc. IEEE* **51**, 80–89 (January 1963).

For an extension of this work which makes clear many of the basic properties of transverse modes in laser resonators. see also A. G. Fox and T. Li, "Computation of optical resonator modes by the method of resonance excitation," *IEEE J. Quantum Electron.* **QE–4**, 460–465 (July 1968).

Many of the concepts and mathematical solutions of transverse modes for periodic beam waveguides or lensguides were also developed quite independently of lasers, in the context of millimeter wave systems, especially in work by G. Gobau as reviewed in a chapter on "Beam Waveguides," in *Advances in Microwaves, Vol. 3*, edited by L. Young (Academic Press, 1968); pp. 67–126.

For further examples of Gobau's important early contributions to propagation in periodic optical and millimeter-wave systems, see, for example, G. Gobau, "On the guided propagation of electromagnetic wave beams," *IEEE Trans.* **AP–9**, 248–255 (May 1961); and G. Gobau and J. R. Christian, "Some aspects of beam waveguides for long distance transmission at optical frequencies," *IEEE Trans.* **MTT–12**, 212–220 (March 1964).

An excellent review of standard stable resonator theory as it developed in the early years of the laser era is given by H. Kogelnik and T. Li, "Laser beams and resonators," appearing both in *Proc. IEEE* **54**. 1312 (October 1966) and *Appl. Opt.* **5**, 1550–1567 (October 1966); see also H. Kogelnik, "Modes in Optical Resonators," in *Lasers: A Series of Advances, Vol. I*, ed. by A. K. Levine (Marcel Dekker, New York, 1966), p. 295. A historical survey of Bell Laboratories work in all aspects of laser optics, with many references, is given by R. Kompfner in, "Optics at Bell Laboratories—optical communications," *Appl. Opt.* **11**. 2412–2425 (November 1972).

A survey of newer types of resonators for high-power lasers is given by A. E. Siegman in, "Unstable optical resonators." *Appl. Opt.* **13**, 353 (February 1974). A somewhat inaccessible reference to Soviet work in resonator theory is L. A. Weinstein, *Open Resonators and Open Waveguides* (Golem Press. Boulder, Colorado, 1969).

Another Soviet reference on laser resonators is Y. Ananiev (or Anan'ev), *Résonateurs Optiques et Problème de Divergence du Rayonnement Laser* (Éditions Mir, Moscow, Russian original 1979. French translation 1982).

Some early references on the mathematical questions involved in nonhermitian resonator integral equations include S. P. Morgan, "On the integral equations of laser theory," *IEEE Trans.* **MTT–11**. 191–193 (May 1963); W. Streifer and H. Gamo, "On the Schmidt expansion for optical resonator modes," in *Quasi Optics*, ed., by J. Fox (Polytechnic Press, Polytechnic Institute of Brooklyn, 1964), pp. 351–365; D. J. Newman and S. P. Morgan, "Existence of eigenvalues of a class of integral equations arising in laser theory," *Bell Sys. Tech. J.* **43** 113–126 (January 1964); J. A. Cochran, "The existence of eigenvalues for the integral equations of resonator theory," *Bell Sys. Tech. J.* **44**, 77–88 (January 1965); and H. Hochstadt, "On the eigenvalue of a class of integral equations arising in resonator theory." *SIAM Rev.* **8** 62 (January 1966).

More recent references include J. A. Arnaud, *Beam and Fiber Optics* (Academic Press, 1976), pp. 122–123 and 175; I. F. Balashov and V. A. Berenberg, "Nonstationary modes of an open resonator," *Sov. J. Quantum Electron.* **5**, 159–161 (August 1975); and A. E. Siegman, "Orthogonality properties of optical resonator eigenmodes," *Optics Comm.* **31**, 369–373 (December 1979).

The problem of finding optical resonator eigenmodes is very closely related to the earlier mathematical problem of shaping a transmitted beam to obtain maximum power transmission between two apertures. Various aspects of this topic are often referred to as "Luneberg apodization problems." since they were posed, and also converted into mathematical eigenvalue problems, in R. K. Luneberg, *Mathematical Theory of Optics* (University of California Press, Berkeley, 1964).

Other and more recent references include A. F. Kay. "Near-field gain of aperture antennas," *IEEE Trans.* **AP-8**, 586–593 (November 1960); G. V. Borgiotti, "Maximum power transfer between two planar apertures in the Fresnel zone," *IEEE Trans.* **AP-14**, 158–163 (March 1966); H. N. Rexroad and B. J. Henderson, "Maximum power-transfer coefficient between two confocal apertures," *J. Opt. Soc. Am.* **59**, 1415–1421 (November 1969); and T. Ueno and T. Asakura, "Apodization for maximum encircled energy with specified over-all transmittance," *J. Optics (Paris)* **8**, 15–31 (1981).

## Problems for 14.3

1. *Higher-order mode suppression during laser turn-on.* A certain laser cavity has a lowest-loss eigenmode $\tilde{E}_{00}$ with eigenvalue $|\tilde{\gamma}_{00}| = 0.9$ and a next-lowest-loss eigenmode $\tilde{E}_{01}$ with eigenvalue $|\tilde{\gamma}_{01}| = 0.8$ (as well as numerous higher-loss eigenmodes). When this laser is first turned on, the unsaturated gain during the initial build-up period is 40% power gain per one-way pass down the laser cavity ($G_1 = |g_1|^2 = 1.4$). How many round trips will it take before the circulating power in the laser cavity has become 99% lowest-order transverse mode, assuming for simplicity that the lowest and next-lowest eigenmodes have equal initial noise amplitudes and that this all takes place during the initial build-up period before gain saturation begins to occur?

2. *Finding the next higher-order transverse mode.* Develop the necessary mathematical formulas and then, using the data from Figure 14.6, find the eigenvalue magnitude $|\tilde{\gamma}_1|$ for the next higher-order transverse mode in this resonator, and its phase angle relative to the lowest-order eigenvalue $\tilde{\gamma}_0$. (Note: The index in this figure is the number of one-way "bounces" in the resonator, rather than the number of two-way round trips.) Hint: Assume that only the $\tilde{\gamma}_0$ and $\tilde{\gamma}_1$ modes are left, and that the complex amplitude of the $\tilde{\gamma}_1$ mode component has become small compared to the $\tilde{\gamma}_0$ mode component.

3. *Higher-order transverse mode beats.* In the previous problem, what will be the beat frequency between the lowest and next lowest-order transverse modes in the resonator of Figure 14.6, assuming that the laser cavity is 1 meter long?

4. *Spectral content of a circulating pulse.* Suppose a "circulating slab" of axial length $\Delta z$ inside a laser cavity of length $L$ has an axial field variation $\cos(\omega_p t - k_p z)$ [or, if you like $\exp[j(\omega_p t - k_p z)]$ within the slab, where $k_p \equiv \omega_p/c$, and where $\omega_p$ (the carrier frequency of the "pancake") is *not* equal to any of the axial mode frequencies $\omega_q = q 2\pi(c/2L)$ of the laser cavity. Suppose gain just equals loss, so that this pulse circulates repeatedly inside the cavity, emitting a short pulse of carrier frequency $\omega_p$ and duration $\Delta t = \Delta z/c$ through the end mirror every $T = 2L/c$ seconds.

It may then seem that this laser is producing output primarily at frequency $\omega_p$, when lasers are supposed to oscillate only at their axial mode frequencies $\omega_q$. Resolve this apparent paradox by a suitable spectral argument. Hint: Consider the frequency spectrum of a single pulse of carrier frequency $\omega_p$; of two such pulses separated in time by $T = 2L/c$; of three such pulses; and so forth. Use some simple pulse shape (e.g., square or gaussian); assume $\omega_p$ is, say, one-third of the way between two axial modes: let $\Delta z$ equal, say, $L/10$; and actually plot the spectral amplitude versus frequency for increasing numbers of pulses.

5. *Output beam characteristics of a multi-transverse-mode laser.* Suppose a laser is oscillating simultaneously in a lowest-order transverse mode that has, say, even symmetry in the transverse $x$ direction and a higher-order transverse mode that has odd symmetry in the same transverse direction. How will the center of gravity of the beam emerging from this laser behave, in the near field and in the far field?

Suppose the output beam is detected by a photodetector large enough to capture all of the energy emerging through the end of this laser. Will this photodetector sense a transverse mode beat? What sort of arrangement will maximize the sensitivity for measuring such beats?

6. *Transverse modes that self-reproduce after several round trips?* The question is sometimes raised: Why does an optical resonator eigenmode $\tilde{E}_{nm}$ have to reproduce itself after only *one* round trip? Could we not have a transverse eigenmode that was self-reproducing in form only after two, or three, or even $k$ round trips?

Discuss this question, explaining why such a "multipass transverse eigenmode" could not be associated with a single axial mode integer, and why such a "multipass eigenmode" would really consist of a mixture of the "true" single-pass transverse eigenmodes. Hint: Consider not only the transverse and longitudinal field expansion inside the laser cavity, but also the space and time dependence of the beam that would come out the output end of a laser cavity oscillating in such a "multipass eigenmode."

# 15

# RAY OPTICS AND RAY MATRICES

Ray optics—by which we mean the geometrical laws for optical ray propagation, without including diffraction—is a topic that is not only important in its own right, but also very useful in understanding the full diffractive propagation of light waves in optical resonators and beams. This chapter, therefore, gives a fairly extensive introduction to ray optics in paraxial optical systems. The following chapter will then give a similar introduction to wave optics in the same systems.

## 15.1 PARAXIAL OPTICAL RAYS AND RAY MATRICES

Ray matrices or "$ABCD$ matrices" are widely used to describe the propagation of geometrical optical rays through paraxial optical elements, such as lenses, curved mirrors, and "ducts." These ray matrices also turn out to be very useful for describing a large number of other optical beam and resonator problems, including even problems that involve the diffractive nature of light. Therefore, we begin the discussion of optical beams and resonators with a detailed review of paraxial ray theory and ray matrices.

### Optical Rays and Ray Transformations

Consider a ray of light—or equally well a particle, such as an electron—that is traveling approximately in the $z$ direction, but with a transverse displacement $r(z)$ from the axis and also a small slope $dr/dz$, as in Figure 15.1. If such a ray propagates in free space from a plane at $z_1$ to a later plane at $z_2 = z_1 + L$, as in Figure 15.2, its input and output ray coordinates will be related by the transformation

$$r_2 = r_1 + L\,dr_1/dz$$
$$dr_2/dz = dr_1/dz. \tag{1}$$

Suppose the same ray passes through a thin lens of focal length $f$ as in the lower part of Figure 15.2. The input and output ray coordinates just before and after

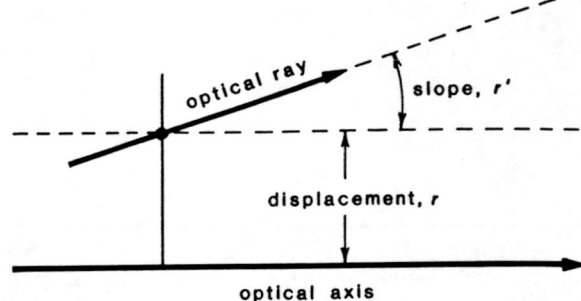

FIGURE 15.1
Definition of an optical ray.

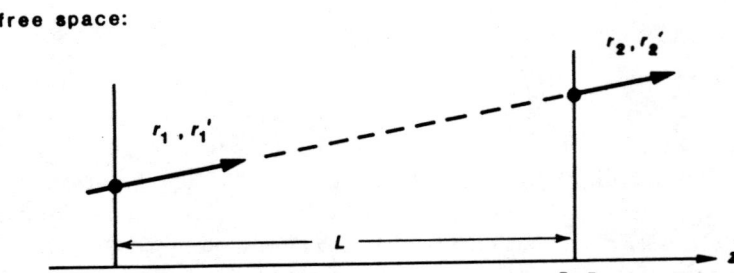

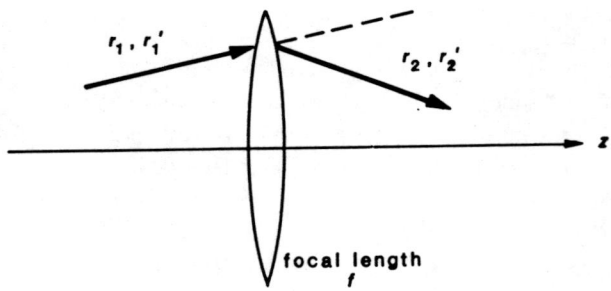

FIGURE 15.2
Optical-ray transformations through free space and through a thin lens.

the lens will then be related by

$$r_2 = r_1$$
$$dr_2/dz = -(1/f)\, r_1 + dr_1/dz.$$

(2)

(Note that we use a sign convention in which a positive value for $f$ means a positive or converging lens.)

Equations 15.1 and 15.2 both give linear transformations between the input and output displacements and slopes of the rays. In rectangular coordinates, of course, these displacements $r$ and slopes $dr/dz$ can represent equally well either the $x$-axis quantities $x$ and $dx/dz$, or the $y$-axis quantities $y$ and $dy/dz$.

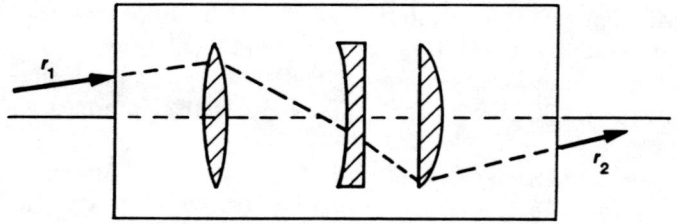

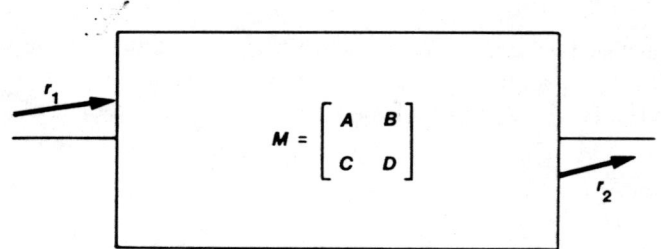

FIGURE 15.3
Example of an overall ray matrix.

### Optical Ray Matrices, or $ABCD$ Matrices

In fact, the change in displacement and slope of an optical ray upon passing through a wide variety of simple optical elements can be written in the same general form as Equations 15.1 and 15.2. One slight additional complexity should be added, however. In writing ray transformations like these, we will simplify many later results if we define the ray slope variable to be, not the actual slope $dr/dz$ of the ray, but rather *this actual slope multiplied by the local index of refraction at the ray position*. Hence, we will define in the most general situation

$$r'(z) \equiv n(z)\frac{dr(z)}{dz} \tag{3}$$

and similarly for $x'(z) \equiv n(z)\,dx(z)/dz$ and $y'(z) \equiv n(z)\,dy(z)/dz$. With these definitions we can connect input and output displacments and slopes in a wide variety of paraxial optical elements by the general form

$$\begin{aligned} r_2 &= Ar_1 + Br'_1 \\ r'_2 &= Cr_1 + Dr'_1. \end{aligned} \tag{4}$$

where we use $r'_1$ and $r'_2$ to denote the modified ray slopes at the input and output planes, and where the coefficients A, B. C. and D characterize the paraxial focusing properties of this element. If we need to, we can refer to the derivatives $dr(z)/dz$ and so forth as the *real slopes* and to the quantities $r'(z)$ and so forth as the *reduced slopes*, in situations where we need to be precise.

It is then natural to write Equation 15.4 in matrix form as

$$\mathbf{r}_2 \equiv \begin{bmatrix} r_2 \\ r'_2 \end{bmatrix} = \begin{bmatrix} A & B \\ C & D \end{bmatrix} \times \begin{bmatrix} r_1 \\ r'_1 \end{bmatrix} \equiv \mathbf{M}\,\mathbf{r}_1, \tag{5}$$

where $M$ is the *ray matrix* for the optical element. Table 15.1 lists the ray matrices for a large number of basic paraxial optical elements, using the actual displacement and reduced slope variables. Note in particular that if we use the generalized definition for the reduced ray slopes, then the bending of a ray trajectory that occurs at a dielectric interface because of Snell's law is automatically taken into account, and the $ABCD$ matrix for a planar dielectric interface is simply the identity matrix.

With the generalized slope definition of Equation 15.3, it is a general property of all the basic elements in Table 15.1 that the ray matrix determinant is given by

$$AD - BC = 1. \tag{6}$$

(If we do not use the reduced slopes, then we have the more cumbersome relation that $AD - BC = n_1/n_2$ where $n_1$ and $n_2$ are the refractive indices at the input and output planes.) Since the determinant of a matrix product is the product of the determinants, Equation 15.6 holds equally well for an arbitrary cascade of optical elements.

### Interfaces and Ducts

The fundamental building blocks for all the paraxial systems of Table 15.1 are *curved dielectric interfaces* and *quadratically varying dielectric media* or "ducts". The general $ABCD$ matrix for a curved interface between two dielectric media can be derived from Snell's law and elementary geometry, and is given in Table 15.1. The corresponding $ABCD$ matrix for a quadratically varying medium can be developed as follows.

First of all, by a "duct" we mean any dielectric medium which has a quadratic transverse variation in its index of refraction, with either a maximum or minimum on axis, as shown in Figure 15.4. We will also extend this concept in later sections to include "complex ducts" in which there may be a quadratic transverse variation of the loss or gain coefficient as well as the real index of refraction.

To analyze ray propagation in a duct, we can consider a ray, or better a light beam of small but finite width, traveling as in Figure 15.5. The inner edge of this beam is at radius $r$, and the outer edge at radius $r + \Delta r$. Suppose the index of refraction $n(r)$ decreases going radially outward from the system axis, so that the inner edge of this light beam is in a region of slightly higher index. The inner edge of the beam then travels more slowly, whereas the outer edge sees a lower index value and travels faster. As a result the beam tends to be continually turned or bent inward toward the axis.

Suppose that the index of refraction in this medium can be written, or at least approximated, in the quadratic form

$$n(r, z) = n_0(z) - \frac{1}{2} n_2(z) \, r^2, \tag{7}$$

where $n_0(z)$ is the variation along the axis, and the parameter

$$n_2(z) \equiv - \left. \frac{\partial^2 n(r, z)}{\partial r^2} \right|_{r=0} \tag{8}$$

is the downward curvature of the index at the axis. Then, within the paraxial approximation a ray traveling through this medium will follow a trajectory given

## TABLE 15.1
Ray Matrices for Paraxial Optical Elements

**(a) "Free space" region, index $n_0$, length $L$**

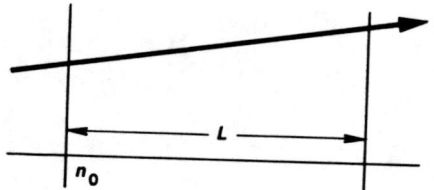

$$\begin{bmatrix} 1 & L/n_o \\ 0 & 1 \end{bmatrix}$$

**(b) Thin lens, focal length $f$**
   $f > 0$ for converging lens

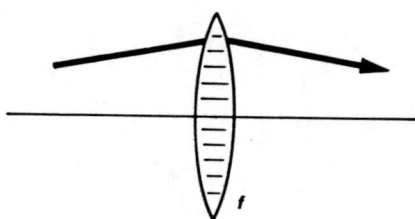

$$\begin{bmatrix} 1 & 0 \\ -1/f & 1 \end{bmatrix}$$

**(c) Curved mirror, radius $R$, normal incidence**
   $R > 0$ for concave mirror

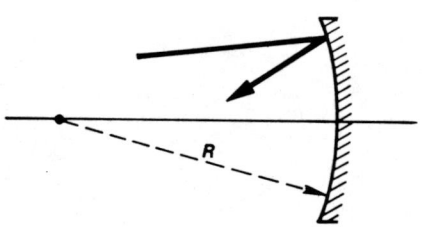

$$\begin{bmatrix} 1 & 0 \\ -2/R & 1 \end{bmatrix}$$

**(d) Curved mirror, arbitrary incidence**
   $R_e = R \cos\theta$ in the plane of incidence ("tangential")
   $R_e = R/\cos\theta$ ⊥ to plane of incidence ("sagittal")

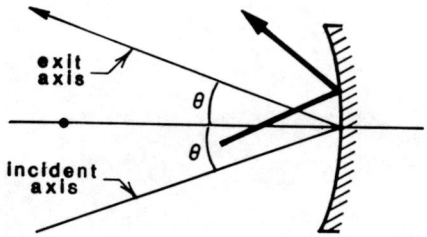

$$\begin{bmatrix} 1 & 0 \\ -2/R_e & 1 \end{bmatrix}$$

### (e) Curved dielectric interface, normal incidence
$R > 0$ for concave surface

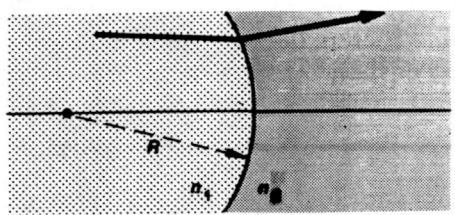

$$\begin{bmatrix} 1 & 0 \\ (n_2 - n_1)/R & 1 \end{bmatrix}$$

### (f) Curved interface, arbitrary incidence, tangential plane
$R > 0$ for concave surface; $n_1 \sin \theta_1 = n_2 \sin \theta_2$

$\Delta n_e = (n_2 \cos \theta_2 - n_1 \cos \theta_1)/ \cos \theta_1 \cos \theta_2$

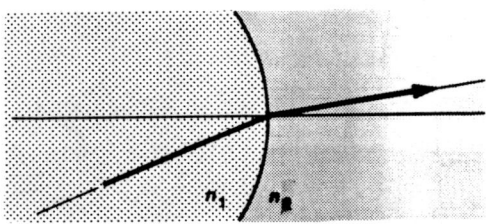

$$\begin{bmatrix} \dfrac{\cos \theta_2}{\cos \theta_1} & 0 \\ \Delta n_e/R & \dfrac{\cos \theta_1}{\cos \theta_2} \end{bmatrix}$$

### (g) Curved interface, arbitrary incidence, sagittal plane
$R > 0$ for concave surface; $n_1 \sin \theta_1 = n_2 \sin \theta_2$

$\Delta n_e = n_2 \cos \theta_2 - n_1 \cos \theta_1$

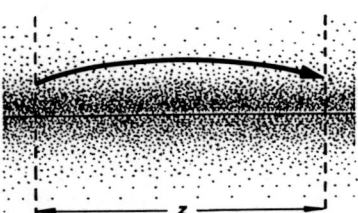

$$\begin{bmatrix} 1 & 0 \\ \Delta n_e/R & 1 \end{bmatrix}$$

### (h) "Duct" (radially varying index and gain)
$n(x) = n_0 - \frac{1}{2} n_2 x^2;\ \gamma^2 \equiv n_2/n_0$

$$\begin{bmatrix} \cos \gamma z & (n_0 \gamma)^{-1} \sin \gamma z \\ -(n_0 \gamma) \sin \gamma z & \cos \gamma z \end{bmatrix}$$

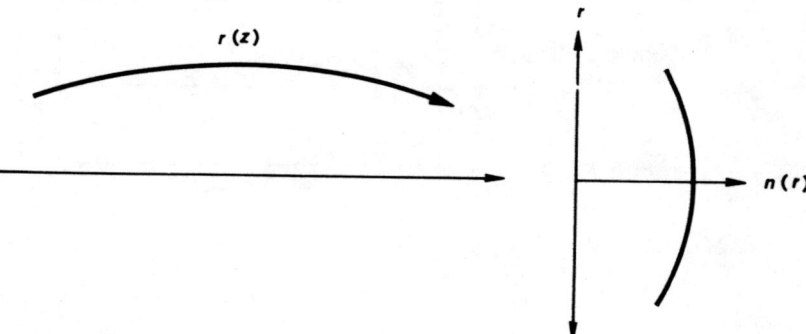

**FIGURE 15.4**
Ray propagation in a "duct."

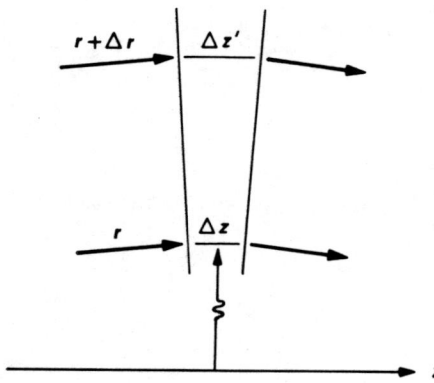

**FIGURE 15.5**
Ray bending in an index gradient.

by the ray propagation equation

$$\frac{d}{dz}\left[n_0(z)\frac{dr(z)}{dz}\right] + n_2(z)\,r(z) = 0. \tag{9}$$

Suppose we define the reduced slope for this ray at any plane, as already discussed in the preceding, by

$$r'(z) \equiv n_0(z)\frac{dr(z)}{dz}. \tag{10}$$

Then we can separate the ray propagation equation (15.9) into the pair of equations

$$\frac{dr(z)}{dz} \equiv \frac{r'(z)}{n_0(z)} \quad \text{and} \quad \frac{dr'(z)}{dz} = -n_2(z)\,r(z), \tag{11}$$

where the first equation is true by definition, and the second accounts for refractive bending in the radially inhomogeneous duct.

### Stable Quadratic Ducts

Ray propagation in real quadratic ducts separates naturally into geometrically *stable* and *unstable* ducts. To show this, let us suppose that the on-axis

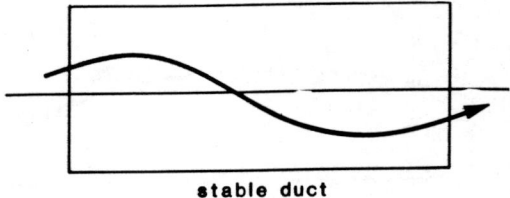

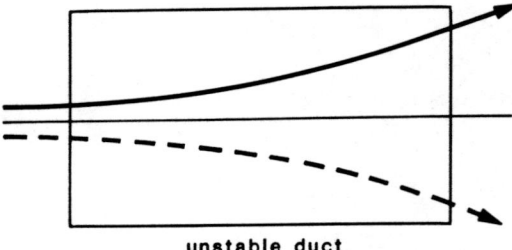

**FIGURE 15.6**
Ray trajectories in geometrically stable and unstable quadratic ducts.

index value $n_0$ and the transverse derivative $n_2$ in Equations 15.11 are both constant with distance. The two ray equations can then be combined to give the single trajectory equation

$$\frac{d^2r(z)}{dz^2} + \frac{n_2}{n_0}r(z) = \frac{d^2r(z)}{dz^2} + \gamma^2 r(z) = 0, \tag{12}$$

where $\gamma$ is given (for positive values of $n_2$) by

$$\gamma^2 = \frac{n_2}{n_0} \quad \text{or} \quad \gamma = \sqrt{\frac{n_2}{n_0}}. \tag{13}$$

The general solution for ray propagation in this kind of paraxial (quadratic) duct becomes

$$r(z) = r_0 \cos \gamma z + \frac{1}{\gamma}\frac{dr_0}{dz}\sin \gamma z$$
$$= r_0 \cos \gamma z + (n_0\gamma)^{-1} r_0' \sin \gamma z, \tag{14}$$

where $r_0$ and $r_0'$ are the initial displacement and (reduced) slope of the ray at $z = 0$.

From Equation 15.14 and its derivative, we can see that the general ray matrix for a duct of length $z$ is

$$M = \begin{bmatrix} \cos \gamma z & (n_0\gamma)^{-1} \sin \gamma z \\ -n_0\gamma \sin \gamma z & \cos \gamma z \end{bmatrix}. \tag{15}$$

A duct with an index maximum on axis and a quadratic variation near the axis will trap optical rays so that they will oscillate periodically back and forth across the centerline of the duct, as shown in the top part of Figure 15.6. We will refer to this as a *stable quadratic duct*.

### Unstable Quadratic Ducts

The same analysis in Equations 15.12 to 15.15 will apply equally well to a medium in which the index of refraction *increases* quadratically going outward from the axis, so that $n_2 < 0$ or $d^2n/dr^2 > 0$. In this situation, however, the value of $\gamma^2$ becomes a negative quantity, and $\gamma$ must be replaced by

$$\gamma^2 = -\left|\frac{n_2}{n_0}\right| \quad \text{or} \quad \gamma = j\sqrt{\frac{1}{n_0}\frac{d^2n}{dr^2}} = j|\gamma|. \tag{16}$$

The general solution analogous to Equation 15.14 then becomes

$$r(z) = r_0 \cosh\gamma z + (n_0\gamma)^{-1} r_0' \sinh\gamma z, \tag{17}$$

and the *ABCD* matrix becomes

$$M = \begin{bmatrix} \cosh\gamma z & (n_0\gamma)^{-1}\sinh\gamma z \\ -n_0\gamma \sinh\gamma z & \cosh\gamma z \end{bmatrix}. \tag{18}$$

Such an "anti-duct," with an index minimum on axis, will diverge (as well as defocus) optical rays. It acts in general in the same way as a thick diverging lens, as shown in the lower part of Figure 15.6.

Ducts thus provide our first illustration of the distinction between *stable ray-propagating systems*, in which rays oscillate periodically back and forth about the ray axis but with bounded excursions; and *unstable ray-propagating systems*, in which rays diverge exponentially outward with distance. We will see many examples of this for more complex types of paraxial focusing systems in later sections.

### Examples of Ducts: Optical Fibers and GRIN Rods

The focusing and ray-trapping properties of stable quadratic ducts are of great practical importance. They provide first of all an idealized model for light propagation in the graded-index optical fibers that are now becoming widely used for long distance optical communications. The simplest type of optical fibers are made up of a uniform core surrounded by a lower-index cladding, as in Figure 15.7, so that the radial index variation is a step-function rather than a smooth quadratic variation. A more detailed waveguide type of analysis is then required to give an accurate description of the modes in fibers having this type of discontinuous index variation.

Many fibers are now being made, however, with a smoothly varying radial profile which more or less approximates a quadratic index variation (Figure 15.7, lower part). The simple results given in the preceding equations will then provide a good first-order approximation to the ray behavior in this kind of fiber, regardless of the actual index variation $n(r)$, provided that the index variation has a quadratic leading term near the axis and provided that the ray trajectories are confined close enough to the axis so that higher-order terms in the radial index variation do not become important. More accurate solutions for other index variations—notably the square-topped or stepped index variations in cladded fibers—are also available but rapidly become more complex.

Optical elements that are of poor optical quality, such as imperfect laser rods and nonlinear optical crystals, may also have unintentional ducts, either stable or unstable, built into them due to local variations in optical index. Laser

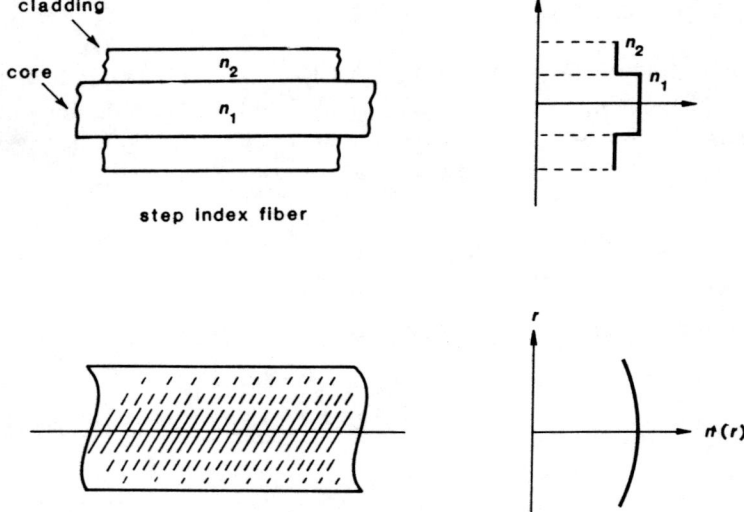

FIGURE 15.7
Examples of step-index and graded-index optical fibers.

oscillations can then be trapped along the stable ducts, and rejected by any unstable "anti-ducts," in such rods. Early laser rods, particularly ruby rods, often exhibited large random index variations and thus random ducting effects across their transverse cross section, leading to poor optical beam quality and possibly to damage at high optical powers. Modern optical rods are generally much better in this regard.

The intense pumping light in solid-state lasers can also cause a temperature rise on the rod axis, which usually produces an increase in index of refraction on axis. The rod as a whole then becomes a duct which acts like a weak positive focusing lens with a pump-power-dependent focal length. Such thermal focusing effects are usually not desirable, and usually limit the oscillation power available from the rod.

Finally, glass rods and fibers with built-in quadratic ducting properties are now commercially manufactured under such trade names as SEL-FOC ("self-focusing") or GRIN ("graded refraction index") rods, and are used as self-focusing laser systems and as specialized lenses for many optical applications.

### Axial Index Variations

We can also consider the situation where there is no transverse variation, or $n_2 = 0$, but there is an axial variation of the index in the medium given by $n_0 = n_0(z)$. The relevant ray equation in this situation is

$$\frac{dr'(z)}{dz} = \frac{d}{dz}\left[n_0(z)\frac{dr(z)}{dz}\right] = 0 \qquad (19)$$

with the solution

$$r(z) = r_0 + r_0' \int_{z_0}^{z} \frac{1}{n_0(z)}\, dz. \qquad (20)$$

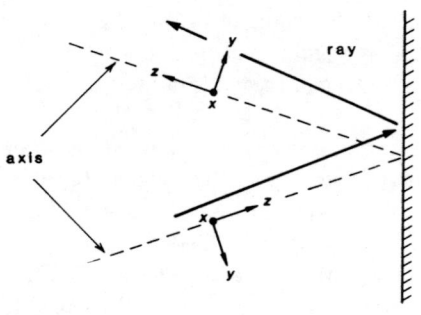 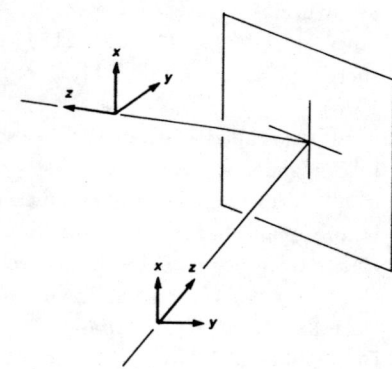

**FIGURE 15.8**
Ray inversion (or coordinate inversion) on reflection.

This gives for the $ABCD$ matrix through a section of length $L$ starting at $z=0$

$$M = \begin{bmatrix} 1 & B(L) \\ 0 & 1 \end{bmatrix} \qquad \text{where} \qquad B(L) \equiv \int_0^L \frac{dz}{n_0(z)}. \tag{21}$$

The ray-bending properties of a segment with an axial index variation are contained in the definition of the reduced slope $r'(z)$.

### Ray Inversion

One additional elementary ray operation that we have not considered yet is *ray inversion* of an optical ray with respect to one or the other of its transverse coordinate axes.

Ray inversion necessarily occurs, for example, in one transverse coordinate or the other whenever an optical ray is specularly reflected from a mirror, as shown in Figure 15.8. If we are to retain a right-handed coordinate system looking in the direction of ray propagation both before and after reflection, the ray displacements and slopes in the planes perpendicular to and lying in the plane of incidence must be related before and after reflection by

$$x = x_0, \; x' = x'_0 \qquad \text{and} \qquad y = y_0, \; y' = -y'_0. \tag{22}$$

The ray matrices along the principal axes can thus be written in the form

$$\boldsymbol{x}_2 = \boldsymbol{I}\,\boldsymbol{x}_1 \qquad \text{and} \qquad \boldsymbol{y}_2 = -\boldsymbol{I}\,\boldsymbol{y}_1, \tag{23}$$

where $\boldsymbol{I}$ is the identity matrix. Ray inversion thus represents one particularly primitive kind of astigmatism in an optical system. Ray inversion also means, among other things, that a ring laser having an odd number of mirrors will have a net overall inversion with respect to one or the other of its axes in one round trip.

### REFERENCES

The paraxial ray matrix approach, though widely used in the laser field, is not yet as widely taught in elementary optics texts. A short survey of this approach, with many useful references, is given by Allen Nussbaum in "Teaching of advanced geometric opics," *Appl. Opt.* **17**, 2128–2129 (July 15 1978). One useful recent book not referenced there is A. Garrard and J.M. Burch, *Introduction to Matrix Methods in Optics* (Wiley, 1975).

The detailed mathematical description of ray propagation and ray bending in an inhomogeneous optical medium with arbitrary spatial variation of the index of refraction is a fairly subtle and complex topic, involving such concepts as eikonal functions, hamiltonian characteristic functions, variational principles, and Euler equations. The classic reference book on the subject is R. K. Luneburg, *Mathematical Theory of Optics* (University of California Press, 1964). Other lengthy discussions can be found in M. Born and E. Wolf, *Principles of Optics* (Pergamon Press, 1959); and in J. A. Arnaud, *Beam and Fiber Optics* (Academic Press, 1976).

A more complex and generalized analysis of ray propagation in ducts with both axial and radial index variations is given in K. Tanaka, "Paraxial theory of rotationally distributed-index media by means of Gaussian Constants," *Appl. Opt.* **23**, 1700–1706 (June 1 1984).

The ray matrices for curved dielectric interfaces at oblique incidence are derived by G. A. Massey and A. E. Siegman, "Reflection and refraction of gaussian light beams at tilted ellipsoidal surfaces," *Appl. Opt.* **8**, 975–978 (May 1969).

Graded-index rods or ducts as discussed in this section are of course essentially equivalent to thick optical lenses. Such graded-index or GRIN rods are now used as lenses in a number of commercially important applications, either singly as fiber optical connectors and medical imaging devices, or in large arrays as imaging systems for photocopying machines. A good series of papers reviewing the technology and applications of graded-index optics can be found in *Appl. Opt.* **21** (March 15 1982), **22** (February 1 1983) and **23** (June 1 1984).

Problems for 15.1

1. *Ray matrix for a curved dielectric interface.* Using Snell's law, derive the ray matrix for a curved interface between two dielectrics.

2. *Ray matrix elements for a curved diffraction grating.* Curved diffraction gratings are occasionally employed as end mirrors for laser cavities, as well as in beam-expanding systems and grating spectrometers. Suppose a curved diffraction grating with radius of curvature $R$ has rulings running in the $y$ direction with grating spacing $d$ in the $x$ direction. An incident beam striking this grating at an angle $\theta_1$ from the normal in the $x, z$ plane will then be diffracted in $N$-th order into angle $\theta_2$ in the same plane given by the grating equation $\sin\theta_1 + \sin\theta_2 = N/d$.

    Show that the ray matrix for reflection from this grating has matrix elements $A = 1/D = M$, $B = 0$, and $C = -2/R_t$ in the tangential or $x, z$ plane, where $M \equiv \cos\theta_2/\cos\theta_1$ is a transverse magnification or beamwidth expansion, and the effective radius of the grating is $R_t \equiv R\cos\theta_1 \cos\theta_2/(\cos\theta_1 + \cos\theta_2)$. Show also that the matrix elements in the perpendicular or sagittal direction are given by $A = D = 1$, $B = 0$, and $C = -2/R_s$ where $R_s \equiv 2R/(\cos\theta_1 + \cos\theta_2)$.

3. *Limiting case.* Show that the limiting case for a short but very strongly focusing duct is a simple thin lens, and give the focal power of the lens in terms of the duct parameters.

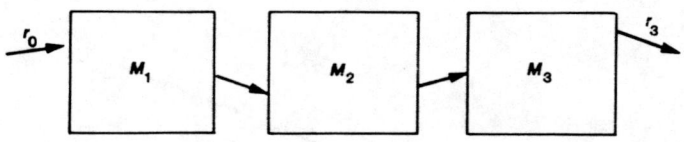

FIGURE 15.9
Ray matrix systems in cascade.

## 15.2 RAY PROPAGATION THROUGH CASCADED ELEMENTS

Let us next look at how rays propagate through cascade optical systems consisting of several different paraxial elements connected together in cascade. It is one of the most important properties of ray matrices that such cascaded paraxial optical elements can be handled simply by matrix multiplying the individual *ABCD* matrices for the individual optical elements, arranged in reverse order.

### Cascaded Ray Matrices

Suppose several optical elements with ray matrices $M_1, \ldots, M_n$—for example, a free-space section, a thin lens, another free-space section, a dielectric interface, and so on—are arranged in cascade as shown in Figure 15.9. The total ray transformation through this cascaded series of elements can then be calculated from the chain multiplication process

$$
\begin{aligned}
r_1 &= M_1 \, r_0 \\
r_2 &= M_2 \, r_1 = M_2 \, M_1 \, r_0 \\
r_3 &= M_3 \, r_2 = M_3 \, M_2 \, M_1 \, r_0,
\end{aligned} \tag{24}
$$

and so on up to the general result

$$ r_n = [M_n M_{n-1} \cdots M_2 M_1] \, r_0 = M_{\text{tot}} \, r_0. \tag{25}$$

The overall or total ray matrix $M_{\text{tot}}$ for this system is thus given by

$$ M_{\text{tot}} \equiv M_n M_{n-1} \cdots M_2 M_1. \tag{26}$$

A single 4-element ray matrix equal to the ordinary matrix product of the individual ray matrices can thus describe the total or overall ray propagation through a complicated sequence of cascaded optical elements. Note, however, that the matrices must be arranged *in inverse order* from the order in which the ray physically encounters the corresponding elements.

### Ray Matrices and Spherical Wave Propagation

Ray matrices and paraxial ray optics provide a general way of expressing the elementary lens laws of geometrical optics, or of spherical-wave optics, leaving out higher-order aberrations, in a form that many people find clearer and more convenient. *Ray optics and geometrical optics in fact contain exactly the same physical content, expressed in different fashion.*

To demonstrate this we can first note that an ideal spherical wave with radius of curvature $R$ can also be viewed as a collection of rays all diverging from a common point, the wavefront's center of curvature $C$ (Figure 15.10). The slope

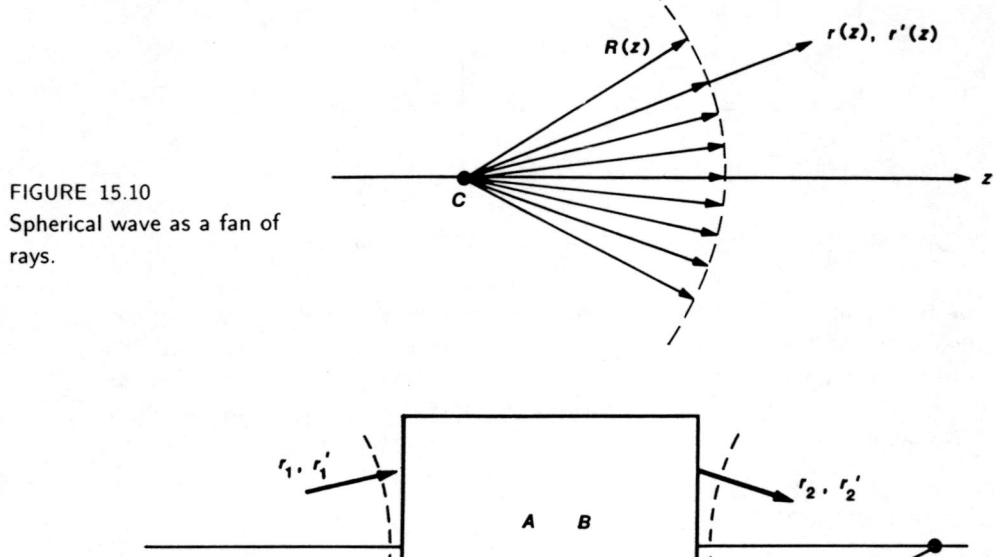

**FIGURE 15.10**
Spherical wave as a fan of rays.

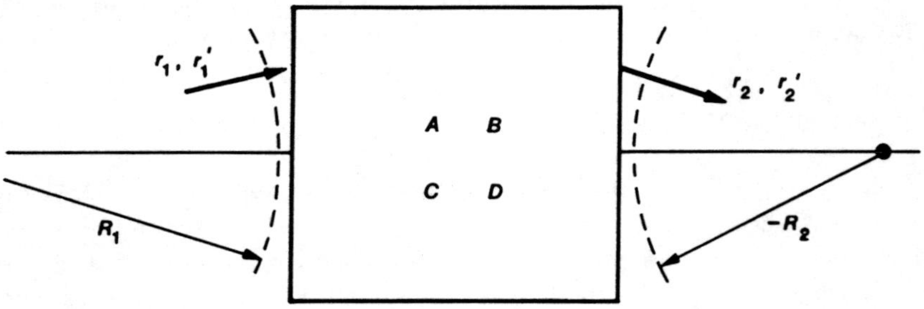

**FIGURE 15.11**
Spherical wave transformation through an arbitrary paraxial system.

and displacement of each of these rays at the plane $z$ where the radius of curvature is $R(z)$—that is, at a distance $R$ from the source point—are then related by

$$r'(z) = n(z)\frac{dr(z)}{dz} = \frac{n(z)r(z)}{R(z)} \quad \text{or} \quad R(z) \equiv \frac{n(z)\,r(z)}{r'(z)}. \tag{27}$$

Equation 15.27 implies a sign convention in which positive $R$ indicates a diverging spherical wave, as drawn, whereas a negative value of $R$ implies a converging spherical wave.

Suppose such a spherical wavefront with radius $R_1$ passes through a paraxial system with ray matrix $ABCD$ as in Figure 15.11. Then the emerging wavefront at the other end of the $ABCD$ system will also be a spherical wavefront with radius $R_2$, which can be calculated from any one of the output rays by writing

$$\frac{R_2}{n_2} \equiv \frac{r_2}{r'_2} = \frac{Ar_1 + Br'_1}{Cr_1 + Dr'_1} = \frac{A(R_1/n_1) + B}{C(R_1/n_1) + D}. \tag{28}$$

(Note that Figure 15.11 shows a converging output wave, which means its radius of curvature $R_2$ would be a *negative* number according to our sign conventions.) More generally, if we define a "reduced radius of curvature" by $\hat{R}(z) \equiv R(z)/n(z)$, then Equation 15.28 in terms of the reduced radii becomes simply

$$\hat{R}_2 = \frac{A\hat{R}_1 + B}{C\hat{R}_1 + D}. \tag{29}$$

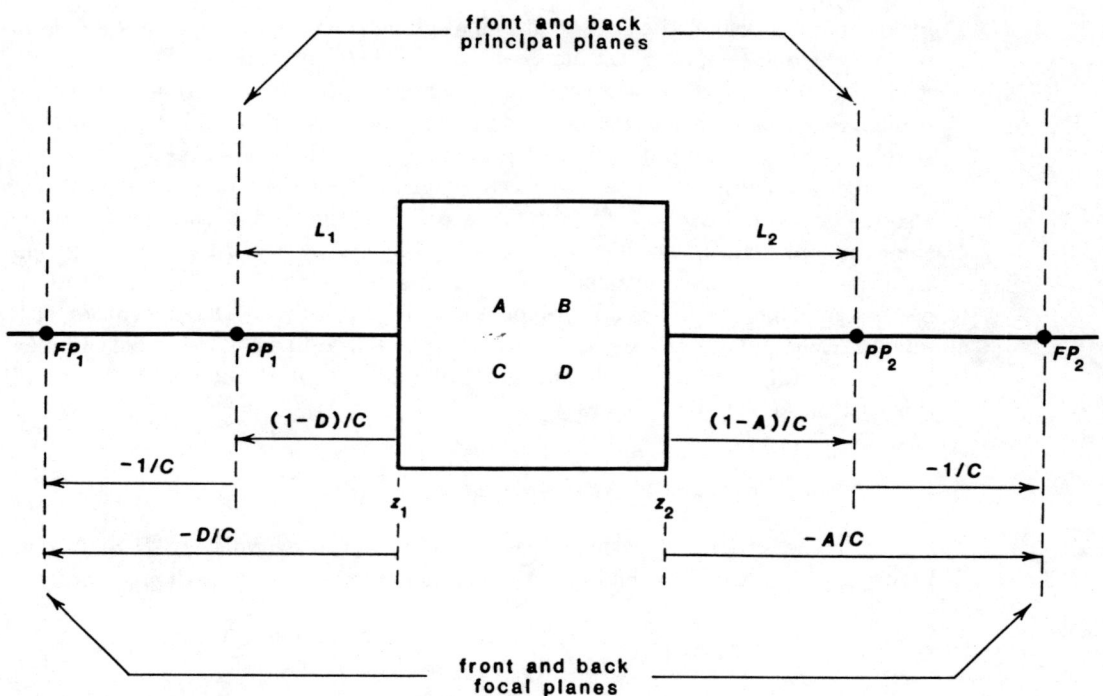

**FIGURE 15.12**
Front and back principal planes and focal planes for an arbitrary $ABCD$ system treated as a compound lens.

This simple but very general connection between $R_1$ and $R_2$, using only the $ABCD$ matrix, will be very important and useful in later sections. It summarizes all of elementary geometrical optics expressed in ray matrix form.

### Thick Lenses and $ABCD$ Matrices

To expand on this last point a bit more, we can note that Equation 15.29 can be manipulated into the alternative form

$$\frac{1}{\hat{R}_2 - L_2} = \frac{1}{\hat{R}_1 - L_1} + \frac{1}{1/C}, \quad (30)$$

with $L_2 \equiv (A-1)/C$ and $L_1 \equiv (1-D)/C$. But this expression is obviously just a slightly generalized form of the usual geometrical optics lens formula. It says that the reduced input and output wave curvatures or image and object distances $\hat{R}_1$ and $\hat{R}_2$ obey the simple lens law for a thin lens of focal length $f \equiv -1/C$, if these quantities are measured from reference planes located at distances $(1-D)/C$ and $(A-1)/C$ behind the input and output planes of the $ABCD$ system.

For simplicity let us consider only the situation where the index of refraction is unity on both sides of the $ABCD$ system, so that $\hat{R} \equiv R$, and the radius $R$ gives the distance to or from the source point for the spherical wave. Then, the two reference planes or *principal planes* for the $ABCD$ system just referred to are located at distances $(1-D)/C$ and $(1-A)/C$ behind and in front of the input and output planes $z_1$ and $z_2$ of the $ABCD$ system itself, as indicated by the points $PP_1$ and $PP_2$ in Figure 15.12. (Note that with our sign convention for radii of curvature, the output principal plane is located a distance $L_2$ behind the output plane $z_2$, or a distance $-L_2 \equiv (1-A)/C$ in front of it.)

If the input and output rays, or spherical waves, are referenced to these principal planes rather than the original reference planes $z_1$ and $z_2$, the overall $ABCD$ system from input to output principal planes then acts exactly like a thin lens with a focal length $f \equiv -1/C$, as given in Equation 15.30. This lens then also has front and back focal points $FP_1$ and $FP_2$ located a distance $f$ outside the principal planes, as indicated in Figure 15.12. Any arbitrary $ABCD$ system with $n_1 = n_2$ is thus equivalent to a thick lens, which can be fully characterized by its two principal planes and its focal length $f$. If $n_1 \neq n_2$ this conclusion still remains true, but the thick lens must be characterized in a slightly more complex fashion by its *principal*, *focal*, and *nodal* planes; see the Problems at the end of this section for details.

### Imaging Properties of $ABCD$ Systems

For the simpler situation of $n_1 = n_2 = 1$, the overall $ABCD$ matrix in Figure 15.12 going from the input to output *principal planes* is then given by

$$M = \begin{bmatrix} 1 & 0 \\ C & 1 \end{bmatrix} \quad \begin{pmatrix} \text{principal plane} \\ \text{to principal plane} \end{pmatrix}. \tag{31}$$

This is the ray matrix for a thin lens with $f = -1/C$. In other words, as we noted in the preceding, the overall $ABCD$ matrix between these planes appears to have an effective length $B = 0$ and a focal power $C$ equivalent to the ABCD matrix itself.

By contrast, the overall $ABCD$ matrix from the input to output *focal planes* is given by

$$M = \begin{bmatrix} 0 & C^{-1} \\ C & 0 \end{bmatrix} \quad \begin{pmatrix} \text{focal plane} \\ \text{to focal plane} \end{pmatrix}. \tag{32}$$

This is the general form of the ray matrix going from focal point to focal point. Note that the apparent length associated with this propagation is $C^{-1} = -f$, even though the actual physical length (for a positive thin lens) is actually $2f$.

More generally, for arbitrary indices, consider an input spherical wave which diverges from an arbitrary *object plane* located at a point $OP$ on the $z$ axis, and is then focused by an arbitrary $ABCD$ system back down to a (real or virtual) *image plane* located at a point $IP$ on the $z$ axis. We can then show that the overall $ABCD$ matrix going from the object plane at $OP$ to the image plane at $IP$ has the general form

$$M = \begin{bmatrix} M & 0 \\ C & 1/M \end{bmatrix} \quad \begin{pmatrix} \text{object plane} \\ \text{to image plane} \end{pmatrix}. \tag{33}$$

Once again the effective length from object plane to image plane is zero, but in the most general situation there will be an *image magnification* $M$ (given in general by $(CR_1 + D)^{-1}$) from any point $r_1$ in the image plane to the corresponding point $r_2$ in the output plane. (Note that because the effective length $B \equiv 0$, all the rays leaving from any input point $r_1$ will pass through the same output point $r_2$.)

A *ray-angle demagnification* given by the $D$ element value of $1/M$ is then necessarily associated with this image magnification $M$. This conclusion repre-

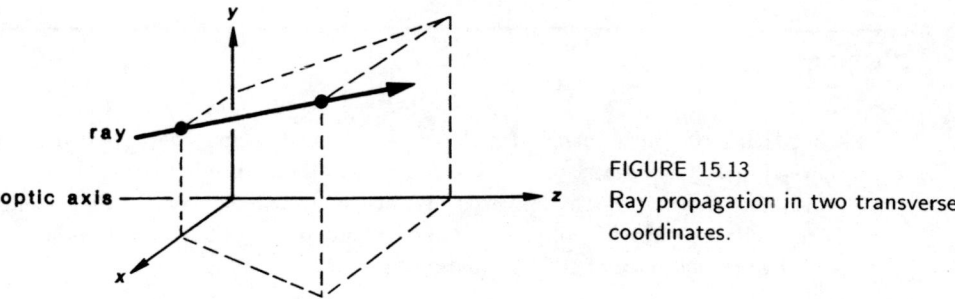

FIGURE 15.13
Ray propagation in two transverse coordinates.

sents, in fact, a paraxial approximation to the more general *sine condition* of optics which says that if a ray leaves a point $r_1$ in an object plane with angle $\theta_1$ and arrives at a point $r_2$ in an image plane with an angle $\theta_2$, these quantities must be related by $n_1 r_1 \sin\theta_1 = n_2 r_2 \sin\theta_2$.

This condition in turn can be given a thermodynamic interpretation: If we collect the blackbody radiation leaving a small area of diameter $r_1$ and temperature $T$ within a cone angle $\theta_1$ and image it, with lateral magnification $M$, so that it is incident within cone angle $\theta_2$ onto another small area of diameter $r_2 = Mr_1$, then this incident radiation must just match the blackbody radiation which the second surface area at the same temperature $T$ would emit back into the same cone angle $\theta_2$. If we take properly into account the difference in blackbody energy densities and velocities in two media with different refractive indices, we can then use the necessity for thermodynamic balance to derive either the more general sine condition, or the ray matrix condition that $AD - BC = 1$.

### Ray Matrices in Astigmatic Systems

When cartesian coordinates are used in an optical system, with propagation primarily in the $z$ direction, then a general ray must be described by its transverse displacements in both the $x$ and $y$ directions (Figure 15.13). For simple optical elements the ray matrix formalism just described then applies separately and independently to both the $x, x'$ and $y, y'$ coordinates. If an overall optical system is rotationally symmetric the same $ABCD$ matrices apply equally to both $x, x'$ and to $y, y'$. If the system contains astigmatic elements, then different $ABCD$ matrices must be used for these elements in the $x$ and $y$ directions, as we will discuss in more detail in a later section.

### Other Ray Matrix Properties

Ray matrices have many other interesting and useful properties and applications which we will introduce in this and later chapters. The properties of real ray matrices in periodic systems, in misaligned ray matrix systems, and in nonorthogonal ray matrix systems (systems with "twist") are discussed in later sections of this chapter. In later chapters we will also show how Huygens' diffraction integral can be written entirely in terms of ray matrix elements; how ray matrix concepts can be extended and all of paraxial optics explained by generalized or complex ray matrices; and how arbitrary ray matrices can be symmetrized, decomposed and/or synthesized by appropriate transformations.

## Problems for 15.2

1. *Evaluating the focal length of a thin lens.* A thin lens may be regarded as two curved dielectric interfaces with vanishingly small distance between them. Using this viewpoint and the ray matrices for a dielectric interface, find the focal length $f$ of a thin lens in terms of the radii of curvature of the two lens surfaces and the index of refraction $n$ of the lens material.

2. *Replacing an arbitrary "black box" ray matrix with a single lens.* An optical black box has various optical elements inside it, producing a given real $ABCD$ matrix from its input plane to its output plane. We want to replace this black box with a box of physical length $L$ containing only a single lens of focal length $f$. Can this be done? What length $L$, focal length $f$, and lens location within $L$ will be required?

3. *Ray matrix of cascaded elements going in the reverse direction.* A collection of optical elements in series has an overall $ABCD$ matrix going in one direction. Find the $ABCD$ matrix going through the same elements in the reverse direction, i.e., assume the direction of the $z$ axis going through these elements is reversed (or equivalently, assume that the whole system is picked up, turned around, and set back down on the same $z$ axis with all the elements now in reverse order).

4. *Evaluating the total ray matrix for a reflection problem.* A ray passes through a collection of optical elements in series having an overall $ABCD$ matrix; bounces off a mirror of radius $R$; and passes back out through the same collection of elements in the reverse direction. What is the total $ABCD$ matrix for the entire round trip?

5. *Replacing an arbitrary ray matrix system with a single mirror.* A certain optical black box has a front entrance plane and various lenses and mirrors inside it, such that a ray entering the entrance plane eventually comes back out through the same plane with a total ray transformation given by a known $ABCD$ ray matrix. Suppose this black box is to be replaced by a single curved mirror of radius $R$ located an appropriate distance $L$ behind (or, if necessary, in front of) the entrance plane of the box. Find the required radius $R$ and position $L$ of the single curved mirror.

6. *Focusing properties of thick-lens ABCD matrices.* Verify the $ABCD$ matrix equations (15.31-15.33) given in the text for transfer between input and output principal planes and between input and output focal planes (for $n_r = 1$), and more generally between object and image planes (for arbitrary $n_r$).

7. *General formulas for an arbitrary thick lens or ABCD system* The focal, principal and nodal planes for an arbitrary thick lens or $ABCD$ system having input and output reference planes $z_1$ and $z_2$ and input and output indices of refraction $n_1$ and $n_2$ are defined by the conditions that:

(1) An input spherical wave emanating from the input focal point $FP_1$ and passing through the $ABCD$ system will emerge as an output plane wave; whereas an input plane wave will emerge as a spherical wave which converges to (or appears to diverge from) the output focal point $FP_2$.

(2) If an input ray $r_1$ which comes from the input focal point $FP_1$, and the output ray $r_2$ parallel to the output axis which it produces, are extended forward or

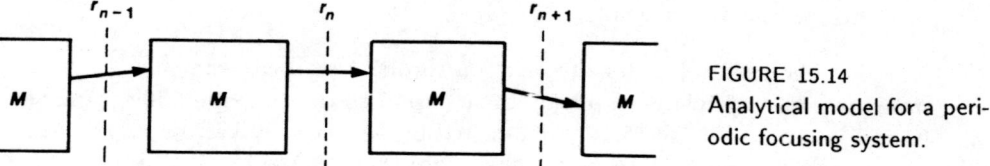

FIGURE 15.14
Analytical model for a periodic focusing system.

backward until they intersect, their intersection point defines the input principal plane $PP_1$. Similarly, the intersection of a parallel input ray $r_1$ and the output ray $r_2$ which it produces defines the output principal plane $PP_2$.

(3) If an input ray with coordinates $r_1, r'_1$ produces a parallel output ray, i.e., $r'_2 = r'_1$, then the line connecting the input and output points $r_1, z_1$ and $r_2, z_2$ crosses the optical axis at the *optical center OC* of the lens. If extensions of the same entering and exit rays are constructed, these rays then intersect the optical axis at the front and back nodal planes $NP_1$ and $NP_2$ of the lens.

(To put this in another way, Ditchburn speaks of any pair of planes which are imaged onto each other as *conjugate planes*, and then says, "The plane conjugate to a plane at an infinitely great distance from the system in a positive direction is called the first focal plane ... In a similar way the second focal plane is conjugate to a plane infinitely distant in the negative direction." Also, "Any ray which before entering the system is directed toward the first nodal point will emerge with its final direction parallel to the original direction and passing through (or coming from) the second nodal point," and finally, the principal planes of a thick lens "...are conjugate planes of unit positive magnification," i.e., any input ray which, when projected, intersects the first principal plane at $r_1 = a$ and any slope $r'_1$ will produce an output ray which intersects the second plane at the same distance $r_2 = a$).

From these definitions, find general formulas for the locations of all these points or planes for an arbitrary $ABCD$ system with $n_1 \neq n_2$, and then illustrate by calculating the actual locations for some representative systems, such as a thick lens with curved front and back faces, or a section of a quadratic duct.

---

## 15.3 RAYS IN PERIODIC FOCUSING SYSTEMS

Perhaps the most interesting and important application of ray matrices comes in the analysis of *periodic focusing systems*, i.e., systems in which the same sequence of elements is repeated many times down a cascaded chain or optical lensguide. An optical resonator can be modeled, as we have already shown in Figure 14.3, by such an iterated periodic focusing system. The eigenvalues and "eigenrays" for such periodic focusing systems play an important role in optical resonator theory, particularly in explaining the stable and unstable properties of optical resonators and lensguides. The stability analysis for periodic optical focusing that we will present here will also apply equally well to periodic particle focusing systems, such as electron beams in periodically focused traveling-wave tubes or in linear accelerators.

### Eigenvalues and Eigenrays

Let the ray matrix for propagation through one period in such a system, from an arbitrary reference plane in one period to the corresponding plane one period later (see Figure 15.14), be denoted by $M$. The ray vectors $r_n$ and $r_{n+1}$ at the $n$-th and $n+1$-th reference planes are then related by

$$r_{n+1} = M\, r_n = M^{n+1}\, r_0, \tag{34}$$

where $r_0$ is the initial ray at the input plane $n = 0$, and $M^{n+1}$ is the matrix for one period raised to the $n + 1$-th power.

Any cascaded matrix problem such as this can best be analyzed by finding the eigenvalues and eigensolutions of the matrix $M$. That is, we look for a set of "eigenrays" $r$ and corresponding eigenvalues $\lambda$ (no connection with optical wavelength $\lambda$) which each individually satisfy the eigenequation

$$M\, r = \lambda\, r. \tag{35}$$

For a 2×2 ray matrix $M$ this is equivalent to the equation

$$[M - \lambda I]\, r = 0 \quad \text{or} \quad \begin{bmatrix} A - \lambda & B \\ C & D - \lambda \end{bmatrix} \begin{bmatrix} r \\ r' \end{bmatrix} = 0, \tag{36}$$

where $I$ is the identity matrix.

Nonzero solutions to Equation 15.36 are possible if and only if the determinant of the matrix in this equation satisfies the relation

$$\begin{vmatrix} A - \lambda & B \\ C & D - \lambda \end{vmatrix} \equiv \lambda^2 - (A + D)\lambda + 1 = 0, \tag{37}$$

where we have used the fact that $AD - BC = 1$. It is convenient to define an "$m$ parameter" for the system, equal to half the trace of the $ABCD$ matrix, or

$$m \equiv \frac{A + D}{2}. \tag{38}$$

The ray matrix eigenvalues are then given by the two values

$$\lambda_a, \lambda_b = m \pm \sqrt{m^2 - 1} \tag{39}$$

which obey the general relationship that

$$\lambda_a \lambda_b \equiv 1. \tag{40}$$

There are also two matching eigenrays $r_a$ and $r_b$, which the reader can calculate for herself, such that

$$M\, r_a = \lambda_a\, r_a \quad \text{and} \quad M\, r_b = \lambda_b\, r_b. \tag{41}$$

The properties of these eigenvalues and eigenrays are fundamental to the theory of stable and unstable optical resonators, as we shall now see.

### Eigenray Expansions

It is a fundamental property of these matrix eigensolutions that any arbitrary ray $r_0$ at the input to the periodic system (or for that matter at any other plane) can always be expanded as a sum of the two eigenrays of the system in the form

$$r_o = c_a r_a + c_b r_b, \qquad (42)$$

where $c_a$ and $c_b$ are suitable expansion coefficients. The ray vector after any number of sections $n$ will then be given by

$$\begin{aligned} r_n = M^n r_0 &= M^n \times (c_a r_a + c_b r_b) \\ &= c_a \times \lambda_a^n r_a + c_b \times \lambda_b^n r_b. \end{aligned} \qquad (43)$$

The propagation of each eigenray is thus specified simply by multiplying it by the corresponding eigenvalue raised to the appropriate power. The eigenrays and their matching eigenvalues therefore contain all the information that is needed to fully describe the propagation of any arbitrary ray in the periodic system.

### Stable Periodic Focusing Systems

All such periodic focusing systems (with purely real ray matrices) can in fact be neatly divided into either *stable* or *unstable periodic systems*, depending only on the properties of the matrix eigenvalues.

Suppose first that the ray matrix for one period has $A$ and $D$ coefficients such that

$$-1 \leq m \leq 1, \quad \text{or} \quad m^2 \equiv \left(\frac{A+D}{2}\right)^2 \leq 1. \qquad (44)$$

In this situation we may write the $m$ parameter as

$$m \equiv \frac{A+D}{2} \equiv \cos\theta, \qquad (45)$$

where $\theta$ is the angle defined by this expression. The eigenvalues of the system can then be written as

$$\lambda_a, \lambda_b = m \pm j\sqrt{1-m^2} = \cos\theta \pm j\sin\theta = e^{\pm j\theta}. \qquad (46)$$

The matrix eigenvalues are thus complex and have magnitude unity. The propagation of any ray in the periodic system then takes the form

$$r_n = c_a r_a \times e^{jn\theta} + c_b r_b \times e^{-jn\theta} = r_0 \cos\theta n + s_0 \sin\theta n, \qquad (47)$$

where $r_0 \equiv c_a r_a + c_b r_b$ is the input ray vector, and $s_0 \equiv j(c_a r_a - c_b r_b)$ is a kind of "input slope vector."

Any periodic focusing system with $|m| \leq 1$ thus represents a *stable periodic focusing system*, analogous to a stable duct. Rays in the system will oscillate back and forth about the axis, as in Figure 15.15, with a maximum excursion determined entirely by the initial ray parameters $r_0$ and $s_0$. The displacement $r_n$ of any ray at successive reference planes down the system will oscillate periodically about the axis in the form

$$r_n = r_0 \cos\theta n + s_0 \sin\theta n, \qquad (48)$$

## CHAPTER 15: RAY OPTICS AND RAY MATRICES

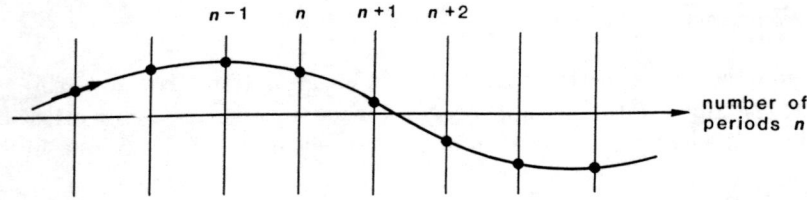

**FIGURE 15.15**
Ray trajectory in a stable periodic system.

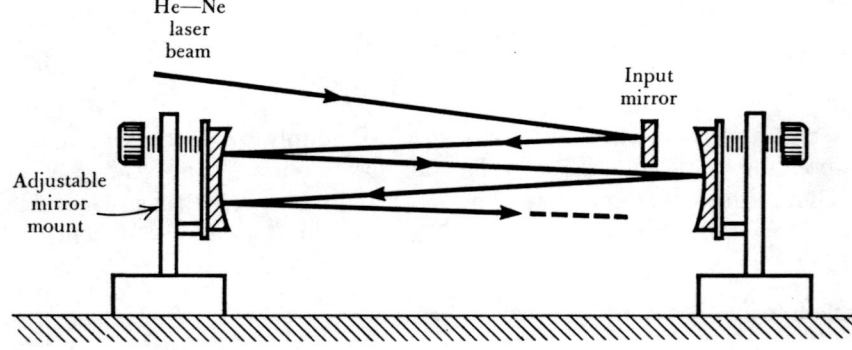

**FIGURE 15.16**
A simple demonstration of a stable periodic focusing system or optical delay line, using a pair of silvered mirrors and a He-Ne laser beam.

where $r_0$ and $s_0$ are the ray initial conditions. Note that it is the index $n$, and not the angle $\theta$, that is the variable which increases with distance down the chain.

Note also that Equations 15.47 and 15.48 only give the displacement $r_n$ as measured at the successive reference planes—they do not say anything about what happens to the ray inside the periodic section between those reference planes. Viewed only at the successive reference planes, however, the ray appears to oscillate about the axis of the periodic focusing system as in Figure 15.15, with an oscillation period equal to $2\pi/\theta$ periods of the periodic focusing system itself.

### Periodic Focusing Demonstration

Any reader who has the opportunity should set up a simple demonstration of such a stable periodic focusing system, using a pair of silvered mirrors perhaps 10 to 15 cm in diameter with a 50 cm to 1 m focal length, as illustrated in Figure 15.16. (Suitable inexpensive mirrors and simple mirror mounts are available from hobby stores or amateur astronomy supply houses.) The beam from a He-Ne laser can be injected at one edge of the resonator, using a small adjustable injection mirror just inside the edge of one of the larger mirrors.

Thoughtful adjustment of the beam injection direction and the mirror spacing and alignment will then lead to various kinds of periodically repeating spot patterns on the end mirrors. A little chalk dust or smoke can make the interlaced beam patterns inside the resonator dramatically visible in a darkened room, although a more effective way to make the beams visible without fouling the mirrors is to attach a few strands of white cord or thin wire to the shaft of a

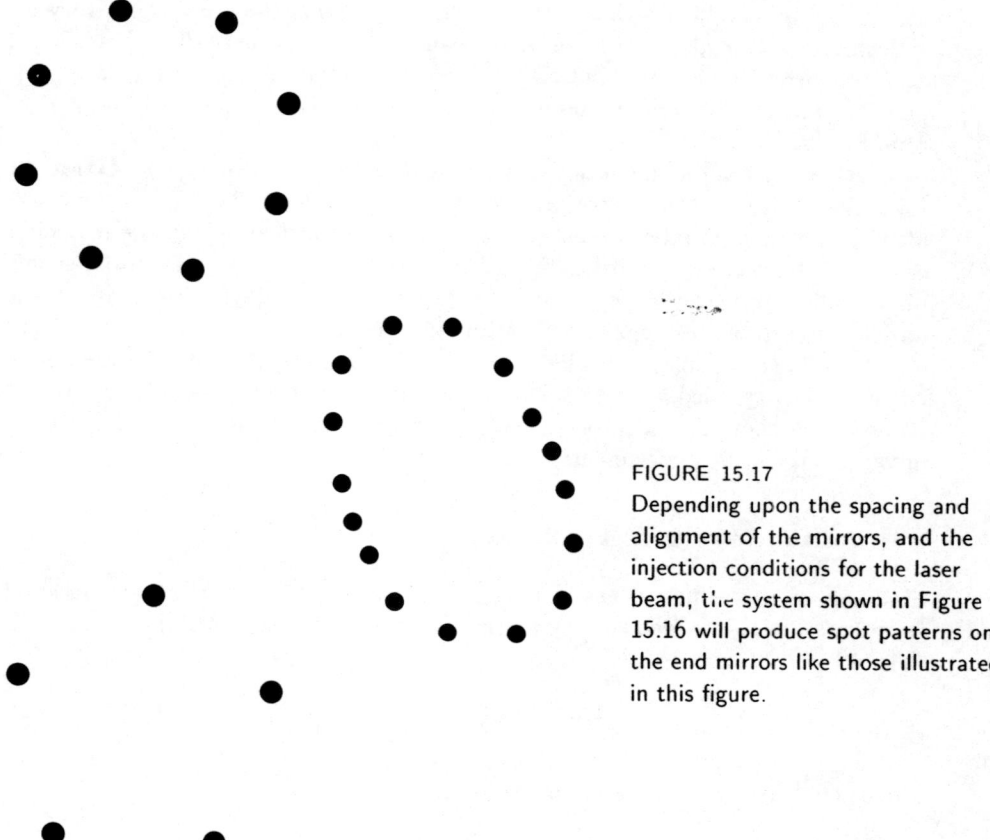

FIGURE 15.17
Depending upon the spacing and alignment of the mirrors, and the injection conditions for the laser beam, the system shown in Figure 15.16 will produce spot patterns on the end mirrors like those illustrated in this figure.

small electric motor so that they sweep transversely across the resonator like a soft buzzsaw.

Note that the periodic solutions derived in Equation 15.47 and 15.48 will apply equally well to both of the transverse displacements $x_n$ and $y_n$, with appropriate (and in general different) initial conditions in each transverse coordinate. The beam in a stable periodic focusing system should thus oscillate sinusoidally about the axis with the same period in both $x$ and $y$ (assuming no astigmatism in the optical system). The oscillations will however in general have different amplitudes and phases in the two directions, depending upon the initial conditions.

But this is just the necessary condition for producing Lissajou figures, except that the Lissajou figures in this situation will be discrete spot patterns at integer values of $n$ rather than continuous line patterns. In the demonstration apparatus, therefore, the successive spots at which the ray strikes either of the end mirrors will trace out a Lissajou pattern with the same frequency or period in the $x$ and $y$ directions. By adjusting the initial beam conditions to vary the phase and amplitude between the $x$ and $y$ oscillations, we can obtain arbitrary circular, elliptical or linear spot patterns (for examples, see Figure 15.17).

Inspection will also show (and we will later verify analytically) that the gaussian laser beam in such a periodic focusing system does not spread due to diffraction as we might expect, even after a large number of round-trip bounces. The same stability conditions that make the ray trajectory stable but oscillatory inside the resonator also make the laser beam spot size be periodically refocused at each mirror. The beam spot size at different points may then oscillate periodically, but it also remains bounded and stable over an indefinite number of round trips.

With proper adjustment we can also catch the laser beam and extract it from the cavity with an extraction mirror (or even with the injection mirror) after any integral number of one-way bounces. The optical delay time in such a cavity is $\sim$6 nsec per round trip for a 1 m long cavity, and with more expensive high-quality mirrors the power loss per bounce can be quite small. Reentrant optical cavities of this type can thus function as optical delay lines. Such delay lines were once seriously considered as potential high-capacity optical memories (with the cavity filled with coded information in the form of very short optical pulses); and they have also been used a number of times as optical delay lines in various scientific experiments.

### Unstable Periodic Focusing Systems

Let us now turn to the opposite example, that of an *unstable periodic focusing system*, in which the ray matrix for one period has instead the property that

$$m^2 \equiv \left(\frac{A+D}{2}\right)^2 > 1 \quad \text{or} \quad |m| > 1. \tag{49}$$

The eigenvalues of the system will then have the values

$$\lambda_a, \lambda_b = m \pm \sqrt{m^2 - 1} = M, 1/M, \tag{50}$$

where $M$ is a "transverse magnification per period," with the property that $|M| > 1$. The ray displacement in this situation will obey the formula

$$r_n = M^n \times c_a r_a + M^{-n} \times c_b r_b = r_0 \cosh \theta n + s \sinh \theta n, \tag{51}$$

where $\theta \equiv \ln M$ and $r_0$ and $s_0$ again represent initial conditions at the start of the periodic system.

The ray displacement $r_n$ in this situation will diverge exponentially with distance down the chain, as shown in Figure 15.18, with the displacements and slopes magnifying by a magnification $M$ in each period. There will also be at first a demagnifying component to the trajectories, decreasing as $1/M$ per section, but this will die out after a few sections. Note that the ray position may also oscillate back and forth across the ray axis in alternate periods, depending on whether the magnification has a value $M < -1$ or $M > +1$. Such unstable periodic focusing systems have an important practical application in the unstable laser resonators we will describe later.

### REFERENCES

Periodic lens waveguides were once of great interest because of what seemed to be their potential for long-distance communications through underground pipes (optical fibers now seem to have made this concept obsolete). Gas lenses and other interesting concepts were invented for use in these optical lensguides. Records of some of these experiments

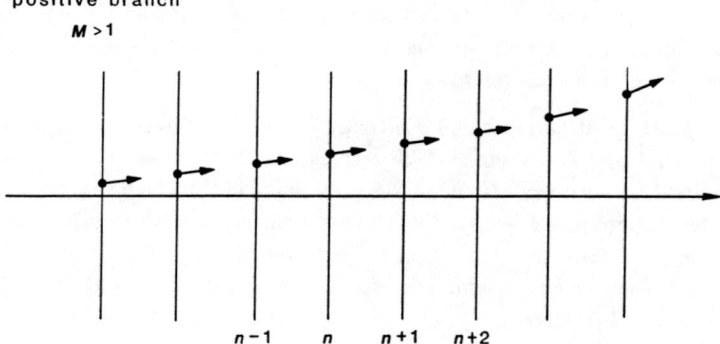
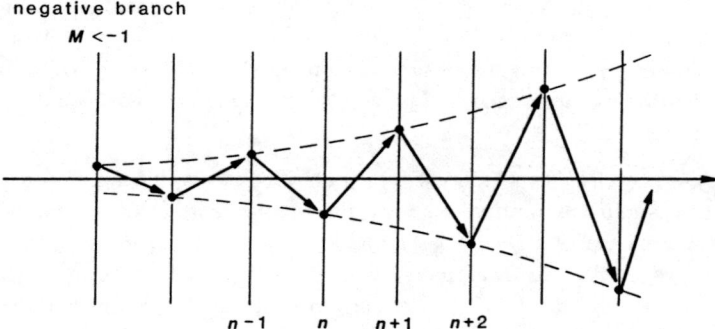

**FIGURE 15.18**
Unstable periodic focusing systems of the "positive-branch" and "negative-branch" types.

can be found in D. Gloge, "Experiment with an underground lens waveguide," *Bell Sys. Tech. J.* **46**, 721–735 (April 1967); and in D. Gloge and W. H Steier, "Pulse shuttling in a half-mile optical lens guide," *Bell Sys. Tech. J.* **47**, 767–782 (May-June 1968).

---

## Problems for 15.3

1. *Properties of the eigenrays in a periodic system.* Calculate the two eigenrays $r_a$ and $r_b$ for a periodic focusing system in terms of the $ABCD$ matrix elements. Note that any physically meaningful ray in the periodic system must be purely real, i.e., must have purely real displacement and slope, yet the eigenvectors $r_a$ and $r_b$ for a stable periodic system are in general complex quantities. How can this be true? Under what conditions (if any) can individual eigenrays be individually or separately excited in the periodic system?

2. *Ray properties of an elementary periodic lensguide.* Calculate the ray eigenvalues and eigenrays for the simplest type of lens waveguide, namely repeated identical convergent lenses of focal length $f$ spaced a distance $L$ apart, using the midpoint between lenses as the reference plane. Use the notation $L = 4f\,(1-\Delta)$, and discuss the mathematical behavior and the physical significance of the eigenvalues and eigenrays as the lens spacing is increased toward the value $L \to 4f$ or $\Delta \to 0$. What would be the optical resonator analog to this limit?

Try repeating this problem working from reference planes located at the midplanes of the lenses (i.e., half the lens focusing power is placed on each side of the reference plane); and compare the eigenvalues at this reference plane to the eigenvalues at the previous reference plane.

3. *Computer plotting of periodic ray positions.* Write a simple computer program to compute and plot (on some suitable plotter or printer) the $x, y$ positions on one end mirror on successive bounces for a ray bouncing through repeated round trips inside a resonator of length $L$ with two identical end mirrors having radii of curvature $R$. Allow for arbitrary initial ray injection conditions and also for astigmatic mirrors, i.e., mirrors having different curvatures $R_x$ and $R_y$ in the $x$ and $y$ transverse directions.

Experiment with different spacings, curvatures, and injection conditions to find the kinds of trajectories the spot will follow around the transverse plane on one end mirror, noting particularly how the spot moves around the mirror from bounce to bounce. (You might also plot side or top views of how the rays bounce in the resonator, or examine the spot patterns at planes inside the resonator other than the end mirror.)

4. *Periodic systems with integer numbers of spots.* Suppose you have set up either the computer simulation outlined in the previous problem, or a working optical delay line model using a He-Ne laser and two identical mirrors with variable spacing. Then you can discover that as you change the spacing between mirrors (with fixed mirror radii $R$), there are certain spacings $L$ for which the beam produces an exactly integral number of spots on each mirror before returning back to the same point where it is injected. (a) Find an expression for the mirror spacings $L_n$ at which there are exactly $n$ spots produced on each mirror, in terms of the radius of curvature $R$ of the two identical mirrors. (b) If the input beam is injected properly the spots on the end mirrors walk around a circular orbit. At any transverse plane in between the mirrors the spots then lie on a circle also, but of smaller diameter than on the end mirrors (the rays lie on a hyperboloid of revolution). Find the ratio between the diameters of the spot circles at the center of the resonator and on the end mirrors.

5. *Alignment procedure for the periodic delay line demonstration.* There is a simple sequence of steps one can follow, using the injected laser beam, to get the optical delay line demonstration initially aligned, with the two mirrors properly aligned to each other, and with the injected beam properly aligned to the resonator. Can you describe how this should be done?

6. *Eigenray solutions for a near-spherical optical resonator.* Find the ray matrix eigenvalues and eigenvectors for a near-spherical resonator (i.e., $R_1 = R_2 \approx 2L$), using the midplane of the resonator as the reference plane. Discuss the physical significance of the results in the limiting situation of an exactly spherical resonator.

7. *Perturbation stability of periodic focusing eigenrays.* Suppose that a ray starts out in a periodic focusing system as primarily one of the eigenrays, say, the $r_a$ eigenray, but with a small perturbation or a small amount of the other eigenray $r_b$ mixed in, so that $r_1 = \alpha_1 r_a + \beta_1 r_b$, with $\beta_1 \ll \alpha_1$. Show that on each successive round trip the relative amount of the $r_b$ component in the ray mixture will grow as $\lambda_b^2$. In other words, show that any small perturbation about either one of the eigensolutions will grow (or decay) with a "perturbation eigenvalue" that is equal to the ray eigenvalue of the other eigensolution squared.

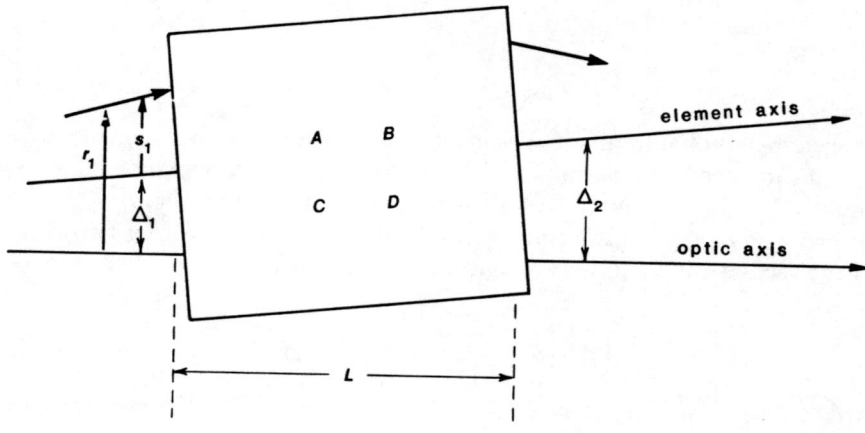

**FIGURE 15.19**
Notation for analyzing a misaligned paraxial optical element.

8. *Ray intersections inside an optical resonator.* In a multiple-pass optical delay line as described in this section, optical rays on different bounces will intersect each other (at least in one transverse dimension) at certain locations inside the cell. Analyze the locations of these intersections, and find the total number of such intersections within a cell as a function of the multiple-pass cell design. Note that beam intersections within such a cell can be significant where nonlinear optical interactions are important, for example, in the multiple-pass Raman gain cells described by B. Perry, et al., "Controllable pulse compression in a multiple-pass-cell Raman laser." *Optics Lett.* **5**, 288–290 (July 1980).

## 15.4 RAY OPTICS WITH MISALIGNED ELEMENTS

The ray matrix formalism we have used thus far assumes that all the paraxial elements are properly aligned and centered with respect to the optical reference axis. What effects will misalignment or transverse misplacement of individual optical elements have on the overall ray matrix performance?

### Analysis of Misaligned Elements

To answer this question let us first consider the effects of misalignment on a single optical element, or perhaps a collection of elements forming a single internally aligned $ABCD$ system. In order to analyze this situation, we must from here on distinguish between the real physical axis (the "true optical axis") of any individual paraxial element or $ABCD$ system, which we will call its *element axis*, and the reference optical axis we use for analyzing the rays in this optical system, which may be arbitrarily chosen, and which we will call the *reference optical axis* or just the *optical axis*, as in Figure 15.19.

Suppose then that the element axis of some arbitrary $ABCD$ system, with overall length $L$, is displaced from the reference optical axis by displacements $\Delta_1$ and $\Delta_2$ at the input and output ends, as in Figure 15.19. The element axis is thus also misaligned in slope with respect to the reference axis by the (small)

## CHAPTER 15: RAY OPTICS AND RAY MATRICES

angle

$$\Delta' \equiv \frac{\Delta_2 - \Delta_1}{L}. \qquad (52)$$

The misalignment of an individual element or collection of $ABCD$ elements with respect to the reference axis can thus be characterized by any two of the three parameters $\Delta_1, \Delta_2, \Delta'$. (Note that $\Delta'$ is a real, not a reduced slope.)

We can also express this misalignment of the paraxial system by two "misalignment vectors" at its input and output ends, as given by

$$\boldsymbol{\Delta}_1 \equiv \begin{bmatrix} \Delta_1 \\ \Delta'_1 \end{bmatrix} \quad \text{and} \quad \boldsymbol{\Delta}_2 \equiv \begin{bmatrix} \Delta_2 \\ \Delta'_2 \end{bmatrix}. \qquad (53)$$

where $\Delta'_1 \equiv n_1 \Delta'$ and $\Delta'_2 \equiv n_2 \Delta'$ are the *reduced* values of the element axis slope at each end. The two misalignment vectors will then be connected by

$$\boldsymbol{\Delta}_2 \equiv \begin{bmatrix} \Delta_2 \\ \Delta'_2 \end{bmatrix} = \begin{bmatrix} 1 & L/n_1 \\ 0 & n_2/n_1 \end{bmatrix} \begin{bmatrix} \Delta_1 \\ \Delta'_1 \end{bmatrix} \equiv \boldsymbol{M}_\Delta \times \boldsymbol{\Delta}_1. \qquad (54)$$

where $\boldsymbol{M}_\Delta$ is shorthand for the $2 \times 2$ matrix in this equation.

The coordinates of any general ray vector *as measured with respect to the arbitrary reference optical axis* we will then continue to denote by $r, r'$ as before, whereas the same ray vector *measured with respect to the element axis* we will denote by $s, s'$. These quantities are then related at the input plane by

$$r_1 = s_1 + \Delta_1 \quad \text{and} \quad r'_1 = s'_1 + \Delta'_1, \qquad (55)$$

and similarly for $r_2$ and $r'_2$. Hence, in vector notation,

$$\boldsymbol{r}_2 = \boldsymbol{s}_2 + \boldsymbol{\Delta}_2 \quad \text{and} \quad \boldsymbol{r}_1 = \boldsymbol{s}_1 + \boldsymbol{\Delta}_1. \qquad (56)$$

(We assume small angles, so that we can simply add the slopes.)

Now, the ray vectors measured with respect to the element axis will transform through the $ABCD$ element in the usual fashion, namely

$$\boldsymbol{s}_2 \equiv \begin{bmatrix} s_2 \\ s'_2 \end{bmatrix} = \begin{bmatrix} A & B \\ C & D \end{bmatrix} \begin{bmatrix} s_1 \\ s'_1 \end{bmatrix} \equiv \boldsymbol{M} \times \boldsymbol{s}_1, \qquad (57)$$

where $\boldsymbol{M}$ is the $ABCD$ matrix for the aligned element(s). However, the input and output displacements and slopes measured with respect to the reference optical axis will now be given, in matrix terms, by

$$\boldsymbol{r}_2 = \boldsymbol{s}_2 + \boldsymbol{\Delta}_2 = \boldsymbol{M}\boldsymbol{s}_1 + \boldsymbol{M}_\Delta \boldsymbol{\Delta}_1 = \boldsymbol{M}\boldsymbol{r}_1 + [\boldsymbol{M}_\Delta - \boldsymbol{M}]\boldsymbol{\Delta}_1 \qquad (58)$$

which we will rewrite in general terms as

$$\boldsymbol{r}_2 = \boldsymbol{M}\boldsymbol{r}_1 + \boldsymbol{E}. \qquad (59)$$

The primary effect of misalignment on a paraxial system is to add to the usual ray matrix transformation what we might call an "error vector" $\boldsymbol{E}$ which is given by

$$\boldsymbol{E} \equiv \begin{bmatrix} E \\ F \end{bmatrix} = [\boldsymbol{M}_\Delta - \boldsymbol{M}]\boldsymbol{\Delta}_1 = \begin{bmatrix} 1 - A & L - n_1 B \\ -C & n_2 - n_1 D \end{bmatrix} \begin{bmatrix} \Delta_1 \\ \Delta' \end{bmatrix} \qquad (60)$$

in terms of the usual $ABCD$ matrix elements and the misalignment quantities $\Delta_1$ and $\Delta'$.

### Three-by-Three Matrix Formalism for Misaligned Systems

These results for a general misaligned paraxial system can be put into a convenient 3×3 matrix form by adding a third dummy element of value unity to each of the ray vectors, and then writing a 3×3 "$ABCDEF$" matrix relation in the form

$$\begin{bmatrix} r_2 \\ r'_2 \\ 1 \end{bmatrix} = \begin{bmatrix} A & B & E \\ C & D & F \\ 0 & 0 & 1 \end{bmatrix} \times \begin{bmatrix} r_1 \\ r'_1 \\ 1 \end{bmatrix}, \qquad (61)$$

where the two additional ray matrix quantities $E$ and $F$ are given by the results derived in Equation 15.60, namely,

$$E = (1 - A)\Delta_1 + (L - n_1 B)\Delta' \quad \text{and} \quad F = -C\Delta_1 + (n_2 - n_1 D)\Delta'. \qquad (62)$$

These 3×3 matrices can then be cascaded, perhaps with the aid of a simple computer program, to handle several such misaligned paraxial elements connected in series.

### Cascaded Misaligned Elements

Suppose several successive optical elements or groups of elements are arranged in cascade, with each element or group of elements having a different degree of (small) misalignment, and hence different $E_i$ and $F_i$ elements, as well as the usual $A_i$, $B_i$, $C_i$, $D_i$ elements. (These individual misalignments are all measured relative to a common reference optical axis passing, in a straight line, through the whole collection.) We can then cascade these 3×3 ray vectors and ray matrices (in reverse order, as usual) to propagate rays through any sequence of cascaded, and individually misaligned, paraxial systems, each with its own $ABCD$ elements and its own distinct $EF$ misalignment elements.

Rather than multiplying and manipulating 3×3 matrices, however, we can analyze the same situation in a more convenient fashion by rewriting Equation 15.61 on the partitioned matrix form

$$\left[ \begin{array}{c} r_2 \\ \hline 1 \end{array} \right] = \left[ \begin{array}{c|c} M & E \\ \hline O & 1 \end{array} \right] \left[ \begin{array}{c} r_1 \\ \hline 1 \end{array} \right], \qquad (63)$$

where $M$ is the usual 2×2 $ABCD$ matrix; $r_1$, $r_2$ and $E$ are 2×1 column matrices; $O$ is a 1×2 row matrix with both elements 0; and 1 is a single "1×1" element. Partitioned matrices of this sort can then be multiplied out analytically by applying the usual rules of matrix multiplication treating each individual submatrix within the partitioned matrix as a fixed element.

Suppose we wish to cascade just two individually misaligned $ABCD$ systems in sequence. The overall 3×3 matrix for the cascaded system can then be calculated from

$$\left[ \begin{array}{c|c} M_{\text{tot}} & E_{\text{tot}} \\ \hline O & 1 \end{array} \right] = \left[ \begin{array}{c|c} M_2 & E_2 \\ \hline O & 1 \end{array} \right] \times \left[ \begin{array}{c|c} M_1 & E_1 \\ \hline O & 1 \end{array} \right]$$

$$= \left[ \begin{array}{c|c} M_2 M_1 & M_2 E_1 + E_2 \\ \hline O & 1 \end{array} \right]. \qquad (64)$$

As a check the reader may want to multiply out the full 3×3 matrices in non-partitioned form to verify that the final result is indeed

$$\left[\begin{array}{c|c} M_{\text{tot}} & E_{\text{tot}} \\ \hline O & 1 \end{array}\right] = \begin{bmatrix} A_2A_1 + B_2C_1 & A_2B_1 + B_2D_1 & A_2E_1 + B_2F_1 + E_2 \\ C_2A_1 + D_2C_1 & C_2B_1 + D_2D_1 & C_2E_1 + D_2F_1 + F_2 \\ 0 & 0 & 1 \end{bmatrix}, \quad (65)$$

or the same as given by the partitioned form.

We see first of all that the 2×2 or $ABCD$ part of the overall cascaded, misaligned system has exactly the same form as the product of the two matrices would have without misalignment, since this part of the product does not depend at all on the misalignment values $E_1, F_1$ or $E_2, F_2$ of the individual elements. To phrase this more generally, *the basic ray matrix properties and paraxial focusing properties of a cascaded system are entirely unchanged by small misalignments of individual elements within the system.*

### Overall Misaligned Systems

These same conclusions obviously remain true even if we cascade an arbitrary number of arbitrarily misaligned paraxial elements. Suppose we propagate an initial ray $r_0$ through $N$ such elements or subsystems, each with an individual misalignment described by an error vector $E_k \equiv [E_k, F_k]$ as referenced to a single straight-line optical axis through the overall system.

The overall transformation through the cascaded system can then be written as

$$r_N = M_{\text{tot}} r_0 + E_{\text{tot}}, \qquad (66)$$

where the overall $ABCD$ matrix is given as usual by the matrix product $M_{\text{tot}} = M_N \cdots M_2 M_1$, and where the cumulative "error vector" through the entire system is given in terms of the error vectors of the individual elements by

$$E_{\text{tot}} = [M_N \cdots M_2] E_1 + [M_N \cdots M_3] E_2 + \cdots + M_N E_{N-1} + E_N. \qquad (67)$$

The overall misalignment elements $E_{\text{tot}}$ and and $F_{\text{tot}}$ for the cascaded system obviously involve the misalignments $E_k, F_k$ of each individual element in the system, as "propagated" through the $ABCD$ matrices of all the subsequent elements in the system. In a cascaded $ABCD$ system with misaligned individual elements, the overall system will thus appear to have a total misalignment $E_{\text{tot}}, F_{\text{tot}}$ that depends in a complicated way both on the misalignment of individual elements and on the transmission of each of these individual misalignments through the individual $ABCD$ matrices of all later elements.

### System Alignment, and the Overall Element Axis

Suppose we do the kind of calculation just outlined, and find the overall misalignment parameters $E_{\text{tot}}$ and $F_{\text{tot}}$ for some particular cascaded system, using some particular arbitrarily chosen reference optical axis that passes in a straight line through the entire system. The preceding results then imply that the overall system acts as if it is a single properly aligned overall system, but one whose overall element axis has end-plane displacements $\Delta_0$ and $\Delta_N$ at its input and output ends like those in Figure 15.20, measured with respect to the reference optical axis that we used in doing all the calculations.

Any system with misaligned individual elements can thus obviously be converted into an effectively aligned overall system, having $E_{\text{tot}} = F_{\text{tot}} = 0$, either by a physical translation and rotation of the overall system to bring its overall element axis into coincidence with the reference optical axis, or equivalently by a redefinition of the reference optical axis to bring it into coincidence with the system's element axis. That is, any overall values of $E = E_{\text{tot}}$ and $F = F_{\text{tot}}$ for the overall system can be canceled out by physically translating the entire system as a unit downward an amount $\Delta_0$ given by

$$\Delta_0 = \frac{(1-D)E - (L-B)F}{(1-A)(1-D) + (L-B)C}, \tag{68}$$

and then physically rotating it toward the system axis, with center of rotation at the input plane, by the angle

$$\Delta' = \frac{CE + (1-A)F}{(1-A)(1-D) + (L-B)C}, \tag{69}$$

where all the quantities $A$, $B$, $C$, $D$, $E$, $F$ and $L$ in these expression are the overall values for the cascaded system. Once this is done the overall system will look perfectly well aligned, despite the individual misalignments of its various internal elements.

### Misaligned Resonators or Periodic Systems

A slightly different viewpoint and approach can also be useful in discussing the ray matrix properties of an optical resonator, or its equivalent iterated periodic focusing system, in the situation where individual optical elements inside the resonator may be misaligned.

Suppose we unfold an optical resonator having one or more misaligned internal elements into an equivalent periodic system. Each individual period of the resulting lensguide, corresponding to one round trip in the resonator, will then have an overall element axis, with respect to which that individual period or round trip will look like an ideal aligned system. *This element axis, however, in general will not come back on itself after one round trip—that is, the element axis in each individual period may be tilted with respect to the reference optical axis running through the repeated sections of the lensguide, so that the element axes in successive periods do not connect to each other.*

Is there then some better or alternative way to define an effective axis in a misaligned resonator or periodic system? To answer this question we might recall that the distinguishing characteristic of the axis in an *aligned* paraxial system is that a ray vector which starts out exactly aligned along the axis always remains exactly on the axis. We might ask therefore if, starting from any given reference plane within a misaligned resonator or periodic system, there will be some unique "axis ray," let us label it by $r_0$, whose displacement and slope (measured with respect to the reference optical axis) will exactly repeat themselves after one period or one round trip through this $ABCDEF$ system?

Such a ray, which self-reproduces after one round trip, is given by the conditions that

$$\boldsymbol{M r_0 + E = r_0} \quad \text{or} \quad r_0 = (\boldsymbol{I - M})^{-1}\boldsymbol{E}, \tag{70}$$

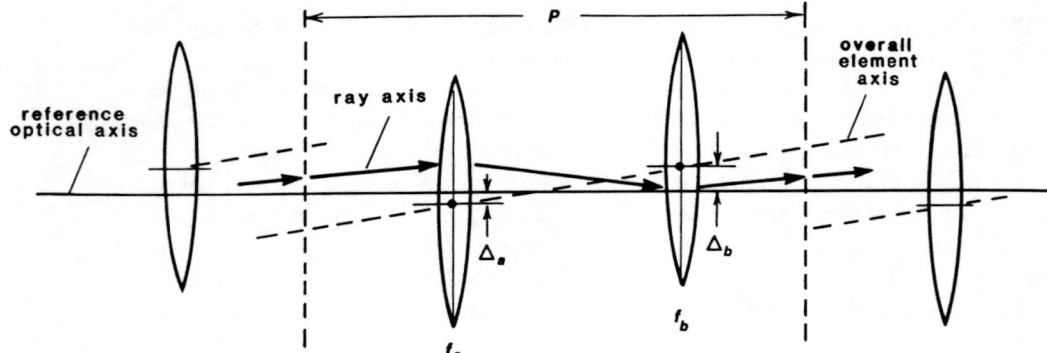

**FIGURE 15.20**
A misaligned periodic focusing system, in which each period of the system has its element axis misaligned with respect to the reference optical axis of the overall system.

where $I$ is the identity matrix, and the $-1$ superscript means the inverse of the matrix within the parentheses. If we carry out the algebra, we can find that the displacement and slope of this "axis ray" are given (at this one particular reference plane) by

$$r_0 \equiv \frac{(1-D)E + BF}{2 - A - D} \quad \text{and} \quad r'_0 \equiv \frac{CE + (1-A)F}{2 - A - D}. \tag{71}$$

It is then easy to show that the transformation of any other input ray $r_1$ through the misaligned system is given by

$$(r_2 - r_0) = M \times (r_1 - r_0), \tag{72}$$

where the $ABCD$ elements are the round-trip elements starting from and coming back to some particular reference plane inside the resonator. This particular ray $r_0$ then represents a kind of misaligned "natural optical axis" for the misaligned periodic system, as observed at this particular reference plane.

The resonator or periodic system becomes in effect a well-aligned $ABCD$ system if the input and output ray coordinates are measured relative to the axis ray $r_0 \equiv [r_0, r'_0]$ at the particular reference plane $z_0$ used to define the $ABCDEF$ matrix elements. If a ray starts around the resonator with input displacement and slope given by $r_0$, it will return to this same position on every successive round trip. Any other ray, however, starting off with different initial values, will oscillate about this ray (or possibly diverge from it) in exactly the stable or unstable periodic fashion described earlier for aligned periodic systems.

### Differences Between the Axis Ray and the Overall Element Axis

We note again that the axis ray for a misaligned resonator or periodic system is not the same in general as the "overall element axis" we discussed a few paragraphs back. The overall element axis through a given collection of misaligned elements is a *straight line* through these elements, as in Figure 15.20, such that if the ray displacements and slopes are measured relative to this axis, the overall system will act like an aligned 2×2 matrix from input to output.

The axis ray through the same collection of elements, by contrast, will consist in general of a series of bent or even curving segments, with respect to which the system again acts like an aligned 2×2 matrix. The axis ray has the property that it comes out parallel to itself after one pass through the system. However, although the axis rays at the input and output planes have the same displacement and slope, and thus are parallel to each other, they do not in general define a single straight line through the system, whereas does the overall element axis does.

In fact, in an optical resonator or periodic system with several individually misaligned elements the axis ray, which acts as the effective optical axis for the periodic system, will trace out a zig-zag course within the $ABCD$ system, shifting or bending from plane to plane within the period or round trip. Moreover, the axis rays going in the forward and reverse directions through a standing-wave resonator may not lie on top of each other (though they must intersect in position, but not necessarily in slope, at the end mirrors); and also the axis ray in a misaligned system may or may not coincide with the element axis of any individual element at the point where it intersects that element. Such an axis ray nonetheless always exists.

### Summary

The overall conclusion of this section is clearly that (small) displacements or misalignments of individual paraxial elements are usually not a serious problem. They can be handled with the extended matrix technique of this section if desired, but in general they do not change the basic focusing or stability problems of a paraxial $ABCD$ system. If we are designing an extended beam transmission system and perhaps wish to know the sensitivity of the overall system alignment to misalignments of individual elements, then the techniques of this section can be very useful. If the problem is merely to design and evaluate the stability and spot size properties of a closed resonator, then misalignment effects can be ignored.

### REFERENCES

The 3×3 matrix technique we have introduced for handling misaligned systems is also briefly described in Appendix B of the book by A. Garrard and J.M. Burch, *Introduction to Matrix Methods in Optics* (Wiley, 1975).

### Problems for 15.4

1. *Error vector for a tilted flat mirror.* What is the error vector $E$ for a flat mirror which is misaligned (i.e., tilted) by a small angle $\theta$ relative to its aligned position (assuming its aligned position is perpendicular to the reference optical axis of the system).

2. *Misaligned optical resonator.* Consider an optical resonator consisting of an aligned flat mirror at the left-hand end; a collection of aligned optical elements having an overal $ABCD$ matrix going in the $+z$ direction from the left-hand mirror to the right-hand mirror; and another planar mirror at the right-hand end which is misaligned by a small tilt angle $\theta$. Find formula for the overall element

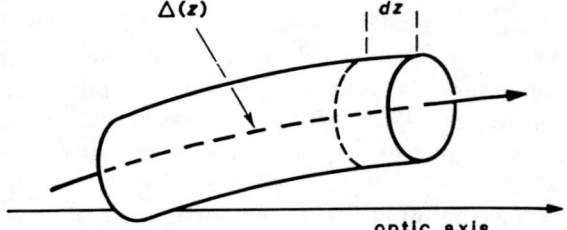

FIGURE 15.21
A curved optical fiber or duct.

axis and axis ray for one round trip in this resonator, or in its equivalent periodic lensguide.

Calculate and sketch the locations of these rays for the specific situation of a resonator of length $L$ with a thin lens of focal length $f$ located at the center of the cavity, for both a stable resonator ($L/4 < f < \infty$) and a positive-branch unstable resonator ($f < 0$).

3. *More misaligned resonators.* Repeat the previous problem assuming the thin lens of focal length $f$ is located just in front of the left-hand mirror, and then just in front of the right-hand mirror. (The stability conditions are different in each of these situations.)

4. *Finding the axis ray in another optical resonator with misaligned elements.* A laser resonator of total length $L$ consists of two intracavity lenses of focal length $f = 2L$ equally spaced between two flat end mirrors. One lens is displaced above the optic axis of the resonator by a small distance $\Delta = \epsilon$; the other is displaced downward by $\Delta = -2\epsilon$. Trace the "axis ray" through this resonator.

## 15.5 RAY MATRICES IN CURVED DUCTS

As still another example of an interesting ray matrix system, consider a quadratic duct as defined previously, in which the transverse index variation $n = n(r)$ is constant with distance, but assume now that this duct is twisted or bent, so that the axis of the duct at any plane $z$ is displaced from a straight reference axis by a small amount $\Delta(z)$ as in Figure 15.21. (This could represent a curved or twisted optical fiber.) What is the $ABCD$ matrix for this curved duct?

### Differential Matrix Analysis

Following the combined approach of the preceding two sections, we can suppose that $M(z)$ represents the 3×3 $ABCDEF$ matrix for such a system from an input plane $z_0$ up to plane $z$, with elements $A(z)$ through $F(z)$. Then from the cascading properties of ray matrices we can write that

$$M(z + dz) = M(dz) \times M(z), \qquad (73)$$

where $M(dz)$ is the ray matrix for the short distance $dz$ from $z$ to $z + dz$.

## 15.5 RAY MATRICES IN CURVED DUCTS

Now, for a thin segment of transversely displaced duct, as in Figure 15.21, this matrix has the form, in the limit as $dz \to 0$, of

$$M(dz) = \begin{bmatrix} 1 & n_0^{-1} dz & 0 \\ -n_0\gamma^2 dz & 1 & n_0\gamma^2\Delta(z)\,dz \\ 0 & 0 & 1 \end{bmatrix}. \tag{74}$$

Multiplying the matrices $M(dz)$ and $M(z)$ together and comparing them term-by-term with the matrix $M(z+dz)$, then gives the differential relations

$$\frac{dA(z)}{dz} = n_0^{-1} C(z), \qquad \frac{dB(z)}{dz} = n_0^{-1} D(z)$$

$$\frac{dC(z)}{dz} = -n_0\gamma^2 A(z), \qquad \frac{dD(z)}{dz} = -n_0\gamma^2 B(z), \tag{75}$$

plus the two additional equations

$$\frac{dE(z)}{dz} = n_0^{-1} F(z) \quad \text{and} \quad \frac{dF(z)}{dz} = -n_0\gamma^2[E(z) - \Delta(z)]. \tag{76}$$

Solving the first four equations, starting from $z_0$, gives the overall $ABCD$ matrix as a function of distance in the form

$$A(z) = D(z) = \cos\gamma(z-z_0), \qquad n_0\gamma B(z) = -(n_0\gamma)^{-1} C(z) = \sin\gamma(z-z_0) \tag{77}$$

which agrees with what we already know from Equation 15.15. The overall $ABCD$ matrix is again unchanged by curvature or misalignment of the duct.

### Effects of Duct Misalignment

The final two equations, which are independent of $ABCD$, however, yield the formal solutions

$$E(z) = \gamma \int_{z_0}^{z} \Delta(z') \sin\gamma(z-z')\,dz'$$

$$F(z) = n_0\gamma^2 \int_{z_0}^{z} \Delta(z') \cos\gamma(z-z')\,dz'. \tag{78}$$

There is one particular situation where these solutions can be quite important. Suppose the axis displacement $\Delta(z)$ in the duct has a natural periodic component with a spatial variation $\Delta(z) = \cos\gamma_1 z$ or $\sin\gamma_1 z$, and suppose that $\gamma_1$ equals or closely matches the natural ray oscillations at $\cos\gamma z$ or $\sin\gamma z$. The integrands in Equation 15.78 will then contain $\cos^2\gamma z$ or $\sin^2\gamma z$ factors which will integrate cumulatively with distance $z$. This then implies that the displacement parameters $E(z)$ and $F(z)$, or in essence the cumulative amount of misalignment in the duct, will grow more or less linearly with distance.

Problems will thus result if the physical curvature or waviness of a duct has a periodic variation that resonates with the natural oscillation period for optical rays about the axis of the duct. The system axis of the duct then seems to diverge by an increasing amount from the physical axis (or element axis) of the duct as we go further down the duct. In more physical terms this means that the periodic oscillations of rays in the duct will appear to grow linearly in amplitude

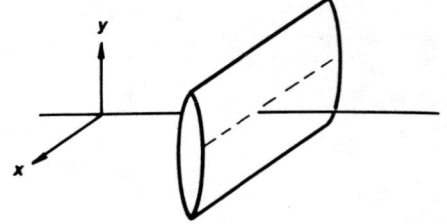

FIGURE 15.22
An astigmatic optical element (cylindrical lens).

with distance, until these rays encounter the edges of the duct, or some other nonlinearity occurs to limit their growth.

If the duct has instead a randomly wavy axis, i.e., with random variations in $\Delta(z)$ along the length of the guide, then the oscillations in off-axis rays will grow as the square root of distance along the guide rather than linearly with the distance $z$. The growth rate for this process will be proportional to the amplitude of the spatial frequency components of $\Delta(z)$ in the immediate vicinity of the natural wave number $\gamma$.

## REFERENCES

The basic differential analysis presented in this section comes from A. Hardy, "Beam propagation through parabolic-index waveguides with distorted optical axis," *Appl. Phys.* **18**, 223–226 (1979).

Though the analysis in this section refers to a continuous parabolic duct with a curved or wavy axis, very similar results will apply to an iterated periodic lensguide having perturbations $\Delta(z)$ in the location of the optic elements from section to section along the guide. Understanding of the growth rate for random perturbations in such periodic systems was of considerable importance in pre-fiber-optics times, when periodic optical lensguides were under serious consideration for long-distance optical communications.

An early analysis of this for the periodic lensguide situation was given by J. Hirano and Y. Fukatsu, "Stability of a light beam in a beam waveguide," *Proc. IEEE* **52**, 1284–1292 (November 1964). A later study was D. W. Berreman, "Growth of oscillations of a ray about the irregularly wavy axis of a lens light guide," *Bell Sys. Tech. J.* **41**, 2117–2132 (November 1965). Another interesting analysis is D. Marcuse, "Physical limitations on ray oscillation suppressors," *Bell Sys. Tech. J.* **45**, 743–751 (May-June 1966).

## 15.6 NONORTHOGONAL RAY MATRICES

We have noted earlier that in optical systems with rotational symmetry the same ray matrices apply equally but separately to the $x, x'$ and the $y, y'$ ray coordinates. In the slightly more complicated situation of optical elements having simple astigmatism, the ray matrices will be different along the $x$ and $y$ coordinates. A thin cylindrical lens having its cylinder axis aligned along the $x$ axis (Figure 15.22), for example, will act as a thin lens with the appropriate $ABCD$ matrix so far as the $y$ transverse coordinate is concerned, but will have no focusing or bending effect on the $x$ displacement of the ray.

## 15.6 NONORTHOGONAL RAY MATRICES

Suppose that an overall optical system contains several such astigmatic elements, but these elements all have their principal axes aligned along the same $x$ and $y$ axes. We can then still analyze the ray behavior in each transverse coordinate separately and independently, using separate $ABCD$ matrices for the $x$ and the $y$ directions. Such an astigmatic system, for example, might even be stable in one coordinate and unstable in the other. Systems having only simple astigmatism, and thus describable by separate and independent ray matrices in two principle planes that are 90° apart, are commonly referred to as *orthogonal systems*.

Systems not having this property are said to be *nonorthogonal*. Nonorthogonal systems in general exhibit one or another kind of "twist" or image rotation, which is more complicated than simple astigmatism, and which in general does not permit the ray matrices to be separated into two separate ray matrices along two orthogonal axes. The ray analysis of nonorthogonal paraxial optical systems has not yet been extensively developed, and we can therefore summarize in this section only a few results concerning such systems.

### General Analysis of Nonorthogonal Ray Optical Systems

It would be useful, for example, to establish the most general forms that the ray matrices of both orthognal and nonorthogonal optical systems can assume if we include such operations as arbitrary astigmatism, image rotation, and image inversion. These questions will not be fully answered in this section, although we will derive some of the general properties of nonorthogonal systems by building up from combinations of elementary ray operations and matrices. We are particularly interested in establishing the conditions under which an optical system will remain orthogonal, so that the system can be described by separate and independent ray matrices along two orthogonal transverse directions.

There are first of all two basically different ways in which we might write the 4×4 matrices needed to describe the ray coordinates in both the $x$ and $y$ transverse coordinates. One way is to organize the ray coordinates in the form of displacements and then slopes, e.g..

$$\begin{bmatrix} x_2 \\ y_2 \\ \hline x'_2 \\ y'_2 \end{bmatrix} = \begin{bmatrix} A_{xx} & A_{xy} & B_{xx} & B_{xy} \\ A_{yx} & A_{yy} & B_{yx} & B_{yy} \\ \hline C_{xx} & C_{xy} & D_{xx} & D_{xy} \\ C_{yx} & C_{yy} & D_{yx} & D_{yy} \end{bmatrix} \begin{bmatrix} x_1 \\ y_1 \\ \hline x'_1 \\ y'_1 \end{bmatrix}, \qquad (79)$$

or in shorthand notation

$$\begin{bmatrix} r_2 \\ \hline r'_2 \end{bmatrix} = \begin{bmatrix} A & B \\ \hline C & D \end{bmatrix} \times \begin{bmatrix} r_1 \\ \hline r'_1 \end{bmatrix}, \qquad (80)$$

where we use the notation in these paragraphs that $r$ and $r'$ are column vectors with elements $r \equiv [x, y]$ and $r' \equiv [n_x\, dx/dz,\, n_y\, dy/dz]$, and $A$, $B$, $C$ and $D$ are all 2×2 matrices.

For an astigmatic but orthogonal system with its principal axes oriented along the $x$ and $y$ directions, all four of these matrices will then be diagonal, i.e., the $xy$ and $yx$ elements that couple between the $x$ and $y$ axes will all be zero. Any rotation of the coordinate system will make these off-diagonal elements nonzero, although for orthogonal systems there will be constraints among the diagonal and off-diagonal elements. A general nonorthogonal system will have off-diagonal elements between the $x$ and $y$ directions that cannot be removed by any coordinate rotation.

Expressing the 4×4 problem in the form of Equation 15.79 has a number of advantages, as discussed for example by Nazarathy (see References). If a superscript $T$ indicates the matrix transpose, then it can be shown (see References) that even in the most general nonorthogonal system these 2×2 matrices must satisfy the constraints

$$AB^T = BA^T, \qquad B^T D = D^T B$$
$$DC^T = CD^T, \qquad C^T A = A^T C \tag{81}$$

as well as

$$AD^T - BC^T = A^T D - B^T C = I, \tag{82}$$

where $I$ is the identity matrix. The last two relations are obviously the nonorthogonal generalizations of the $AD - BC = 1$ relation for orthogonal 2×2 ray matrices. There are potentially sixteen elements in the general 4×4 ray matrix, but as a result of these six relations there are only ten independent elements (as also pointed out by Arnaud).

We can also show, following Nazarathy, that with this form for the 4×4 matrices the general form of the Huygens-Fresnel integral that we will introduce in a later chapter can be put into the very beautiful form

$$\tilde{u}_2(\mathbf{r}_2) = \frac{j}{|B|^{1/2}\lambda} \int_{-\infty}^{\infty} \tilde{K}(\mathbf{r}_2 . \mathbf{r}_1) \, \tilde{u}_1(\mathbf{r}_1) \, d\mathbf{r}_1, \tag{83}$$

where $|B|^{1/2}$ is the square root of the determinant of the $B$ matrix, and $\tilde{K}$ is the exponential part of the Huygens' kernel given by

$$\tilde{K}(\mathbf{r}_2, \mathbf{r}_1) \equiv \exp\left[-j\frac{\pi}{\lambda} \left(\mathbf{r}_1 \cdot B^{-1} A \cdot \mathbf{r}_1 - 2\mathbf{r}_1 \cdot B^{-1} \cdot \mathbf{r}_2 + \mathbf{r}_2 \cdot DB^{-1} \cdot \mathbf{r}_2\right)\right]. \tag{84}$$

with $B^{-1}$ being the inverse of the $B$ matrix. This form of Huygens' integral is then equally valid for orthogonal or nonorthogonal systems.

### Alternative Matrix Notation

An alternative notation to Equation 15.79 for ray systems in two transverse dimensions is to organize the coordinates and matrix elements in the form

$$\begin{bmatrix} x_2 \\ x'_2 \\ y_2 \\ y'_2 \end{bmatrix} = \begin{bmatrix} A_{xx} & B_{xx} & A_{xy} & B_{xy} \\ C_{xx} & D_{xx} & C_{xy} & D_{xy} \\ A_{yx} & B_{yx} & A_{yy} & B_{yy} \\ C_{yx} & D_{yx} & C_{yy} & D_{yy} \end{bmatrix} \begin{bmatrix} x_1 \\ x'_1 \\ y_1 \\ y'_1 \end{bmatrix}. \tag{85}$$

## 15.6 NONORTHOGONAL RAY MATRICES

As a shorthand notation we will write this equation in the partitioned matrix form

$$\begin{bmatrix} \boldsymbol{x}_2 \\ \hline \boldsymbol{y}_2 \end{bmatrix} = \begin{bmatrix} \boldsymbol{M}_{xx} & \boldsymbol{M}_{xy} \\ \hline \boldsymbol{M}_{yx} & \boldsymbol{M}_{yy} \end{bmatrix} \begin{bmatrix} \boldsymbol{x}_1 \\ \hline \boldsymbol{y}_1 \end{bmatrix}, \qquad (86)$$

where $\boldsymbol{x}$ and $\boldsymbol{y}$ are the ray vectors in the $x$ and $y$ coordinates, respectively; $\boldsymbol{M}_{xx}$ and $\boldsymbol{M}_{yy}$ are the ordinary $2\times 2$ ABCD matrices applying to the $x$ and $y$ directions; and $\boldsymbol{M}_{xy}$ and $\boldsymbol{M}_{yx}$ are the cross-matrices between the $x$ and $y$ directions. We will pursue some applications and consequences of this alternative matrix arrangement in the remainder of this section.

### Rotated Astigmatic Optical Systems

Most of the difficulties in nonorthogonal systems arise from questions of rotation, where the term rotation can mean either *coordinate system rotation* or *actual image rotation* of a ray bundle by arbitrary angles about the direction of propagation. Let us therefore examine in some detail the analytical effects that arise from such rotations.

For example, we might begin by organzing the $4\times 4$ ray matrix for an astigmatic but still orthogonal system, aligned along its principle axes, in the form

$$\begin{bmatrix} x_2 \\ x'_2 \\ y_2 \\ y'_2 \end{bmatrix} = \begin{bmatrix} A_x & B_x & & \\ C_x & D_x & & \\ & & A_y & B_y \\ & & C_y & D_y \end{bmatrix} \begin{bmatrix} x_1 \\ x'_1 \\ y_1 \\ y'_1 \end{bmatrix}, \qquad (87)$$

where we will follow the convention that any elements not written are zero. The $x$ and $y$ quantities in this situation are entirely uncoupled.

At any position $z$ we can always make a coordinate rotation from our original $x_1, y_1$ coordinates to a set of axes $x_2, y_2$ which are rotated about the $z$ axis by an angle $\theta$ (Figure 15.23). This is done analytically by applying the general rotation matrix

$$\begin{bmatrix} x_2 \\ x'_2 \\ y_2 \\ y'_2 \end{bmatrix} = \begin{bmatrix} \cos\theta & & \sin\theta & \\ & \cos\theta & & \sin\theta \\ -\sin\theta & & \cos\theta & \\ & -\sin\theta & & \cos\theta \end{bmatrix} \begin{bmatrix} x_1 \\ x'_1 \\ y_1 \\ y'_1 \end{bmatrix}, \qquad (88)$$

where subscript 1 refers to the ray coordinates measured in the old coordinate system and subscript 2 refers to the same ray measured in the new (rotated) coordinate system. We can then write this in shorthand notation as

$$\begin{bmatrix} \boldsymbol{x}_2 \\ \hline \boldsymbol{y}_2 \end{bmatrix} = \begin{bmatrix} C_\theta & S_\theta \\ \hline -S_\theta & C_\theta \end{bmatrix} \begin{bmatrix} \boldsymbol{x}_1 \\ \hline \boldsymbol{y}_1 \end{bmatrix}, \qquad (89)$$

where $C_\theta$ and $S_\theta$ (with suitable subscripts) represent the cos and sin of the rotation angle, with each of these understood to be multiplied by the identity matrix which is not written out. Rotation in the opposite direction simply reverses the sign of $S_\theta$.

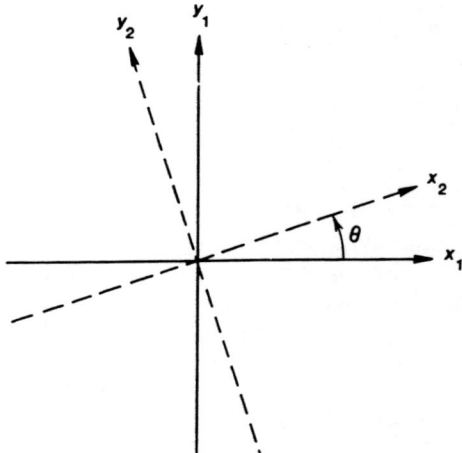

FIGURE 15.23
Coordinate system rotation.

Suppose an orthogonal astigmatic element is physically rotated about the $z$ axis by an arbitrary angle $\theta$, as in Figure 15.23, and that we wish to describe the ray propagation through this element written in the original or unrotated coordinate system. To pass a ray through this rotated element analytically using the original $x_1, y_1$ axes, we must transform from our original axes into the rotated principal axes of the element; propagate through the element using the $ABCD$ matrices along its principal axes; and then rotate back to our original axes by a rotation of amount $-\theta$. If we carry out this procedure, the ray matrix of the rotated astigmatic element written in the original $x, y$ coordinate axes is the cascade product

$$\left[\begin{array}{c|c} C_\theta & -S_\theta \\ \hline S_\theta & C_\theta \end{array}\right] \times \left[\begin{array}{c|c} M_{xx} & \\ \hline & M_{yy} \end{array}\right] \times \left[\begin{array}{c|c} C_\theta & S_\theta \\ \hline -S_\theta & C_\theta \end{array}\right] \tag{90}$$

which can be manipulated into the form

$$\left[\begin{array}{c|c} C_\theta^2 M_{xx} + S_\theta^2 M_{yy} & S_\theta C_\theta (M_{xx} - M_{yy}) \\ \hline S_\theta C_\theta (M_{xx} - M_{yy}) & S_\theta^2 M_{xx} + C_\theta^2 M_{yy} \end{array}\right]. \tag{91}$$

An orthogonal system rotated to an arbitrary angle $\theta$ will thus have a 4×4 matrix of this general form. In particular we can deduce that *in an orthogonal but arbitrarily rotated system, the upper right and lower left 2×2 blocks may not be zero, but they will always be identical,* as illustrated in Equation 15.91.

### Two Rotated Elements in Cascade

Suppose next that two individually orthogonal but astigmatic elements or systems are arranged in cascade, and are rotated to arbitrary angles $\theta_1$ and $\theta_2$ about the $z$ axis (see Figure 15.24), with element #1 passed through first. The overall ray matrix of these cascaded elements is then the matrix product of two rotated matrices of the type given in Equation 15.91, with appropriate subscripts to identify the first and second systems (e.g., $S_{\theta_1} \equiv \sin\theta_1$ for the first element; $M_{xx,1}$ is the $x$-axis ray matrix of the first element in its own principal axes;

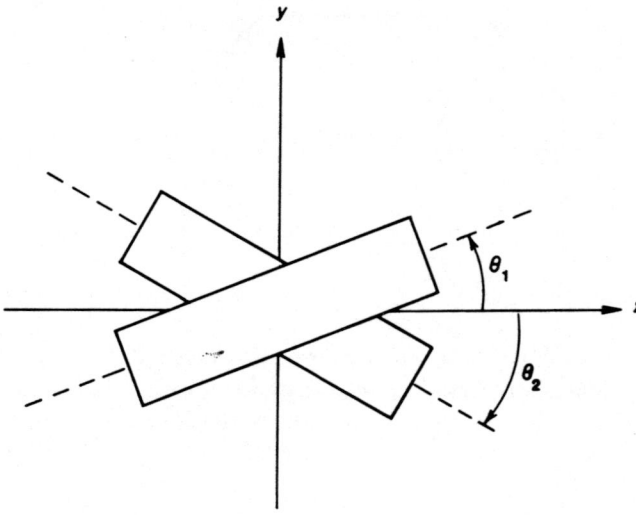

FIGURE 15.24
Rotated astigmatic optical elements.

$M_{yy,2}$ is the $y$-axis ray matrix of the second element in its own principal axes; and so forth).

The overall matrix product that results from carrying out this multiplication is lengthy and not particularly transparent. But suppose this overall product is written in the shorthand form

$$\left[\begin{array}{c|c} M_{xx} & M_{xy} \\ \hline M_{yx} & M_{yy} \end{array}\right] = \left[\begin{array}{c} \text{overall } 4\times 4 \\ \text{matrix product} \end{array}\right]. \qquad (92)$$

In this situation the 2×2 $M_{xx}$ and $M_{yy}$ matrices are no longer necessarily correct $ABCD$ matrices by themselves, but are merely the upper left and lower right blocks of the overall 4×4 matrix, whereas the $M_{xy}$ and $M_{yx}$ are cross matrices between the $x$ and $y$ coordinates.

Now, if this overall cascaded system is to be an orthogonal system, then the upper right and lower left blocks must be identical, i.e., $M_{xy} = M_{yx}$, in the same way as in the rotated orthogonal system of Equation 15.91. All of these blocks are complicated functions of the rotations $\theta_1$, $\theta_2$ and the individual system matrices. It can be shown, however, after some algebra, that the upper right and lower left blocks of the cascade product of Equation 15.91 will differ by the amount

$$M_{xy} - M_{yx} = \sin(\theta_2 - \theta_1)\cos(\theta_2 - \theta_1)(M_{xx,1} - M_{yy,1})(M_{xx,2} - M_{yy,2}). \qquad (93)$$

We can deduce from this that a cascaded system of two rotated astigmatic elements will in general be orthogonal only if (i) $\theta_2 - \theta_1 = 0°$ or $90°$, which means the two elements have relative rotations such that their principal planes coincide; or else if (ii) $M_{xx,1} = M_{yy,1}$, or $M_{xx,2} = M_{yy,2}$, which means that one or the other of the cascaded systems is not astigmatic (e.g., is rotationally symmetric).

To phrase this in the opposite sense, we can conclude that, except for these very special situations, *an optical system having cascaded astigmatic elements rotated at arbitrary angles will in general not be orthogonal.* Such a system will not have any pair of transverse coordinates separated by 90° with respect to which a ray can be analyzed by separate and independent $ABCD$ matrices.

**FIGURE 15.25**
Image inversion in a Dove prism.

### Image Rotation

Paraxial optical systems of the most general form can also exhibit *image rotation* in addition to inversion and astigmatism. Image rotation means that the displacement and slope of a ray on passing through an element are actually rotated in the $x, y$ plane in the manner given analytically by the general 4×4 rotation matrix given in Equation 15.88.

We introduced the coordinate rotation notation given above at first to represent simply a purely mathematical transformation of coordinates. In simple situations we may rotate the $x, y$ coordinate system by an angle $\theta$, perhaps in order to line up the coordinate system with the principal axes of an astigmatic element. We may then rotate the coordinate system back by $-\theta$ to the original axes further along the $z$ axis, after passing through the astigmatic element.

However, there are also optical systems which accomplish genuine *physical rotation* of the ray position even with respect to fixed coordinate axes. This image rotation is also given analytically by the same rotation matrix using $C_\theta$ and $S_\theta$ as given in Equation 15.88, but with the rotation operation now viewed as operating on the rays with respect to fixed coordinate axes. Such image rotation systems often also contain one or more image inversions. A beam passing a partially rotated Dove prism is one simple example of this type. In such a system the rotation matrix only operates once—there is no "reverse rotation" later on.

### Nonplanar Ring Resonators

The concepts of coordinate rotation versus image rotation become particularly indistinguishable for a twisted or nonplanar ring resonator (see Figure 15.26). When rays bounce off a mirror at other than normal incidence, as in any ring resonator, it is most natural to use transverse coordinate axes that lie in the plane and perpendicular to the plane of incidence defined by the ray axes just before and after reflection. This is particularly desirable when reflecting off spherical mirrors at other than normal incidence, since the effective radius of curvature of the mirror becomes $R\cos\theta_0$ for rays in the plane of incidence and $R/\cos\theta_0$ for rays perpendicular to the plane of incidence, where $\theta_0$ is the angle between the incident direction and the normal to the mirror.

Analyzing the ray propagation in going around a twisted or nonplanar ring then requires repeated coordinate rotations just before each mirror, in order to bring the transverse $x, y$ axes into agreement with the plane of incidence and reflection of the optical rays on that particular mirror. For a twisted ring, these rotations at each mirror may or in general may not sum to zero net rotation after a complete round trip.

We can then view this situation either as a set of sequential coordinate transformations which do not bring the final coordinate axes back in alignment with the initial axes after one round trip; or alternatively we may view this as a phys-

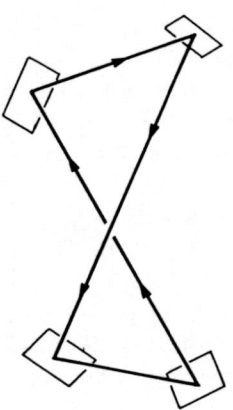 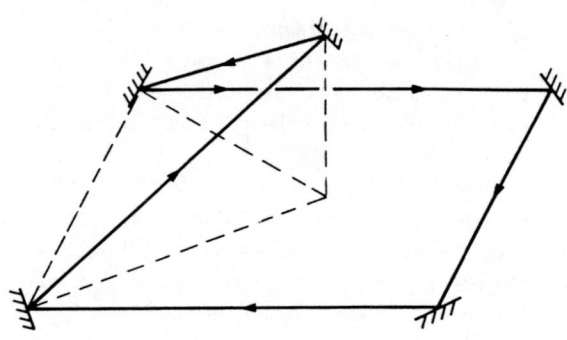

FIGURE 15.26
Nonplanar ring resonators.

ical rotation of the image or of the ray vectors as seen in the original transverse coordinates after one round trip. The result either way is a net nonzero rotation of the ray coordinates in one round trip. An image rotation plus an orthogonal system in cascade will have a net 4×4 matrix in one of the two forms

$$\left[ \begin{array}{c|c} C_\theta M_{xx} & S_\theta M_{xx} \\ \hline -S_\theta M_{yy} & C_\theta M_{yy} \end{array} \right] \quad \text{or} \quad \left[ \begin{array}{c|c} C_\theta M_{xx} & S_\theta M_{yy} \\ \hline -S_\theta M_{xx} & C_\theta M_{yy} \end{array} \right], \qquad (94)$$

depending on whether the rotation or the astigmatic element comes first. Systems with image rotation are clearly not orthogonal.

### Summary

The analysis of general nonorthogonal systems, i.e., those having image rotation, inversion, and/or cascaded and rotated astigmatic elements, thus becomes significantly more complicated than for the simple 2×2 ray matrix. Arnaud and others have shown, for example, that the most general 4×4 ray matrix has just ten independent elements out of the sixteen total elements. A general nonorthogonal system can also be separated into independent $x$ and $y$ coordinates in a particular *nonorthogonal* set of $x$ and $y$ axes, i.e., a set of transverse coordinates that are not at 90° to each other. Systems with image rotation generally also rotate the electric field polarization of a real optical wave, leading to added complexities for the polarization eigenmodes of such a resonator.

We will not explore any of these properties of nonorthogonal optical systems further in this text, and the remainder of our discussions in the following chapters will apply only to orthogonal astigmatic systems, with separable and orthogonal $x$ and $y$ axes.

### REFERENCES

The theory of ray propagation, paraxial optical propagation, and gaussian beam propagation in nonorthogonal systems has been treated by a number of authors using ray matrix, eikonal, Huygens' integral, and differential operator methods. An important early reference is S. A. Collins, Jr., "Lens-system diffraction integral written in terms of matrix optics," *J. Opt. Soc. Am.* **60**, 1168–1177 (September 1970).

Other references from this same period include E. E. Bergmann, "Optical resonators with paraxial modes," *Appl. Opt.* **11**, 113–119 (January 1972); and Y. Suematsu and H. Fukinuki, "Matrix theory of light beam waveguides," *Bull. Tokyo Inst. Technol.* **88**, 33–47 (1968).

Extensive discussions have also been given by J. A. Arnaud in a series of papers, including "Degenerate optical cavities," *Appl. Opt.* **8**, 189–195 (January 1969); "Degenerate optical cavities. II: Effects of misalignments," *Appl. Opt.* **8**, 1909–1917 (September 1969); "Gaussian light beams with general astigmatism," (with H. Kogelnik) *Appl. Opt.* **8**, 1687–1693 (August 1969); "Nonorthogonal optical waveguides and resonators." *Bell Sys. Tech. J.* **49**, 2311–2347 (November 1970): and "Mode coupling in first-order optics," *J. Opt. Soc. Am.* **61**, 751–758 (June 1971).

The matrix results using the first of the two formalisms outlined in this chapter come from M. Nazarathy. *Operator Methods in First Order Optics*, D. Sc. Dissertation, Technion, Israel Institute of Technology (1982).

Other relevant references, particularly on nonplanar ring resonators, include S. I. Zavgorodnava, V. I. Kuprenyuk, and V. I. Sherstobitov, "Unstable resonator with field rotation," *Sov. J. Quantum Electron.* **7**, 787–788 (June 1977); G. B. Al'tshuler, et al., "Analysis of misalignment sensitivity of ring-laser resonators," *Sov. J. Quantum Electron.* **7**, 857–859 (July 1977); and F. Biraben, "Efficacite des systemes unidirectionnels utilisables dans les lasers en anneau," *Optics Comm.* **29**, 353–356 (June 1979), which proposes the use of a nonplanar ring to rotate the plane of polarization for an optical isolator.

Russian theorists have published extensively on field rotation in resonators and nonplanar rings, including papers by V. I. Kuprenyuk and V. E. Sherstobitov, "Calculations on the mirror system of an unstable resonator with field rotation," *Sov. J. Quantum Electron.* **10**, 449–453 (April 1980): M. M. Popov, "Resonators for lasers with unfolded directions of principal curvatures." *Optics and Spectrosc.* **25**, 213–217 (1968): and "Resonators for lasers with rotated directions of principal curvatures," *Optics and Spectrosc.* **25**, 170–171 (1968).

See also E. F. Ishchenko and E. F. Reshetin, "Sensitivity to misalignment of an optical ring resonator with a focusing element." *Optics and Spectrosc.* **46**, 202–207 (February 1979); and Yu. A. Anan'ev, V. I. Kuprenyuk, and V. E. Sherstobitov, "Properties of unstable resonators with field rotation. I. Theoretical principles," *Sov. J. Quantum Electron.* **9**, 1105–1110 (September 1979). and D. A. Goryachkin et al., "II. Experimental results," *Sov. J. Quantum Electron.* **9**, 1110–1114 (September 1979).

Recent Soviet work on nonplanar rings is reported by Yu. lD. Golyaev, et al., in "Spatial and polarization characteristics...," and "Temporal and spectral characteristics of radiation from a cw neodymium-doped garnet laser with a nonplanar ring resonator," *Sov. J. Quantum Electron.* **11**, 1421–1426 and 1427–1432 (November 1981).

Fundamental formulas for the rotation of an image upon reflection from a plane mirror are given in D. A. Berkowitz, "Design of plane mirror systems," *J. Opt. Soc. Am.* **55**, 1464–1467 (November 1965).

## Problems for 15.6

1. *Image rotation in a Dove prism.* Using ray matrices, show that when you physically rotate the Dove prism of Figure 15.25, the image transmitted through the prism rotates twice as fast as the prism itself.

# 16
## WAVE OPTICS AND GAUSSIAN BEAMS

A more accurate treatment of optical beams and laser resonators must take into account diffraction and the wave nature of light. Practical laser beams are almost always well enough collimated even under worst conditions, however, that we can describe their diffraction properties using a scalar wave theory, and working in the paraxial wave approximation.

In this chapter, therefore, we introduce the paraxial wave analysis and the equivalent Huygens-Fresnel integral approach for optical beams in free space. We also introduce the lowest and higher-order gaussian mode solutions of these equations as a widely useful set of "normal modes of free space."

The Hermite-gaussian or Laguerre-gaussian modes which we introduce in this chapter are exact and yet mathematically convenient solutions to the paraxial wave equation in free space. They also provide very close (though not quite exact) approximations for the transverse eigenmodes of stable laser resonators with finite diameter mirrors. Gaussian beams are therefore very widely used in analyzing laser beams and related optical systems. Our approach in this chapter is to focus primarily on the mathematical derivation of these modes, whereas in the following chapter we summarize most of the important practical properties of gaussian beams in considerable detail.

## 16.1 THE PARAXIAL WAVE EQUATION

One fundamental way of analyzing free-space wave propagation, using a differential approach, is through the *paraxial wave equation*, which we can derive once again here in the following fashion.

### Derivation of the Paraxial Wave Equation

Electromagnetic fields in free space (or in any uniform and isotropic medium) are governed in general by the scalar wave equation

$$\left[\nabla^2 + k^2\right] \tilde{E}(x,y,z) = 0, \tag{1}$$

## 16.1 THE PARAXIAL WAVE EQUATION

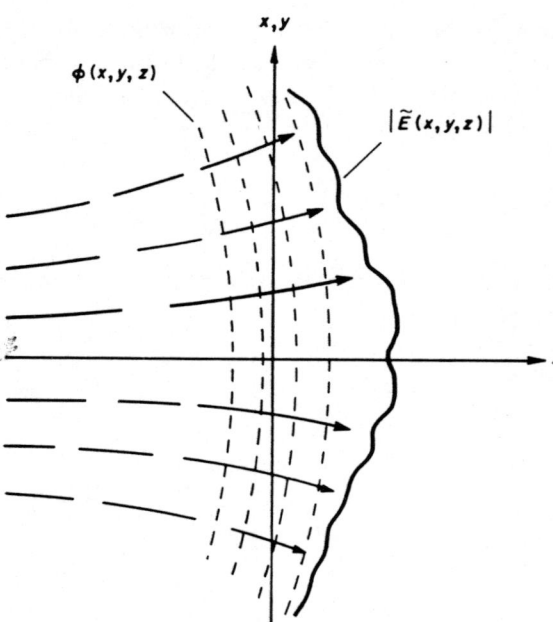

FIGURE 16.1
Transverse amplitude and phase variation of a paraxial optical wave.

where $\tilde{E}(x, y, z)$ is the phasor amplitude of a field distribution that is sinusoidal in time. We will be concerned in this section with optical beams propagating primarily along the $z$ direction, so that the primary spatial dependence of $\tilde{E}(x, y, z)$ will be an $\exp(-jkz)$ variation. This $\exp(-jkz)$ variation has a spatial period of one wavelength $\lambda$ in the $z$ direction.

In addition, for any beam of practical interest the amplitude and phase of the beam will generally have some transverse variation in $x$ and $y$ which specifies the beam's transverse profile, as shown in Figure 16.1; and this transverse amplitude and phase profile will change slowly with distance $z$ due to diffraction and propagation effects. Both the transverse variations across any plane $z$, however, and especially the variation in beam profile with distance along the $z$ axis, will usually be slow compared to the plane-wave $\exp(-jkz)$ variation in the $z$ direction for a reasonably well-collimated beam.

It is then convenient to extract the primary $\exp(-jkz)$ propagation factor out of $\tilde{E}(x, y, z)$, by writing each relevant vector component of the field (such as $E_x$ or $E_y$) in the form

$$\tilde{E}(x,y,z) \equiv \tilde{u}(x,y,z)e^{-jkz}, \qquad (2)$$

where $u$ is a complex scalar wave amplitude which describes the transverse profile of the beam. Substituting this into the wave equation 16.1 then yields, in rectangular coordinates, the reduced equation

$$\frac{\partial^2 \tilde{u}}{\partial x^2} + \frac{\partial^2 \tilde{u}}{\partial y^2} + \frac{\partial^2 \tilde{u}}{\partial z^2} - 2jk\frac{\partial \tilde{u}}{\partial z} = 0. \qquad (3)$$

Now, we emphasize once again that with the $\exp(-jkz)$ dependence factored out, the remaining $z$ dependence of the wave amplitude $\tilde{u}(x, y, z)$, is caused basically by diffraction effects, and this $z$ dependence will in general be slow compared not only to one optical wavelength, as in $\exp(-jkz)$, but also to the transverse variations due to the finite width of the beam. This slowly varying dependence of $\tilde{u}(x,y,z)$ on $z$ can be expressed mathematically by the *paraxial approximation*

$$\left|\frac{\partial^2 \tilde{u}}{\partial z^2}\right| \ll \left|2k\frac{\partial \tilde{u}}{\partial z}\right| \quad \text{or} \quad \left|\frac{\partial^2 \tilde{u}}{\partial x^2}\right| \quad \text{or} \quad \left|\frac{\partial^2 \tilde{u}}{\partial y^2}\right|. \tag{4}$$

By dropping the second partial derivative in $z$, we thus reduce the exact wave equation 16.3 to the *paraxial wave equation*

$$\frac{\partial^2 \tilde{u}}{\partial x^2} + \frac{\partial^2 \tilde{u}}{\partial y^2} - 2jk\frac{\partial \tilde{u}}{\partial z} = 0. \tag{5}$$

More generally we may write this paraxial wave equation as

$$\nabla_t^2 \tilde{u}(\mathbf{s}, z) - 2jk\frac{\partial \tilde{u}(\mathbf{s}, z)}{\partial z} = 0, \tag{6}$$

where $\mathbf{s}$ refers to the transverse coordinates $\mathbf{s} \equiv (x,y)$ or $\mathbf{s} \equiv (r,\theta)$, depending on what coordinate system (rectangular or cylindrical) we elect to use, and $\nabla_t^2$ means the laplacian operator operating on these coordinates in the transverse plane. This equation will be the primary governing equation for all the analysis of this and the following several chapters.

### Paraxial Wave Propagation: Finite Difference Approach

The paraxial wave equation can also be turned around and written in the form

$$\frac{\partial \tilde{u}(\mathbf{s}, z)}{\partial z} = -\frac{j}{2k}\nabla_t^2 \tilde{u}(\mathbf{s}, z). \tag{7}$$

This equation can then be integrated forward in the $z$ direction in order to compute the forward propagation and diffraction spreading of an arbitrary paraxial optical beam. That is, we can employ any suitable numerical differentiation and integration algorithms, first to evaluate the transverse derivative $\nabla_t^2 \tilde{u}(\mathbf{s}, z)$ at a given plane $z$, and then to step forward to a new plane $z + \Delta z$. We can thus accomplish numerical forward propagation of an arbitrary optical wavefront, making sure to use adequate numbers of sampling points in both the transverse and longitudinal directions.

This numerical approach, sometimes referred to as the "finite difference approach," has been applied to practical beam propagation problems by several workers. For almost any free-space beam propagation problem that we may consider, however, the integral formulation that we will consider in the next section is probably a better choice for numerical calculations, because of the much greater computational efficiency of fast Fourier transforms that can be employed.

### Validity of the Paraxial Approximation

The paraxial wave equation in either of the above forms is fully adequate for describing nearly all optical resonator and beam propagation problems that arise with real lasers. As perhaps the simplest but most effective way to confirm

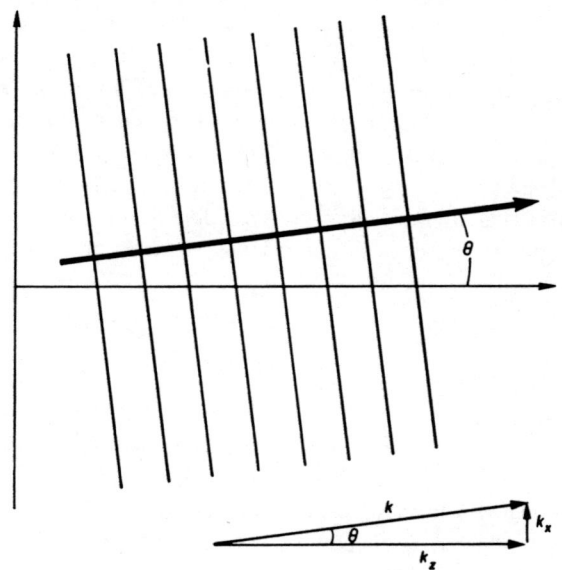

FIGURE 16.2
A plane wave traveling at a small angle $\theta$ to the optic axis.

this, and also to illustrate the physical limitations of the paraxial approximation, we can consider the following argument.

Any optical beam can always be viewed as being made up of a superposition of plane wave components traveling at various angles to the $z$ axis. (We will discuss this kind of plane wave expansion more rigorously in a later section.) Consider, for example, a plane wave component $\tilde{E}(x, z)$ traveling at angle $\theta$ to the $z$ axis in the $x, z$ plane, as shown in Figure 16.2. (For simplicity we consider only one transverse coordinate.) The axial and transverse variations of this plane wave component are then given by

$$\tilde{E}(x, z) = \exp[-jkx \sin \theta - jkz \cos \theta] = \tilde{u}(x, z)e^{-jkz}. \tag{8}$$

The exact form for the reduced wave amplitude $\tilde{u}(x, y, z)$, and its approximate form within the paraxial approximation, then become

$$\tilde{u}(x, z) = \exp[-jkx \sin \theta + jkz(1 - \cos \theta)] \approx \exp\left[-jk\theta x + jk\frac{\theta^2 z}{2}\right]. \tag{9}$$

The normalized first and second derivatives of $\tilde{u}(x, z)$ in the transverse direction then take on the values

$$-j\frac{2k}{\tilde{u}}\frac{\partial \tilde{u}}{\partial z} = +2k^2(1 - \cos \theta) \approx k^2\theta^2$$

$$\frac{1}{\tilde{u}}\frac{\partial^2 \tilde{u}}{\partial x^2} = -k^2 \sin^2 \theta \approx -k^2\theta^2. \tag{10}$$

However, the second derivative in the $z$ direction takes on the form

$$\frac{1}{\tilde{u}}\frac{\partial^2 \tilde{u}}{\partial z^2} = -k^2(1 - \cos \theta)^2 \approx -\frac{k^2\theta^4}{4}. \tag{11}$$

This particular derivative is smaller than either of the preceding terms by the ratio $\theta^2/4$ (with $\theta$ measured in radians)—a ratio which will be $\ll 1$ so long as $\theta$ is $\leq 1/2$ radian.

We can conclude that so long as all (or at least most) of the plane wave components making up any optical beam are traveling at angles $\theta \leq 0.5$ rad, or $\theta$ less than about $30°$, the $\partial^2 \tilde{u}/\partial z^2$ terms will be at least an order of magnitude smaller than either of the other two terms, in agreement with the basic paraxial approximation. Paraxial optical beams can thus be focused or can diverge at cone angles up to $\approx 30°$ before significant corrections to the paraxial wave approximation become necessary.

## REFERENCES

Corrections to the paraxial optical theory derived in this section do become significant when beams are focused so tightly, or diverged so rapidly, that local wavefronts become tilted by more than about 30° to the beam axis. The next higher-order extensions that are then required are discussed in M. Lax, W.H. Louisell, and W.B. McKnight, "From Maxwell to paraxial optics," *Phys. Rev. A* **11**, 1365–1370 (1975).

Their approach is extended and simplified by L.W. Davis, "Theory of electromagnetic beams," *Phys. Rev. A* **19**, 1177–1179 (March, 1979); and the resulting next-order correction term for gaussian optical beams is discussed by G.P. Agrawal and D.N. Pattanayak, "Gaussian beam propagation beyond the paraxial approximation," *J. Opt. Soc. Am.* **69**, 575–578 (April 1979).

## 16.2 HUYGENS' INTEGRAL

Another equally valid and effective way of analyzing paraxial wave propagation, but now using an integral approach, is to employ *Huygens' principle, expressed in the Fresnel approximation*. We can derive this alternative approach to paraxial beam propagation as follows.

### Spherical Waves, and the Fresnel Approximation

Let us first note that one very general solution to the exact wave equation, which corresponds physically to a uniform spherical wave diverging from a source point $r_0$ (Figure 16.3) may be written in the form

$$\tilde{E}(r; r_0) = \frac{\exp[-jk\rho(r, r_0)]}{\rho(r, r_0)}, \tag{12}$$

where $\tilde{E}(r; r_0)$ means the field at point $r$ due to a source at point $r_0$, and where the distance $\rho(r, r_0)$ from the source point $s_0, z_0$ to the observation point $s, z$ is given by

$$\rho(r, r_0) \equiv \sqrt{(x - x_0)^2 + (y - y_0)^2 + (z - z_0)^2}. \tag{13}$$

We emphasize that a spherical wavefunction in this form is an *exact* solution to the *full* scalar wave equation of the previous section.

Consider, however, a situation in which the source point $x_0, y_0, z_0$ for this wave is located somewhere not too far off the $z$ axis; and suppose we only wish

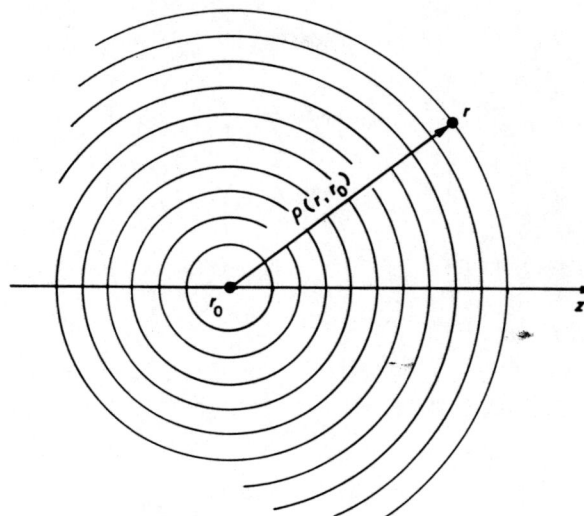

FIGURE 16.3
A general spherical wave.

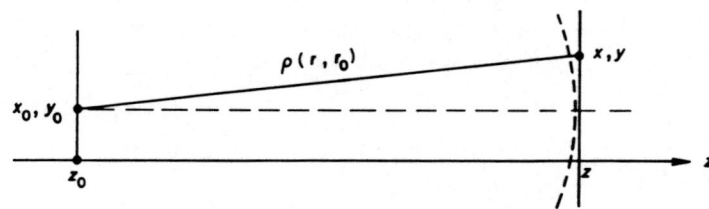

FIGURE 16.4
Fresnel approximation to the spherical wave of Figure 16.3.

to write the resulting field distribution $\tilde{E}(x,y,z)$ or $\tilde{u}(x,y,z)$ of this spherical wave on some transverse plane $x,y$ that is farther along the $z$ axis, for values of $x$ and $y$ that are also not too far off the axis, as in Figure 16.4. The *Fresnel approximation* to diffraction theory says that if we expand the distance $\rho(\boldsymbol{r},\boldsymbol{r}_0)$ in Equation 16.13 in a power series in the form

$$\rho(\boldsymbol{r},\boldsymbol{r}_0) = z - z_0 + \frac{(x-x_0)^2 + (y-y_0)^2}{2(z-z_0)} + \cdots, \qquad (14)$$

then we can drop all terms higher than quadratic in this expression, at least in writing the phase shift factor $\exp[-jk\rho(\boldsymbol{r},\boldsymbol{r}_0)]$ (We will examine the validity of this assumption in more detail a bit further on.) In the $1/\rho$ denominator of Equation 16.12, on the other hand, we will drop even the quadratic terms, and replace $\rho(\boldsymbol{r},\boldsymbol{r}_0)$ by simply $z - z_0$.

The spherical wave of Equation 16.12 is then converted, in this Fresnel approximation, into what we might call a "paraxial-spherical wave" given by

$$\tilde{E}(x,y,z) \approx \frac{1}{z-z_0} \exp\left[-jk(z-z_0) - jk\frac{(x-x_0)^2 + (y-y_0)^2}{2(z-z_0)}\right], \qquad (15)$$

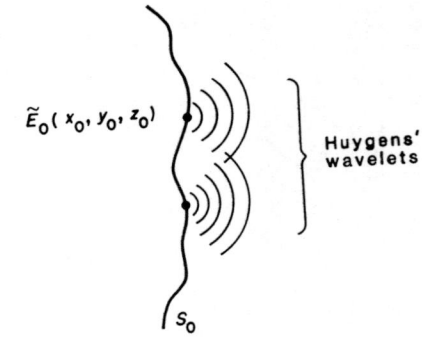

FIGURE 16.5
Huygens' principle.

or

$$\tilde{u}(x,y,z) = \frac{1}{z-z_0}\exp\left[-jk\frac{(x-x_0)^2+(y-y_0)^2}{2(z-z_0)}\right]. \qquad (16)$$

These expressions approximate the spherical wave by a quadratic phase variation as observed on a transverse plane $x, y$ located at a distance $z - z_0$ away from the source point along the $z$ axis, as shown in Figure 16.4. The "paraxial-spherical wave" given by this Fresnel approximation is, as the reader can verify, an *exact analytical solution to the paraxial wave equation rather than to the exact wave equation.*

### Huygens' Integral

We can next connect these spherical-wave and paraxial-spherical-wave ideas to Huygens' integral.

Huygens' integral originated as an intuitive physical principle, which was only later put into more rigorous mathematical terms. This principle says in physical terms that if we are given an incident field distribution $\tilde{E}_0(x_0, y_0, z_0)$ over some closed surface $S_0$, we may regard the field at each point on that surface as the source for a uniform spherical wave or "Huygens' wavelet" which radiates from that point on the surface, as illustrated in Figure 16.5. The total field at any other point $s, z$ inside, or beyond, the surface $S_0$ can then be calculated by summing the fields of all these Huygens' wavelets coming from all the points on the surface $S_0$.

Huygens' intuitive ideas concerning this principle were put into more formal mathematical form, first by Fresnel and Kirchoff, and later by Rayleigh and Sommerfeld. The general idea is that each of the Huygens' wavelets should be viewed as a spherical wave with a form like Equation 16.12, leading to Huygens' integral equation in the form

$$\tilde{E}(s,z) = \frac{j}{\lambda}\int\int_{S_0}\tilde{E}_0(s_0,z_0)\frac{\exp[-jk\rho(\boldsymbol{r},\boldsymbol{r}_0)]}{\rho(\boldsymbol{r},\boldsymbol{r}_0)}\cos\theta(\boldsymbol{r},\boldsymbol{r}_0)\,d\boldsymbol{S}_0, \qquad (17)$$

where $\rho(\boldsymbol{r},\boldsymbol{r}_0)$ is the distance between source and observation points as defined earlier. In this formulation $d\boldsymbol{S}_0$ is an incremental element of surface area at point $s_0, z_0$ on the surface $S_0$, and the factor $\cos\theta(\boldsymbol{r},\boldsymbol{r}_0)$ is an "obliquity factor" which depends on the angle $\theta(\boldsymbol{r},\boldsymbol{r}_0)$ between the line element $\rho(\boldsymbol{r},\boldsymbol{r}_0)$ and the normal to the surface element $d\boldsymbol{S}_0$.

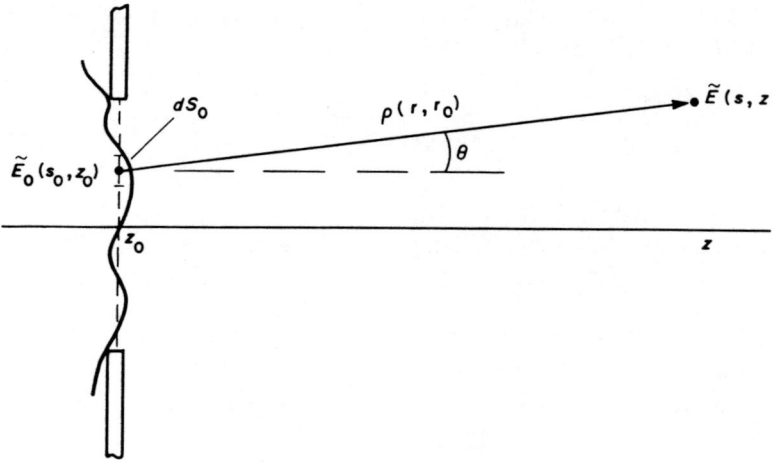

**FIGURE 16.6**
Geometry for evaluating Huygens' integral in the Fresnel approximation.

Slightly different forms for the obliquity factor in Equation 16.17 are predicted by the Kirchoff-Fresnel or the Rayleigh-Sommerfeld approaches to diffraction theory, although in either situation this factor goes to unity if the angle $\theta$ is limited to small values. The $j/\lambda$ factor in front of Huygens' integral is a normalization factor which comes out of a more detailed approach to the theory, and which is necessary in order to get the correct near-field and far-field dependence from Huygens' integral. (We will also see another physically meaningful explanation for this factor in the following chapter.)

### Fresnel Approximation to Huygens' Integral

Suppose now that we are given the input field distribution or beam profile $\tilde{E}_0(x_0, y_0, z_0)$ of a paraxial optical beam across an input transverse plane at $z = z_0$, as in Figure 16.6, and we wish to calculate the output beam profile $\tilde{E}(x, y, z)$ across another plane at distance $z = z_0 + L$, accurate to within the paraxial degree of approximation. We can then make this calculation by using Huygens' principle in the form just given, except that we can replace the exact spherical wave $\exp[-jk\rho]/\rho$ from Equation 16.12 with a "paraxial-spherical wave" in the form derived in Equation 16.15.

If we do this, we obtain *Huygens' integral in the Fresnel approximation*, as given by

$$\tilde{E}(x,y,z) \approx \frac{je^{-jk(z-z_0)}}{(z-z_0)\lambda} \int\int \tilde{E}_0(x_0, y_0, z_0) \exp\left[-jk\frac{(x-x_0)^2 + (y-y_0)^2}{2(z-z_0)}\right] dx_0 dy_0. \tag{18}$$

We have assumed here that the distance $z - z_0$ between input and output planes in Figure 16.6 is large enough so that the angle of the line element connecting source and observation points is always $\leq 0.5$ rad. Hence we can approximate the obliquity factor $\cos\theta$ in Equation 16.17 by unity, as well as using the Fresnel approximation in the exponent.

This formulation of Huygens' integral can equally well be written for the reduced wavefunction $\tilde{u}(x, y, z)$ in the form

$$\tilde{u}(x, y, z) = \frac{j}{L\lambda} \int\int \tilde{u}_0(x_0, y_0, z_0) \exp\left[-jk\frac{(x-x_0)^2 + (y-y_0)^2}{2L}\right] dx_0 dy_0, \tag{19}$$

where $L \equiv z - z_0$ is again the distance from input to output plane, and the integrations are over the transverse input plane located at $z = z_0$. We will most often use Huygens' integral in this second form, with the on-axis or plane-wave phase shift $\exp(-jkL)$ or $\exp[-jk(z-z_0)]$ omitted, since it is the transverse variation of $\tilde{u}(x, y)$ that is usually of primary interest. The plane-wave phase factor must be brought back into the calculation if resonant frequencies or total phase shifts are to be calculated.

Validity of the Fresnel Approximation

The reader can verify that the "paraxial spherical wave" given in Equations 16.15 or 16.16 satisfies exactly the paraxial wave equation, just as the exact spherical wave in Equation 16.12 satisfies the exact wave equation. Any field expression for $\tilde{u}(x, y, z)$ given by Huygens' integral in the Fresnel approximation, as given in Equation 16.19 with $L = z - z_0$, will therefore satisfy the paraxial wave equation exactly, and vice versa. *Huygens' integral and the paraxial wave equation represent exactly the same mathematical (and physical) approximations.*

There are, however, some subtleties in the physical interpretation of the Fresnel approximation which it is useful to understand. Suppose we wish to calculate the forward propagation of a paraxial beam through a distance $L$, from an input wave $\tilde{u}_1(x_1, y_1)$ at plane $z = z_1$ to an output wave $\tilde{u}_2(x_2, y_2)$ at plane $z = z_2 = z_1 + L$. Suppose also that this beam is confined to a width $\approx 2a$, i.e., $\tilde{u}_1$ and $\tilde{u}_2$ have negligible values outside of $-a \leq x, y \leq a$.

Now the Fresnel approximation of Equations 16.14 to 16.19 assumes that the quartic and higher-order terms in the expansion of the exponent $e^{-jk\rho}$ can be dropped, because their contribution to the complex exponent will be small compared to, say, $\pi/2$. If we consider either of the transverse coordinates $x$ or $y$, this condition on the next higher-order terms seems to require that

$$\left|\frac{k(x_2-x_1)^4}{4L^3}\right| \approx \left|\frac{2\pi}{\lambda}\frac{(2a)^4}{4L^3}\right| \leq \frac{\pi}{2}, \tag{20}$$

since by assumption $|x_2 - x_1| \leq 2a$. This condition can be rewritten in the alternative form

$$\frac{L}{2a} \geq \left(\frac{2a}{\lambda}\right)^{1/3}. \tag{21}$$

In any normal situation, where the beam width is large compared to a wavelength, or $2a \gg \lambda$, this condition apparently says that Huygens' integral in the Fresnel approximation can only be applied to calculate forward beam propagation over lengths that are significantly greater than the beam diameter, or $L \gg 2a$. There seems to be an inconsistency in this limitation, however, in that the paraxial wave equation 16.7 of the previous section can obviously be applied over arbitrarily short forward steps in $z$, and the paraxial wave equation and the

Huygens-Fresnel integral are supposed to be mathematically equivalent. Is the condition in Equation 16.21 really a limitation on the use of Huygens' integral in the Fresnel approximation?

This question can be answered (in the negative) as follows. Suppose the optical wavefront $\tilde{u}(x_1, y_1, z_1)$ at the input plane $z_1$ is a freely propagating paraxial beam, without any sharp discontinuities or aperturing at or near that particular plane. Then, *the effective sources for this beam are really located at source points $x_0, y_0, z_0$ that are located far behind (or, for a converging beam, far ahead) of either of the planes $z_1$ or $z_2$*. We can then clearly apply the Huygens-Fresnel integral to calculate the wave propagation from plane $z_0$ to plane $z_1$, and we can equally well apply this integral to calculate the propagation from plane $z_0$ to plane $z_2$. But the inherent cascading properties of the Huygens-Fresnel integral (which the reader may want to verify; see the Problems) then imply that the Huygens-Fresnel integral in exactly the same form must also be valid over the distance $z_2 - z_1 = L$, even if this distance is too small to satisfy Equation 16.21, and in fact even if $L \to 0$.

The essential physical point here is that *the paraxial or Fresnel approximation is a physical property of the optical beam, not a mathematical property of the Huygens-Fresnel formulation*. The paraxial wave equation 16.7 and also the Huygens-Fresnel integral 16.19 can be applied over arbitrarily short distances $L$ *if* the optical beam itself is truly paraxial.

### Sharp-Edged Aperture Effects

We can give a more physical interpretation to this assertion about paraxial beams—which may seem somewhat confusing at first—by the following illustration.

Suppose for example that an input wave $\tilde{u}_1(x_1, y_1)$ does have some sharp discontinuity in its wavefunction at the input plane $z_1$, either in amplitude or phase, such as would be caused by a hard-edged aperture or by a discontinuous phase step in the input plane $z_1$, say, at radius $x = a$. We will see later that such discontinuities appear to act in effect as sources for quasi spherical diffraction waves or "edge waves" which appear to radiate from the discontinuous edges of the aperture. The criterion of Equation 16.21 must then be applied, not only to the kernel of Huygens' integral, but also to these "edge waves" as seen at any plane a distance $L$ away from the aperture plane. Huygens' integral in the Fresnel approximation can thus *not* be used for distances closer than $L/2a \approx (2a/\lambda)^{1/3}$ beyond such an aperture.

But in fact, *neither* Huygens' integral in the Fresnel approximation *nor* the paraxial wave equation can be applied to this sort of sharp-edged diffraction situation over distances $L$ shorter than given by the preceding condition. The Huygens-Fresnel integral cannot be used because this would violate the Fresnel approximation for those diffracted wavelets which appear to be scattered from the edges of the aperture. The paraxial wave equation cannot be applied accurately in this region, because the rapid changes in $\tilde{u}(x, z)$ with both $x$ and $z$ near the sharp aperture edges violate the approximations inherent in the paraxial wave equation.

The fundamental point, then, is that *paraxial methods, however they may be formulated, can only be applied to paraxial beams*, and a beam diffracted by a sharp-edged aperture does not again become paraxial until we move far enough past the aperture to satisfy Equation 16.21.

## Huygens' Integral in One Dimension

Huygens' integral may be rewritten in the general form

$$\tilde{u}(s, z) = \iint \tilde{K}(r, r_0) \, \tilde{u}_0(s_0, z_0) \, ds_0, \tag{22}$$

where $r_0$ and $r$ indicate points on the input and output transverse planes; $\tilde{K}(r, r_0)$ is shorthand for the Huygens kernel of Equation 16.19; and $ds_0$ indicates an element of area on the input plane. In rectangular coordinates the Huygens-Fresnel kernel separates into a product kernel in the form

$$\tilde{K}(r, r_0) = \tilde{K}_1(x - x_0) \times \tilde{K}_1(y - y_0), \tag{23}$$

where the one-dimensional kernel $\tilde{K}_1$ has the form

$$\tilde{K}_1(x - x_0) = \sqrt{\frac{j}{L\lambda}} \exp\left[-j\frac{\pi(x - x_0)^2}{L\lambda}\right]. \tag{24}$$

If the wavefunction $\tilde{u}(s, z)$ is also separable in $x, y$ coordinates, the entire integral can be separated into two one-dimensional integrals of the form

$$\tilde{u}(x, z) = \sqrt{\frac{j}{L\lambda}} \int \tilde{u}_0(x_0, z_0) \exp\left[-j\frac{\pi(x - x_0)^2}{L\lambda}\right] dx_0 \tag{25}$$

and the same for $\tilde{u}(y, z)$ and $\tilde{u}_0(y_0, z_0)$. We will frequently write such integrals in only one transverse dimension, in order to simplify the mathematical expressions.

Note that the $j/L\lambda$ factor in front of the three-dimensional Huygens' integral 16.19 reduces to $\sqrt{j/L\lambda}$ if there is only one transverse dimension. If only one transverse coordinate is included, the Huygens' wavelet is in essence a cylindrical wave rather than a spherical wave. Hence its amplitude decreases with distance as $1/\sqrt{\rho}$ or $1/\sqrt{L}$, rather than as $1/\rho$. The phase shift of $\pi/2$ due to the $j$ factor in two dimensions also reduces to $\pi/4$ in one dimension. This phase factor represents in essence an initial phase shift of the Huygens' wavelet compared to the actual field value at the input point. We will see later that this corresponds to a well-known "Guoy phase shift" associated with any wave passing through a focus, or coming from a small enough source point.

## REFERENCES

For an excellent introduction to scalar diffraction theory, with earlier references, see Chapter 3 of J. W. Goodman, *Introduction to Fourier Optics* (McGraw-Hill, 1968). Much more extensive discussions will also be found in M. Born and E. Wolf, *Principles of Optics* (Pergamon Press, 1959).

---

Problems for 16.2

1. *Solid angular spread from a uniformly illuminated aperture.* A **transmitting aperture** of total area $A_0$ transmits a collimated wavefront with total **power $P_0$** having (ideally) a uniform intensity distribution over the aperture. Using Huygens' integral, show that the transmitted intensity on axis in the far-field (as $z \to \infty$)

## 16.3 GAUSSIAN SPHERICAL WAVES

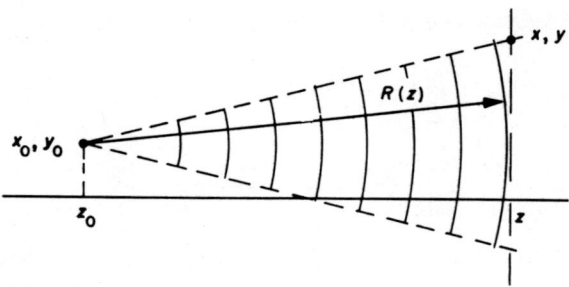

FIGURE 16.7
Spherical waveform coming from a real point source.

will be given in all cases by $I_z = P_0/(z\lambda)^2$, entirely independent of the shape of the aperture. What does this say about the approximate spread in solid angle of the far-field beam diverging from an aperture having total area $A_0$ but arbitrary shape? Can you give a simple argument why a uniform intensity distribution is optimum?

2. *Cascade properties of the Huygens-Fresnel integral.* Suppose we use the operator notation $\tilde{u}(x_1, z_1) = \tilde{K}(x_1 - x_0; z_1 - z_0) \star \tilde{u}(x_0, z_0)$ as shorthand to indicate the operation of propagating an optical field $\tilde{u}(x, z)$ from plane $z_0$ to plane $z_1$ using the one-dimensional Huygens-Fresnel integral 16.19. Verify that the cascade of two such steps over successive distances $z_0$ to $z_1$ and then $z_1$ to $z_2$ is the same as a single step over the combined distance $z_0$ to $z_2$, independent of the length of the individual steps. That is, verify that $\tilde{K}(x_2 - x_0; z_2 - z_0) \star \tilde{u}(x_0, z_0) \equiv \tilde{K}(x_2 - x_1; z_2 - z_1) \star \tilde{K}(x_1 - x_0; z_1 - z_0) \star \tilde{u}(x_0, z_0)$ independent of the spacings between $z_0$, $z_1$ and $z_2$.

---

## 16.3 GAUSSIAN SPHERICAL WAVES

Our next important step is to derive the analytical form for a gaussian spherical wave, or a so-called "gaussian beam" in free space; and then to show that this gaussian beam is a very useful exact solution to the paraxial wave equation, or to Huygens' integral in the Fresnel approximation.

### Paraxial Spherical Waves

Consider a uniform spherical wave diverging from a source point located at $x_0, y_0, z_0$, and observed at an observation point $x, y, z$, as in Figure 16.7. If the axial distance $z - z_0$ between the source and observation points is sufficiently large compared to the transverse coordinates $x_0, y_0$ and $x, y$, then the field distribution produced by this wave at point $x, y$ on the plane located at distance $z$ can be written, using the paraxial approximation, in the form

$$\tilde{u}(x,y,z) = \frac{1}{z-z_0} \exp\left[-jk\frac{(x-x_0)^2+(y-y_0)^2}{2(z-z_0)}\right]$$
$$= \frac{1}{R(z)} \exp\left[-jk\frac{(x-x_0)^2+(y-y_0)^2}{2R(z)}\right], \quad (26)$$

where $R(z) = z - z_0$ gives the radius of curvature of the spherical wave at plane $z$. The phase variation $\exp[-j\phi(x,y,z)]$ across a transverse plane at fixed $z$ for such a paraxial spherical wave with radius of curvature $R(z)$ thus has the *quadratic* form

$$\phi(x,y,z) \equiv k\frac{(x-x_0)^2+(y-y_0)^2}{2(z-z_0)} = \frac{\pi}{\lambda}\frac{(x-x_0)^2+(y-y_0)^2}{R(z)}. \quad (27)$$

The radius of curvature $R(z)$ of the wave at plane $z$ can be written in a more general form as

$$\tilde{R}(z) = R_0 + z - z_0, \quad (28)$$

with $R_0$ being the value at the earlier plane $z_0$ (and with $R_0 = 0$ if the earlier plane is the source plane, as in the present situation). As such a spherical wave propagates forward to any other plane $z$, the radius of curvature of the wave thus increases linearly with distance.

Note that with our sign convention, a value of $R > 0$ indicates a diverging or expanding wave, whereas $R < 0$ indicates a converging wave moving inward toward the source point. Note also that if the wave is propagating in some dielectric medium, $k$ and $\lambda$ are the values in that medium, not in free space. This quadratic phase variation of course represents only a paraxial or Fresnel approximation to the true surface of a sphere, so that this form will have a sizable phase error if we move far enough out from the optic axis.

### Introducing Complex Source Point Coordinates

A paraxial spherical wave in the form given in Equation 16.26 cannot by itself be a very useful analytical form for a real physical beam, however, because the amplitude of the spherical wave does not fall off with transverse distance from the axis. Such a wave instead extends out to infinity in the transverse direction, and carries infinite energy and power across the transverse plane (as well as having large deviation from a true sphere far off the axis).

A very simple way to overcome these difficulties can be developed, however, as follows. Let us first note that the spherical wave expressions in Equations 16.26–16.28 satisfy the paraxial wave equation, or the Huygens-Fresnel integral, exactly for any arbitrary choice of the source point coordinates $x_0, y_0, z_0$. That is, these coordinates are simply constant parameters, which cancel out identically when the spherical wave expression is put into the paraxial wave equation or the Huygens-Fresnel integral. What then will happen if we explore the possibility of employing *complex values* for these source point coordinates?

In particular, suppose that for simplicity we set $x_0$ and $y_0$ to zero, but that we convert the axial location $z_0$ of the source point into a *complex number*, by subtracting from it an arbitrary complex quantity which we will call $\tilde{q}_0$. That is, we replace the purely real value $z_0$ in the spherical wave expression by the

## 16.3 GAUSSIAN SPHERICAL WAVES

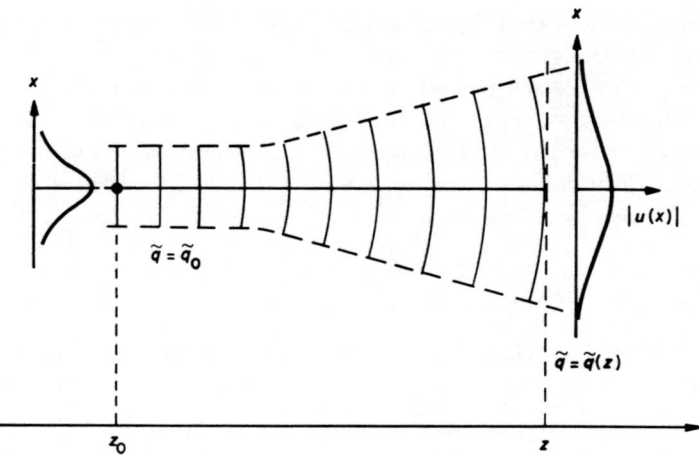

**FIGURE 16.8**
Gaussian-spherical wave from a complex point source.

complex value $z_0 - \tilde{q}_0$. This amounts to the same thing as replacing the radius of curvature $R(z) = R_0 + z - z_0$ by the complex quantity $\tilde{q}(z) = z - (z_0 - \tilde{q}_0) = \tilde{q}_0 + z - z_0$. [We introduce a new notation $\tilde{q}(z)$ in place of $R(z)$ because this quantity will turn out to be a complex generalization of the purely real spherical-wave radius of curvature $R(z)$]. Note that the value of this new "complex radius" at the source plane $z = z_0$ is just $\tilde{q}(z_0) = \tilde{q}_0$.

The spherical wave diverging from this complex source point is now written, following the same form as Equation 16.26, as

$$\tilde{u}(x,y,z) = \frac{1}{z - z_0 + \tilde{q}_0} \exp\left[-jk\frac{x^2 + y^2}{2(z - z_0 + \tilde{q}_0)}\right]$$
$$= \frac{1}{\tilde{q}(z)} \exp\left[-jk\frac{x^2 + y^2}{2\tilde{q}(z)}\right], \qquad (29)$$

where the complex radius $\tilde{q}(z)$ is given by

$$\tilde{q}(z) = \tilde{q}_0 + z - z_0 \qquad (30)$$

in direct analogy to the expression 16.28 for $R(z)$.

But if $\tilde{q}(z)$ is complex, then we can separate the exponent in Equation 16.29 into real and imaginary parts, by first separating the quantity $1/\tilde{q}(z)$ into real and imaginary parts in the form

$$\frac{1}{\tilde{q}(z)} \equiv \frac{1}{q_r(z)} - j\frac{1}{q_i(z)}, \qquad (31)$$

and then writing the spherical wave expression in the form

$$\tilde{u}(x,y,z) = \frac{1}{\tilde{q}(z)} \exp\left[-jk\frac{x^2 + y^2}{2q_r(z)} - k\frac{x^2 + y^2}{2q_i(z)}\right]. \qquad (32)$$

(Note that $q_r$ and $q_i$ as defined here are not in general the real and imaginary parts of $\tilde{q}(z)$, but rather are defined by Equation 16.31.)

The resulting exponent for this complex-source-point beam now has both an imaginary quadratic transverse variation, corresponding to a quadratic phase front, or a spherical wave with a real radius of curvature; and also a purely real quadratic transverse variation, which gives a gaussian transverse amplitude profile or amplitude variation, with a transverse fall-off determined by the imaginary part of $1/\tilde{q}$. Both of these variations are contained in the complex radius of curvature $\tilde{q}$ through Equation 16.31.

### Gaussian-Spherical Beams

At this point let us convert this result into the standard notation which is very widely used in the laser field, by rewriting it in the form

$$\tilde{u}(x,y,z) = \frac{1}{\tilde{q}(z)} \exp\left[-jk\frac{x^2+y^2}{2\tilde{q}(z)}\right]$$
$$\equiv \frac{1}{\tilde{q}(z)} \exp\left[-jk\frac{x^2+y^2}{2R(z)} - \frac{x^2+y^2}{w^2(z)}\right], \quad (33)$$

where $R(z)$ is now the radius of curvature and $w(z)$ the so-called "gaussian spot size" of this solution. Equation 16.33 gives the *lowest-order spherical-gaussian beam solution in free space*. We can view this solution if we like as being simply a paraxial-spherical wave diverging from a complex source point, which is located a complex distance $z - z_0 + \tilde{q}_0$ rather than a real distance $z - z_0$ behind the observation plane $z$.

It is very important to note here that, although we use the same notation for $R(z)$ as before, *this radius of curvature can no longer be calculated by the formula $R(z) = R(z_0) + z - z_0$* from Equation 16.28. Rather, the radius of curvature $R(z)$ and the spot size $w(z)$ of the wave at any plane $z$ are derived from the complex radius $\tilde{q}(z)$ by the definition that

$$\frac{1}{\tilde{q}(z)} \equiv \frac{1}{R(z)} - j\frac{\lambda}{\pi w^2(z)}. \quad (34)$$

The variation of the complex radius of curvature $\tilde{q}(z)$ with distance is then determined by the formula

$$\tilde{q}(z) = \tilde{q}_0 + z - z_0. \quad (35)$$

The fundamental propagation law for all such gaussian beams in free space is entirely contained in this simple relation for $\tilde{q}(z)$.

### Summary

The primary result of this section, then, is that replacing the purely real radius of curvature $R(z)$ for a spherical wave coming from a real source point $z_0$ with a *complex radius of curvature* $\tilde{q}(z)$, or a *complex source point* $(z_0 - \tilde{q}_0)$, converts the paraxial-spherical wave solution given in the opening paragraphs of this section into the *gaussian-spherical wave solution* given by Equations 16.29–16.35. This gaussian-spherical wave solution is still an exact mathematical solution to either the paraxial wave equation or the Huygens-Fresnel integral. Now, however, it has in addition the physically desirable properties that its amplitude falls off smoothly and rapidly with distances from the $z$ axis; it carries finite total power across the beam cross section; and it also remains complex gaussian in profile at all later planes $z$.

This gaussian-spherical solution, together with various higher-order Hermite-gaussian or Laguerre-gaussian extensions, will prove to be extraordinarily useful in the analysis of optical resonators and beams. This gaussian-spherical wave solution and its higher-order extensions for paraxial beams in free space can in fact be derived in at least four different ways, by using:

(i) The complex source point derivation used in the present section; or
(ii) A differential equation approach based on the paraxial wave equation, which we will use in the following two sections; or
(iii) Direct substitution of the spherical-gaussian solution into the Huygens-Fresnel integral; or
(iv) A plane-wave expansion approach, which we will introduce in several later sections.

The complex-source-point approach we have used here is, despite its simplicity, both subtle and entirely rigorous. We will explore the physical properties of this gaussian mode solution, and its higher-order extensions, in great detail in the following chapter.

## REFERENCES

A very good summary of much of the early work on gaussian beams and resonator modes is given by H. Kogelnik and T. Li, "Laser beams and resonators," *Appl. Optics* **5**, 1150–1567 (October 1966).

The concept of a gaussian beam as a spherical wave emanating from a complex-valued source point has been developed in numerous places in the literature. See, for example, G. A. Deschamps, "Gaussian beam as a bundle of complex rays," *Electron. Lett.* **7**, 684–685 (1971); or M. Couture and P-A. Belanger, "From gaussian beam to complex-source-point spherical wave," *Phys. Rev. A* **24**, 355–359 (July 1981).

---

Problems for 16.3

1. *Changes in wavefront curvature on reflection from a curved mirror.* The phase variation across a transverse plane (i.e., at constant $z$) for a diverging spherical wave with radius of curvature $|R|$ traveling in the $+z$ direction has the form $\exp[-jkx^2/2|R|]$. What will be the transverse variation of this same wave (a) after reflecting off a planar mirror set up normal to the $z$ axis? (b) After reflecting off a concave spherical mirror with radius of curvature $R_m = |R|$? (c) After reflecting off a concave spherical mirror with the same radius?

2. *Using a reversed coordinate system.* In the previous problem, suppose that in writing the fields of the reflected wave we decide to use a coordinate system in which the $z$ direction is reversed, i.e., we write the reflected fields as $\tilde{E}(x, z')$ where $z' = -z$. What will be the answers to the same questions?

3. *Spherical waves and circular interference rings.* When we work with coherent light sources it is commonplace to observe circular interference rings or "bulls-eye patterns" in the beam coming from a laser resonator or from some subsequent

optical system. Suppose that a set of such concentric rings with maxima at radii $r_n$ are observed in the output beam pattern from an optical system, and the hypothesis is that these rings must result from interference between two spherical waves coming from different source points within the system. Analyze what the radial spacing of such interference rings would be and discuss how, if you were able to measure a set of such radii $r_n$ in the output beam, you might be able to determine where one or both of the source points were located within the system.

4. *Complex transverse source point coordinates.* Demonstrate that if we introduce complex values for one or both of the transverse source point coordinates—say, a small imaginary value for the transverse coordinate $x_0$—as well as a complex value for $z_0$, the result is to produce a tilted (and possibly transversely displaced) gaussian-spherical beam which travels at a small angle to the $z$ axis in the $x, z$ plane.

---

## 16.4 HIGHER-ORDER GAUSSIAN MODES

The gaussian beam expression we introduced in the previous section is really only the lowest-order solution in an infinite family of higher-order solutions to the same Huygens' integral, or the same free-space paraxial wave equation. The higher-order solutions to the same equations can take the form either of Hermite-gaussian functions in rectangular coordinates, or of Laguerre-gaussian functions in cylindrical coordinates. These higher-order gaussian modes are of considerable importance both in practical lasers and in optical beam analyses. Hence we reproduce their mathematical derivation in some detail in this section.

### Lowest-Order Mode: The Differential Approach

A slightly more formal way of deriving the expressions we have just given for a lowest-order gaussian beam is to assume a trial solution to the paraxial wave equation of the form

$$\tilde{u}(x, y, z) = A(z) \times \exp\left[-jk\frac{x^2 + y^2}{2\tilde{q}(z)}\right], \quad (36)$$

where $A(z)$ and $\tilde{q}(z)$ are initially assumed to be unknown. If we substitute this trial solution into the paraxial wave equation, we obtain

$$\left[\left(\frac{k}{2}\right)^2 \left(\frac{d\tilde{q}}{dz} - 1\right)(x^2 + y^2) - \frac{2jk}{\tilde{q}}\left(\frac{\tilde{q}}{A}\frac{dA}{dz} + 1\right)\right] A(z) = 0. \quad (37)$$

Now, the only way in which this equation can be satisfied for all $x$ and $y$ is to set both of the differential expressions inside the large square brackets to zero. We then have the two differential equations

$$\frac{d\tilde{q}(z)}{dz} = 1 \quad \text{and} \quad \frac{dA(z)}{dz} = -\frac{A(z)}{\tilde{q}(z)}, \quad (38)$$

whose solutions are given by

$$\tilde{q}(z) = \tilde{q}_0 + z - z_0 \quad \text{and} \quad \frac{A(z)}{A_0} = \frac{\tilde{q}_0}{\tilde{q}(z)}. \tag{39}$$

These correspond exactly to the complex-source-point expressions derived in the preceding section.

### Higher-Order Solutions in Rectangular Coordinates

We will now show how this same trial solution approach can be extended to find higher-order Hermite-gaussian eigensolutions $\tilde{u}_{nm}(x, y, z)$ or Laguerre-gaussian eigensolutions $\tilde{u}_{pm}(r, \theta, z)$ to the paraxial wave equation.

We will derive first the higher-order Hermite-gaussian solutions to the wave equation in rectangular coordinates. In rectangular coordinates the elementary solutions can be separated into products of identical solutions in the $x$ and $y$ directions, i.e.,

$$\tilde{u}_{nm}(x, y, z) = \tilde{u}_n(x, z) \times \tilde{u}_m(y, z), \tag{40}$$

where $\tilde{u}_n(x, z)$ and $\tilde{u}_m(y, z)$ have the same mathematical form. We can therefore find the solutions in only one rectangular coordinate and bring in the other coordinate by analogy.

The paraxial wave equation in one transverse coordinate reduces to

$$\frac{\partial^2 \tilde{u}_n(x, z)}{\partial x^2} - 2jk \frac{\partial \tilde{u}_n(x, z)}{\partial z} = 0. \tag{41}$$

As one way of looking for higher-order solutions, let us now write a more general trial solution for the wave amplitude $\tilde{u}(x, z)$ in the form

$$\tilde{u}_n(x, z) = A(\tilde{q}(z)) \times h_n\left(\frac{x}{\tilde{p}(z)}\right) \times \exp\left[-jk\frac{x^2}{2\tilde{q}(z)}\right], \tag{42}$$

where $\tilde{q} = \tilde{q}(z)$ is the same as in the preceding; $A(\tilde{q})$ and $h_n(x/\tilde{p})$ are initially unknown functions; and $\tilde{p} = \tilde{p}(z)$ is a distance-dependent scaling factor in the argument of $h_n$.

Substituting this form into the paraxial wave equation, and assuming that $\tilde{q}(z)$ will continue to obey the propagation rule $d\tilde{q}/dz = 1$, then converts the paraxial wave equation into a differential relation for $h_n(x/\tilde{p})$, namely

$$h_n'' - 2jk\left[\frac{\tilde{p}}{\tilde{q}} - \tilde{p}'\right]xh_n' - \frac{jk\tilde{p}^2}{\tilde{q}}\left[1 + \frac{2\tilde{q}}{A}\frac{dA}{d\tilde{q}}\right]h_n = 0, \tag{43}$$

where $h_n'$ and $h_n''$ mean the first and second derivatives of $h_n$ with respect to its total argument, e.g., $h_n'(y) = dh_n(y)/dy$, and $\tilde{p}' \equiv d\tilde{p}(z)/dz$. But, Equation 16.43 is very similar to the standard differential equation for the Hermite polynomials $H_n(x/\tilde{p})$, which has the form

$$H_n'' - 2(x/\tilde{p})H_n' + 2nH_n = 0. \tag{44}$$

The two equations for $h_n$ and $H_n$ will in fact become the same if we can find solutions for $\tilde{p}(z)$ and $A(\tilde{q})$ which satisfy simultaneously the two conditions

$$2jk\left[\frac{\tilde{p}}{\tilde{q}} - \frac{d\tilde{p}}{dz}\right] = \frac{2}{\tilde{p}} \quad \text{or} \quad \frac{d\tilde{p}}{dz} = \frac{\tilde{p}}{\tilde{q}} + \frac{j}{k\tilde{p}}, \qquad (45)$$

and

$$\frac{-jk\tilde{p}^2}{\tilde{q}}\left[1 + \frac{2q}{A}\frac{dA}{d\tilde{q}}\right] = 2n \quad \text{or} \quad \frac{2q}{A}\frac{dA}{d\tilde{q}} = \frac{2jnk\tilde{p}^2}{\tilde{q}} - 1. \qquad (46)$$

Now, there are at least two, and probably many different ways in which Equations 16.45 and 16.46 can be solved, with each solution leading to a different family of higher-order Hermite-gaussian solutions. We will describe in this section one family of such solutions, which we will refer to as the "standard" Hermite-gaussian solutions. In the following section we will describe another alternative set of solutions which we will refer to as the "elegant" solutions.

### The "Standard" Hermite Polynomial Solutions

The set of Hermite-gaussian solutions that we will derive in this section are by far the most widely used set of such solutions, as well as being the closest to simple physical solutions in ordinary stable lasers. However, this set is also perhaps the most complicated and inelegant approach from a mathematical viewpoint. This standard approach to Hermite polynomial solutions is obtained by assuming that the scale factor $\tilde{p}(z)$ in the function $h_n(x/\tilde{p})$ will be purely real, and in fact will be related to the gaussian spot size $w(z)$ in the form

$$\frac{1}{\tilde{p}(z)} \equiv \frac{\sqrt{2}}{w(z)}. \qquad (47)$$

As motivation for this approach we can note that if this is valid then the higher-order solutions of Equation 16.42, namely

$$\tilde{u}_n(x,z) = h_n\left(\frac{\sqrt{2}x}{w(z)}\right)\exp\left[\frac{-jkx^2}{2R(z)} - \frac{x^2}{w^2(z)}\right], \qquad (48)$$

will have the same normalized shape at every transverse plane $z$. That is, these functions will change in transverse scale like $w(z)$, and will acquire spherical curvature $R(z)$, but their amplitude profiles will remain unchanged in shape at any plane $z$.

We must first verify that this form for $\tilde{p}(z)$ will satisfy the differential equation 16.45. Since the spot size $w(z)$ is related to $\tilde{q}(z)$ by

$$\frac{1}{\tilde{q}(z)} = \frac{1}{R(z)} - j\frac{\lambda}{\pi w^2(z)} \qquad (49)$$

we can use this to obtain

$$\frac{1}{w^2(z)} = \frac{jk}{4}\left[\frac{1}{\tilde{q}(z)} - \frac{1}{\tilde{q}^*(z)}\right] = \frac{jk}{4}\frac{\tilde{q}^*(z) - \tilde{q}(z)}{\tilde{q}(z)\tilde{q}^*(z)}. \qquad (50)$$

We can also note that the formulas for $\tilde{q}(z)$ imply the useful relations that

$$d\tilde{q}^*/dz = d\tilde{q}/dz = 1 \quad \text{and hence} \quad \tilde{q}^* - \tilde{q} = \tilde{q}_0^* - \tilde{q}_0. \qquad (51)$$

## 16.4 HIGHER-ORDER GAUSSIAN MODES

The reader can then verify that the form for $\tilde{p}(z)$ given in Equation 16.47 does indeed satisfy Equation 16.45.

The simplest way to satisfy the equation for $A(q)$ is then perhaps to use the definition of $1/\tilde{q}$, and the fact that $d\tilde{q}^* = d\tilde{q}$ to rewrite Equation 16.46 as

$$\frac{dA}{A} = -\frac{1}{2}\frac{d\tilde{q}}{\tilde{q}} + \frac{n}{2}\left(\frac{d\tilde{q}^*}{\tilde{q}^*} - \frac{d\tilde{q}}{\tilde{q}}\right). \qquad (52)$$

Integrating this equation then yields

$$A(\tilde{q}) = A_0 \times \left(\frac{\tilde{q}_0}{\tilde{q}(z)}\right)^{1/2} \times \left(\frac{\tilde{q}_0}{\tilde{q}_0^*}\frac{\tilde{q}^*(z)}{\tilde{q}(z)}\right)^{n/2}. \qquad (53)$$

A complete set of properly normalized higher-order Hermite-gaussian mode functions for a beam propagating in free-space are thus given, in one transverse dimension, by

$$\tilde{u}_n(x, z) = \left(\frac{2}{\pi}\right)^{1/4}\left(\frac{1}{2^n n! w_0}\right)^{1/2}\left(\frac{\tilde{q}_0}{\tilde{q}(z)}\right)^{1/2}\left[\frac{\tilde{q}_0}{\tilde{q}_0^*}\frac{\tilde{q}^*(z)}{\tilde{q}(z)}\right]^{n/2}$$
$$\times H_n\left(\frac{\sqrt{2}x}{w(z)}\right)\exp\left[-j\frac{kx^2}{2\tilde{q}(z)}\right], \qquad (54)$$

where the $H_n$'s are the Hermite polynomials of order $n$, and $\tilde{q}(z)$ and $w(z)$ are exactly the same as for the lowest-order gaussian mode.

### The Guoy Phase Shift

The most compact and efficient way of writing the higher-order Hermite-gaussian eigenmodes is as in Equation 16.54, using the ratios of $\tilde{q}(z)$ and $\tilde{q}^*(z)$ raised to appropriate powers. This form can also be converted, however, to a more commonly used form involving the real spot size $w(z)$ and a phase angle $\psi(z)$, as follows.

Let us associate a magnitude and especially a phase angle $\psi(z)$ with the complex $\tilde{q}$ parameter at any plane $z$ by writing

$$\frac{j}{\tilde{q}} = \frac{\lambda}{\pi w^2}\left[1 + j\frac{\pi w^2}{R\lambda}\right] \equiv \frac{\exp[j\psi(z)]}{|\tilde{q}|}, \qquad (55)$$

so that the phase angle $\psi = \psi(z)$ is given at any plane $z$ by

$$\tan\psi(z) \equiv \frac{\pi w^2(z)}{R(z)\lambda}. \qquad (56)$$

We have included the factor of $j$ in Equation 16.55 because it will be convenient later on to have $\psi(z) = 0$ at the "waist" of a gaussian beam, where the spot size $w$ is finite but the radius of curvature $R$ becomes infinite.

If we use this definition, we can then show (after some algebra) that the first part of the $\tilde{q}_0/w_0\tilde{q}(z)$ normalization factor in Equation 16.54 can be written as

$$\frac{1}{w_0}\frac{\tilde{q}_0}{\tilde{q}(z)} = \frac{\exp[j(\psi(z) - \psi_0)]}{w(z)}, \qquad (57)$$

where $\psi_0 \equiv \psi(z_0)$ is the initial value of $\psi(z)$ at $z = z_0$. The lowest-order gaussian-spherical wave (Equation 16.33) may then be written in the alternative form

$$\tilde{u}_0(x,z) = \left(\frac{2}{\pi}\right)^{1/4} \sqrt{\frac{\exp j\left[\psi(z) - \psi_0\right]}{w(z)}} \exp\left[-j\frac{kx^2}{2\tilde{q}(z)}\right]. \tag{58}$$

In other words, the factor $(1/w_0) \times (\tilde{q}_0/\tilde{q}(z))$ contains the necessary $1/w(z)$ normalization factor in front of the gaussian beam expression, along with an added phase shift term given by $\psi(z) - \psi_0$.

For all higher-order Hermite-gaussian modes we must also include the additional factors given by the term

$$\left[\frac{\tilde{q}_0}{\tilde{q}_0^*}\frac{\tilde{q}^*(z)}{\tilde{q}(z)}\right]^{n/2} \equiv \exp\left[jn\left[\psi(z) - \psi_0\right]\right] \tag{59}$$

appearing on the right-hand side of Equation 16.54. This factor gives rise to a pure phase shift. With this factor included, the higher-order Hermite-gaussian mode functions of Equation 16.54 can be written in the alternative form

$$\begin{aligned}\tilde{u}_n(x) = \left(\frac{2}{\pi}\right)^{1/4} &\sqrt{\frac{\exp[-j(2n+1)(\psi(z) - \psi_0)]}{2^n n! w(z)}} \\ &\times H_n\left(\frac{\sqrt{2}x}{w(z)}\right) \exp\left[-j\frac{kx^2}{2R(z)} - \frac{x^2}{w^2(z)}\right].\end{aligned} \tag{60}$$

We will discuss the physical significance of the so-called "Guoy phase shift term" $\psi(z)$ in the following chapter.

### Hermite-Gaussian Mode Expansions

The Hermite-gaussian functions $\tilde{u}_n(x,z)$ we have derived here provide a complete basis set of orthogonal functions characterized by a single complex parameter, the complex $\tilde{q}_0$ parameter at any arbitrary reference plane $z_0$. (We will discuss the physical significance of this parameter in the following chapter.) These functions obey the orthonormality condition

$$\int_{-\infty}^{\infty} \tilde{u}_n^*(x,z) \tilde{u}_m(x,z)\, dx = \delta_{nm}, \tag{61}$$

independent of either $z$ or of $\tilde{q}_0$. They can thus be used as a basis set to expand any arbitrary paraxial optical beam $\tilde{E}(x,y,z)$ in the form

$$\tilde{E}(x,y,z) = \sum_n \sum_m c_{nm} \tilde{u}_n(x,z) \tilde{u}_m(y,z) e^{-jkz}. \tag{62}$$

If we multiply both sides of this by $u_n^*(x,z) u_m^*(y,z)$ and integrate across the full cross section, we can find that the expansion coefficients $c_{nm}$ are given by

$$c_{nm} = \int_{-\infty}^{\infty} \int_{-\infty}^{\infty} \tilde{E}(x,y,z) u_n^*(x,z) u_m^*(y,z)\, dx\, dy. \tag{63}$$

The coefficients $c_{nm}$ will depend upon the arbitrary choice of $\tilde{q}_0$ at $z_0$. There is thus in general no unique or necessary way of choosing the waist size $w_0$ or waist

## 16.4 HIGHER-ORDER GAUSSIAN MODES

location $z_0$ for the basis set to expand a given beam pattern $\tilde{E}(x,y)$. We may attempt to choose these parameters to given an expansion which best fits some other physical constraint in the problem, or which gives an expansion that best fits the actual field $\tilde{E}(x,y,z)$ with the smallest number of terms.

### Astigmatic Mode Functions

As we noted earlier, the Hermite-gaussian function $\tilde{u}_n(x,z)$ in one dimension corresponds essentially to a cylindrical wave, and hence has a normalization factor of $1/\tilde{q}^{1/2}(z)$ or $1/w^{1/2}(z)$ rather than $1/\tilde{q}(z)$ or $1/w(z)$. The overall normalization factor for the field function $\tilde{u}_{nm}(x,y,z)$ is then the product of the individual normalization functions for $\tilde{u}_n(x,z)$ and $\tilde{u}_m(y,z)$.

There is no fundamental reason in fact why the complex beam parameters $\tilde{q}_{0x}$, $w_{0x}$ and $z_{0x}$ associated with the functions $\tilde{u}_n(x,z)$ in the $x$ transverse coordinate cannot have entirely different values from the corresponding parameters $\tilde{q}_{0y}$, $w_{0y}$ and $z_{0y}$ associated with the functions $\tilde{u}_m(y,z)$ in the $y$ direction. The gaussian beam solutions can thus be converted, where it seems necessary or useful, into a set of somewhat more general astigmatic gaussian beam modes by replacing $\tilde{q}(z)$, and the related quantities $w(z)$ and $\psi(z)$, by separate values for the $x$ and $y$ coordinates wherever these quantities appear in the Hermite-gaussian solutions.

Note in particular that only half of the fundamental Guoy phase comes from each transverse coordinate. The Guoy phase shift for a uniform cylindrical wave through a focus is only 90° rather than 180°, and the phase factor in Huygens' integral in one transverse dimension is $\sqrt{j}$ rather than $j$.

### Cylindrical Coordinates: Laguerre-Gaussian Modes

An alternative but equally valid family of solutions to the paraxial wave equation can be written in cylindrical rather than rectangular coordinates. These solutions are in general the *Laguerre-gaussian solutions* of the form

$$\tilde{u}_{pm}(r,\theta,z) = \sqrt{\frac{2p!}{(1+\delta_{0m})\pi(m+p)!}} \frac{\exp j(2p+m+1)(\psi(z)-\psi_0)}{w(z)}$$

$$\times \left(\frac{\sqrt{2}r}{w(z)}\right)^m L_p^m\left(\frac{2r^2}{w(z)^2}\right) \exp\left[-jk\frac{r^2}{2\tilde{q}(z)} + im\theta\right]. \tag{64}$$

In these solutions the integer $p \geq 0$ is the radial index and the integer $m$ is the azimuthal mode index; the $L_p^l$ functions are the generalized Laguerre polynomials; and all the other quantities $\tilde{q}$, $w$ and $\psi$ are exactly the same as in the Hermite-gaussian situation.

These solutions are written using the "standard" transverse scaling $r/\tilde{p}(z) = \sqrt{2}r/w(z)$ that we used for the Hermite-gaussian solutions in Equation 16.47 of this section. An alternative set of complex Laguerre-gaussian cylindrical solutions using the complex scaling $r/\tilde{p}(z) = \sqrt{jkr^2/2\tilde{q}(z)}$ which we will introduce in the following section could equally well be developed. In either situation these modes have cylindrical symmetry, with modes having circles of constant intensity in the radial direction and an $e^{im\theta}$ variation in the azimuthal direction. Alternatively linear combinations of the $\pm m$ terms can be formed to give $\cos m\theta$ and/or $\sin m\theta$ variations, leading to $2m$ nodal lines running radial outward from the mode axis. Laguerre-gaussian exhibit the same Guoy phase shift as the rectangular modes.

The Laguerre-gaussian solutions provide an equally general but alternative basis set to the Hermite-gaussian solutions for expanding an arbitrary optical beam $\tilde{u}(r,\theta,z)$ in free space (provided we are knowledgeable about generalized Laguerre polynomials). Since both the Hermite-gaussian and Laguerre-gaussian functions form complete sets, we must be able to expand any Hermite solution in terms of the Laguerre functions and vice versa.

The Laguerre-gaussian functions will perhaps be more convenient for problems having a large amount of cylindrical symmetry, and will probably not provide the most convenient set for expanding any real optical beam having substantial astigmatism between $x$ and $y$ axes. In real lasers the Brewster windows and any other tilted surfaces or distorted elements usually provide a small but inherent rectangular symmetry to the laser cavity. Real lasers, therefore, overwhelmingly elect to oscillate in near-Hermite-gaussian rather than near-Laguerre-gaussian modes. Experiments with very carefully aligned gas lasers having internal mirrors and no Brewster windows have, however, clearly demonstrated oscillations in Laguerre-gaussian modes with higher-order radial and azimuthal symmetry.

## REFERENCES

A summary of the standard Hermite-gaussian mode functions and some of their properties is given in A. E. Siegman and E. A Sziklas, "Mode calculations in unstable resonators with flowing saturable gain. I: Hermite-gaussian expansion," *Appl. Optics* **13**, 2775–2792 (December 1974).

More detailed discussions of the vector properties of free-space beam modes can be found in, for example, L. W. Davis and G. Patsakos, "TM and TE electromagnetic beams in free space," *Optics Lett.* **6**, 22–23 (January 1981) and "Comment on 'Representation of vector electromagnetic beams'," *Phys. Rev. A* **26**, 3702–3703 (December 1982), and the references cited therein.

A very useful reference source for orthogonal polynomials and almost all other special functions is M. Abramowitz and I. A. Stegun, *Handbook of Mathematical Functions* (Dover Publications, Inc., New York, 1965).

---

### Problems for 16.4

1. *Intensity contours for a higher-order Hermite-gaussian mode.* Consider as an example the $\tilde{u}_{22}(x,y)$ higher-order gaussian mode in rectangular coordinates. Develop a formula and write a computer program to trace out the constant-amplitude contours in the $xy$ plane for this mode, and make an isoamplitude contour map or model.

2. *Finding the mode content of an arbitrary optical beam (research problem).* Suppose you have an optical beam made up of an arbitrary, unknown combination of lowest and higher-order Hermite-gaussian modes. How could you *experimentally* separate this beam into its individual Hermite-gaussian components, with each component being directed into a separate single-mode optical fiber?

   So far as I know, no good solution to this problem has yet been given. The problem of identifying and measuring the mode content of an arbitrary optical beam is addressed, however, in M. A. Golub, *et al.*, "Synthesis of spatial filters for investigation of the transverse mode composition of coherent radiation," *Sov. J. Quantum Electron.* **12**, 1208–1209 (September 1982).

3. *Recursion relation for Hermite-gaussian modes* A standard recursion relation for the Hermite polynomials $H_n(x)$ is $H_{n+1}(x) = 2xH_n(x) - 2nH_{n-1}(x)$. Using this relation derive a similar recursion relation for the normalized Hermite-gaussian functions $\tilde{u}_n(x)$ introduced in this section.

---

## 16.5 COMPLEX-ARGUMENT GAUSSIAN MODES

The "standard" Hermite-gaussian solutions developed in the preceding section are the most commonly used form for the Hermite-gaussian eigenmodes, and the form most often given in the laser literature. They represent, among other things, the set of Hermite-gaussian modes that are the closest approximation to the actual higher-order modes of finite-mirror stable resonators. These modes are perhaps somewhat inelegant mathematically, however, in that we have a complex scaling factor in the gaussian function but only a real scaling factor in the Hermite polynomials. As a result, Equation 16.54 is somewhat messy, since the inelegant combinations of $\tilde{q}$ and $\tilde{q}^*$ values must be carried along in all the normalization factors.

There are, however, many other alternative choices for $\tilde{p}(z)$ and $A(\tilde{q})$ that will satisfy the differential equations 16.45 and 16.46 for these quantities derived in the preceding section. To illustrate this we will develop one of these alternative families of solutions in this section. The motivation behind this particular solution is to use the same *complex* scaling factor, that is, the quantity $\sqrt{jkx^2/2\tilde{q}}$, as the argument both in the gaussian exponent and in the Hermite polynomial functions.

### The "Elegant" Hermite Polynomial Solutions

To do this we will define the complex scale factor $\tilde{p}$ in the functions $h_n(x/p)$ of the previous section not by $1/\tilde{p} = \sqrt{2}/w$ as in Equation 16.47, but by

$$\frac{1}{\tilde{p}(z)} \equiv \sqrt{\frac{jk}{2\tilde{q}(z)}}. \tag{65}$$

The reader can verify that this choice will also satisfy the differential equation 16.45, and that the differential condition 16.46 for $A(\tilde{q})$ then takes on the significantly simpler form

$$\frac{\tilde{q}}{A}\frac{dA}{d\tilde{q}} = -\frac{n+1}{2}. \tag{66}$$

If we solve this, the Hermite-gaussian eigenfunctions then take on the alternative form

$$\hat{u}_n(x,z) = \hat{u}_0 \left[\frac{\tilde{q}_0}{\tilde{q}(z)}\right]^{n+1/2} H_n\left(\sqrt{\frac{jkx^2}{2\tilde{q}(z)}}\right) \exp\left[-j\frac{kx^2}{2\tilde{q}(z)}\right]. \tag{67}$$

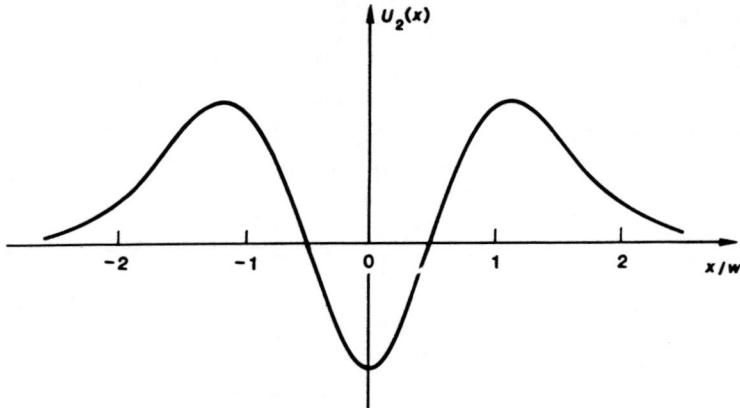

**FIGURE 16.9**
The "standard" (real-argument) Hermite-gaussian function $\tilde{u}_n(x)$ for $n = 2$.

This alternative solution puts the same complex argument $\sqrt{jkx^2/2\tilde{q}(z)}$ into both the Hermite polynomial and the gaussian exponent, and is much simpler than the "standard" solutions given in Equation 16.54.

This alternative set of Hermite-gaussian solutions represents an equally valid complete set of solutions to the paraxial wave equation in free space. This alternative set is more compact and elegant, with some interesting analytical properties, but it is also perhaps less directly useful physically. For example, whereas these alternative functions $\tilde{u}_n(x,z)$ still form a mathematically complete set, they are no longer orthogonal to each other in the usual sense. Rather the alternative functions $\tilde{u}_n$ are *biorthogonal* to a set of adjoint functions $v_n$ given by

$$\hat{v}_n(x,z) = H_n\left(\sqrt{\frac{-jk}{2\tilde{q}^*}}\, x\right), \tag{68}$$

(no gaussian factor) with the orthogonality relation now being

$$\int_{-\infty}^{\infty} \hat{u}_n(x,z)\, \hat{v}_n^*(x,z)\, dx = c_n \delta_{nm}, \tag{69}$$

where $c_n$ is an appropriate normalization constant.

### Properties of the "Elegant" Solutions

The lowest-order or $n=0$ and $n=1$ members of the "standard" and the alternative or "elegant" sets of Hermite-gaussian functions given in Equations 16.54 and 16.67 are indistinguishable from each other, since they consist only of the gaussian exponential, or of this exponential multiplied by $x$. There are, however, significant differences between the higher-order modes in the two sets.

The next higher-order function $n = 2$, for example, uses the Hermite polynomial $H_2(x) = 4x^2 - 2$, so that the "standard" solution $\tilde{u}_2(x)$ has the form

$$\tilde{u}_2(x,z) = \text{const} \times \left[\frac{4x^2}{w^2} - 1\right] \exp\left[-j\frac{kx^2}{2\tilde{q}}\right], \tag{70}$$

## 16.5 COMPLEX-ARGUMENT GAUSSIAN MODES

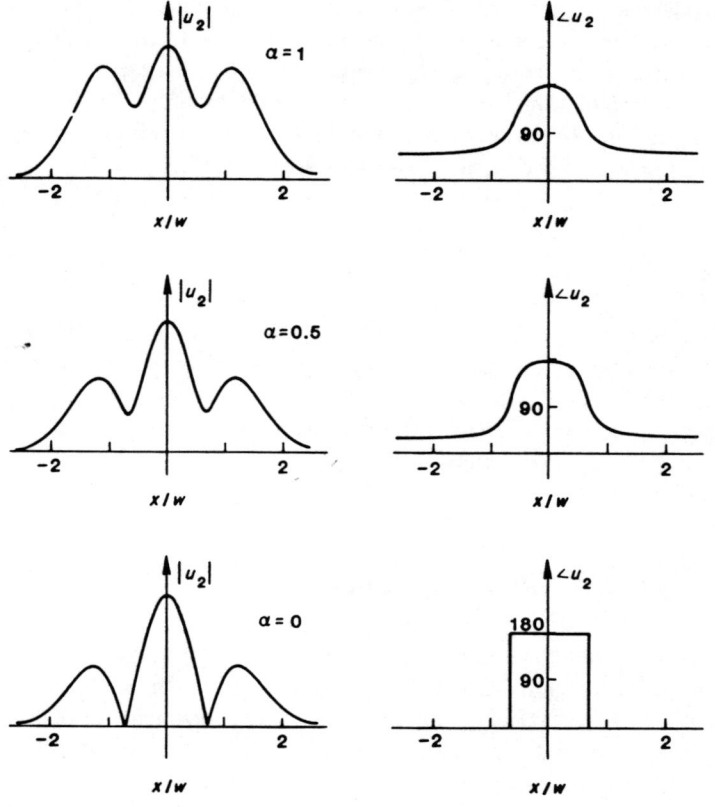

**FIGURE 16.10**
The "elegant" (complex-argument) Hermite-gaussian function $\hat{u}_n(x)$ for $n = 2$ and for different values of the parameter $\alpha \equiv \pi w^2/R\lambda$.

whereas the "elegant" solution $\hat{u}_2(x)$ becomes

$$\hat{u}_2(x) = \text{const} \times \left[\frac{jkx^2}{\tilde{q}} - 1\right] \exp\left[-j\frac{kx^2}{2\tilde{q}}\right]$$

$$= \text{const} \times \left[\frac{2(1+j\alpha)x^2}{w^2} - 1\right] \exp\left[-j\frac{kx^2}{2\tilde{q}}\right], \tag{71}$$

where $\alpha = \pi w^2/R\lambda$. The two functions are different, though somewhat similar, even for $\alpha = 0$; and for nonzero values of $\alpha$ the complex argument in the polynomial causes the usual nulls in the function to be filled in, as illustrated in Figure 16.10, and also produces an additional phase variation across the beam which is not purely spherical in form, as Figure 16.10 also shows. For $n \geq 2$ the amplitude and phase patterns of the higher-order alternative modes also change in shape with propagation distance $z$ (unlike the "standard" modes) because of the change in $\tilde{q}(z)$ with distance.

Either family of Hermite-gaussian solutions, 16.54 or 16.67, is equally valid as a general basis set for analytic expansions of arbitrary optical beams. Stable laser resonators with spherical mirrors and neligible beam aperturing will generally have real eigenmodes that are much closer to the "standard" or $\sqrt{2}x/w$ family of Hermite-gaussian modes, and hence this form for the Hermite-gaussian solutions is much more widely used in the laser literature. More general complex paraxial systems including soft gaussian apertures, such as we will discuss later, do lead to Hermite modes with complex arguments, like the "elegant" or $\sqrt{jkx^2/2\tilde{q}}$ family, and this type of solution is now being more extensively considered. In all situations we can develop astigmatic mode solutions with different fundamental parameters in the $x$ and $y$ coordinates if this seems useful.

## REFERENCES

The complex Hermite solutions described in this section were introduced in A. E. Siegman, "Hermite-gaussian functions of complex argument as optical-beam eigenfunctions," *J. Opt. Soc. Am.* **61**, 1093–1094 (September 1973).

More general complex free-space solutions are also given by R. Pratesi and L. Ronchi, "Generalized gaussian beams in free space," *J. Opt. Soc. Am.* **67**, 1274–1276 (September 1977).

## 16.6 GAUSSIAN BEAM PROPAGATION IN DUCTS

Gaussian beams in free space always remain gaussian, but do of course spread outward due to diffraction effects as they propagate. In a medium with a quadratic transverse variation of index of refraction, however, it becomes possible to trap and propagate a particular confined gaussian beam which neither spreads nor contracts with distance. We have already discussed the ray-trapping properties of graded-index optical waveguides or ducts. Such ducts are also of substantial practical and analytical interest in gaussian beam optics as well as in ray optics. The two topics are, in fact, essentially identical in concept and in results.

### Gaussian Beam Propagation in Ducts

Suppose again that the index of refraction $n(r)$ in a duct has a radial (or transverse) variation given by

$$n(r) = n_0 - \frac{1}{2}n_2 r^2. \tag{72}$$

The wave equation 16.1 in a medium with a quadratic transverse variation such as this must then be expanded to the form

$$\left[\nabla^2 + \omega^2 \mu\epsilon[1 - n_2(x^2 + y^2)]\right] \tilde{E}(x, y, z) = 0. \tag{73}$$

Converting this to the paraxial approximation in the same form as used in deriving Equation 16.5 then gives

$$\left[\nabla_{xy}^2 - k^2 n_2(x^2 + y^2) - 2jk\frac{\partial}{\partial z}\right] \tilde{u}(x, y, z) = 0. \tag{74}$$

## 16.6 GAUSSIAN BEAM PROPAGATION IN DUCTS

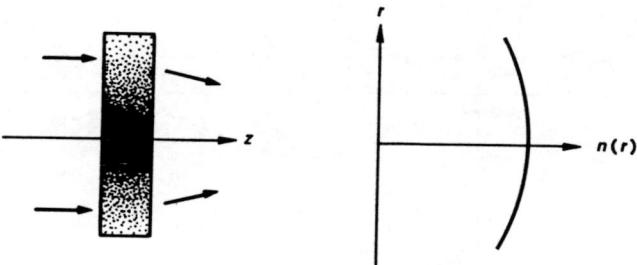

**FIGURE 16.11**
A short section of a duct.

The reader can verify that a stable solution to this equation is given by the confined or trapped gaussian beam

$$\tilde{u}(x,y,z) = \tilde{u}_0 \exp\left[-\frac{x^2 + y^2}{w_1^2} + j\frac{\lambda z}{w_1^2}\right], \tag{75}$$

where the spot size $w_1$ is given by

$$w_1^2 = \frac{\lambda}{\pi\sqrt{n_2}}. \tag{76}$$

This solution thus represents a *stable trapped gaussian eigenmode of fixed diameter in the waveguide or duct*. Note that the quadratic exponent for this wave is purely real, i.e., the wavefronts in this guided beam are exactly plane waves. Higher-order modes with Hermite-gaussian or Laguerre-gaussian form can also propagate in the same duct.

### Physical Interpretation

One way of understanding this confined mode is the following. In a duct with an index variation like Equation 16.72, each axial segment of length $\Delta z$ is like a thin lens with focal length $f = 1/n_2\Delta z$, as illustrated in Figure 16.11. For a gaussian beam with spot size $w_1$ as given in the preceding, the effects of diffraction spreading in each unit length are just canceled by this focusing effect, so that the beam size remains constant.

Note again that the steady-state gaussian spot size $w_1$ varies inversely as the strength $n_2$ of the transverse index variation—the stronger the focusing the smaller the steady-state beam profile that will be propagated in the duct. This gaussian eigenmode also acquires a small added phase shift per unit length, over and above the $e^{-jkz}$ factor, that is expressed by the $+j\lambda z/\pi w_1^2$ term. This indicates that the $z$-directed propagation constant in the duct, call it $k_d$, is not the on-axis value of $k$ in the medium but rather a guided-wave value given by

$$k_d = k - \lambda/\pi w_1^2. \tag{77}$$

The added phase shift $\psi(z)$ associated with the $-\lambda z/\pi w_1^2$ per unit length in the duct is just given by $d\psi(z)/dz = \lambda/\pi w_1^2 = 1/z_R$ where $z_R$ would be the Rayleigh range for the guided beam without the ducting effects. This is in fact exactly the same as the derivative $d\psi(z)/dz$ exactly at the waist for the Guoy effect which we will discuss in the following chapter.

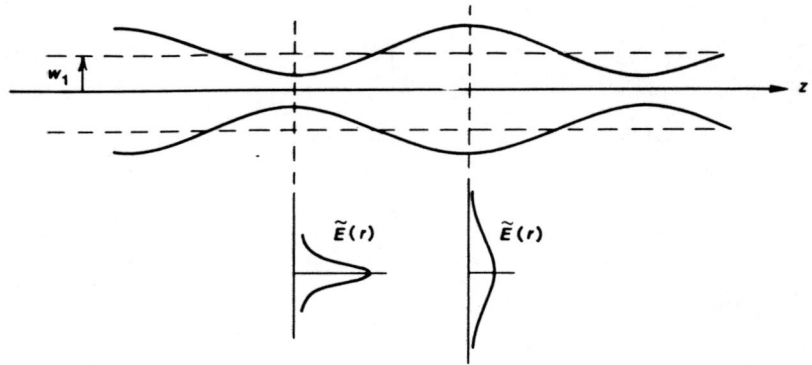

FIGURE 16.12
Beam scalloping in a quadratic duct.

### Beam Oscillations or Beam Scalloping in Ducts

Suppose we introduce into such a duct a gaussian beam which does not match the gaussian eigenmode of the duct, either in spot size or in wavefront curvature. Suppose the input beam is initially smaller than the steady-state spot size $w_1$. The diffraction spreading for this smaller beam will then be larger than for the steady-state spot size, whereas the refocusing produced by each unit length of the duct will be the same. The spot size will therefore begin to grow, and the gaussian beam will begin to spread with distance.

As soon as its spot size becomes larger than the steady-state value, however, the opposite condition will prevail, and the beam will be refocused again. An input beam with a spot size $w(z)$ either larger or smaller than the steady-state value $w_1$ will thus oscillate or scallop periodically inward and outward about the steady-state value in sausage-like fashion as it propagates down the guide, as shown in Figure 16.12.

This gaussian beam behavior is very much analogous to the oscillatory behavior of rays trapped in a stable duct, and to first order the beam oscillations will have the same oscillation period as the stable ray solutions we derived earlier. As an equally valid alternative explanation, we can say that the mismatched input beam excites in the duct a mixture of the lowest and higher-order eigenmodes of the duct. These eigenmodes then propagate with slightly different phase velocities, since the added phase shift term $\lambda z/\pi w_1^2$ is different for the lowest-order and for the higher-order eigenmodes. The periodic scalloping behavior then represents a beating phenomena in the axial direction as the higher-order modes propagate and interfere with different phases at difference distances along the duct.

### Ducts in Real Systems

The paraxial wave solutions in a duct will be exactly gaussian only if the index variation about the axis is exactly quadratic—and a purely quadratic decrease of the index with radius cannot continue indefinitely. The Hermite-gaussian solutions for propagation in a duct are thus generally approximations to the real modes, although they will be quite good approximations if the index variation is in fact approximately quadratic out to at least a few spot sizes $w_1$ of the trapped gaussian beam (as is quite often the situation). If this approximation fails, or if the index variation is other than quadratic, then other mathematical forms for the trapped modes must be sought; and there is a very large published literature on such trapped modes, especially of course for the modes in simple optical fibers which have step variations rather than continuous variations in index of refraction.

Transverse index variations leading to duct-like effects occur naturally on a random basis in many situations—for example, in laser rods, in inhomogeneous optical materials, in optical (or radio-wave) propagation through the atmosphere, or in sound-wave propagation through the oceans. Poor quality laser rods in the early days of lasers, in fact, often exhibited highly irregular transverse beam patterns, with many small spots across the face of the rod, due to strong ducting effects at random locations across the rod cross-section. More controlled ducting effects are also now deliberately produced to create waveguiding effects in optical fibers or in laser rods manufactured under such trade names as GRIN (graded refractive index) or SEL-FOC rods. Note that the spot size $w_1$ of the trapped beam in a duct will almost always be much smaller than the half-width $1/n_2^{1/2}$ of the duct itself, unless the duct is extremely small. Hence it will only be the quadratic or $n_2 \equiv -\partial^2 n/\partial r^2$ variation of the index close to the axis that will usually be important, even if the index profile no longer remains purely quadratic at larger radii.

Larger ducts can also trap an entire family of higher-order Hermite-gaussian or Laguerre-gaussian modes such as we have analyzed in the previous chapter. We will develop a more general theorem of gaussian ducts in Chapter 20, including both quadratic index and loss variations, by using a complex *ABCD* matrix approach

## REFERENCES

Two examples of more general treatments of optical ducting and guided waves with arbitrary index profiles are C. N. Kurtz and W. Streifer, "Guided waves in inhomogeneous focusing media, Part I and Part II," *IEEE Trans.* **MTT–17**, 11–15 and 250–253 (January 1969 and May 1969); and W. J. Firth, "Propagation of laser beams through inhomogeneous media," *Optics Commun.* **22**, 226–230 (August 1977).

Many of the optical properties and applications of graded-index or GRIN laser rods and lenses are surveyed in a Special Issue of *Appl. Optics* **21** (March 15 1982).

## Problems for 16.6

1. *Practical criteria for trapping a gaussian beam in a duct.* To express the criterion for trapping of a gaussian beam in an index duct in a more visible form, suppose that the index variation across a duct in given by the formula $n(x) = n_a + \Delta n \exp[-(x/a)^2]$ where $n_a$ is the background value in the duct; $\Delta n$ is the (small) peak value of the transversely varying part of the index; and $a$ is the $1/e$ spot size for this index variation. Show that the spot size $w$ of the gaussian beam that will be trapped in this duct is then given by

$$\frac{w}{a} \approx \left(\frac{n_a}{2\Delta n}\right)^{1/4} \times \left(\frac{\lambda}{\pi a}\right)^{1/2},$$

with the obvious condition that $w$ must be somewhat smaller than $a$ for the gaussian beam theory to be a reasonable approximation.

2. *Lensing or ducting effects in a saturated laser amplifier.* In an experiment carried out at the Bell Telephone Laboratories, a powerful gaussian laser beam was passed through a laser amplifier, and certain interesting power-dependent focusing effects were observed. The incident laser beam was powerful enough to cause at least partial saturation of the amplifying transition in the laser amplifier, and because the signal fields are strongest at the center of a gaussian incident beam, the degree of saturation was largest on the axis of the amplifier tube, decreasing radially outward. When the saturation took place, it was found that the amplifier tube began to act as a (weak) lens, with the sign of this lens effect (convergent or divergent) depending on whether a weak probing beam used to observe lens effect had a frequency slightly above or slightly below the atomic center frequency.

   Explain the physical causes of this effect, and in particular whether the lens should be convergent or divergent above and below the laser center frequency. (For simplicity, assume that the gain profile is uniform across the amplifier-tube cross section before saturation, and that the atomic transition is homogeneously broadened.)

3. *Higher-order eigenmodes in ducts.* Find the higher-order Hermite-gaussian modes in a quadratic duct—that is, find the higher-order solutions to the paraxial wave equation given in this section. (Hint: The standard analysis given in quantum-mechanics texts for the quantum wavefunctions of a harmonic oscillator in a quadratic potential well may be useful.)

## 16.7 NUMERICAL BEAM PROPAGATION METHODS

Hermite-gaussian modes, particularly the lowest-order gaussian beams, provide extremely useful tools for analyzing optical beam propagation in simple situations, especially in low-loss stable optical resonators, and in other situations where the physical modes of the problem are close to Hermite-gaussian in character, and where the effects of diffraction by hard-edged apertures are negligible or entirely absent. We can even, if necessary, expand any arbitrary optical beam as a summation of Hermite-gaussian modes; and then calculate the propagation of the arbitrary beam by calculating the propagation of the individual Hermite-gaussian modes.

There are many situations in real optical systems, however—such as unstable resonator problems, for example—where aperture diffraction effects play a major role. We must then treat the effects of edge diffraction, and analyze the propagation of beams with rather arbitrary and irregular amplitude and phase profiles. Numerical calculation methods play a large role in such analyses. The most efficient numerical methods then generally center around the use of Huygens' integral together with various "fast transform" numerical methods. In this section we will review briefly some of the analytical and numerical tools that become important in handling these more messy situations that arise in real-world problems.

### Paraxial Wave Propagation: The Finite Difference Approach

We have already mentioned the possibility of calculating the forward propagation of an arbitrary optical beam by writing the paraxial wave equation in the form

$$\frac{\partial \tilde{u}(s,z)}{\partial z} = -\frac{j}{2k}\nabla_t^2 \tilde{u}(s,z), \tag{78}$$

and then integrating this equation forward numerically using so-called finite difference methods, first to calculate the transverse derivative of the known wavefunction at one plane, and then to step forward in $z$ to the next plane.

This finite-difference approach to paraxial wave propagation can be of some practical usefulness for calculating beam propagation through inhomogeneous regions, such as perturbed atmospheres or problems involving thermal blooming or an inhomogeneous laser medium. Even in such inhomogeneous situations, however, fast transform methods, properly applied, are probably still superior.

### Huygens' Integral: Fourier Transform Interpretation

Huygens' integral provides another straightforward way to propagate an arbitrary optical wavefront from an input plane at $z_0$ to any later plane $z$. In doing this, it is particularly useful to note that the one-dimensional Huygens' integral written in rectangular coordinates has exactly the form of a convolution integral.

That is, Huygens' integral as given in Equation 16.25 has exactly the form of a convolution (in $x$) of the input field $\tilde{u}_0(x_0, z_0)$ with a spherical wavefunction $\exp[-j\pi x^2/(z-z_0)\lambda]$ in the form

$$\tilde{u}(x,z) = \tilde{u}_0(x_0) * \exp[-j\pi x_0^2/(z-z_0)\lambda], \tag{79}$$

where the symbol $*$ indicates the convolution operation. (We leave out the constant in front for simplicity.)

The convolution of two functions, as in Equation 16.79, can be accomplished, however, according to Fourier transform theory, by ($i$) calculating the Fourier transform of each function individually; ($ii$) multiplying together the two Fourier transforms point by point to get a product transform; and then ($iii$) inverse Fourier transforming this product transform to get the desired convolution.

Efficient fast Fourier transform algorithms then provide a very practical way of doing the required transforms in order to numerically convolve two arbitrary functions such as Equation 16.79 on a digital computer. (In doing Huygens' integral calculations in this fashion, the spherical function corresponding to the kernel in Huygens' integral must be transformed only once and then stored.) This approach is generally by far the most efficient way to evaluate the Huygens-Fresnel integral numerically, in order to calculate the propagation and diffraction spreading of an arbitrary optical beam in a numerical calculation, if the work is to be done in rectangular coordinates.

### Alternative Fourier Transform Approach

As a slightly different approach, we can also rewrite Huygens' integral for one transverse dimension in the form

$$\tilde{u}(x,z) = \exp\left(\frac{-j\pi x^2}{L\lambda}\right) \sqrt{\frac{j}{L\lambda}} \int_{-\infty}^{\infty} \tilde{u}_0'(x_0, z_0) \times \exp[j(2\pi/L\lambda)xx_0]\, dx_0, \quad (80)$$

where $L \equiv z - z_0$, and $\tilde{u}_0'(x_0, z_0)$ is a modified input function given by

$$\tilde{u}_0'(x_0, z_0) \equiv \tilde{u}_0(x_0, z_0)\, e^{-j\pi x_0^2/L\lambda}. \tag{81}$$

But Equation 16.80 simply has the mathematical form of a Fourier transform between the variables $x$ and $x_0$. In this form, therefore, the Huygens-Fresnel propagation integral appears as a single (scaled) Fourier transform between the input and output functions $\tilde{u}_0$ and $\tilde{u}$. This transform is applied, however, to the modified function $\tilde{u}_0(x_0, z_0) \exp(-j\pi x_0^2/L\lambda)$, and is followed by multiplication by another factor of $\exp(-j\pi x^2/L\lambda)$.

Applying a fast Fourier transform algorithm directly to the evaluation of Equation 16.80 provides another related but different way to do the same propagation calculation, using now a single Fourier transform. However, this transform is now applied to a more complex input function, because of the additional spherical wave factor $\exp(-j\pi x_0^2/L\lambda)$. The total amount of numerical work seems to come out about the same for either approach in most practical situations.

### Fourier Transforms and Gaussian Beams

Huygen's integral in the Fresnel approximation thus has the mathematical form either of a convolution of the input wavefront against a spherical wavefunction, or of a Fourier transform of the input wavefront multiplied by a spherical wavefront. Suppose we put a gaussian-spherical beam as the input into either of these mathematical forms. The reader should then know, or learn, that *the convolution of a gaussian function with another (possibly complex) gaussian always gives still another gaussian*. A gaussian beam passing through the convolution process of Equation 16.79 will thus always come out again gaussian, as we already know.

To express this same point in an alternative form, the Fourier transform of a gaussian function is always another gaussian transform (and in fact the Fourier transform of a Hermite-gaussian function is always another Hermite-gaussian function of the same order). All of the gaussian beam properties we have derived earlier and will discuss later are thus deeply imbedded in the self-transforming properties of generalized complex gaussian functions, and their higher-order Hermite-gaussian extensions.

### Paraxial Plane Waves and Transverse Spatial Frequencies

The mathematical procedures that are employed in evaluating the Huygens-Fresnel integral by Fourier transform methods can be given a simple and graphic physical interpretation in terms of *an expansion of the optical beam in a set of infinite plane waves traveling in slightly different directions*. The Fourier transforms that are calculated in the convolution procedure correspond, in fact, to the transformation of the beam profile into a "spatial frequency" domain, or into a

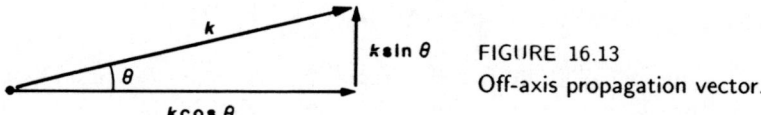

FIGURE 16.13
Off-axis propagation vector.

"k-vector space" of plane waves traveling at different directions with respect to the z axis, as illustrated earlier in Figure 16.2.

Because this spatial-frequency or $k$-vector viewpoint is very graphic and because it may give useful physical insights into the nature of beam propagation, we will rederive this plane wave description in some detail in the following paragraphs, even though the final results we obtain will be entirely identical to what we have already obtained merely by applying Fourier theorems to the Huygens-Fresnel integral.

To carry out this derivation, we consider as fundamental building blocks a set of infinite plane waves of the form

$$\tilde{u}_{pw}(x,y,z) \equiv \exp[-j\mathbf{k}\cdot\mathbf{r}] = \exp\left[-j(k_x x + k_y y + k_z z)\right]. \tag{82}$$

The propagation vector $\mathbf{k} = (k_x, k_y, k_z)$ for any such plane wave traveling at angles $\theta_x$ and $\theta_y$ with respect to the $z$ axis in the $x, z$ and $y, z$ planes has transverse components which we can write (see Figure 16.13) as

$$k_x = k\sin\theta_x \equiv 2\pi s_x \quad \text{and} \quad k_y = k\sin\theta_y \equiv 2\pi s_y. \tag{83}$$

The quantities $s_x \equiv (k/2\pi)\sin\theta_x \approx \theta_x/\lambda$ and $s_y \equiv (k/2\pi)\sin\theta_y \approx \theta_y/\lambda$ are then the transverse spatial frequencies along the $x$ or $y$ axes for a plane wave traveling at a small angle $(\theta_x, \theta_y)$ away from the $z$ axis toward the $x$ or $y$ directions.

The longitudinal $k$-vector component for a given plane-wave component can then be written, using the paraxial or Fresnel approximation, in the form

$$\begin{aligned} k_z &= \sqrt{k^2 - k_x^2 - k_y^2} \approx k - (k_x^2 + k_y^2)/2k \\ &= k - \pi\lambda(s_x^2 + s_y^2). \end{aligned} \tag{84}$$

Each individual plane wave component, characterized by its angles $\theta_x$ and $\theta_y$, or by its spatial frequencies $s_x$ and $s_y$, will then have a $z$ propagation given by

$$\tilde{u}_{pw}(x,y,z) = \tilde{u}_{pw}(x,y,0) \times \exp[-jkz + j\pi\lambda(s_x^2 + s_y^2)z]. \tag{85}$$

Each plane-wave component thus travels with a slightly different propagation constant in the $z$ direction, given (within the paraxial approximation) by the $\pi\lambda(s_x^2 + s_y^2)z$ factor.

### Expansion as a Distribution of Plane Waves

We then assume that an arbitrary paraxial optical beam $\tilde{u}(x,y,z)$ can be expanded into a plane-wave or spatial-frequency expansion in the form

$$\tilde{u}(x,y,z) = \iint \tilde{U}_{pw}(s_x, s_y, 0) \times \tilde{u}_{pw}(x,y,z)\, ds_x\, ds_y$$

$$= \iint \left[\tilde{U}_{pw}(s_x, s_y, 0) e^{-jkz + j\pi\lambda(s_x^2 + s_y^2)z}\right] \times e^{-j2\pi(s_x x + s_y y)}\, ds_x\, ds_y$$

$$= \iint \tilde{U}_{pw}(s_x, s_y, z) \times e^{-j2\pi(s_x x + s_y y)}\, ds_x\, ds_y, \tag{86}$$

where $\tilde{U}_{pw}(s_x, s_y, 0)$ gives the complex amplitude at plane $z = 0$ of each plane-wave component in the beam having spatial frequencies $s_x, s_y$, or traveling at angles $\theta_x \approx \lambda s_x$ and $\theta_y \approx \lambda s_y$. In writing the second and third lines we have made use of the fact that each such plane-wave component propagates forward in $z$ with a differential propagation constant given by

$$\tilde{U}_{pw}(s_x, s_y, z) = \tilde{U}_{pw}(s_x, s_y, 0) \times e^{-jkz + j\pi\lambda(s_x^2 + s_y^2)z}, \tag{87}$$

so that each component of this spatial-frequency distribution rotates in phase by a slightly different amount as the beam propagates forward.

At any arbitrary input plane $z = z_0$ the beam intensity pattern can thus be written as

$$\tilde{u}(x, y, z_0) = \iint \tilde{U}_{pw}(s_x, s_y, z_0) \times e^{-j2\pi(s_x x + s_y y)}\, ds_x\, ds_y. \tag{88}$$

But this expression is exactly a two-dimensional Fourier transform between the transverse spatial coordinates $x, y$ and the spatial frequencies $s_x, s_y$. If we know the input field distribution $\tilde{u}(x, y, z_0)$ at plane $z_0$, therefore, we can evaluate the spatial frequency distribution $\tilde{U}_{pw}(s_x, s_y, z_0)$ by carrying out the inverse Fourier transformation given by

$$\tilde{U}_{pw}(s_x, s_y, z_0) = \iint \tilde{u}(x, y, z_0) \times e^{+j2\pi(s_x x + s_y y)}\, dx\, dy. \tag{89}$$

Having transformed $\tilde{u}(x, y, z_0)$ into the spatial-frequency domain, we can then propagate this spatial-frequency distribution forward (or for that matter backward) to any other plane $z$ by multiplying it by the phase shift factor

$$\tilde{U}_{pw}(s_x, s_y, z) = \tilde{U}_{pw}(s_x, s_y, z_0) \times e^{-jk(z-z_0) + j\pi\lambda(s_x^2 + s_y^2)(z-z_0)}. \tag{90}$$

The field distribution $\tilde{u}(x, y, z)$ at the second plane can then be evaluated from the second Fourier transformation

$$\tilde{u}(x, y, z) = \iint \tilde{U}_{pw}(s_x, s_y, z) \times e^{-j2\pi(s_x x + s_y y)}\, ds_x\, ds_y. \tag{91}$$

Propagating any arbitrary paraxial wavefunction from plane $z_0$ to plane $z$ in free space is thus carried out using two Fourier transformations plus one simple multiplication step. Note also that the $e^{-jk(z-z_0)}$ term is just the on-axis or plane-wave phase shift, whereas the $e^{j\pi\lambda(s_x^2 + s_y^2)(z-z_0)}$ factor gives the differential phase rotation of each individual spatial frequency component.

This whole approach, from Equation 16.86 to Equation 16.91, is nothing more than a physical reinterpretation of the convolution or double-Fourier-transform approach to the evaluation of Huygens' integral which we discussed in connection with Equation 16.79. Looking at it in spatial-frequency terms, however, emphasizes again the importance of the small-angle or Fresnel approximation and the quadratic spatial-frequency dependence which this produces in all the exponents.

### Huygens' Integral in Cylindrical Coordinates

Huygens' integral 16.19 for free space can also be written in cylindrical coordinates $(r, \theta, z)$, leading to the result

$$\tilde{u}(r,\theta) = \frac{j}{L\lambda} \int_0^\infty r_0 \, dr_0 \int_0^{2\pi} \tilde{u}_0(r_0, \theta_0) \times \qquad (92)$$
$$\exp\left\{-j\left(\frac{\pi}{L\lambda}\right)\left[r^2 + r_0^2 - 2rr_0\cos(\theta - \theta_0)\right]\right\} d\theta.$$

Suppose the wavefunction has $m$-th order azimuthal symmetry, so that we can separate the radial and azimuthal variables in the form

$$\tilde{u}(r,\theta) = \tilde{u}_m(r) \times e^{\pm jm\theta} \qquad (93)$$

for both $u$ and $\tilde{u}_0$. Huygens' integral then reduces to the simpler form

$$\tilde{u}_m(r) = \frac{2\pi j^{m+1}}{L\lambda} \int_0^\infty r_0 \tilde{u}_0(r_0) e^{-j(\pi/L\lambda)(r^2+r_0^2)} J_m(2\pi r r_0/L\lambda) \, dr_0, \qquad (94)$$

where $J_m$ is the $m$-th order Bessel function.

Huygens' integral in this situation takes the form of a Fourier-Bessel transform, more commonly called a Hankel transform. A quasi "fast Hankel transform" is then available for the carrying out of numerical propagation and diffraction calculations on optical beams in cylindrical coordinates.

### REFERENCES

One representative example of the finite-difference method for numerical beam calculations can be found in P.B. Ulrich and J. Wallace, "Propagation of collimated pulsed laser beams through an absorbing atmosphere," *J. Opt. Soc. Am.* **63**, 8 (1973).

The plane-wave or spatial-frequency approach to beam propagation has been discussed, among many other references, in L. M. V. Camargo and I. Palocz, "A new Fraunhofer zone and some of its applications," *Proc. IEEE* **60**, 149 (January 1972).

The fast Fourier transform was first extensively applied to optical beam and resonator calculations by E. A. Sziklas and A. E. Siegman, "Diffraction calculations using fast Fourier transform methods," *Proc. IEEE* **62**, 410 (March 1974), and "Mode calculations in unstable resonator with flowing saturable gain. II. Fast Fourier transform method," *Appl. Opt.* **14**, 1873–1889 (August 1975). See also M. M. Johnson, "Direct application of the fast Fourier transform to open resonator calculations," *Appl. Opt.* **13**, 2326–2328 (October 1974): and A. E. Siegman, "How to compute two complex even Fourier transforms with one transform step," *Proc. IEEE* **63**, 544 (March 1975).

Optical resonator and beam propagation calculations using the standard fast Fourier transform algorithm (sometimes called the Cooley-Tukey algorithm) can be speeded up still further by a newer and even faster FFT algorithm developed by S. Winograd, "On computing the discrete Fourier transform," *Math. of Comp.* **32**, 175–199 (January 1978). This approach is followed, for example, by D. Heshmaty-Manesh and S. C. Tam, "Optical transfer function calculation by Winograd's fast Fourier transform," *Appl. Opt.* **21**, 3273–3277 (September 15 1982).

A "fast Hankel transform" algorithm for carrying out beam calculations in cylindrical coordinates is outlined (in a primitive form) in A. E. Siegman, "Quasi fast Hankel transform," *Optics Lett.* **1**, 13–15 (March 1977); and applied in a more sophisticated form in S-C. Sheng and A.E. Siegman, "Nonlinear optical calculations using fast transform methods: Second harmonic generation with depletion and diffraction," *Phys. Rev. A* **21**, 599–606 (February 1980).

For other ways of calculating Hankel transforms, see A. V. Oppenheim, G. V. Frisk, and D. R. Martinez, "Computation of the Hankel transform using projections," *J. Acoust. Soc. Am.* **68**, 523–529 (1980); S. M. Candel, "Dual algorithms for fast calculation of the Fourier-Bessel transform," *IEEE Trans.* **ASSP-29**, 963–972 (October 1981); S. M. Candel, "An algorithm for the Fourier-Bessel transform," *Comp. Phys. Commun.* **23**, 343–353 (1981); and P. K. Murphy and N. C. Gallagher, "Fast algorithm for the computation of the zero-order Hankel transform," *J. Opt. Soc. Am.* **73**, 1130–1137 (September 1983).

## Problems for 16.7

1. *Center of gravity of a paraxial optical beam.* Given an arbitrary paraxial optical beam $\tilde{u}(x, z)$ in free space, prove that the "center of gravity" $\bar{x}(z)$ of this beam as defined by

$$\bar{x}(z) \equiv \frac{\int_{-\infty}^{\infty} x |\tilde{u}(x,z)|^2 \, dx}{\int_{-\infty}^{\infty} |\tilde{u}(x,z)|^2 \, dx}$$

travels in a straight line in the $x, z$ plane. (Hints: Use the wave equation and differential identities; or a plane wave expansion and Fourier transform theorems; or Huygens' integral; or a Hermite-gaussian expansion and the Hermite-gaussian recursion relation.)

What physical interpretation can you give to the formula for the slope of this line? (Each of the preceding techniques in fact gives different and interesting insights into this question.)

2. *Second moment of a paraxial optical beam.* Extend the calculations of the previous problem, using a plane wave expansion, to calculate the second moment and the standard deviation $\langle (x - \bar{x})^2 \rangle$ of an arbitrary paraxial beam as a function of propagation distance. Again, give physical interpretations of the quantities involved, and see if you can develop an uncertainty relation between the standard deviations of the beam in real ($x$) space and in spatial frequency space.